CISM COURSES AND LECTURES

The series presents lecture notes, monographs, edited works and proceedings in the field of Mechanics, Engineering, Computer Science and Applied Mathematics.
Purpose of the series is to make known in the international scientific and technical community results obtained in some of the activities organized by CISM, the International Centre for Mechanical Sciences.

INTERNATIONAL CENTRE FOR MECHANICAL SCIENCES

COURSES AND LECTURES - No. 397

APPLICATION OF NUMERICAL METHODS TO GEOTECHNICAL PROBLEMS

PROCEEDINGS OF THE
FOURTH EUROPEAN CONFERENCE ON
NUMERICAL METHODS IN GEOTECHNICAL ENGINEERING
NUMGE98
UDINE, ITALY
OCTOBER 14-16, 1998

EDITED BY

ANNAMARIA CIVIDINI
TECHNICAL UNIVERSITY (POLITECNICO), MILANO, ITALY

Springer-Verlag Wien GmbH

Organized by:
European Regional Technical Committee ERTC7 on
NUMERICAL METHODS IN GEOTECHNICAL ENGINEERING
of the International Society for Soil Mechanics and Foundation Engineering.

in cooperation with
CISM, International Centre for Mechanical Sciences, Udine
Department of Structural Engineering, Politecnico di Milano
IACMAG, Intern. Assoc. for Computer Methods and Advances in Geomechanics
AGI, Associazione Geotecnica Italiana (Italian Geotechnical Association)

This volume contains 509 illustrations

Originally published by CISM, Udine in 1998.
SPIN 10699974

In order to make this volume available as economically and as rapidly as possible the authors' typescripts have been reproduced in their original forms. This method unfortunately has its typographical limitations but it is hoped that they in no way distract the reader.

ISBN 978-3-211-83141-0 ISBN 978-3-7091-2512-0 (eBook)
DOI 10.1007/978-3-7091-2512-0

PREFACE

The intent of the NUMGE98 Conference, held in Udine (Italy) during October 1998, is to provide a forum for exchange of ideas and for discussion on the various aspects of the analysis of Geotechnical Problems through Numerical Methods. Senior and young researchers, scientists and practicing engineers from European and overseas countries were invited at the Conference, to share their knowledge and experience.

The Conference is the fourth in a series that began informally in 1986 and continued, every four years, with the sponsorship of the European Technical Regional Committee ERTC7 of the International Society for Soil Mechanics and Foundation Engineering (ISSMFE).

The proceedings include the papers selected from a review of the abstracts and of the full length manuscripts submitted to the Conference.

The selected papers address a broad spectrum of geotechnical problems, including tunnels and underground openings, shallow and deep foundations, slope stability, seepage and consolidation, partially saturated soils, geothermal effects, constitutive modelling, etc. Depending on the nature of the addressed problem, the papers have been allocated in one of the following groups:

1. *Interpretation of field measurements and of small scale tests;*
2. *Laboratory tests and constitutive modelling;*
3. *Numerical algorithms and theoretical aspects;*
4. *Miscellaneous.*

The effort of the ERTC7 members, and of the colleagues serving as "National Contacts", in publicizing this meeting and in organizing the participation of delegates from their countries is gratefully acknowledged.

I would like to thank K.J. Bakker, A. Bolle, M. Boulon, M. Dolezalova, P. Fritz, M. Gryczmanski, G. Houlsby, B. Lehane, J.M. Couto Marques, Ph. Mestat, Y. Murzenko, S. Nordal, G. Pati, O. Ravaska, H. Schweiger, I.M. Smith for their active interest in the conference.

My thanks also to G. Bianchi (secretary general of CISM), to D. Sterpi (technical secretary of the Conference), to R. Contro (head of the Department of Structural Engineering of the Politecnico di Milano) and to my colleague and friend M.A. Parisi, for their scientific and technical assistance.

Annamaria Cividini
Udine, October 1998

Previous NUMGE Conferences:

1st Stuttgart (Germany)	*1986,*	*chaired by Peter Gussman*
2nd Santander (Spain)	*1990,*	*chaired by Cesar Sagaseta*
3rd Manchester (UK)	*1994,*	*chaired by Ian M. Smith*

CONTENTS

NUMGE98 Conference Chairperson:
Annamaria CIVIDINI

Conference Secretariat:
Donatella STERPI

Department of Structural Engineering
Technical University of Milan (Politecnico di Milano)
Piazza Leonardo da Vinci, 32 - 20133 Milano, Italy

INVITED LECTURES

ON THE ROLE OF ANALYTICAL SOLUTIONS FOR THE EVALUATION OF SOIL DEFORMATION AROUND TUNNELS

C. Sagaseta
University of Cantabria, Santander, Spain

ABSTRACT: The role of analytical solutions in the evaluation of stress and displacement fields around tunnels is commented. The paper is divided into two parts. In the first part, ground loss and tunnel distortion are evaluated from elastic and elastic-plastic solutions, starting from the basic case of a circular tunnel at infinite depth and with uniform and isotropic initial stresses, and by introducing the effect of uneven stresses and finite depth. The second part deals with the analysis of distant deformations. Available solutions are used to derive surface and deep displacements. The results of observations in actual cases are used to assess the applicability of the solutions presented.

1 INTRODUCTION

The use of analytical solutions has been reduced in the last decades due to the large development of numerical methods. Finite element analyses in 2 or 3 dimensions, with non-linear soil behaviour, simulation of construction phases, and coupled deformation and flow are now possible in a portable computer, with commercially available programs.

Under these conditions, the old analytical solutions, requiring drastic simplifications of the problem geometry and stress-strain properties are no longer a usual tool. However, they can play still an important role for several reasons: first, they are the reference benchmarks for the results of numerical analyses; they are useful for sensitivity analysis and to identify problem variables; and finally, they provide in some cases simple useful results, either directly or with some empirical corrections (see, for instance, Booker, 1988).

In any case, analytical solutions, used in conjunction to numerical methods and interpretation of actual observations form the basis for the prediction of performance in engineering problems. This is particularly evident in the present case of soil deformation around tunnels. This problem can be divided into two parts, of very different nature.

On one hand, the construction of the tunnel produces a deformation in its immediate vicinity. Its magnitude and distribution are strongly dependent on the soil mechanical behaviour and the tunnel construction process. These factors are difficult to consider, given the three-dimensional character of the problem, the non-linearity and time-dependency of soil deformability, and the nature of construction details. This poses a serious problem, and even the most refined numerical analyses require some empirical correction factors.

This deformation attenuates with the distance to the tunnel. The distribution of soil movements depends only on a limited number of factors. The kinematical constrains of the problem are strong, the strain level is moderate or small, and the problem geometry is relatively simple. As a result, the pattern of soil displacements at some distance from the tunnel does not vary significantly between different cases.

These two aspects: near field and far field deformations, are addressed separately in the following paragraphs.

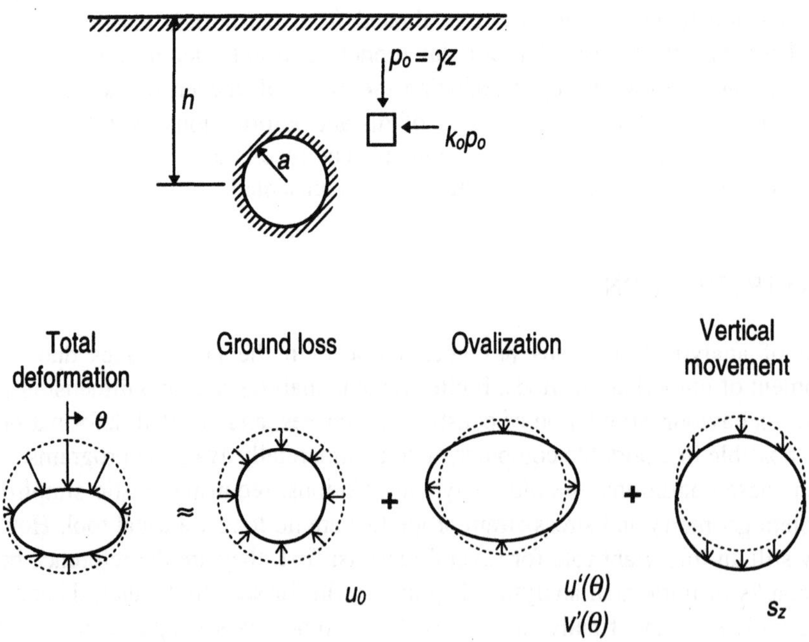

Fig. 1. Principal components of deformation of the tunnel wall

2 TUNNEL DEFORMATION

In the general case of a circular tunnel of radius a (diameter $D=2a$) at finite depth, h, uneven initial stresses ($k_o \neq 1$) and stress gradient with depth, the displacements of the tunnel wall are non-uniform and with radial and tangential components ($u(\theta), v(\theta)$). This deformation can be considered as the sum of several fundamental modes (Fig. 1).

First, a pure radial contraction, u_o,, which can be expressed as an average radial convergence ($\varepsilon = u_o/a$) or as a relative ground loss (both will be used in the following):

$$\varepsilon_s = \frac{\Delta S}{S_o} = \frac{\int u(\theta)ad\theta}{\pi a^2} = \frac{2\pi a u_o}{\pi a^2} = 2\frac{u_o}{a} = 2\varepsilon \qquad (1)$$

The second term is an ovalization or distortion of the tunnel without change of section (no ground loss). Although the displacements have radial (u') and tangential (v') components, the radial one is predominant. The distortion is usually taken positive as depicted, i.e., with vertical shortening. It is defined by:

$$\delta = \frac{u'_{max}}{a} \qquad (2)$$

The third component is a downward movement, s_z, with no distortion. It will be represented by its value divided by the tunnel radius (s_z/a).

In the following, the ovalization and settlement terms will also be defined by their values relative to the average radial convergence, and it will be denoted as:

$$\rho = \frac{\delta}{\varepsilon} = \frac{u'_{max}}{u_o} \quad ; \quad \eta = \frac{s_z/a}{\varepsilon} = \frac{s_z}{u_o} \qquad (3)$$

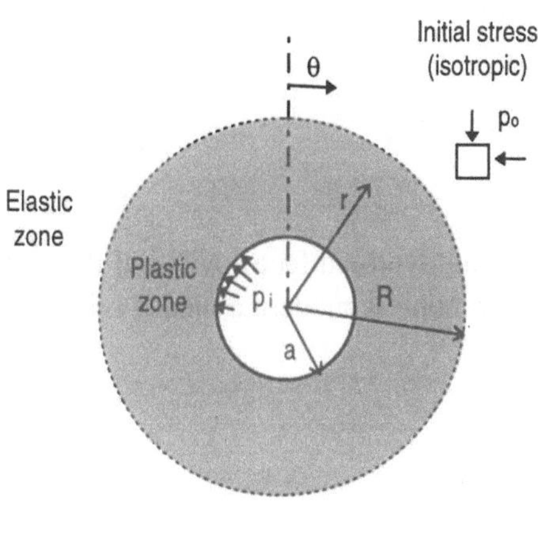

Fig. 2. Basic case. Plastic zone

2.1 Basic case. Axial symmetry

The simplest case is a circular tunnel of radius a, infinite length, at infinite depth in a soil initially subjected to uniform isotropic compression, p_o, ($k_o=1$). A uniform pressure, p_i ($p_i \leq p_o$), acts at the tunnel wall (Fig. 2).

The problem is one-dimensional, all the variables being independent of the polar angle, θ, and the axial coordinate, z. A closed form solution is possible, even in the elastoplastic range. The stress changes are maximum at the tunnel wall, so the plastic zone is a concentric annulus around the tunnel.

Of the modes of deformation of Fig. 1, only the first one exists. The ground loss for an elastic (G,μ) soil, is:

$$\varepsilon_s = 2\frac{u_o}{a} = \frac{p_o - p_i}{G} \tag{4}$$

- Purely cohesive case (ϕ=0)

The case of elastoplastic Tresca soil ($c=c_u$, $\phi=0$) represents the situation of undrained tunnel excavation in clay. Elastic incompressibility requires a Poisson's ratio $\mu=1/2$.

Peck (1969) introduced the use of the overload factor, N:

$$N = \frac{p_o - p_i}{c_u} \tag{5}$$

This factor has been universally adopted in tunnel design as a qualitative measure of the mobilized soil strength. Its meaning comes from the consideration of a plastic deformation taking place due to an active pressure, p, against a resisting pressure, q. The general expression for the limit load for a weightless soil is:

$$p = q\,N_q + c\,N_c \tag{6}$$

For the present case, the active pressure is p_o and the resisting pressure is p_i. For $\phi=0$, $N_q=1$, and N can be identified as N_c. However, some confusion exists about the meaning of the inner pressure, p_i. Originally, the overload factor was used for the analysis of tunnel face stability, and the pressure p_i was accordingly taken as the inner pressure at the tunnel face, which is zero except for closed face tunnelling (air pressure, slurry or EPB shields). However, when it is used for the analysis of the tunnel cross-section (Clough and Schmidt, 1981; Uriel and Sagaseta, 1989), p_i must be the inner pressure at the tunnel wall, i.e., the pressure taken by the lining. This second definition will be used here.

The extension of the plastic zone, R, results:

$$\frac{R}{a} = e^{\frac{N-1}{2}} \qquad (N \geq 1) \tag{7}$$

This result is shown in the upper curve in Fig. 3. For an overload factor $N\leq 1$, all the soil is in the elastic range. For N between 1 and 3 the plastic zone is moderate, and it becomes very large for N approaching 5-6. The loss of ground is:

$$\varepsilon_s = \begin{cases} \dfrac{p_o - p_i}{G} = \dfrac{1}{I_r} N & \text{if } N \leq 1 \\ \dfrac{1}{I_r} e^{N-1} & \text{if } N \geq 1 \end{cases} \tag{8}$$

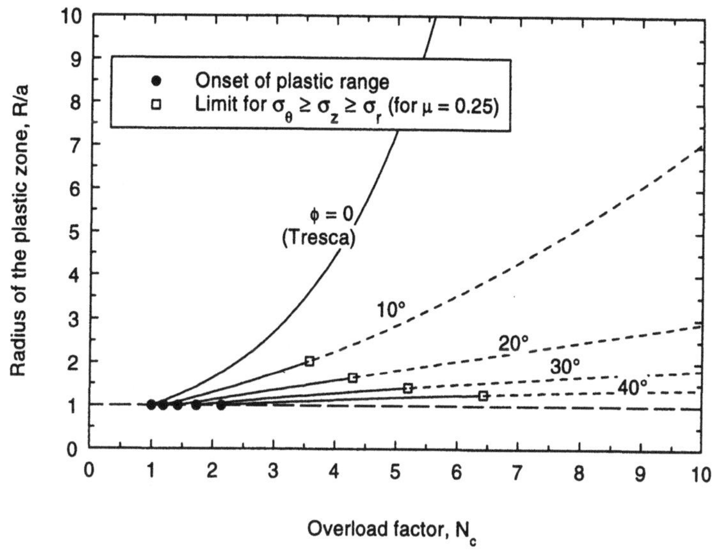

Fig. 3. Basic case. Extension of the plastic zone

where I_r is the so-called "rigidity index" (shear stiffness to strength ratio):

$$I_r = \frac{G}{c_u} \tag{9}$$

The overload factor is not bounded in theory. However, in real soils the undrained shear strength c_u is related to consolidation pressure, and hence to p_o. In the less favourable cases (unlined tunnel in a normally consolidated soil with water level at the ground surface), N can take values up to 8 or 10. Regarding the rigidity index, the usual range is 50 to 200.

The ground loss (8) is plotted in Fig. 4 (upper curve) in logarithmic scale for a typical value of I_r=100. The ground loss is less than 1% in the elastic range, increasing up to 10% for N between 1 and 3, and reaching the collapse of the tunnel cavity for values of N of about 5 to 6 (for $N \geq 4$, large strains become significant; if they are considered in the analysis, then the curve tends asymptotically to 100% ground loss for $N \to \infty$).

A noticeable feature in Fig. 4 is that the slope of the line (log ε_s - N) in the plastic range does not depend on the soil properties, but it is always equal to 1 (i.e., an increase of N by one unit means to multiply the ground loss by e=2.71). A variation of I_r implies a pure vertical translation of the curve.

- *Cohesive-frictional case* (c, ϕ, ν)

The solution for the general case with cohesion c friction angle ϕ, with Mohr-Coulomb criterion, and dilatancy angle ν, is more complex and of less direct applicability.

There are several reasons for this. First, the number of parameters increases from two (G,c_u) to five (G,μ,c,ϕ,ν), and this makes the presentation of the solution more cumbersome. Furthermore, the incompressibility condition in the case ϕ=0 has several implications that simplify the solution, and that are lost in the general frictional case.

On the other hand, the presence of volumetric strains in the soil imply that the ground loss decreases with the distance to the tunnel, and only a fraction reaches the ground surface. Hence, the coincidence of the ground loss at the tunnel and the settlement volume no longer coincide. In real cases, only the latter of these two parameters, is measured, and this decreases the practical interest of evaluation of the ground loss at the tunnel.

Besides these reasons (or perhaps due to them), there is no universal agreement about the use of non-dimensional parameters similar to the overload factor or rigidity index defined above for the undrained case. As a consequence, the expressions for stresses and displacements given in the literature use different parameters and notation. Their results are not always coincident, sometimes due to printing errata, and in other cases to the

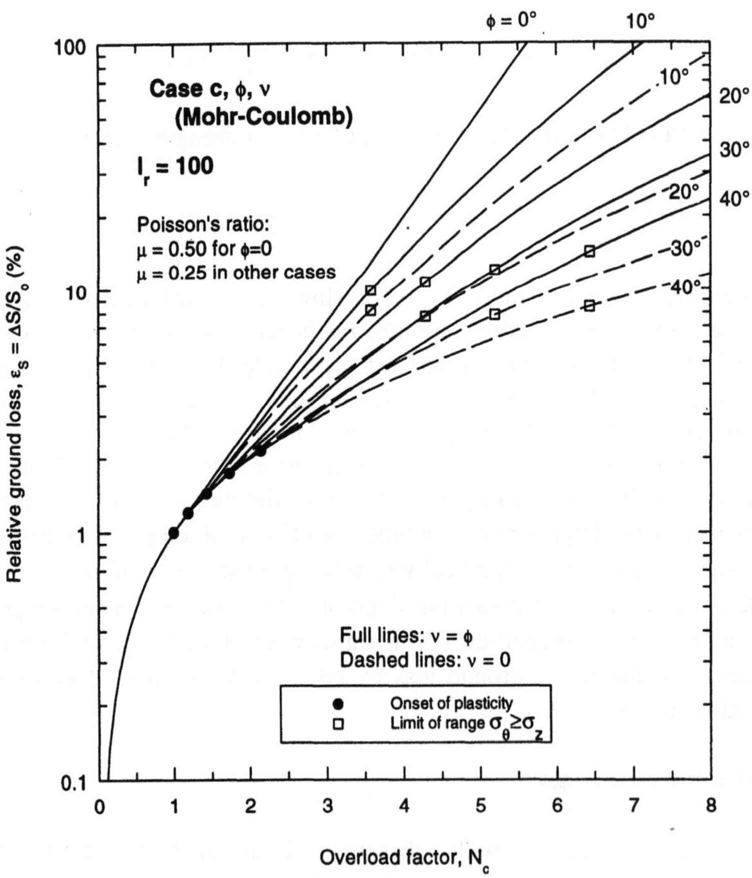

Fig. 4. Relative ground loss. Case c, ϕ, ν.

consideration of different simplifying assumptions, not always stated explicitly.

The non-dimensional parameters must fulfil two main requirements: i) they must not be arbitrary, but based on physical considerations; and ii) it is desirable that they tend to the expressions defined above when the friction angle tends to zero.

From the application of the Theory of Plasticity, the limit equilibrium equation (6) can be written in the most general case as:

$$p + c \cot\phi = (q + c \cot\phi) N_q \tag{10}$$

It is then convenient to use the following stress variables:

$$\begin{aligned} s_o &= p_o + c \cot\phi \\ s_i &= p_i + c \cot\phi \end{aligned} \tag{11}$$

By comparing eqs. (6) y (10), it is obtained:

$$N_c = (N_q - 1)\cot\phi \quad ; \quad N_q = 1 + N_c \tan\phi \tag{12}$$

$$\begin{aligned} N_q &= \frac{s_o}{s_i} = \frac{p_o + c \cot\phi}{p_i + c \cot\phi} \\ N_c &= \left(\frac{s_o}{s_i} - 1\right)\cot\phi = \frac{p_o - p_i}{c + p_i \tan\phi} \end{aligned} \tag{13}$$

It follows from this that any of the factors N_c or N_q above can be used as load parameter. N_q gives generally more simple expressions, but the factor N_c has the advantage of reducing to eq. (5) when $\phi \to 0$, and hence it is a direct generalisation of the overload factor, N. So, in the following both N_c and N_q are used, having in mind that they are linked by (12).

As for the rigidity index, I_r, it can be generalised accordingly as:

$$I_r = \frac{G}{c + p_i \tan\phi} \tag{14}$$

which also coincides with (9) for $\phi=0$.

The onset of plasticity is given by the Mohr-Coulomb at the tunnel wall:

$$\sigma_\theta - \sigma_r = (\sigma_\theta + \sigma_r)\sin\phi + 2c \cos\phi \tag{15}$$

from where the following critical load factors result:

$$N_{qe} = \frac{1}{1 - \sin\phi} \quad ; \quad N_{ce} = \frac{\cos\phi}{1 - \sin\phi} \tag{16}$$

which reduce to $N_{qe}=N_{ce}=1$ when $\phi=0$.

The radius R of the plastic zone is then:

$$\frac{R}{a}=\left(\frac{N_q}{N_{qe}}\right)^{\frac{1-sin\phi}{2sin\phi}} \tag{17}$$

which reduces to (7) for $\phi\rightarrow 0$.

R/a is plotted in Fig. 3 against N_c for different values of ϕ. For constant cohesion c, the friction reduces noticeably the extension of the plastic zone.

Equation (15) assumes implicitly that the intermediate principal stress acts along the tunnel axis, i.e., in the direction of plane strain:

$$\sigma_\theta \geq \sigma_z \geq \sigma_r \tag{18}$$

It can be easily proved that this always happens for incompressible soil (ϕ=0, μ=1/2). However, in the general case, condition (18) only holds for moderate values of the load factor. The limit values N_{ez} and N_{cz} are:

$$\frac{N_{qz}}{N_{qe}}=1+\frac{sin\phi}{1-2\mu} \quad ; \quad \frac{N_{cz}}{N_{ce}}=\frac{2(1-\mu)}{1-2\mu} \tag{19}$$

where N_{qe} and N_{ce} are defined by (16).

Beyond this limit (i.e., for $N_q>N_{qz}$ or $N_c>N_{cz}$), σ_z becomes the major principal stress, σ_θ the intermediate one and σ_r the minor one, and so failure takes place in the meridian planes (rz). The analytical solution is still possible, but it is necessary the simultaneous consideration of all the stress and strain variables and equations, and the algebra involved increases enormously. The interest of this case is almost purely academic. In Fig. 3 and in the following it is not solved, and only the existence of the limit (19) is pointed out.

Under these conditions, the general expression for the elastoplastic ground loss is:

$$\varepsilon_s=\frac{s_o}{G}(1-2\mu)\left[\left(1+\frac{sin\phi}{1-2\mu}\right)\left(\frac{R}{a}\right)^{1+m}-1\right]-\frac{s_i}{G}\frac{1-2\mu+sin\phi\, sin\nu}{1-sin\phi\, sin\nu}\left[\left(\frac{R}{a}\right)^{k+m}-1\right] \tag{20}$$

where:

$$k=\frac{1+sin\phi}{1-sin\phi} \quad ; \quad m=\frac{1+sin\nu}{1-sin\nu}$$

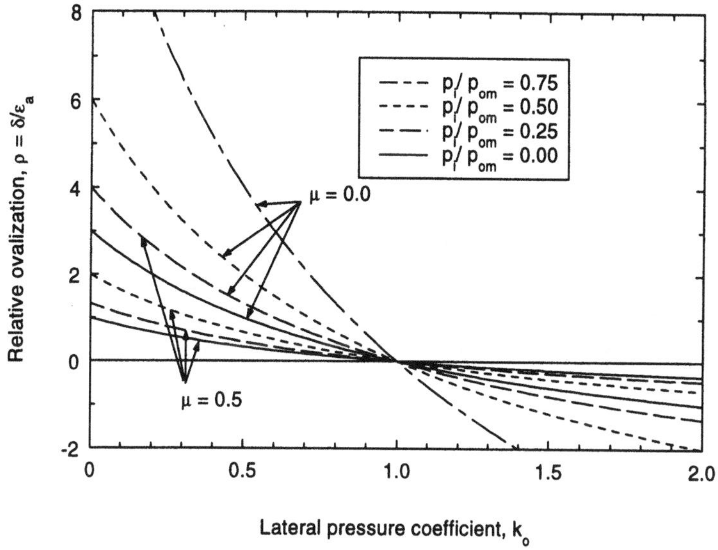

Fig. 5. Relative ovalization in Kirsch's solution

It is usual to find simplified versions of equation (20). The most common assumption is to neglect the elastic component of strain in the plastic region. The resulting expression is:

$$\varepsilon_s = \begin{cases} \dfrac{1}{I_r} N_c & \text{if } N_c \le N_{ce}, \left(N_q \le N_{qe}\right) \\ \dfrac{1}{I_r} N_{ce} \left(\dfrac{N_q}{N_{qe}}\right)^{\frac{1-\sin\phi \sin\nu}{\sin\phi(1-\sin\nu)}} & \text{if } N_q \ge N_{qe}, \left(N_c \ge N_{ce}\right) \end{cases} \tag{21}$$

Eqs. (21) and (20) give very similar results. Sometimes, further simplifications are made, the most frequent being the extreme assumptions of zero plastic volume change (ν=0), or associative behaviour ($\nu=\phi$). The results of eq. (21) are plotted in Fig. 4 for these two extreme hypotheses.

– *Other cases*

The relative simplicity of the above solution allows for the consideration of other cases. So, Duncan Fama (1993) provides solutions for strain softening material, with a gradual decrease of the strength with plastic strain. For application to rock masses, there are also solutions with the criterion of Hoek y Brown (1980).

2.2 Solutions without axial symmetry

In this section, the solutions for the cases mentioned above (finite depth, uneven initial stresses and stress gradient with depth) are considered. In these cases the problem has two dimensions, and the solution is restricted to the elastic regime. However, the elastic range is relevant in actual cases, as it will be shown later.

– *Anisotropic initial stresses ($k_o \neq 1$)*

The problem with initial stresses being different in vertical (p_o) and horizontal directions ($k_o p_o$) ($k_o \neq 1$), maintaining the assumption of infinite depth, was solved by Kirsch (1898) and can be found in any textbook of Elasticity. However, in its original presentation, the solution was given for the case of stresses acting on a medium with a pre-existing hole. For application to tunnelling, the interest is focused to the excavation of the cavity in a pre-stressed medium. The stresses are the same in both cases, but the displacements are different (in the pre-existing hole case, they do not vanish with the distance to the tunnel). Pender (1980) has pointed out this fact and given the correct solution.

The radial, u, and tangential, v, displacements at a distance r from the tunnel axis are:

$$\frac{u}{a} = \frac{p_o}{2G}\left\{\frac{1+k_o}{2}\frac{a}{r} + \frac{1-k_o}{2}\left[4(1-\mu)\frac{a}{r} - \left(\frac{a}{r}\right)^3\right]\cos 2\theta\right\} - \frac{p_i}{2G}\frac{a}{r}$$

$$\frac{v}{a} = \frac{p_o}{2G}\frac{1-k_o}{2}\left[2(1-2\mu)\frac{a}{r} + \left(\frac{a}{r}\right)^3\right]\sin 2\theta \qquad (22)$$

The associated ground loss is:

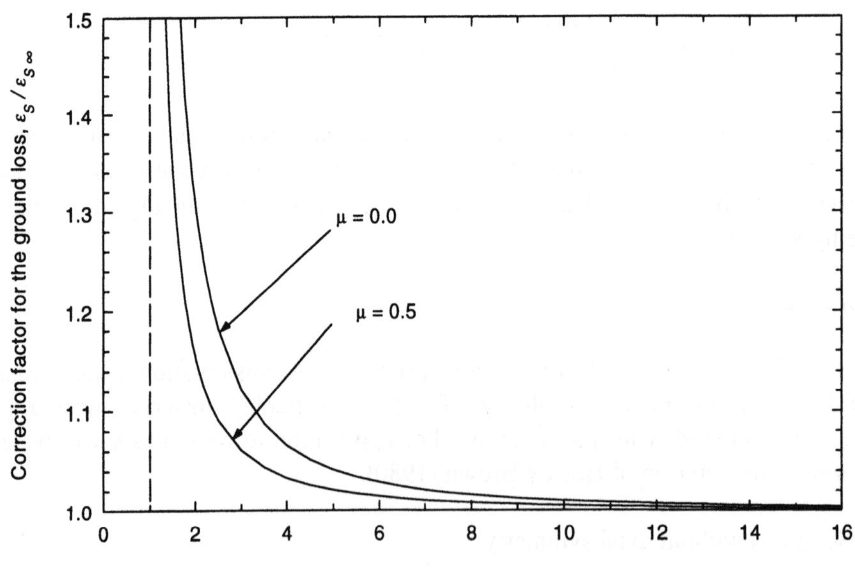

Fig. 6. Free surface. Influence on ground loss, from Verruijt (1996) solution

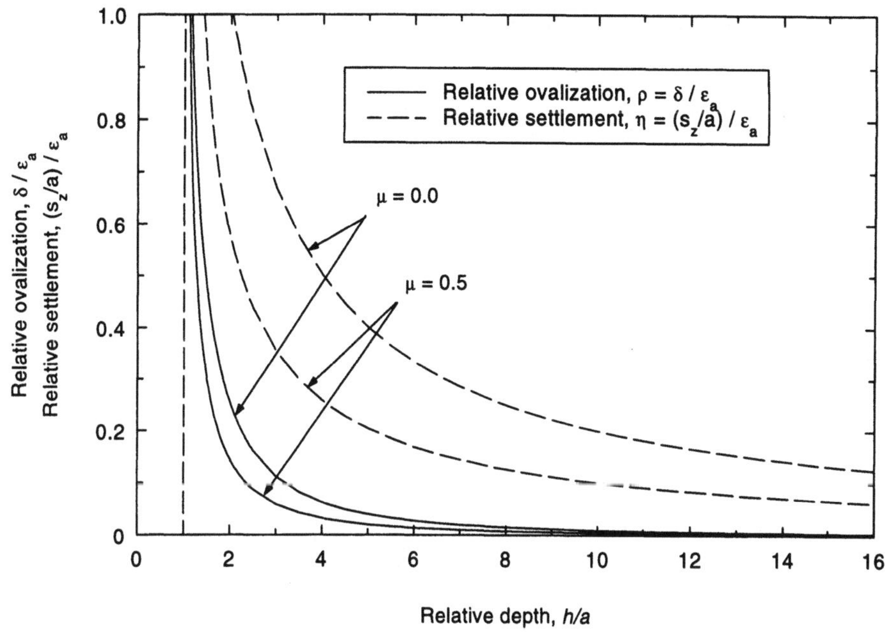

Fig. 7. Free surface. Relative ovalization and settlement from Verruijt (1996) solution

$$\varepsilon_s = 2\frac{u_o}{a} = \frac{1}{G}\left(\frac{1+k_o}{2}p_o - p_i\right) \tag{23}$$

which coincides with eq. (4), taking the mean initial pressure: $p_{om}=p_o(1+k_o)/2$.

The tunnel ovalization (2) results:

$$\delta = \frac{u'_{max}}{a} = \frac{p_o}{2G}\frac{1-k_o}{2}(3-4\mu) \tag{24}$$

The ovalization is independent from the inner pressure, p_i (provided that it is uniform). The relative ovalization, $\rho=\delta/\varepsilon$, is plotted in Fig. 5. For k_o = 0.5 to 2.0, it varies between -1.0 and +1.0, except for very high ratios p_i/p_o (ground loss tends to zero for $p_i=p_o$).

– *Finite depth*

Mindlin (1940) obtained the elastic solution for the stresses in the case of near surface tunnel, with initial stress gradient with depth. His results indicate that the influence of the surface on the stress distribution is not significant for depth-to-diameter ratios over 2 or 3.

Regarding the deformations, there is a recent analysis by Verruijt (1996), using complex variable techniques. The boundary condition of the stress free surface is considered, but keeping the assumptions of isotropic (k_o=1), and uniform initial stresses (no gradient with depth). The tunnel wall displacements are:

$$\frac{u}{a}=\frac{p_o-p_i}{2G}\left[1+2(1-\mu)\frac{\cos\theta}{\frac{h}{a}-\cos\theta}\right]$$
$$\frac{v}{a}=\frac{p_o-p_i}{2G}2(1-\mu)\frac{\sin\theta}{\frac{h}{a}-\cos\theta} \tag{25}$$

This means a ground loss:

$$\varepsilon_s=\frac{p_o-p_i}{G}\left[1+2(1-\mu)\frac{2\frac{h}{a}\left(\frac{h}{a}-\sqrt{\left(\frac{h}{a}\right)^2-1}\right)-1}{1-\frac{h}{a}\left(\frac{h}{a}-\sqrt{\left(\frac{h}{a}\right)^2-1}\right)}\right] \tag{26}$$

This expression is a generalisation of (4), and the factor in brackets denotes the influence of the surface on the ground loss. It takes only moderate values, being less than 1.10 for *h/a*>3, increasing abruptly for extremely shallow tunnels (Fig. 6).

The ovalization is:

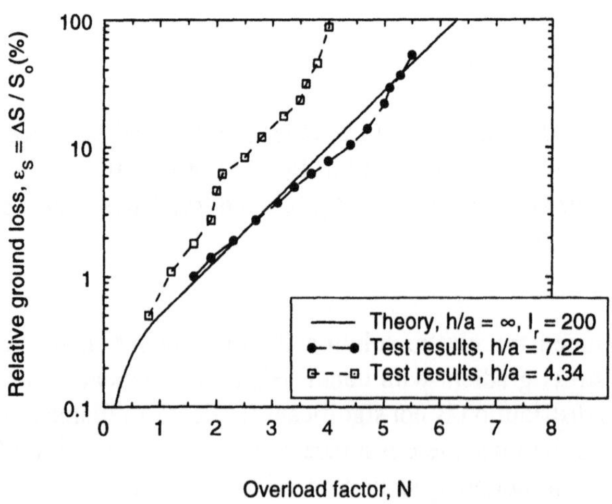

Fig. 8. Ground loss in a tunnel in clay. Centrifuge tests (Mair et al., 1981)

$$\delta = \frac{p_o - p_i}{2G}(1-\mu)\frac{1}{\left(\frac{h}{a}\right)^2 - 1} \tag{27}$$

The relative ovalization, ρ, is below 0.1 for the usual range of h/a (Fig. 7).

The presence of the free surface implies that the radial displacements are different at the crown (θ=0) and invert ($\theta=\pi$). So, a third component of the tunnel deformation emerges: a uniform downward movement, s_z, shown in Fig. 1. It can be calculated from (25) as:

$$\frac{s_z}{a} = \frac{p_o - p_i}{2G}(1-\mu)\frac{2\frac{h}{a}}{\left(\frac{h}{a}\right)^2 - 1} \tag{28}$$

The value of this component relative to the average radial convergence, $\eta=(s_z/a)/\varepsilon$, is also shown in Fig. 7. It ranges from 0.2 to 0.6 for the usual cases. It must be noticed that this component is the only responsible for the settlement of the crown being larger than the invert heave, as usually observed in real tunnels. For a relative ovalization ρ=0.2 and a relative settlement η=0.4, the crown settlement results twice as large as the invert heave.

2.3 Application

Starting from the basic, symmetric case, the above described solutions consider some of the

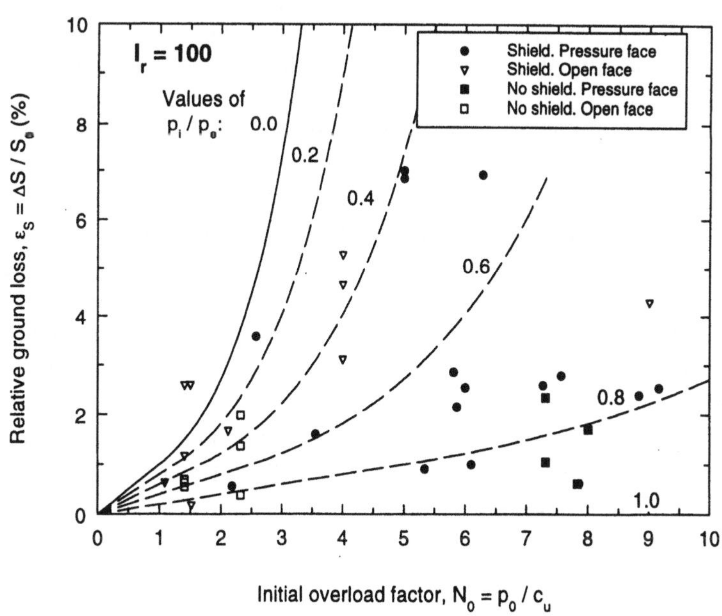

Fig. 9. Tunnels in clay. Measured and calculated ground loss

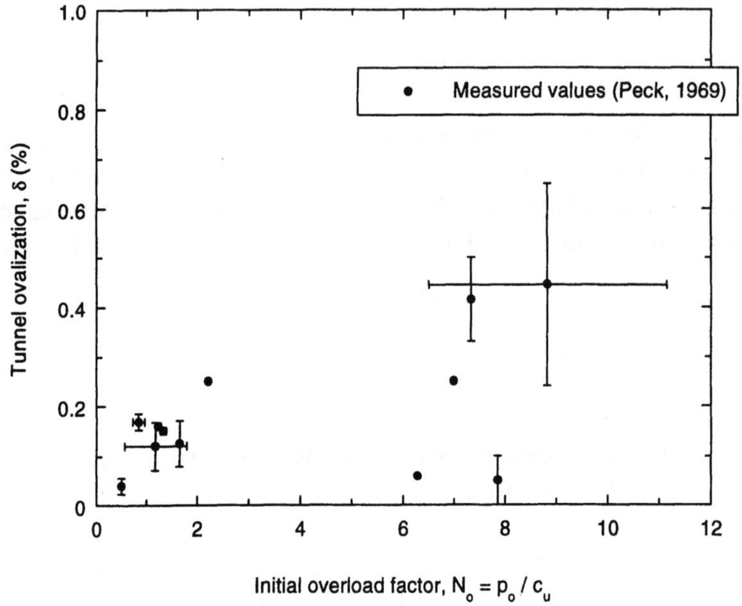

Fig. 10. Tunnel ovalization. Data from actual cases (from Peck, 1969)

factors controlling tunnel deformation: plastic strains, uneven initial stresses and presence of the free surface. However, these factors are taken separately, and there is no solution, for instance, for shallow tunnels in the plastic range. Some kind of assessment is needed of their applicability, using the results of observations in actual cases or numerical methods.

Fig. 8 shows the results of centrifuge tests of a circular tunnel in clay (Mair et al., 1981), in which the pressure difference (p_o-p_i) is increased progressively until collapse. Two tests are presented, with different h/a ratios. In the initial part of the curves, near the elastic range, both tests give similar values of the ground loss. This indicates a relatively small effect of the free surface, in agreement with Fig. 6. For the plastic range, the deeper case (h/a = 7.22) follows closely the theoretical curve (the slope of this curve is fixed, as it was pointed out in Fig. 4). However, in the shallower case (h/a = 4.34) the measured ground loss is clearly higher than the theoretical value.

For tunnels in clay, the overload factor N is often used as reference to analyse the measured ground loss. However, as commented above, for a proper comparison between theoretical and measured values, p_i must be considered as the pressure at the tunnel wall, i.e., at the ground-lining interface. This pressure is not known. So, it must be taken as variable, expressing the solution in terms of the initial overload factor, N_o, i.e., the value of N for zero inner pressure (N_o=p_o/c_u):

The ground loss (8) can then be written in the form:

$$\varepsilon_s = \begin{cases} \dfrac{1}{I_r} N_o(1-\dfrac{p_i}{p_o}) & if \;\; N_o(1-\dfrac{p_i}{p_o}) \le 1 \\ \dfrac{1}{I_r} e^{N_o(1-\frac{p_i}{p_o})} & if \;\; N_o(1-\dfrac{p_i}{p_o}) \ge 1 \end{cases} \tag{29}$$

Eq. (29) is plotted in Fig. 9 for an average rigidity index (I_r=100) and different values of the pressure ratio, p_i/p_o, together with the results of actual observations. It is clearly seen that closed face tunnels give lower ground losses, at higher lining pressures. For conventional excavation, without shield, there are only cases for N_o slightly over 2. The range for open face shields extends to N_o=4, and beyond this limit there are only cases with pressurised face. There are some remarkable cases: four points of conventional (no shield) tunnelling, with N_o >7, corresponding to Chicago Metro, excavated with air pressure in 1940 (Peck, 1969); there is also a case of open shield in a soil with N_o=9 with moderate ground loss (4.3%), in a section of San Francisco BART (Peck, 1969).

There are fewer data on measurements of tunnel ovalization. Fig. 10 shows cases compiled by Peck (1969) for tunnels in clay. Only a qualitative analysis is possible, because the measured distortion includes the influence of lining stiffness, which has not been considered in the theoretical solutions given above. The distortion tends to increase with N_o. Comparing with Fig. 9, the order of magnitude of the distortion is about one tenth of the ground loss. This is in general agreement with the theoretical results of Fig. 5 and Fig. 7.

3 DISTANT DEFORMATION

For the distribution of far field movements, the influence of construction details and tunnel precise geometry is not so important, due to attenuation with distance. As a result, soil deformation follows relatively simple patterns. On the other hand, the influence of the soil surface becomes of primary importance, and must be considered in any realistic analysis.

3.1 Empirical methods

The above two reasons explain the success of empirical methods to fit the patterns of soil deformation. The use of the error curve, proposed by Peck (1969) and Schmidt (1969), has become the most efficient tool to fit the final transverse profiles of surface settlements:

$$s_z(x) = s_o \, e^{-\frac{x^2}{2i^2}} \tag{30}$$

where $s_z(x)$ is the settlement at a distance x from the centreline, s_o is the (maximum) settlement for x=0, and i is the abscissa of the point of inflection of the curve. At this inflection point, s_z= 0.61s_o. The volume of settlements is given by the integral of (30) as:

$$\Delta V_s = \sqrt{2\pi} \, i \, s_o \cong 2.5 \, i \, s_o \tag{31}$$

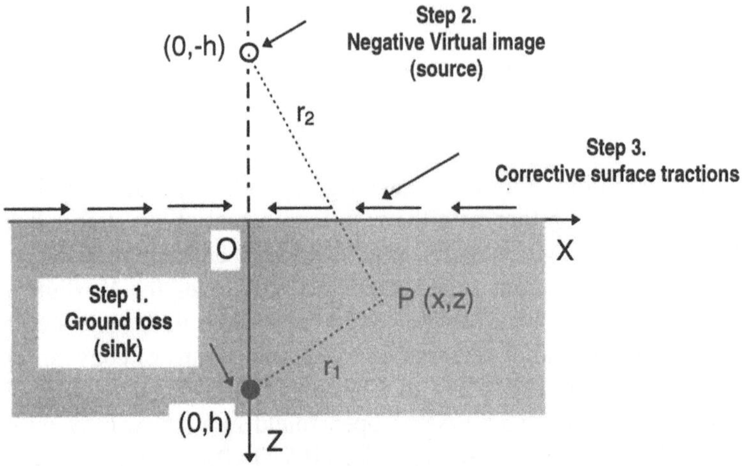

Fig. 11. Free surface. Virtual image technique (Sagaseta, 1987)

In clays, this volume is assumed to be equal to the ground loss at the tunnel, and it is also given as relative to the tunnel cross-section *($\Delta V_s/S_o$)*.

The success of this approach is based on its capability for reproducing the actual patterns of settlement profiles, but there is no theoretical reason demonstrating that settlement distribution is governed by expression (30). Some further extensions of the error curve have been introduced for horizontal displacements, and movements near to face.

3.2 Theoretical solutions

– *Ground loss at finite depth*

A theoretical approach to this problem was proposed by Sagaseta (1987), based on solutions for incompressible irrotational fluid flow (Fig. 11). First, the undrained ground loss in an infinite space is considered, reducing the tunnel to a point sink (Step 1), with the conditions of incompressibility and spherical symmetry determining a radial field of displacements, decreasing with the distance to the sink. The surface is considered by using a virtual image technique (Step 2), combined with corrective surface tractions (Step 3), for which elastic solutions for the half space are used. It is easily proved that the displacements at the surface are twice those that would occur if the sink was in an infinite space.

The final displacement field due to a ground loss ε_s (radial deformation ε) is given by:

$$s_x = -\varepsilon a^2 \left[x\left(\frac{1}{r_1^2} + \frac{1}{r_2^2} \right) - 4xzz_2 \frac{1}{r_2^4} \right]$$
$$s_z = -\varepsilon a^2 \left[z_1\left(\frac{1}{r_1^2} + \frac{1}{r_2^2} \right) - 4x^2 z \frac{1}{r_2^4} \right] \qquad (32)$$

where $z_1=(z-h)$, $z_2=(z+h)$, and r_1 and r_2 are the distances to the sink and its image, respectively (Fig. 11).

At the soil surface (z=0):

$$s_x(x)=\varepsilon a^2 \frac{2}{h}\frac{\bar{x}}{1+\bar{x}^2}$$
$$s_z(x)=\varepsilon a^2 \frac{2}{h}\frac{1}{1+\bar{x}^2} \qquad (\textit{with: } \bar{x}=\frac{x}{h}) \tag{33}$$

These expressions imply that the displacement vector at the surface is addressed towards the tunnel centre. The complex variable solution by Verruijt (1996) has confirmed that only for extremely shallow tunnels (h/a<2), there is a significant departure from this.

– *Tunnel ovalization*

Referring to the deformation components at the tunnel wall, step 1 in the analysis (Fig. 11) implies a pure radial ground loss, and steps 2 and 3 give a significant downward movement, but only a very small ovalization. Recently, Verruijt and Booker (1996) have presented an extension of the above analysis including the effect of the ovalization. For the sink in the infinite space, Kirsch's solution (22) is used, neglecting the third order terms, $(a/r)^3$. The displacements are written in terms of the radial tunnel deformation, ε, and the ovalization, δ, instead of p_o, p_i and k_o. In this way, δ is taken as a basic input parameter, as the ground loss, regardless its origin (uneven stresses, different support conditions at the crown and at the sides, or plastic deformations that imply maximum movements at the crown).

For μ=1/2, the displacement field due to the tunnel ovalization, δ, is:

$$s_x = \delta a^2 \left[x\left(\frac{x^2-z_1^2}{r_1^4} + \frac{x^2-z_2^2}{r_2^4} \right) - 4hxz\frac{x^2-3z_2^2}{r_2^6} \right]$$
$$s_z = \delta a^2 \left[z_1\left(\frac{x^2-z_1^2}{r_1^4} + \frac{x^2-z_2^2}{r_2^4} \right) - 4hzz_2\frac{3x^2-z_2^2}{r_2^6} \right] \tag{34}$$

These displacements must be added to (32) to give the total displacement field. The resulting movements at the surface are:

$$s_x(x)=-\varepsilon a^2 \frac{2}{h}\frac{\bar{x}}{1+\bar{x}^2}\left(1+\rho\frac{1-\bar{x}^2}{1+\bar{x}^2}\right)$$
$$s_z(x)= \varepsilon a^2 \frac{2}{h}\frac{1}{1+\bar{x}^2}\left(1+\rho\frac{1-\bar{x}^2}{1+\bar{x}^2}\right) \tag{35}$$

where $\rho=\delta/\varepsilon$ is the relative tunnel ovalization (3). For ρ=0, the tunnel cavity contracts horizontally and vertically by the same amount (no ovalization), and expression (35) reduces to (33). The horizontal convergence is less than the vertical one for $\rho>0$, and becomes negative (the tunnel expands horizontally) for $\rho>1$.

The transversal settlement troughs given by (33) are wider than usually measured in actual cases (Schmidt, 1988; Sagaseta, 1988). The ovalization $\rho>0$ tends to reduce this lateral spreading, thus giving a better reproduction of actual measurements.

– *Soil compressibility. Plastic strains*

The volumetric strains in the plastic range (soil dilatancy) can also contribute to give settlement profiles more concentrated than in the undrained, constant volume case. In order

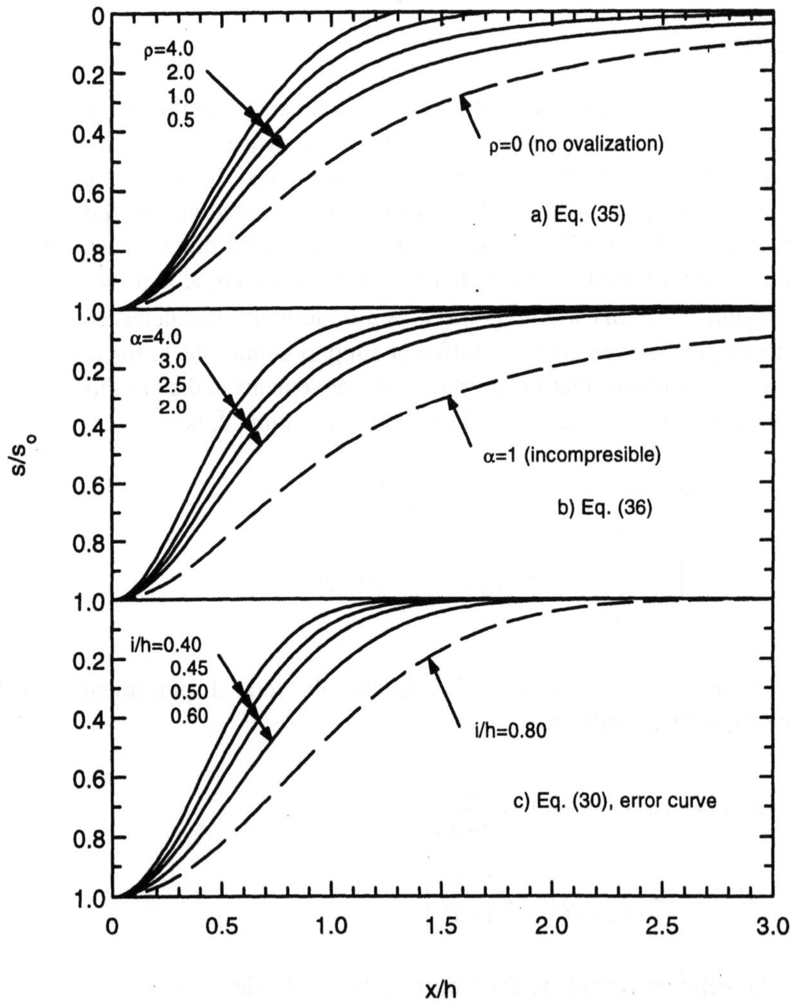

Fig. 12. Normalised surface settlement troughs

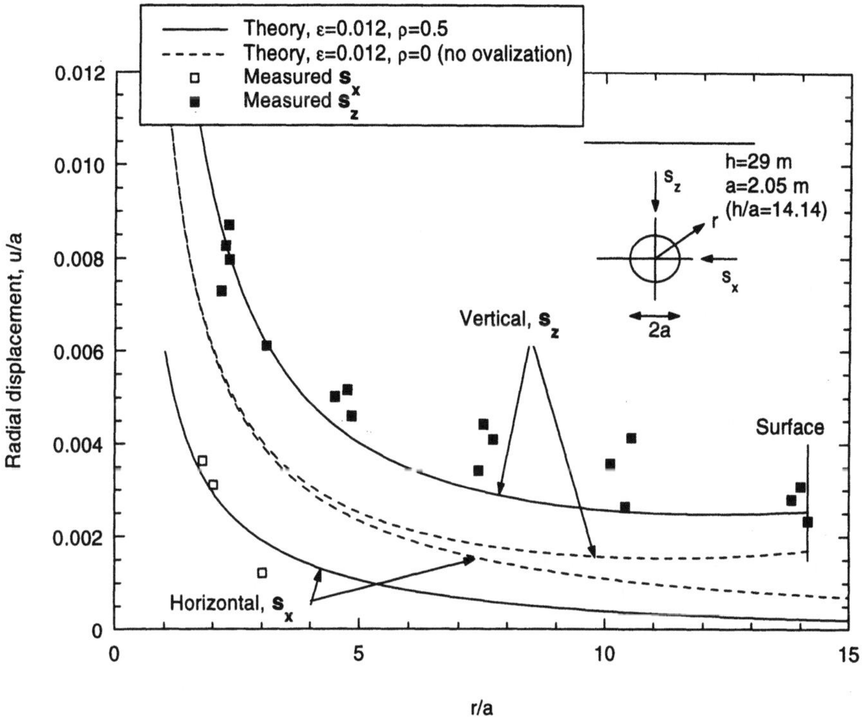

Fig. 13. Tunnel in London clay. Measured inner radial movements along vertical and horizontal axes (Attewell and Farmer, 1974).

to include these effects, the use of an exponent in the denominator of the settlement profile equation (33) was proposed (Sagaseta, 1988; Oteo and Sagaseta, 1996). This was based on the fact that in non-elastic medium, the displacements attenuate with a power of the distance, $O(1/r^{\alpha})$, $\alpha > 1$:

$$s_z(x) = s_o \frac{1}{\left(1+\bar{x}^2\right)^{\alpha}} \quad ; \quad s_x(x) = s_o \frac{\bar{x}}{\left(1+\bar{x}^2\right)^{\alpha}} \tag{36}$$

3.3 Application

In Fig. 12 the three above expressions (30, 35 and 36) for the settlement troughs are compared. They give very similar results when appropriate values are taken for their respective parameters, i/h, ρ and α. The range for i/h in actual cases is from 0.4 to 0.5, which corresponds to values of α from 2 to 4 and ρ from 0.5 to 4.

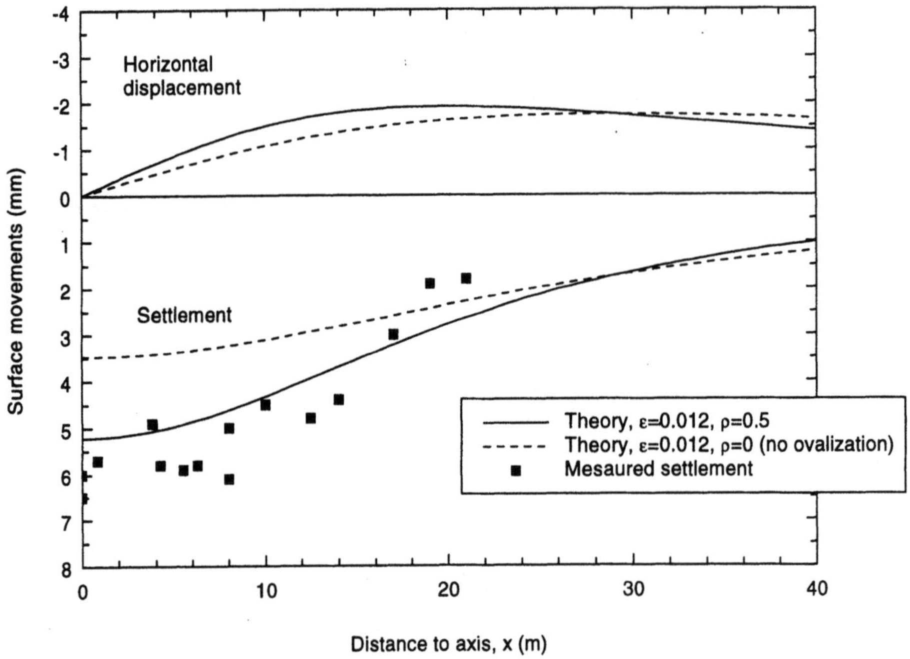

Fig. 14. Tunnel in London clay (Attewell and Farmer, 1974). Surface settlements

The expressions (34) have been applied to a well documented case of shield tunnelling in London clay (Attewell and Farmer, 1974). The tunnel has 4.10 m diameter at a depth of 29 m. Fig. 13 shows inner soil movements: vertical displacements above the tunnel, between the crown and the surface, and horizontal displacements at the tunnel sides. These data were analysed by Mair and Taylor (1992) to illustrate the applicability of cavity contraction solutions. As can be seen, the theoretical solutions closely match the observations, for a convergence, $\varepsilon = 0.012$ (relative ground loss 2.4%), and a relative ovalization, ρ, of 0.5. For comparison, the theoretical curves for no ovalization (ρ=0), with the same ground loss are shown. As it can be seen, the effect of the tunnel ovalization is to increase the settlements above the crown and to reduce movements at the sides. Fig. 14 shows the predictions for surface displacements for these parameters, together with measured settlements.

For soils other than clay, there are not so many documented cases. In Fig. 15 a case is shown of a tunnel excavated by the New Austrian Tunnelling Method (NATM) in Lisbon, Portugal, with a cross-section area of 100 m², at a depth of 25 m. The soil above the tunnel is dense, partially cemented Miocene sand.

The surface settlement profile has been fitted using the above three equations (30, 35 and 36). The three curves are almost identical, with a volume of settlements of about 0.7% of the excavated area, a maximum settlement of about 25 mm and *i/h* (defined for s/s_{max}=0.61), in the range 0.40-0.46. For eq. (35), the fitting parameters are $\varepsilon = 0.0024$ and

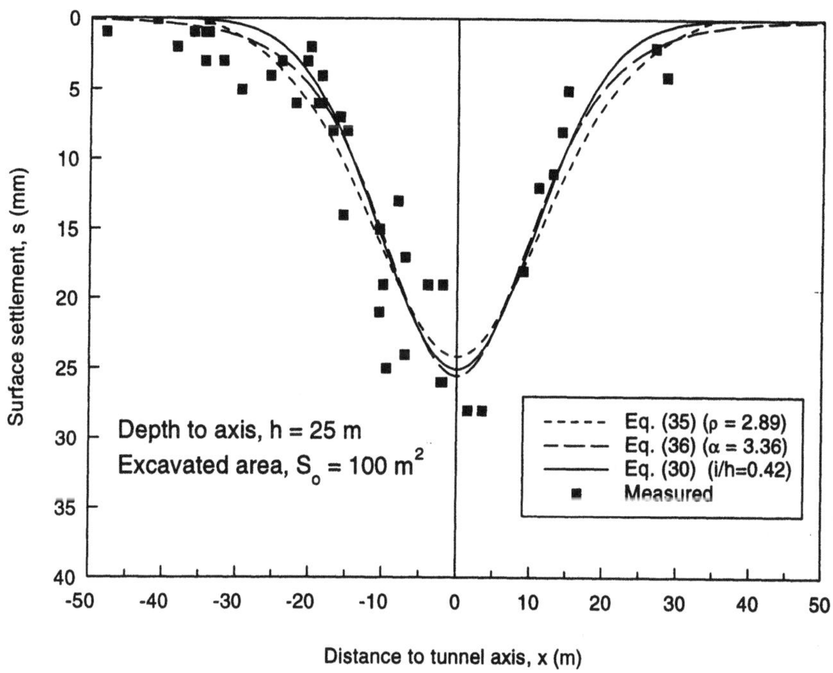

Fig. 15. Tunnel in Miocene sand (Sagaseta et al., 1999). Surface settlements

ρ=2.89. The total integral of the settlement profile is actually 0.48% of S_o with a volume of settlements for x<35 m of 0.72%, and an outside heave of 0.24%.

This value of the relative ovalization is abnormally high, and it does not seem to be realistic. This is possibly due to the influence of non-elastic deformations (shear strain concentrations, dilatancy, volume changes) mentioned above.

So, in cases in granular soils, a combination of the above factors (ovalization and soil compressibility) is needed, and this is being the subject of subsequent work.

4 CONCLUSIONS

Some analytical solutions for the deformation around tunnels have been reviewed. Some of them are old and classical, and some others are very recent. In any case, the aim of the review has been to show the potential applicability of these simple solutions to an apparently very complex problem.

As stated in the introductory comments, analytical methods do have still an important place, as reference benchmarks for the results of numerical analyses, for identification of problem variables, and as a framework for presentation and analysis of observations in actual cases. In some instances they can give by themselves simple useful results, as it has been shown in the preceding paragraphs.

5 REFERENCES

Attewell, P.B. and I.W. Farmer (1974). Ground deformations resulting from shield tunnelling in London clay. Canadian Geot. J. **11**:3,380-395.

Booker, J.R. (1988). Application of analytic and semi-analytic techniques to geotechnical analysis. Proc. 6th. Int. Conf. on Numer. Methods in Geomech. Innsbruck **1**,23-36.

Clough, G.W. and B. Schmidt (1981). Design and performance of excavations and tunnels in soft clay. Chapt. 8 in *'Soft Clay Engineering'* (E.W. Brand and R.P. Brenner, eds.). Elsevier. Amsterdam.

Duncan Fama, M.E. (1993). Numerical modelling of yield zones in weak rock.. Chapt. **2**:2 in *'Comprehensive Rock Engineering'*, (J.A. Hudson, ed.). Pergamon Press. London.

Hoek, E. and E.T. Brown (1980). *Underground Excavations in Rock.* Inst. of Mining and Metallurgy. London.

Mair, R.J., M.J. Gunn and M.P. O'Reilly (1981). Ground movements around shallow tunnels in soft clay. Proc. 10th Int. Conf. Soil Mech. Found. Eng. Stockholm. **1**,323-328.

Mair, R.J. and R.N. Taylor. (1992). Prediction of clay behaviour around tunnels using plasticity solutions. *Predictive Soil Mechanics. Wroth Memorial Symposium* (G.T. Houlsby and A.N. Schofield, eds.). Oxford, U.K., 449-463.

Mindlin, R.D. 1940. Stress distribution around a tunnel. Trans. ASCE, **105**,1117-1140.

Oteo, C.S. and C. Sagaseta (1996). Some Spanish experiences on measurement and evaluation of ground displacements around urban tunnels. *Geotechnical Aspects of Underground Construction in Soft Ground* (R.J. Mair and R.N. Taylor, eds.). London, 731-736.

Peck, R.B. 1969. Deep excavations and tunnelling in soft ground.. Proc. 7th ICSMFE. Mexico. State-of-the-Art Volume, 225-290.

Pender, M. J. (1980). Elastic solutions for a deep circular tunnel. Géotechnique **30**,216-222.

Sagaseta, C. (1987). Analysis of undrained soil deformation due to ground loss. Géotechnique **37**:3,301-320.

Sagaseta, C. (1988). Author's reply. Géotechnique **38**:4,647-649.

Sagaseta, C., J.M. Sánchez-Alciturri, C. González, A. López, P. Gómez and R. Pina (1999). Soil deformations due to the excavation of two parallel underground caverns. Proc. 12th Eur. Conf. Soil Mech. Found. Engg. Amsterdam. In press.

Schmidt, B. (1969). Settlements and ground movements associated with tunnelling in soil. Ph.D. Thesis. Univ. of Illinois.

Schmidt, B. (1988). Discussion. Géotechnique **38**:4,647.

Uriel, A.O. and C. Sagaseta (1989). Selection of parameters for underground construction. Proc. XIIth. Int. Conf. Soil Mech. Found. Engng. R. Janeiro, Brasil. Gen. Report, Session 9.9. **4**,2521-2551.

Verruijt, A. (1996). Complex variable solutions of elastic tunnelling problems. Geotechnical Laboratory, Delft University of Technology.

Verruijt, A. and J.R. Booker (1996). Surface settlements due to deformation of a tunnel in an elastic half plane. Géotechnique **46**:4,753-757.

PARALLEL COUPLED ANALYSES IN GEOTECHNICAL ENGINEERING

I.M. Smith
University of Manchester, Manchester, UK

ABSTRACT: Geotechnical materials are multi-phase combinations of solid, fluid and gas. The simplest mathematical formulation of their behaviour is based on Biot-type equations coupling the properties of the solid grains to those of an incompressible porefluid (water). Progress in the large scale computation of geotechnical situations depends on the calculation processes being parallelisable. This paper shows that coupled calculations of importance in geotechnical engineering can be parallelised, and evaluates the performances of different algorithms for solving the governing equations.

1 INTRODUCTION

The processing speed attainable by single processors increases gradually year by year and may ultimately be limited only by physical consideration. But whatever this speed, carrying out calculations in parallel offers the possibility of increasing it many-fold, at present up to a factor of about 500.

Parallel processing of engineering computations has been possible in principle for about 20 years, but lack of suitable hardware and software has led to the decade of the 1980s being called the "lost decade for parallel processing". Only in the 1990s has significant progress been made.

It is not the intention of the present paper to dwell on considerations of hardware and of software, for example by comparing shared memory, and distributed memory systems. What is important for geotechnical engineers is that the computations described have been carried out on a variety of computer systems ranging from "supercomputers" to clusters of "workstations". The latter are particularly significant since they fall within the budgets of small- to medium-sized engineering organisations.

The method of calculation described involves explicit message-passing using MPI. This has been shown to be portable and has run successfully on the following systems: Cray T3D, Cray T3E, Cray J932, IBM SP2, Silicon Graphics Origin 2000, Silicon Graphics Workstation Cluster.

All programs and libraries are written in Fortran 90 making full use of the features of that language, Smith (1995).

The paper begins with a brief review of the governing differential equations describing coupled geotechnical problems and their semi-discretisation in space using finite elements. It is assumed that "large" problems will tend to involve three space dimensions. In the time domain, implicit solutions are used exclusively with consequent unconditional stability properties of the algorithms.

The options of direct versus iterative solution of the descretised equations are briefly discussed but the paper considers only the latter option. The solution algorithms compared in the paper are the preconditioned conjugate gradient method (PCG) and the minimum residual method without and with preconditioning (MINRES and PMINRES respectively). These methods are compared in the solution of a variety of two- and three-dimensional problems of interest in geotechnical engineering. Looking to the future the generalised minimum residual method (GMRES) is also discussed and evaluated.

2 THE BIOT EQUATIONS

Assuming for the moment a porous elastic solid whose pores are filled with an incompressible fluid (water), the well known Biot equation is:

$$\frac{K'}{\gamma_w}\left[k_x\frac{\partial^2 u_w}{\partial x^2}+k_y\frac{\partial^2 u_w}{\partial y^2}+k_z\frac{\partial^2 u_w}{\partial z^2}\right]=\frac{\partial u_w}{\partial t}-\frac{\partial p}{\partial t} \tag{1}$$

where K' is the effective elastic bulk modulus of the solid, u_w is the excess water pressure, p the mean total stress, γ_w the unit weight of water and k_x, k_y, k_z the D'Arcy permeabilities in the three coordinate directions.

Equilibrium in the absence of body forces dictates that:

$$\begin{aligned}&\frac{\partial \sigma_x'}{\partial x}+\frac{\partial \tau_{xy}}{\partial y}+\frac{\partial \tau_{zx}}{\partial z}+\frac{\partial u_w}{\partial x}=0\\&\frac{\partial \tau_{xy}}{\partial x}+\frac{\partial \tau_{yz}}{\partial z}+\frac{\partial \sigma_y'}{\partial y}+\frac{\partial u_w}{\partial y}=0\end{aligned} \tag{2}$$

$$\frac{\partial \tau_{zx}}{\partial x} + \frac{\partial \tau_{yz}}{\partial y} + \frac{\partial \sigma_z'}{\partial z} + \frac{\partial u_w}{\partial z} = 0$$

where σ_x' etc are the effective stresses ($\sigma_x - u_w$) for example.

The elastic constitutive law is

$$\underline{\sigma}' = \underline{D}\,\underline{\varepsilon} \tag{3}$$

where $\underline{D}$ is the conventional elasticity matrix expressed in terms of effective Young's modulus E' and Poisson's ratio ν' (Smith and Griffiths, 1998). The stress and strain vectors, $\underline{\sigma}'$ and $\underline{\varepsilon}$ are $\{\sigma_x'\sigma_y'\sigma_z'\tau_{xy}\tau_{yz}\tau_{zx}\}^T$ and $\{\varepsilon_x\,\varepsilon_y\,\varepsilon_z\,\gamma_{xy}\,\gamma_{yz}\,\gamma_{zx}\}^T$ respectively.

Continuity dictates that:

$$\frac{\partial q_x}{\partial x} + \frac{\partial q_y}{\partial y} + \frac{\partial q_z}{\partial z} = -\frac{d}{dt}\left(\frac{\partial u}{\partial x} + \frac{\partial v}{\partial y} + \frac{\partial w}{\partial z}\right) \tag{4}$$

where u,v,w are the displacements in the coordinate directions and q_x,q_y,q_z the volumetric flow rates per unit volume.

These flow rates are related to the excess porewater pressure by:

$$\begin{Bmatrix} q_x \\ q_y \\ q_z \end{Bmatrix} = \frac{1}{\gamma_w}\begin{bmatrix} k_x & 0 & 0 \\ 0 & k_y & 0 \\ 0 & 0 & k_z \end{bmatrix}\begin{Bmatrix} \partial u_w / \partial x \\ \partial u_w / \partial y \\ \partial u_w / \partial z \end{Bmatrix} \tag{5}$$

leading to equation (4) being written as:

$$\frac{k_x}{\gamma_w}\frac{\partial^2 u_w}{\partial x^2} + \frac{k_y}{\gamma_w}\frac{\partial^2 u_w}{\partial y^2} + \frac{k_z}{\gamma_w}\frac{\partial^2 u_w}{\partial z^2} + \frac{d}{dt}\left(\frac{\partial u}{\partial x} + \frac{\partial v}{\partial y} + \frac{\partial w}{\partial z}\right) = 0 \tag{6}$$

In a conventional "displacement" formulation σ' and ε are eliminated in terms of u,v and w leading to the correct number of equations in the "primitive" variables u,v,w and u_w (four degrees of freedom per node in a finite element discretisation).

The well-known LBB (Ladyzhenskaya, Babuska, Brezzi) condition encourages the use of different orders of interpolation for displacement and pressure variables as follows:

$$\begin{aligned} u &= \underline{N}_s\,\underline{u} \\ v &= \underline{N}_s\,\underline{v} \\ w &= \underline{N}_s\,\underline{w} \\ u_w &= \underline{N}_w\,\underline{u}_w \end{aligned} \tag{7}$$

In this paper, $\underline{N}_s$ are tri-quadratic over 20-node cuboids or "bricks" while $\underline{N}_w$ are tri-linear over the 8 corner nodes of the bricks.

After semi-discretisation in space the elemental matrix equations to be solved are:

$$\underline{KM}\,\underline{r} + \underline{C}\,\underline{u}_w = \underline{f} \qquad (8)$$

$$\underline{C}^T \frac{d\underline{r}}{dt} - \underline{KP}\,\underline{u}_w = \underline{0}$$

where $\underline{r}$ collects together all the displacements u,v,w.

Here $\underline{KM}$ is the conventional "element stiffness matrix" and $\underline{KP}$ the conventional Laplacian "element fluid matrix". For poro-elastic solids, $\underline{KM}$ contains E' and ν' and for incompressible laminar porefluids $\underline{KP}$ contains k_x, k_y, k_z. Matrix $\underline{C}$ is a function of the interpolating polynomials only and does not contain any material parameters.

Note that $\underline{KM}$ can in principle be any material stiffness matrix, symmetrical or unsymmetrical, linear or nonlinear. Also in $\underline{KP}$, k_x, k_y, k_z can be nonlinear.

3 DISCRETISATION IN TIME

Equations (8) are coupled algebraic/differential equations conventionally discretised by differencing the second equation. In this paper only implicit discretisations are considered. For example "fully implicit" differencing leads to the recurrence:

$$\begin{bmatrix} \underline{KM} & \underline{C} \\ \underline{C}^T & -\Delta t\underline{KP} \end{bmatrix} \begin{Bmatrix} \underline{r}_1 \\ \underline{u}_{w1} \end{Bmatrix} = \begin{bmatrix} \underline{0} & \underline{0} \\ \underline{C}^T & \underline{0} \end{bmatrix} \begin{Bmatrix} \underline{r}_0 \\ \underline{u}_{w_0} \end{Bmatrix} + \begin{Bmatrix} \underline{f} \\ \underline{0} \end{Bmatrix} \qquad (9)$$

where subscripts "1" and "0" refer to successive time levels separated by the interval Δt.

For nonlinear problems, the incremental statement of equation (9) is

$$\begin{bmatrix} \underline{KM} & \underline{C} \\ \underline{C}^T & -\Delta t\underline{KP} \end{bmatrix} \begin{Bmatrix} \Delta\underline{r} \\ \Delta\underline{u}_w \end{Bmatrix} = \begin{Bmatrix} \Delta\underline{f} \\ \Delta t\,\underline{KP}\,\underline{u}_w \end{Bmatrix} \qquad (10)$$

4 DIRECT VERSUS ITERATIVE SOLUTION

Equations (9) and (10) are linearised simultaneous algebraic equations applicable to all elements in a finite element mesh. "Large" problems might arbitrarily be considered to be those for which the total number of unknowns in the mesh exceeds 100,000. Traditional methods of solution of (9) and (10) over a whole mesh have involved the assembly of all the element equations into a "global" system of equations. For a typical mesh involving 100,000 unknowns, the half-band-width of the resulting assembled global system matrix might be about 6000. This matrix would occupy 0.6 Gwords or roughly 5Gb of memory - at the limit of the largest currently available computers.

A forward-looking strategy is therefore unlikely to consider assembling banded or skyline matrices, necessary in a direct solution method due to "fill-in" during the factorisation stage. It is also the case that "direct" methods are not particularly suited to vectorisation or parallelisation.

As an alternative, iterative methods have been widely researched in recent years. The algorithms for all of these follow a similar pattern, typified by the conjugate gradient method shown below.

To solve $\underline{A}\,\underline{x} = \underline{b}$ first calculate $\underline{p}^0 = \underline{r}^0 = \underline{b} - \underline{A}\,\underline{x}^0$ where $\underline{r}^0$ is the "residual" or error for a first trial $\underline{x}^0$, and then conduct k steps of the process

$$\underline{u}^k = \underline{A}\underline{p}^k\ ,\ \alpha^k = \frac{\left(\underline{r}^k\right)^T \underline{r}^k}{\left(\underline{p}^k\right)^T \underline{u}^k}\ ,\ \underline{x}^{k+1} = \underline{x}^k + \alpha^k \underline{p}^k\ ,\ \underline{r}^{k+1} = \underline{r}^k - \alpha^k \underline{u}^k\ ,$$

$$\beta^k = \frac{\left(\underline{r}^{k+1}\right)^T \underline{r}^{k+1}}{\left(\underline{r}^k\right)^T \underline{r}^k}\ ,\ \underline{p}^{k+1} = \underline{r}^{k+1} + \beta^k \underline{r}^k \tag{11}$$

until the difference between $\underline{x}^{k+1}$ and $\underline{x}^k$ is "sufficiently" small. In the above, $\underline{u}$,$\underline{p}$ and $\underline{r}$ are vectors of length NEQ, the number of equations to be solved, while α and β are scalars.

The main part of the solution process can be seen to consist of a matrix-vector multiplication

$$\underline{u}^k = \underline{A}\,\underline{p}^k \tag{12}$$

followed by a series of operations on vectors. Equation (12) can be solved in two main ways, each of which conserves storage. Either $\underline{A}$ can be assembled in sparse form, with pointers to its non-zero entries, or the operation called for by equation (12) can be carried out "element-by-element" in which case $\underline{A}$ is never assembled at all, and only the individual element matrices on the left hand side of equations (9) and (10) are involved in the matrix-vector multiplication. For the kind of elements used in this paper, these element matrices are of size 68 x 68.

Whichever of the two ways of solution described above is implemented, storage requirements will be an order of magnitude less than for the assembly method, say 500 Mb for a 100,000 equation problem.

In the results which follow, the "element-by-element" option has been used, for two reasons. Firstly the matrix-vector multiplication vectorises very well and secondly there is no "domain decomposition" to be done (partitioning of $\underline{A}$) but rather a "domain composition" in which individual elements, or blocks of element, are built up for the calculation of equation (12). This process is relatively easy to parallelise as shown below.

It is however an open question as to which way of calculating equation (12) would be optimal in any given vector/parallel environment.

5 ITERATIVE ALGORITHMS

The conjugate gradient method, including preconditioning, is well known to engineers but a considerable volume of literature has been developed over the past 10 years or so on alternative algorithms, in particular MINRES, BiCGSTAB and GMRES - all potentially with preconditioning. Texts by Kelley (1995) and Greenbaum (1997) give details.

Strictly speaking, CG should only be used when matrix $\underline{A}$ is symmetric and positive definite. Inspection of equations (9) and (10) shows than an $\underline{A}$ matrix assembled from these element submatrices would not possess this property due to the negative terms associated with $\underline{KP}$. However, preconditioning to form a PCG algorithm is known to be very effective. The simplest kind of preconditioning involves the inverse of the diagonal of the $\underline{A}$ matrix and this yields a preconditioned $\underline{A}$ which is positive definite.

It is one of the aims of the present paper to investigate how well this diagonally preconditioned method works for practical coupled problems in geotechnical engineering.

All successful methods minimise the error during the iterative process - $\underline{r}$ in equations (11) - in some sense or other. The CG (PCG) method minimises the "A-norm" of the error, given by

$$\| e \|_A = \{e . \underline{A} . e\}^{1/2} \tag{13}$$

and so the absolute error may never actually be known. The MINRES (or PMINRES) process, recommended for symmetric but indefinite systems such as equations (9) and (10) minimises the "Euclidean norm" or "2-norm" of the error given by

$$\| e \|_2 = \left\{ \{\sum_n |e_i|^2\}^{1/2} \right\} \tag{14}$$

and therefore more truly generates a solution with a minimised error directly related to the input tolerance.

The MINRES algorithm (after Kelley, 1995) is listed below in Fortran 90 for the solution of

$$\underline{A}\,\underline{xMR} = \underline{f} \tag{15}$$

for a starting guess of $\underline{x0}$.

```
!           initialisation phase
        cjiters = 0
        v = .0; v_hat = f - matmul (A, x0); beta = norm (v_hat)
        c = 1. ; c_old = 1. ;  s = .0;   s_old = .0
        w = .0; w_old = w; eta = beta
        xMR = x0; norm_rMR = beta    ; norm_r0 = beta
             do
                   cjiters = cjiters + 1 ; converged = norm_rMR/norm_r0 < cjtol
```

```
                if(cjiters==cjits.or. converged) exit
!          Lanczos phase
                 v_old =v
                 v = v_hat/beta; Av = matmul (A , v); alpha = dot_product (v,Av)
                 v_hat = Av – alpha*v –beta*v_old
                 beta_old = beta ; beta = norm (v_hat)
!       QR factorisation
                 c_oold = c_old; c_old = c; s_oold = s_old; s_old = s
                 r1_hat  = c_old*alpha – c_oold*s_old*beta_old                  (16)
                 r1       =sqrt (r1_hat**2 + beta**2)
                 r2       =s_old*alpha+c_oold*c_old*beta_old
                 r3       =s_oold*beta_old
!       Givens rotation
                 c1 = r1_hat/r1
                 s  = beta/r1
!       Update phase
                 w_oold = w_old   ; w_old = w
                 w = (v – r3*w_oold-r2*w_old) /r1
                 xMR = xMR + c*eta*w ; norm_rMR = norm_rMR*abs(s)
                 eta = -s*eta
        end do
        if (abs (c) > epsilon (small) )  then
         xGAL=xMR-s*eta*w/c
         norm_rGAL = norm_rMR/abs(c)
        end if
```

Computational costs and memory requirements are considered in more detail below but in essence CG and MINRES are very comparable in the amount of work they involve i.e. a matrix-vector multiplication followed by vector operations. In the author's implementation MINRES uses somewhat more memory for vectors – 12 @ NEQ-long vectors compared with CG's 9.

For some geotechnical engineering applications, KM in equations (9) and (10) is not symmetrical – for example when the constitutive D matrix arises from non-associated elastoplasticity. In such cases, as explained by Greenbaum (1997) for example, error minimisation is much more complicated than it was for symmetric systems. It can be shown that in exact arithmetic the CG and MINRES algorithms lead to an optimal approximation in order (NEQ) operations in addition to the matrix-vector multiplication with no additional storage of intermediate approximation vectors. In practice, see below, results with sufficient accuracy for engineering purposes are attained in far less than NEQ iterations.

For non-symmetric systems the options are either to perform extra work and use extra storage to find the optimal approximation as in the GMRES (generalised minimal residual) algorithm or to settle for a non-optimal approximation. Methods of the latter type include BiCG, CGS, QMR, BiCGSTAB, restarted GMRES, hybrid GMRES and several others. Of these BiCGSTAB is possibly the most promising but the author has not coded it so far.

Full GMRES has been coded and the implications of using it for geotechnical engineering problems are considered later.

6 ALGORITHM COSTS IN TIME AND MEMORY

The computational cost of diagonal preconditioning in the conjugate gradient method is negligible and so CG and PCG have essentially the same cost in time and memory. Similarly, MINRES costs essentially the same in time, but the memory resources are somewhat increased.

Preconditioning in the minimum residual method (PMINRES) does however incur significant extra costs in computer time and in memory : 17 @ NEQ-long vectors compared to 12 in MINRES and 9 in PCG in the author's implementations. It also becomes necessary to store some information from previous iterations but this is not too serious a limitation.

The extra computational costs associated with PMINRES arise from the need to precondition the updated norm. This involves two extra matrix-vector multiplications (Kelley, 1995) and so preconditioning would need to lead to significantly reduced iteration counts to be competitive.

For unsymmetric systems, GMRES actually involves less storage for the element matrices and the associated NEQ-long vectors than PCG. However, significant extra storage is involved during the iteration process as an NEQ-long vector is accumulated with each iteration. Since it can be difficult to find good preconditioners for nonsymmetric systems, large iteration counts tend to arise for "large" problems, say several thousand iterations. This is obviously significant, involving 1.6 Gb for 2000 iterations and NEQ of 100,000.

7 STRATEGY FOR PARALLEL COMPUTATIONS

Shared memory systems (SGI Origin for example) may attempt to parallelise automatically and HPF is an attempt at portable parallelisation which the author has investigated (Smith and Pettipher, 1997). However the main technique considered in this paper involves explicit message-passing using MPI. The essential features are described by Smith and Wang (1998) in the context of elastoplastic solid mechanics, and the same techniques carry over directly to the PCG and MINRES algorithms described above for coupled problems. GMRES presents extra difficulties associated with the vectors accumulating during iterations and has not been successfully parallelised yet by this author. Algorithms like BiCGSTAB may well be more easily parallelisable which will tend to counteract their slower convergence rates compared to GMRES.

Table 1 shows the parallel performance of these types of algorithm on a Cray T3D. Points to notice are that something like 25% of the peak performance of such a system was achieved in the intensively computational parts of the code and that a speedup of a factor of 40 was achieved in going from 8 to 512 processors. Even if the 8-processor version performed at single processor speed, a very pessimistic assumption, a minimum speed up of about 50 was achieved.

		Serial EL98	8	16	32	64	128	256	512
Total	Wall	3584	493	257	141	80	52	50.1	52.0
Bcast.	Wall		8.1	8.3	9.1	10.0	11.7	18.1	24.6
Local g	Wall		6.4	2.6	1.9	1.7	3.4	11.4	15.8
Load inc.	Wall		478	246	130	68	37	21	11.9
	Spdup		1.0	1.9	3.7	7.0	12.9	22.8	40.2
	(max)			(2)	(4)	(8)	(16)	(32)	(64)
PCG	Wall		459	236	125	66	36	20	11.5
Solver:	Mflp/s		504	979	1852	3505	6446	11550	20110
	% Peak		42%	41%	39%	37%	34%	30%	26%
Peak	Mflp/s	133	1200	2400	4800	9600	19200	38400	76800

Table 1. Parallelisation statistics, Cray T3D, MPI

8 COMPARISON RESULTS

Although problems with complex geometries have been solved (eg Smith, 1996) attention is concentrated here on "benchmark" problems in two- and three-dimensions. The two-dimensional case involves a 10x10 mesh and only about 1000 unknowns. In the "ideal" case the elements are all square but the effects of refining the mesh by creating thinner horizontal layers near the draining surface is investigated. The system of equations becomes less well conditioned as the surface elements are elongated, giving a measure of the effect of this ill-conditioning on the iteration counts.

The three-dimensional case involves a cube of 10x10x10 elements and about 15,000 equations. The top surface of the cube is drained and the other faces undrained. A vertical surface load is applied to the four elements at one corner of the mesh, thus representing a square footing on a 20x20x20 grid.

Results are shown in Table 2. From these the following can be concluded:

Timestep	**1**	**2**	**3**	**4**	**5**	**6**	**7**	**8**	**9**	**10**	**11**	**12**	**13**	**14**	**15**	**16**	**17**	**18**	**19**	**20**	**21**
CG	421	427	409	380	374	381	370	376	374	369	367	357	357	306	299	299	299	306	295	289	292
PCG	117	109	113	26	104	109	106	53	111	90	62	12	47	68	52	42	62	46	55	58	26
MINRES	454	461	460	465	464	467	466	467	470	467	462	449	457	455	455	454	454	452	454	452	452
PMINRES	409	406	417	412	409	404	404	400	400	408	405	351	358	346	332	324	326	307	289	304	304
PCG(3-d)	788	709	582	289	618	479	575	571	730	699	236	348	285	288	128	266	199	295	332	104	193

Table 2. Iterations to convergence for various methods on 2-d and 3-d benchmark problems

1. Preconditioning (diagonal in this case) is very effective in the CG method for Biot-type problems. In this case the improvement in run time was about 5.
2. MINRES has comparable performance to CG (although it is true that it can be difficult to compare relative accuracies).
3. Preconditioning MINRES does not lead to comparable savings to those achieved for CG. However this test problem may not be critical and PMINRES stands as an alternative to PCG if necessary in future.
4. As has been found for 3-d solid elasto-plastic problems (Smith and Wang, 1998), PCG gains in effectiveness as problem sizes increase, especially in 3-d. The iteration counts for the 3-d benchmark problem with 15,000 unknowns were only 6 times those for the 2-d problem with less than 1000 unknowns. The 3-d problem was being solved with an average of about 400 PCG iterations per timestep. Although this is greater (perhaps by a factor of about 2) than the average iteration count in a comparable uncoupled elastoplastic problem, the efficiency of parallelism means that solutions of this kind are perfectly practicable.

In all of the above results (as in elasto-plasticity) it pays to use the solution at the end of a timestep as the initial guess for the next timestep.

The effects of element aspect ratio (equation ill-conditioning) were evaluated in the 2-d analysis. As the aspect ratios of the surface elements increased from 1 through 2 to 10 the PCG iterations to convergence increased in the proportions 70/81/112.

It can be seen that ill-conditioning in this case increases the number of iterations to convergence but not dramatically so.

REFERENCES

KELLEY C.T. (1995). Iterative Methods for Linear and Nonlinear Equations.

GREENBAUM A. (1997). Iterative Methods for Solving Linear Systems. SIAM Press, Philadelphia.

SMITH I.M. (1995). Programming in Fortran 90. John Wiley, Chichester.

SMITH I.M. (1996). Computation of Large Scale Viscoplastic Flows of Frictional Materials in 'Dynamics of Complex Fluids' eds M.J. Adams, R.A. Mashelker and A.R. Rennie, Royal Society.

SMITH I.M. & M.A. PETTIPHER (1997). The Development of an MPP Implementation of a Suite of Finite Element Codes. Proc. HPC Conference, Vienna. In "Lecture Notes in Computer Science" No. 1225, pp 400-409, Springer.

SMITH I.M. & A. WANG (1998). Analyses of Piled Rafts, IJNAMG, 22, 7.

SMITH I.M. & D.V. GRIFFITHS (1998). Programming the Finite Element Method, 3rd ed., John Wiley, Chichester.

1. INTERPRETATION OF FIELD MEASUREMENTS AND OF SMALL SCALE TESTS

THE CALCULATIONS ON HAARAJOKI TEST EMBANKMENT WITH THE FINITE ELEMENT PROGRAM PLAXIS 6.31

A. Aalto, R. Rekonen and M. Lojander
Helsinki University of Technology, Espoo, Finland

ABSTRACT: The Finnish National Road Administration has organised an international competition to calculate the behaviour of the Haarajoki test embankment. This paper concentrates on calculations with the finite element program PLAXIS. The evaluation of the calculation parameters is discussed. The main interest is put on preconsolidation pressure and the Poisson's ratio. The subsoil is slightly overconsoldiated and therefore the exact value of the preconsolidation pressure is very important. The choice of Poisson's ratio has a great effect on the stress paths and displacements. The results of the calculations are compared with the measurements of the test embankment.

INTRODUCTION

The test embankment was instrumented and constructed during July and August 1997 (FinnRa, 1997). The measurements of the settlement plates, piezometers and inclinometers are also available in the internet (www.tieh.fi/pailas/).

Several oedometer tests (Incremental Loading and Constant Rate of Strain) and triaxial tests (Consolidation test, undrained and drained shear) were carried out in the laboratory of Soil mechanics and Foundation Engineering at the Helsinki University of Technology.

Calculations by several different methods were made before the construction of the embankment (Näätänen et al, 1997). After the first measurements some modifications of the parameters have been done. The modification of the permeability and the elastic parameter (kappa) is discussed in the article of Näätänen et al (1998).

The profile of the preconsolidation pressure of Haarajoki is typical for the Finnish clay layers. The upper clay layer, dry crust, is heavily overconsolidated. Under the dry crust the pre-consolidation pressure is diminishing towards the minimum value. In deeper layers the pre-consolidation pressure profile is similar to the overburden pressure profile.

The value of the Poisson´s ratio is commonly determined from the results of drained triaxial tests. In the calculation program PLAXIS it is necessary to use the real elastic value of the Poisson´s ratio. The effect of the change of the value of the Poisson´s ratio for one layer (2...5m) from 0.1 to 0.35 is shown in the article.

2. SOIL PROPERTIES

The Haarajoki embankment is 2.9 meters high, 100 meters long and the width at the top is 8 meters. The geometry of the test embankment can be shown in the paper by Näätänen & al 1998. The fill material is gravel with a density of 21 kN/m^3. The embankment is lying on layers of overconsolidated clay and silt with an altogether depth of 20 meters. The ground water table is at the surface and the excess pore pressures are between –3 and 10 kPa have been measured in this area.

The subsoil is divided in to 9 geotechnical layers for the calculations. The layers are shown in Table 1. On the top there is a 2 meters thick layer of dry crust and under that a 20 meters thick layer of fat clay (the content of clay size particles > 50%). The water content of the fat clay varies between 67 to 112 % depending on the depth. The bulk density varies from 14 to 16 kN/m^3 and the specific gravity varies from 2.73-2.79 g/cm^3. The undrained (undisturbed) shear strength is between 15 and 42 kPa and was determined by a fall cone test and field vane test. The initial water content in the fat clay layer is almost the same as the liquid limit or above it. The organic content is between 1.4 and 2.2 % in the depth of 3-13 meters. The classification properties of the ground are shown in figure 1.

Table 1 shows the calculation parameters used in PLAXIS. These parameters are based on oedometer tests and triaxial tests. The preconsolidation pressure (σ'_p) and the in-situ effective vertical stress (σ'_{vo}) is shown in figure 2. The calculation of σ'_{vo} is based on the bulk density values, the measured ground water level and the excess pore pressure values. After the first calculations the permeabilities and preconsolidation pressures have been re-evaluated. The first permeability values were calculated directly from the oedometer consolidation coefficient values, but after some corrections they were multiplied by 5 and the new values are shown in table 1.

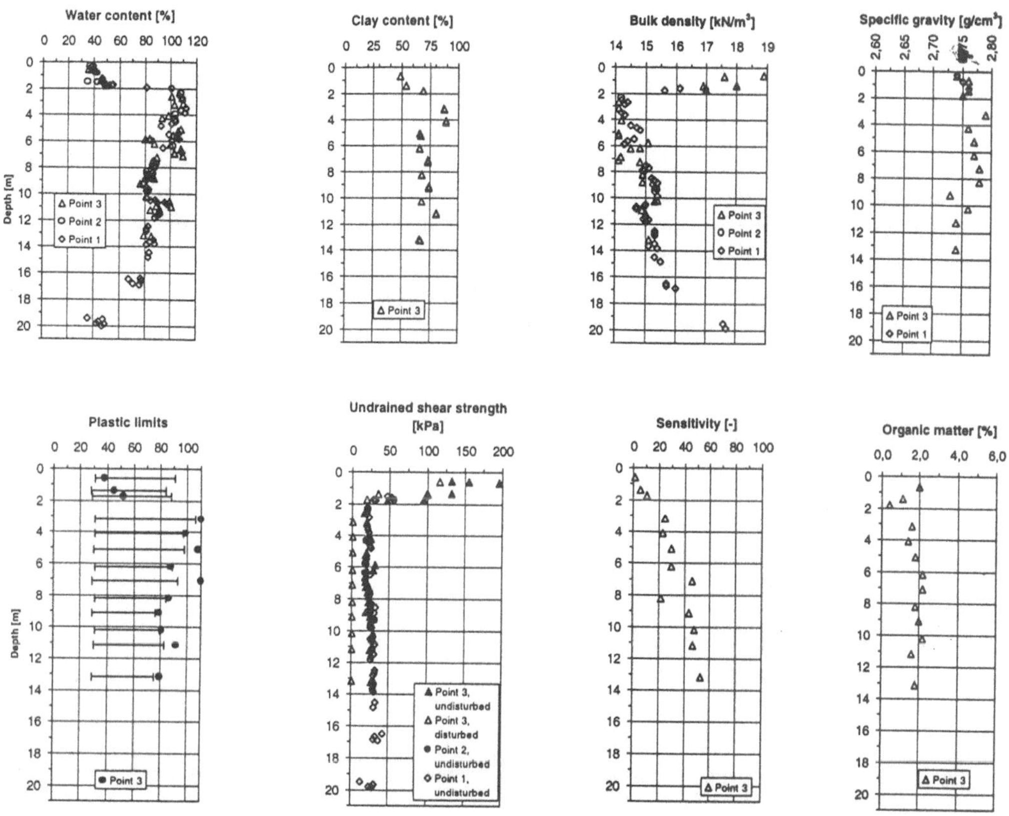

Figure 1. The Haarajoki test embankment. The classification properties.

layer no	Depth m	γ kN/m^3	M	ϕ' °	ν	κ^* $\kappa/(1+e_o)$	λ^* $\lambda/(1+e_o)$	k_x m/d	k_y m/d	e_0	$K_0^{OC}= K_0^{NC}\times OCR^{\sin\phi}$	OCR/POP
1	0-1	17	1.50	36.9	0.35	0.0029	0.042	6.5×10^{-3}	6.5×10^{-3}	1.4	1.0	/ 76.5
2	1-2	17	1.50	36.9	0.35	0.0054	0.042	6.5×10^{-3}	6.5×10^{-3}	1.4	1.0	/ 60.0
3a	2-3	14	1.15	28.8	0.18	0.0072	0.277	7.8×10^{-4}	6.5×10^{-3}	2.9	0.7	/ 41.4
3b	3-4	14	1.15	28.8	0.18	0.0072	0.277	7.8×10^{-4}	6.5×10^{-3}	2.9	0.7	/ 30.0
3c	4-5	14	1.15	28.8	0.18	0.0072	0.277	7.8×10^{-4}	6.5×10^{-3}	2.9	0.7	/ 18.7
4	5-7	14	1.10	27.7	0.10	0.0074	0.489	6.0×10^{-4}	4.3×10^{-4}	2.8	0.7	1.45 /
5	7-10	15	1.07	27.0	0.10	0.0064	0.197	6.9×10^{-4}	3.5×10^{-4}	2.3	0.7	1.54 /
6	10-12	15	1.07	27.0	0.28	0.0116	0.384	1.3×10^{-3}	6.5×10^{-4}	2.2	0.7	1.64 /
7	12-15	15	1.15	28.8	0.28	0.0084	0.341	1.3×10^{-3}	6.5×10^{-4}	2.2	0.7	1.66 /
8	15-18	16	1.50	36.9	0.28	0.0167	0.160	4.0×10^{-3}	5.6×10^{-4}	2.0	0.55	/ 33.5
9	18-22,2	17	1.50	36.9	0.28	0.0038	0.042	4.0×10^{-3}	8.0×10^{-3}	1.4	0.45	1.0 /

Table 1. The Haarajoki test embankment. Material Parameters.

3. CALCULATIONS

3.1 CALCULATIONS WITH THE FINITE ELEMENT PROGRAM PLAXIS 6.31

The geometry is modelled by using a finite element mesh with 780 6-noded triangular elements (each triangular element contains three stress points). The shape of the test embankment was assumed symmetric and only half of it was included in the model. The problem was handled in a plane-strain state.

The consolidation analysis is based on the Biot's theory where only the primary consolidation is allowed as well as Darcy's law for fluid flow.

The material model used for the clay layers was the isotropically hardening elasto-plastic Modified Cam-Clay Model (MCCM). In this model the yield surface (f = 0) represents an ellipse in the p'–q –plane (equation 1) (PLAXIS, 1995).

$$f = \frac{q^2}{M^2} + p'(p' - p_c) \tag{1}$$

q	deviatoric stress
p'	mean effective stress
p_c	preconsolidation stress
M	slope of critical state line / shape factor for ellipse

In order to obtain the correct shear strength parameter M should be based on the friction angle ϕ' according the equation 2.

$$\sin \phi' = \frac{3M}{6 + M} \tag{2}$$

In PLAXIS the overconsolidated material can be modelled by using either the Over-Consolidation Ratio (OCR) or the Pre-Overburden Pressure (POP) (equations 3 a and 3 b).

$$OCR = \frac{\sigma'_p}{\sigma'_{v0}} \tag{3a}$$

$$POP = \sigma'_p - \sigma'_{v0} \tag{3b}$$

σ'_p	vertical preconsolidation stress
σ'_{vo}	in-situ effective vertical stress

The stress state of the clay layers below the Haarajoki test embankment is shown in figure 2. The other curve describes the in-situ vertical stress (σ'_{vo}) and the other vertical preconsolidation stress (σ'_p) based on oedometer tests.

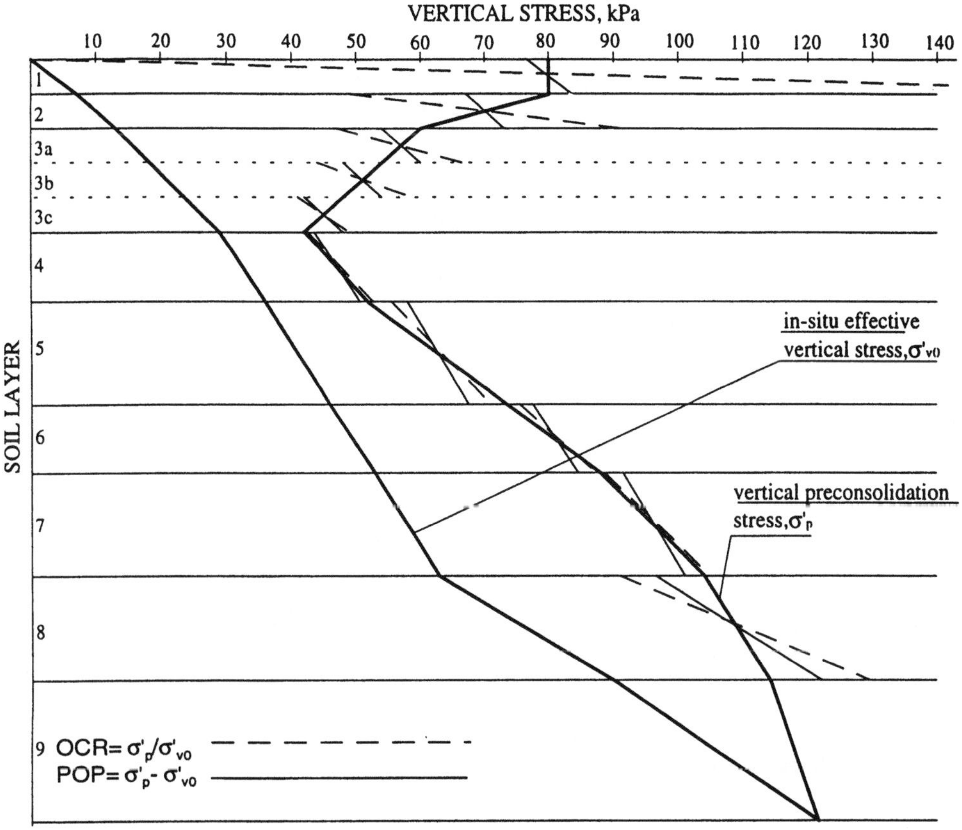

Figure 2. The Haarajoki test embankment. The In-situ vertical stress (σ'_{vo}) and vertical preconsolidation stress (σ'_p). The use of OCR / POP in PLAXIS.

The thinner lines (solid and dashed) in the preconsolidation stress-curve represent the vertical preconsolidation stress state in the PLAXIS calculations (OCR / POP). It can be seen from the three upper soil layers that there are difficulties to describe real preconsolidation stress state by using the POP or the OCR when σ'_p decreases with depth. The use of OCR / POP requires the curves of σ'_{vo}- and σ'_p to be parallel.

To get as close as possible to the real value of σ'_p either OCR or POP has been chosen at every layer. Note the 3rd layer has been divided into three different parts.

3.2 THE EFFECT OF THE POISSON'S RATIO

In the Modified Cam-Clay Model of PLAXIS the Poisson's ratio (ν) is a purely elastic parameter. The Poisson's ratios for the different layers were chosen from the reload curves of the triaxial tests. The Poisson's ratio of the 3rd layer (ν=0,18) is presented in figure 3.

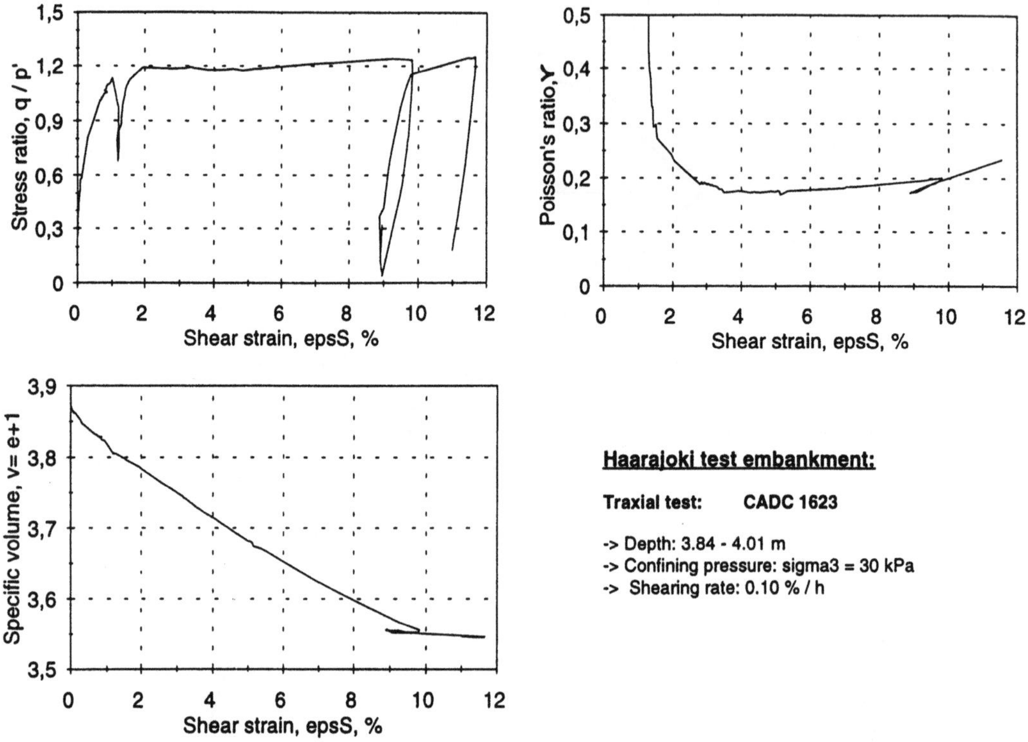

Figure 3. The Haarajoki test embankment. The choice of the Poisson's ratio from the reload curve of the triaxial test CADC 1623.

The effect of the Poisson's ratio has been studied in the third layer (depth: 2-5m) where the relation between the total load after construction of the embankment ($\sigma'_{vo}+\Delta\sigma'_{emb}$) and the vertical preconsolidation stress (σ'_p) is quite remarkable.

The calculated settlements after fifty months, that is four years and two months are, presented in figure 4 a. The Poisson's ratio of 0.35 (Trialem 7) gives nearly six times larger vertical displacements than value of 0,1 (Trialem 5) in the 3rd layer. Also the mobilized horizontal stresses are about two times higher in the case where the Poisson's ratio is 0.35 and its effects on the stress path are shown in figure 4 b (mean effective stress, p' is increasing).

The stress path caused by the Poisson's ratio ν=0,18 (Trialem 6, used in final calculations) is shown in figure 5 a. Figure 5 b shows the stress paths in the 5th layer for the Trialem 5 and the Trialem 7 calculations.

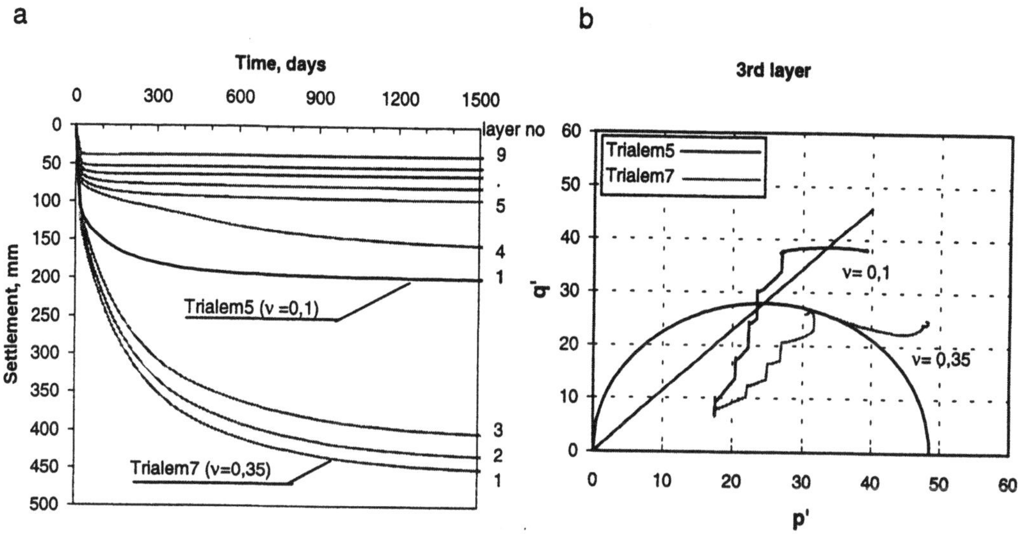

Figure 4. The Haarajoki test embankment. a) The calculated settlements and b) stress paths when using the Poisson's ratio of 0.1 and 0.35 in the 3rd soil layer (Trialem5 and Trialem7).

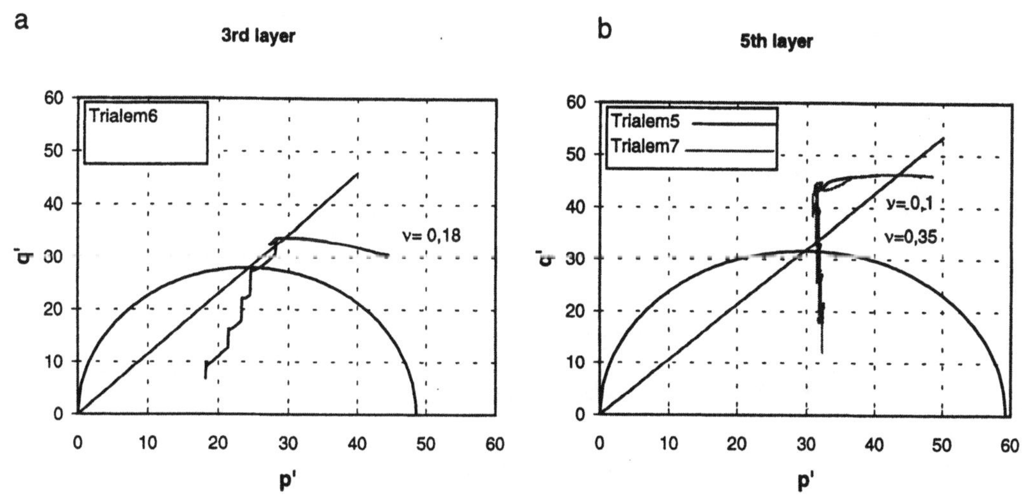

Figure 5. The Haarajoki test embankment. Stress paths a) in the 3rd layer (ν= 0.18) and b) in the 5th layer when the Poisson's ratio of ν= 0.1 and ν= 0.35 in the 3rd layer has used.

3.3 RESULTS

The construction of a 2.9-meter high test embankment was done in six layers and the real construction schedule was taken into account. The graphical presentation of the calculated settlements at the centreline of the embankment during the first 5 years are shown in figure 6. The measured values after 9 months are also presented in the same figure.

The total pore pressure (calculated and measured) in the four different depths at the centreline are presented in figure 7.

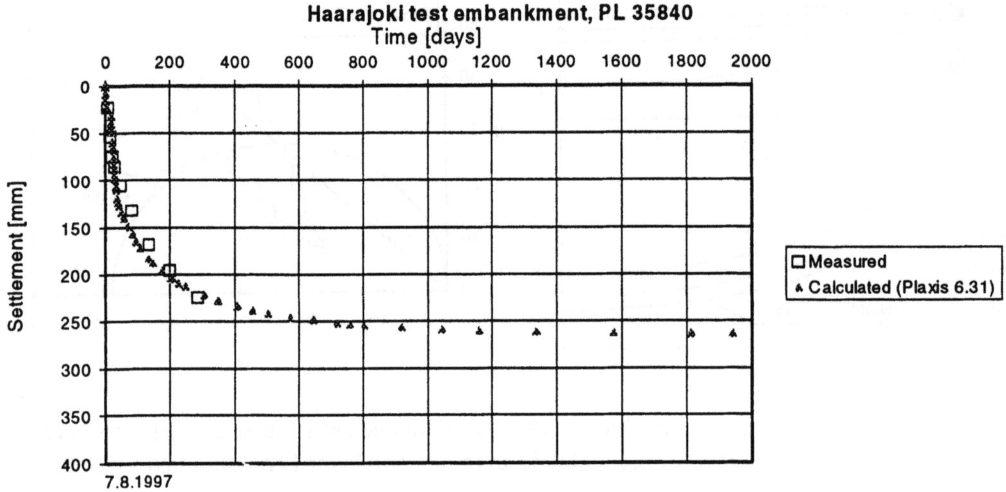

Figure 6. The Haarajoki test embankment. The calculated and measured settlements.

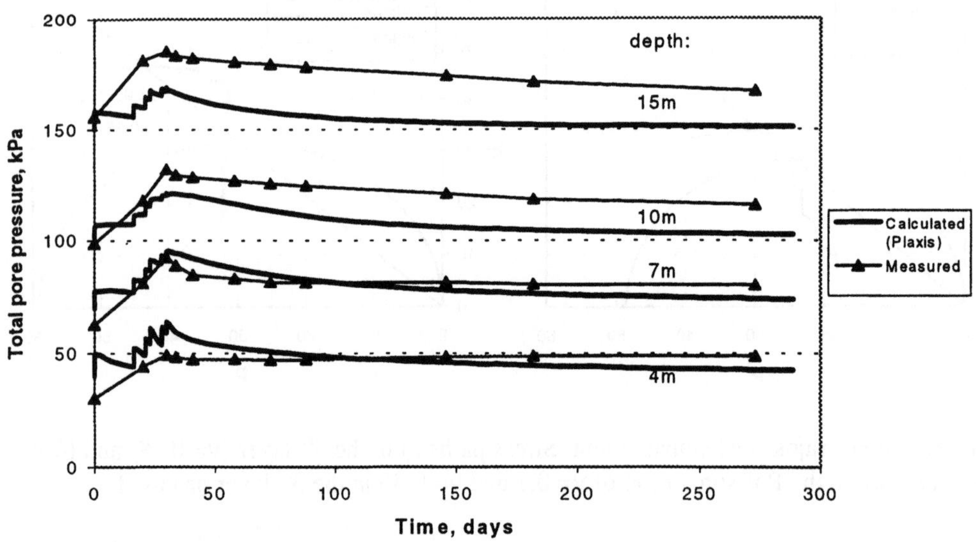

Figure 7. The Haarajoki test embankment. The calculated and measured total pore pressure in depths 4, 7, 10 and 15 m at centreline.

The horizontal movements of the soil layers are measured by inclinometers at 9 and 4 meters from the cross sectional centreline of the embankment. The calculated and measured values after 9 months are compared in figure 8.

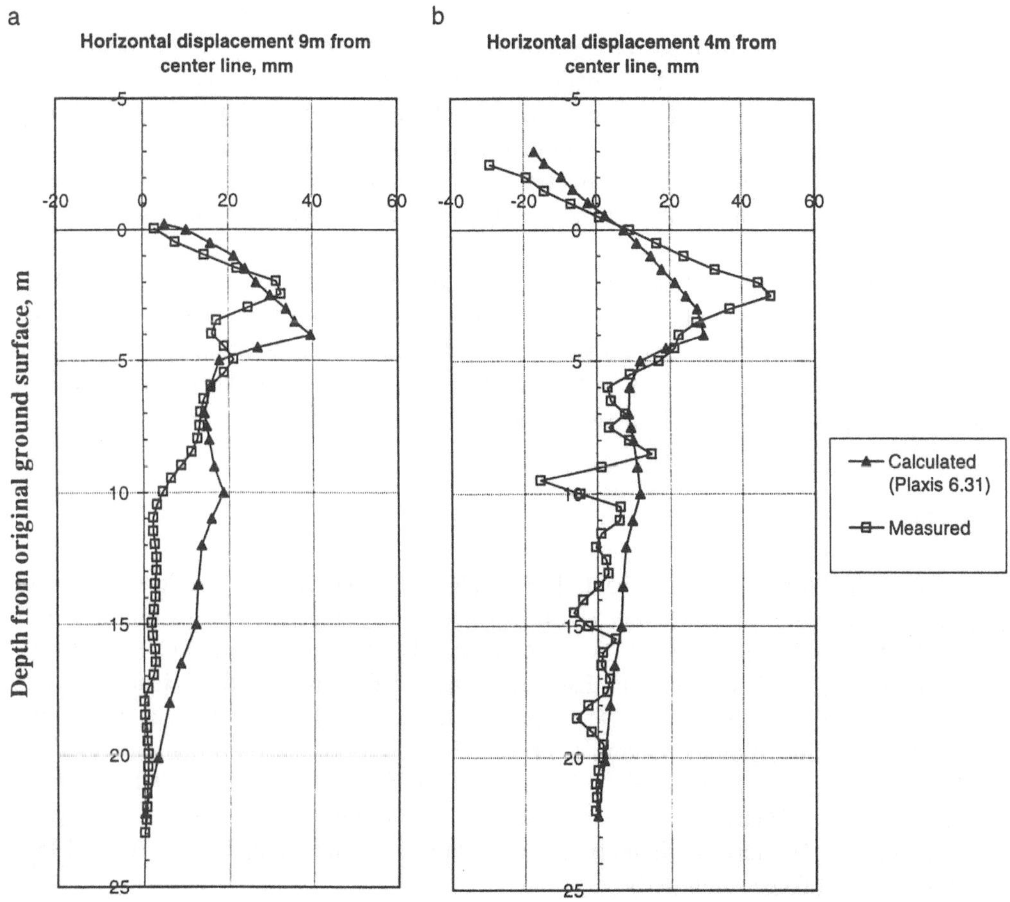

Figure 8. The Haarajoki test embankment. The calculated and measured horizontal movements of the soil layers at a) 9 and b) 4 meters from the centreline of the embankment.

4. CONCLUSIONS

The curves of calculated and measured settlements at centreline during 9 months coincide together very well. However the shape of the measured curve seems to be steeper than calculated curve and the real settlements could probable be continued over calculated values. The same estimation can be done from the values of the pore pressures. The predicted values of excess pore pressure at different depths are quite close to zero whereas primary consolidation process is still going on according to measured pore pressure.

The measured total pore pressures just after construction exceed 11-17 kPa higher than calculated values in depth of 10-15 m. One reason for this is in-situ initial excess pore pressures 3-10 kPa, which can not be modelled in PLAXIS. And when comparing the observed and the calculated settlements of the embankment it should be taken into consideration that it is not possible to model the secondary consolidation with PLAXIS.

Of course all parameters needed in material models play an important role, but a quite remarkable influence of the Poisson's ratio can be noticed in the Modified Cam-Clay Model of PLAXIS. Other critical parameter when using consolidation analysis is in-situ permeability, which can not really be modelled in laboratory conditions, but better in-situ methods should be found out.

The use of the Over-Consolidation Ratio (OCR) or the Pre-Overburden Pressure (POP) for modelling vertical preconsolidation stress (σ'_p) in overconsolidated materials causes some inaccuracy compared with real in-situ stress state. Because of that it is important to divide the thick layers to thinner sections in the soil layers where the vertical preconsolidation stress-curve (σ'_p) goes toward the in-situ vertical stress-curve (σ'_{vo}).

REFERENCES:

FINNISH NATIONAL ROAD ADMINISTRATION (FinnRA), (1997). Competition to calculate settlements at the Haarajoki test embankment. Competition programme.

NÄÄTÄNEN A., PUUMALAINEN N., SAARELAINEN K., AALTO A., LOJANDER M. & VEPSÄLÄINEN P. (1997). Estimation of settlement of the Haarajoki test embankment. Proceedings of the 6th Finnish Mechanical Days, September 5-6, 1997, Oulu, Finland. Eds. Aalto, J., Salmi, T. pp. 295-310.

NÄÄTÄNEN A., VEPSÄLÄINEN, P. & LOJANDER M. (1998). Finite element calculations on Haarajoki test embankment. Proceedings of the 4th European Conference on Numerical Methods in Geotechnical Engineering, Udine, Italy, 14-16 October 1998. (To be submitted)

PLAXIS Finite Element Code for Soil and Rock Analyses (1995). Version 6. P.A. Vermeer, R.B.J. Brinkgreve (eds.). A.A. Balkema. Rotterdam 1995.

TUNNEL MODELLING IN ANALOGICAL SOIL: COMPARISON OF EXPERIMENTAL AND COMPUTATIONAL RESULTS OBTAINED WITH A FINITE DIFFERENCE PROGRAM

I. Al Abram, D. Dias, P. Mathieu and R. Kastner
U.R.G.C. Géotechnique, Institut National des Sciences Appliquées de Lyon, Villeurbanne, France

ABSTRACT: The aim of this work is to obtain a better simulation of soil movements due to reconstruction of shallow tunnels, dug with pressurised shield tunnelling boring machine, inducing in the soil mass, successive deconfining, reconfining cycles.

The reduced scale experimentation is realised on a two-dimensional model with the analogical Taylor-Schneebeli material consisting of metallic rods of 3 to 5 mm diameter.

The experiments have been analysed with digital image techniques, and the complete displacement and strain fields were obtained within the entire soil mass. The numerical simulation is done with the explicit finite difference program, FLAC V3.30. by ITASCA. Two different constitutive laws for the analogical soil have been tried: a elastic-plastic law (Mohr-Coulomb) and a constitutive law developed by our laboratory using a non-linear, irreversible law before failure (Duncan type) and a Mohr-Coulomb flow rule after failure. These constitutive laws have also been written for a purely 2D material and numerical results and will be compared with classical results obtained with the 3D model using strain plane analysis. The law parameters were determined from biaxial and oedometer tests done with the analogical soil. Experimental and numerical displacement field within the entire mass as well as the settlement of the soil surface are compared.

1. INTRODUCTION:

According to experimental results by Ollier (1997), the measurement of soil movements due to pressurised shield tunnelling (Fig. 1) shows successive deconfining, confining phases of the soil mass. These different cycles are due to overcutting, to the conicity of the boring machine, to grouting and grout consolidation.... For these reasons, we decided to reproduce, with a small scale model, these different phases in order to better understand the influence of a tunnel boring and to measure the induced soil movements.

A deconfining then a confining phases will be studied with the experimental model.

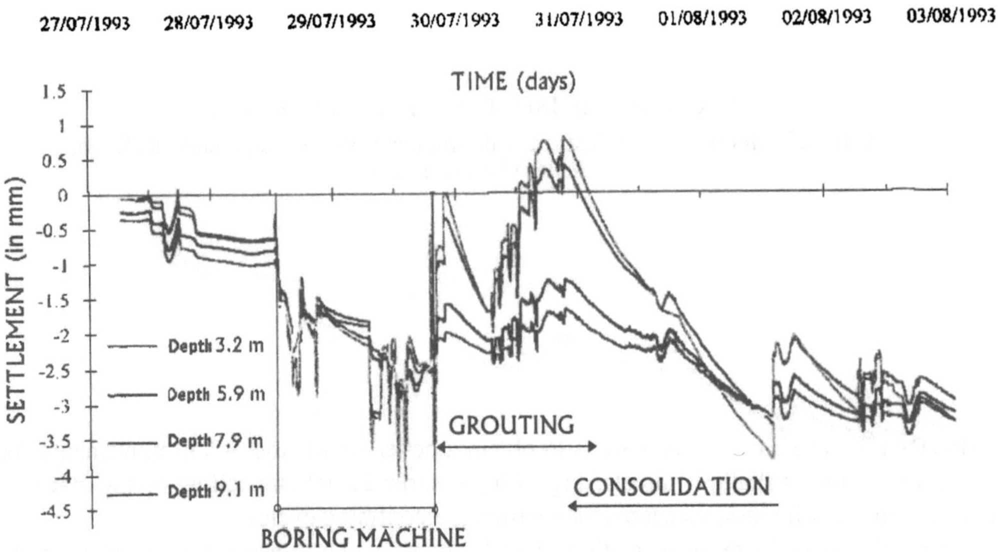

Fig. 1 : Soil movements due to tunnel boring (pressured shield machine), OLLIER (1997)

2. SMALL SCALE EXPERIMENTAL MODEL

2.1 Model description

The experimental model is composed of different parts:

- a rigid frame (2m wide, 1.5m high) made of thick structural steel sections. It is possible to reduce or increase the diameter: minimum diameter: 197 mm, maximum: 212 mm).
- the analogical soil made of a mixture of small steel rods, 60mm long, with 3 different diameters: ϕ=3mm (34%), ϕ=4mm (35%), ϕ=5mm (31%) (percentages are given by weight).

The specific weight of the mixture is: γ=65kN/m^3.

Fig. 2: Small scale experimental model

The displacement field in the soil mass is obtained frame digital pictures using correlation technique (Sifasoft program, developed by Morestin F., LMSO, INSA de Lyon). Strain field, iso-values ... are obtained with Deplac program, developed in our laboratory (Al Abram [1]). Results of several experiments are presented elsewhere (Al Abram [2] et [3]).

2.2 Loading plot:

A decreasing diameter phase (ΔD=-2.9%, percentage of initial diameter) and then a increasing diameter phase (ΔD=+2.9%) are done as shown in Fig. 3.

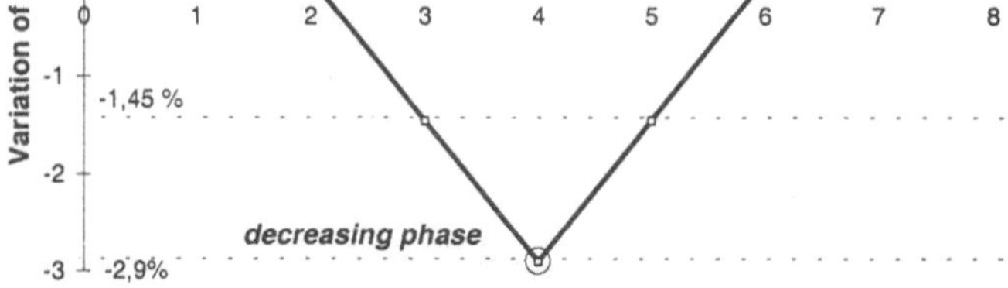

Fig. 3: Loading plot

3. NUMERICAL CALCULATIONS:

3.1 Constitutive laws:

We use two constitutive elastic-plastic laws for our calculations:

- Mohr-Coulomb law (M-C): In this law, before failure, Young's Modulus E, and Poisson's ratio ν are constant within calculation. But we use a different Young's modulus according to the simulated phase and the modulus increases with depth according to the formula: $E = (E_{ch} ou E_{ur}).P_a((K_0\gamma h)/P_a)^{(n_{ch} ou\, n_{ur})}$, with: P_a, atmospheric pressure; E_{ch} and n_{ch} are constants for the increasing phase; E_{ur}, n_{ur} constants for the decreasing phase; K_0 is earth pressure coefficient at rest.
- Mohr-Duncan law (M-D): In this law, before failure, we use the hyperbolic law described by Duncan (1984), taking in account the non-linear, non elastic stress-strain behaviour (with differentiation of primary loading and unloading-reloading behaviour).

In these two laws, the behaviour after failure (Mohr-Coulomb criterion) will be plastic (non associated flow rule).

3.2 Geotechnical parameters values for M-C et M-D constitutive laws

These parameter values have been determined using both biaxial tests done by Kastner in (1982) and oedometer tests done by Al Abram [3]. The different tests were considered as plane strain cases. We just give here final values used in our calculations:

a) M-C law: E_{ch}=560, n_{ch}=0.4; E_{ur}=1450; n_{ur}=0.8; ν=0.35; φ = 21°

b) M-D law: K_{ch}=560, K_{ur}=1450; K_b=600; m=n=0.4; R_f=0.65; φ=22°; Δφ=2°.

The dilatancy angle is: ψ=4° for both laws.

Fig. 4 shows that in an oedometer test, the modulus has a higher value during unloading than during primary loading.

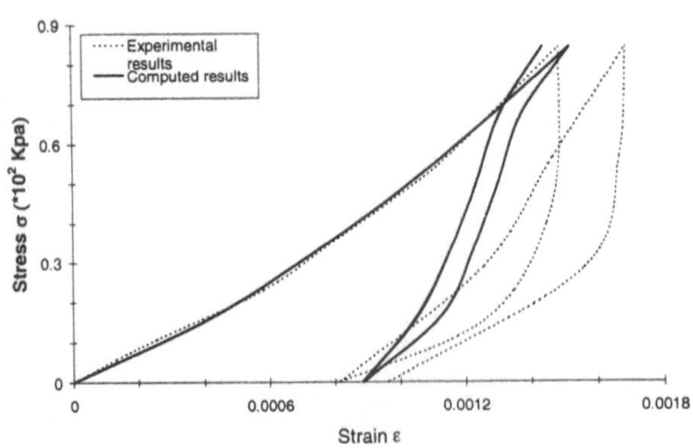

Fig. 4: Oedometer test: experimental results and computed results (M-D law) for two loading-unloading cycle

3.3 Mesh and Boundary conditions:

The tunnel mesh is a ring, 206 mm external, 10 mm thick and purely elastic, (Fig. 5-a).
The mesh geometry is given on Fig. 5-b. The mesh length is bigger than the frame length to take in account the material-frame interaction. On both lateral sides, no (horizontal or vertical) displacements will be allowed.
Initial vertical stresses are equal to: $\sigma_v=\gamma.h$ and horizontal stresses: $\sigma_h=K_0.\gamma.h$, where: h is the depth, $K_0 = 0.64$ and $\gamma=65$ kN/m^3.

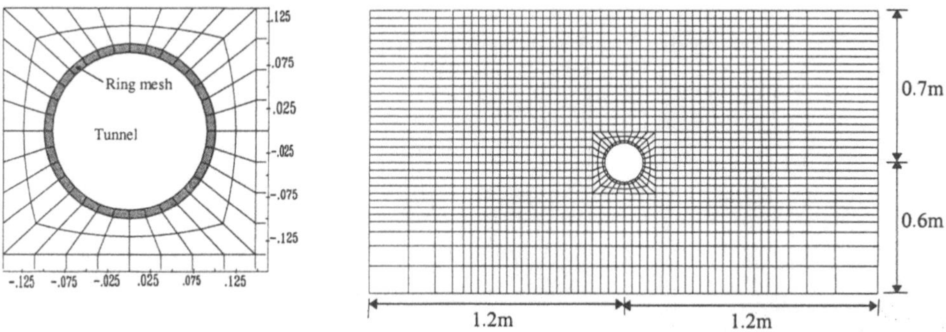

a) Mesh around tunnel

b) the whole Mesh

Fig. 5:Mesh used for computation with FLAC program.

3.4 Loading

The experimental displacements measured at the tunnel-soil contact have been imposed to the ring. The Fig. 6 shows the displacement values corresponding to the end of the decreasing phase (ΔD=-2.9%).

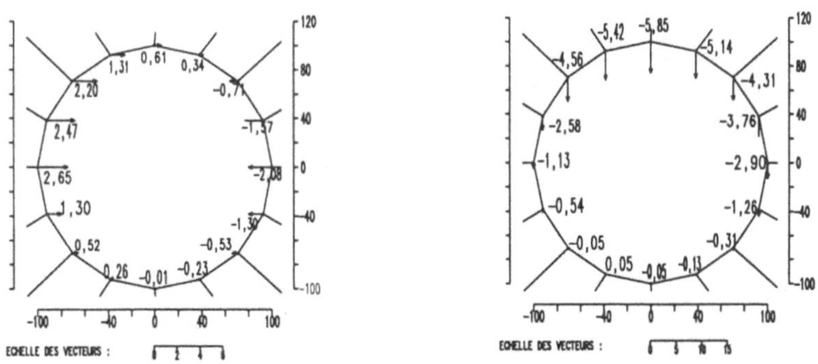

a) horizontal displacement **b) vertical displacement**

Fig. 6: Displacements imposed at the tunnel-soil contact, end of decreasing phase (ΔD=-2.9%)

4. Comparison of experimental results and 3-D plane strain computed results

4.1 End of decreasing phase

We can make the following statements for this phase:

- From tunnel top to the surface, the value of M-D computed vertical displacement varies rapidly close to tunnel and then becomes constant when going to surface. With M-C computed values, this variation is slightly more regular (Fig. 7).

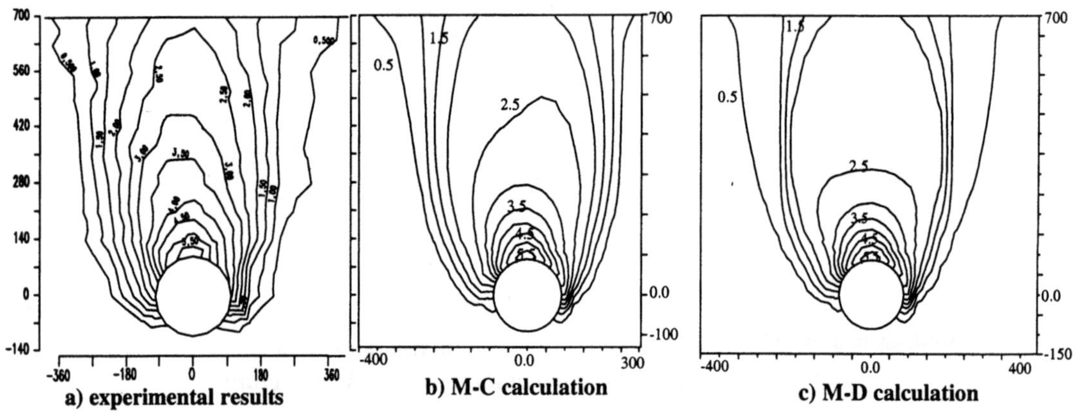

Fig. 7: Experimental and computed displacements in mm, end of decreasing phase

- The M-D computed settlement for an horizontal distance bigger than D (tunnel diameter) has an higher value than M-C calculation (Fig. 7 and Fig. 8). The centre part of the experimental and computed settlement profiles are almost identical.

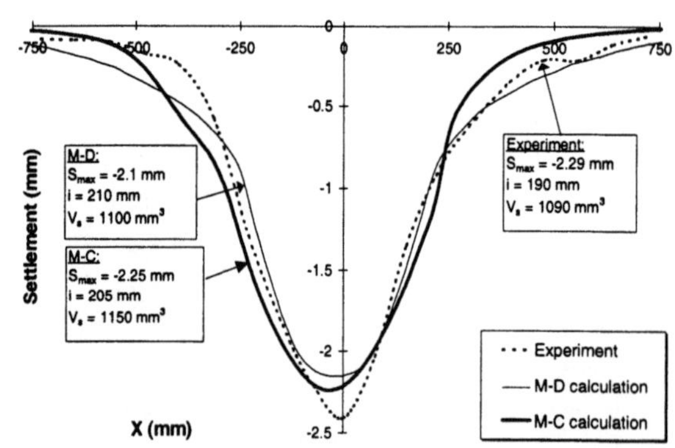

Fig. 8: Settlement profiles, end of decreasing phase, comparison of experimental and M-C and M-D calculations

- The spacing between maximum shear strain iso-values is smaller with M-D calculation than with M-C calculation (Fig. 9).

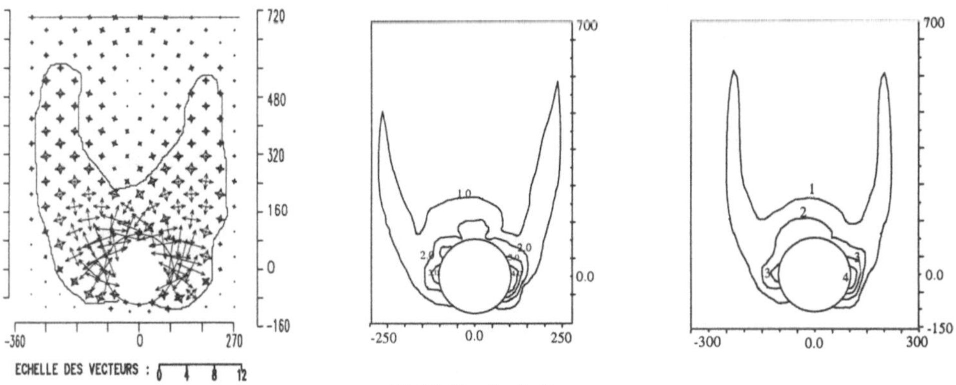

a) Experiment **b) M-C calculation** **c) M-D calculation**

Fig. 9: Maximum shear strain in %, end of decreasing phase experimental, M-C and M-D calculations

- The M-D computed failure zone is more symmetrical and narrower than the M-C computed failure zone (Fig. 10) as we noticed for the maximum shear strain (Fig. 9).

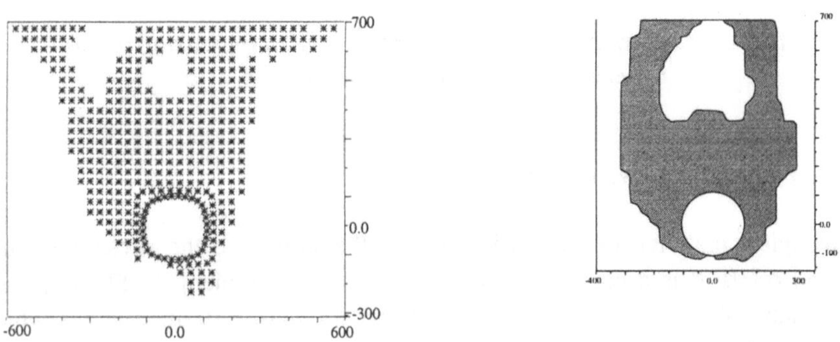

a) Failure zone, M-C calculation **b) Failure zone, M-D calculation**

Fig. 10: Comparison of failure zones computed with the two laws

4.2 End of increasing phase (ΔD=2.9%)

We can observe that the M-C computed vertical displacement field (Fig. 11) as well as the vertical surface displacement profiles (Fig. 12) are broader (and closer than experiment) than the M-D ones.

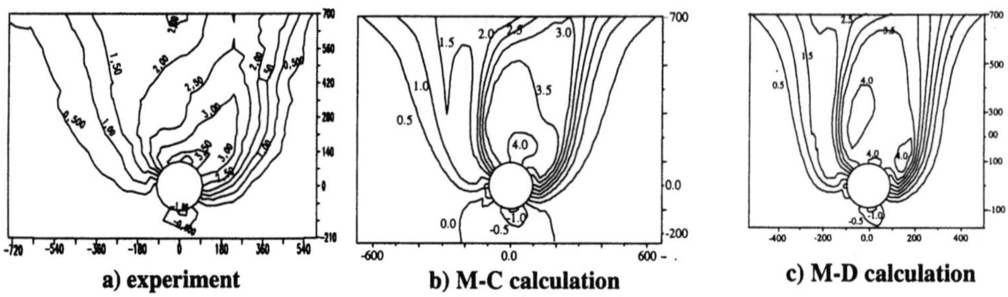

Fig. 11: Displacement computed with both constitutive laws: increasing phase

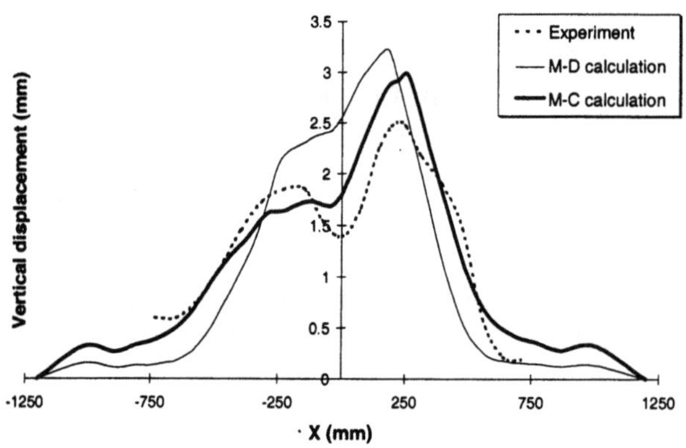

Fig. 12: Vertical surface displacement profile, end of increasing phase Experiment, M-C and M-D calculation

5. Calculation with a true two dimensional 2D-law

In our analogical material, in the third direction, the strain and the stress are equal to zero: $\varepsilon 3=0$ and $\sigma 3=0$. It is a plane strain and a plane stress problem. In FLAC program, strain-stress relationships have been rewritten except for plastic-strain calculus which is already two-dimensional.

In order to find the 2D-law parameter values, we have to consider differently the biaxial and oedometer tests:

1. **with M-C law**, we have: $\nu_{(2D)} = 2\nu_{(3D)}$ et $K_{0(2D)} = \nu_{(2D)}$

   The slope of the deviatoric curve of the biaxial test is equal to the Young's modulus $E_{(2D)}$, (in 3D plane strain, this slope was equal to: $E_{(3D)} / (1-\nu_{3D}^2)$).
2. **with M-D law**: we obtained the following values: K_{ch}= 640, K_{ur}=2000, n=0.4, K_b=750, m=0.4, K_0 = 0.8, the other parameters: φ, $\Delta\varphi$, R_f and ψ are not changed.

The Fig. 13 shows displacement results computed with true 2D laws (end of decreasing phase).

Comparing these results and the previous results obtained in a 3D strain plane case, we can make the following statements:

a) <u>M-C calculation</u>: the vertical displacement iso-values are almost identical in 2D and in 3D. The difference between the 2 settlement profiles (2D and 3D) is less than: 0.1mm. We can explain this with the fact that the strain are mainly plastic.

b) <u>M-D calculation</u>: the vertical displacement fields in 2D or 3D have the same width. We have a better distribution of the iso-values in 2D than in 3D. The maximum settlement computed in 2D is slightly less than the one computed in 3D (and less than the experimental one).

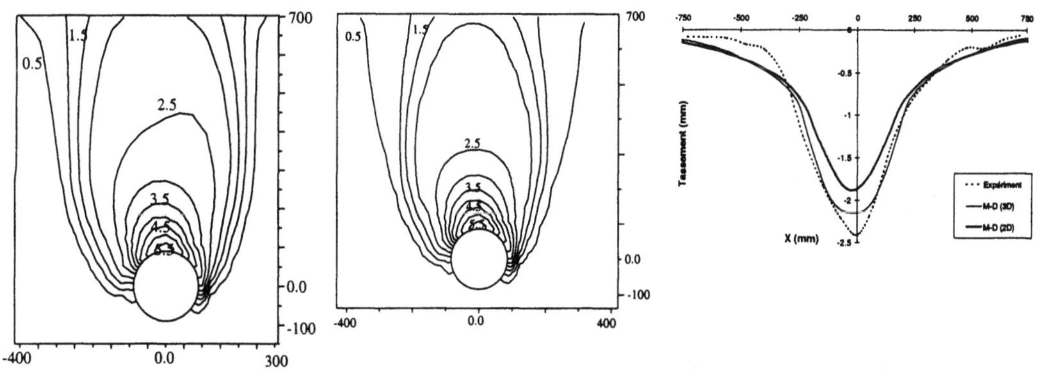

a) M-C calculation (2D) **b) M-D calculation (2D)** **c) Settlement profiles**

Fig. 13: Vertical displacement iso-values and vertical surface displacement profiles computed with M-C and M-D two-dimensional laws (end of decreasing phase: ΔD=-2.9%)

6. Conclusions

Two elasto-plastic constitutive laws (M-C and M-D) written for a three-dimensional plane strain case and for a true two-dimensional case have been used to compute our experiment on analogical soil, simulating soil movements due to tunnel construction:

- It is possible to determine the displacement field in the whole soil mass in such an experiment. The results can be used to validate a computation technique.
- We obtain good agreement comparing experimental and computed results.
- The calculations show that the strain are mainly plastic in the zone where the soil moves. The results are not significantly different using M-C or M-D law, 2D or 3D, because the flow rule formulation is exactly the same in the 4 cases.

- In case of increasing diameter phase, the surface displacement profile is twice wider than in case of decreasing phase. This phase corresponds to the grouting phase. So the width of a too strong grouting phase could be twice bigger than of a settlement phase. Nevertheless, it is difficult to project our two-dimensional experiment results on a real tunnel construction site with the variety of existing soils.
- The biaxial tests we have used, were realised at very high confining pressure (compared to stresses in our experiment) and without unloading measurement. So to improve the rheology of our analogical soil, we are now carrying new tests in our laboratory.

REFERENCES:

AL ABRAM I., MATHIEU Ph., KASTNER R. et al. [1] (1998): Mesures de déplacements par Imagerie Numérique par un modèle réduit simulant le creusement de tunnel. Colloque photomécanique 98, 14, 15 et 16 April à Marne la Vallée, 1998, 65-72.

AL ABRAM I., MATHIEU Ph., KASTNER R. [2] (1998): Déplacements dans un modèle réduit bidimensionnel de creusement de tunnel, comparaison mesures et calculs. Congrès Universitaire de Génie Civil, Les ouvrages dans leur Environnement, Reims: AUGC, -481p.

AL ABRAM I. [3] (1998): Etude sur modèle réduit bidimensionnel du champ de déplacement induit par le creusement d'un tunnel à faible profondeur; interaction avec les ouvrages existants. PhD Thesis, sustained at URGC-Géotechnique of l'INSA de Lyon.

DUNCAN J.M., SEED R.B., WONG K.S. et al. (1984): *FEADAM/* a computer program for Finite Element Analysis of Dams. Research report No. SU/GT/84-03. Department of Civil Engineering Stanford University , 14 p.

KASTNER, R. (1982): Excavations profondes en site urbain. PhD Thesis: INSA-Lyon and Claude Bernard-Lyon 1; 203-208.

OLLIER C. (1997): Etude expérimentale de l'interaction sol-machine lors du creusement d'un tunnel peu profond par tunnelier à pression de boue, PhD Thesis sustained at URGC-Géotechnique of l'INSA de Lyon.

3-D DISCRETE ELEMENT ANALYSIS OF RIGID STRUCTURES RESTING ON NON-LINEAR SOIL

C. Athanasiu
NOTEBY AS Consulting Engineers, Oslo, Norway

T. Guttormsen
SAGA PETROLEUM ASA, Oslo, Norway

ABSTRACT:

A 3-D discrete element method is developed as a simple tool to analyse the contact stresses between rigid structures (such as off-shore subsea structures, cylindrical skirt foundations or suction anchors, caissons and so on) and the surrounding soil as well as soil bearing capacity under applied loads. The structure is divided into a number of small discrete elements, defined by their position, orientation and geometry, rigidly connected at the structure centre. The soil-structure interaction is modelled by non-linear springs. The method is incorporated into a PC, Fortran program - DISEL. The structure geometry, the displacements and the resulting contact stresses at discrete element centres are displayed graphically by a post-processor Visual Basic program. The use of the program in designing and analysing the subsea structures is exemplified with a levelling analysis for a North Sea Template Foundation Structure in order to estimate the required levelling forces.

1 INTRODUCTION

The study of soil-structure interaction and distribution of contact stresses between structural parts and surrounding soil is a very important step in designing the offshore «non-piled» foundations such as skirted foundations, subsea structures, anchors and so on.

This paper presents a simple discrete element analysis program, **DISEL.** It performs 3-D soil structure interaction analysis of rigid structures in contact with soil. The structure is divided into a number of small, rigid elements defined by their position and orientation, and contact area with the soil. The soil - structure contact is modelled by integrated force-displacement relationships (non-linear springs) in contrast to the constitutive, stress-strain relationship used in finite element analysis.

2 METHOD OF SOLUTION

The structure is divided into a number of small discrete elements, all rigidly connected at structure centre. One or more discrete elements can be assembled in a macroelement as shown in Fig.1. The position of a macroelement is specified by the coordinates of its centre, x_o, y_o and z_o. The macroelement centre, O, and two of its local axes, u and w, lay in a vertical plane that makes the angle β with x-axis. The local axis, w, makes an angle θ with horizontal as shown in Fig.1. Other types of macroelements such as rectangular plate,

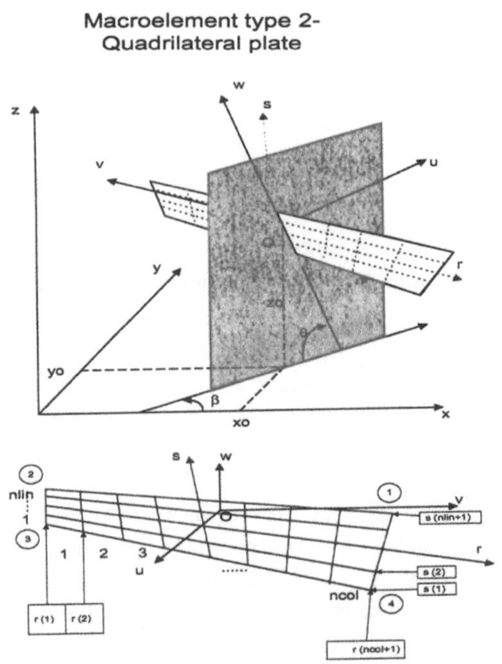

Fig.1 Quadrilateral macroelement

cylindrical surface or circular plate can be specified as shown in Fig.2. The global coordinates x,y,z ,β and θ are automatically generated from the coordinates of macroelements.

The soil reactions on each element are defined by three forces one normal to the element, F_u, and two in the plane of the element, F_v and F_w and three moments M_u, M_v and M_w. The element force vector in the local coordinate system is thus defined by $\{F\}_l^T=\{F_u\ F_v\ F_w\ M_u\ M_v\ M_w\}$. As the discrete elements are small, the element moments are neglected and

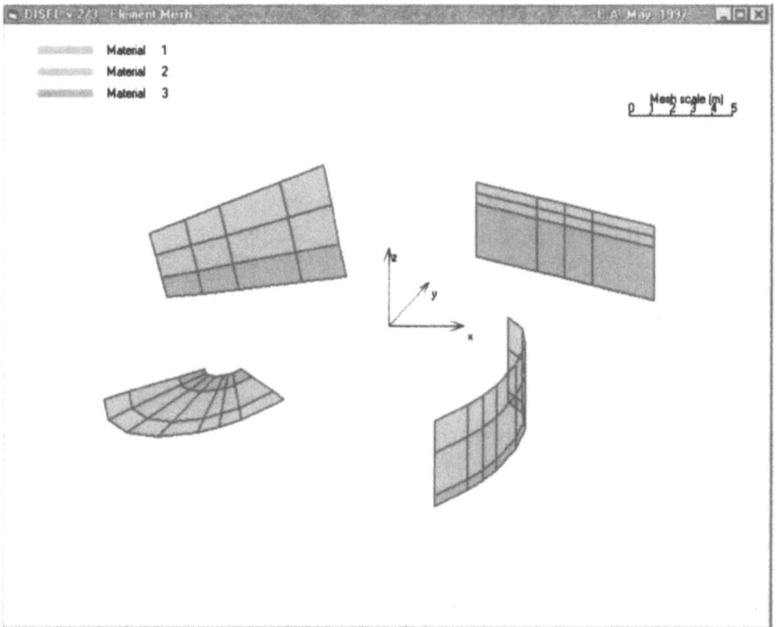

Fig.2. Different types of macroelements.

the contact soil-structural element is modelled by three displacement springs (no rotational springs). The discrete element forces in the global system are expressed in terms of those in local system as:

$$\{F\}_g = [T]\cdot\{F\}_l \tag{1}$$

where $\{F\}_g^T=\{F_x\ F_y\ F_z\ M_x\ M_y\ M_z\}$and [T] is the transformation matrix. The transformation equation can also be written to obtain element displacement vectors in local system from structure displacement vector defined in global coordinate system:

$$\{\delta\}_l = [T]^T\cdot\{u\} \tag{2}$$

The contact between soil and structural discrete element is modelled by non-linear springs. The vector of local element forces can thus be expressed in terms of element local displacements as:

$$\{F\}_l = [K]\cdot\{\delta\}_l \tag{3}$$

where [K] is the element spring stiffness matrix . Substituting $\{\delta\}_l$ from eq.(2) in (3) the local element force vector is obtained in terms of global displacements:

$$\{F\}_l = [K]\cdot[T]^T\cdot\{u\} = [B]\cdot\{u\} \tag{4}$$

where [B] is the force-displacement matrix. Using eq.(2), the global force vector can be expressed in terms of global displacements:

$$\{F\}_g = [T]\cdot[B]\cdot\{u\} \tag{5}$$

The structure equilibrium requires that the sum of all element forces equals the applied forces on the structure:

$$\sum\{F\}_g = \left(\sum[T]\cdot[K]\cdot[T]^T\right)\cdot\{u\} = [K_s]\cdot\{u\} \tag{6}$$

where $[K_s]$ is the stiffness matrix of the entire structure. As there is only one rigid structure, the eq.(6) represents a system of 6 linear equations, which is solved for structure displacements, {u}. The soil non-linearity is accounted for by a variable secant modulus technique in which the stiffness is gradually adjusted to match the specified stress-displacement curves until convergence is obtained. The element displacements can then be computed from eq.(2) and element contact forces from eq.(4). The element forces are tranformed in global system and the equilibrium is checked by summing up the element forces in global system.

3 STRESS-DISPLACEMENT RELATIONSHIPS

The stiffness of contact springs is calculated from the stress-displacement relationships for normal and shear directions to the element. These stress-displacement curves can either be read as input data or automatically generated by the program.

The automatic generation is based on the assumption that the stress-displacement stiffness of a rigid element (foundation) in contact with soil varies with the displacement in the same way as the average secant shear modulus along a possible failure surface varies with strain.

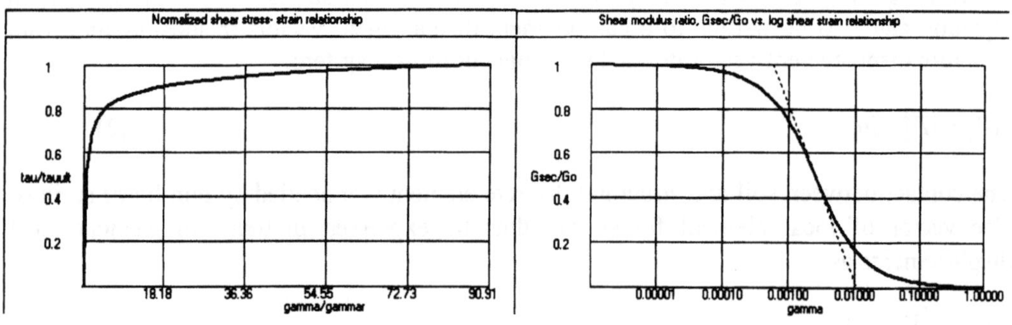

Fig.3. Undrained shear stress-strain relationship for N.C. clays

The procedure is described for undrained behaviour of N.C. clays but the principles are also valid for drained behaviour of friction materials. The stress - strain curve (Fig.3) is expressed as the shear secant modulus ratio degradation with strain:

$$\frac{G_{sec}}{G_o} = 1 - c_1 \cdot a\tan(e^{c_2 \cdot \log(\frac{\gamma}{\gamma_i})}) \qquad (7)$$

where : G_{sec} and G_o are the secant and initial shear moduli respectively; c_1 and c_2 are constants related to the slope, β_o, at inflection point on G_{sec}/G_o vs. logγ curve (Fig.3) and to G_{ult}/G_o :

$$c_1 = \frac{1 - \frac{G_{ult}}{G_o}}{a\tan(e^{c_2 \cdot \log(\frac{\gamma_{ult}}{\gamma_i})})} \quad \text{and} \quad c_2 = -\frac{2 \cdot \beta_o}{c_1} \qquad (8)$$

γ_r is the reference shear strain ($\gamma_r = s_u/G_o$); γ_i is the inflection point shear strain; and γ_{ult} is the ultimate (failure) shear strain. The same relation is assumed for stress-displacement curves (the strain, γ, is replaced by displacement , δ, and the shear moduli by the stress-displacement stiffnesses, K_o and K_{sec}).This will be illustrated for horizontal and vertical plates as follows.

Horizontal plates transfer shear in the horizontal plane. The ultimate shear is ,$\tau_{ult} = s_{u\,DSS}$ at the depth of the element. The stress - displacement curve has an initial stiffness defined as (Wolf, 1997):

$$K_{sho} = \frac{8 \cdot G_o \cdot \sqrt{\frac{A}{\pi}}}{(2 - \nu)} \qquad (9)$$

where A is the area of superelement and ν is the Poisson's ratio of the soil. The reference displacement is calculated from initial stiffness:

$$\delta_r = \frac{s_{uDSS}}{K_{sho}} \qquad (10)$$

Assuming that the ultimate secant stiffness ratio $K_{shult}/K_{sho} = G_{ult}/G_o$,

$$\delta_{ult} = \frac{s_{uDSS}}{K_{sho} \cdot \frac{G_{ult}}{G_o}} = \delta_r \cdot \frac{G_o}{G_{ult}} \qquad (11)$$

For normal stress reactions on horizontal plates the initial stiffness of the superelement is calculated as:

$$K_{no} = \frac{4 \cdot G_o \cdot \sqrt{\frac{A}{\pi}}}{(1-\nu)} \qquad (12)$$

The ultimate normal stress is equal to the bearing capacity of the plate: $\sigma_{ult} = N_c * s_{uDSS} + q$.

Vertical plates can transfer both shear and normal stresses. The shear stress - displacement curves are defined in the same way as for horizontal plates. In normal direction, an outward horizontal movement of the plate will mobilise soil reaction from in situ «at rest» horizontal stresses to passive earth pressure. The ultimate normal stress is thus $\sigma_{ult} = (1\text{-}k_o)\cdot p_o' + \kappa \cdot s_{uP}$. A horizontal movement from the soil (negative) will reduce the horizontal stress from in situ «at rest» horizontal stresses to active earth pressure. The ultimate normal stress for negative movements is thus $\sigma_{ult} = \kappa \cdot s_{uA} - (1\text{-}ko) \cdot po'$. The initial stiffness has the same expression as for horizontal plates in normal direction.

4 LEVELLING ANALYSIS OF A TEMPLATE FOUNDATION

The foundation soil beneath a North Sea Template bottom frame can locally be washed away during drilling operations. The washing out may cause uneven settlements of the structure with the risk of loading the subsea tree flowline connector. In order to immediately secure the situation uplift forces must be applied by attaching buoayancy elements to the structure lifting trunnions in a levelling operation. The estimation of the required levelling forces by a soil-structure interaction analysis using the program **DISEL** is illustrated in this section.

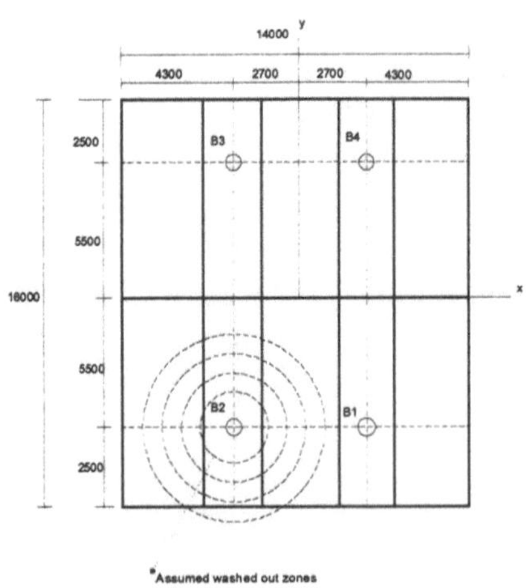

Fig.4. The plan of bottom frame

The structure is modelled by a series of stiff discrete elements rigidly connected to the centre of the structure. The elements within the assumed wash out zone are assigned material "0" (i.e. the elements are desactivated and there are no soil reactions on these elements). The plan of bottom frame structure with the four wells, B1 to B4, showing the assumed "washed out" zone is presented in Fig. 4. It consists of 6 skirt walls parallel with y-axis and of 3 skirt walls paralell to x-axis as shown in Fig.5. The skirt walls are corrugated (corrugation factor is 1.363) and stiffened with a 10 mm thick and 500 mm high steel band on both sides (corrugation factor for bands is 2.0). The bands are placed between elevations –1.0 m and –1.5 m. The skirt walls have different lengths varying between 1.5m and 2.4m. There are 6 skirt compartments with fixed mud mats, with a foundation area of 25.6 m^2 each (Fig.6).

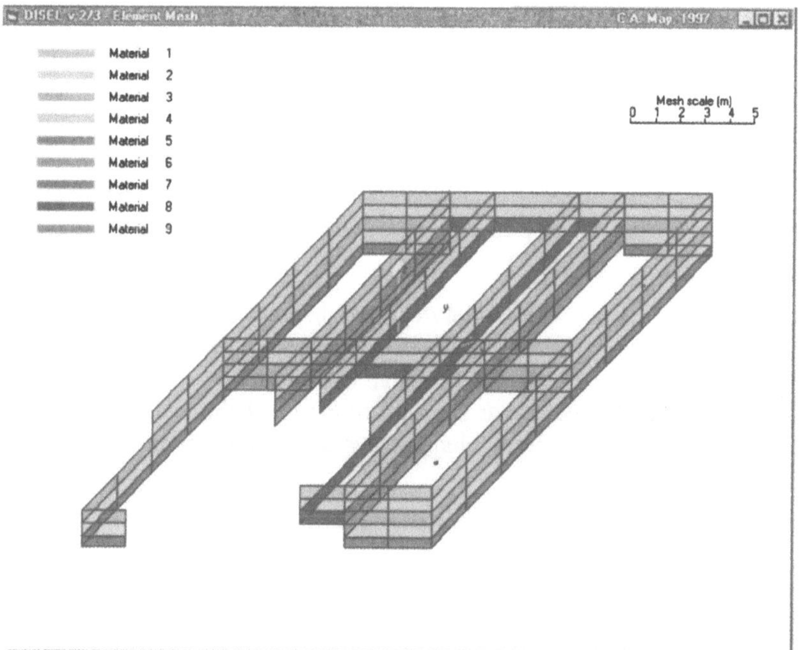

Fig.5. Vertical skirt walls

The soil conditions at the site consist of 0.4m sand layer. Underneath this layer there is a soft clay layer extending down to 12m depth. The undrained, active (compression) shear strength, s_{uA}, is represented by "soft soil" profile, Fig. 7. "Stiff soil" profile is also considered in a parametric study of the required uplift forces. Direct shear strength, s_{uDSS}, and passive shear strength, s_{uP}, are taken as $0.8 \cdot s_{uA}$ and $0.75 \cdot s_{uA}$ respectively. The shear strength along vertical plane is taken as $s_{uDSS} \cdot k_o$ with the coefficient of earth pressure at rest, $k_o = 0.6$ in assumption "A" and $k_o = 1$ in assumption "B". It is assumed that the relative displacement soil / structural element required to fully mobilize the soil shear strength is about 12 mm, corresponding to approximately 0.5 % shear strain. The same displacement is

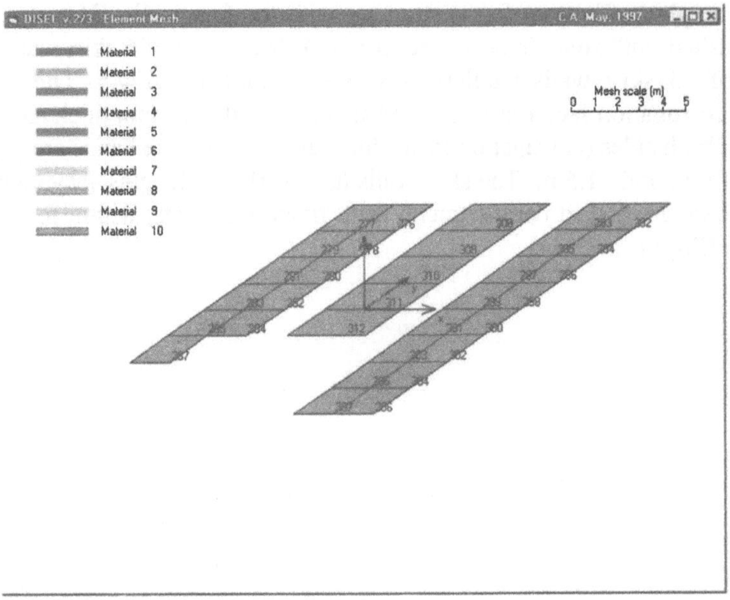

Fig.6 Horizontal mud mat discrete elements

assumed for mobilization of active and passive earth pressure on vertical skirt elements. It is also assumed that a residual shear strength of 80 % of the peak value is reached when relative displacement soil / structural element is greater than 300 mm.

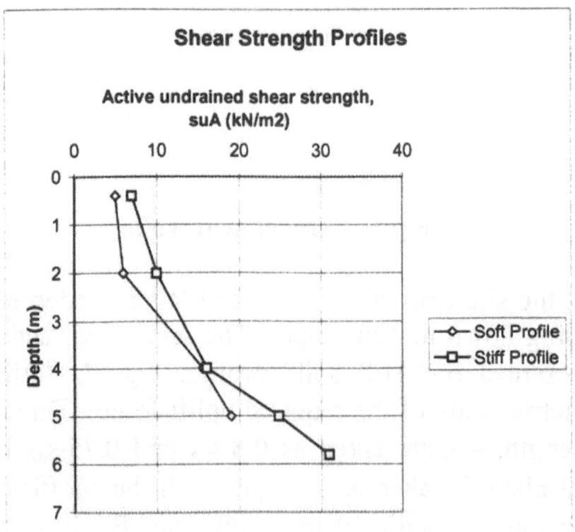

Fig.7 Shear strength profile

The required vertical displacement to reach vertical capacity on horisontal elements is taken equal to the displacement required to reach active/passive earth pressure. The stress-displacement relation for the soil-structure contact depends on element depth, corrugation factor and on the ability of the structure to mobilize soil response in different directions. Different stress-displacement relations are considered as different "materials" in which the structure elements are embeded. The different material numbers assigned to different discrete elements are shown in Figs. 6 and 7.

The levelling analysis is a typical loading/unloading problem, which requires modelling of hysteretical behaviour upon reversal of the load. Consequently, the "as-is" forces were applied in a first run and the corresponding displacements and mobilized soil reactions obtained. Upon the reversal of the load, the upward shear stresses mobilized during the "as-is" loading will be reduced and for elements with large displacements downward shear stresses will be mobilized up to shear strength as shown skematically in Fig.8.

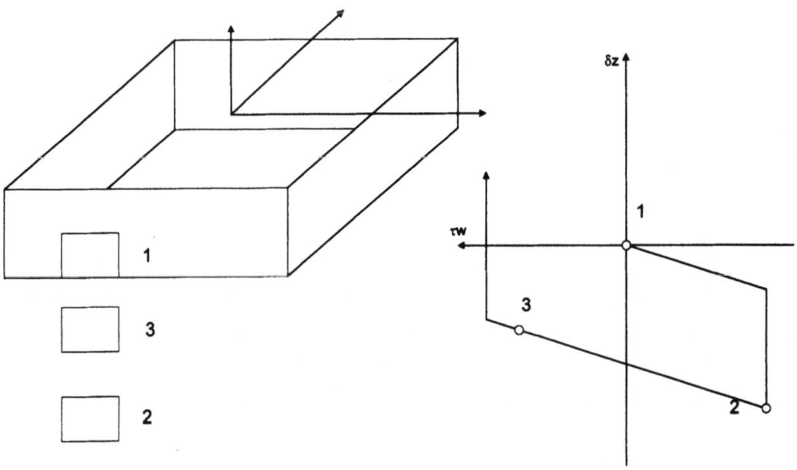

Fig.8. Mobilization of shear stresses upon reversal of the load.

Based on this approach the equivalent shear strength for a second run is taken as the sum of the existing shear stresses plus the shear strength in the opposite direction. A second run is then performed where uplift forces (vertical forces at the centre of the "washed out " side, i.e. 8m excentricity about x-axis) are applied. The origin of the stress-displacement relation is moved in point 2, Fig.8, and the upward movement of the bottom frame at well heads calculated directly from the displacements obtained in the second run. Fig.9 shows the upward movements at the well heads for different applied uplift forces.

The calculated required force and moment to obtain an upward target movement of 24 mm at well head B2 are Fz=2000kN and Mx=16000kNm, respectively.

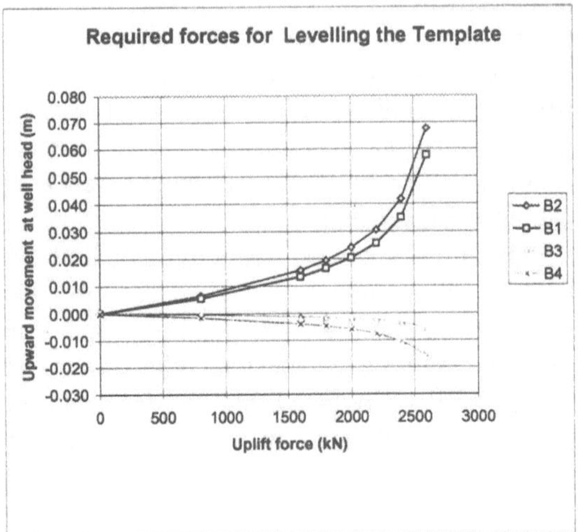

Fig.9. The relation between applied uplift forces and resulting upward displacements

5 CONCLUSIONS

The discrete element analysis of soil-structure interaction incorporated into the program DISEL is a versatile tool for the design of rigid structures in contact with soil. It combines the relative simple, user friendly, input with 3-dimensional model of the structure enabling the designer to make parametric studies and to estimate the structure displacements and capacity. As any other analysis tool, it can not, however, substitute the engineering judgement and the results depend on the ability and experience of the user.

6 AKNOWLEDGEMENTS

The permission from SAGA PETROLEUM ASA – Norway to illustrate the use of DISEL with the analysis of a North Sea Template Structure, is greatefully aknowledged.

7 REFERENCES

Wolf,.J.P. -1997- Spring-dashpot-mass models for foundation vibrations. Earthquake Engineering and Structural Dynamics, vol.26, 931-949.

NOTEBY A/S-1993- DISEL – User's Manual. Report 39153 No.1.

BACKANALYSIS OF SHEETPILE WALL TEST KARLSRUHE (1993) APPLYING INVERSE ANALYSIS

K.J. Bakker
Delft University of Technology, Delft, The Netherlands
and
Public Works Department "Rijkswaterstaat", Utrecht, The Netherlands

ABSTRACT: The Finite Element modelling of sheetpile walls has been evaluated in the light of the measurements of the 1993 sheet-pile wall in Karlsruhe. The method applied is a simplified version of the Maximum Likelihood approach, as used by Ledesma (1989), applying the Inverse analysis equations and FEM analysis subsequently. A reasonable fit for stresses and displacements was found, including the force deformation curve for the strut, which was not a part of the fit. The soil stiffness based on the laboratory test result seemed to have under estimated the in situ stiffness, as observed, largely.

INTRODUCTION

Figure 1 The construction pit at the test site

In 1993 at the test-site Hochstetten near Karlsruhe, a sheet-pile wall test was performed. The test was organised by the University of Karlsruhe in co-operation with the Dutch Centre for Research and Codes; CUR (Gouda). In advance a prediction contest was held. The test itself, and the prediction results where published by von Wolffensdorfer

(1997). The back-analyses included used by him focused among other things on the material model used; hypoplastic model. The best fit for the parameters was found, as far as could be observed, based on trial and error.

Here in this paper, Bayesian analysis (Ledesma 1989), is used to fit parameters for a Finite Element analysis with PLAXIS, see Vermeer (1995). For the material model the hard-soil model was chosen; a stress dependent stiffness, and hyperbolic stress-strain relation between strain and deviatoric stress in the elastic range, a distinction between primary loading and unloading/reloading, and failure according to the Mohr-Coulomb theory.

TEST EXECUTION AND MEASUREMENTS

The test, was performed in sandy soil, and was heavily instrumented. The test was carried out from the end of may to the begin of June 1993. the final loading was carried out on the 8th of June.

As the ground water level was 5.5 m below soil surface, it has to be considered that the sand showed some apparent cohesion due to suction. The test itself was performed executing the following stages of construction, see table 1:

Table 1

Stage	Stage	Description
0		Initial conditions
1		Excavation up to 1.00 m.
2		excavation to -1.75 m.
3		Installation of the struts and pretension to 4.29 kN/m.
4		Excavation to -3.00 m
5	I	Excavation to -4.00 m.
6	II	Excavation to -5.00 m
7	III	Surface load (in order to reduce the effect of the apparent cohesion).
8	IV	Release the strut length up to 'failure'.

Preliminary to the test, after that the instrumented sheet piles where placed but before excavation, horizontal soil stresses where measured, see Fig 2. According to von Wolfferdorff; *"the initial horizontal stress as observed are quite in disagreement with 'as expected' distributions, but nevertheless have to be considered accurate as the measurement was observed four times independently, and showing a coherent view"*. One of the critical things to be predicted was the deformation, (of the strut) at soil failure. As there might be a dispute whether this deformation is well defined, here a comparison between strutforce and deformation will be made. In Fig 6 a comparison between back-analysis and measurement is given. As one can observe, only minor deformations of the wall lead to diminishing values of the strutforce (and apparently to the soil loading of the wall).

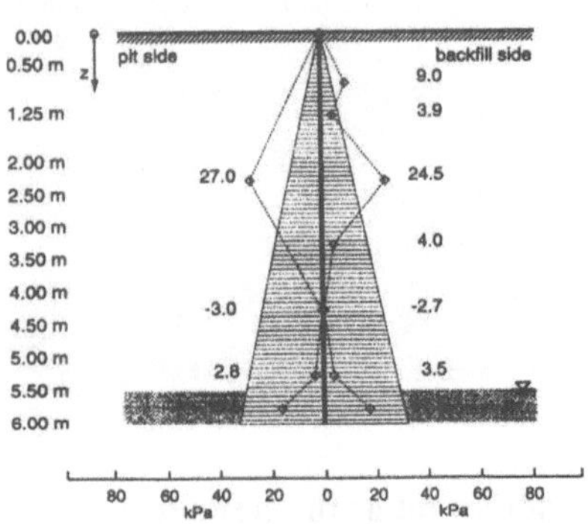

Figure 2 Measured earth pressures

PREDICTION CONTEST

Among other predictors the Civil Engineering Division of the "Rijkswaterstaat" made two predictions. One with an engineering model based on a Subgrade Reaction Model, the other one with a Finite Element Model; PLAXIS. The prediction was discussed in a paper by Bakker & Beem in the former conference in Manchester (1994).

An elaborate description of the predictions and of the test results, was given by von Wolfferdorff (1994), and presented at a Workshop held in Delft at Delft University, October 6 and 7, 1994. One of the characteristic results presented; here repeated in Fig 4 and 5, is a comparison between all the predictions, and the FEM predictions; PLAXIS, for stage III of the test, (When the pit is excavated, and after placing the "water surcharge" load), showing the bandwidth in predictions. Looking at these pictures one is tempted to derive an estimate for the standard deviation of models, estimating this from the predicted bending moments by; $\sigma \approx (\frac{\max(M)-\min(M)}{4})$. It must be considered however that although all the predictors where based on the same set of parameters, the transformation between, bare geotechnical survey data, and model parameters could be and will have been diverse. Therefor a large part of the standard deviation found would thus have to be attributed to the parameters, and not to the model itself.

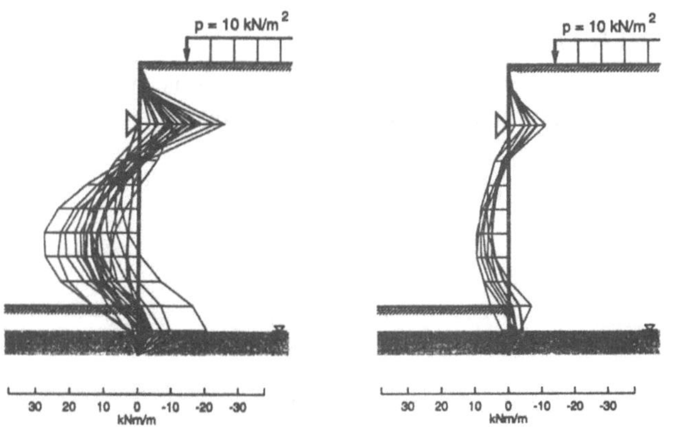

Figure 3 Prediction results stage III, after excavation and surcharge load; left; 4a), all predictions FEM. right; 4b) PLAXIS results

THEORY FOR THE BACK ANALYSIS

In order to perform a postdiction for Karlsruhe sheet piling test, "Inverse analysis" Ledesma (1989), Nova (1995), was applied.

In this theory, to begin with, an explicit model relating parameters; **x**, and postdiction results; $\mathbf{f}^c$ (where ^{c}, stands for 'calculation'), has to be available;

$$\mathbf{f}^c = \mathbf{M}(\mathbf{x}) \qquad \textbf{(1)}$$

The results of which (the post diction), might be evaluated in relation to measurements; $\mathbf{f}^t$, (where ^{t}, stands for test). Both $\mathbf{f}^c$ and $\mathbf{f}^t$ are assumed to be vectors here, with a length $\boldsymbol{n}$; the number of measurements taken in consideration. Here only a limited number of measurements will be used to fit the parameters; e.g. a maximum bending moment, a strut force and/or a maximum deformation, for a number of successive steps in the excavation, i.e. the engineering parameters being used in the evaluation against construction criteria.

The measurements being taken in consideration and the calculation results of the model might be ordered in vectors according to;

$$f_i^t = (f_1^c, f_2^c, \ldots\ldots, f_n^c)^T \tag{2}$$

and

$$f_i^c = (f_1^c, f_2^c, \ldots\ldots, f_n^c)^T \tag{3}$$

After Ledesma, (1989), it is assumed that the probability distributions of the prior information of the parameters and the measurements are multivariate Gaussian;

$$P(\mathbf{x}) = \left|\mathbf{C}_\mathbf{x}^0\right|^{-1/2} (2\pi)^{-m/2} \exp\left[-\frac{1}{2}(\mathbf{x} - \langle\mathbf{x}\rangle)^\mathrm{T} \left(\mathbf{C}_\mathbf{x}^0\right)^{-1} (\mathbf{x} - \langle\mathbf{x}\rangle)\right] \tag{4}$$

and

$$P(\mathbf{f}^c) = \left|\mathbf{C}_\mathbf{f}\right|^{-1/2} (2\pi)^{-n/2} \exp\left[-\frac{1}{2}(\mathbf{f}^t - \mathbf{f}^c)^\mathrm{T} \left(\mathbf{C}_\mathbf{f}\right)^{-1} (\mathbf{f}^t - \mathbf{f}^c)\right] \tag{5}$$

Where;

$\mathbf{C}_\mathbf{x}^0$ is the covariance matrix, based on the available 'a priori' information.
$\mathbf{C}_\mathbf{f}$ measurements covariance matrix
$\langle\mathbf{x}\rangle$ " a priori" estimated value's of parameters, e.g. the mean value's
$\mathbf{f}^t$ the measured variable values
m is the number of parameters evaluated
n is the number of measurements
$()^\mathrm{T}$ is used to indicate a transpose

If the measurements and the 'a priori' estimates for the parameters are independent, the likelihood of a combination of a priori parameters and measurements is assumed according to;

$$L(\mathbf{x}) = k\, P(\mathbf{x}) P(\mathbf{f}^c) \tag{6}$$

where k is an arbitrary constant.

The most likely combination of parameters to fit the measurements can be found, solving the minimum of the natural logarithm, which yields the same optimum, as the latter function is monotone. Therefor an additional function S is postulated to be minimised;

$$S = -\ln L(\mathbf{x}) \tag{7}$$

Which written out yields:

$$S=\left(\mathbf{f}^{\mathbf{t}}-M(\mathbf{x})\right)^{T}\mathbf{C}_{\mathbf{f}}^{-1}\left(\mathbf{f}^{\mathbf{t}}-M(\mathbf{x})\right)+\left(\mathbf{x}-\langle\mathbf{x}\rangle\right)^{\mathrm{T}}\left(\mathbf{C}_{\mathbf{x}}^{0}\right)^{-1}\left(\mathbf{x}-\langle\mathbf{x}\rangle\right)+ \\ +\tfrac{1}{2}\ln\left|\mathbf{C}_{\mathbf{f}}\right|+\tfrac{1}{2}\ln\left|\mathbf{C}_{\mathbf{x}}^{0}\right|+\tfrac{n}{2}\ln(2\pi)+\tfrac{m}{2}\ln(2\pi)-\ln k \tag{8}$$

If the error structure of the measurements and parameters is considered to be fixed, only the first to terms of the equation have to be considered in the minimisation process, the other terms being constants;

$$S^{*}=\left(\mathbf{f}^{\mathbf{t}}-M(\mathbf{x})\right)^{T}\mathbf{C}_{\mathbf{f}}^{-1}\left(\mathbf{f}^{\mathbf{t}}-M(\mathbf{x})\right)+\left(\mathbf{x}-\langle\mathbf{x}\rangle\right)^{\mathrm{T}}\left(\mathbf{C}_{\mathbf{x}}^{0}\right)^{-1}\left(\mathbf{x}-\langle\mathbf{x}\rangle\right) \tag{9}$$

It is assumed that here that the results of the numerical analysis; $\mathbf{f}^{\mathbf{c}}$ may be expanded using a linear Taylor's expansion according to;

$$\mathbf{f}^{\mathbf{c}}=\mathbf{f}_{0}^{\mathbf{c}}+\frac{\partial\mathbf{f}^{\mathbf{c}}}{\partial\mathbf{x}}\Delta\mathbf{x}=\mathbf{f}_{0}^{\mathbf{c}}+\mathbf{A}\Delta\mathbf{x} \tag{10}$$

Combination of equations 9 and 10 lead to;

$$S^{*}=\left(\mathbf{f}^{\mathbf{t}}-\mathbf{f}_{0}^{\mathbf{c}}-\mathbf{A}\Delta\mathbf{x}\right)^{T}\mathbf{C}_{\mathbf{f}}^{-1}\left(\mathbf{f}^{\mathbf{t}}-\mathbf{f}_{0}^{\mathbf{c}}-\mathbf{A}\Delta\mathbf{x}\right)+\left(\mathbf{x}-\langle\mathbf{x}\rangle\right)^{\mathrm{T}}\left(\mathbf{C}_{\mathbf{x}}^{0}\right)^{-1}\left(\mathbf{x}-\langle\mathbf{x}\rangle\right) \tag{11}$$

Because we intend to improve the solution with respect to trial values of the parameters $\mathbf{x}$, (related to the trial values of $\mathbf{f}^{\mathbf{c}}$); $\mathbf{f}_{0}^{\mathbf{c}}$. If we use the notation $\Delta\mathbf{f}=\mathbf{f}^{\mathbf{t}}-\mathbf{f}_{0}^{\mathbf{c}}$, equation 11, yields;

$$S^{*}=\left(\Delta\mathbf{f}-\mathbf{A}\left(x-x^{tr}\right)\right)^{T}\mathbf{C}_{\mathbf{f}}^{-1}\left(\Delta\mathbf{f}-\mathbf{A}\left(x-x^{tr}\right)\right)+\left(\mathbf{x}-\langle\mathbf{x}\rangle\right)^{\mathrm{T}}\left(\mathbf{C}_{\mathbf{x}}^{0}\right)^{-1}\left(\mathbf{x}-\langle\mathbf{x}\rangle\right) \tag{12}$$

Equation 12 can be minimised, differentiating by $\mathbf{x}$;

$$\frac{\partial S^{*}}{\partial x}=-\mathbf{A}^{\mathrm{T}}\mathbf{C}_{\mathbf{f}}^{-1}\Delta\mathbf{f}-\mathbf{A}^{\mathrm{T}}\mathbf{C}_{\mathbf{f}}^{-1}\mathbf{A}\,\mathbf{x}^{\mathbf{tr}}+\mathbf{A}^{\mathrm{T}}\mathbf{C}_{\mathbf{f}}^{-1}\mathbf{A}\,\mathbf{x}+\mathbf{C}_{\mathbf{x}}^{-1}\mathbf{x}-\mathbf{C}_{\mathbf{x}}^{-1}\langle\mathbf{x}\rangle=0 \tag{13}$$

Rearranging the equation in a dependant part with the unknown parameters $\mathbf{x}$ on the left side, and the a priori information; trial values and a priori values of the unknowns on the right hand, yields;

$$\left(\mathbf{A}^{\mathrm{T}}\mathbf{C}_{\mathbf{f}}^{-1}\mathbf{A}+\mathbf{C}_{\mathbf{x}}^{-1}\right)\mathbf{x}=\mathbf{A}^{\mathrm{T}}\mathbf{C}_{\mathbf{f}}^{-1}\left(\Delta\mathbf{f}+\mathbf{A}\,\mathbf{x}^{\mathbf{tr}}\right)+\mathbf{C}_{\mathbf{x}}^{-1}\langle\mathbf{x}\rangle \tag{14}$$

Equation 14 is the general form for the Maximum Likelihood formulation for back-analysis. If the a priori information is not taken in consideration, the solution simplifies to;

$$\left(\mathbf{A}^{\mathrm{T}}\mathbf{C}_{\mathbf{f}}^{-1}\mathbf{A}\right)\mathbf{x}=\mathbf{A}^{\mathrm{T}}\mathbf{C}_{\mathbf{f}}^{-1}\left(\Delta\mathbf{f}+\mathbf{A}\,\mathbf{x}^{\mathbf{tr}}\right) \tag{15}$$

Finally, if the error structure matrix is the identity; = one, the more common form of the least squares formulation is obtained;

$$\left(\mathbf{A}^{\mathrm{T}}\mathbf{A}\right)\mathbf{x}=\mathbf{A}^{\mathrm{T}}\left(\Delta\mathbf{f}+\mathbf{A}\,\mathbf{x}^{\mathbf{tr}}\right) \tag{16}$$

FINITE ELEMENT MODELLING

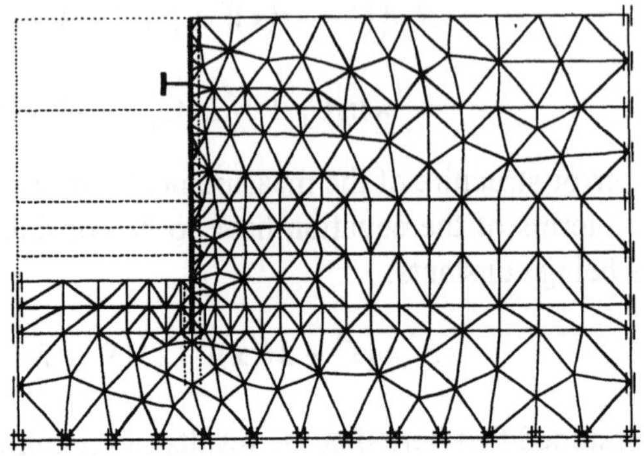

Figure 4 Finite Element model for post diction

The FEM analyses both for the prediction as well as for the postdiction were performed with PLAXIS, the prediction with version 4.5, and the postdiction with version 7. The test is modelled in plane strain. The mesh for the postdiction is given in Fig. 4. The mesh displayed is the mesh at a certain stage of construction, i.e. the soil elements in the pit are removed yet. In the initial situation a level soil surface is modelled. In order to improve the analysis, all stages of the test are modelled and analysed subsequently.

The shortening of the struts in the final stage of the analysis was performed by removing the strut in a staged construction phase, up to the point in the analysis that the soil yields.

PARAMETERS AND MEASUREMENTS TO EVALUATE

After the test comparing measurements and predictions, discussion focused on 1) the soil stiffness; i.e. for small strains, 2) Apparent cohesion due to suction, 3) initial stresses due to the installation procedure. Here the following considerations where made;

Friction angle; apparently the soil is 'stronger' than anticipated in the predictions. The bending moments and strut-forces are largely over estimated in the predictions Therefor, in the back-analysis to begin with a friction angle at failure (the top value, instead of at $\varphi_m(3\%)$), will be assumed, i.e. $\varphi = 42^\circ$.

Apparent cohesion; In the back-analysis by von Wolffensdorfer (1996), it is mentioned that for the top-layer of 1.5 m approximately, a capillary underpressure of approx. 13 kPa is active, leading to an apparent cohesion of $C_{uns} = 13 \tan(42) = 11.7$ kPa.

Elasticity of the soil; In the prediction by Beem & Bakker (1994), the Mohr-Coulomb model with a G_{50} was used. With this approach, the unloading of the soil, with a much stiffer behaviour was disregarded. In the back-analysis, the PLAXIS 'hard-soil' model; with an hyperbolic strain hardening relation acc. to Duncan & Chang, 1970) is applied. The hard soil model, see Vermeer & Brinkgreve (1995), identifies a Initial Young's modulus; E_i and unloading-reloading Young's modulus E_{ur}. As a trial value, the modulus from the Triaxial-test results is $E_{50} \approx 2(1+\upsilon)G_{50}^{ref} \approx 35000$ is used. Subsequently the Cone-penetration results have been looked at, with respect to the emperic relation that $E \approx (3 \text{ to } 5)\, q_c$. Based on that 5 layers with a different stiffness have been distinguished..

For the unloading reloading modulus, according to the a priori data set for the test, the"Platten-druckversuch"; the load plate test, the stiffness ratio, for unloading reloading is 1.6. As the initial stiffness, E_i, assuming an hyperbolic shape for the hardening curve is twice the value at E_{50}, the Young's modulus for unloading reloading E_{ur} is assumed to be 1.6*2 = 3.2 times E_{50}. The young's moduli used are gathered in Table 2

Initial stresses; The earth pressure measurements, in advance of the excavation, see Fig 2 indicate that in the upper zone, 2.0 m an increased horizontal stress is active. Approximately twice the value acc. to Jaky; $K_0 \approx (1-\sin\varphi)$, was observed. Whereas below 3.5' soil surface, the horizontal stresses seem to contradict with plasticity theory, as for active failure; $K_0 \approx \frac{1-\sin\varphi}{1+\sin\varphi} - \frac{c}{\sigma_v}\cos\varphi$.

For a depth of 3.5 m. e.g. with a $\sigma_v \approx 3.5 * 16.5 \approx 58\,kPa$, a soil friction of approximately $\varphi \approx 42^0$ and a cohesion of $c \approx 5kPa$ this would lead to a minimum value of $K_0 \approx 0.14$. The observed K_0 value of appr. 0.0 suggests a Cohesion of appr. 15 kPa which is considered to be unrealistic. In the postdiction analysis, a value for K_0 for the below 3.5 m of 0.2 is used , whereas the value for the undeeper layers was being considered a free parameter in the optimisation.

Table 2 Soil data used as initial values in the back analysis

Layer top	γ_d	γ_n	φ	ψ	c	c_{depth}	Ref C_{depth}	E_{50}^{ref}	E_{ur}	υ
+. MSL	kN/m³	kN.m³	[°]	[°]	kPa	kPa	m	kPa	kPa	[-]
+0.00	16.9	-	42.0	12.0	11.7	-	-	65000	208000	0.3
- 1.25	16.5	-	42.0	12.0	11.7	-2.52	-1.25	65000	208000	0.3
- 3.50	16.5	-	42.0	12.0	11.7	-2.52	-1.25	35000	112000	0.3
- 4.50	16.5	-	42.0	12.0	1	-	-	70000	224000	0.3
- 5.50	16.5	19.0	42.0	12.0	1	-	-	35000	112000	0.3

BACK ANALYSIS

The measurements taken in consideration for the back-analysis, are a subset of the total measurements. This subset of characteristic measurements, such as anchor force, bending moment, and maximum deformation is evaluated for several stages of construction. The measurements are: M2(2.0), which stands for the bending moment in the second stage of excavation, at the height 2.0 m below the top, M6(1.0), M6(3.0), F6, where F stands for the strutforce, U7(0.0); the displacement in stage 7 at the top, U7(3.10), M8(2.0), M8(3.0) and finally F8.

The solution of equation 14 and 15 demands that a covariance matrix for the measurements is established. This would not be necessary for the plain least squares approach acc. to equation 16, which implicitly assumes a standard deviation for the measurements of 1.0.

In order to weigh the importance of the measurements, and to do this in a way not too subjective, here it was assumed that measurements are independent, off diagonal terms are assumed to be zero, whereas a variance; $\sigma_i/\mu_i = 0.1$ is adopted. Above that error's

(Δf_i) with respect to bending moments and strutforces are weighed heavier in comparison than deformations, for a factor 5.

OBSERVATIONS OF THE BACK-ANALYSIS

The back analysis was started, with the weighed Least Square approach acc. to equation 15 The method itself was applied by extracting the gradients assembled in the A matrix, from the EEM model only once, consecutively improving the solution iteratively, updating the trial value for a next step using the result of the former LS analysis. The assumption implicitly made here was that the derivatives of the model, for the reach in consideration are not too strongly depending from the position of the model in the solution space. The necessity to do so was that the FEM model could only be handled menu driven so that the extraction of the derivatives for the matrix A could in practice not be made automatically.

This procedure appeared to be reasonably stable, giving convergence in approximately 15 steps. Within this process, equation 14 was solved using Mathcad ©.
One of the final results of this analysis was that the Ratio for the Young's modulus came out nearly four times as high as the value extracted from the triaxial test results, presented in table 2. After the LS result was derived the Maximum Likelihood formulation according to equation 14 was used in order to try to improve the result. It came out very soon that the procedure without updating matrix A, did not yield a good converge. The improved solution is given in table 3, though it has to be mentioned that only, 2 or 3 convergent iterations could be made, depending on the relaxation factor applied; the smaller, the more steps, the less the improvement derived. After that divergence appears. The a priori information seems to yield higher cohesion values.

A comparison of back-analysis results and measurements is given in table 3.

Table 3 Comparison Measurements and back analysis

		Prediction	Measurement	Back-analysis Least squares	Back-analysis Maximum Likelihood
Bending moment stage 2	M2(2.00)	pm	2.26	1.93	1.99
	M6(1.00)	-5.8	-4.41	-5.159	-4.96
Field moment stage 6	M6(3.00)	5.38	2.2	1.778	1.65
Strut force stage 6	F6	23.36	28.64	29.68	28.38
Field displacement stage 6	u6(3.00)	3.51	2.99	2.637	2.49
Head moment stage 7	M7(1.25)	-6.78	-5.06	-6.234	6.03
Field moment stage 7	M7(3.00)	6.72	2.76	2.138	1.99
Strut force stage 7	F7	30.07	33.72	34.86	33.91
Top displ. Stage 7	U7(0.00)	-0.586	5.15	2.86	2.90
Field displ. Stage 7	U7(3.00)	7.19	3.4	3.27	3.14
Ultimate bending moment	M8(2.00)	5.99	4.67	3.919	3.87
Ultimate bending moment	M8(3.00)	12.2	3.41	9.41	9.44
Ultimate strutforce	F8	10.0	4.22	3.035	3.38
Objective function J				72.5	61.17

The terminology is, that M stand for bending moment, U for displacements, and F for the strutforce. The parameters derived in the subsequent analyses are listed in table 4,

Table 4 Post diction results

	k0(1)	K0(2)	φ	ψ	R(δ)	C(1)	C(2)	R(E)
Prediction	0.38	0.38	38	5	0.66	5	5	1.0
Least squares	2.85	0.6	42.7	4.7	1	6.35	5.5	4.25
Maximum Likelihood	2.92	0.59	42.5	6.0	1	7.2	5.85	4.1

where

$K_0(1)$ = the K_0 in the upper soil layer until 1.25 m. deep
$K_0(2)$ = the K_0 in the soil layer between 1.25 m. and 3.5 m. deep
C(no) = The cohesion in soil layer (no) (numbering is top downwards)
R() – the Ratio with respect to 1) δ; wall friction and 2) E; Young's modulus, (for all soil layers)
φ and ψ are varied for all soil layers

Finally in Figure 5, The soil pressure, bending moments and displacements are displayed for stage 7 . The displacement anchor-force plot is given in Fig 6. As one can observe for the ultimate values of stresses; bending moments, a reasoble agreement hase been derived whereas the distribution indicates that the soil loading, in situ, is acting on a higher level than in the model. Apparently the stiffness of layer 3 is not conform the assumptions acc. to table 2. The distribution of stiffnesses could therefor be optimized which was not a part of this analysis. With respect to displacements; the actual stiffnes of the strut is appr. 30 % less, yielding the disagreement for the hor. displacements caused by the swing.

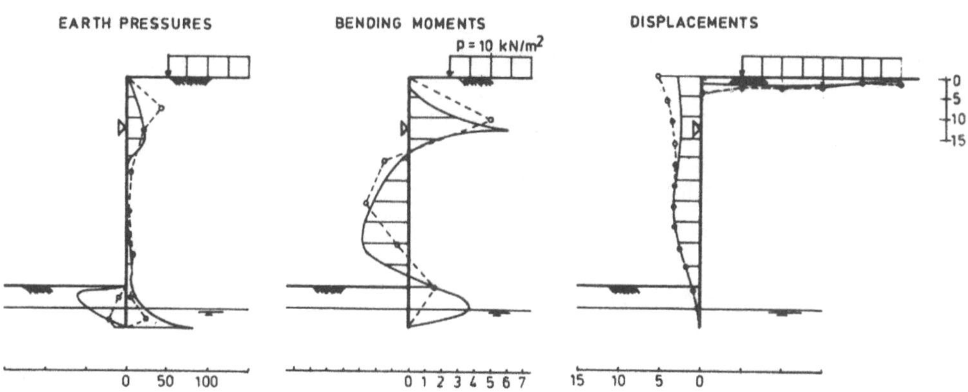

Figure 5 Comparison of postdiction results and measurements

CONCLUDING REMARKS

A reasonable fit of the parameters has been derived. Apart from aspects such as the importance of initial stresses, and the underestimated influence of under-pressure in the soil, it appeared that in this case the stiffness based on triaxial cell tests strongly underestimated the observed behaviour. With respect to wall friction; the common used value of 2/3 of the soil friction underestimes practice.

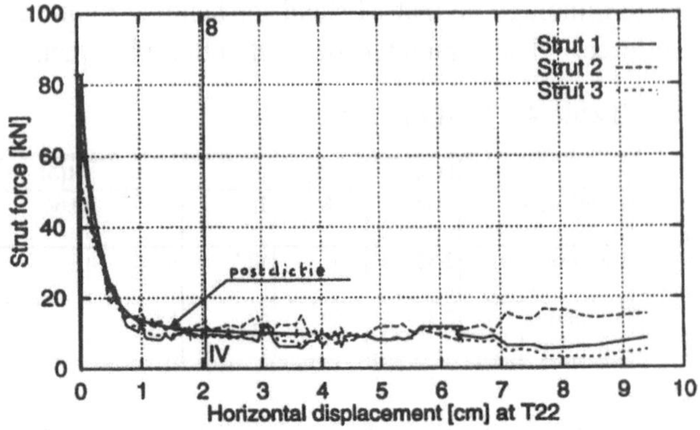

Figure 6 Strut force as a function of deformations.

It is thought too, that the description of small strain behaviour is largely improved by the Hard soil model. For convergence of the Maximum Likilihood analysis, an update of the gradient matrix A seems to be necessary to derive convergence.

REFERENCES

BAKKER, K.J. & BEEM R.C.A. (1994)
Modelling of the sheet pile wall test in Karlsruhe 1993
Proc. 3the European conf. on Numerical methods in
Geotechnical Engineering; ECONMIG 94, Manchester/UK/7-9-sept.

DUNCAN J.M.., CHANG C.Y. (1970)
Nonlinear Analysis of Stress and Strain in Soil,
ASCE J. of the Soil Mech. and Found. Div. Vol 96

NOVA R., BOULON M.,GENS A., (1995)
Numerical modelling and information technology, Theme Report 6,
In: The Interplay between Geotechnical Engineering and Engineering Geology,
Proc. XI European Conf. on Soil Mechanics and foundation Engineering, Copenhagen

LEDESMA, A. GENS, A., ALONSO E.E. (1989)
Probabilistic backanalysis using a maximum likelihood approach
XII ICSMFE, Rio de Janeiro.

VERMEER, P.A. and BRINKGREVE, (1995)
PLAXIS, Finite Element Code for Soil and Rock Analysis
Balkema, Rotterdam.

WOLFFENSDORFER, P.A. von (1994)
Workshop Sheet Pile Test Karlsruhe
Delft University, October 6 and 7, 1994

WOLFFENSDORFER, P.A. von (1995)
Vervormungsprognosen von Stutzkonstruktionen.
Habilitationsschrift, Universitat Fridriciania Karlsruhe

NUMERICAL SIMULATION OF NON LINEAR BEHAVIOR OF LATERALLY LOADED PILE

S. Belkhir and D. Levacher
Le Havre University, Le Havre, France

W. Hamadeh
BCEOM, Montpellier, France

S. Mezazigh
Le Havre University, Le Havre, France

ABSTRACT

Due to the recent research advances in understanding mechanisms which govern the deep behavior foundations, now the vertical piles can be used to support correctly lateral loads. The paper concerns the development of a computer program for studying non linear behavior of lateral loaded piles taking into account the shear stress at the sand interface layers.

Some examples issued from the use of the numerical code are compared with experimental results from centrifuge modelling loaded piles tests.

The influence that a shear stress at the soil on the behavior laterally loaded piles was also studied.

KEY WORDS : Piles, Lateral loaded, Deep foundation, Shear stress, Numerical model, Centrifuge model-piles.

INTRODUCTION

Deep piles foundations, which are widely used in construction of coastal or offshore structures, are in a growing development. Improvements are made in the design and for technical innovations in construction piles. The weak performance of soils given to the constructors, and the importance of the construction dimensions, are at the origin of this development. Nowadays, the theoretical approaches are insufficient to take in account all

the parameters of the soil-pile interaction. The design codes used do not take the shear stress at the soil into account. Moreover, the obtaining of experimental data on site is more difficult and more expensive. So, studying the influence of different parameters linked to the soil-pile interaction, led many researchers to use model-piles.
Referred to the bibliography, only few non-linear design models relative to the behavior laterally piles are proposed to the case of that concern simple configurations.

The paper concerns a design code, based on the matrix transfer method which is applied to the laterally loaded piles, taking in account the shear stress at the sand. Then, the performances of the developed design code is illustrated by some examples. The theoretical results (used the PHRI method) are compared with the experimental results obtained by tests on the models of piles in centrifuge.

NUMERICAL MODELLING

Pasternak (1954) has proposed a behavior soil model taking in account the interaction of shear stress between the springs of the Winkler model (1867) (Fig. 1). This is obtained by linking the springs to an incompressible layer that may support only tangential deformation caused by the shear stress. The shear stress layer is assumed isotropic depending on a shear stress modulus Gs.

The governing equation is given under the following relationship :

$$p = K_h y + G_s \nabla^2 y \qquad [1]$$

with the operator

$$\nabla^2 = \frac{\partial}{\partial x^2} + \frac{\partial}{\partial z^2}$$

In the case of the study of the behavior of a lateral loaded pile, the lateral displacement is only function of the depth z, then the equation [1] becomes :

$$p = K_h \, y + G_s \frac{d^2 \, y}{dz^2} \qquad [2]$$

This model is interesting, since it takes into account the interaction of shear stress between the springs and the formulation of the soil reaction behavior is function of the subgrade reaction modulus and of the shear stress modulus. This model is used for the study of the behavior of the lateral loaded piles which takes into account the shear stress at the soil.

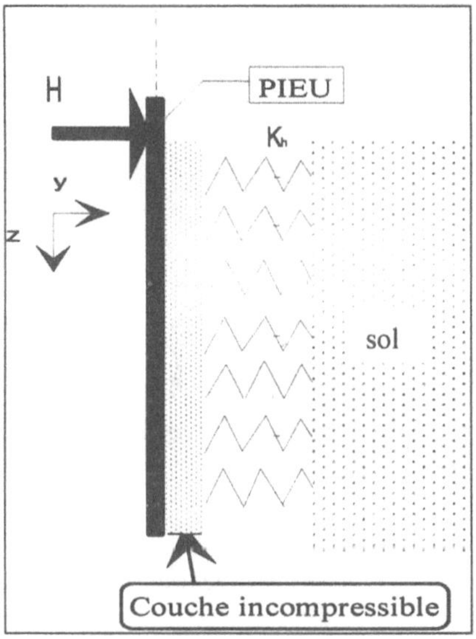

Fig. 1 : Pasternak model.

Seed and al. (1986) have expressed the maximal shear stress modulus of sand according to the density index (I_D) and the effective unit weight (γ'), under the form :

$$G_{max} = 21,7\,(K_2)_{max}\,P_a\left(\frac{\sigma'_m}{P_a}\right)^{0,5} \quad [3]$$

$$\sigma'_m = \gamma'.z$$

Seed and Idriss (1970) have estimated the modulus $(K_2)_{max}$ as follows :

$$(K_2)_{max} = 3,5\,(I_D)^{2/3} \quad [4]$$

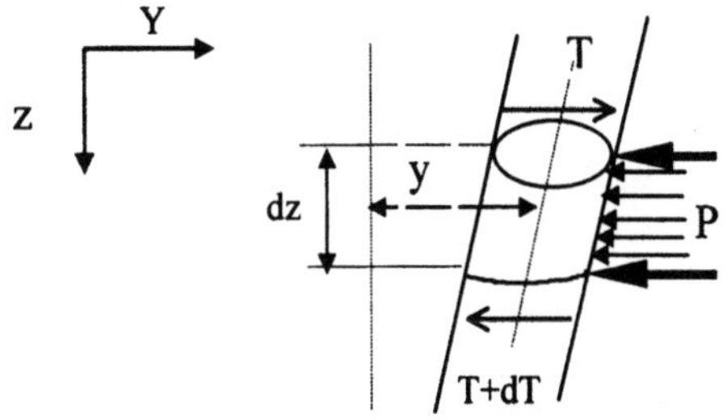

Fig. 2 : An element of pile.

The equilibrium equation of the submitted pile section to a distributed loading p(z) (Fig. 2) is :

$$\frac{dT}{dz} = -P(z) \qquad [5]$$

$$T = \frac{dM}{dz} \qquad [6]$$

$$M = EI\,\frac{d^2y}{dz^2} \qquad [7]$$

From the relationships [5], [6] and [7], we obtain the equilibrium equation of the pile element :

$$EI\frac{d^4y}{dz^4} + P(z) = 0 \qquad [8]$$

The soil reaction taking into account the shear stress is written under the form :

$$P = p.B = E_s y + P_0 + G_s.B\frac{d^2(y)}{dz^2} \qquad [9]$$

with :
P_0 : the initial soil reaction (kN/m)
E_s : subgrade reaction modulus (kPa)
B : frontal width or diameter of the pile (m).

The differential equation of the problem is written as follows :

$$EI\,\frac{d^4y(z)}{dz^4} + G_s.B.\frac{d^2y(z)}{dz^2} + E_s y(z) = -P_0 \qquad [10]$$

The solution of the equation without the second member, where $E_s \neq 0$, is:

$$y = C_1 e^{\frac{z}{z1}} cos\left(\frac{z}{z_2}\right) + C_2 e^{\frac{z}{z1}} sin\left(\frac{z}{z_2}\right) + e^{-\frac{z}{z1}} cos\left(\frac{z}{z_2}\right) + C_4 e^{-\frac{z}{z1}} sin\left(\frac{z}{z_2}\right) \qquad [11]$$

with

$$z_1 = \sqrt{\frac{4}{-\frac{G_s.B}{EI} + \sqrt{\frac{4E_s}{EI}}}} \quad and \quad z_2 = \sqrt{\frac{4}{\frac{G_s.B}{EI} + \sqrt{\frac{4E_s}{EI}}}}$$

The numerical resolution of the differential equation, taking into account the shear stress with the soil behavior laws introduced in the model, is undertaken by the transfer matrix method (Gery and al., 1973).

The process of reduction is a mixed method which takes as unknown factors both forces and displacements. It consists of the values of forces and compatibility equilibrium of displacement along these different elements. This method behaved to a quick and suitable solution.

At the beginning, the pile is cut into sections along its center line. The first passage is made along the pile to define the Transfer Matrix for all the sections.

By calling $[K_n]$ the transfert matrix , whose terms are function of the length of the section and the rigidity EI, knowing the vector $\{V_g\}$ to the high extremity of a section, we can to determine the vector $\{V_d\}$ of the low end thanks to the next relationship :

$$\{V_d\}=[K_n]x\{V_g\} \quad [12]$$

with

$$\{V_d\} = \begin{Bmatrix} y_d \\ w_d \\ M_d \\ T_d \\ 1 \end{Bmatrix} \quad and \quad \{V_g\} = \begin{Bmatrix} y_g \\ w_g \\ M_g \\ T_g \\ 1 \end{Bmatrix}$$

In the case where $E_s \neq 0$ the transfer matrix is written as follows :

$$[K_n] = \begin{bmatrix} k_1 & k_2 & k_3 & k_4 & y_p(l) \\ k_5 & k_6 & k_7 & k_8 & w_p(l) \\ k_9 & k_{10} & k_{11} & k_{12} & M_p(l) \\ k_{13} & k_{14} & k_{15} & k_{16} & T_p(l) \\ 0 & 0 & 0 & 0 & 1 \end{bmatrix} \quad [13]$$

with

k_i : function of (EI, ℓ, E_s, G_s)

y_p, w_p, M_p,et T_p : function of $(EI, \ell, E_s, G_s$ and $P_0)$

In the case where $E_s=0$ the transfer matrix is simplified as follows :

$$[K_n] = \begin{bmatrix} 1 & l & \frac{l^2}{2EI} & \frac{l^3}{2EI} & -\frac{Pol^4}{24EI} \\ 0 & 1 & \frac{l}{EI} & \frac{l^2}{2EI} & -\frac{Pol^3}{6EI} \\ 0 & 0 & 1 & l & -\frac{Pol^2}{2EI} \\ 0 & 0 & 0 & 1 & -\frac{Pol}{EI} \\ 0 & 0 & 0 & 0 & 1 \end{bmatrix} \qquad [14]$$

The design code called CAPELA is developed under FORTRAN 90. CAPELA gives the design of displacement and effort of the single lateral loaded pile, with taking into account the shear stress at the soil.

The validation and the performance of the design code is made by comparison the obtained results with experimental test results on centrifuge model-piles.

TESTS ON MODEL-PILES IN CENTRIFUGE

A series of tests on lateral loaded model-piles in a centrifuge has been performed at the L.C.P.C of Nantes by Mezazigh (1995). The aim was the determination of the effect of the proximity of a slope on the P-y reaction curves. Fifty-nine laterally loaded pile tests in a sand have been undertaken. The model-piles that were used, are considered as long and flexible. Parameters taken into account in these tests are : the slope, the distance of the pile from the crest of the slope and the soil density. For the test comparison, only the results for the single pile installed in a horizontal soil-surface (without slope), submitted to the lateral loading are considered. In the study, the characteristics of model-piles and P-y reaction curves of the tests (Mezazigh, 1995) are used as data for the design code CAPELA.

The used soil is the sand, a clean siliceous material (sand of Fontainebleau) commonly used during tests on site or in laboratory. It is a sand which approximately 70% of grains have a diameter inferior to 200 µm. In the presented examples of design made with PILATE and CAPELA, the soil has a unit weight of approximately 16.19 kN/m^3 for a dense sand and approximately 15 kN/m3 for a lightly dense sand what corresponds to respectively to a density index of 0.81 and 0.63.

The model-piles is an aluminum tube AU4G with an external diameter of 18 mm, a thickness of 1.5 mm and a length of 380 mm. The centrifuge acceleration 40 g allows simulate of a prototype pile of 720 mm diameter and of 15.2 m length . The modulus of rigidity is 514 MN.m^2.

The pile instrumentation allows to undertake displacement measurements at the top of the model-pile and bending moments derived from the strain gages installed along the pile for each level of loading.

The deformations and pressures along the pile are obtained by a double integration and a double derivation of obtained bending moments curves.

VALIDATION AND ANALYSIS OF RESULTS

The comparative study has been made with the experimental results using the data of the centrifuge model-piles tests, to validate the design code CAPELA . The different data are based on mechanical and geometrical pile characteristics and on soil characteristics of the centrifuge tests. The results of comparison show a great concordance between them (Fig. 3 et 4). The difference between deflections at the top and the maximal bending moments obtained by the design code and centrifuge tests has not exceeded 10%. Experimental moments and deflections curves are compared with the obtained theoretical curves (used the curves P-y of PHRI method) by CAPELA, without taking into account the shear stress (Fig. 5). The gap between the maximal moments obtained by this comparison has exceeded 14%. Taking into account the shear stress has reduced the gap between results obtained by CAPELA and the centrifuge tests (the new gap between the maximal moments obtained is of 5%) (Fig. 6). When we take into account the shear stress in the design, the theoretical results obtained by CAPELA are near with the experimental results.
Thus, the influence of shear stress at the soil on the behavior of the lateral loaded pile can be explained clearly.

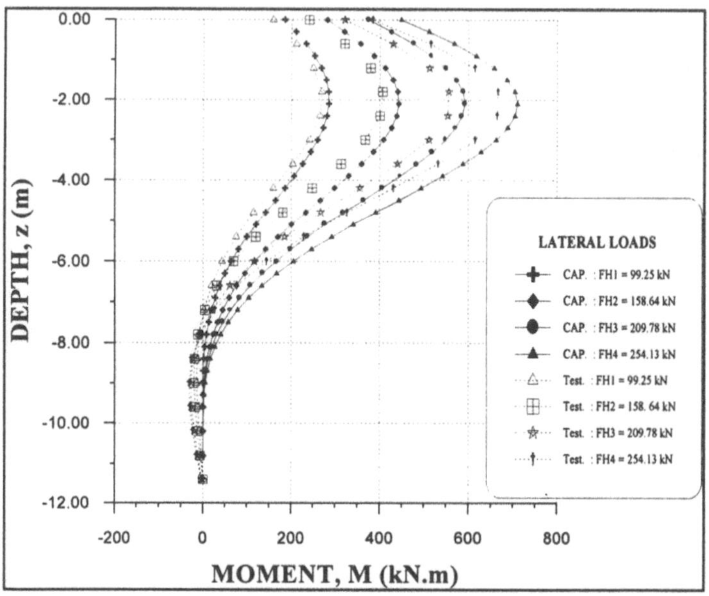

Fig. 3 : *Comparison between bending moments obtained by CAPELA and* centrifuge tests.

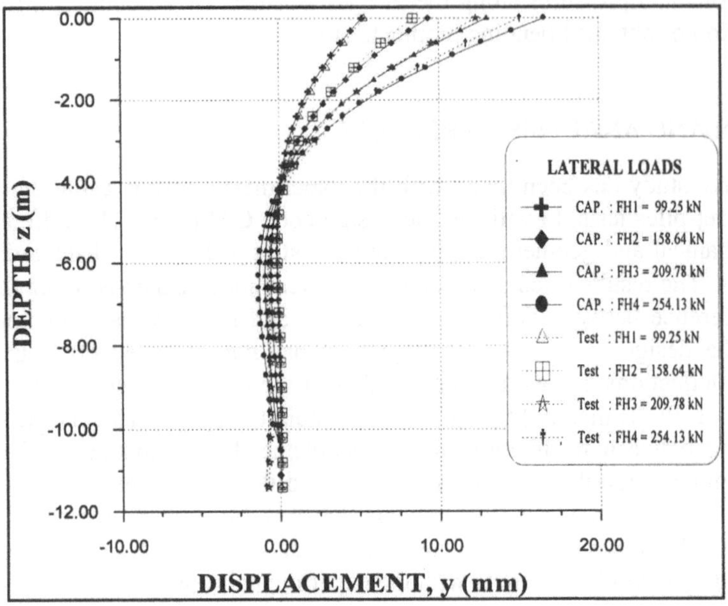

Fig. 4 : *Comparison between pile deflections obtained by CAPELA and* centrifuge tests.

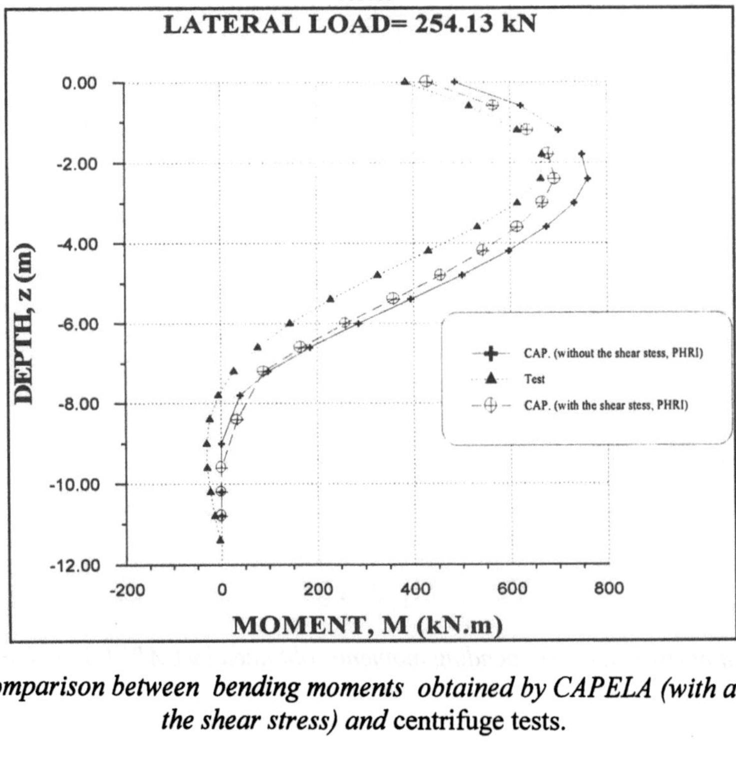

Fig. 5 : *Comparison between bending moments obtained by CAPELA (with and without the shear stress) and* centrifuge tests.

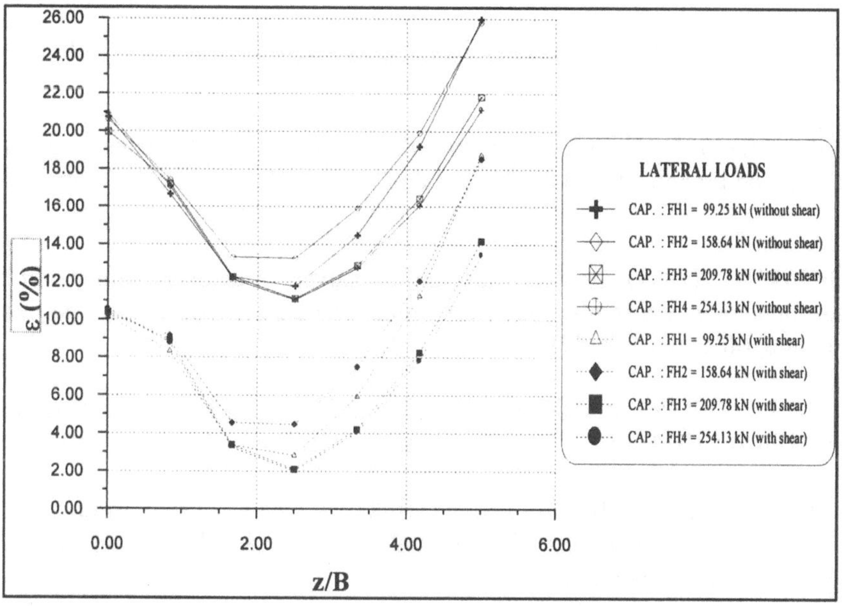

Fig. 6 : *Comparison between bending moments obtained by CAPELA (with and without taking into account the shear stress) and* centrifuge tests.

CONCLUSION

The code CAPELA is useful to design the loaded lateral pile. The design code is based on the method of the subgrade reaction modulus and supported on a suitable design method that is a transfer matrix . Furthermore, this code has been endowed with a suitable menu of data introduction and a convivial use.

The quick and simple `` transfer matrix '' method allows computation of the distribution of moments and deflection along a lateral loaded pile. This method is of interest for the design of the pile itself.

A comparative study of results obtained by CAPELA and tests on model-piles in the centrifuge has been undertaken. This study on illustrated examples has a double objective: the validation of the design code CAPELA, and the usefulness of the taking into account of the effect of shear coupling is shown.

The analysis method, can be developed for the design methods of the lateral loaded pile, in different configurations and environmental conditions.

REFERENCES

GERY P.M., CALGARO J.A. (1973). Les matrices transfert dans le calcul des structures. p.263. Eyrolles, Paris.

MEZAZIGH S.(1995). Etude expérimentale des pieux chargés latéralement : proximité d'un talus et effet de groupe. Thèse de doctorat de l'Université et de l'Ecole Centrale de Nantes, 217p.

PASTENAK P.L. (1954). Òn a new method of analysis of an elastic foundation by means of two foundation constants. Gosudarstvennoe Izdatelstro Liberatuni po Stroitelstvui Arkhitekture, Moscow, 355-421

SEED H.B ET IDRISS I.M. (1970). Soil moduli and damping factors for dynamic response analysis. Report UCB/EERC - 70/10, University of California, Berkeley, 180-193.

SEED H.B., WONG R.T., IDRISS I.M. ET TOKIMATSU K. (1986). Moduli and damping factors for dynamic analyses of cohesionless soils. ASCE Journal of Gotechnical Engineering, 112, , 1016-1032.

WINKLER E. (1867). Die lehhre von der elastizitat und festigkeit. Dominicus, Prague.

THREE-DIMENSIONAL STRUCTURAL ANALYSES AND DESIGN OF SEGMENTED TUNNEL LINING AT CONSTRUCTION STAGE

C.B.M. Blom and H.C.W. Duurland
Holland Railconsult, Utrecht, The Netherlands

G.P.C. Oosterhout
TNO-Bouw, Delft, The Netherlands

P.S. Jovanovic
Holland Railconsult, Utrecht, The Netherlands

ABSTRACT: This paper deals with the structural analyses of the segmented tunnel lining using three-dimensional finite element model. With this model it is possible to observed stresses and deformations which occur in concrete lining during the construction stage (assembling) and latter at serviceability stage. Most of the conventional models of tunnel lining assume structural loads as full overburden at the serviceability stage and the resulted sectional forces, bending moments and deformations are used for the lining design. Loads from the construction stage when assembling the segments are essentially different from the loads at serviceability stage. Huge monitoring programme of the tunnel lining in Netherlands by the Second Heinenoord tunnel is still running. Hypotheses based on the conclusions from measurement data and the results of three-dimensional FEM analyses confirmed each other.

1. INTRODUCTION

The lining design of tunnels driven by TBM is a well known subject all over the world. There are few theories developed to predict tunnel lining behaviour that differ from each other concerning structural system. These models vary from simple single ring to the sophisticated Finite Element Models. Most of the theories assume that loading condition comes exclusively from the serviceability stage. As it could be seen from the results of measurement data from The Second Heinenoord tunnel the construction stage is more important for tunnel lining design than the serviceability stage. With the three-dimensional Finite Element Method (FEM) model, developed by Holland Railconsult, it is possible to predict stresses and displacements of the tunnel lining, as well for the construction stage as for the serviceability stage.

2. PILOT PROJECT - THE SECOND HEINENOORD TUNNEL

The Second Heinenoord Tunnel has been driven by TBM (diameter 8,35m) with precast concrete tunnel segments as ground support. The total length of the tunnel is 1350m with 950m as bored tunnel. The Ministry of Public Works decided, 1992, to name two projects The Second Heinenoord Tunnel and the Botlek Railway Tunnel as pilot projects in The Netherlands. The aim of these projects is to obtain knowledge and to get experience with tunnels driven by TBM in Dutch soft soil.

One part of complete research program deals with measurements of the tunnel lining behaviour. For this purpose are the two rings equipped (TNO-Dutch research Institute) with measurement devices such as vibration gauges, stress cells and displacement devices. Measurements have been done during the construction stage and are still running. Available measurement data have been collected and evaluated by TNO and Holland Railconsult. Some observations and conclusions led to certain hypotheses which could be used as a basement for the structural analyses and design of tunnel lining. These hypotheses are presented in this paper.

3. MEASUREMENTS AND HYPOTHESES

3.1. HYPOTHESIS 1 - ASSEMBLING STRESSES

It was concluded from de results of measurement data that the internal stresses introduced in to the lining during the assembling of segments remain in the concrete sections. The total stresses of the lining in the serviceability stage are strongly influenced by the assembling stresses. Figure 1 and figure 2 shows (for one segment, angle 160 ref. vertical as) the distribution of axial and tangential normal forces during the first hours after assembling of the second ring equipped with measurement devices.

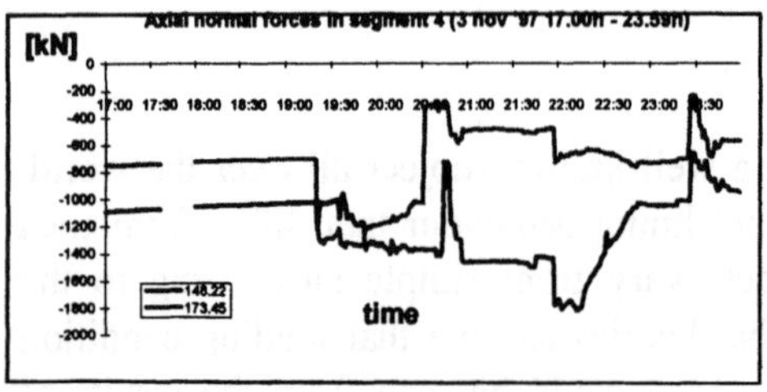

Fig. 1: Axial forces after assembling

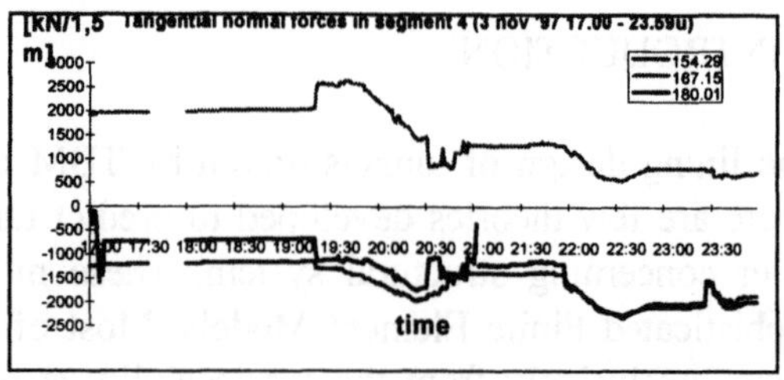

Fig. 2: Tangential forces after assembling

Figure 3 shows the measured normal forces distributed over the perimeter during the first days after assembling of the ring equipped with devices. Stresses and deformations have been measured at the moment that the ring equipped with devices was in the shield (between 5 p.m. and 7.20 p.m.). At the moment that the ring leave the shield during the tunnel driving additional loads occurred from grout and excentricity of jack forces. This can be seen on figure 1 where the axial normal forces increased caused by the influence of jacks. It is obvious (see figure 2) that the stresses are introduced in the ring direct after assembling when the ring is still in the shield and the tangential normal forces remain constant within few hours. This means that the initial stresses introduced to the ring are caused by the assembling process. The totally other screen of the stresses could be observed when the ring comes out the shield. At that moment the ring is loaded with external forces from the grouting, ground and water and the stress redistribution is changed. On figure 3 is given the redistribution of tangential normal forces over the perimeter through the longer period where also the assembling stresses could be seen (date: 3/11/97 at 5 p.m.)

3.2. HYPOTHESIS 2 : PLANE STRESSES

The results of measurements have shown that the distribution of tangential stresses around the key segment is not uniform. Figure 3 indicates that the absolute value of the tangential stresses in the key segment is low. Hypothetically speaking this could be provoked by the fact that the stresses are locally transmitted through the edges of the segments or that they are migrating from one ring to another (see fig.4). Two hypotheses have been established to explain these phenomena and both of them indicate the fact that the key segment is not properly connected with the other neighbour segments:

- non accurate assembling of the segments during the construction stage
- deformation difference between the de rings. Rings that are already assembled are loaded with external loads (grouting, ground, water). Geometric imperfection, joint position, key segment, friction in the joints.

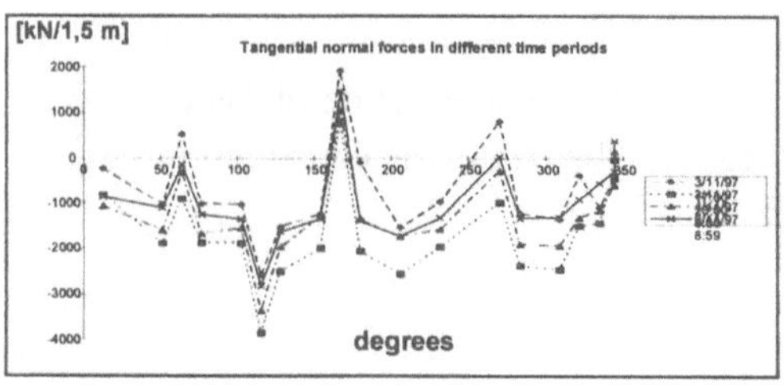

Fig. 3: Tangential forces in time

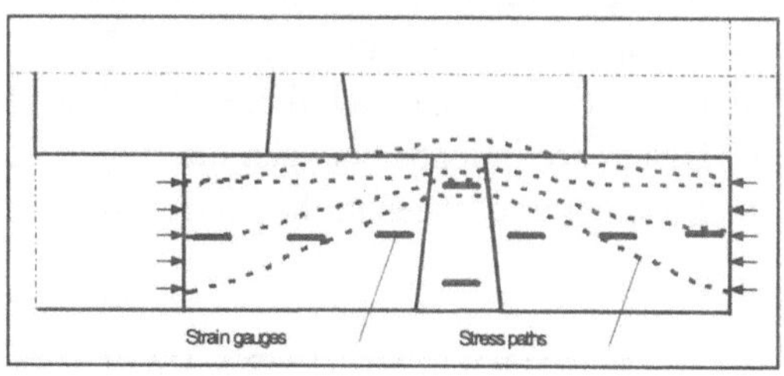

Fig. 4: Hypothetical stress paths

3.3. HYPOTHESIS 3: AXIAL NORMAL FORCES

The axial normal forces in the tunnel lining develop mainly because of the influence from the TBM jack forces. During the observation of the first ring equipped with devices it was noticed that some sort of stress relaxation occurs. The same phenomenon has been observed by the second ring with devices but here is the relaxation shown through decreasing of the axial normal forces at the top of the tunnel and stagnation at the rest of the lining. On figure 5 are both phenomena screened. Please compare the redistribution on 3/11/97 at 11 p.m. and on 6/11/98 at 8:59 a.m. (decrease interval 300°-350°). The possible explanation of the phenomenon that occur by the second ring with devices could be find on the existence of the second mechanism beside the first one:

- The first mechanism deals with the stress relaxation caused by the lining-ground interaction. This can be seen through the balance between axial normal forces in the lining and the total jack forces from TBM. The average value of jack forces shows slight decrease in time and notable decrease of axial normal forces.
- The second mechanism deals with eccentricity of the axial normal force's sum (bending moment of the tunnel as a complete section). This bending moment shows, during the first days after assembling, significant change of values. The position of the result force from TBM is close to the neutral line and after the few days sinks more down. This means that the value of axial normal forces at the top of the tunnel decrease and those at the bottom increase. The increasing of forces could be seen as a compensation of the decreasing caused by relaxation.

3.4. CONCLUSIONS BASED ON THE THREE HYPOTHESES

According to the measurement data set, first perceptions, observations and hypotheses it is obvious that the stresses in the concrete lining segments are initialised by the following factors:

- Ring's assembling
  a) Eccentric placing of the jacks on the segments
  b) imperfection of segment's dimensions
  c) non accurate assembling
- Excentricity of the jack forces (bending moment and driving correction of TBM)
- Difference in loads between the ring in the shield and the rest in the construction stage
- Load circumstances in the serviceability stage

3.5. PREDICTIONS VERSUS MEASUREMENTS

Conventional structural design models of the tunnel lining assume only serviceability loads. The influences of assembling are almost ignored and only the loading combination from serviceability stage has been shown as decisive when dimensioning lining cross-section.

The assumption that the assembling occur under stress-free conditions is contradictory with measurements. Initial stresses have been observed during the construction phase when assembling. Furthermore, the conventional models recognise only plane stress state (stresses with uniform distribution in a cross-section) in thc scgmcnt's cross-section. Totally different stress distributions have been observed, close to the key segment, during the observation of the measurement data. Moreover, the conventional models are not taking into account the influence of axial stresses in the lining and on the other hand the influence from these stresses can be seen in measurement data. It is obvious that these stresses in axial direction initialise stresses in other directions. Because of the reasons mentioned here there is a discrepancy between the conventional models and measurements.

4. THREE-DIMENSIONAL FEM MODEL HOLLAND RAILCONSULT® [2]

To explain the hypotheses based on measurements using conventional models seems to be very difficult. Because of this reason and because of obvious large difference between the theory and practice it was necessary, for Holland Railconsult, to design a three-dimensional model which could be of use for both stages, construction and serviceability stage.

This 3D FEM model had to fulfil following requirements:

- Ring assembling: excentric placing of the jack on the segment
- Excentric sum of the axial normal forces and the case without forces
- Difference in loading condition between the ring in the shield and the rest during the construction stage. Rings behind the TBM could be loaded with ground and water in comparison with the ring in the shield which has no external loads from ground and water (only jack forces)
- Ground and water loads (optionally dead weight) in serviceability stage
- Imperfection of segments and accuracy when assembling could be implied too but the attention has been paid on the items mentioned here above.

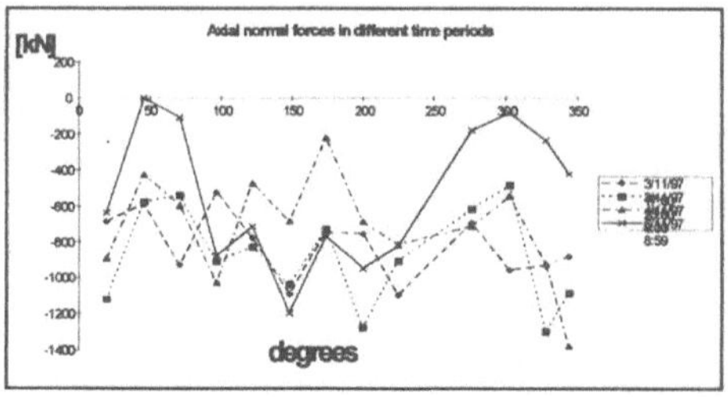

Fig. 5: Axial forces in time

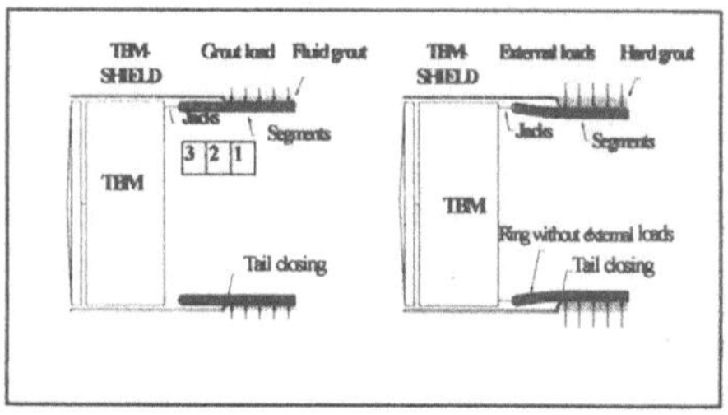

Fig. 6: "Horn-effect"

The requirements on geometry are as follows:

- Possibility to model more than one ring, each ring should consist of loose segments

- Possibility to model a key segment
- longitudinal joint should be placed in the middle of the neighbour ring
- possibility of applying various material properties to the joints

4.1. MODEL DESCRIPTION

The three-dimensional finite element model is made by Holland Railconsult using software package ANSYS. There are three rings designed with in total 8100, eight nodes solid brick elements. Each node has three degrees of freedom. Ground properties are shown applying spring-damper elements (in total 1418) at three quarter of perimeter (after Prof.Dr.H.Duddeck[5]). Interaction between the segments could be observed through the behaviour of the contact elements placed in the ring joint (in total 576) and in longitudinal joints (in total 842). These elements are capable to transmit only pressure and friction between segments.

4.2. RELATED TO HYPOTHESIS 1 - "HORN" EFFECT

The presented three-dimensional finite element model deals with the step wise process from construction stage to serviceability stage that is not possible to be observed with conventional models. Observing the deformations of the segments at construction stage it is obvious that the first ring (out of shield) has been deformed in radial direction at the top and the bottom (Uy) under the external load conditions (ground and water). These deformations are transmitted to the second ring which hangs on the third ring in shield. The deformations Uy from ring 1 are transmitted to ring 3. Because of the radial displacements of ring 3 at the top and bottom some sort of "egg" form could be seen. The Ring 3 is not supported by ground as rings 2 and 1. From the left to right side of ring 3 the radial deformations increase (Ux). This radial increase transmits from ring 3 to ring 2 and finally ring 1.

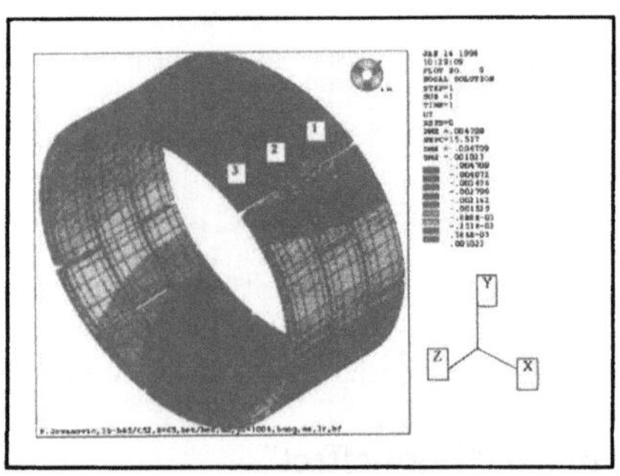

Fig. 7: Displacement in "Y" direction

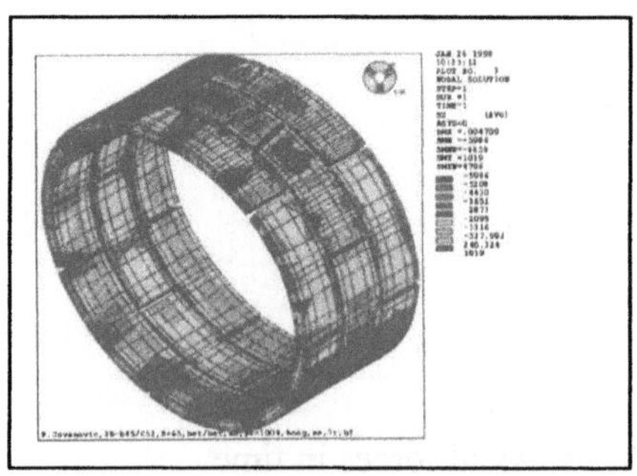

Fig. 8: Axial stresses when assembling

Radial displacements (see fig.7) of ring 3 at the top and bottom are smaller in comparison with radial displacement of ring1. Radial displacements at the sides of ring 3 are larger than at ring 1. The system diameter of ring 3 is bigger than the ring 1. The jacks are placed at ring 3 and they are transmitted to ring 1 with eccentricity. Hereby the diameter of ring 3 is increasing again because of the jacks influence (comparison with second order effects). The fact that the diameter of ring 3 is bigger than ring 2 and this one is bigger than diameter of ring 1 it seems that the structure is getting some sort of horn form (horn-effect). Because of this "horn"-effect the jack forces are introduced to the segments with eccentricity where these segments must follow the corresponding movements and rotations in all directions. This is a reason why the local stress concentrations appear in joints (see fig 8).

4.3. RELATED TO HYPOTHESIS 2: STRESS CONCENTRATIONS AROUND THE KEY SEGMENT

Damages on the tunnel segments occur quite often during construction phase. Some of them are taking place at joints or very close to the joints. Analysing the results of three-dimensional FEM calculations, for case of perfect assembling, stress concentrations have been observed. If these stress concentrations reach a value that is higher than the ultimate limit strength of the concrete section, damage will occur. Conventional lining models assume plain-stress distribution in a longitudinal concrete cross section. Because of this fact it is possible to apply beam elements when modelling. Measurements at the Second Heinenoord Tunnel show the complete disturbance of the plain-stress distribution in a segments cross section and especially near the key segment. Two possible explanations of these phenomena are analysed:

- Non accurate assembling
- Difference in deformations between the ring in the shield and the rest of the lining

Presented three-dimensional FEM assume perfect and accurate assembling stage. Figure 9 shows the ring 3 with stress distribution in tangential direction (Sx) at the construction stage. Difference between the construction stage (assembling phase) and serviceability stage could be seen through load distribution on the lining. To obtain a boundary load condition at the construction stage it is assumed that the grout hardening process behind the TBM is finished. The external load from grout, ground and water when assembling are acting on the first two rings where the third ring receive only jack forces (TBM). First two rings are supported by ground and the third ring is not. The serviceability stage assumes total load combination on all rings with complete ground support (conventional models).

4.3.1. STRESS PATH - CASE 1

Figure 9 shows a non-uniform stress redistribution under the circumstances of perfect assembling that is contrary to the assumptions used by conventional models. Higher stress values occur at the narrow side of a key segment in comparison to the wide side because of the joints geometry. Actually, stresses are not able to flow through the whole cross section but they will try to follow a path as close as possible to the segment surface. In this case the narrow part of a key segment will receive higher stress values because these stresses have to pass a shorter way from one segment to the other through key segment. On the other wide side of a key segment the distance between two segments is longer so the stresses develop through the whole thickness of the key segment but with lower stress values. This means that there is no plain-stress distribution in a cross-section.

4.3.2. STRESS PATH - CASE 2

Figure 10 shows the stress path at the longitudinal joint between the key and neighbour segments. The stress paths can be recognised both on figure 4 and figure 9 where the measured and calculated data coincide. The explanation of stress increase at the segments corners could be found in deformation differences between the rings when assembling because corners of the segments are moving and rotating in all directions as a result of complex structure - loads constellation and the "horn" effect.

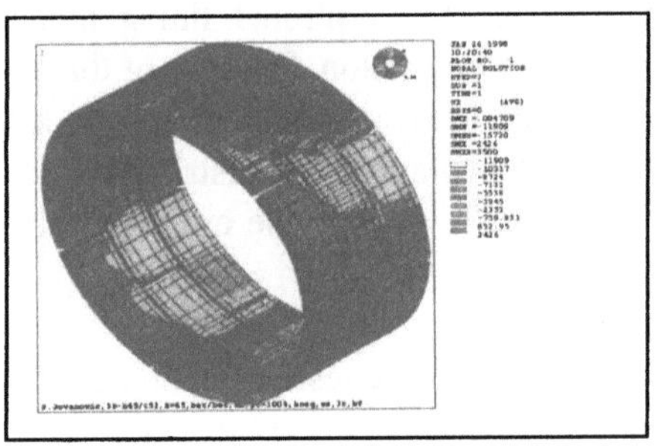

Fig. 9: Sresses in "X" direction

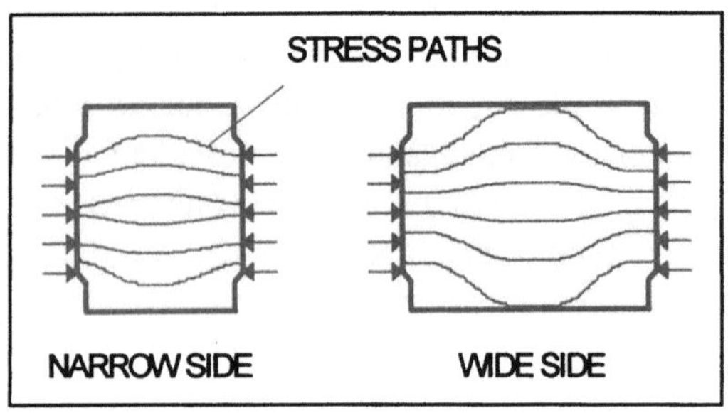

Fig. 10: Key segment - Stress paths

4.4. THE AXIAL STRESSES AND BEDDED BEAM EFFECT

Conventional design models of tunnel lining are not taking into account the axial stress contribution. Mostly, models are defined by one ring or two coupled rings with coupling stiffness which are quite difficult to determine. This coupling stiffness is directly related to the joint geometry. Usually these coupled forces should be taken by the copping system producing full shear stresses. The influence of the area near the copping that could possibly take over the shear stresses through friction (if the axial forces are available) is neglected. However, this is not neglected in the presented three-dimensional FEM model.

Measurements data at The Second Heinenoord Tunnel indicate that the eccentricity of the axial normal forces lead to the stress relaxation at the place where the pressure reaches the lowest value (i.e. mostly at the top of the lining). This relaxation have been observed in a longer time period and it still remains unchangable. Moreover, it gives definitely an important impact to the lining where the friction forces in combination with axial normal forces can develop and determine the complex structure - load's mechanism.

The excentricity of the total axial normal force is automatically assumed in the three-dimensional FEM model. This could be seen on Figure 11 where the displacements in axial direction are shown. It is obvious that the bottom of the lining is pushed more backwards than the top and the friction between the rings reach the maximum value.

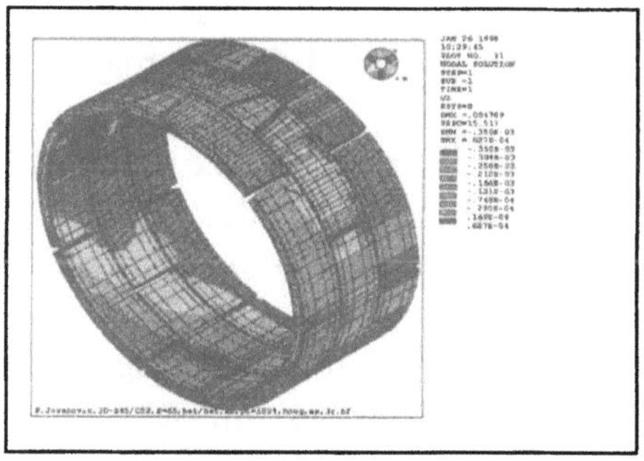

Fig. 11: Displacement in axial direction

5. CONCLUSIONS

The loads from construction stage should be taken into account when dimensioning. These loads have a definitely very important influence on general tunnel lining behaviour. Three dimensional FEM model (Holland Railconsult®) is capable to predict the influences of geometrical non-linearity and load combinations on the complex behaviour of a bore tunnel structure and at the same moment gives a possibility to predict stress distribution in the construction stage when assembling tunnel segments. These stresses remain in the lining cross section in time determining the final stresses distribution in the serviceability stage. The stress concentrations that could lead to the partial damage of the structure and at the same time shortening of tunnel's durability can be easily found and observed using this model. Hypotheses based on the conclusions from measurement data and the results of three-dimensional FEM analyses confirmed each other.

6. RECOMMENDATIONS

The further research on complex space behaviour of the tunnel structure should be held on with three-dimensional FEM models. The future model analyses have to be verified through appropriate experimental data and measurements in situ as well as in a laboratories. Furthermore, the conventional models which are usually used in engineering practice, should fulfil the requirements obtained from these verifications. The influence of the imperfection of dimensions and the non-accurate assembling on assembling stresses should be investigated. The results of this study could be applied as a requirements to the segments production and assembling procedures (i.e. protocol of assembling).

The problem of assembling stresses could be solved through reduction of "horn"-effects. The difference in deformations between the ring in a shield and the rest of the lining must be minimal. This minimising can be obtained applying following measures:

- Applying support structure. Usually named "Reformer", can be used to keep the circle form of the rings which are already placed behind the shield.
- Applying longer shield tail. More rings could be assembled within the shield. The distance between the deformed rings behind the shield and not deformed ring in the shield will be longer.
- Applying thicker lining. When he stiffness of the lining increase than the deformation differences between the rings will decrease. Thicker lining provides better stress distribution within the segments and because of this the influence of axial stresses on tunnel lining behaviour will be reduced.
- Applying steel-fibre concrete. The stress concentrations caused by assembling process could lead to cracks in segments. Steel-fibre concrete can absorb these stresses.
- Applying controlled grouting. With special grout consistency, considering the time of grout hardening, the influence of external loads on the lining deformations behind the shield could be controlled.

The tunnel lining behaviour is defined by number and type of joints. It is necessary to investigate the influence of different materials which are usually applied in joints when assembling (strips, water seals, geometry of joints, etc.). These materials could provoke unpredicted tunnel lining behaviour. One of the research programs, which is running now in Holland Railconsult, is to determine the influence of different materials in joints on tunnel lining behaviour.

REFERENCES:

BLOM, C.B.M. & G.P.C.OOSTERHOUT (1997). The Second order evaluation tunnel structure The Second Heinenoord Tunnel, part 1 en 2, Gouda

JOVANOVIC, P.S. (1996). Three-dimensional analyses of segmented tunnel lining, Holland Railconsult, Utrecht

COB (1997), News letter No.3, Gouda,

COB (1997), News No.10, Gouda,

DUDDECK, H. (1980). Empfehlungen zur Berechnung von Tunneln im Lockergestein, Die Bautechnik No.10, 349-356

AXIALLY LOADED PILE
NUMERICAL MODELS VS. EXPERIMENTAL DATA

E. Cabella and R. Passalacqua
University of Genoa, Genoa, Italy

ABSTRACT:

In the paper the performances of two numerical models are compared through the backanalysis on experimental data. The first numerical model has been implemented by the authors and, under many aspects, can be regarded as a simplified algorithm; the second one is an "open code", commercially available. The experimental data come from medium size models performed in a laboratory facility.

1. INTRODUCTION

In the present work two numerical models have been used to backanalyze a set of directly observed data, these being some of the results of an experimental program devoted to study the behaviour of axially loaded, medium-size pile models.

The main objective is to assess the performances of two programs by using as a benchmark a set of experimental data, these gained from a series of tests conducted within a laboratory facility where the relevant parameters are strictly under control.

2. THE NUMERICAL MODELS, GENERAL DESCRIPTION.

The two programs used in the present work are substantially different: the first one is a multi-purpose, commercially available code (FLAC®) which allows the user's definition of constitutive laws and the computed data manipulation in a wide variety of ways; the second one, named PILE, has been implemented directly by the authors and can simulate exclusively a pile-granular soil interaction.

2.1 THE COMMERCIAL CODE.

The commercial code is a two-dimensional, explicit difference based, program (Cundall 1976) originally developed for geotechnical and mining engineering problems. The explicit Lagrangian calculation scheme and a mixed discretization zoning technique (Marti & Cundall 1982) are used to ensure that plastic collapse and flow are modelled very accurately.

Each element of the discretization grid could behave according to either a linear or a non-linear stress-strain prescribed law. The code also contains a powerful built-in programming language which offers a way to manipulate the introduced and calculated data and also the opportunity to implement a user's defined constitutive model.

The drawbacks of using this approach instead of the classic Finite Element Method (Zienkiewicz 1977) lies in the need to perform the analysis through small timesteps and in the necessity to take into account of a damping effect (Cundall 1982).

Other advantages allowed by this program are, for instance, the opportunity of selecting the most appropriate among numerous built-in constitutive models, the possibility to create interface elements in order to simulate planes along which slip and/or separation can occur and plane strain, plane stress or axisymmetric geometry modes. The program offers a wide range of capabilities especially suited to solve geotechnical problems and complex geometry in mechanics. Particularly, it permits the modelling of steady and flowing groundwater (with fully coupled consolidation), of visco-elastic behaviour, has a structural elements library and dynamic analysis capability.

2.2 THE AUTHORS' CODE.

The PILE program is specifically devoted to simulate exclusively a pile-soil interaction. It has been developed deepening its roots into specific and related works due many Contributors pertaining to the same Academic Structure of the authors (Berardi 1961, Dalerci 1971, Berardi et al. 1973,Dalerci & Del Grosso 1978, Berardi et al. 1981, Passalacqua 1986).

The numerical model analyse the axisymmetric geometry of a cylindrical pile embedded into a non-homogeneous, nonlinearly elastic soil, under drained conditions; the axial load is transferred to the surrounding soil both by skin friction, this as a function correlating the lateral shear with the amount of the relative pile-soil slip, and by the vertical stress under the pile tip. The computational procedure takes into account of the initial stress conditions existing into the soil, of its variations as induced by the pile installation and, particularly, of the confining stress development applied from the soil to the pile due to the stress state induced from the pile to the soil. The state of displacement is computed regarding to a continuously variable modulus of deformation.

The incremental analysis allows to obtain, for every load step, the pile relative and absolute displacements along its axis, its elastic strain, the mobilised skin friction distribution and the state of stress within the surrounding soil.

In its present version a parametric algorithm has been introduced in order to better discretize the problem geometry in any specific case, by keeping the width/height ratio of the soil elements as close as possible to unity. In fact, the former version allowed to examine a dominion locked by a structural element subdivision into 10 segments and then by reaching horizontally, into the surrounding soil, at a distance from the pile axis equal to 5 times the pile diameter divided into 13 parts, and vertically down to a depth from ground level equal to 1.5 times the pile length, this divided into 15 equally spaced parts. This caused ill-mannered griddings when dealing with piles of high slenderness, having in such cases the discrete soil-elements very thin and long.

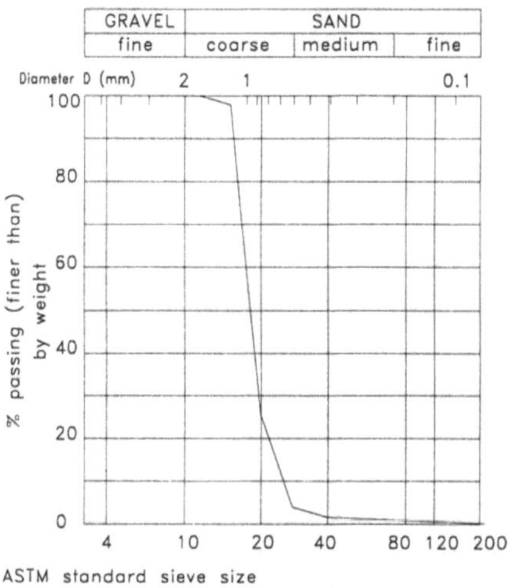

Fig. 1 Grain size distribution of the soil used in the tests.

Ticino Sand		
G_s	2.65	g/cm^3
$\gamma_{d\,min}$	1.386	g/cm^3
$\gamma_{d\,max}$	1.703	g/cm^3
D_{10}	0.65	mm
D_{50}	0.93	mm
D_{60}	0.97	mm
D_{100}	1.70	mm
$C = D_{60}/ D_{10}$	1.49	#
e_{min}	0.550	#
e_{max}	0.905	#
Dominant mineral: Quartz (≈30%)		

Table 1. Principal characteristics of the granular soil.

3. EXPERIMENTAL MODELS.

Using a sand spreader for the reconstitution of granular soil models (Passalacqua 1991) the authors performed a set of tests on axially loaded, medium scale, pile models. The use of this facility ensure the repeatability of the soil model with respect to the relevant geotechnical characteristics, in order to have always a significant set of tests (Seed et al. 1975).

Three different pile models, having diameters 40, 50 and 63 mm, were tested. They had fixed length equal to 900 mm, thus resulting in length to diameter ratios (L/D) 15, 19 and 24. A natural granular soil (see Fig. 1 and Table 1) was repeatedly deposited at need, attaining each time one of the two chosen relative densities: 45 and 70 %.

The pile models have been internally and externally instrumented in order to asses the reading of the most important geotechnical quantities.

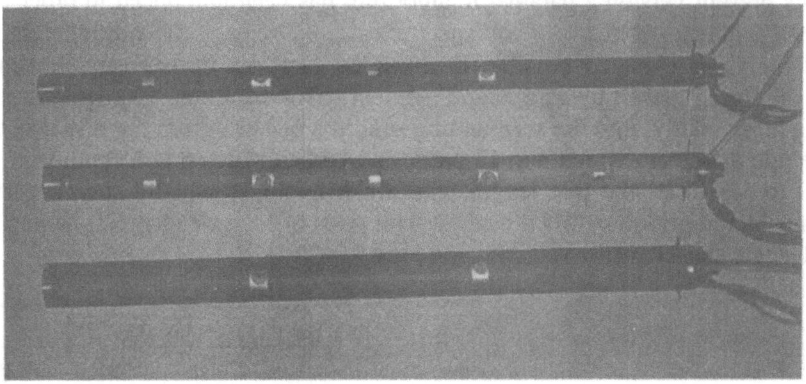

Fig. 2. Assembled pile models.

3.1 PILE MODELS INSTRUMENTATION.

Each pile model (Fig. 2) was equipped with a miniaturised point load cell, specifically designed and calibrated for that model.

The pile models were designed as formed by three parts, in order to be assembled in place during the reconstitution of the soil model around them; a transducer, sensitive to the local shear at the external surface of the pile, has been provided within every one of the two coupling elements. The three parts are made by a plastic tube and each one of them has its mid-axial section strain gaged with a full-bridge wheatstone circuit: in such way the tube itself behaves as an axial load sensor. A compression load cell and a set of three LVDTs were mounted at the pile top to measure the total applied load and the pile settlements during the load test.

The described pile models were used in a group of tests to investigate parametrically the axial behaviour of these structural elements. The external surface of a specific series was grinded at a prescribed roughness, in order to mobilise a different friction level.

The installation procedure used in the case of the referred tests aims to reproduce the behaviour of friction piles having a very slight contribution from the point bearing capacity. The pile tip in fact was open and the installation mode (with a constant velocity of penetration) was performed through continuous removal of the intaking soil. This method simulates an unplugged, open-ended pile. The instrumentation set up permits the continuous readings of: the soil-pile shear stress, the axial load distribution and the tip load, during both the installation phase and the static load test.

4. THE NUMERICAL MODEL IMPLEMENTED WITH THE COMMERCIAL CODE.

The numerical model is based upon an axisymmetric geometry with the pile schematised by fifty zones with elastic behaviour. Between the pile and the soil an interface has been generated to represent the surface between the real elements.

The soil is modelled by a regular grid of about 1800 zones. The model cover an area of 10 • R wide and 50 • R high (R max pile radius). A rigid boundary condition is defined at the extremities of the soil model.

The elements that represents the pile uses a simple linear elastic isotropic model, fully justified by the real behaviour of the material and by the magnitude of the axial load applied to the pile models.

The interface elements introduced are governed by a Coulomb shear-strength criterion; the maximum shear force is:

$$F_{smax} = c \cdot L + \tan\phi \cdot F_n \tag{1}$$

where c is the cohesion along the interface, L is the effective contact length, and ϕ is the friction angle of the interfaced surfaces. In addition to this a normal and shear elastic stiffness exists between two planes which may contact, and a tension bond is present in order to simulate the separation resistance of the interface.

The soil models used here have, mainly, a simple elasto-plastic behaviour with a Mohr-Coulomb failure envelope, while some are non linear elastic models. The Mohr-Coulomb model uses a non associated shear flow rule and the tensile flow rule is associated; the failure envelope is defined as:

$$f^s = \sigma_1 - \sigma_3 \cdot N_\phi + 2 \cdot c \cdot \sqrt{N_\phi} \tag{2}$$

where: $$N_\phi = \frac{1+\sin\phi}{1-\sin\phi}$$

The non linear elastic models use an initial constant modulus or a modulus defined by the principal stress according to Lade (Lade 1988):

$$E = M \cdot p_a \cdot \left[\left(\frac{I_1}{p_a} \right)^2 + R \cdot \frac{J_2'}{p_a^{\,2}} \right]^\lambda \tag{3}$$

where: $$R = 6 \cdot \frac{1+\nu}{1-2\cdot\nu}$$

$$I_1 = \sigma_1 + \sigma_2 + \sigma_3 \ , \ I_2 = \sigma_1 \cdot \sigma_2 + \sigma_2 \cdot \sigma_3 + \sigma_3 \cdot \sigma_1 \ , \ J_2' = \frac{I_1^2}{3} - I_2$$

p_a = atmospheric pressure

This modulus is then modified according to an hyperbolic decay model based on the principal stress difference:

$$2 \cdot t = \sigma_1 - \sigma_3 \tag{4}$$

acting at the relevant spot. The non linear law for elastic behaviour is based on:

$$\sigma_1 - \sigma_3 = \frac{\varepsilon_1}{\dfrac{1}{E_i} + \dfrac{\varepsilon_1}{Y}} \tag{5}$$

with: E_i = initial Young's modulus

Y = maximum value of $\sigma_1 - \sigma_3$

in one type of analysis and on the hyperbolic model defined by Duncan and Chang (Duncan & Chang 1970) in a second type. The tangent modulus is defined, in the Duncan and Chang material as:

$$E_t = (1 - R_f \cdot SL)^2 \cdot K \cdot p_a \cdot \left(\frac{\sigma_3}{p_a}\right)^n \tag{6}$$

with: $SL = \dfrac{(\sigma_1 - \sigma_3)}{(\sigma_1 - \sigma_3)_f}$

K,n = model parameters (Duncan et al. 1980)

$R_f = \dfrac{(\sigma_1 - \sigma_3)_f}{(\sigma_1 - \sigma_3)_{ult}}$

In case of an unloading step an unloading-reloading modulus number is used instead of the normal (K) modulus number, and the following relationship is used:

$$E_t = K_{ur} \cdot p_a \cdot \left(\frac{\sigma_3}{p_a}\right)^n \tag{7}$$

5. THE NUMERICAL MODEL IMPLEMENTED BY THE AUTHORS.

The following assumptions have been adopted to implement the PILE program:

- pile material obey laws of the elasticity theory and is characterised by the Young's modulus E_p,
- the soil response to the applied loads, in terms of effective stress under fully drained conditions, is considered assuming a linear Mohr-Coulomb's strength envelope, this having a cohesion intercept c'=0 and an effective angle of shearing resistance φ' (cohesionless material),
- a transfer law, which correlates the tangential stress to the relative displacements W at the pile-soil interface, is introduced in the model as a function of the soil properties and of the actual radial effective stress σ'_r, the latter depending on the axial load applied to the pile.

The pile-soil interaction problem, when the former is subjected to the external load P, may be schematically described as follows:

- every generic pile element "i" having length L/n is subjected to a radial stress $\sigma'_r(z_i)$ to which corresponds a mobilised shaft friction $\tau(z_i)$, being: L the pile length, "n" its divisor and "z" the absolute depth from ground level,
- the axial load on the considered pile segment is P_i,
- the vertical displacement of the pile segment is η_i and the corresponding vertical displacement of the soil adjacent to that pile segment is S_i, thus leading to a relative pile-soil interface displacement $W_i = \eta_i - S_i$,
- the axial force transmitted to the soil through the pile tip is P_B.

Under such assumptions the solution of the discretized problem may be obtained by solving "n" compatibility equations (8) and one equilibrium equation (9), leading to the determination of the "n" unknown forces P_i along the pile axis and of the tip load P_B;

namely, for i = 1..n:

$$\eta_i = W_i + S_i \tag{8}$$
$$P = P_B + \Sigma_i X_i \tag{9}$$

where:

$$X_i = \pi \cdot D \cdot L \cdot \tau(z_i) / n \tag{10}$$

from which follows:

$$P_i = P_{i-1} - X_i. \tag{11}$$

The solution of the problem requires the knowledge of the existing relationship between τ and W and of the friction coefficient on the pile-soil interface tgδ. Following previous experiences reported in literature (Kezdi 1957, Potyondy 1961, Berardi 1961, Coyle & Reese 1966, Coyle & Sulaiman 1967, Clough & Duncan 1971) the following hypotheses are adopted:

$$tg\delta(W) = W \cdot tg\delta / W_C \quad \text{when } W \le W_C \tag{12}$$
$$tg\delta(W) = tg\delta \quad \text{when } W > W_C \tag{13}$$

and assuming that the critical relative displacement W_C falls within the range 0.01÷0.5 cm, valid for a cohesionless material. In sands is generally argued that W_C decreases as relative density and pile smoothness increase. Under these assumptions, the equation (10) may be rewritten as follow:

$$X_i = \pi \cdot D \cdot L \cdot tg\delta \cdot W_i \cdot (\sigma'_{ho,i} + \sigma'_{r,i}) / (n \cdot W_C) \tag{14}$$

where $\sigma'_{ho,i}$ represents the effective radial stress after the pile installation but before the application of any external load and $\sigma'_{r,i}$ corresponds to the effective radial stress induced by the load P applied to the pile.

In order to write in explicit form the equations (10), the absolute displacements η_i and S_i are evaluated by directly integrating the closed form solution which gives the stress-strain state due to a force applied at the interior of a Boussinesq half-space (Mindlin 1936). This integration is performed only once, for the real geometry to be analysed and for unitary loads, thus obtaining a set of influence coefficients to be used during all the computational process.

While explicitating eq. (10), a clear distinction must also be made between the pile segments for which the $W_i \ge W_C$ condition is verified, corresponding to a full shaft friction mobilisation, from those where $W_i < W_C$, this meaning that only a part of the ultimate shaft friction is mobilised.

The resulting system of simultaneous equations formed by (9) and (10), the latter in their explicit form as described, is solved by means of an iterative process in order to satisfy the compatibility conditions for the shaft friction as expressed through the relationships (12) and (13).

During the computational process, performed through incrementally stepped load, the soil mass is characterised by a stress dependent deformability modulus E' and a constant Poisson coefficient ν'. In order to perform the computations as described above, an average soil deformability, representative of the discretized soil cylinder, is considered. Therefore the value of E' adopted during each computation step is the average performed upon the single values pertaining to each discretized soil element, this weighted with respect to the induced vertical stress σ'_{zz} acting on that element.

An hyperbolic stress-strain decay criterion (Duncan & Chang 1971) is adopted, leading to the following relationship for the definition of the tangential modulus value:

$$E_t' = E_i' \cdot (1-f_d)^2 \quad (15)$$

where f_d is a coefficient of mobilisation for the ultimate soil strength, defined by:

$$f_d = (T-T_0) / (T_{MAX} - T_O) \quad (16)$$

being: T the actual stress level in the considered soil element,
T_0 the initial stress level in that same element,
T_{MAX} the stress level at failure.

The stress level T is defined, according to Lade (1977, 1978), by means of:

$$T = (I_1^3 / I_3 - 27) \cdot I_1^q \quad (17)$$

as a function of the first and the third stress invariant, while T_{MAX} and the exponent q are specific sand properties to be measured from appropriate triaxial tests.
The use of equation (17) allows to take into account of the complete stress state (σ_1', σ_2' and σ_3') and of the resulting principal planes rotations for every soil element.
The initial soil deformability E_i' is computed, before each calculation step, according to Janbu (1963) by the empirical relationship:

$$E_i' = m \cdot p_a \cdot (\sigma_3' / p_a)^\alpha \quad (18)$$

in which the value of σ_3' resulting from the previous computational step is introduced.

The algorithm, as schematically described above, was coded in a structured FORTRAN 90 program by using a PC-platform compiler/linker capable of addressing the extended memory up to 4 gigabytes.

As a raw indication, a complete PILE run considering n = 100 subdivisions for the structural element (to which corresponds a surrounding soil mass discretization into 1750 elements) and 30 load steps is executed in 45 seconds on a Pentium class PC (133 Mhz) having 16 Mbytes RAM.

6. NUMERICAL VS. EXPERIMENTAL RESULTS COMPARISONS AND CONCLUSIONS

The numerical computations are compared with the experimental data in the following Figure 3. Due to the lacking of space only the data for pile dia. 63mm into a D_R = 70% deposit are reported, being in any case the others consistent with these.

Regarding the commercial code, it can be seen that the best agreements are reached when both the hyperbolic and the Duncan-Chang model is adopted but only if using, in each case, different moduli during the virgin load and unload-reload conditions. The consequent computational procedure thus needs a first phase to simulate the pile installation, a second one to reach the unloaded condition and then the load test simulation phase. This full process involve a computing time close to six hours on a Pentium PC (166 Mhz).

The PILE program performs rather correctly in the case of the smooth piles simulations; for rough piles it exhibit a clear disagreement and this fact suggest that some improvement have to be implemented in it such as, for instance, the emulation of the contrasted dilatancy at the pile-soil interface which surely is highly mobilized in those cases. The computing time is negligible, as said before, and will become a great advantage when pile groups will be implemented through such model. An other improvement will regard the implementation of a suitable model for cohesive materials.

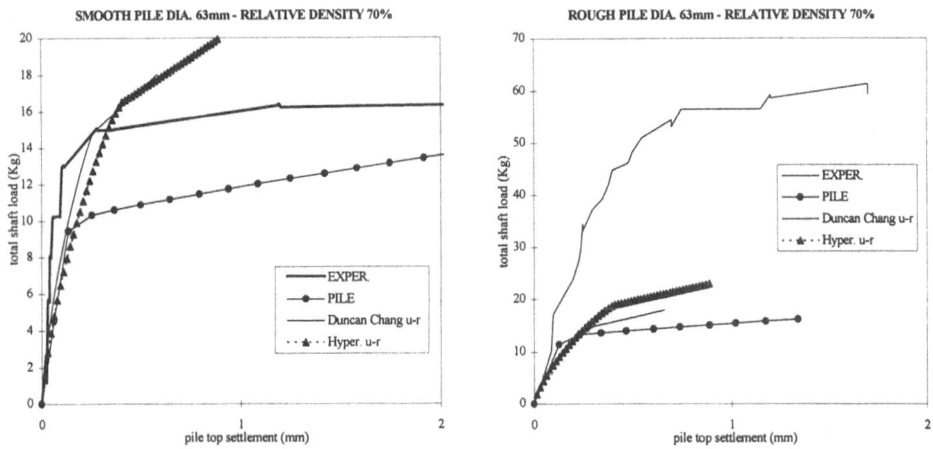

Fig. 3 Numerical versus experimental results comparisons.

7. ACKNOWLEDGEMENTS.

This research had its financial backing part from a MURST (Italian Ministry for University, Scientific and Technological Research) 40% quote grant and part from a University of Genoa, Athenaeum Type, fund.

Thanks are due to Professor G. Berardi for his continuous stimulating support and to whom have to be acknowledged the original formulation of the PILE algorithm.

Finally, thanks are due to the lab technician G. Bertoloni for his precious contribution during the design and the setting up of the experimental tests.

REFERENCES

BERARDI G. (1961). Sul comportamento del palo di fondazione in terreno incoerente. Atti dell'Università di Pisa, n° 85, Pisa, Italy

BERARDI G., DALERCI G., & A. LA MAGNA (1973). Stato di tensione-deformazione, in fase di esercizio, del palo in terreno incoerente. Proc. XIth Conf. of Italian Geotechnical Association (AGI), Theme 1: Deep Foundations - publ. n.r 3, Milan, Italy

BERARDI G., LA MAGNA A., & R. PASSALACQUA (1981). Pile embedded into a nonlinearly elastic medium. Proc. Xth ICSMFE, Session 8, vol. 2, pp. 619-624, Stockholm, Sweden

CLOUGH G. W. & J. M. DUNCAN (1971). Finite element analyses of retaining wall behaviour. ASCE Journ. SMF Div., 8583, Dec. pp. 1657-1673

COYLE H. M. & L. C. REESE (1966). Load transfer for axially loaded piles in clay. ASCE Journ. SMF Div., 4702, Mar. pp. 1-26

COYLE H. M. & I. H.SULAIMAN (1967). Skin friction for steel piles in sand. ASCE Journ. SMF Div., 5590, Nov. pp. 261-278

CUNDALL P. A. (1976). Explicit finite difference methods in Geomechanics. Proc . Conference on Numerical Methods in Geomechanics, Blacksburg, Virginia, 1: 132-150

CUNDALL P. A. (1982). Adaptive density-scaling for time-explicit calculations. Proc. 4th Int. Conference on Numerical Methods in Geomechanics, 1: 23-26, Edmonton, Canada

DALERCI G. (1971). Sulla piastra circolare rigida caricata simmetricamente profondamente immersa in un semispazio elastico. Atti dell'Istituto di Scienza delle Costruzioni, Università di Genova, Vol. 5, pp. 75-81, Genova, Italy

DALERCI G. & A. DEL GROSSO (1978). Analisi del comportamento non lineare del complesso palo-terreno mediante l'uso di modelli ad elementi finiti. Atti dell'Istituto di Scienza delle Costruzioni, Università di Genova, Serie IV, pubbl. 13, Genova, Italy

DUNCAN J. M. & C-Y. CHANG (1970). Non linear analysis of stress and strain in soils. ASCE Journ. of Soil Mechanics, 96 (SM5), pp. 1629-1653

DUNCAN J. M., P. BYRNE, K. S. WONG & P. MABRY (1980). Stregth stress-strain and bulk modulus parameters for finite element analyses of stresses and movements in soil masses. University of California, College of Engineering, Rep. No. UCB/GT/80-01 Berkeley, U.S.A.

JANBU N. (1963). Soii compressibility as determined by oedometer and triaxial tests. European Conf. of Soil Mech. & Foundation Eng.rg, Vol. 1 pp. 19-25, Wiesbaden, Deutschland

KEZDI A. (1957). Bearing capacity of piles and pile groups. Proc. 4th Int. Conf. of Soil Mech. & Foundation Eng.rg, Vol. 2, London, England

LADE P. V. (1977). Elasto-plastic stress-strain theory for cohesionless soil with curved yield surfaces. Int. Journ. Solids and Structures (13) pp. 1019-1035, November

LADE P. V. (1988). Model and parameters for the elastic behaviour of soils. Proc. of Numerical Methods in Geomechanics, pp. 359-364, Innsbruck, Austria

MARTI J. & P. A. CUNDALL (1982). Mixed discretization procedure for accurate solution of plasticity problems, Int. Journal Numerical Methods and Analytical Methods in Geomechanics, 6: 129-139

MINDLIN R. D. (1936). Force at a point in the interior of a semi-infinite solid. Physics, May, Vol. 7, pp. 195-202

PASSALACQUA R. (1986). Behaviour of a single pile subjected to lateral loading: experimental versus numerical model. Proc. 3rd Int. Conf. on Computational Methods and Experimental Measurements. Vol. 2, pp. 673-687, Porto Carras, Greece

PASSALACQUA R. (1991). A sand spreader used for the reconstitution of granular soil models, Soils and Foundations, 2: 175-180.

POTYONDY J. G. (1961). Skin friction between various soils and construction materials. Géotechnique, Vol. XI, pp. 339-353, London, England

SEED H. B., J. P. MULILIS & C. K. CHAN (1975). The effect of method of sample preparation on the cyclic stress-strain behaviour of sands. Rep. E. E. R. C. No. 75-18, Coll. of Eng., University of California.

ZIENKIEWICZ O. C. (1977). The finite element method, 3rd Edition, London, McGraw-Hill.

A NUMERICAL ANALYSIS OF SOME STRUCTURAL AND GEOTECHNICAL PROBLEMS OBSERVED AT VAT PHU MONUMENT SITE

A. Cividini, G. Gioda and D. Sterpi
Polytechnic of Milan, Milan, Italy

ABSTRACT: The relevant geotechnical and structural problems characterizing Vat Phu monument site (Lao People's Democratic Republic) are briefly outlined, on the basis of a preliminary site visit and of the performed in situ investigation. After recalling the historical background of the site, the main engineering aspects to be considered in the possible restoration of its monuments are outlined. Then, the results of some numerical analyses are presented, which aim at obtaining an insight into the causes that induced the observed severe damages of the ancient structures. The numerical results lead to some conclusions on the possible provisions which can be adopted for the stabilization and restoration of the site.

1. INTRODUCTION

Vat Phu [Pichard, 1997] is located in the southern region of Lao People's Democratic Republic, in the vicinity of the town of Champasak, 5 km west of the Mekong river. Historical evidence shows that this zone had a prominent role during the growth of the ancient Khmer kingdom. In fact, a wide walled city existed in this area from the 5th century AD. Most of the Hindu temple complex of Vat Phu was erected in its present form during the 11th century AD at the foot of the Phu Kao mountain, formerly known as Lingaparvata. The remains of several religious buildings are present in the zone, as well as channels, causeways, artificial ponds (baray) and stone quarries used for building the monuments. Quite probably Vat Phu was connected to the ancient city of Angkor, seat of the Khmer empire, by a 250 km long road.

Due the relevant archeological potential of this site, and to his significance for the history of mainland Southeast Asia between 5th and 13th centuries, in the past years various projects concerning Vat Phu have been financed by the United Nation Educational, Scientific and Cultural Organization (UNESCO). They aim at planning, and implementing, the archeological and architectural conservation of this site, within a wider environmental program for the protection of the whole area. Part of this work is presently carried out by the Lerici Foundation (Rome).

In the framework of this activity, a site inspection has been carried out by a group of researchers of the Technical University (Politecnico) of Milan to investigate the possible causes of damages of the monuments and to work out guidelines for their restoration. In particular, the hydraulic, geotechnical and structural problems characterizing the site have been analyzed. In the following, only those having structural and geotechnical nature will be briefly discussed. Further details can be found in [Cividini & Gioda, 1998].

2. GEOTECHNICAL OVERVIEW OF THE SITE

From the geotechnical view point Vat Phu area can be subdivided into three main portions: the lower zone, roughly flat, where two main buildings, known as north and south "quadrangles", are located (cf. fig.1); the slope, with the terraces and the stairway, (fig.2) and the upper zone, hosting the Hindu shrine, (fig.3) bounded on its east side by the slope and on its west side by the upward, almost vertical, rock wall of Lingaparvata.

The superficial soil in the lower zone apparently consists of a man made fill the thickness of which increases from north to south. The rectangular quadrangles are founded on this fill. They have a length of about 67m (in the east-west direction) and are 44m wide. Each side, or "gallery", of these buildings consists of two parallel vertical walls, about 3m high (fig.4). Presently, no roof exists on the galleries, but it is possible that originally they had a wooden roof covered with tiles.

The walls rest on a massive "base", located above ground level, having a height of about 2m. A shallow, continuous foundation exists below the base. In some places, this foundation is partially exposed by the erosion caused by the water running on the ground surface during the rainfalls. The four galleries encompass a central courtyard having an elevation close to, or higher than that of the external ground.

Blocks of sandstone and laterite were mainly used for erecting the quadrangles, but it was not known whether the inner part of their base consists of rock blocks or of a soil fill.

Various portions of the walls are markedly tilted towards the central part of the galleries, as shown in fig.5, regardless their internal or external location with respect to the courtyard. Some portions of the walls collapsed due to their exceedingly large inclination

From the present aspect of these structures it can be guessed that their failure depends on several reasons, which can be subdivided into "random" and "systematic" causes. Among the random causes, one can consider the growth of vegetation and local settlements of the

Fig. 1. Lower part of Vat Phu site with north (left) and south (right) quadrangles.

Fig. 2. View of the slope.

Fig. 3. Front part of the shrine, close to the slope.

Fig. 4. Internal view of one of the galleries.

Fig. 5. Tilting of the walls of one of the galleries.

soil, which could depend on the non homogeneous extension of the bearing layers.

It should be considered, however, that the systematic inward tilting of the walls cannot be induced solely by random actions. They should rather depend on causes having a systematic nature. Some of them could be related to the features of the method of construction adopted by the Khmer, like e.g. the absence of cementing material between the stones. Also the use of exceedingly shallow foundations can be one of these causes. In fact, the settlements could partially depend on the erosion of the water running on the ground surface during the heavy rainstorms characterizing the region. Another systematic cause of failure could be represented by the particular mechanical and physico-chemical properties of the materials used for walls and base. For instance, the limited shear strength of laterite [Bawa, 1957; Townsend, 1985], and its tendency to loose strength when subjected to wetting/drying cycles, could induce a progressive deformation of the zones with high stress concentration.

A third and crucial, from the geotechnical view point, factor influencing the tilting of the walls is related to the behaviour of the foundation soil. This is apparently a man made fill of natural soils which could not have been properly compacted. The settlements of this fill, perhaps induced also by the removal of its fine particles by the action of seeping water, could be one of the causes of the observed damages of the quadrangles.

Finally, the tilting of the walls could have been increased during time by the action of severe external loads, the origin and amplitude of which deserve further investigation.

The central part of Vat Phu site consists of a slope (fig.2), within which a variety of rock elements are present with dimensions varying from few tens of centimeters to some meters. The major ones belong to the natural slope, while smaller blocks are parts of failed structural elements erected by the Khmer. In fact, the original shape of the slope was modified through the construction of two series of terraces. These are located on the north and south sides of the central stairway connecting the lower zone to the upper shrine.

It is likely that the construction of the terraces required an increase of the total volume of soil constituting the slope. This, in turn, led to the reduction of its global factor of safety.

The terraces located, in particular, on the north portion of the slope are heavily damaged, and many parts of them collapsed. This could be caused by an extremely slow "creep" movement of the slope. The movement, perhaps induced by the increment of weight occurred during the construction of the terraces, can be influenced by the pore water pressure build up taking place during the rainy season, and might be locally emphasized by the non homogeneous distribution of the shear strength of the slope.

Erosion of the surface soil, due to the action of the rainfalls, is present at various locations of the terraces. This is an additional cause of the observed failures.

Settlements and failures are present also in various parts of the stairway. These seem related to the mentioned creep movement of the slope and to the induced marked compressive stresses in the structure of the stairway.

The third part of Vat Phu site, referred to as upper zone or upper terrace, is located about 60 m above the lower zone. This is a flat ground limited on its east side by the terraced slope and on its west side by an almost vertical rock wall which extends upwards, reaching an elevation much higher than that of the upper terrace.

The Hindu shrine is a relatively small rectangular building (32m by 8m) having its main dimension oriented in the east-west direction. Its front portion, close to the slope, consists mainly of outer sandstone blocks covering internal brick walls, while only bricks were used for its back portion.

The severe damages of the shrine where probably induced by large differential settlements of its foundations, and perhaps also by the impact of rock blocks falling from the vertical rock wall. The differential settlements of the two main sides of the building have different distributions and values. This led to appreciable deformation, and failure, of the structural members of the shrine.

It is likely that the upper zone originally consisted mainly of rock fragments and blocks fallen from the rock wall, and that it was not suitable for the construction of the shrine. Probably the Khmer leveled out this terrace with the same soil fill used in the lower zone. As a consequence, the foundations of the shrine rest on a highly non homogeneous mixture of rock elements and soil, characterized by a marked spatial variation of its strength and compressibility. This non homogeneity is a possible cause of the observed differential settlements. In addition, the creep movement of the slope can have some influence also on the upper zone and, hence, could be a second cause of the settlements.

3. EXPERIMENTAL INVESTIGATION

In situ and laboratory tests have been planned to improve the available geotechnical information. The in situ investigation was completed during the site visit, and consists of the following steps: topographical and GPS survey for assessing the position of some relevant points; boring tests for determining the sequence of soil layers and for recovering soil and rock samples; cone penetration tests for assessing the in situ mechanical properties of the foundation soil; drillings executed within the galleries of the quadrangles in order to characterize the material constituting the inner part of their base.

The laboratory tests on the recovered soil and rock samples, and on laterite samples from the quadrangles, has not being completed yet. Here only the findings of the in situ investigation are summarized.

Six drillings (WP) and 15 cone penetration tests (P) were carried at the locations shown in fig.6. Two boring tests (WP05 and WP06) were carried out within the base of the north quadrangle, to identify the material constituting its internal part.

The marked non homogeneity of the soil in Vat Phu area does not allow for a schematic interpretation of the borings. It can be observed, however, that an upper layer is generally present containing vegetation, like roots or fragments of wood, and in some cases fragments of bricks. The thickness of this first layer is of the order of 50 cm, or less. Below it, up to the maximum explored depth of about 5 m, different kinds of soils are found that consist mainly of sand and gravel, sandy silt, silty sand and clayey silt, with the presence of sandstone elements. Fragments of laterite or bricks, used as construction materials by the Khmer, are also found below the upper layer, e.g. at a depth of about 1.5 m in WP04. This supports the hypothesis that shrine and quadrangles were at least partially founded on a soil fill.

The two borings within one of the galleries showed that the internal part of the base, starting from an elevation of about 2m above the level of the surrounding ground, can be schematically subdivided into three main layers: an upper layer of soil, having a thickness of 30÷40 cm, that consists of particles transported by water and wind; a layer of stones, with thickness of 50÷60 cm, that represents the original pavement of the gallery; a soil fill, down to the elevation of the surrounding ground.

The presence of a relatively large quantity of soil within the base of the gallery involves that the stiffness of this structural component is affected by a non negligible variation along its section. In fact, the internal part of the base is likely to be more deformable than the stone assemblage constituting its outer portion.

This non homogeneous stiffness distribution could be one of the causes of the inward settlements of the gallery walls. To check this hypothesis, some numerical analyses have been performed, which will be discussed in a subsequent Section.

The penetration tests (P, in fig.6) were performed by means of a portable, hand driven penetrometer similar to, but smaller than, the Dutch Cone. This apparatus was chosen because the operator can bring it up to the shrine without appreciable effort. The penetration of the conical tip is produced by the blows of a hammer, hand driven by the operator. A

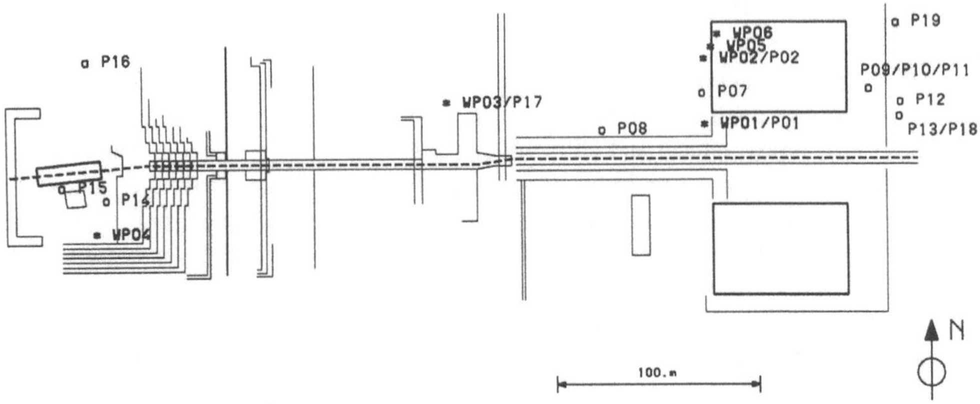

Fig. 6. Location of the performed drillings (WP) and cone penetration tests (P).

small computer, connected to a series of transducers, works out the value of the mechanical resistance q_c of the soil below the cone during penetration. Unfortunately, this apparatus showed severe limits when used in a non homogeneous deposit, since the presence of even small boulders prevented further advancing of its conical tip.

4. NUMERICAL ANALYSES

Some numerical analyses, concerning the walls of the quadrangles and the slope, have been carried out to get an insight into the mechanisms that led to the observed damages and failures. These calculations are based on values of the relevant mechanical characteristics estimated on the basis of the results of the in situ investigation.

It was previously observed that the inward inclination of the walls of the quadrangles depends on various causes, among which the non homogeneity of their base and the possible action of severe external loads were mentioned. One of the possible hypotheses on the nature of the external loads is that they depend on strong winds during the rainy seasons.

A preliminary attempt to evaluate quantitatively the effects of such a kind of lateral load has been based on the finite element analysis of a section of the gallery of the north quadrangle. The calculations were carried out in plane strain conditions adopting the mesh shown in fig.7, which consists of four node, quadrilateral isoparametric elements. The centerline of the gallery was assumed as axis of symmetry and, consequently, only half of the gallery and of the underlying base was discretized.

The finite element code SoSIA2, for Soil Structure Interaction Analysis [Cividini & Gioda, 1992], was adopted for the analyses. They were carried out in terms of effective stresses assuming an elasto-plastic behaviour of the materials. The yield surfaces for both soil and stone assemblage depend on the frictional and cohesive components of the shear resistance. A tension cut-off was introduced to account for the negligible tensile strength of these materials. The plastic flow rule is non associative, considering that the angle of dilatancy is in general markedly lower than the friction angle.

Even though the adopted program can consider strain softening effects [Sterpi, 1997], this feature was not used in these preliminary analyses. In fact, a quantitative evaluation of the relevant parameters can be obtained only after completing the laboratory tests.

The performed finite element analyses consist of three main steps. The first one concerns the evaluation of the in situ stress state before the quadrangles were erected. This was done considering a unit value of the coefficient of earth pressure at rests and adopting a linear variation of the vertical effective stress with depth. In this first analysis, only the lower part of the mesh, discretizing the foundation soil, was "activated". Then, the construction of the base of the quadrangle and of its walls was simulated by activating the elements belonging to the upper part of the mesh and by adding the contribution of their own weight.

The stress and strain states obtained at the end of this second step represent the initial condition of the structure, before the external loads or other kind of actions are applied to it.

These results show that only a minor permanent deformation develops within the base and in its foundation under the structure own weight. As expected, a moderate tilting of the wall is also observed towards the central part of the gallery.

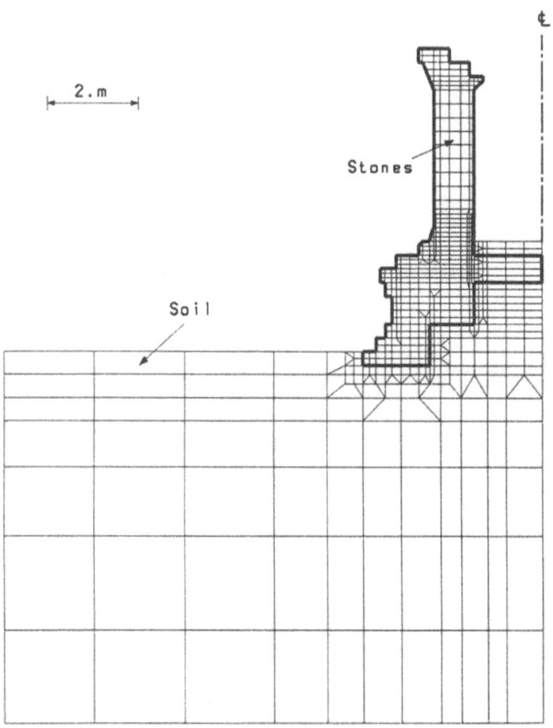

Fig. 7. Finite element mesh for one of the walls of the north quadrangle.

During the third, and final, part of the analysis the external side of the wall was subjected to a uniformly distributed horizontal pressure, increased by small increments, which represents the action exerted by a possible strong wind. Two calculations were performed. The first one refers to the case in which no "erosion", or damages due to weathering, are present in the wall before the application of the external load. On the contrary, erosion is present in the second case and is simulated by removing some elements from the lower part of the internal side of the wall, before applying the wind pressure. In both cases the external load has been increased until the collapse of the wall is closely approached.

During the loading process, tensile vertical stresses tend to develop within the finite elements located in the lower part of the wall, which cannot be supported by the stone assemblage. When this condition is detected at the end of a loading increment, the corresponding element is removed from the mesh before initiating the next increment. This can be viewed as an approximated but effective procedure for estimating the development of a tensile crack at the base of the wall under an increasing horizontal load.

As an example, fig.8 shows the contour lines of the square root of the second invariant of the deviatoric plastic strains for the "eroded" cases in the limit equilibrium state. The horizontal segment at the base of the wall represents the mentioned tensile crack.

The diagrams in fig.9 report, for the two mentioned cases, the variation of the maximum horizontal displacement of the top of the wall with increasing horizontal pressure.

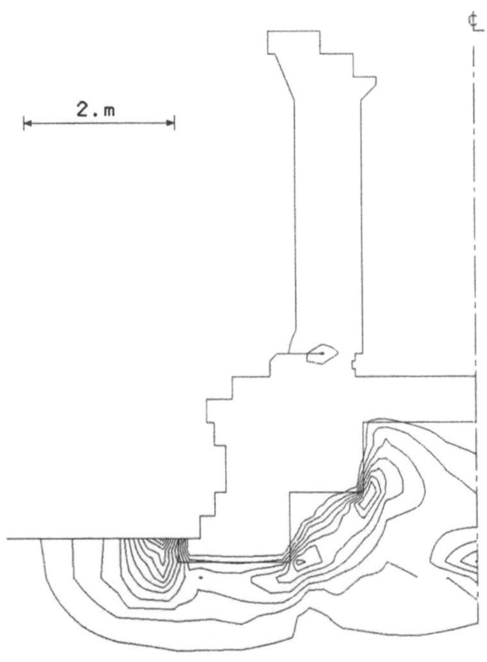

Fig. 8. Contour lines of deviatoric plastic strains after base erosion, under a lateral pressure of 160 kg/m^2 (min.value=0.05%; Δ=0.05%; max.value=0.60%).

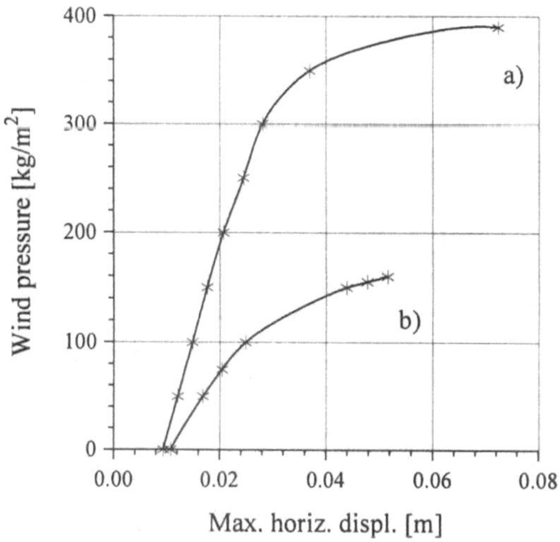

Fig. 9. Variation of the horizontal displacement of the top of the wall with the lateral pressure: a) before base erosion; b) after erosion.

From the results of the finite element analyses it can be concluded that the non homogeneous distribution of soil and rock within the base of the quadrangles is one of the initial causes that led during time to the inward tilting of their walls. This tilting increases under the action of a lateral pressure simulating the effect of a strong wind, until a collapse condition is attained. Note, in fact, that the plastic strains in the foundation soil never reached particularly high values (cf. fig.8). This indicates that a collapse condition due to excessive loading of the foundation is far from being reached.

Finally, the diagrams in fig.9 show that even a relatively small weakening of the inner part of the walls could lead to an appreciable reduction of the overall resistance of the structure with respect to horizontal loads.

It has been previously observed that the slope between the lower and upper zones could be subjected to a slow creep movement. This could be one of the causes of the damages observed both along the slope and in the upper terrace, where the shrine is located.

It should be considered, in fact, that in many instances this kind of movement is associated to the presence of relatively weak soils within the slope that leads, in turn, to values of its global factor of safety close to 1. Since layers of silty sand and sandy silt, with a limited shear resistance, have been found during the in situ investigation, an experimental evidence exists that supports the mentioned hypothesis.

To reach a conclusion about the possible occurrence of a creep movement, it was decided to evaluate the stability of the shallow part of the slope by a series of two dimensional stability analyses, which concerns its vertical cross section through the central stairway. They were performed by program STARS, for the STability Analysis of Reinforced Soils [Cividini, 1995], which can handle slopes having a general shape, also in the presence of systems of one dimensional, linear reinforcements (like e.g. tie backs, geotextile sheets, etc.).

The calculations are based on Rosenbrock algorithm, for the direct search minimization of non linear functions [Himmelblau, 1972], that finds among the class of possible failure surfaces chosen by the user (e.g. those having circular or log spiral shape, those meeting the slope surface at a chosen point, etc.), the one fulfilling a given requirement on the value of the factor of safety FS (e.g. FS=1; minimum value of FS; etc.). At each step of the iterative procedure, the factor of safety for the current shape of the sliding surface is evaluated through the modified Bishop method [Bishop, 1954].

The stability analyses were carried out considering a wide range of variation of the overall shear resistance of the slope. In particular, the friction angle varies between 15° and 30°. In fact, a precise estimation of this parameter is not straightforward, due to the non homogeneous composition of the soil and to the presence of a variety of rock elements within the slope that lead to sharp variations of its local strength. The calculations where carried out assuming a circular shape of the sliding surface. They were limited to the shallow part of the slope, close to the ground surface, since no evidence of deep movements was noticed during the site inspection.

The numerical results are summarized in fig.10 which shows, for the above mentioned values of the friction angle, the sliding surfaces that correspond to collapse (i.e. factor of safety FS=1) and that meet the ground surface at the base of the 5th terrace. These data indicate that the shallow, upper portion of the slope is close to an unstable condition, even for relatively high values of its overall frictional resistance, and that the consequent failure mechanism could involve part of the upper zone, where the shrine is situated.

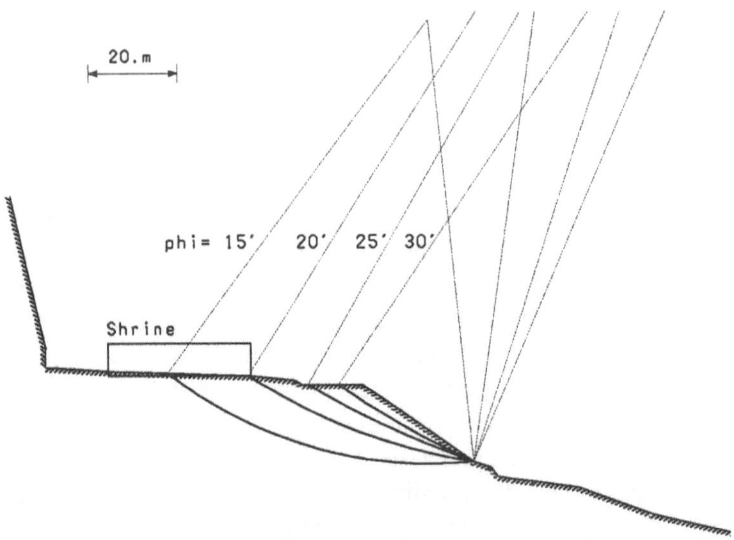

Fig. 10. Failure surfaces (factor of safety FS=1 for different values of the friction angle ϕ (vertical pressure due to the shrine $q=40\ kN/m^2$).

5. CONCLUDING REMARKS

The site inspection and the experimental and numerical results lead to some conclusions on the geotechnical and structural problems characterizing Vat Phu area. On this basis, some possible geotechnical provisions could be suggested for the architectural restoration of the site.

The action of the water running on the ground surface during the heavy rainfalls led to erosion phenomena in various parts of the slope, as well as in the soil surrounding the foundations of the quadrangles. This action, and the associated increase of the pore water pressure, is one of the causes of the extended failures of the terraces, in particular of those located on the north side of the stairway.

Another cause of these damages is a slow creep movement of the slope that is probably induced, from the one hand, by the mentioned increase of pore water pressure and, from the other hand, by the weight of the soil fill used by the Khmer for constructing the terraces.

Since it is not possible to "unload" the slope by removing part of the soil fill, or to "reinforce" it by the construction of heavy retaining structures, the only possible way for improving its stability reduces to the execution of proper hydraulic works able to reduce the effect of the rainfalls and the consequent build up of the pore pressure.

The major damages suffered by the "quadrangles" are related to the exceedingly large inward inclination of their walls that developed during time. This, in turn, has two relevant causes, namely: the settlement of the internal part of the base, which consists of a soil fill and is more deformable than the outer part of the base made of laterite blocks, and the action of lateral loads on the walls, such as those due to the wind.

Even though differential settlements are present in some places, caused by the non homogeneity of the foundation soil or by the erosion of water, they can be considered as a secondary cause of damage, with respect to the previous ones.

It should be also pointed out that many parts of the quadrangles failed because the Khmer apparently did not use any cementing material between the stones of walls and foundations.

The above considerations show that the hydraulic provisions alone, aimed at reducing the erosion of water, are not sufficient to improve the stability of the quadrangles. The structural and geotechnical remedial should, either, drastically reduce the possible level of tensile forces within the walls and in their foundation, or, increase their tensile resistance.

A temporary provision of the first kind, involving the use of bracing for stabilizing the walls of the galleries, is relatively straightforward. A possible provision of the second kind would involve the anastylosis of the quadrangles. In this case a continuous, reinforced concrete foundation can be prepared for the structure, prior to its re-assembling, with a marginal cost compared with that of the entire operation. During re-assembling, the shear and tensile resistance between adjacent stones could be increased by connecting them through internal metal pins, which would not have any appreciable impact from archeological and architectural view point.

The Hindu shrine located in the upper terrace presents differential settlements much higher than those of the lower quadrangles. In this case the possibility of planning the anastylosis of the shrine, or at least of its stone portions, should be carefully considered. In fact, it should not be overlooked that the severe damages to its external and internal structures would increase the difficulties of a proper underpinning of the foundations and that, in the present conditions, a mere restoration of the foundation would not be effective and would not appreciably reduce the rate of degradation of this structure.

ACKNOWLEDGEMENTS

The authors are indebted to Patrizia Zolesi and Mauro Cucarzi, of Lerici Foundation (Rome), and to Phakhanxay Sikhanxay, of the Ministry of Information and Culture of Lao People's Democratic Republic, who made this investigation possible through the financial support of the United Nation Educational, Scientific and Cultural Organization (UNESCO). The advice of Giulio Ballio, Silvio Franzetti and Pierre Pichard is gratefully acknowledged.

REFERENCES

BAWA K.S. (1957). Laterite soils and their engineering characteristics, J. Soil Mech. Found. Div., Proc. ASCE, Vol.83, No.SM4.

BISHOP A.W. (1954). The use of the slip surface in the stability analysis of slopes, Geotechnique.

CIVIDINI A. (1995). Limit equilibrium methods for the stability analysis of reinforced earth structures (in Italian), Proc.19th Italian Congress on Soil Mechanics, Pavia.

CIVIDINI A. & G. GIODA (1992). Finite element analysis of direct shear tests on stiff clays, Int. J. Numer. Anal. Methods in Geomechanics, Vol.16.

CIVIDINI A. & G. GIODA (1998). Geotechnical aspects of the restoration of Vat Phu site, Internal Report prepared for Lerici Foundation, Department of Structural Engineering, Politecnico di Milano, May 1998.

HIMMELBLAU D.M. (1972) Applied non-linear programming, McGraw-Hill, New York.

PICHARD P. (1997). The conservation of Vat Phu temple, UNESCO Assignment Report FIT/536/LAO/70, Bangkok.

STERPI D. (1997). Influence of the strain localization phenomenon on the stability of underground openings (in Italian), Ph.D. Thesis, Department of Structural Engineering, Politecnico di Milano.

TOWNSEND F.C. (1985). Geotechnical characteristics of residual soils, J. of Geotechnical Engineering, Vol.111, No.1.

NON-LINEAR ANALYSES OF AXIALLY AND LATERALLY LOADED BORED PILES

J. Kos
Czech Technical University, Prague, Czech Republic

ABSTRACT: The proposed method of pile analyses is based on a beam-on-Winkler's foundation model. However, the non-linear subgrade reaction-deflection relationship is not based on predicted p-y or t-z curves. Subgrade reactions are calculated as the result of the synthesis of three models, i.e. the continuous beam on elastic supports as the basic model of the system, the elastic half-space for stress calculations, the Mohr-Coulomb failure law and the soil bilinear elastic-perfectly plastic stress-strain relationship. The breaking point of the bilinear elastic relationship is calculated by using the Mohr-Coulomb failure law with reduced (at rest) strength parameters, which are derived from Jaky's coefficient of earth pressure at rest. The cyclic loading and the pile group effect can be solved by this incremental iterative method.

1 INTRODUCTION

The evolution of the proposed method has been initiated by several back analyses of horizontally loaded large diameter piles based on the beam-on-Winkler's foundation model. Resultant values of modules of the subgrade reaction have been even more then fifty times higher than soil modules of deformability (an analogy of Young's modulus including plastic deformations) at low load levels. Considering that a value of the modulus of the subgrade reaction can be calculated as the quotient of a modulus of deformability and a pile diameter (Terzaghi 1955), there are two possibilities: the modulus of soil deformability can be considerably higher than its typical value or an active (deformable) zone can be considerably smaller than the pile diameter, i.e. 0,95 m in this case. The modulus of the subgrade reaction has been remarkably non-linear. The solution of this problem has been found in the theory of so-called structural strength of soils. According to this theory, critical (measurable) deformations of soils (particulate materials) start after overcoming the shear resistance of statistically important number of particle contacts - the soil structure falls. Before this moment most of particles cannot move relative to each other and the soil is many times stiffer. The soil behaviour is elastic. The proposed method uses standard soil parameters - γ, ϕ, c, E_{def} - preferred in foundation engineering in the Czech Republic.

2 STRUCTURAL STRENGTH OF SOILS

The structural strength is a property of any particulate material. The soil with the structural strength undergoes theoretically no deformation within a stress interval between so-called active (lowest) and passive (highest) stresses at rest. The values of these stresses can be calculated as Rankine's stresses with reduced shear strength parameters (Myslivec 1976). The method is based on the idea that critical deformations of soils start after overcoming the shear resistance of statistically important number of particle contacts (Lambe and Whitman 1969). At this moment the Mohr-Coulomb strength parameters are partly mobilised. The values of these partly mobilised parameters have been derived from the assumption that the active stress calculated with these reduced parameters is equal to the stress at rest from Jaky's equation:

$$\sigma_{0a} = \sigma_z \cdot \tan^2\left(\frac{\pi}{4} - \frac{\phi_{rst}}{2}\right) = \sigma_z \cdot \left(1 - \sin\phi_{pk}\right) \tag{1}$$

$$\phi_{rst} = \arcsin\left(\frac{\sin\phi_{pk}}{2 - \sin\phi_{pk}}\right) \tag{2}$$

$$c_{rst} = c_{pk}\frac{\tan\phi_{rst}}{\tan\phi_{pk}} \tag{3}$$

where σ_{0a} = active (lowest) stress at rest, σ_z = vertical pressure, φ_{pk} (φ_{rst}) = peak (at rest) internal friction angle of soil, c_{pk} (c_{rst}) = peak (at rest) cohesion of soil.

The active (σ_{0a}) and passive (σ_{0p}) horizontal stresses at rest can be then calculated as Rankine's limit stresses with partly mobilised strength parameters:

$$\sigma_{0a} = \sigma_z \cdot \tan^2\left(\frac{\pi}{4} - \frac{\phi_{rst}}{2}\right) - 2 \cdot c_{rst} \cdot \tan\left(\frac{\pi}{4} - \frac{\phi_{rst}}{2}\right) \tag{4}$$

$$\sigma_{0p} = \sigma_z \cdot \tan^2\left(\frac{\pi}{4} + \frac{\phi_{rst}}{2}\right) + 2 \cdot c_{rst} \cdot \tan\left(\frac{\pi}{4} + \frac{\phi_{rst}}{2}\right) \tag{5}$$

The structural strength of the horizontally loaded soil is characterised by the stress interval between the actual (original at load beginning) horizontal stress at rest and the active stress at rest when the soil is horizontally unloaded, or between the first one and the passive pressure at rest when the soil is overloaded.

These assumptions have been extended. At first, the relationship between horizontal and vertical principal stresses (Eqs. 4 and 5) may be generalised for principal stresses of any direction, i.e. the state of soil at rest can be tested by Mohr-Coulomb's failure law with partly mobilised strength parameters at rest. When the actual stress exceeds any of stress at rest limits, the interval of the structural strength is pulled by this stress modifying the interval extent. Methods based on these assumptions can model the soil hysteresis when the load is cyclic.

3 MODEL OF HORIZONTALLY LOADED PILES

The method for the solution of horizontally loaded piles is based on the beam-on-Winkler's foundation model. The static scheme of a continuous beam on elastic supports has been used for the numerical solution of the general equation:

$$E \cdot I \cdot \frac{d^4 w}{dx^4} + k \cdot d \cdot w = p \tag{6}$$

where E = Young's modulus of elasticity of the pile material, I = pile flexural stiffness inertia, w = lateral deflection, k = modulus of subgrade reaction, d = pile diameter, p = applied transverse loading at point x.

Winkler's model is used only for the calculation of soil-structure interaction. The final behaviour of its elastic springs (elastic supports of the model continuous beam) is non-linear elastic-perfectly plastic. However, the model is not based on predicted p-y curves as usual. The stiffness of Winkler's springs is calculated as the stiffness of the elastic half-space with the structural strength as the result of the synthesis of three partial models in the iterative process.

The first partial model - the continuous beam on linear elastic Winkler's supports - is used for the calculation of internal forces and lateral deflections of the pile and also for the calculation of horizontal pile-soil contact stresses.

The second partial model - the Boussinesq elastic half-space - is used for the computation of stress distribution in the soil around the pile. However, the computer time saving simplified model is used now. This model consists of perfectly separated thin horizontal elastic layers. It allows only a horizontal distribution of stresses. The stress distribution in the vertical direction has been ignored due to its insignificance.

The third partial model - the soil with the structural strength - is used for the calculation of the modulus of the subgrade reaction and the continuous beam support stiffness. The soil with the structural strength undergoes very small deformation within the stress interval (σ_{0a}; σ_{0p}). The soil is characterized by the modulus of elasticity with many times higher value than the value of the modulus of deformability (including plastic deformations) characterizing the soil when the actual stress is outside the interval (σ_{0a}; σ_{0p}). This is the bilinear model simplification. In the reality, of course, the change of the soil modulus is continuous. Despite this, thanks to a lot of calculation points where the state of the soil is tested, the model behaviour is continuous, too. The evaluation of the subgrade reaction modulus is based on the requirement of the equivalent deformability of Winkler's medium and the soil medium with the structural strength. It can be expressed by these equations:

$$w = \frac{1}{E_d} \cdot \int_0^{L_a} (\sigma_{ol} - \sigma_s) \cdot dx = \frac{1}{E_d} \cdot \sigma_{ol,c} \cdot L_W \quad (7)$$

$$\int_0^{L_a} (\sigma_{ol} - \sigma_s) \cdot dx = \sigma_{ol,c} \cdot L_W \quad (8)$$

$$k = \frac{E_d}{L_W} = \frac{E_d \cdot \sigma_{ol,c}}{\int_0^{L_a} (\sigma_{ol} - \sigma_s) \cdot dx} \quad (9)$$

where w = lateral deflection, E_d = design modulus of deformability, σ_{ol} = overloading horizontal stress, σ_s = structural strength of soil (σ_{0p} - $\sigma_{0,or}$ or σ_{0a} - $\sigma_{0,or}$, $\sigma_{0,or}$ = original stress at rest), $\sigma_{ol,c}$ = overloading contact stress, L_a = length of active (deformable) zone, L_W = length of Winkler's spring.

The highest value of the modulus of the subgrade reaction is restricted by the condition:

$$k \le E_d / t \quad (10)$$

where t = thickness of disturbed (by technology) contact soil without the structural strength. By this way the maximum (elastic) stiffness of Winkler's spring can be limited when the actual stress is within the structural strength interval.

The following computation approach has been applied. The structure is analysed as the continuous beam on elastic supports (1st model). The initial stiffness of elastic supports has to be predicted from Eq. (10). The results of the continuous beam analysis are contact stress changes in the elastic supports. The elastic half-space (2nd model) is loaded by means of these contact stress changes. Its (reduced) state of stresses is computed. It is assumed in the soil deformation computation that at a stress decrease under the soil structural strength value, the soil undergoes very small (theoretically none) deformation (3rd model). New values of continuous beam supports stiffness are calculated from Eq. (9), i.e. the quotient of the contact stress change and the soil deformation. Through the repeated analysis of the continuous beam on elastic supports and by the elastic supports stiffness modification a state is reached when all deformations of elastic supports of the continuous beam correspond to deformations of the non-linear half-space of the soil with the structural strength - values of the modulus of the subgrade reaction are correct. At this moment the extreme values of contact stresses are tested.

The values of active and passive (Weissenbach 1962) stresses should not been exceeded. In the opposite case the elastic support becomes plastic, i.e. its stiffness is very small (theoretically equal to zero) and the continuous beam is loaded by the resultant of the active or passive stress. By this the static scheme of the structure has been changed. That is why the iterative process has to be repeated. Finally, internal forces, lateral deflections and contact stresses of the pile are calculated.

The pile load-deflection relationship is calculated by the incremental method. It means that the described iterative process is repeated many times because of one hundred or more calculated load increments. Nevertheless, the computer time is relatively short, usually less then 15-20 s on a common PC computer.

4 MODEL OF VERTICALLY LOADED PILES

The principles of the method are similar to those which have been used for the solution of horizontally loaded piles. The main model is the pile on linear elastic springs which is used only for the pile-soil interaction solution. The stiffness of the model spring is calculated in the incremental iterative process to be equal to the stiffness of the soil medium with the structural strength. This is why the final behaviour is non-linear. There are two types of model springs. The first one is the classic Winkler's spring (discussed above) supporting the pile base. The second one is the elastic cantilever adjacent to the pile shaft. The cantilever, loaded only by the shear stress not by a bending moment, is the result of the simplification of the adjacent soil medium. There have been some ideas leading to this simplification.

The shear stress is the most effective in supporting the pile shaft. It rapidly decreases in the increasing distance from the pile shaft. In the distance equal to the pile radius, the skin friction increment value is reduced to 50 %. It means that the soil medium with the structural strength undergoes practically no deformation outside the coaxial cylinder with the radius equal to the pile diameter. The shear modulus value is equal to 33÷40 % of the value of Young's modulus, however, the pile and the adjacent "soil pier" are many times longer than the pile radius. So that the shear elastic cantilevers are many times stiffer than the coaxial soil pier. This soil pier is reinforced by the pile. Before the skin friction is fully mobilised, the ratio of pile and adjacent soil normal stresses is similar to the ratio of Young's modules due to displacement compatibility. A typical pile normal stress change induces only a small normal stress change of the soil pier. This is why the vertical normal stress change in the soil is not taken into account as a second-rate influence. The adjacent soil is simulated only by elastic cantilevers to transfer the shaft load to the surrounding rigid soil where stresses being inside the structural strength interval are dissipated practically without deformations. However, this simplification is acceptable only for the single pile solution. In close arranged pile groups the soil surrounding the pile has to transfer the full shaft load by the vertical stress. When it exceeds the structural strength, the settlement increases rapidly or when the settlement is limited, the shaft bearing capacity decreases, especially in the upper part of the shaft. The proposed method tests whether the area of surrounding soil is large enough to solve the pile as single. In the opposite case the total settlement is calculated as the sum of the pile settlement relative to the soil and the soil settlement. However, the real settlement is bigger because of the elastic cantilever stiffness decrease due to the structural strength reduction. The more accurate method is being developed.

The shear structural strength calculation rather differs from the normal structural strength calculation mentioned above. The calculation is based on the equation:

$$\tau_{\max} = \sqrt{\left[\tfrac{1}{2}\left(\sigma_z + \sigma_x\right)\cdot \sin\phi + c\cdot\cos\phi\right]^2 - \tfrac{1}{4}\left(\sigma_z - \sigma_x\right)^2} \qquad (11)$$

where τ_{max} = maximum vertical shear stress (shear strength) at the moment of soil failure, σ_z (σ_x) = vertical (horizontal) normal stress, φ = friction angle of soil, c = cohesion of soil.

The structural strength shear stress is calculated from Eq. (11) using partly mobilised Mohr-Coulomb strength parameters from Eqs. (2) and (3). When the mobilised shear stress is less than this limit value, the soil is in the elastic state, i.e. many times stiffer than typical.

The equation (11) is also used for the calculation of the skin shear strength, however, with peak Mohr-Coulomb strength parameters. The equation:

$$\tau_{\max} = c + K_0 \cdot \sigma_z \cdot \tan\phi \qquad (12)$$

is used only for the calculation of the skin shear strength reduced by the pile installation technology and accepted when it is less then the strength calculated from Eq. (11).

The graphic interpretation of Eq. (11) for $\phi = 32^{\circ}$, c = 14 kPa, σ_z = 100 kPa and variable σ_x is in Fig. 1. It can be seen that results from equations (11) and (12) are equal only when $\sigma_x \cong (1 - \sin\phi)\, \sigma_z$. For higher values of the horizontal stress the skin resistance from Eq. (12) should be considerably reduced.

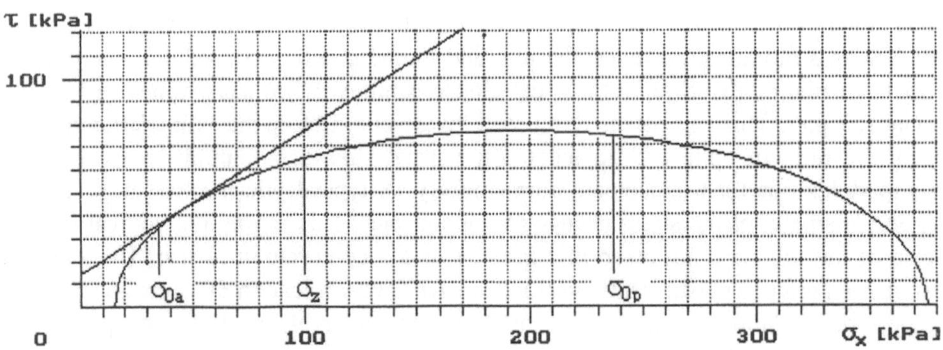

Fig. 1. Lateral pressure-skin resistance relationship

The iterative process of the elastic springs stiffness refinement is similar to the one for horizontally loaded piles described above. There are also three partial models, i.e. the pile on shear elastic springs for the pile-soil interaction calculation, the model for shear stress distribution - the elastic cantilever of constant height and width equal to the perimeter of the cylinder on which the shear stress is calculated - and the same elastic cantilever as the third model for the calculation of the vertical settlement:

$$w = \frac{1}{G} \cdot \int_0^{L_a} (\tau_{ol} - \tau_s) \cdot dx = \frac{1}{G} \cdot \int_0^{L_a} \left(\frac{r}{r+x} \tau_{ol,c} - \tau_s \right) \cdot dx \tag{13}$$

where w = vertical settlement, G = shear modulus of soil, τ_{ol} ($\tau_{ol,c}$) = overloading (contact) shear stress, τ_s = structural strength shear stress, L_a = length of elastic cantilever (active zone), r = pile radius, x = distance from pile skin.

The description of the horizontally loaded model can be used for the illustration of the process after formal changes of normal springs and stresses onto shear springs and stresses.

5 HORIZONTAL PRESSURE OF FRESH CONCRETE

The skin resistance reaches its maximum value at the certain depth and do not grow with its next increase. It is explained as the effect of soil volumetric strains. In the proposed models the limitation of the bored pile skin resistance is considered to be an effect of the limitation of the fresh concrete lateral pressure due to the silo effect of the borehole. The equilibrium between the fresh concrete weight and the skin resistance is solved by the numerical iterative process in certain levels of the borehole (middles of model pile computation elements).

The influence of different coefficient K_0 of the fresh concrete has been studied. The concrete lateral pressure has been calculated as Rankine's active stress (Eq. 4) for 10 values of the party mobilised concrete friction angle ϕ varying from 0,1 ϕ to 1,0 ϕ with increment 0,1 ϕ, where $\phi = 4^\circ$, c = 0 kPa, $\gamma = 23$ kNm^{-3}. The fresh concrete skin friction has been then calculated from Eq. (11), where $\phi = 4^\circ$. The study of this problem is illustrated in the left side of Fig. 2, where the right straight line represents the concrete hydrostatic pressure, the second line from the right represents the concrete active pressure (without friction, see Eq. (11) and Fig. 1). It can be seen that there are practically no differences among stresses for mobilised concrete friction angles less than 0,5 ϕ. The initial soil contact stress is equal to the concrete pressure. The skin resistance is then calculated from Eq. (11) with soil strength parameters - the right side of Fig. 2, where the right straight line represents the concrete hydrostatic pressure, the second line represents the drilling mud pressure, the left curve represents the lateral contact pressure and the right curve represents the skin resistance in the soil of $\gamma = 18{,}5$ kNm^{-3}, $\phi = 32^\circ$ and c = 14 kPa. The skin resistance has been calculated from Eq. (11) with respect to favourable effects of radial and tangential stresses around the circular hole of a relatively small diameter.

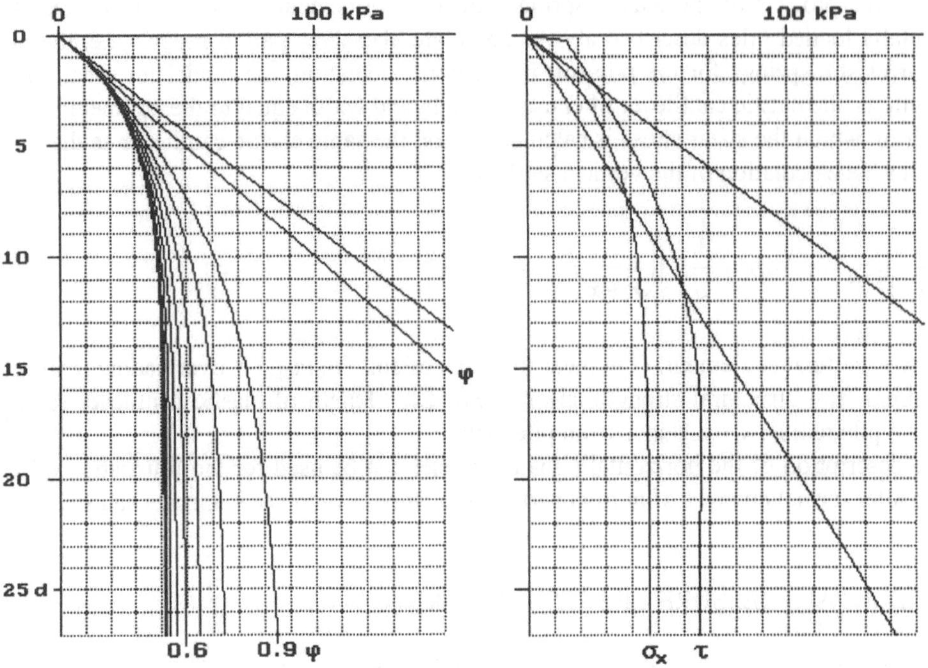

Fig. 2. Lateral stresses of the fresh concrete and the skin resistance along the pile shaft

6 VERIFICATION OF MODELS

The proposed models have been verified by comparison of calculated and experimental load-displacement relationships.

The example of the horizontally loaded pile verification has been based on experimental results published by Wittke et al (1974). The horizontal load-displacement relationship (Fig. 3) has been calculated at the ground level (1,2 m below the pile head). The published soil parameters have been used (E_d = 45 MPa - between Young's and the constrained modules). The computer time of 325 load increments has been 25 s (1. cycle - 200 increments - 6 s) on 133 MHz PC.

The example of the vertically loaded pile verification has been based on experimental results published by Van Impe (1985). The test pile was the Atlas-Franki auger pile of length 13,5 m and diameter approx. 0,46 m. The exact soil parameters have not been published, thus the values typical for soil classes described on the test site in Ghent, i.e. clayey sand and sandy clay, have been tested (back analysis): $\gamma = 18{,}5\ \mathrm{kNm}^{-3}$, $\phi' = 32°$, c' = 14 kPa, E_d = 7 MPa and 28 MPa (below the pile base in the medium dense to dense sand).

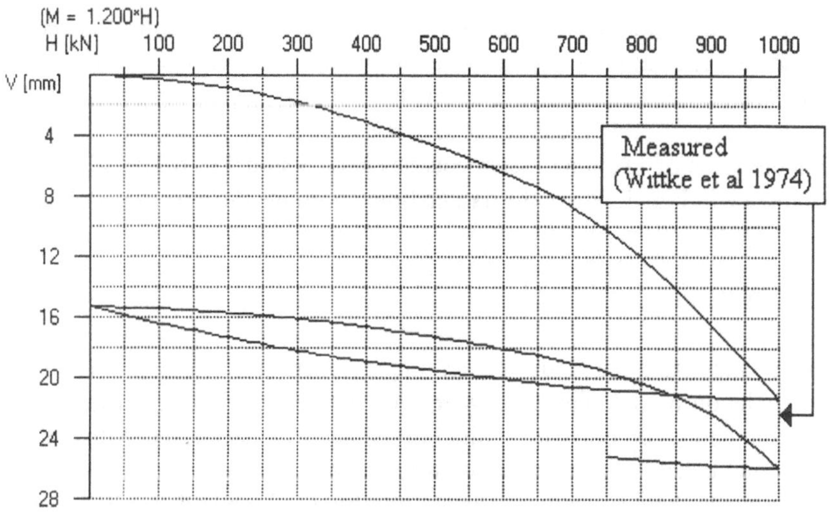

Fig. 3. Horizontal load-displacement relationship

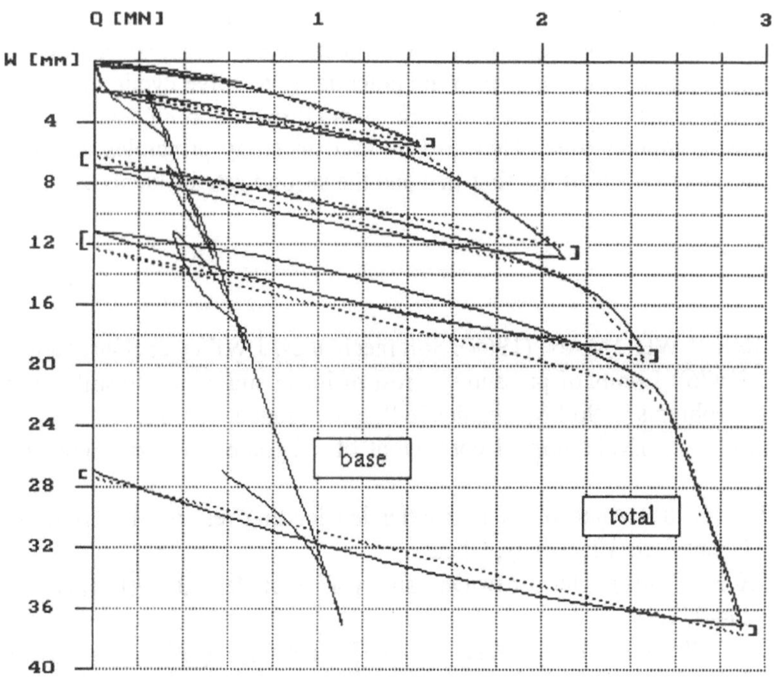

Fig. 4. Vertical load-settlement relationship

The reduction of the contact pressure by the silo effect has not been supposed in this case of the soil heavily compressing technology, $K_{0c} = 1{,}0$ (hydrostatic pressure) has been used for the fresh concrete. The resultant K_{0s} in the adjacent soil has been $K_{0s} = K_{0c} \cdot \gamma_c / \gamma_s = 23 / 18{,}5 = 1{,}24$. The relationship in Fig. 1, calculated with the same soil parameters illustrates that $K_0>1$ cannot considerably increase the skin resistance.

The computer time of 180 load increments has been 6 s on 133 MHz PC. The results of the verification are in Fig. 4, where the dashed line represents the experimental data.

7 CONCLUSIONS

The proposed calculation methods are an improvement of the well known beam-on-Winkler's foundation model. The used system of the elastic springs stiffness refinement leads to the simulation of the pile in the particulate soil half-space with the structural strength. The final behaviour is non-linear elastic-perfectly plastic. In comparison with p-y or t-z curve methods, there are certain advantages: Proposed methods are based on standard soil parameters, which are preferred in foundation engineering in the Czech Republic. Their values are well known and extensively tabulated. Thus, the preliminary calculations of piles can be based on this wide knowledge database without soil testing. Time effects can be introduced by using drained or undrained strength parameters. The good accuracy and the high speed of calculations make it possible to use these methods for parametric studies. Even statistical analyses of the input data variation influence seem to be executable on today's computers. This methods have also the potentiality to solve cyclic loading and pile group effects because of the stress path and the stress distribution calculation, with good accuracy and short computer time. They can be used to study principles of the pile-soil interaction and to select effects of the first-rate and the second-rate influence.

REFERENCES

LAMBE T.W. & R.V.WHITMAN. (1969). Soil mechanics. J.Willey & Sons, New York.

MYSLIVEC A. (1976). Limits of pressures at rest of loose and cohesive soils, (in Czech). Proc. Actual Problems in Soil Mechanics in Prague, Prague, 1: 5-17.

TERZAGHI K. (1955). Evaluation of coefficient of subgrade reaction. Géotechnique 5, 4: 297-326.

VAN IMPE W.F. (1985) Analysis of four extended loading test results on Atlas-Franki auger piles. Ghent State University, Ghent.

WEISSENBACH A. (1962). Der Erdwiderstand vor schmalen Druckflächen. Die Bautechnik 39, 204-211.

WITTKE W., J.SPANG, W.RODATZ & S.SEMPRICH. (1974). Bemessung von horizontal belasteten Grossbohrpfählen nach der Methode Finiter Elemente. Der Bauingenieur 49, 6: 219-226.

FINITE ELEMENT ANALYSIS OF PILE FOUNDATIONS SUBJECT TO PULL-OUT

S.P.G. Madabhushi and S.K. Haigh
Cambridge University, Cambridge, UK

ABSTRACT:

Pile behaviour under conventional downward loading and lateral loading has been extensively investigated. Pile behaviour when subjected to pull-out loads, however, is yet to be understood fully. In this paper we investigate the behaviour of piles in sand under tensile pull-out loading using finite element analysis. The variation of pull-out resistance with soil density, pile size and speed of pull is investigated. To impart a degree of confidence in the finite element results, centrifuge modelling of the same problem was carried out. The results from both these modelling techniques are compared. The implications for the finite element modellers of using simplified constitutive models are discussed. Modelling of interfaces between sand and structure when high degrees of relative movement occur, is considered.

Keywords: piles & piling; pull-out loading; finite element modelling; centrifuge modelling.

1. INTRODUCTION

Use of piles in both on-shore and off shore structures has become the most popular foundation technique. Pile foundations are often constructed to transfer structural dead and live loads through unstable ground to a solid stratum. The behaviour of these piles acting in compression has been widely studied using numerical modelling, centrifuge modelling and full-scale testing. Recent studies investigated the behaviour of piles under lateral loading as in the case of a bridge abutment, Ellis (1997). On the other hand, upward tensile loads may be applied to some of these piles owing to the action of horizontal wind or wave forces on the structure. The behaviour of the piles under these 'pull-out' loads is much less understood.

In this paper, work undertaken at Cambridge University into the behaviour of these tension piles in sand using the techniques of finite element analysis and centrifuge modelling will be described. The results from the finite element analysis are compared to the centrifuge test data to give an insight into the force-displacement behaviour of these piles. The variation of pull-out resistance with soil density, pile size and speed of pull is investigated.

In sandy soils the resistance of these piles to pull-out comes from two major sources, skin friction between pile and soil and suctions generated at the base of the pile. Both of these effects are greatly influenced by the generation of excess or suction pore pressures as the soil suffers shear and volumetric strains. Suctions are generated at the base of the pile owing to the opening up of a void as the pile moves up. At the sides of the pile, undrained shearing of the soil when the pile is pulled quickly will result in excess pore pressure generation in loose soils and suction pore pressures being generated in dense soils. These pore pressures will alter the effective stress state of the soil and will hence have a great impact on the skin friction component of the pull-out resistance.

2. NUMERICAL ANALYSIS

Numerical analysis was carried out using the ABAQUS finite element analysis program. This was used to investigate load-displacement behaviour for the pile pull-out along with pore pressure and shear stress distribution for various combinations of pile size, soil density and saturation state.

2.1 MESH DESIGN

The pile-soil system was modelled in the finite element analysis as an axi-symmetric problem. The finite elements used are 8-noded axi-symmetric elements, with porous properties for the soil elements. The pile elements, however, are modelled as solid elements. The finite element mesh used in the analysis is shown in Fig. 1. The element size is biased towards the pile in order to give most data in the region of greatest interest, i.e. close to the pile. The limits of the mesh were at a radius of 10m and 10m below the base of the pile, as this was the prototype size of the physical models tested in the centrifuge (see Sec.3). It was seen that enlarging the mesh domain has little effect on the analysis.

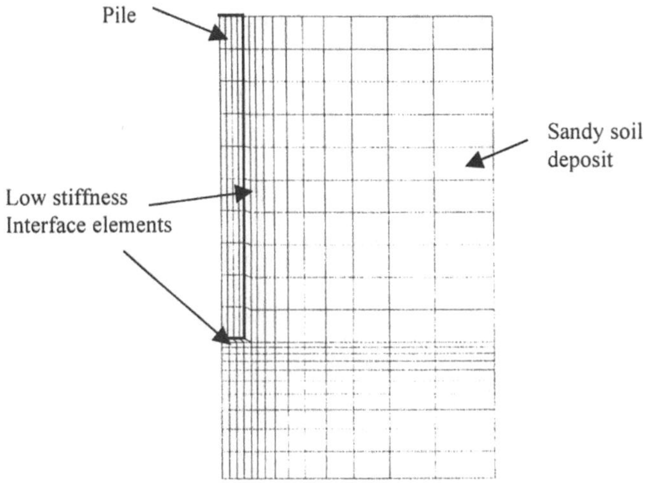

Fig. 1. Finite Element Mesh

In order to allow true behaviour to be modelled, slip must occur on the sides of the pile, and a void must open up at the base. For the void at the pile tip to open, duplicate nodes were employed on the pile and soil in the initial analysis, giving no attachment between the two materials. This caused numerical instability before achieving initial equilibrium, as the soil at the base of the pile yields under the high horizontal effective stress, with no vertical stress present across the interface. In saturated tests, the problem is exemplified by the fact that pore pressures cannot act across the interface, so base suctions do not show an effect on pull-out capacity.

The problem was hence analysed using a layer of interface elements at the pile tip with a greatly reduced axial stiffness. This allows forces and pore pressures to act across the boundary, whilst giving very limited resistance to movement from tensile soil stiffness. This procedure works well until failure but adequate post-failure behaviour cannot be correctly modelled this way. The analysis can be improved by use of "Gap" elements, which have an infinite compressive axial stiffness but zero tensile axial stiffness. These allow a negligibly thin layer of elements along the interface to model void opening. These will be incorporated in future analysis.

At the pile shaft interface, a layer of elements was incorporated with a reduced friction angle. This was an attempt to model the fact that the friction angle between sand and pile material is much lower than the internal friction angle ϕ of the sand. This models the correct frictional resistance offered by the soil on the pile shaft but ignores the reduction of the contact surface area between pile and soil as the pile suffers significant axial displacements. As the nodes on pile and soil are connected, the contact area modelled remains constant and large shear strains are imposed on the boundary layer elements. Thus once any significant movement has occurred, the modelling breaks down, as the situation being modelled is incorrect. These differences between the actual and modelled pile behaviours are shown in Fig. 2. The pre-failure behaviour of the pile under pull-out loading is however accurately modelled by the finite element analysis.

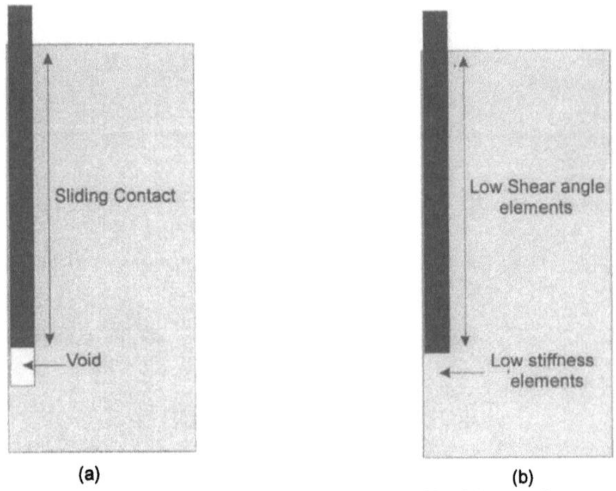

Fig. 3. a) True boundary conditions. b) Modelled boundary conditions.

2.2 CONSTITUTIVE MODEL

The analysis used the Drucker-Prager (1952) elasto-plastic constitutive model for the sand. This model gives a good approximation to Mohr-Coulomb soil behaviour, although the Drucker-Prager model defines the yield surface as being a circle in the deviatoric π plane, rather than a hexagon, assuming failure to be independent of the intermediate principle stress. A friction angle ϕ of 30° was taken for the soil outside the boundary layer, and a friction angle of 20° for the boundary layer. A value of K, defining the ratio between tensile and compressive triaxial strengths, of 1 was used.

The Yield function is defined by:

$$F = t - d - p \tan \beta \tag{1}$$

where:

$$t = \frac{1}{2} q \left[1 + \frac{1}{K} - \left(1 - \frac{1}{K} \right) \left(\frac{r}{q} \right)^3 \right] \tag{2}$$

The Drucker-Prager model cannot deal with the excess or suction pore pressures generated during shearing of loose or dense soils as they approach the Critical State. The generation of these pore pressures has a great effect on the skin friction experienced by the pile, as they change the effective stress state of the soil around the pile. This excess pore pressure/suction generation will be modelled using a more complex constitutive models for the soil, such as the Pastor-Zienkiewicz (1985) Mark III model, in the future. The finite element analysis is however able to handle the generation of base suction due to upward pile displacements.

2.3 BOUNDARY CONDITIONS AND LOADING

The nodes at the base of the soil, 10m below the pile tip, were fixed in the radial direction and the nodes at the side, 10m from the axis of the pile, were fixed in the vertical direction. The pore pressure was fixed at zero at the surface of the sand on the outside of the mesh.

A ramped vertical displacement was applied to the node at the centre of the pile top to simulate pull-out. For fast tests, this was a displacement of 10cm in 20 seconds, whereas for the slow tests it was a displacement of 10cm in 10 minutes.

3. CENTRIFUGE MODELLING

Centrifuge modelling is based on the principle of modelling correct prototype stresses and strains in scale models. As soil is a highly non-linear material, only if this condition is satisfied will correct prototype behaviour be observed in the centrifuge tests.

This is accomplished by testing scale models in the enhanced gravity field of a geotechnical centrifuge. The behaviour of 1:N scaled model in the N times gravity field of the centrifuge is related to the behaviour of the prototype by means of a set of scaling laws. A brief summary of the scaling laws is presented in Table 1; these can be derived by dimensional analysis, as was done by Langhaar (1951), or by considering governing equations, Schofield (1980).

Parameter	model / prototype	Dimensions
Length	1/N	L
Acceleration	N	LT^{-2}
Velocity	1	LT^{-1}
Time (dynamic)	1/N	T
Stress	1	$ML^{-1}T^{-2}$
Force	$1/N^2$	MLT^{-2}
Mass	$1/N^3$	M
Seepage velocity	N	LT^{-1}
Time (seepage)	$1/N^2$	T
Force	$1/N^2$	MLT^{-2}

Table 1: Centrifuge scaling laws. Schofield (1980)

A series of twelve 1:100 scale models of piles in a sand deposit were tested using the CUED 10m beam centrifuge running at a nominal acceleration of 100g, the design and operation of which are described by Schofield (1980).

Each test package consisted of four models of a pile in sand contained within an 850mm tub. The package is shown in Fig. 3. Each model comprises of a plastic tub full of sand, with an aluminium pile with an embedded length of 23cm on the axis of the tub. The tub extends 10cm below the base of the pile, and has an internal diameter of 20cm, thus modelling a cylinder of soil 20m in diameter and 33m deep at prototype scale. The top of the pile is connected to an electrical actuator via a load cell and two pairs of links, which allow movement in two degrees of freedom to the top of the pile to compensate for any small, non-verticality of the pile.

The models were saturated under vacuum with either water or silicone oil, or were dry depending on the situation being modelled. Silicone oil is used as pore fluid in order to correct the anomaly between scaling laws for dynamic and seepage velocities seen in Table 1, Madabhushi (1994).

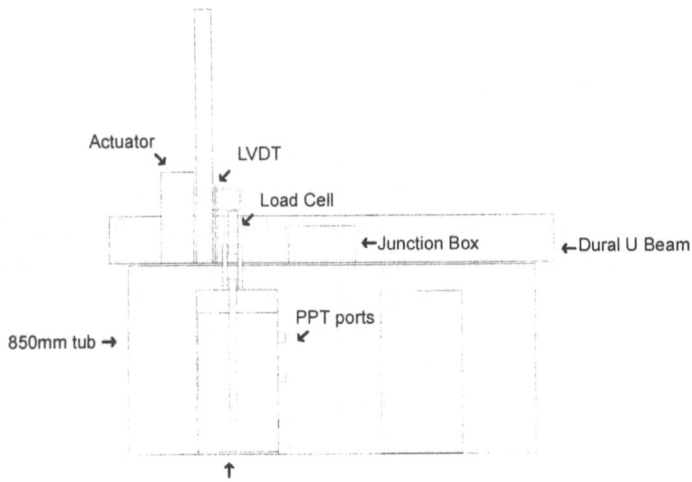

Fig 3. Centrifuge Package

Three Druck PPTs, (Pore pressure transducers), fitted with porous bronze stones, allow the pore pressures at the base and at two positions up the side of the pile, and an LVDT on the actuator allows the position of the pile to be monitored throughout the test.

4. FINITE ELEMENT ANALYSIS RESULTS

The ABAQUS finite element package was used to investigate force-displacement behaviour of the pile, and the distribution of pore pressures and shear stresses in the surrounding soil at failure. All the tests predicted force-displacement behaviour with the general features of Figure 4, showing an initial linear section to failure and linearly increasing post-failure load with displacement. The post failure behaviour is not reliable for the reasons discussed earlier.

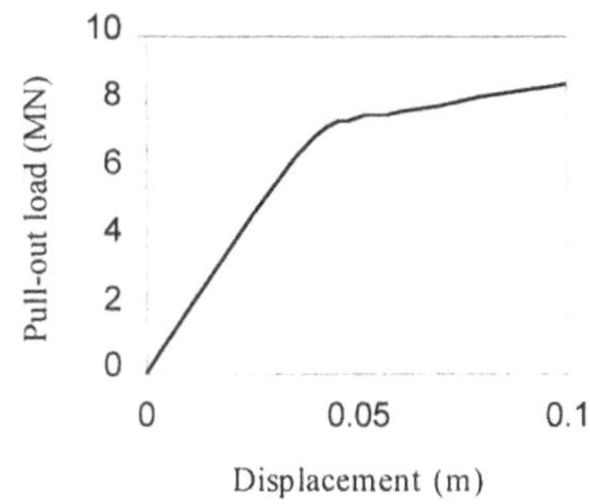

Fig. 4: Predicted force-displacement behaviour of a 19mm pile in dense dry sand

Plots of pore pressure at failure are seen in Fig. 5, showing suctions generated at the base of the pile owing to the void opening up and suctions generated up the side of the pile owing to relaxation of the horizontal stress in the soil as the pile moves upwards.

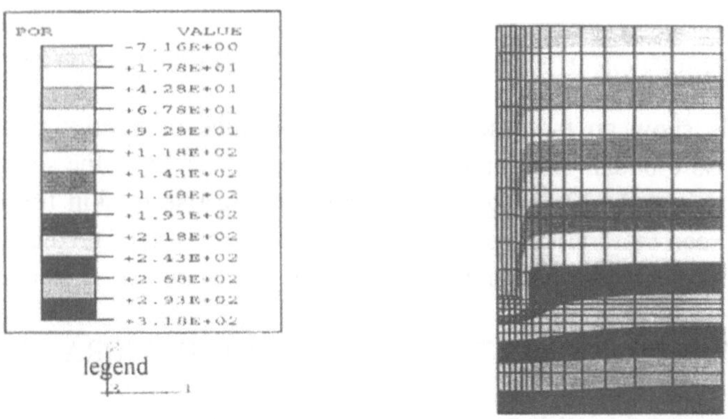

Fig.5 Pore pressure distribution at failure in loose sands.

4.1 COMPARISON OF RESULTS FOR SATURATED AND DRY SANDS

Dry tests show significantly higher resistance to pull-out than do saturated tests. This is due to an increase by a factor of approximately two in the effective earth pressures at any point in the model in dry tests. This increases by a factor of two the skin friction experienced by the pile.

In the saturated tests, a further contribution to pull-out resistance is provided by suctions generated at the base of the pile as it moves. The absolute pore pressures at the base were shown to have fallen to approximately zero at failure, corresponding to a suction of 230 kPa in the tests carried out. This suction contributes to an increase of around 15% in pull-out resistance. The results from all the analyses carried out are shown in Table 2.

Diameter	Length	Density	Rate of Pull-out	Saturation	Predicted force from F.E. analysis	Observed force from centrifuge tests
1.9 m	25 m	Loose	Fast	Water	4.5 MN	2.24 MN
1.9 m	25 m	Dense	Fast	Water	5.45MN	5.95 MN
0.95 m	23 m	Loose	Fast	Water	1.75 MN	2.67 MN
0.95 m	23 m	Dense	Fast	Water	2.6 MN	3.13 MN
1.9 m	23 m	Loose	Fast	Dry	6.4 MN	9.24 MN
1.9 m	23 m	Dense	Fast	Dry	8.0 MN	9.35 MN
0.95 m	23 m	Loose	Fast	Dry	3.2 MN	3.67 MN
0.95 m	23 m	Dense	Fast	Dry	3.9 MN	7.9 MN
1.9 m	23 m	Loose	Fast	Silicone Oil	4.1 MN	5.0 MN
1.9 m	23 m	Dense	Fast	Silicone Oil	5.0 MN	7.3 MN
1.9 m	23 m	Loose	Slow	Silicone Oil	3.8 MN	5.58 MN
1.9 m	23 m	Dense	Slow	Silicone Oil	4.5 MN	7.66 MN

Table 2: Prototype scale ultimate pull-out forces from finite element analysis and the centrifuge tests

4.2 COMPARISON OF RESULTS FOR DENSE AND LOOSE SANDS

Dense models show an increase in pull-out resistance of approximately 20% over that shown by loose models. The change in buoyant unit weight of the soil between these two cases is approximately 17.5%, giving a corresponding rise in skin friction experienced. It can hence be seen that in this case the difference between forces predicted is almost entirely due to the increase in effective pressure in the soil due to increased soil density.

4.3 COMPARISON OF RESULTS FOR DRAINED AND UNDRAINED TESTS

Drained tests showed a lower pull-out resistance than did undrained tests. The suctions seen at the base were reduced to 50 kPa for the slow tests, and this 180 kPa drop in suction corresponds almost exactly to the fall in pull-out resistance.

5. RESULTS FROM CENTRIFUGE TESTS

The centrifuge test results gave agreement to within about 10-15% for the forces required to achieve pull-out in each of the cases modelled. The tests however gave a great deal more in formation about post-failure behaviour than could be seen from the finite element results

5.1 COMPARISON OF RESULTS FOR LOOSE AND DENSE TESTS

As loose sands shear around the boundary of the pile, they contract to critical state, causing excess pore pressures to be generated in undrained events, lowering the effective stress and hence the skin friction experienced by the pile. Conversely, with dense sands, suction pore pressures are generated and increased skin friction is seen.

This is illustrated by Figure 6, showing load-displacement curves for loose and dense sands, in which it can be seen that the pile in dense sand shows a much more sustained resistance to pull-out, owing to the presence of suctions around the pile. These pore pressures were observed by the PPTs situated at the side of the pile.

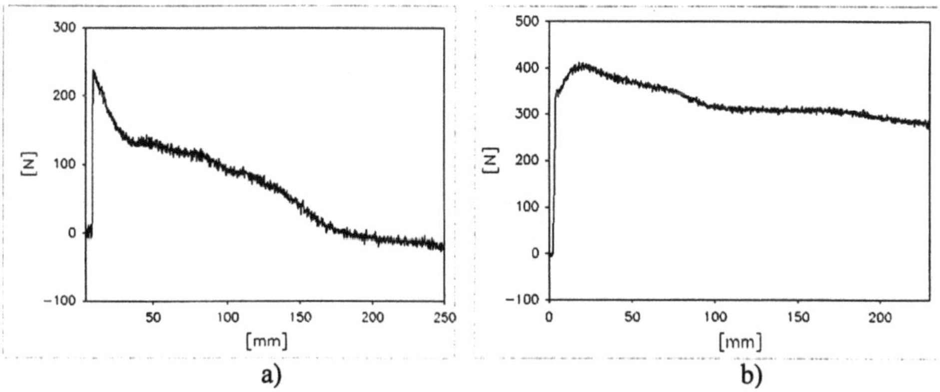

Fig. 6 Load-displacement plots for piles in a) Loose sand, b) Dense sand.

5.2 COMPARISON OF RESULTS FOR DRAINED AND UNDRAINED TESTS

Very little difference was noted between drained and undrained test results from the centrifuge. In no test was the suction predicted by F.E. analysis generated at the base of the pile recorded. It has been suggested that this is due to movement of the base PPT during model preparation, although it may also be that this base pore pressure dissipates rapidly owing to a fast drainage path at the pile-sand interface.

6. COMPARISON OF RESULTS

Agreement was found to within 15% for the forces seen between ultimate resistances predicted for piles by numerical and physical modelling methods. The behaviour up to failure was well predicted by the numerical code, but post-failure, little useful information could be gained on the behaviour, owing to shortcomings of the boundary modelling and constitutive model used. Efforts are underway to improve the finite element modelling of these aspects. This is felt to be particularly important, as the centrifuge test data on loose and dense sands show completely different post-failure behaviour and it would be important for the numerical code to capture this aspect.

7. IMPLICATIONS TO FINITE ELEMENT ANALYSIS

From the analyses presented in this paper the following implications may be arrived at;

- In situations where shear is a major element of soil behaviour, the constitutive model must be able to predict the expansive or dilatant behaviour of the soil and the consequent pore pressure changes in undrained tests, if adequate model behaviour is to be predicted. This is a limitation of simple soil models but can be implemented with use of more complex constitutive models.
- The handling of soil-structure boundaries plays a significant role in the behaviour of numerical models of soil-structure interaction problems, especially when significant movement occurs between the two substances. Slip contact and the opening of voids are two complications which need to be dealt with if behaviour is to be modelled accurately once significant movement has occurred.

REFERENCES

DRUCKER, D.C. & PRAGER, W., (1952) Soil Mechanics and Plastic Analysis or Limit Design. Quarterly of Applied mathematics, Vol. 10., pp. 157-165.

ELLIS, E.A., (1997), Soil-structure interaction for full-height piled bridge abutments constructed on soft clay, PhD Thesis, Cambridge University.

LANGHAAR, H.L., (1951), Dimensional Analysis and theory of Models, John Wiley, New York.

MADABHUSHI S.P.G, (1994), Effect of pore fluid in dynamic centrifuge modelling, Proc. Centrifuge 94, Balkema, Rotterdam, pp 127-132.

PASTOR M. & ZIENKIEWICZ, (1985), A simple model for transient soil loading in earthquake events Part II non-associative model for sands. International Journal for numerical and analytical methods and geomechanics, Volume 9.

SCHOFIELD, A.N., (1980), Cambridge geotechnical centrifuge operations, Geotechnique, Vol. 20, pp. 227-268.

SIMULATING INFILTRATION AND SUCTION EVOLUTION IN UNSATURATED SOIL SLOPES

J.R. McDougall and I.C. Pyrah
Napier University, Edinburgh, UK

ABSTRACT
Infiltration has been frequently cited as a trigger for instability in unsaturated slopes. This paper seeks to illustrate how an unsaturated flow model interprets such phenomena and the process of suction evolution by simulating a series of rainfall events infiltrating an idealised homogeneous slope. Numerical moisture profiles indicate that infiltration following the onset of a rainfall event can be characterised in relation to two extreme patterns of infiltration. On the one hand, where rainfall rates are low and initial hydraulic conductivity is high then infiltration occurs with minimal moisture accumulation. In contrast, where infiltration exceeds the ability of the soil to redistribute it then moisture accumulates and infiltration proceeds as a distinct wetting front.

1.0 INTRODUCTION

In soils where the water table is at some depth, slope stability assessment must deal with potential failure surfaces which lie, at least partly, in the unsaturated zone. Conventional techniques of analysis using only positive pore water pressures may then reveal factors of safety of less than unity. A generalised approach to slope stability analysis (Fredlund & Barbour, 1992), seeks to overcome such anomalies by defining soil water pressures and associated shear strength continuously within both saturated and unsaturated soil regions.

In this paper, the numerical interpretation of infiltration in unsaturated slopes, and its influence on soil water suctions, is examined.

2.0 UNSATURATED FLOW MODEL

The unsaturated flow model uses Richards equation and is solved using a numerical approach which is now well documented (e.g. Neumann, 1975; Thomas & Rees, 1990). It is based on a Galerkin finite element formulation for spatial discretisation and a finite difference scheme for the time domain which is implemented via linear quadrilateral isoparametric elements; this is more fully described in McDougall (1996). Of central importance to any unsaturated flow model are the fundamental soil water relations; these are, (a) the soil moisture characteristic which relates soil suctions to volumetric moisture content and, (b) the relationship between hydraulic conductivity and moisture content, commonly referred to as the relative permeability function.

The soil moisture characteristic used here is that proposed by van Genuchten (1980),

$$\theta_e = \left[\frac{1}{1+(\alpha\psi)^n}\right]^m \qquad [1]$$

where ψ is the negative soil water pressure or suction, α, m & n are parameters which control the shape of the soil moisture characteristic and θ_e is the effective volumetric moisture content (dimensionless) defined by,

$$\theta_e = \frac{\theta - \theta_r}{\theta_s - \theta_r} \qquad [2]$$

where θ is the moisture content; θ_r is the residual moisture content, i.e. the lowest moisture content obtainable in the field; and θ_s is the saturation moisture content which is equivalent to the porosity under conditions of full saturation (all dimensionless).

The unsaturated hydraulic conductivity, $K(\theta)$, is defined using parameters already established for the soil moisture characteristic, i.e.,

$$K(\theta) = K_{sat}\sqrt{\theta_e}\left[1-\left(1-\theta_e^{1/m}\right)^m\right]^2 \qquad [3]$$

where K_{sat} is the saturated permeability.

	Sand	Clay
α	0.50	0.06
n	2.0	2.0
m	2.0	2.0
θ_s	0.375	0.375
θ_r	0.05	0.10

Table 1. van Genuchten soil water parameters

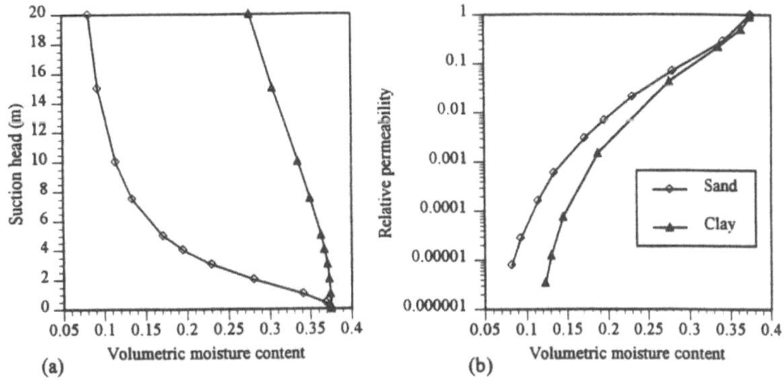

Fig.1 Soil water relations for sand and clay soils, after van Genuchten (1980); (a) soil moisture characteristic; (b) relative permeability function

3.0 MODELLING INFILTRATION IN HOMOGENEOUS, UNSATURATED SLOPES

Two different soil types were represented in the numerical tests and are distinguishable by their respective soil water relations as shown in Fig.1. One soil is of the clay-type and is characterised by a gradual reduction in moisture content as soil suctions are applied; the other has the more rapid desorption characteristics of a sand. In practice these two soils would have saturated permeabilities which differ by three to five orders of magnitude. Here, an identical saturated permeability of 1×10^{-4}m/s [i.e. 8640mm/day] has been used for both soil types to emphasise the influence of the soil water relations on the infiltration process.

3.1 Geometry, rainfall & initial conditions

The geometry and finite element discretisation of the slope is shown in Fig.2. Investigations by Pradel & Raad (1993) indicate that rainfall induced slips are commonly planar failures triggered by suction attenuation and occur about 3-4ft (1-1.2m) below the slope surface. These simulations have therefore been configured to monitor soil suctions at three locations on such a failure surface, as depicted in Fig.2.

Rainfall events of 86.4, 432 & 864mm/day were applied for a 24 hour period to an upper section of the slope surface using Neumann, i.e., prescribed flux type boundary conditions. Infiltration is not permitted through the lower portion of the slope to avoid boundary interference. The slope is initially in hydrostatic equilibrium with a horizontal phreatic surface prescribed along the base. All other boundaries have no-flow conditions. The test coding system uses a prefix letter to denote the type of soil water relations, (C for clay or S for sand); a number to denote the rainfall rate in millimetres per day; and suffix codes A, B & C when referring to the suction monitoring locations shown in Fig.2.

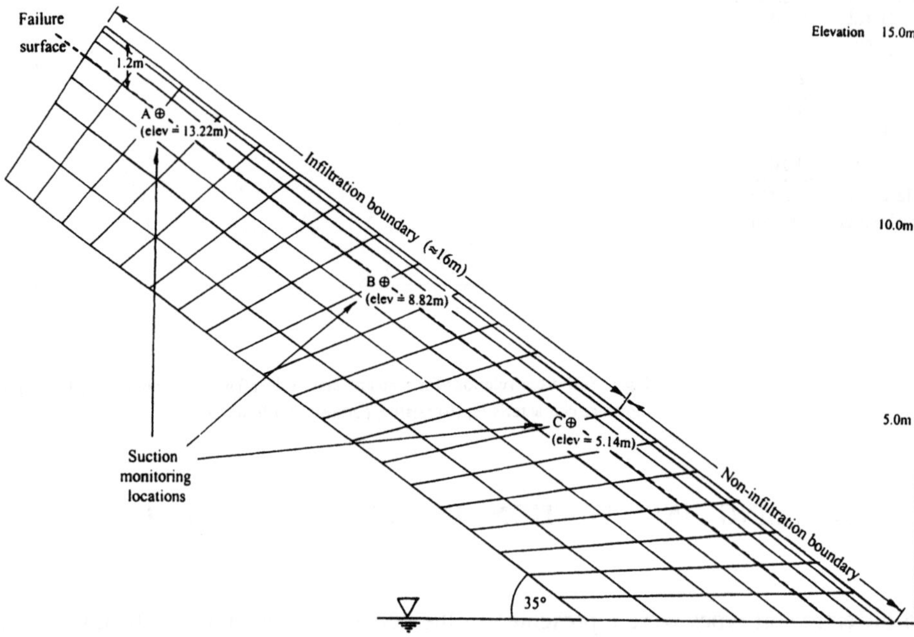

Fig.2 Geometry, finite element mesh and boundary conditions for slope tests

3.2 Results - clay soil water relations

Figure 3 presents the patterns of suction evolution at the monitoring locations during each of the three rainfall events with the clay soil water parameters. The hydrostatic condition means that soil suctions at each monitoring location have initial values corresponding to their elevation above the phreatic surface. Rainfall/infiltration then leads to reductions in

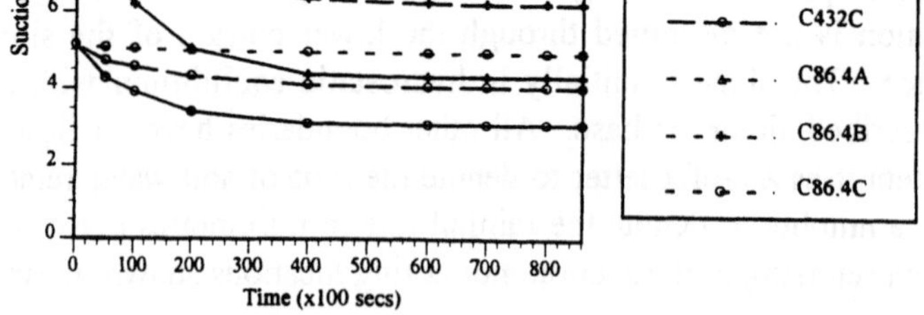

Fig.3. Evolution of suctions at monitoring locations; A,B & C - clay-soil water relations

soil suctions which in Fig.3 appear to vary directly with the magnitude of the initial suction and intensity of the rainfall event, i.e., greatest attenuations occur under high initial suction and high rainfall conditions. This type of suction response has been noted in experimental observations reported by Lim, Rahardjo, Chang & Fredlund (1996).

3.3 Results - sand soil water relations

The same infiltration tests run with the sand-type soil water parameters produced the suction evolution curves in Fig.4. In all cases, soil suctions attenuate more slowly and by larger amounts than is the case for the clay-type parameters. However, the patterns of attenuation, especially those associated with the larger rainfall events (i.e., Run S864 & S432) appear to fall to a common suction value. Once again a similar pattern of suction response is reported by Lim et al (1996) for heavy rainstorms on a bare earth slope.

4.0 INTERPRETATION OF INFILTRATIVE STATES: ANALYTICAL & NUMERICAL METHODS

In numerical flow models, infiltration is handled by prescribed boundary conditions which introduce moisture into the flow domain. Under Neumann type boundary conditions, the rate of moisture transfer is specified and in an unsaturated soil inevitably leads to an increase in the moisture content. Strictly, any flux must be accounted for in relation to both hydraulic gradient and conductivity which, in the case of unsaturated soils, are functionally related to the prevailing moisture content. It is these functional relationships which confer a high degree of non-linearity upon the unsaturated flow model and militate against the use of simple analytical solution methods. Nevertheless, there are certain idealised states for which the hydraulic regime can be assumed or could be empirically established. In such cases, direct reference to the soil water relations can adequately

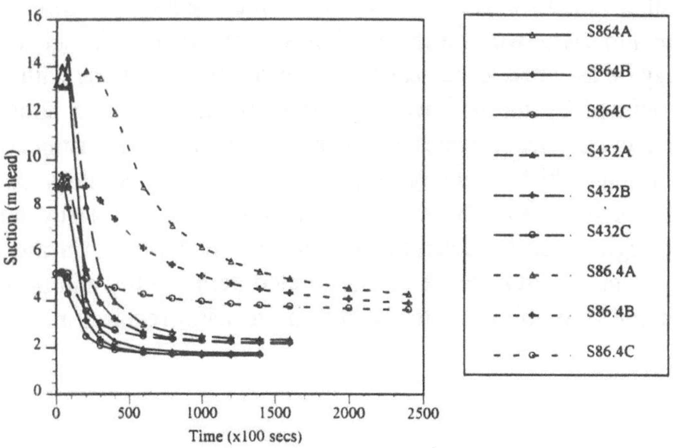

Fig.4. Evolution of suctions at monitoring locations, A, B & C - sand soil water relations

describe the hydraulic status and provide a useful reference framework for the interpretation of infiltration processes. These are; (a) under hydrostatic or zero gradient conditions and (b) steady vertical infiltration.

4.1 Hydrostatic moisture profile

The hydrostatic condition defines the moisture profile to which a soil tends in the absence of any prescribed fluxes. The exact range of moisture contents depends on the soil moisture characteristic and the value of prescribed suctions but must comply with a zero hydraulic gradient throughout. Since the elevation gradient is by definition unity, the pressure gradient under hydrostatic conditions is -1 and initial suctions are thus equivalent in magnitude to their elevations above the phreatic surface.

4.2 Steady vertical infiltration with unit gradient

Assuming Darcy type flow, Kisch (1959) shows that the variation of pore water pressure with elevation during steady state vertical infiltration is given by,

$$\frac{dp}{dz} = \frac{q}{k(\theta)} - 1 \qquad [4]$$

where p is the pore water pressure head (m), q is the flux velocity (m/s or mm/day), z is the vertical coordinate increasing upwards (m) and $k(\theta)$ is used to denote the unsaturated hydraulic conductivity (m/s or mm/day). Equation [4] shows that when $q = 0$, i.e. the hydrostatic condition, the pressure head gradient is -1; when $q < k(\theta)$ the pressure gradient lies between -1 and 0; and when $q = k(\theta)$, the pressure gradient is zero. In the latter case, moisture contents are constant and the hydraulic gradient is unity. Such a condition is generally recognised to occur during prolonged steady infiltration, for example, from ponded rainfall wherein a transmission zone of constant moisture content extends down from the surface.

The impact of a rainfall event upon the hydraulic regime of a soil profile can be interpreted in relation to these two limiting equilibria. For example, initial conditions with little or no flow may be assumed to be near or at hydrostatic conditions with the particular hydraulic regime defined by the soil water parameters and elevation above a phreatic surface. A rainfall event then initiates a process of hydraulic adjustment to the new boundary flux conditions which, if the rainfall event is of sufficient duration, produces a steady infiltrative state in at least the near-surface locations. For steady infiltration the resulting hydraulic regime can be determined using equation [4] or, in the case of unit hydraulic gradient, from the equivalence of unsaturated permeability and prescribed flux. However, the transition from initial to steady infiltrative states requires more sophisticated solution methods.

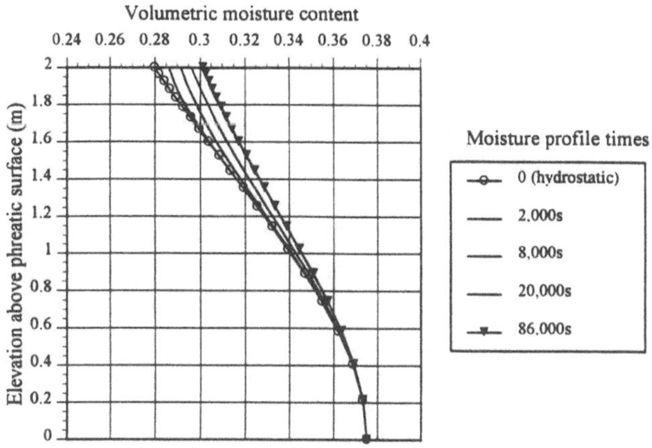

Fig. 5 Transient infiltration moisture profiles generated by 43.2mm/day rainfall rate on 2m soil column with saturated permeability of 864mm/day and initially in equilibrium with phreatic surface 2m below upper surface (test 5-43.2/864/2)

4.3 Numerical simulation of transient vertical infiltration

Consider the simulation of infiltration into a vertical soil column; infiltration events are applied to the upper surface using Neumann type boundary conditions whereas the lower surface is free draining with a prescribed water pressure. The soil water relations are those of the sandy soil shown in Fig.1. Figure 5 shows the predicted moisture profiles at various stages of a 43.2 mm/day rainfall event applied to a 2 m column which is initially in hydrostatic equilibrium with upper / lower soil water pressures of -2m / 0m. Infiltrating moisture readily travels down the column causing only a small increase in the initially high moisture contents.

In contrast, Fig.6 shows a quite different moisture response. Here, the rainfall event

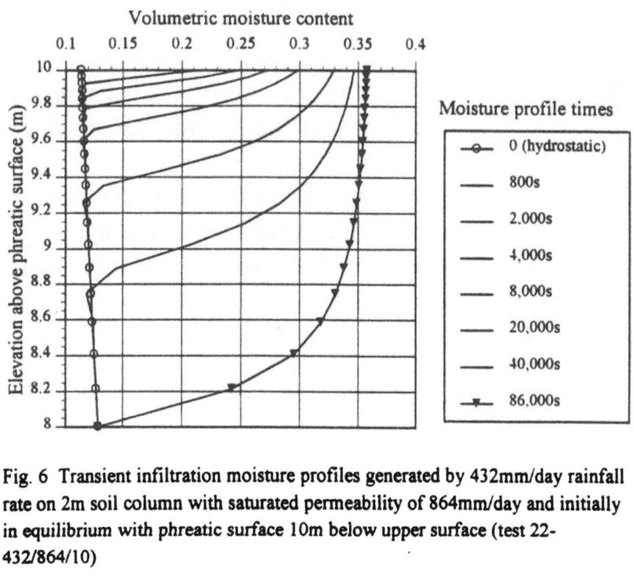

Fig. 6 Transient infiltration moisture profiles generated by 432mm/day rainfall rate on 2m soil column with saturated permeability of 864mm/day and initially in equilibrium with phreatic surface 10m below upper surface (test 22-432/864/10)

is 432 mm/day and it is applied to a column whose upper / lower soil water pressures are -10m / -8m. There is a relatively rapid and significant increase of moisture levels in the near surface zones before penetration to depth occurs.

A consideration of the hydraulic conditions existing within the soil at the start of the rainfall event provides some physical rationale for the numerical interpretation. In the case of the column at -2m / 0m surface pressures, initial moisture contents (θ = 0.28) and associated relative permeabilities are relatively high. The newly introduced moisture is readily redistributed down through the soil column. At the higher suction pressures, i.e. -10m / -8m, initial moisture contents are less (θ = 0.112) but the corresponding relative permeabilities can be seen by reference to Fig.1 to be lower by several orders of magnitude. The newly introduced moisture cannot easily move down through the soil so accumulates in the near surface region. As infiltration continues, moisture contents and the prevailing hydraulic conductivity in the near surface region increase and the infiltration process advances down through the soil profile as a distinct wetting front.

The infiltration rates used in the tests described above have been chosen to highlight these different patterns of transient infiltration. However, the patterns of infiltration shown in Fig.5 are extremes; a range of tests at more moderate infiltration rates and initial conditions were run and produced patterns of infiltration lying between the two extremes.

4.4 Effect of different patterns of infiltration on soil water suctions

The results of these tests are significant because the mechanism of hydraulic adjustment, which differs fundamentally at the two extreme infiltration patterns, has implications for soil suction evolution in unsaturated slopes. For example, where minimal moisture accumulation occurs, the ease with which newly introduced moisture is redistributed means the soil water suction regime remains under the influence of the initial hydraulic conditions. In contrast, where rainfall is high in relation to the unsaturated hydraulic conductivity and a distinct wetting front occurs, moisture is constrained to accumulate in the near surface region. Eventually, increases in moisture content extend beyond the immediate sub-surface region to establish a transmission zone and the wetting front moves down into the soil. The saturation of the upper soil profile leaves a suction regime that is determined by the infiltration characteristics and largely independent of any initial suction values.

The extent to which transient flow phenomena fully develop will be dependent on the history, uniformity and persistence of prescribed boundary fluxes and in general, numerical solution methods provide the only means of tackling more complex boundary condition sequences. However, it is instructive to re-examine the original slope tests in the light of the foregoing discussion.

5.0 RE-ASSESSMENT OF SLOPE INFILTRATION TESTS

Figure 4 revealed that for test runs S864 & S432, the reported suction attenuation at locations A, B & C converged to common values. Such behaviour is consistent with the concept of an advancing wetting front modifying the hydraulic regime in accordance with transmission zone characteristics.

Run ID	Soil type	Rainfall mm/day	Unsaturated permeability mm/day	Run ID	Soil type	Rainfall mm/day	Unsaturated permeability mm/day
S864A	Sand	864	0.426	C864A	Clay	864	1097
S864B	Sand	864	2.514	C863B	Clay	864	2307
S864C	Sand	864	24.11	C864C	Clay	864	4200

Table 2. Initial hydraulic variables at suction monitoring locations in slope infiltration tests

Table 2 contains a summary of the initial hydraulic variables for the suction monitoring locations and rainfall events. In the case of the sand soil water relations the unsaturated hydraulic conductivities range from 0.426 mm/day to 24.11 mm/day for suction monitoring locations A to C respectively. A rainfall event of 864 mm/day therefore exceeds the initial hydraulic conductivity at the slope monitoring locations by factors ranging from 2028 to 35 respectively. Other rainfall events are similarly calculated and in all cases exceed the initial hydraulic conductivity but under the lower rainfall rate (S86.4), the fall to a common suction value is less pronounced.

In the case of the clay soil water relations, Table 2 shows that even the most extreme rainfall event has a flow rate which is less than the initial hydraulic conductivity; minimal moisture accumulation occurs in all tests. Soil suctions will reflect the magnitude of the rainfall event and are related to the initial hydraulic condition. However, the magnitude of suction reductions is dependent on the particular soil water relations. Under the clay-type soil water parameters, suction reductions (Fig.3) are noticeably less than those occurring under the sand-type soil (Fig.4) but can hardly be described as minimal. Inspection of the soil moisture characteristic for the clay soil in Fig.1 shows that suction response is sensitive to changes in moisture content at all moisture contents.

Given that the initial hydraulic regime is fundamental to the pattern of infiltration, the initial groundwater conditions are clearly important to the accurate simulation of suction attenuation. With a horizontal phreatic surface, a variation in initial suctions along the slope can exist and their subsequent differential evolution gives rise to a complex variation of soil strength.

6.0 CONCLUSIONS

This numerical investigation of infiltration into unsaturated soil has revealed a range of transient infiltration patterns lying between the extremes of (a) a distinct wetting front and

(b) minimal moisture accumulation. The former condition occurs when the prescribed influx exceeds the ability of the soil to redistribute it. In the latter condition, the background permeability is high and newly introduced moisture is rapidly removed.

Where the accumulation of moisture is minimal, suctions communicate via the soil moisture characteristic with their initial hydraulic state. Infiltration which inundates a soil profile as a pronounced wetting front does so in advance of a transmission zone in which the soil moisture status is defined in relation to the surface flux. Soil suctions are thus independent of the initial hydraulic condition. When the initial hydraulic conditions give rise to non-uniform suctions along potential shear surfaces the impact of different infiltration phenomena on unsaturated slopes may be complex.

7.0 REFERENCES

FREDLUND, D.G. & BARBOUR, S.L. (1992). Integrated seepage modelling and slope stability analyses: A generalized approach for saturated/unsaturated soils. *In* Geomechanics and water engineering in environmental management. *Edited by* R.N.Chowdhury, A.A.Balkema, Rotterdam.

KISCH, M. (1959). The theory of seepage from clay-blanketed reservoirs, Geotechnique, 9: 9-21.

LIM, T.T., RAHARDJO, H., CHANG, M.F. & FREDLUND, D.G. (1996). Effect of rainfall on matric suctions in a residual soil slope. Canadian Geotechnical Journal, 33, pp 618-628.

McDOUGALL, J.R. (1996). Application of variably saturated flow theory to the hydraulics of landfilled waste: A finite element solution. Ph.D. thesis, University of Manchester/Bolton Institute, U.K.

NEUMAN, S.P. (1973). Saturated-unsaturated seepage by finite elements. Journal of Hydraulics Division, American Society of Civil Engineers, Vol.99, HY12, pp 2233-2250.

PRADEL, D.P. & RAAD, G. (1993). Effect of permeability on surficial stability of homogeneous slopes. Journal of Geotechnical Engineering, American Society of Civil Engineers, Vol.119, No.2, pp 315-332.

THOMAS, H.R. & REES, S.W. (1990). Modelling field infiltration into unsaturated clays. Journal of Geotechnical Engineering, American Society of Civil Engineers, Vol.116, No.10, pp 1483-1501.

VAN GENUCHTEN, M.T. (1980). A closed form equation for predicting the hydraulic conductivity of unsaturated soils. Soil Science Society of America Journal, Vol.44, pp 892-898.

FINITE ELEMENT CALCULATIONS ON HAARAJOKI TEST EMBANKMENT

Näätänen, P. Vepsäläinen and M. Lojander
Helsinki University of Technology, Espoo, Finland

ABSTRACT: The Finnish National Road Administration has organised an international competition to calculate the behaviour of the Haarajoki test embankment. Half of the embankment was constructed on vertically drained area and the other half on natural deposits without any ground improvement. The embankment lies on over 20 meters of overconsolidated clay and silt and it has been well instrumented. The Laboratory of Soil Mechanics and Foundation Engineering at Helsinki University of Technology has been calculating the behaviour of this embankment with several methods. This paper concentrates on calculations with Sage-CRISP. Existing measurements have been used to verify and re-evaluate the soil parameters. The calculations are compared with the field measurements and the validity of finite element calculations is discussed. Especially, the modelling of the vertically drained area has been discussed.

1. INTRODUCTION

The Finnish National Road Administration has organised an international competition to calculate the behaviour of the Haarajoki test embankment (FinnRA, 1997). The embankment was constructed as a part of a noise barrier during July and August 1997. The embankment was instrumented with settlement plates, piezometers, inclinometers, extensometers and total stress gauges and the measurements are going to be carried on for two years. The embankment is a part of a larger project called Road Structures Research Programme. As a part of this project, the behaviour of this embankment has been analysed at the Laboratory of Soil Mechanics and Foundation Engineering. These calculations have been done outside the competition, because some of the tests had been conducted in the laboratory and due to this fact the Laboratory was not allowed to participate. The first calculations were done before any measurements were available and both classical and finite element methods have been used (Näätänen & al., 1997). All the programs used are either commercially available, like EMBANKCO (Användarhandbok, 1994), Z-Soil (1995), PLAXIS (1995) and Sage-CRISP (1995), or are used by geotechnical design consultants like, RAKPA and KONSOL (Vepsäläinen, 1990). The aim was specifically to test the applicability of the programs, which are being used in geotechnical design. In this paper only revised Sage-CRISP calculations are presented. These calculations have been performed with re-evaluated parameters after studying the measured data from the first 6 months.

Although quite a lot of experience has already been gained on modelling the behaviour of embankments on soft ground with elasto-plastic isotropic hardening models and finite element method, problems still arise when designing embankments on vertically drained areas. Generally, the existence of vertical drains complicates the modelling, and the simulation is extra sensitive to the choice of consolidation parameters. In this case the parameter decision was eased by the reference section available: the section of embankment on the non-improved ground. The field data from that side has been used to calibrate and re-evaluate some of the parameters, so that the inaccuracy in the modelling of the vertically drained section can no more be attributed to the material parameters.

For practical purposes it is more relevant to model the behaviour of a full embankment than just a single drain. For most embankments plane strain conditions exist, but when considering the effects of vertical drains the conditions around each drain are more close to axial symmetry. Special consideration has to be dedicated on how to match this combined effect in a numerical simulation.

2. GEOMETRY AND GROUND CONDITIONS

The embankment lies on layers of overconsolidated clay and silt with a combined depth of 20 meters. More details on soil classification data can be found in the paper by Aalto & al (1998). The subsoil has been divided in to nine geotechnical layers for the calculations. The ground water table is assumed to be at the ground level, although –3 to 10 kPa excess pore pressures have been measured on this area. The embankment is 2.9 meters high and 100 meters long, and the width at the top is 8 meters, as shown in Figure 1. Half of the embankment is constructed on vertically drained area and the other half on natural deposits without any ground improvement. The construction was done in 0,5 meter stages using a gravel fill with a density of 21 kN/m^3.

3. MATERIAL PARAMETERS AND CALCULATION METHODS

Revised calculations were done with two finite element programs, namely Sage-CRISP and PLAXIS. This paper only the Sage-CRISP calculations are shown while the latter can be found in Aalto & al (1998).

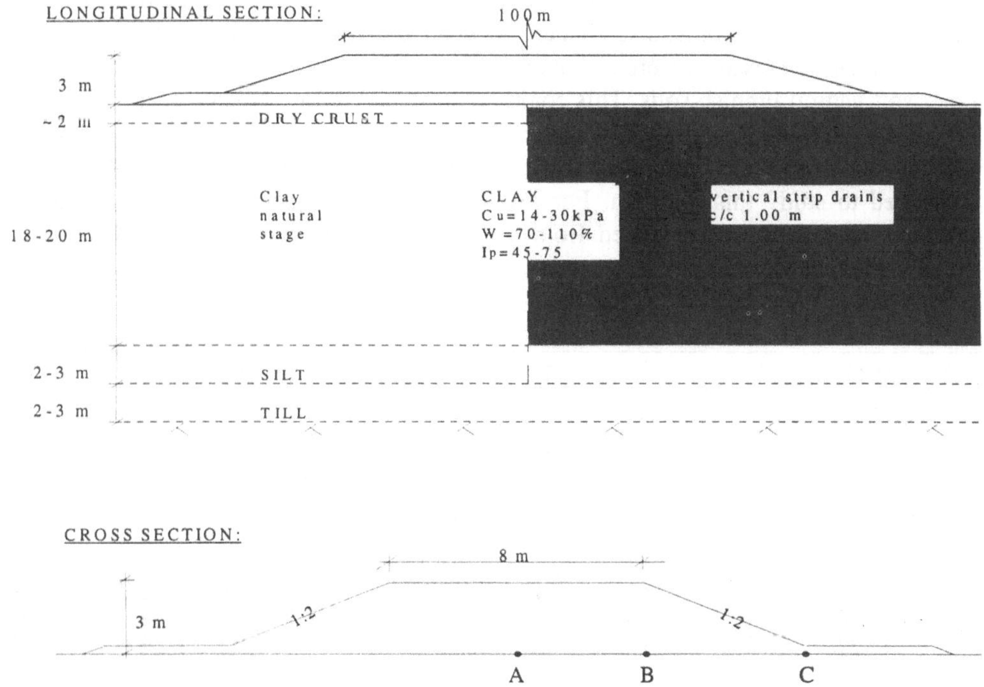

Figure 1. The Haarajoki test embankment. Embankment geometry.

The parameters used for calculations are shown in Table 1. These parameters are based on over 50 oedometer tests and 16 triaxial tests. In addition to vertical samples some horizontal samples have also been studied with oedometer. The preconsolidation pressure assumed for the simulations is shown in Figure 2. The measured excess pore pressures have been taken into account when determining the preconsolidation pressures. After comparing the first calculations (Näätänen & al, 1997) with the field measurements after the first 6 months, the following parameters have been re-evaluated: the values for κ, permeability and preconsolidation pressures. The settlement speed in previous calculations (Näätänen & al. 1997) seemed much too slow compared to the field observations. The first permeability values were calculated directly from the oedometer test results. Leroueil & al. (1990) have done some comparison between coefficient of consolidation determined in the laboratory and deducted from observations of settlement of embankments. According to these comparisons the in situ values are 3 to 200 times the laboratory values, and the average multiplier is 20. A factor of 5 was chosen to multiply the originally derived values for permeability for the calculations presented here. The laboratory values and the values applied for the simulations are shown in Table 2.

The values of κ, the slope of the unloading-reloading line in v:lnp' plane, are now determined from oedometer test results from the beginning of the recompression line. For overconsolidated layers, this elastic parameter plays an important role. For the first calculations the preconsolidation pressures for some layers were based on triaxial tests. There seemed to be some discrepancy between the triaxial and oedometer tests when preconsolidation pressures were considered. The author's have had similar experience with other soft natural Finnish soils. This could be attributed to rate-effects, or it may be that in the case of extremely soft sensitive soils the triaxial set-up (rubber etc) 'strengthens' the sample and hence systematically over predicts the preconsolidation pressure. More research is needed to study this problem. For revised calculations the isotropic preconsolidation pressure value p_c were calculated from the oedometer test results and OCR values. The procedure is shown in Figure 2.

layer	Depth m	γ kN/m^3	M	ϕ °	ν	κ	λ	e_0	e_{cr}
1	0-1	17	1.5	36.9	0.38	0.007	0.1	1.4	1.78
2	1-2	17	1.5	36.9	0.38	0.013	0.1	1.4	1.76
3	2-5	14	1.15	28.8	0.1	0.028	1.08	2.9	6.68
4	5-7	14	1.1	27.7	0.1	0.028	1.86	2.8	9.51
5	7-10	15	1.07	27	0.1	0.021	0.65	2.3	4.84
6	10-12	15	1.07	27	0.28	0.037	1.23	2.2	7.22
7	12-15	15	1.15	28.8	0.28	0.027	1.09	2.2	6.63
8	15-18	16	1.5	36.9	0.28	0.05	0.48	2	3.98
9	18-22.2	17	1.5	36.9	0.28	0.009	0.1	1.4	1.84

Table 1. Haarajoki test embankment. Material parameters.

layer	Depth	k_x $*10^{-4}$ m/d	k_y $*10^{-4}$ m/d	k_x $*10^{-4}$ m/d	k_y $*10^{-4}$ m/d
		laboratory values		*Estimated field values*	
1	0-1	13	13	65	65
2	1-2	13	13	65	65
3	2-5	1.56	1.3	7.8	6.5
4	5-7	1.21	0.86	6	4.3
5	7-10	1.38	0.69	6.9	3.5
6	10-12	2.59	1.3	13	6.5
7	12-15	2.59	1.3	13	6.5
8	15-18	8	1.12	40	5.6
9	18-22.2	80	80	80	80

Table 2. Haarajoki test embankment. Values for permeability.

The triaxial test on layer 4 showed that the shape of the initial yield locus is a rotated ellipse rather than the isotropic ellipse in Modified Cam Clay (MCC) model. This is related to the anisotropic depositional history. More accurate results could be achieved with anisotropic soil model. Since none of the programs available had suitable anisotropic model, the isotropic hardening models have been used for subsoil. For layers 1 and 2 Schofield model and for layers 3 to 9 MCC are chosen. The embankment has been modelled with an elastic-perfectly plastic Mohr-Coulomb model.

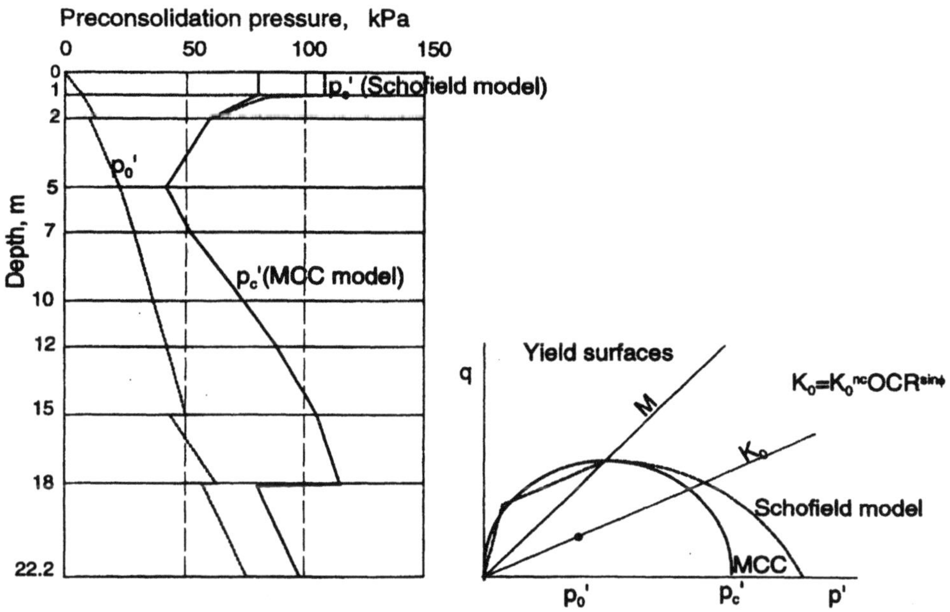

Figure 2. Haarajoki test embankment. In situ hydrostatic pressures and preconsolidation pressures.

3. CALCULATIONS

3.1 CROSS SECTION PL 35840

The cross section PL 35840 is situated on the natural clay and silt deposits. A finite element mesh with 575 8-noded quadrilateral elements is used to model the subsoil and the embankment. The symmetry has been taken into account, and only a half of the problem is modelled. Settlement results (see www.tieh.fi/pailas/) show that this assumption has been correct. The construction was done in 0,5 meter layers and the real construction schedule has been simulated in calculation. The top and the bottom of the mesh have been considered permeable. The calculations have been taken up to ten years to find out the final consolidation settlement. The embankment has been constructed using 35 increments and 120 time steps have been used to model the consolidation progress.

The calculation results for the first 9 months have been compared with measurements in Figure 3. The final consolidation settlement of 44 cm is reached less than 10 years. In Figure 3 the amount of settlement caused by construction of first layer of fill, three centimetres, has been subtracted from the calculated values because the settlement plates were levelled for the first time only after the first 0,5-meter gravel layer was constructed. Horizontal movements have been measured with inclinometers at points B and C (Figure 1). Both the calculated and the measured values are presented in Figure 4. Excess pore pressures at the centreline are presented in Figure 5.

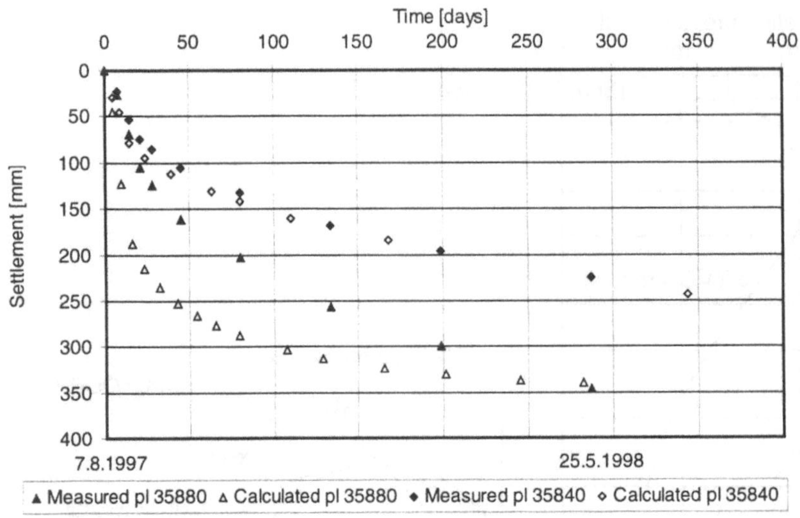

Figure 3. Haarajoki test embankment. Calculated and measured settlements for 9 months.

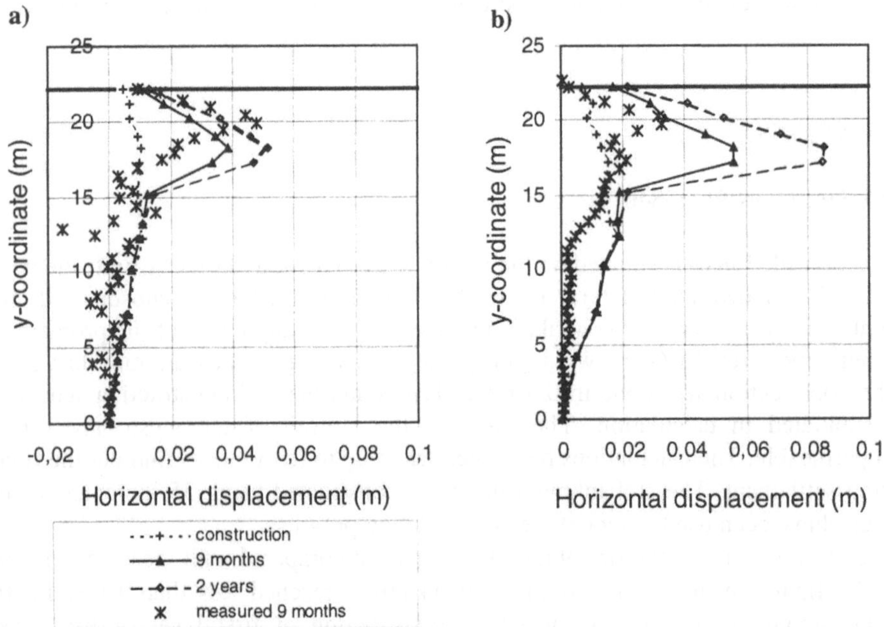

Figure 4. Haarajoki test embankment, cross section PL 35840. Horizontal displacements. a) 4 meters from the centreline b) 9 meters from the centreline.

The prediction of excess pore pressures is not as satisfactory as the predicted displacements. This can partly be explained by the fact that the pore pressures were not measured just before construction but two months earlier. So these measurements had to be used as the reference pressures. The calculated change in excess pore pressures however, corresponds quite well with the measurements at depths 7 and 10 meters (Figure 5). The fact that there were excess pore pressures (of to 10 kPa) initially before construction, may complicate the comparison. The calculation is in fairly good agreement with measurements and the parameters used seem therefore adequate for the study of vertically drained area.

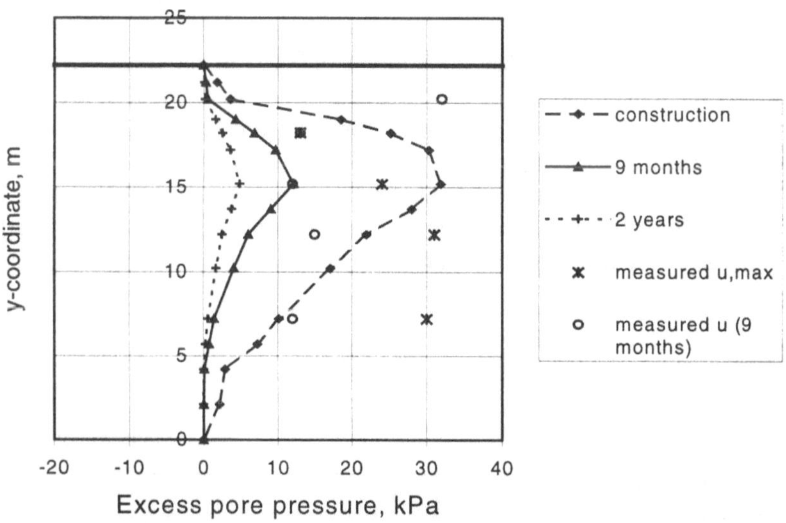

Figure 5. Haarajoki test embankment, cross section PL 35840. Excess pore pressures at the centreline.

3.2 CROSS SECTION PL 35880

Cross section PL 35880 is situated in the middle of the part of embankment constructed on vertically drained area. Vertical drains are 15 meters long and they were installed in a square grid with one-meter spacing. The equivalent radius of geodrains is 0.034 meters. The permeability was assumed to be five times the laboratory value, as on the other side of the embankment, and it is kept constant during the consolidation process.

The 2-D modelling for vertically drained area was performed following the procedure by Hird & al (1995). The geometric matching was done according the Equation:

$$\frac{B}{R}=\left\{\frac{3}{2}\left[\ln\left(\frac{R}{r_s}\right)+\left(\frac{k}{k_s}\right)\ln\left(\frac{r_s}{r_w}\right)-\left(\frac{3}{4}\right)\right]\right\}^{1/2} \qquad (1)$$

where B is the half-width of the plane strain unit cell; R, r_w and r_s are the radius of the axisymmetric unit cell, the drain and the smear zone, respectively and k and k_s are the horizontal permeabilities of the undisturbed and smeared soil. In the plain strain cell the soil has the same permeability as the undisturbed soil in the axisymmetric cell and no smear zone need to be explicitly represented. Hird & al (1994) presented also alternative matching procedures. The matching can as well be achieved by manipulating the permeability of the soil instead of the cell geometry. By combining these two matching procedures a combined matching can be adopted to gain further control over the mesh. Hird & al. (1994) has shown that the results of geometry and permeability matching were inseparable. A significant well resistance can also be taken account with independent matching rules.

According to Equation (1) with a smear zone radius of 0.065 m and a k/k_s value of 5 the drain spacing for 2-D model is 3 meters. Since no special drainage elements were available, the drains have been modelled by setting the excess pore pressure zero along the boundaries where drains are situated. The well resistance was not taken into account. The finite element mesh is shown in Figure 6. The embankment construction has been modelled with 79 increments and the consolidation procedure is followed up to five years by using 130 time steps. The time-settlement curve for 9 months is presented in Figure 3. Again, 3.8

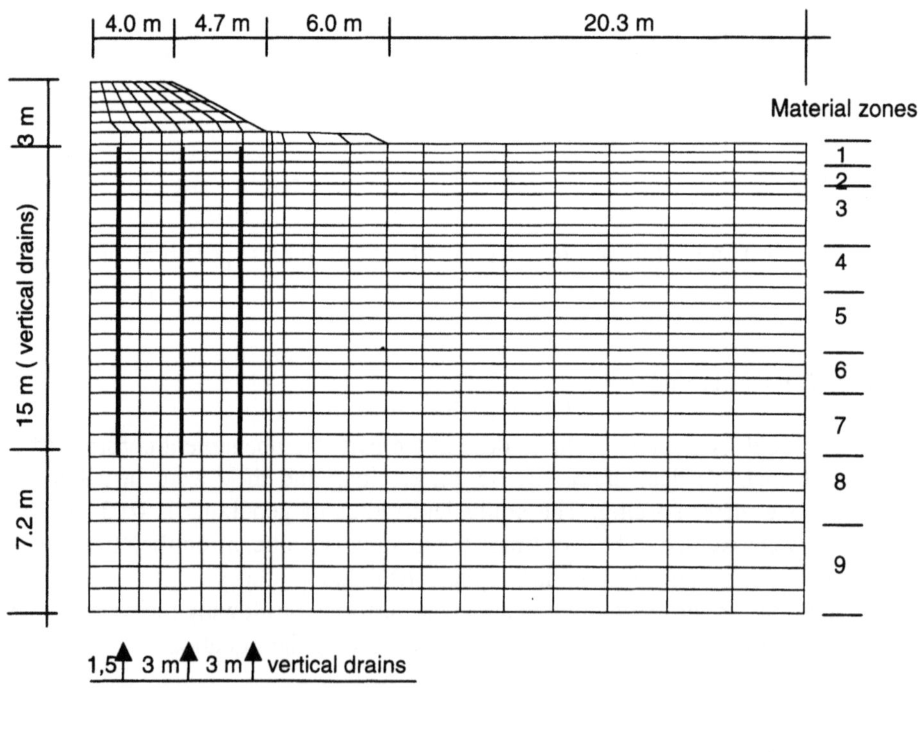

Figure 6. Haarajoki test embankment, cross section PL 35880. Finite element mesh.

centimetres have been taken off from the calculation results, because the settlement plates were levelled after the first 0.5-meter layer was constructed.

Horizontal movements were again measured at points B and C (Figure 1). After 9 months underneath the point B the maximum measured displacement was 40 mm at depth 2.7 m and below the point C 27 mm at depth 1.97 m. In the calculation the maximum horizontal displacements were 71 mm and 96 mm, underneath point B and C respectively, at depth 4 m. The pore pressure measurements 0,5 meters aside from the centreline show that at the depth of 7.2 m there are still about 5.5 kPa, and at the depth of 2.2 m 7 kPa excess pore pressures. Calculations showed only 2 kPa maximum excess pore pressures after 9 months.

For the vertically drained area the effect of secondary consolidation may be significant. That has not been taken into account though, because there was no creep model in the program.

4. CONCLUSIONS

- The computed time-settlement behaviour for first nine months at the centreline in cross section PL 35840 is very satisfactory.
- In cross section PL 35880 the measured and calculated settlements are not in equally good agreement. The calculations overestimate the settlement rate at the beginning. This could be due to the fact that, it is quite difficult to take into account the way the vertical drains start to function at the field. In the calculations the drains were modelled by a free excess pore pressure boundary condition, and they started functioning at once when the boundary condition was introduced.
- The prediction of excess pore pressures is not as satisfactory. This can partly be explained by the fact the reference pressures had to be estimated from values measured two months before construction. The fact that according to the measurements there were excess pore pressures initially before construction, may complicate the comparison. The measurements are not able to show clearly the excess pore pressures caused by construction.
- As the clay is overconsolidated, a proper prediction of elastic deformations is extremely important. In this case isotropic elastic behaviour was assumed, this may not be the case in the field. Care is also needed when the parameters for selected models are determined.
- Due to depositional processes many soft soils are anisotropic. This affects not only the orientation of initial yield curve, but also the permeabilities in different directions. The anisotropic properties should be taken into account in determining the parameters and modelling soft soil behaviour.
- The most problematic parameter is inevitably the permeability. A lot of research is still needed on how to determine the permeability reliable in all directions. An in situ method would be most desirable.
- The procedure suggested by Hird & al (1995) seems straightforward and is easily adopted for design purposes. Problems that arise are attributed to parameter

determination. More studies are needed for estimating the radius of the smear zone and, the permeabilities within.

While the vertical drains speed up the consolidation process, the settlement due to the secondary consolidation can be significant. For modelling the vertically drained areas a practical secondary consolidation model would be required.

REFERENCES:

AALTO A., (1998). The calculations on Haarajoki test embankment with the finite element program Plaxis 6.1. Proc. of 4th European Conference on Numerical Methods in Geotechnical Engineering, Udine, Italy, 14-16 October 1998. (To be submitted)

ANVÄNDARHANDBOK (1994). Program Embankco. Version 1.02. Program för sättningberäkning för vägbank på finkornig jord (in Swedish).

FINNISH NATIONAL ROAD ADMINISTRATION (FinnRA), (1997). Competition to calculate settlements at the Haarajoki test embankment. Competition programme.

HIRD C.C., PYRAH I.C., RUSSELL D. & CINICIOGLU F., (1995). Modelling the effect of vertical drains in two-dimensional finite element analyses of embankment on soft ground. Canadian Geotechnical Journal, 32: 795-807.

LEROUEIL S., MAGNAN, J-P. & TAVENAS, F., (1990). Embankments on Soft Clays. (English translation, D. Muir Wood.) Ellis Horwood, Chichester, England.

NÄÄTÄNEN A., PUUMALAINEN N., SAARELAINEN K., AALTO A., LOJANDER M. & VEPSÄLÄINEN P. (1997). Estimation of settlement of the Haarajoki test embankment. Proceedings of the 6th Finnish Mechanical Days, September 5-6, 1997, Oulu, Finland. Eds. Aalto, J., Salmi, T. pp. 295-310.

PLAXIS (1995) Finite Element Code for Soil and Rock Analyses. Version 6. P.A. Vermeer, R.B.J. Brinkgreve (eds.). A.A. Balkema. Rotterdam 1995.

Sage CRISP (1995) User Guide and Technical Reference Guide. Version 3.02 Update. 1995.

VEPSÄLÄINEN, P., (1990). Computers in Geomechanics. RIL 157-2, Geomechanics II. Handbook. Helsinki 1990, RIL. pp. 339-422. (In Finnish)

ZSOIL V 3.11(1995). User's guide. Zace Services Ltd, Lausanne.

SOME FACTORS AFFECTING THE PREDICTIONS OF GROUND MOVEMENTS AROUND AN EXCAVATION IN STIFF CLAY

N.D. Pierpoint
Mott MacDonald, London, UK

C.C. Hird
University of Sheffield, Sheffield, UK

ABSTRACT: The paper describes a sensitivity study of predictions of excavation behaviour in a heavily overconsolidated clay using a non-linear cross-anisotropic elastic model for the soil. The model is experimentally based and is capable of including non-linearity of stress-strain behaviour, dependence of stiffness on both stress level and stress path direction, and changes in stiffness caused by changes in the stress path direction. Analyses are reported of an instrumented full-scale trial excavation and computed displacements are compared with the observed behaviour. Firstly, the effect of including anisotropy is quantified, by comparing analyses with isotropic and anisotropic soil models. The effects of modelling the recent stress history and changes in the stress path direction during excavation are then investigated. Finally, analyses with linear and non-linear stress-strain behaviour are compared to isolate the effect of including a strain-dependent stiffness.

1. INTRODUCTION

A ground investigation at Elstow, Hertfordshire, UK, was conducted in the mid-1980s to establish the suitability of the site for the construction of underground repositories for low level nuclear waste (Pierpoint, 1996). The ground investigation included the construction and monitoring of a full scale trial excavation within the Oxford Clay to evaluate the ground response to relatively large stress changes representative of the final construction. The excavation was heavily instrumented to monitor horizontal movement, vertical movement, and pore water pressures in the surrounding ground. A plan and transverse section of the excavation, including the location of instrumentation is shown in Fig. 1. The soil profile consists of approximately 1 m of superficial Head deposits overlying stiff to very stiff heavily overconsolidated Oxford Clay to a depth of about 14 m, being weathered over a depth of about 3 m. Underlying the Oxford Clay the Kellaways beds provide a gradual and rather variable transition to generally harder limestone and mudstone strata. The upper bed, the Kellaways Sand, consists of 4-4.5 m of variable fine to medium sands and sandy clays. The lower Kellaways bed is the Kellaways Clay which consists of 1.5-2 m of a very stiff, thinly laminated closely to very closely fissured clay.

A laboratory research programme was undertaken primarily to investigate the stress-strain characteristics of the Oxford Clay. The laboratory testing programme was designed to provide information about the non-linearity of stiffness and the degree of anisotropy at small strains (<0.1%).

A constitutive model was developed to make a Class C prediction (Lambe, 1973) of the ground behaviour in response to the construction of the trial excavation. The constitutive model was based on the form of cross-anisotropic elasticity proposed by Graham & Houslby (1983) and also featured a non-linear stiffness response which varied with incremental strain energy, stress history, stress path direction, and stress level. Subsequently, an attempt was made to optimise the input parameters through back analysis. As part of this a series of parametric studies was carried out to investigate the sensitivity of the results to the various features within the model. The results of this study are reported here.

2. CONSTITUTIVE MODEL AND MATERIAL PARAMETERS

In order to make a genuine prediction, the numerical modelling had to be completed before the field data were interpreted and, in the time available, only a relatively simple model could be employed. Simplicity in numerical models has been strongly advocated (Wroth and Houlsby, 1985), particularly when they are required as a tool for engineering analyses.

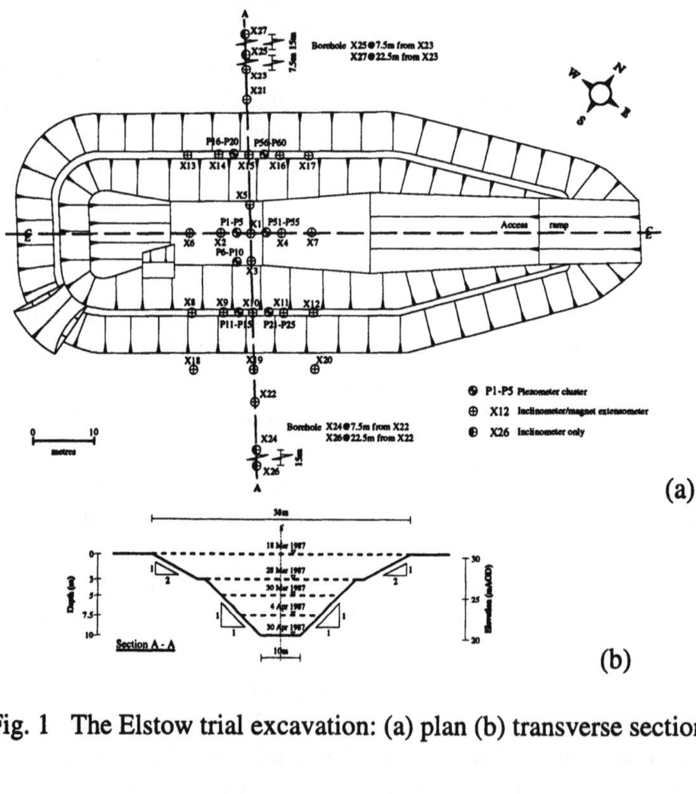

Fig. 1 The Elstow trial excavation: (a) plan (b) transverse section

Nevertheless, this model incorporated both the non-linearity and anisotropy observed in the laboratory tests (Hird and Pierpoint, 1994). The behaviour of the Oxford Clay was modelled using a non-linear cross-anisotropic elastic model (Graham and Houlsby, 1983). The model was implemented in the finite element program CRISP (Britto & Gunn, 1987). Details of the numerical analysis were given by Pierpoint (1996) and Hird and Pierpoint (1997).

The incremental effective stress-strain behaviour of soil may be described by equation 1:

$$\begin{bmatrix} \delta\varepsilon_v \\ \delta\varepsilon_s \end{bmatrix} = \begin{bmatrix} \dfrac{1}{K'} & \dfrac{1}{J'_{qv}} \\ \dfrac{1}{J'_{ps}} & \dfrac{1}{3G'} \end{bmatrix} \cdot \begin{bmatrix} \delta p' \\ \delta q \end{bmatrix} \tag{1}$$

For an elastic implementation the stiffness matrix is symmetric, and the constitutive model therefore requires the incremental parameters G', K' and J', where $J' = J'_{qv} = J'_{ps}$. The incremental strain energy, U, was chosen as a suitably generalised parameter through which to incorporate non-linearity into the stiffness parameters. The input parameters for this model were entirely derived from the experimental programme, and consist of two, complementary, non-linear stiffness relations, for increasing and decreasing stress levels respectively.

The stiffness was dependent on the current stress path direction through a similar approach to that of Jardine *et al.* (1991), in which regions of broadly similar behaviour were assigned similar stiffness parameters. The need to discretize specific zones within a finite element model is removed in this case by allowing the model to calculate appropriate stiffness parameters for a soil element based on the stress increment direction. For the undrained analyses considered here, stress path directions were defined in the same way as in the laboratory triaxial tests where different parameters were determined depending on whether the soil element was deforming in compression or extension.

When a reversal in the stress path direction was detected, stiffnesses were simply reassigned as if the soil was following a path from its original starting point and the reference strain energy was reset to zero.

The underlying strata were modelled as linear cross-anisotropic elastic materials. The stiffness of the underlying strata was high compared to that of the Oxford Clay.

3. NUMERICAL MODELLING OF THE ELSTOW EXCAVATION

The soil profile was sufficiently horizontally uniform for the excavation to be treated as symmetric about the centreline. Plane strain conditions were assumed although such behaviour was only confirmed subsequently, once the instrumentation data had been examined (Pierpoint, 1996). The mesh configuration (Fig. 2) comprised linear strain finite elements. The mesh discretisation was naturally based on providing more elements where the largest changes in the ground stresses and strains were expected.

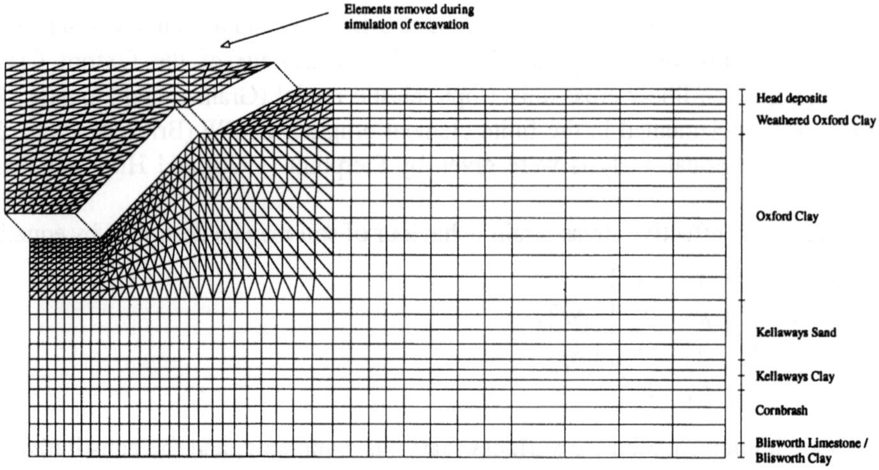

Fig. 2 Finite element mesh showing elements removed during excavation

The overconsolidation of the Oxford Clay has been brought about by the removal of approximately 500 m of overburden (Jackson and Fookes, 1974). Given the high degree of overconsolidation it was assumed that the Oxford Clay at Elstow was close to a state of passive failure. The horizontal effective stresses were therefore calculated based on an earth pressure coefficient calculated for this limit state. A hydrostatic pore water pressure distribution was assumed from a water table depth of 0.9 m. The Oxford Clay, Kellaways Clay and Blisworth Strata were modelled as undrained, by specifying a relatively high value of bulk stiffness for the pore water (2 GPa). The Kellaways Sand and the lower limestone strata were assumed to be fully drained.

Interpretation of the instrumentation data has been presented by Pierpoint (1996) and the initial prediction has been described by Hird and Pierpoint (1997). In the following sections the sensitivity of the model predictions is established, and the analysis results are compared with the instrumentation data. The instrumentation data are interpreted along section lines of the instruments shown in Fig. 1. In this figure the longitudinal section line of instruments on the centreline is termed section F, sections G and H are 4.5 m from the centreline, sections I and J are 12.5 m from the centreline, sections K and L are 21 m from the centreline, and sections M and N are 26 m from the centreline.

The instrumentation data were monitored during excavation and subsequently for a period of 3 months. This monitoring period has been combined into a number of ranges, 'A' to 'S', for which comparisons with the numerical analyses can be made. The end of excavation is termed 'I' and the end of monitoring is termed 'S'.

4. INFLUENCE OF ANISOTROPY

The sensitivity of the model prediction to assumptions regarding the degree of anisotropy was investigated. The principal simplification of the experimental results was the implementation of cross-anisotropy into an elastic model by combining the two coupling stiffness parameters, J'_{qv} and J'_{ps}, into a single parameter J'. This reduction is performed using the factor β such that $J' = J'_{ps} + \beta(J'_{qv} - J'_{ps})$. In the genuine prediction (analysis A01) this factor is determined for each element automatically from the current stress path direction, e.g. for a constant p' stress path β would be calculated as 1.0 and J' is set to J'_{qv}.

The first stage of the sensitivity analysis therefore consisted of varying the non-linear variation of J' with strain energy between the two limits defined by J'_{qv} and J'_{ps}. This involved varying the factor β so that for analyses A04, A05, and A06, β = 0, 0.5, and 1.0 respectively. Additionally, the cross-anisotropic aspect of the model was removed by setting J' to a suitable high value (analysis A07). The off-diagonal coupling stiffness terms in equation 1 were then effectively zero, making the soil response isotropic.

The instrumentation data for both horizontal and vertical movements covering the period following the completion of excavation to the end of the monitoring period are presented.

The measured ground response in terms of vertical displacement is compared with the cross-anisotropic and isotropic responses of the non-linear model in Fig. 3. The analyses A04, A05 and A06 are not included in this figure for clarity but would have plotted close to those of A01. Contrasting the isotropic and anisotropic model predictions, little difference between the predictions is seen. The main differences were seen adjacent to the excavation sides where horizontal stress changes had a dominating influence. This resulted in a deeper settlement trough for the isotropic prediction, although the responses converged at the excavation base as shown in Fig. 3 and away from the excavation where stress changes were more one-dimensional.

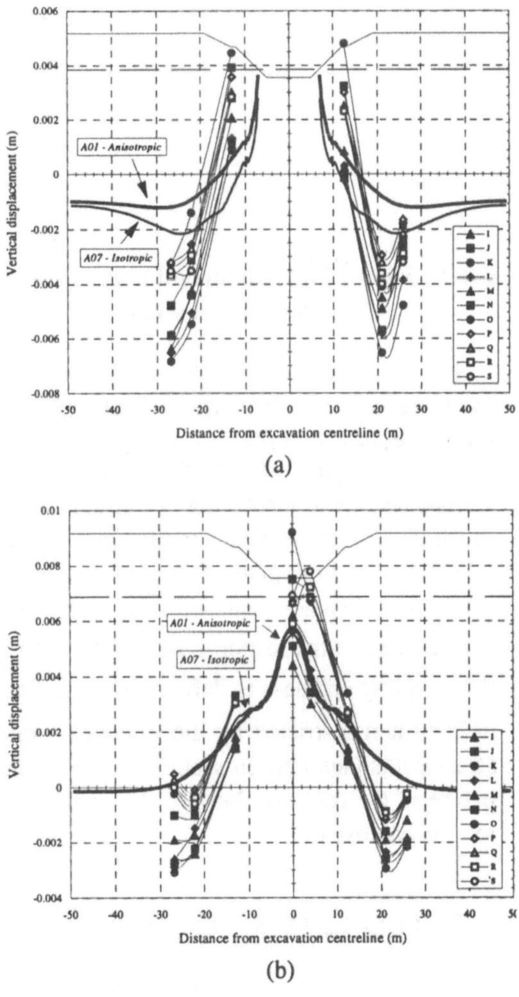

Fig. 3 Prediction of vertical displacement at (a) 8 m and (b) 14 m depth

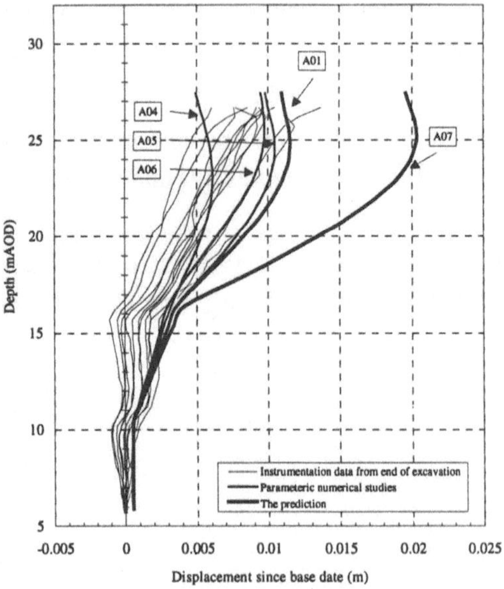

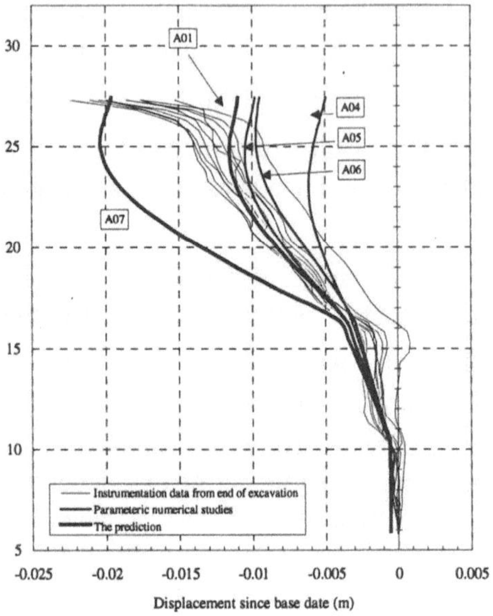

Fig. 4 (a) Prediction of horizontal displacement at section I (b) Prediction of horizontal displacement at section J

For horizontal displacements, the influence of incorporating cross-anisotropy was far more marked than for vertical displacements and also considerably improved the ability of the model to match the instrumentation data. The form of the predicted response shown in Fig. 4 was typical of the other instrument locations: very small displacements below the Kellaways strata, increasing but still small displacements within the Kellaways strata, and much larger displacements in the Oxford Clay.

For the genuine prediction (analysis A01), the stress path direction dependent β factor resulted in displacements near the upper bound of the range from $\beta = 0 \rightarrow 1$. This is as expected since the response is attributable to the constant p' stress path response in which the coupling stiffness is defined by J'_{qv}. The undrained stress paths in the Oxford Clay are defined and restricted by the imposed requirement of constant volume deformation. The stress paths (see later in section 5) are inclined to the left of vertical due to the anisotropic nature of the response but are still reasonably near vertical and therefore result in a factor relatively close to 1.0.

5. INFLUENCE OF STRESS HISTORY AND STRESS PATH DIRECTION

The effect of excavation on the stresses nearby is shown for a particular analysis in Fig. 5 which shows actual stress path responses for 6 elements at key points around the excavation. All the stress paths have a consistently negative slope which is indicative of a

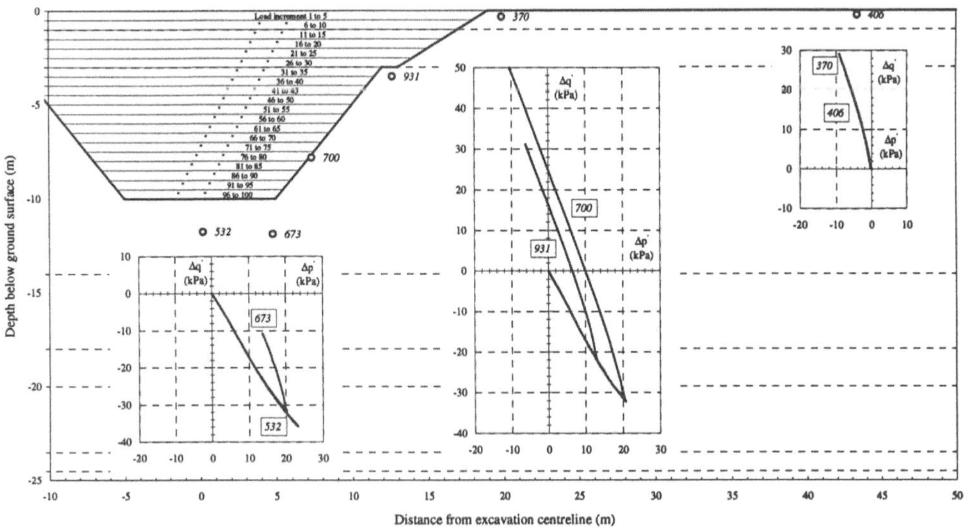

Fig. 5 Predicted stress changes during the process of excavation

cross-anisotropic response for soil that is stiffer horizontally than vertically. The elements at the base of the excavation experience a continual fall in deviator stress, whilst those beyond the excavation crest experience a continual increase in deviator stress. At the excavation sides, however, elements are subject to a change in the loading direction as the excavation passes their elevation.

Incorporated into the model implementation is the ability to interrogate the current stress path direction for each element and also to determine whether there has been a change in the stress path direction from the previous load increment. At the start of an analysis a directionally dependent stiffness response is attached to each soil element depending on the loading direction incurred in the first loading increment. If a stress path reversal subsequently takes place, a different response is substituted. In the excavation analyses, a reversal was deemed to have taken place if there had been a reversal in the direction of the change in deviator stress as defined in triaxial stress space. For the undrained analyses considered here the change in effective mean normal stress is constrained by the degree of anisotropy.

Although the model implementation allocates the stiffness definitions dynamically as the excavation analysis proceeds, the effect of using a single (fixed) set of parameters was investigated. Fig. 6 shows horizontal and vertical displacement contours, contrasting analyses in which the stiffness definition was dependent on the stress path direction with analyses in which the stiffness characteristics were fixed. If a stiffness relation determined purely from the compression data is used (Fig 6b), the heave at the base of the excavation was 30-40% less than when the lower stiffness associated with unloading is taken into account (Fig 6a), although the general form of the displacement response was recovered. If only the extension data are used, the predicted displacements are clearly inaccurate (Fig 6c).

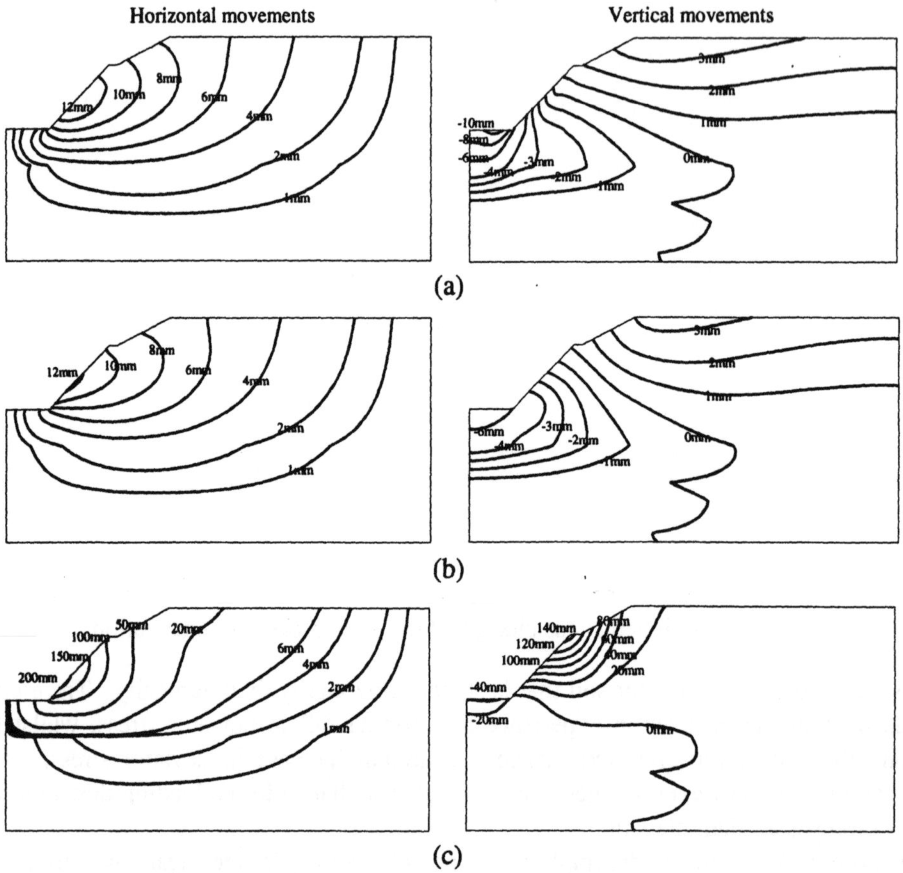

Fig. 6 Predicted ground movements using a stiffness relation (a) determined from stress path direction (b) for compression stress paths (c) for extension stress paths.

6. INFLUENCE OF NON-LINEARITY

During the back analysis optimisations were attempted for linear elastic analyses using both isotropic and cross-anisotropic stiffnesses. Fig. 7 shows the horizontal and vertical displacements respectively comparing a number of linear isotropic elastic analyses to the genuine prediction. A range of G' was taken to cover the range encountered in the experimental programme, and analyses are shown for $G' = 10,35$, and 60MPa. The resulting displacements vary considerably between analyses although, interestingly, the $G' = 35$MPa analysis produces a response very similar to the non-linear cross-anisotropic elastic analysis proposed as the genuine prediction (A01). A further analysis was carried out in which either a low ($G' = 10$MPa) or a high ($G' = 60$MPa) stiffness was assigned to elements of Oxford Clay depending whether they were considered to be in extension or compression respectively. The high stiffness value was assigned to elements experiencing a reversal in the loading direction during excavation. This analysis, labelled $G' = 10/60\text{MPa}$, is also shown in Fig. 7 and produces a similar response to the genuine prediction. Clearly, there are a number of combinations of parameters that can produce similar predictions for a variety of models. A similar conclusion was reached by Ng and Lings (1995) who obtained suitable predictions of a retaining wall problem using either a linear or a non-linear model.

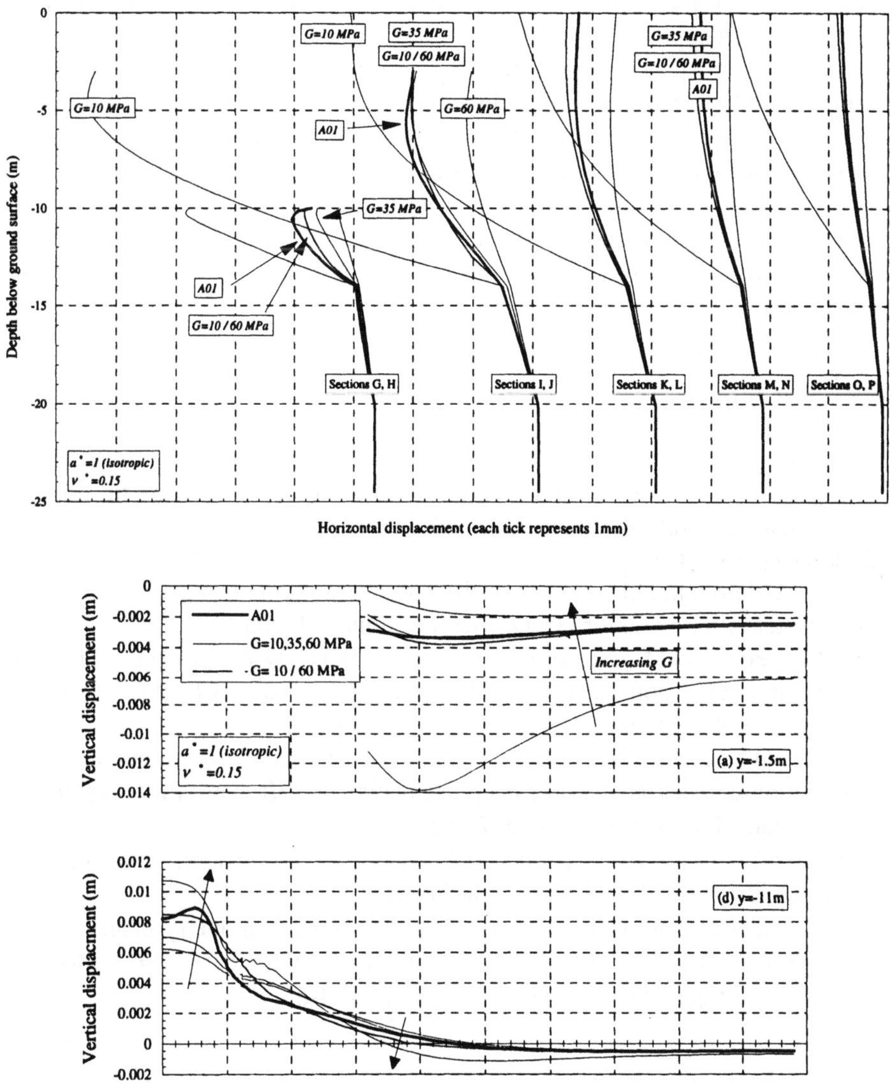

Fig. 7 Comparison of the non-linear cross-anisotropic model prediction with a series linear isotropic analyses

8. CONCLUSIONS

The importance of the cross-anisotropy in predicting the ground response of a stiff, heavily overconsolidated clay has been demonstrated. Horizontal displacements at the sides of an excavation were overpredicted by a factor of up to two if the anisotropic nature of the stiffness response was neglected.

Although the non-linearity of stiffness is important, in this case the non-linear predictions could be represented by equivalent linear analyses in back analysis. The key difference is that an appropriate stiffness and degree of anisotropy for a linear analysis may only be found in retrospect, whereas the non-linear cross-anisotropic stiffness variation may be obtained directly from suitable experimental data.

The material anisotropy has been shown to have at least as great an effect on the ground displacements than the non-linearity of the stiffness response.

Changes in stress path direction during construction occur even for such an apparently monotonic loading condition as considered here. Changes in stress path direction significantly affect the stiffness response and their effect should be included in the modelling process.

9. REFERENCES

Hird, C.C. and Pierpoint, N.D. (1994). A non-linear anisotropic elastic model for overconsolidated clay based on strain energy. Proc. 3rd European Conference on Numerical Methods in Geotechnical Engineering, Manchester, UK. Published by Balkema, Rotterdam: 67-74.

Hird, C.C. and Pierpoint, N.D. (1997). Stiffness determination and deformation analysis for a trial excavation in Oxford Clay. Géotechnique, 47: 665-691.

Jardine, R.J., Potts, D.M., St. John, H.D. and Hight, D.W. (1991). Some practical applications of a non-linear ground model. Proc. 10th ECSMFE, 'Deformation of Soils and Displacements of Structures', Florence: 223-228.

Ng, C.W.W. and Lings, M.L. (1995). Effects of modelling soil nonlinearity and wall installation on back-analysis of deep excavation in stiff clay. Proc. ASCE Journal of Geotechnical Engineering, 121: 687-695.

Pierpoint (1996), The prediction and back analysis of excavation behaviour in Oxford Clay. Ph.D. Thesis, University of Sheffield, UK.

Wroth, C.P. and Houlsby, G.T. (1985). Soil mechanics - property characterisation and analysis procedures. Proc. 11th International Conference on Soil Mechanics and Foundation Engineering, San Francisco, USA, 1: 1-55.

10. ACKNOWLEDGEMENT

The work was supported by an Engineering and Physical Sciences Research Council studentship in collaboration with Mott MacDonald.

A NUMERICAL SIMULATION OF A FULL SCALE IN-SITU EXPERIMENT ON THE HYDROMECHANICAL BEHAVIOUR OF A CLAY BASED BACKFILL

S.W. Rees and H.R. Thomas
University of Wales, Cardiff, UK

F. Bernier
SCK/CEN, Mol, Belgium

ABSTRACT

The results a numerical simulation of an in situ test, performed in the HADES underground laboratory (Mol, Belgium), are presented. The experiment (known as BACCHUS 2) demonstrates an installation procedure for a clay based backfill material made of a mixture of high density Boom clay pellets and Boom clay powder. The results achieved have been used as one of a series of validation exercises designed to assess the performance of a thermo-hydro-mechanical computer code.

1. INTRODUCTION

The engineering problem addressed here is concerned with the disposal of high level nuclear waste within a deep geological repository. One of the concepts under consideration requires the design of a multi-barrier containment system to mitigate radionuclide migration to the biosphere. Within this overall framework, the performance of highly impermeable engineered barriers which will directly surround the waste canister is of importance. In particular, such barriers, which are occasionally referred to as 'buffers', may rely on the use of low permeability expansive clays - possibility mixed with other materials. However, the overall performance of the system will depend on many factors, not least on the nature of the host material. Analysis of the problem requires an understanding of the hydraulic and mechanical interaction of the materials involved. In addition chemical and thermal effects will also require consideration. However, these are not included within the scope of this work.

The paper summarises a recent numerical simulation, performed by the University of Wales Cardiff, of an in-situ test, performed by SCK/CEN, located in the HADES underground laboratory (Mol, Belgium). In particular, the work presented is one of a series of validation exercises designed to assess the performance of thermo-hydro-mechanical computer codes. The model employed here was developed at the University of Wales Cardiff and is known as COMPASS (Code for Modelling Partly Saturated Soil) [Thomas and He, 1995].

2. DESCRIPTION OF THE IN-SITU EXPERIMENT

The in-situ test BACCHUS 2 (BACkfilling Control experiment with Hydration for the Underground Storage of radioactive waste) was designed to demonstrate and optimise an installation procedure for a clay based backfill material and to study its hydration behaviour. The backfill material was made of a mixture of high density Boom clay pellets and Boom clay powder. Total stress, pore water pressure and water content measurements were performed both in the backfill material and in the surrounding clay massif.

BACCHUS 2 was installed in a borehole at a location of between 11 m and 14 m below the gallery floor level (some 222 m below ground surface). The detailed design of the experiment is shown in Figure 1. It can be seen that the experiment consists of a central filter surrounded by the backfill. The backfill is in direct contact with the saturated host clay.

Figure 1 shows the location and range of instruments installed in the backfill. The sensor codes given in the figure are; PW for pore water pressure and PT for total stress. C, F and V indicate that the sensor is installed respectively on the Central tube, the bottom Flange, or on a Vertical plate. BACCHUS 2 was installed in such a way that most of its sensors were orientated towards the piezometers located in the host clay.

In addition to the above, thermocouples were also installed in the backfill. Together with a heater, this allowed measurement of thermal conductivity to take place.

Pore water pressure and total stress measurements were also made in the surrounding host clay. Instruments were located in small diameter boreholes drilled adjacent to the main experiment. Absolute values of ground pressures thus obtained were considered unreliable. However, it was thought that the relative ground pressure changes were realistic. Pressure levels had reached steady state before the installation of BACCHUS 2.

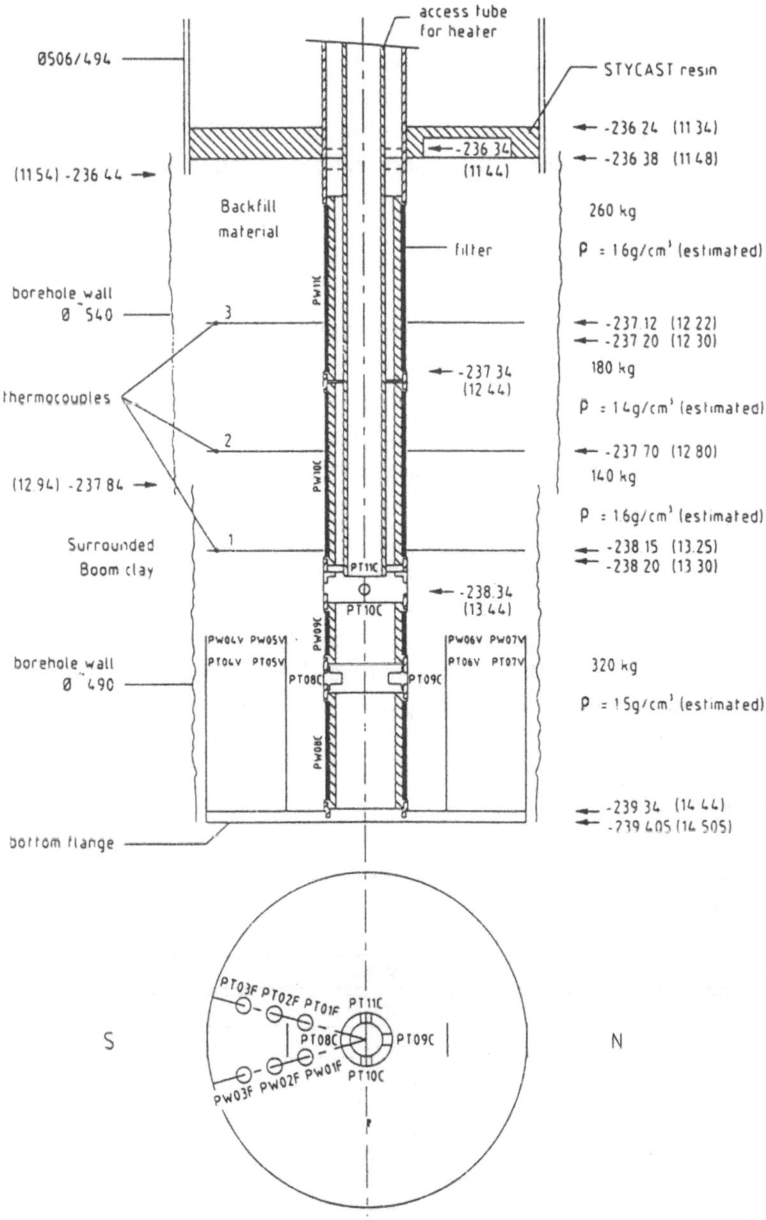

Fig. 1 Experimental layout of BACCHUS 2

The experiment commenced on 01.07.93 with natural hydration of the backfill material. This proved to be a very slow process and as a result, after 516 days into the experiment, an 'artificial' hydration phase was initiated on November 25, 1994. This was achieved by injecting water, from a reservoir installed in the gallery, through the filters in the central tube directly into the backfill. The injection pressure corresponded to the height of a column of water between the filter and the gallery, i.e. between 11 m and 15 m, and was maintained to day 611. When the saturated state was reached in the backfill, the hydraulic circuit was closed (at day 624). Further details of the experiment can be found in EUR 16860 [1996].

3. DESCRIPTION OF COMPASS

The development of the theoretical basis of COMPASS has been achieved in a gradual step-by-step manner. Thomas and Sansom [1995] extended an early version of the model to include air transfer, in a fully coupled manner. Recent work has seen the development of the software to include the stress/strain behaviour of unsaturated soils [Thomas and He, 1995]. Validation of the code has been addressed on a number of occasions [Thomas *et al* 1996; Thomas and Li, 1997; Volckaert *et al* 1996].

COMPASS has been developed to solve a wide range of problems involving unsaturated soils. The primary variables to be determined are; pore water pressure, pore air pressure, temperature and displacements. The mechanical behaviour of the soil is described by an elasto-plastic approach in this work [Thomas and He, 1998]. For this application only the hydro-mechanical aspects of the model are employed.

4. MATERIAL PROPERTIES

Natural Boom Clay: The data presented in Table 1, for drained saturated conditions, were provided by SCK/CEN.

Material Property	**Value**
Young's modulus	300 MPa
Poisson's ratio	0.125
Friction angle	18°
Cohesion	300 kPa
Dilation angle	0°
Porosity	0.39
Storage coefficient	10-5 m-1
Vertical permeability	2.1·10-12 m/s
Horizontal permeability	4.5·10-12 m/s

Table 1 Some material properties for natural Boom clay

In addition to the above parameters, experimental data was also available regarding the water retention curve of the clay powder compacted at the same dry density as the host clay (17 kN/m^3) and at the in situ temperature (about 22 °C). The water retention curve shown in Figure 2 has been employed to represent the host clay. The hydraulic conductivity relationship used is shown in Figure 3. Both relationships are based on the available measured data.

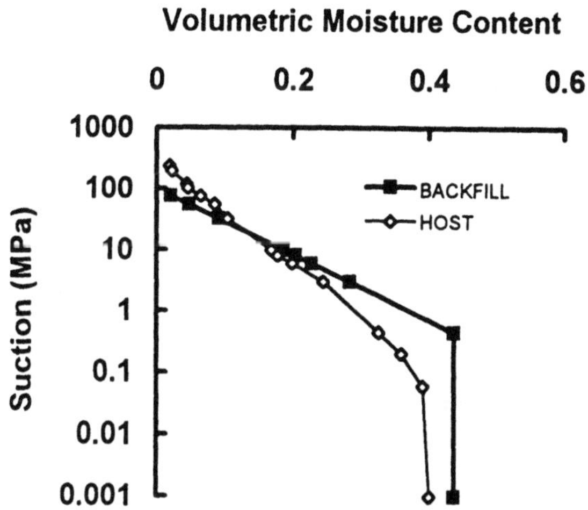

Fig. 2 Water Retention Curves

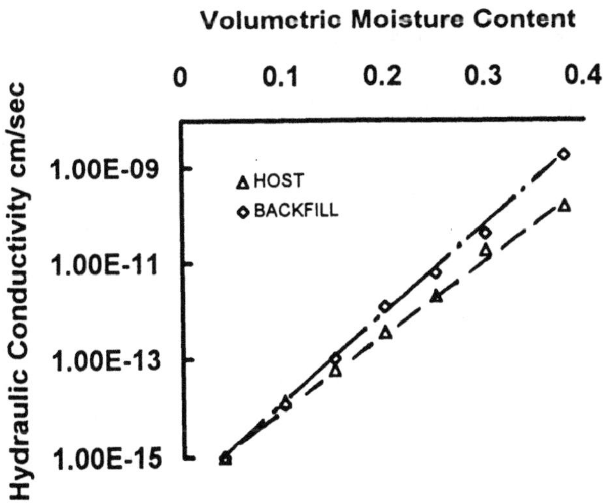

Fig. 3 Hydraulic Conductivity Relationships

Engineered Backfill: The backfill material is a 50/50 mixture of Boom clay powder and high density Boom clay pellets. The high density pellets are produced by compaction of dry Boom clay powder using industrial equipment. The equipment produces pellets with a dry density of about 2.1 g/cm^3. The pellets are about 2 cm long and 0.5 cm thick. The mean dry density of the mixture is about 1.5 to 1.6 g/cm^3.

Some of the basic mechanical properties of Boom clay powder were provided by SCK/CEN [Volckaert *et al,* 1996]. These include: Liquid limit 55.9 %; Plastic limit 29.2 %; Plasticity index 26.7 %; Specific gravity of particles 2.70.

The water retention curve and hydraulic conductivity relationship used for the backfill are also shown in Figures 2 & 3 respectively. The water retention curve yields an initial suction of approximately 55 MPa for the backfill, based on an emplacement water content of 0.03 (by weight). The elasto-plastic soil properties used to represent the backfill are summarised in Table 2.

SYMBOL	PROPERTY	VALUE
$\lambda(0)$	Slope of the virgin compression line for saturated conditions.	0.158
κ	Elastic stiffness parameter for changes in net mean stress.	0.01
$\lambda(s)$	Stiffness parameter dependent on suction.	variable
κ_s	Elastic stiffness parameter for changes in suction.	0.03
λ_s	Stiffness parameter for changes in suction for virgin states of the soil.	0.052
β	Parameter controlling the rate of increase of soil stiffness with suction.	12.5 (MPa^{-1})
r	Parameter defining the maximum soil stiffness.	0.87
k	Parameter describing the increase in cohesion with suction.	0.6
v_o	Initial specific volume.	1.932
p_o*	Preconsolidation stress for saturated conditions.	150 kPa
p_c	Reference stress.	10 kPa

Table 2 - Elasto-plastic material properties for the backfill

5. PRELIMINARY NUMERICAL SIMULATION

Preliminary assessment of the problem revealed that a rigorous analysis is likely to require the inclusion of material non-homogeneity (density variations) to emulate the nature of the measured response. Given the relatively limited amount of measured data available for the specification of the hydraulic and mechanical material properties it was decided, in the first instance, to conduct a one-dimensional analysis of the problem. This was achieved by employing a sector to represent a horizontal plane at the mid-elevation of the backfill. The solution was then determined via an axisymmetric formulation of the problem.

The hydraulic problem was solved in the above manner and the results were then utilised within the constitutive model to determine the average swelling pressure evolution in the backfill. This was achieved by direct integration of the stress/strain model for the average suction changes in the backfill (obtained directly from the hydraulic analysis).

The sector was discretised using 108 quadrilateral (8 node) elements. The mesh was designed to provide very fine sub-divisions near the interface between the host clay and the backfill. Earlier work indicated that mesh refinement is important for this problem and may cause practical difficulties for a fully coupled two-dimensional analysis of the problem. A variable timestep size was permitted during the analysis, starting with a timestep size of 0.25 days and subject to a upper limit of 15 days.

It was known that prior to any excavation, in the virgin host clay, the total stress at the depth of the gallery was approximately 4500 kPa and the pore pressure was approximately 2200 kPa. The initial state of the Backfill material is summarised by the following parameters: Degree of saturation 10.37 %; Water content 3 % (by weight); Density 1.55 g/cm^3; Suction 55 MPa (based on Figure 2).

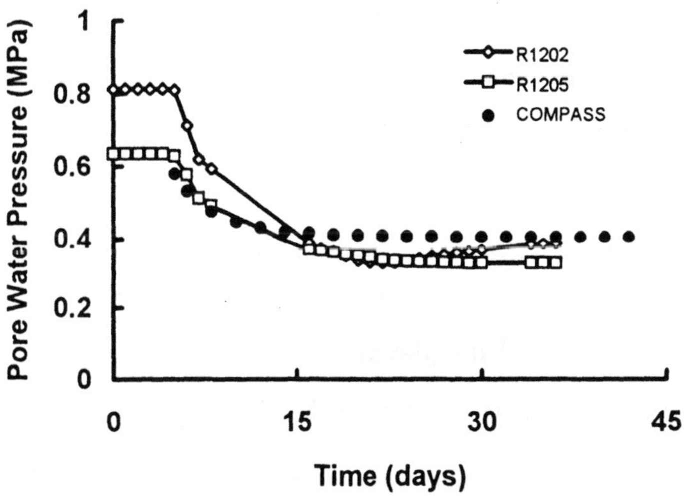

Fig. 4 Pore pressure variation at 0.75 m into host clay

The first stage of the numerical simulation involved the analysis of the 'open' borehole to obtain the 'drawdown' of pore water pressures that occurred prior to the emplacement of the backfill. This was achieved by running an uncoupled hydraulic analysis of the problem. At the outer edge of the host clay a pore water pressure boundary condition of approximately 0.6 MPa was applied. Nodes at the borehole wall were held at a zero pore pressure. The analysis was run for a period of 40 days in accordance with the testing schedule. Typical results obtained during this stage are shown in Figure 4. The figure shows a comparison between measured and predicted pore water pressure variation over the 40 day period. The results represent the variation that occurred in the host clay at a horizontal distance of 0.75m from the centreline of the experiment. A reasonable agreement between measured and predicted results can be seen to have been achieved.

The second stage of the analysis involved inclusion of the backfill material at an emplacement water content of 0.03 (suction approx. 55 MPa). This section of the analysis was run for the period to day 516. The artificial hydration phase of the experiment was then included via the use of a fixed pore pressure boundary at the inner backfill surface providing an equivalent pressure head of 13 m of water. The artificial hydration ended at day 611 after which the water injection boundary was removed and a natural hydration phase continued through to the end of the analysis.

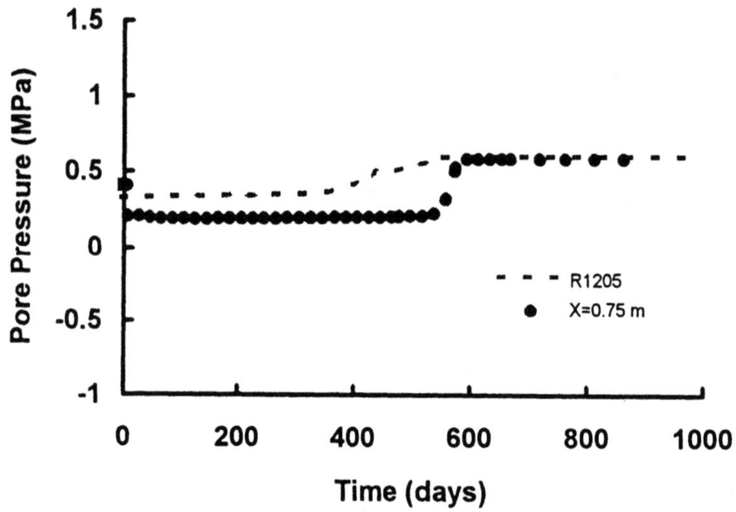

Fig. 5 Pore pressure variation at 0.75 m into host clay - hydration phase.

Typical results achieved during the hydration phase are presented in Figure 5. The location considered is as described above for Figure 4. In general, the model appears to show a more abrupt increase in pore water pressure than that measured. However, the overall change in pore pressure has been predicted with acceptable accuracy.

The hydraulic interaction between the backfill and the host clay is an important feature of this problem. The work presented here assumed full contact between the backfill and the host clay on emplacement - this results in a very large hydraulic gradient across the interface (i.e. the backfill is at an initial suction of around 55 MPa and the host clay is saturated). This aspect of the problem is the subject of ongoing investigation at the Cardiff School of Engineering.

The results of the corresponding elasto-plastic stress analysis are shown in Figure 6. It can be seen that the model has produced a good prediction of the final swelling pressure generated in the backfill. The predicted results are in reasonable agreement with the measured data for the first 500 days of the simulation. The prediction then shows a more rapid increase in swelling pressure than that measured. This feature of the prediction is linked to the nature of the predicted hydraulic response of the backfill. The model predicted a sharp change in suction at around 520 days - this translates to a rapid increase in swelling pressure. Overall, it is concluded that the constitutive modelling approach adopted is capable of predicted the magnitude of swelling pressure developed in the experiment.

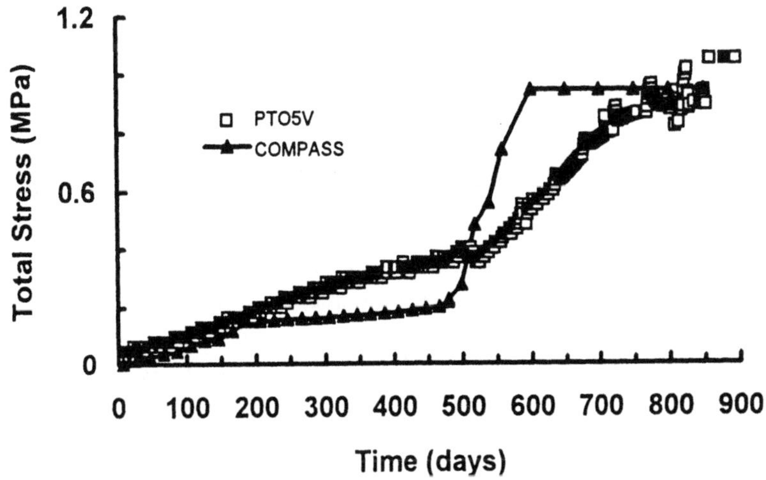

Fig. 6 Evolution of swelling pressure in backfill

6. CONCLUSIONS

The results of a numerical simulation of a full-scale in-situ test, performed at an underground nuclear waste storage research laboratory, have been presented. The in-situ test was designed to demonstrate and optimise an installation procedure for a clay based backfill material and to study its hydration behaviour. The work presented is one of a series of validation exercises designed to assess the performance of thermo-hydro-mechanical computer codes. In particular, a model developed at the University of Wales Cardiff, known as COMPASS, has been shown to be capable of predicting the general trends and behaviour associated with the test. Results obtained illustrate that the model is able to predict the overall pore water pressure response within the host clay with reasonable accuracy. An elasto-plastic approach was used to calculate the development of swelling pressure within the backfill during the hydration process. Overall these results were found to be in good agreement with the variations measured in-situ. Further work appears necessary regarding the precise nature of the hydraulic interaction between the backfill and the host material.

ACKNOWLEDGEMENT: The work forms a component of an EC funded research programme known as CATSIUS CLAY (Calculation and Testing of Behaviour of Unsaturated Clay - Contract No. F14WCT950003). The financial support of the EC is gratefully acknowledged.

7. REFERENCES

EUR 16860 EN/FR. 1996 - Demonstration of the in situ application of an industrial clay-based backfill material (Démonstration de la mise en place in situ d'un matériau de remplissage industriel à base d'argile) (Bacchus 2). Report. Directorate-General. Science, Research and Development. ISBN 92-827-7761-8, pp209.

THOMAS. H. R. AND SANSOM, M. R. (1995) "Fully coupled analysis of heat, moisture and air transfer in unsaturated soil". ASCE, Journal of Eng. Mech., 121(3) 392-405

THOMAS, H. R. AND HE, Y. (1995) "Analysis of coupled heat, moisture and air transfer in a deformable unsaturated soil." Geotechnique, 45(4) 677-689.

THOMAS, H. R., HE, Y., SANSOM, M. R. AND C. L. W. LI (1996) "On the development of thermo-mechanical-hydraulic behaviour of unsaturated soils." Eng. Geology 41, 197-218.

THOMAS, H. R. AND LI, C. L. W. (1997) "An assessment of heat and moisture transfer in unsaturated soil." Geotechnique, 47(10), 113-131.

THOMAS, H. R. AND HE, Y. (1998) "Modelling the behaviour of unsaturated soil using an elasto-plastic constitutive relationship". To appear in Geotechnique, 1998.

VOLCKAERT, G., BERNIER, F, ALONSO, E., GENS, A., SAMPER, J., VILLAR, M., MARTIN-MARTIN, P.L., CUEVAS, J., CAMPOS, R., THOMAS, H., IMBERT, C., AND ZINGARELLI, V., (1996), "Thermal-hydraulic-mechanical and geochemical behaviour of the clay barrier in radioactive waste repositories (model development and validation)", Final Report on CEC Contract Nos. F12W-CT90-0033 and FI2W-CT91-0102, EUR 16744 EN Pub. European Commission.

AN ELASTOPLASTIC ANALYSIS OF 3-D GROUND MOVEMENTS

Y. Riou and P. Chambon
École Centrale de Nantes, Nantes, France

ABSTRACT: This paper describes numerical modelling of the deformation of sand caused by the reduction of pressure at tunnel face. This study refers to a laboratory test carried out on a centrifuge with measurements of surface settlements and displacements in soil mass. A series of 3D finite element calculations using three elastoplastic stress-strain constitutive models (Mohr Coulomb, Vermeer and Nova models) is performed. The results of this analysis are compared with experimental findings. If numerical results show good conformity with displacements observed at tunnel face, we note some limitations of the physical modelling and the numerical calculation when achieved in a standard way. These limitations in the two approaches are due to simplifications inducing different deformation mechanisms and show that experimental tests and numerical modellings are highly connected and should be carried out simultaneously.

1. INTRODUCTION

The increasing demand of new infrastructure coupled with the technical breakthrough in the shield tunnelling design has led to a great number of shallow tunnels constructed in urban areas. The common problem associated with tunnelling in such areas, arises from the disturbances on the surrounding structures caused especially by settlement on the surface.

In order to gain some insight into the effects of tunnel driving, a three dimensional numerical study predicting deformations, is carried out. To test the effectiveness of the simulation, these results are compared with measurements on a centrifuge test. The present work reports on soil movements generated by a reduction of pressure applied by the shield toward the tunnel face. Three soil models are examined in this study, covering a range of those commonly used in geotechnical analysis :

- an elasto-perfectly plastic model :

In this case, the plastic response is represented by Mohr Coulomb criterion,

- a hardening model characterised by a single yield surface for a given stress state :

The model refers to Nova model originally used to describe behaviour of sand (Nova and Wood, 1979). Both volumetric strain and deviatoric plastic strains occurs as hardening parameter in the potential function.

- a hardening model characterised by two yield surfaces:

Vermeer model is used here (Vermeer, 1978). The two yield surfaces are :

  - a cone shaped surface around space diagonal introducing deviatoric strains
  - a spherical yield surface introducing volumetric strain.

We note that all these constitutive models propose various yield surfaces. The Mohr Coulomb surface depends on major and minor principal stresses. The Vermeer model refers to the Matsuoka yield surface depending on the third invariant. Contribution of the intermediate principal stress is not so negligible. This intermediate stress plays the same part as other principal stresses in the Nova model. For the three models, the non-associated cone-flow rules are simple plastic potential functions expressed in terms of p (mean stress) and q (deviatoric stress).

The version of the two last models used in this study, considers a potential function depending on p and q. As a result, the direction of plastic strain in the deviatoric plane is the same. However, in p-q plane, this direction is based on three different ways. In order to save CPU time, elastic strain is controlled by linear Hooke's law, and assumed, for convenience, to be isotropic. The volumetric strain of the three models is not restricted at failure. For all models, we consider an isotropic hardening. The numerical calculation was performed with the finite element program Cesar-LCPC.

2. TESTING PROCEDURE

A 1/55 small-scale physical model was designed to idealise the three-dimensional tunnel heading. This model consists of a half longitudinal tunnel section. A detailed description of the test carried out in the LCPC geotechnical centrifuge is given by Chambon (Chambon, 1997). A brief summary of the model is reported below. The principal components of the tunnel model are as follow (Fig. 1.) :

- a strong half box filled with Fontainebleau sand,

Dry sand is poured dry into the model container by pluviation through air.

- a 3mm thick semi circular aluminium pipe : 260 mm length with an outer diameter of

90 mm (495 cm prototype),

- a thin rubber membrane at the tunnel face,

This membrane represents the interface between tunnel inside and ground.

- a vertical Perspex window, in the plane of symmetry,

It permits observation of the tunnel face and sand mass during the centrifuge flight.

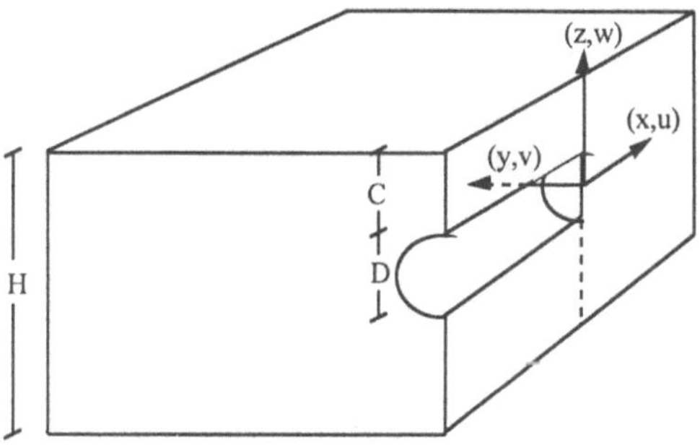

dimensions of the simulated prototype :

. cover	C	4m95
. outer diameter	D	4m95
. total height	H	13m75

Figure 1: Geometric lay-out of the tunnel

Internal air pressure is applied to the tunnel heading at 1g. This temporary tunnel support pressure is equal to the total overburden stress in the sand at the tunnel axis level. The centrifuge is then accelerated in steps to the predetermined test acceleration of 55 g. For each step, the internal air pressure is increased accordingly to the approximated overburden stress at the tunnel axis level. After centrifugation, the internal air pressure is gradually decreased at a uniform rate to simulate tunnel excavation process. Subsequent displacements are observed on the following items as shown in figure 2.

- horizontal displacements in front of the tunnel face : points A and B, (LVDT Transd.)
- vertical and horizontal displacements in the sand near the tunnel face (point C), (digital image processing),
- ground surface settlements at the centre line, (LVDT Transducers)

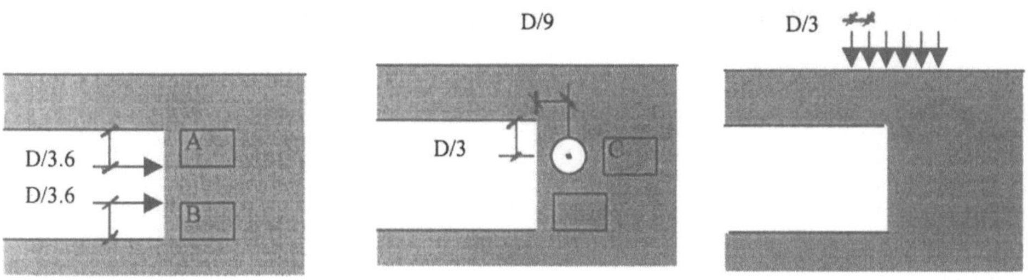

Transducers at open face

mark in the soil

transducers at surface

Figure. 2 : Set up of equipment used for measuring the displacement

3. FINITE ELEMENT CALCULATION

3.1 *Programme and calculation scheme*

The FEM simulations of the effect of unloading are carried out by performing 3D analysis. The soil mass is subdivided into 2936 elements with 6 nodes. The initial stress state is characterised by a geostatic stress state with : $K_0 = \nu / (1-\nu)$ ν : Poisson's ratio

In order to simulate the physical model, this value of K_0 corresponds to a condition of gravity force and no lateral displacement in elastic behaviour. The pressure at front, which balances initial stresses, is then replaced by a uniform pressure. This imbalance leads to an initial stress state slightly different from the geostatic stress state. The subsequent displacement is reset to zero before the reduction of uniform pressure. The tunnel lining is assumed to have an infinite stiffness. The nodes at the interface soil-lining can only move in the longitudinal direction of the tunnel, except at the open face where all the nodes are fixed. All the results fitted in this document are achieved by the initial stress process. The convergence test is performed by checking the global residual force norm and the applied force norm. In this analysis, the tolerance is, at least, 0.005. Each unloading is applied incrementally in about 15 steps. Generally, good convergence is observed up to a support pressure of 8 kPa.

3.2 *Properties of material*

The behaviour of soil is modelled as Mohr Coulomb, Vermeer, Nova material. The values of parameters are determined from three triaxial compression tests, at constant lateral stress, carried at different cell pressures : 75 kPa, 100 kPa, et 150 kPa. Parameter sets were obtained using specific procedures suggested by authors of constitutive relationships, on particular experimental data (constant volumetric strain, locus of peaks ...). These values were then adjusted by optimal simulation over all data. These values used in finite element calculations are given in table 1. We note that the value of Mohr Coulomb's stiffness is the secant stiffness for the initial mean deviatoric stress at the open face.

E MPa	ν	C kPa	φ	ψ
72	0.27	0	34°	12°

Mohr Coulomb parameters

E MPa	ν	B_0	L_0	M	l	D	m	μ
170	0.27	0.00084	0.00086	0.83	0.013	O.462	0.16	1.18

Nova parameters

E MPa	ν	β	ε_0^c	ε_0^e	φ_p	φ_{cv}
170	0.27	0.22	0.0035	0.005	35°	23°

Vermeer parameters

Table 1 : Set of soil parameters

3.3 *Results of the finite element analysis*

Horizontal displacement to the tunnel face

Observed and predicted axial displacements at tunnel face (point A) are shown in figure 3.

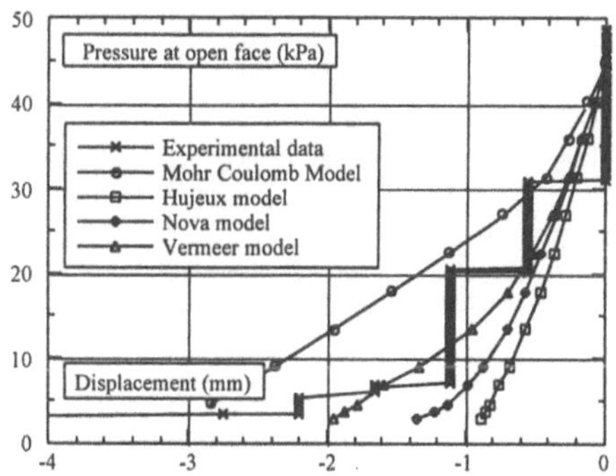

Figure 3 : Observed and predicted axial displacement at the tunnel face (point A : under cover of tunnel)

The results obtained suggest the following :

The predicted displacements are roughly in agreement with the observed displacements. The measured range of displacements is within the bounds of calculated displacements. However, it can be noticed that Mohr Coulomb model overestimates volume loss, in comparison with experimental data and simulations from other models. There is a contradiction between this softer behaviour and the stiffer behaviour in simulation of triaxial tests. So, explanation of such a behaviour does not lie on a wrong assessment of Young modulus. It cannot be due to a different volumetric behaviour from the other models either. Indeed, as it will be observed later, this model causes a greater volumetric strain likely to prevent movement of the soil towards the open face. The reason of greater volume loss refers to the fact that plastification is delayed in the Nova model in comparison with the Vermeer model, due to the shape of yield surface. In accordance with this principle, the plastification in Vermeer model occurs later than in Mohr Coulomb model. The discrepancy between models is all the more significant since, in this case, the stress path is more representative of a passive compression than of an active one, particularly in the second phase of unloading (Fig. 4). So, the relative position of yield surface in different models, when depicted in the deviatoric plane, is found in the behaviour of the soil near the open face. This feature does not constitute the only reason of the softer behaviour of the soil when modelled by the Mohr Coulomb failure criterion. In fact, Mohr Coulomb model presents plastic potential function expressed in terms of minor and major principal stresses while the two other models resort to mean stress (p) and deviatoric stress (q). That leads obviously to various plastic strain tensor orientation and to a different deformation mechanism. The results of numerical analysis show that the Mohr Coulomb model increases strain in the direction normal to the open face and consequently causes a greater volume loss. This plastic flow is linked to the relaxation of the intermediate stress (the horizontal stress in transversal section) which occurs at the same rate that for minor principal stress (stress normal to the open face). The predicted stress paths in diagram p-q and in deviatoric plane, given at figure 4, are in accordance with this interpretation.

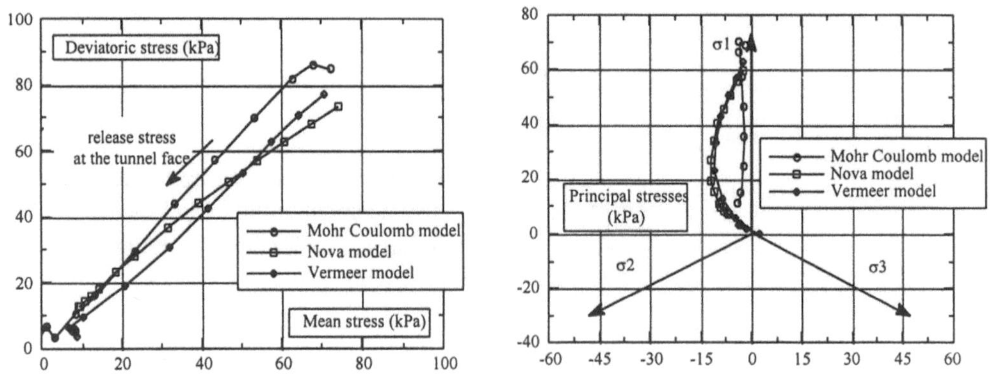

Figure 4 : Stress paths of point A during the decreasing front pressure.

In a general way, the rate of decreasing pressure at failure is overestimated with regard to the centrifuge test. But the Nova model shows, just before failure, a behaviour in conformity with the experimental measurements, better than the Vermeer model. The volume loss increases sharply when the rate of release stress reaches 85 %. When a stress patch reaches the yield locus, the Vermeer model causes numerical instability due to yield surface expressed in terms of the third stress invariant. When we replace the Matsuoka Nakaï yield surface of the Vermeer model by the Drücker Prager yield surface, we obtain the best numerical results. But unrealistic results can be obtained if no care is paid to the friction angle (Riou, 1997).

Vertical and horizontal displacement in soil mass

Figure 5 shows the axial and vertical displacements ahead of tunnel face (point C). Due to the small values of these measures and their scattering, we have reported maximum, mean and minimum values. It should be noted that the physical analysis induces mainly a collapse at point C. The vertical displacement reaches 2mm when the support pressure is 10 kPa. This settlement is produced with barely perceptible horizontal displacement toward the tunnel. The experiment indicates a horizontal displacement up to 1mm, at point A, 60 cm from the open face. It suggests that an internal failure surface occurs in the centrifuge model, and that formation of a shear band starts at the crown of the tunnel face. This strain mechanism at the open face has been observed on previous tests.

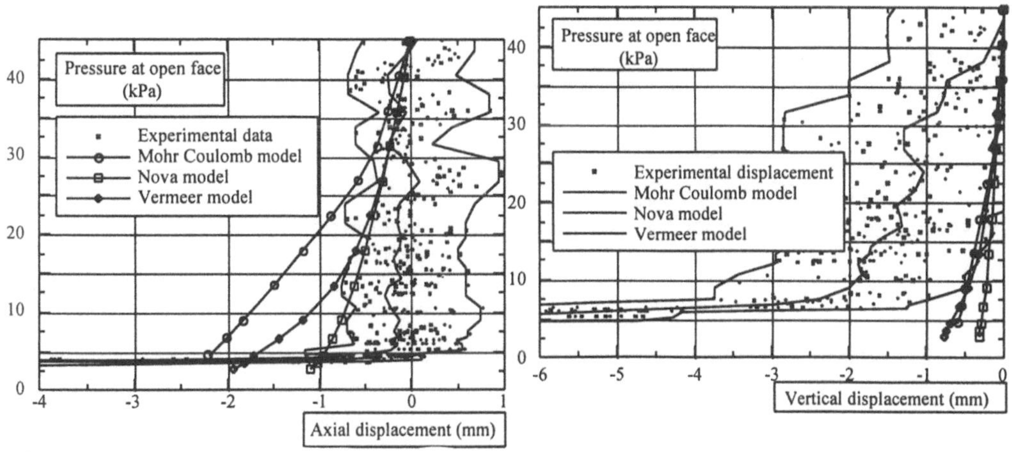

Figure 5 : Axial and vertical displacements at point C.

Remark :
The digital image processing needed, for satisfactory precision, to bring the camera closer to the physical model. This requirement prevented us from having a general view of displacement ahead the front. So, this experiment does not provide confirmation of this assumption in the failure mechanism.

Calculations show a deformation field inconsistent with experimental results in term of magnitude of horizontal and vertical displacements. The simulation proposes a spatial distribution of mainly horizontal displacement, with a very small damping of this displacement over the first 55 cm ahead the front. This small damping is more obvious for the Vermeer model which creates a larger plastic zone in the vicinity of the face.

Settlement at surface

Face displacements produce surface settlements of similar magnitude : figure 6. This observation suggests that no reduction of vertical displacement occurs in the cover. Such a result supports the assumption of a rigid body collapse mechanism. Nevertheless, some questions arise from reliability of those values. Indeed, if equilibrium volume amounts need to be reached (volume loss at tunnel face Vf and settlement trough Vs), we have to assume a highly localised face collapse phenomena (Fig. 7).

. $Vf = \pi.D^2/4.U_{max}$

that is, for a pressure at the face, ranging from 40 kPa to 20 kPa : $Vf = 0.01\ m^3$

. $Vs = \pi.Lc^2/4.V_{max}$

Lc : length of surface area where transducers are located

That is, for pressure at the face, ranging from 40 kPa to 20 kPa : Vs = 0.01 m^3
Assuming more classical shapes :
. Volume of settlement trough with bell shaped profile

$$Vs = 2.i^2.\pi.V_{max}$$

. Volume loss at open face with spheric deformation

$$Vf = \pi.U_{max}.(0.75.D^2 + U_{max}^2)./6$$

the volume of settlement trough is four times greater than the volume loss.

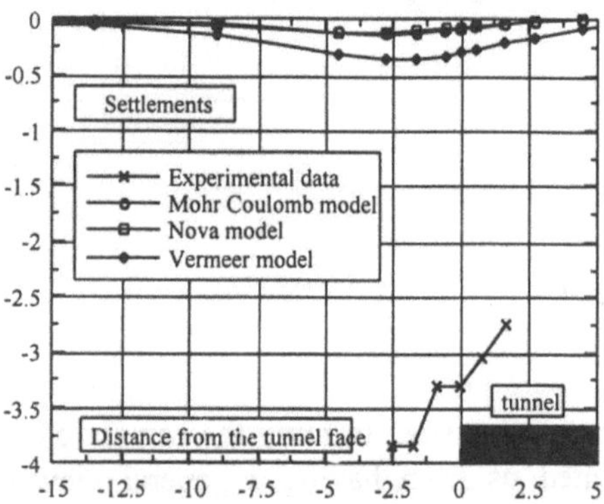

Figure 6 :
Surface settlement over centreline of tunnel
Physical model and numerical models.

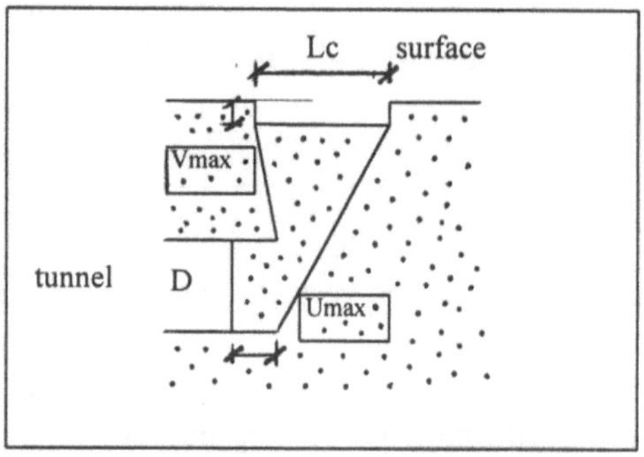

Figure 7 : Schematic failure mechanism giving equilibrium in volumes

In spite of plausible collapse, this imbalance between volumes seems unrealistic and can be the result of a measurement defect (global settlement during the fly of the centrifuge, precision of measurement ...). We notice here that a 1 mm prototype makes up 0.02 m in the centrifuge test. Consequently, the substantial gap between calculated and measured settlements will not be discussed in the following section. Attention will be focussed on analysis of different calculated results.

In figure 8, a discrepancy can be seen in the settlement profile with different models. In order to explain this result, we have plotted, in figure 9, the evolution of V_s and V_f during the decreasing pressure. the Mohr Coulomb model exhibits dilative behaviour whereas the Vermeer model shows a tendency to contract. With the Nova model, the soil is close to deforming at a constant volume. The divergence from the Mohr Coulomb model and the Vermeer model is due to the particular stress path ahead of the tunnel face. During this phase, the horizontal stress decreases, while vertical stress remains constant and then decreases during the arch effect. A passive compression triaxial test is reproduced in figure 10, with the same parameters defined in the compression triaxial test. Shearing leads, with the Vermeer model, to a more dilatant behaviour. This is due to the fact that, in the compression triaxial test, the Mohr Coulomb model integrates all the plastic strains into a deviatoric mechanism, and these strains do not exist in the passive compression triaxial test. However, this argument cannot explain the less contracting behaviour in the Nova model with regard to the Vermeer model (figure 8). Indeed, for the passive compression triaxial test, when the two minor stresses are decreasing, the volumic strain of the Nova model is smaller than that of the Vermeer model (figure 10). The less contracting behaviour of the Nova model, for this configuration, lies in the fact that there is a delay in the development of the plasticity due to the yield surface expressed in term of p and q.

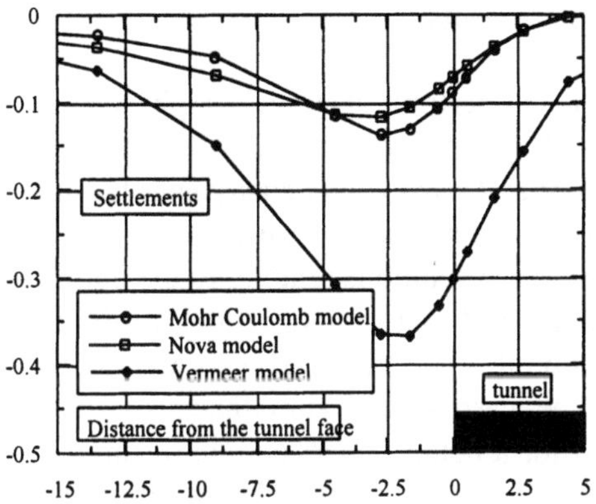

Figure 8 :
Surface settlement over centreline
Physical model and numerical models.

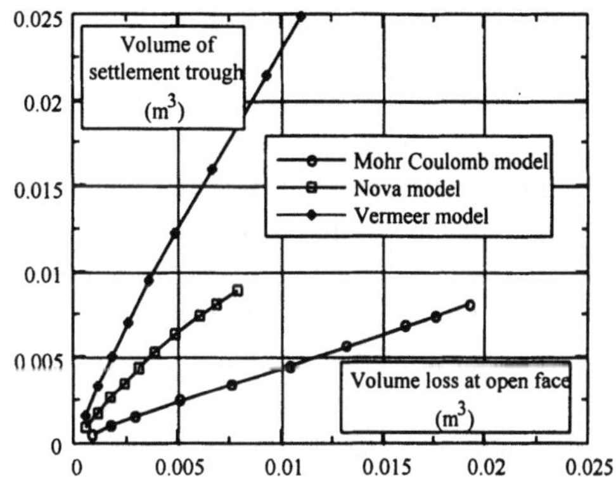

Figure 9 :
Comparison of volume loss during decreasing pressure at the tunnel face

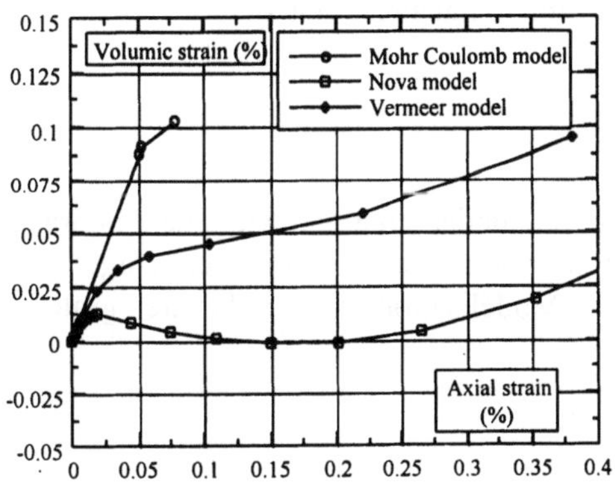

Figure 10 :
Compression triaxial test - volumetric strain

4. CONCLUSION

The principal conclusions of this comparison are as follows :

- the calculated axial displacements at the tunnel face show a reasonable agreement with the measurement. Experimental results are bracketed between the FEM results obtained with the three constitutive relationships,

- the numerical results at the open face change according to any form of yield surface and plastic potential. The displacements can be as sensitive to the constitutive model as to the values of parameters,
- each constitutive model of soil shows different settlement troughs according to the representation of passive triaxial test.
- Calculations show that the deformation field is inconsistent with experimental results in terms of magnitude of horizontal and vertical displacements.

Previous sections have shown all the difficulties to perform accurately a physical and a numerical model of volume loss at tunnel face. Experimenters are faced with sensitive subjects : precision in measurements, preparation of soil mass (low density near the side of centrifuge box linked to pluviation process), friction between the sand and the Perspex window. The numerical modelling calls for compromise solutions relating to the number of nodes, to the identification of parameters, to the type of constitutive model, to the initial stress. Comparison of results leads to the conclusion that experimental tests and numerical modellings are highly connected and should be carried out simultaneously. Further physical and numerical investigations are needed for a better understanding of tunnel face under decreasing pressure. New centrifuge tests are needed to answer the questions arising from half space configuration. Particular care should be paid to the settlement trough and to deformation of tunnel face. Regarding numerical studies, new calculations should be conducted with more relevant values of parameters and a new constitutive model restricting volumetric strain. Moreover, It should be carried out in accordance with a centrifuge test. So, the accelerated spinning phase with locking of tunnel in the vertical direction should be considered in order to control the initial stress state.

ACKNOWLEDGEMENTS :

The work described in this paper was carried out under contract with the Laboratoire Central des Ponts et Chaussées (LCPC). The guidance and advice from Centrifuge Centre (Nantes) and the Section de Rhéologie et Modélisation des Sols are gratefully acknowledged.

REFERENCES

Chambon P. (1997) Souterrains et forage en site urbain., *Rapport interne, GEO, Ecole Centrale de Nantes. Lab. de Génie Civil de Nantes-Saint Nazaire (LGCNSN)*

Mestat Ph. (1992) Sur quelques développements du module de résolution en comportement non linéaire (MCNL) – Progiciel CESAR – LCPC, *Annales de l'ITBTP, n°509, décembre 1992*

Nova R & D.M. Wood (1979). A constitutive model for sand in triaxial compression, *Vol.3, n°3, Int. Journal. for Numerical and analytical Methods in geomechan., pp 255-278.*

RiouY., Nicolas Ch., Abdallah N. (1997), Modélisation numérique d'une section de tunnel en service : incidence du modèle de comportement,, *mini symposium AFPC AFREM, GEO'97, Aussois*

Vermeer P.A. (1978) A double hardening model for sand, *Geotechnique, 28, 1978, pp 413-433*

PARTICLE SIZE EFFECT OF ANCHOR PROBLEM WITH GRANULAR MATERIALS

T. Sakai and Erizal
Ehime University, Matsuyama, Ehime, Japan

T. Tanaka
Meiji University, Kawasaki, Kanagawa, Japan

ABSTRACT In this study, the difference in the scale effect on shallow anchors in dense Toyoura sand, Soma sand and Leighton-Buzzard sand was evaluated by comparing the experimental result with the results of finite element analysis. A shear band thickness was introduced in this finite elemet analysis, so as to evaluate the particle size effect due to shear banding. In the experimental result, up to 10cm of anchor diameter (D), there was clear evidence of a scale effect in Toyoura sand, but there was a negligible scale effect in Leighton-Buzzard sand. In the finite element analysis, in the case of evaluating the scale effect using D, an effect was observed for D over 5cm in Toyoura sand, over 20cm in Soma sand and over 30cm in Leighton-Buzzard sand. Next, in the case of evaluation on using D/d_{50} (d_{50}: mean particle diameter), within the range 350 to 1000 for D/d_{50}, there was clear evidence of the particle size effect. For D/d_{50} over 1000, the trend of the scale effect was similar in all cases. The results indicate that it is necessary to consider the particle size effect when evaluating the scale effect due to progressive failure.

1 INTRODUCTION

The scale effect of a footing problem was discussed by de Beer(1965), who showed that the bearing capacity factor generally decreased with an increase in the footing width. The scale effect gave rise to problems in the transformation of the model test result to a prototype. Many studies of the scale effect have been carried out using model tests and analysis (Yamaguchi et al., 1977, Ovesen, 1980, Tatsuoka et al., 1997). The experimental studies of the scale effect are carried out using both a centrifugal and a conventional 1g test. The causes of the scale effect are the progressive failure and crushing of sand particles under a high confining pressure. Since the anchor problem in the conventional 1g test can be carried out at a low confining pressure, crushing of sand particles may be negligible.

The scale effect of the anchor problem has been investigated by some investigators (Dickin, 1988, Tanaka and Sakai, 1993). Ovesen (1981) studied the anchor problem using the centrifugal test and observed no scale effect. Hutchinson (1982) studied it in the conventional 1g test with dense Leighton-Buzzard sand, and also observed no scale effect. However, Sakai and Tanaka (1998) studied the scale effect with dense Toyoura sand and compared the experiment results with those of analysis, and showed clear evidence of the scale effect. Leighton-Buzzard sand and Toyoura sand differ in grain size. It is important to evaluate the scale effect caused by the effect of particle size (particle size effect) due to progressive failure.

Sutherland et al. (1988) stated that the finite element analysis for cohesionless soil showed unsatisfactory results. The failure of a sand mass is usually progressive and linked to the development of a narrow shear band of localized deformation. Tanaka et al. (1997) presented a finite element analysis in which the shear band thickness was introduced into the constitutive equation as a characteristic length. The results of this finite element analysis were very close to the experimental results of footing (Tatsuoka et al.,1997), anchor (Sakai and Tanaka,1994) and retaining wall problems (Tanaka et al., 1997).

In the present study, we attempt to explain the difference in the scale effect on a shallow circular anchor in dense sand, by comparing experimental results with those of finite element analysis. Sand samples used in the test were Toyoura sand and Soma sand. Data for Leighton-Buzzard sand presented by Hutchinson (1982) were also evaluated.

2 EXPERIMENTAL PROCEDURE

Figure 1 shows the testing apparatus. This apparatus consisted of a cylindrical container and an anchor pullout machine. The cylindrical container fabricated from 10mm-thick hard vinyl chloride was 59cm in diameter and 50cm in height. The anchors were flat circular 0.5cm-thick steel plates with diameters (D) of 2.5cm, 5cm, 10cm and 15cm. The pullout load was measured using a load cell connected to the anchor rod. The anchor was moved up using a D.C. motor until the shear band passed completely into the sand mass. The tests were displacement controlled. The anchor pullout rate was approximately 0.3mm/min.

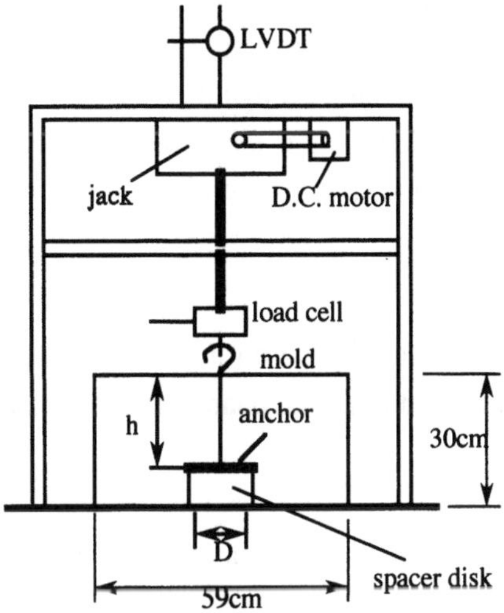

Fig.1 Testing apparatus

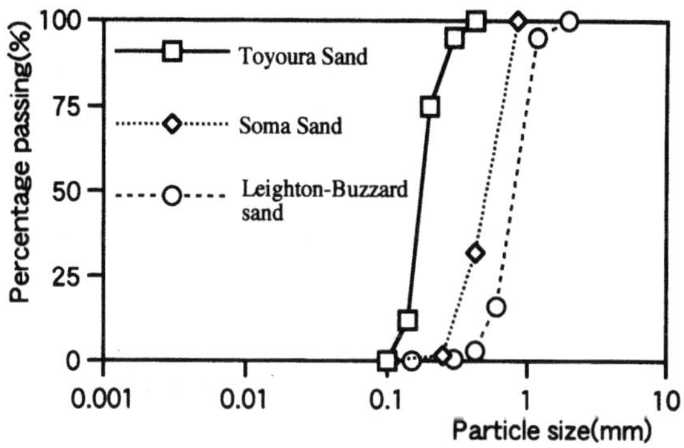

Fig.2 Particle size distributions

The sand samples used in the test were Toyoura sand and Soma sand (grade No.4). Toyoura sand is the Japanese standard sand and Soma sand has also been used in laboratory and model testing in Japan. The particle distributions are shown in Figure 2. The results for Leighton-Buzzard sand (grade 1630), which is the standard sand in the UK. are also plotted in Figure 2. The mean particle diameters (d_{50}) of Toyoura sand, Soma sand and Leighton-Buzzard sand are 0.16mm, 0.45mm and 0.8mm, respectively. The sand beds were prepared by pluviating air-dried sand. The dry densities of sand beds with Toyoura sand and Soma sand were 1.60g/cm^3 (D_r=90%) and 1.57g/cm^3 (D_r=91%), respectively. In order to observe shear band propagation in sand mass, the apparatus which divided the mold and anchor into halves was used. Its sidewall was made of 1cm-thick glass plate and thin horizontal colored sand layers were placed at intervals of approximately 1cm in sand mass adjacent to the sidewall. All the tests were performed in h/D=2 (h : anchor depth).

3 NUMERICAL METHODS

The finite element analysis has been carried out by Tanaka et al. (1997). In this analysis, shear band thickness was introduced into the constitutive equationas as a characteristic length. A constitutive model for nonassociated strain hardening-softening elastoplastic material was introduced. A yield function of the Mohr-Coulomb type and a plastic potential function of the Drucker-Prager type were employed. The element employed for the analysis was a 4-noded Lagrange-type element with one point integration. A dynamic relaxation method with the return mapping algorithm was applied to the integration algorithms of the elastoplastic constitutive equation including the shear band effect.

In the return mapping algorithm (Ortiz and Simo, 1986), the elastically predicted stresses (σ_A) are relaxed onto a suitably updated yield surface (σ_B). A change in stresses can cause an associated change in the elastic strains given by

$$\varepsilon^e = [D]^{e-1}(\sigma_B - \sigma_A), \quad (1)$$

where $[D]^e$ is the elastic matrix. Since the total strain does not change during the relaxation process, the plastic strain change is balanced by an equal and opposite change in the elastic strains;

$$S\varepsilon^P = -\varepsilon^e = -[D]^{e-1}(\sigma_B - \sigma_A). \quad (2)$$

The ratio S was incorporated into the elastoplastic constitutive model as a characteristic length scale. S is the ratio of the shear band area to the finite element area.

$$S = \frac{F_b}{F_e}, \quad (3)$$

where F_b is the area of the shear band in one element and F_e is the total area of one element. The ratio was approximately given by

$$S = \frac{W}{\sqrt{F_e}}, \quad (4)$$

where W is the thickness of the shear band.

The unit weight (γ_d), residual friction angle (ϕ_r), Poisson's ratio (ν) and initial shear modulus (G_0) are chosen to be $\gamma_d = 15.7\mathrm{kN/m^3}$, $\phi_r = 35°$ and $\nu = 0.3$, $G_0 = 78400\mathrm{kPa}$ for Toyoura sand (Tatsuoka et al., 1986a and 1986b).

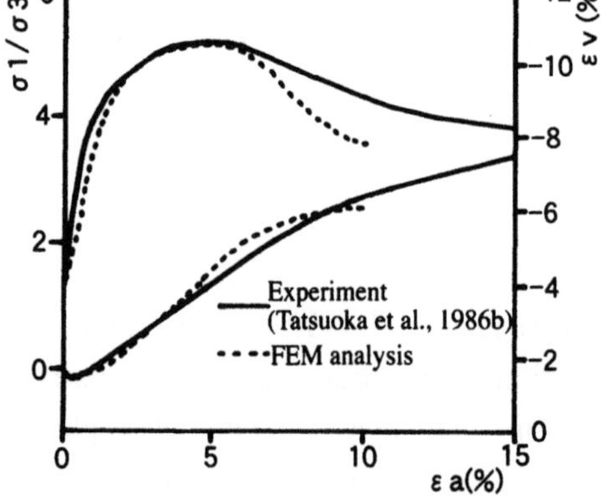

Fig.3 Simulated stress-strain-volume chang relationship

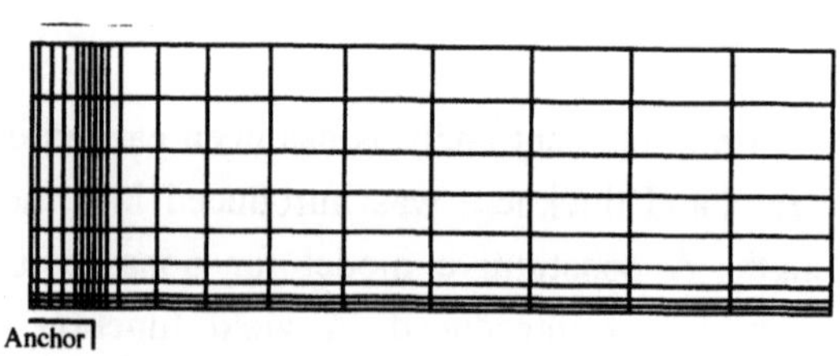

Fig.4 Finite-element mesh (fine mesh, elements of 189 and nodal points of 220)

Vardoulakis et al. (1981) reported that the shear band thickness was 20 times the mean particle diameter, and Sakai (1997) also reported that the shear band thickness in glass beads was 20 times the particle diameter. Then, the thickness of the shear band was estimated to be 20 times the mean particle diameter.

The confirmation of the results of the triaxial compression test by the finite element method using one element was carried out employing the material properties.The calculated stress-strain-volume change relationship when $\sigma_3 = 98$kPa is shown in Figure 3. The finite element mesh (fine mesh, elements of 189 and nodal points of 220) used for the analysis is shown in Figure 4. All the analyses were conducted under axisymmetric problem.

4 EXPERIMENTAL AND NUMERICAL RESULTS

Figures 5 and 6 show the relationship between the dimensionless breakout factor (N_q) and dimensionless displacement (s/D; D: anchor diameter, s: displacement) curves in Toyoura sand and Soma sand. Both tests show that Nq generally decreases with an increase in D. There is clear evidence of the scale effect. The peak N_q appears approximately at the same s/D for both types of sand. Figure 7 shows the theoretically calculated and experimentally observed uplift resistance-displacement curves for Toyoura sand. A calculated result obtained by using coarse finite element mesh (see Figure 8, elements of 75 and nodal poins of 96) is also plotted in this figure. It shows that the calculated result of peak load obtained by fine mesh is very close to the experimental result.

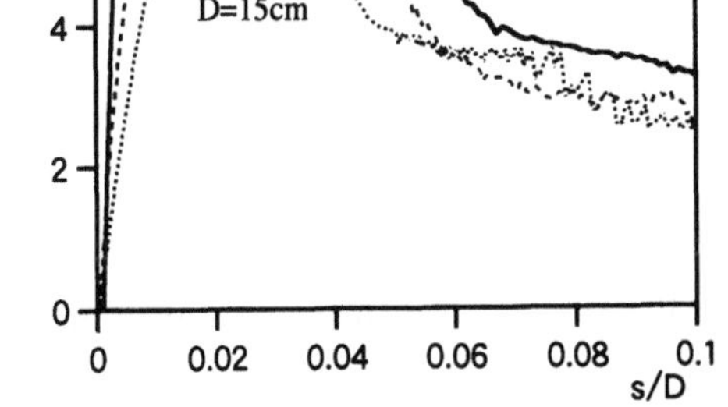

Fig.5 Experimental Nq-s/D curves (Toyoura sand)

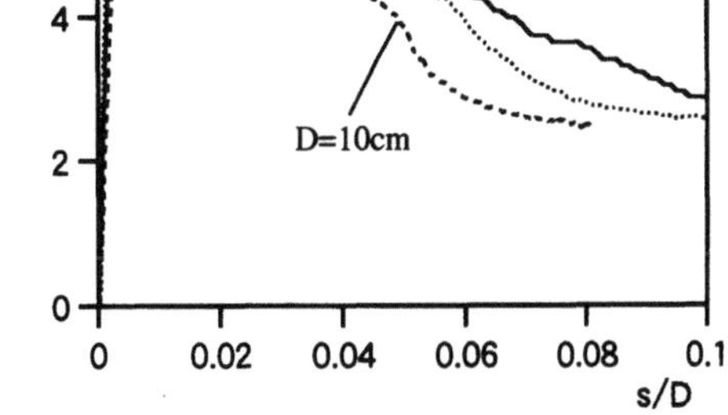

Fig.6 Experimental Nq-s/D curves (Soma sand)

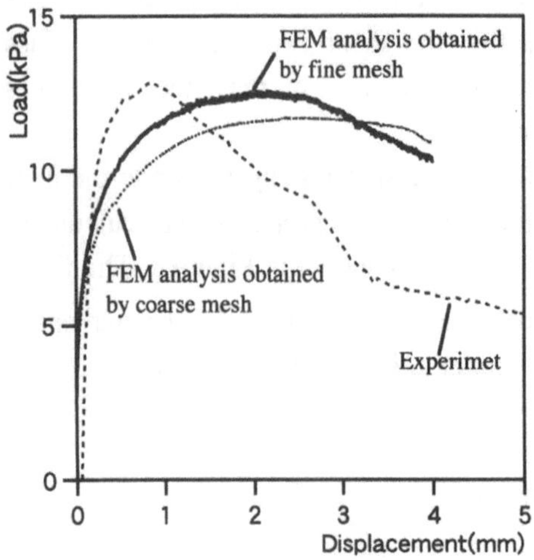

Fig.7 Calculated and experimental load-displacement curves(D=5cm)

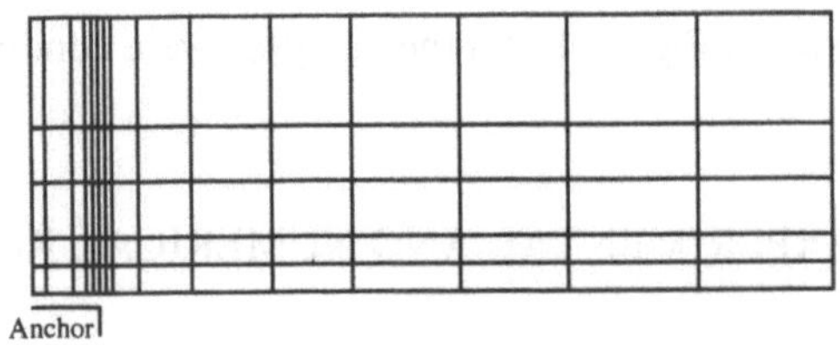

Fig.8 Finite-element mesh (coarse mesh, elements of 75 and nodal points of 96)

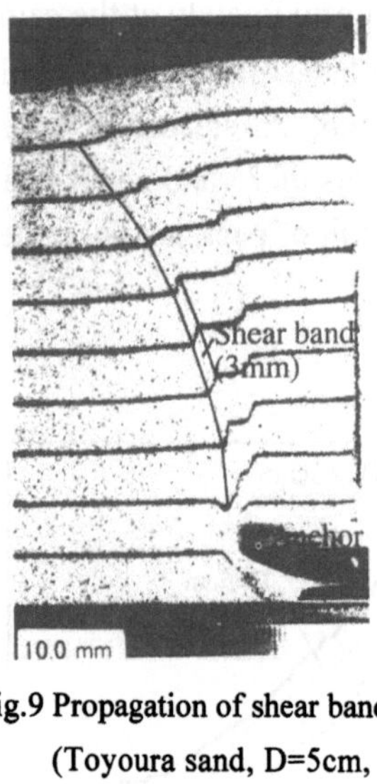

Fig.9 Propagation of shear band (Toyoura sand, D=5cm, disp.=10mm)

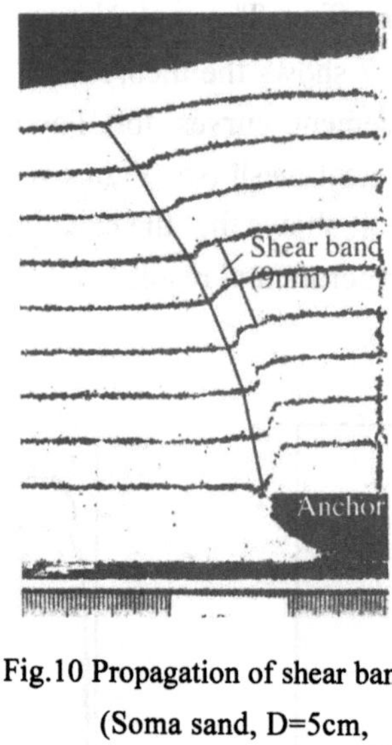

Fig.10 Propagation of shear band (Soma sand, D=5cm, disp.=10mm)

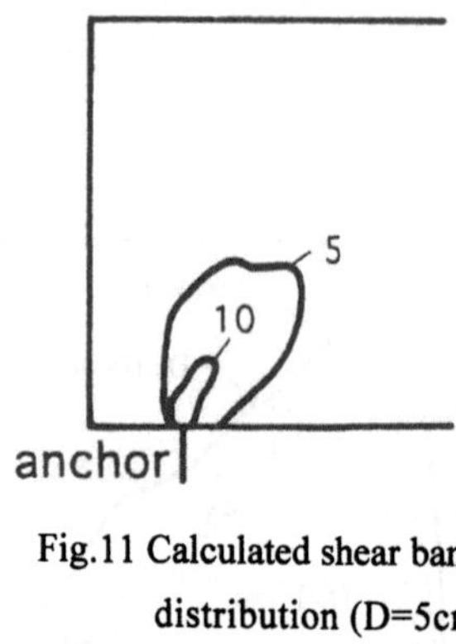

Fig.11 Calculated shear band distribution (D=5cm, disp.=3mm, unit %)

Figures 9 and 10 show the propagation of the shear band in Toyoura sand and Soma sand at a displacement of 10mm, respectively. Since the symmetrical shear band developed from the edges of the anchor plate, the photographs were taken from the left side. The direction of primary shear band development for the two types of sand are similar. The interior angles between the direction of primary shear band propagation and the horizontal line are approximately 65 degrees. Figure11 shows the calculated shear strain distribution in Toyoura sand at a displacement of 3mm. The shear strain shown in the figures is the apparent maximum shear strain averaged at the element level. The direction of the localized narrow zone is calculated to be approximately the same as the direction of the shear band observed experimentally.

The experimental results for Toyoura sand and Soma sand are illustrated in Figure12, in which N_q is plotted as a function of D. The results for Leighton-Buzzard sand, reported by Hutchinson (1982), are also plotted in this figure (Dr=91%). For Toyoura sand, there is clear evidence of the scale effect on the value of Nq. However, for Leighton-Buzzard sand, there is no clear evidence of the scale effect.

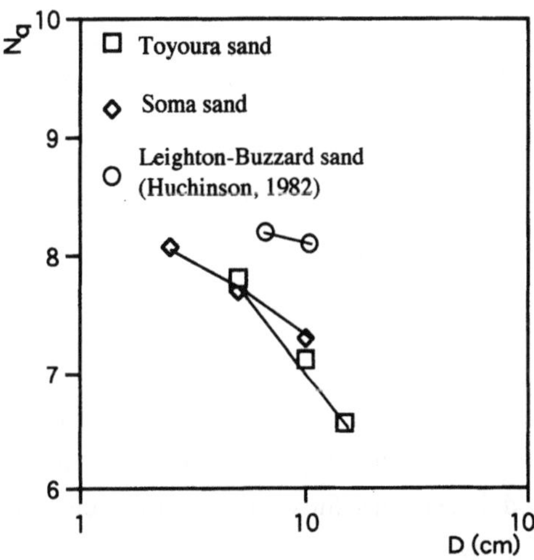

Fig.12 Relationship between Nq and D obtained experimentally

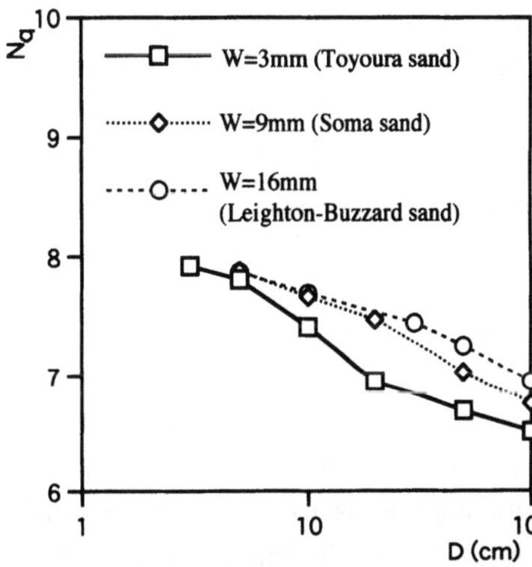

Fig.13 Relationship between Nq and D obtained by calculation

Consider the results of finite element analyses, presented in Figure13 in which Nq is plotted as a function of D. These analyses were carried out employing the material properties of Toyoura sand, changing only the thickness of the shear band (W). In the analyses, since the thickness of shear band was known to be 20 times the mean particle diameter, W was estimated to be 3mm for Toyoura sand, 9mm for Soma sand and 16mm for Leighton-Buzzard sand. In all cases, the scale effect of the anchor problem can be determined by finite element analysis. For Toyoura sand, the scale effect is evident for D over 5cm. For Soma sand and Leighton-Buzzard sand, it was shown that there is clear evidence of the scale effect for D over 20cm and over 30cm, respectively. Therefore, the range of D in which the scale effect is apparent differs according to particle size.

It is important to evaluate the scale effect using the parameter d_{50}/D. A very similar parameter was identified by Steenfelt (1982) in an investigation of the scale effect in model tests of footing. The reciprocal of d_{50}/D was used purely for convenience. The finite element results are shown in Figure 14, in which Nq is plotted as a function of D/d_{50}. In the D/d_{50} range from 350 to 1000, there is clear evidence of the particle size effect. Above D/d_{50} of 1000, the trend of the scale effect was similar for all the cases.

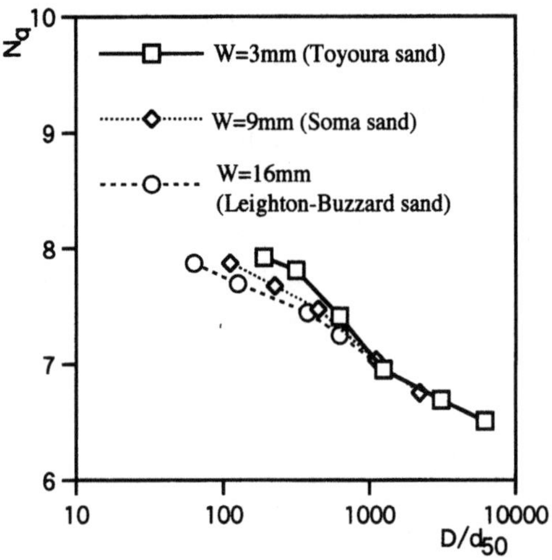

Fig.13 Relationship between Nq and D/d_{50} obtained by calculation

5 CONCLUSION

We evaluated the scale effects observed in the behavior of a shallow anchor with dense sand of different particle sizes (Toyoura sand, Soma sand and Leighton-Buzzard sand), by comparing the results of the conventional 1g model test with those of the finite element analysis. The conclusions can be summarized as follows.

1) The results of finite element analysis showed good agreement with experimental results, and the scale effect was observed in both theoretical and experimental results.

2) The conventional 1g model test revealed that the scale effect of each sand was different. Although there is clear evidence of the scale effect in Toyoura sand, there is negligible scale effect in Leighton-Buzzard sand up to 10cm of D.

3) The finite element analysis of evaluating the scale effect using D, clearly revealed the scale effect when D was above 5cm in Toyoura sand, above 20cm in Soma sand and above 30cm in Leighton-Buzzard sand. The range of D in which the scale effect is apparent differs according to particle size. Next, in the case of evaluating the scale effect using D/d_{50}, within the D/d_{50} range from 350 to 1000, there was clear evidence of the particle size effect. For

D/d_{50} over 1000, the trend of the scale effect was similar in all the cases.

These results indicated that it is necessary to consider the particle size effect when evaluating the scale effect due to progressive failure.

REFERENCE

de BEER E.E. (1965). Bearing capacity and settlement of shallow foundations on sand. Proc. of Symp. held at Duke University.

DICKIN E.A. (1988). Uplift behavior of horizontal anchor plates in sand. ASCE, J. Geotech. Enrg., Vol.114: 1300-1317

HUTCHINSON N.J. (1982). Groups of plate anchors in sand. Int. Research Report CE1/1982, Dept. of Civil Eng., University of Glasgow

ORTIZ M. & J.C. SIMO (1986). An analysis of a new class of integration algorithms for elastoplastic constitutive relations. Int. J. Numer. Meth. in Eng., Vol.23: 353-366

OVESEN N.K. (1980). The use of physical models in design. Design Parameters in Geotechnical Eng,. BGS, Panel Discussion in Session 9: 319-323

OVESEN N.K. (1981). Centrifugal tests of uplift capacity of anchors. Proc. 10th. Int. Conf. on SMFE: 717-722

SAKAI T. (1997). A study of a particle size effect of a trap-door problem with glass beads. Int. Symp. on Deformation and Progressive Failure in Geomechanics, Nagoya: 145-150

SAKAI T. & T. TANAKA (1994). Finite element analysis of progressive failure and scale effect of anchor problem in dense sand. Int. Conf. on Computational Methods in Structural and Geotech. Engrg., Hong Kong, 4: 1283-1288

SAKAI T. & T.TANAKA (1998). Scale effect of a shallow circular anchor in dense sand. Soils and Foundations, Vol.38: 93-99

STEENFELT J.S. (1982). Scale effects on observations in particulate media. Euromech. Colloquium Quality of Mech. Observations: R1-R4

SUTHERLAND H.B., T.W. FINLAY & M.O. FADL(1988). Uplift resistance of soils. Geotechnique, Vol.38: 493-516

TANAKA T., H. MORI & M. KIKUCHI (1997). Finite element elastoplastic analysis of underground structures and retaining walls. 9th. Int. Conf. on Computer Methods and Advances in Geomechanics, Wuhan, 2: 1465-1470

TANAKA T. & T. SAKAI (1993). Progressive failure and scale effect of trap-door problems with granular materials. Soils and Foundations, Vol.33: 11-22

TATSUOKA F., S. GOTO, T. TANAKA, K. TANI & Y. KIMURA (1997). Particle size effects on bearing capacity of footing on granular material. Int. Symp. on Deformation and Progressive Failure in Geomechanics, Nagoya: 133-138

TATSUOKA F., S. GOTO and M. SAKAMOTO (1986b). Effects of some factors on strength and deformation characteristics of sand at low pressures. Soils and Foundations, Vol.26: 105-114

TATSUOKA F., M. SAKAMOTO, T. KAWAMURA and S. FUKUSHIMA (1986a). Strength and deformation characteristics of sand in plane strain compression at extremely low pressures. Soils and Foundations, Vol.26: 65-85

VARDOULAKIS I., B. GRAF & G. GUDEHUS (1981). Trap-door problem with dry sand: a statical approach based upon model test kinematics. Int. J. Numer. and Anal. Methods in Geomech., Vol.5: 57-78

YAMAGUCHI M., T. KIMURA & N. FUJII (1977). On the scale effect of footings in dense sand. 9th Int. Conf. on SMFE.: 795-798

SEISMIC SITE EFFECTS ON THE FRENCH RIVIERA
ANALYSIS BY THE BOUNDARY ELEMENT METHOD

J.-F. Semblat and P. Dangla
Laboratoire Central des Ponts et Chaussées, Paris, France

A.-M. Duval
CETE Méditerranée, L.R.P.C., Nice, France

ABSTRACT: The analysis of site effects through numerical approaches is interesting since it gives quantitative results on both following parameters : the amplification level and the location of its highest values. It is then possible to compare or improve (with) the experimental investigations concerning this phenomenon. In this paper, site effect is investigated considering boundary element method (in the frequency domain) and a pure plane SH wave as the seismic loading. The specific site considered is located in the center of Nice on the french Riviera. The influence of frequency and incidence is analyzed. In a second part, pure P-waves and SV-waves are considered. Shear waves (SH and SV) give higher amplification factors than pressure waves. The thickness of the layer, its general shape as well as the seismic wave type involved have a great influence on the maximum amplification factor and the frequency for which it occurs. These results are in good agreement with experimental ones.

1 SITE EFFECTS ON THE FRENCH RIVIERA

1.1 *Amplification of seismic motion*

The amplification of seismic waves in some specific sites is sometimes important. Reflections and scattering of seismic waves near the surface strengthen the effects of earthquakes. It was the case for the terrific Mexico 1985 earthquake : maximum acceleration was around 0.28 g at sixty kilometers from the source and 0.2 g in the center of Mexico located much further (400 kilometers). In the town itself, seismic motion was amplified up to a factor of 60 compared to the bedrock because of the thick clay deposit under the city. The detection and estimation of site effects is therefore very important and the main goal is to avoid, in the design of structures, buildings involving resonant features close to that of the soil surface layers. If the seismic motion amplification is supposed to be strong, the local seismic response of soils must be analyzed to precisely determine the characteristics of the reference earthquake used for the design of structures.

Seismic wave propagation can be investigated using experimental techniques (Duval 96a, 96b, Heitz 92, Semblat 98a) or various numerical methods (Bard 83, Bonnet 95, Dangla 89, Semblat 97, 98b). This paper investigates the amplification of seismic waves (site effects) through experimental measurements (real earthquake and microtremor) and numerical models based on the boundary element method.

1.2 *Seismic measurements in Nice*

In Nice, areas having a high population density are, like in many other towns, located above alluvial soil layers. Experimental measurements of site effects have been performed near the railway station of Nice. This area involves an alluvial deposit corresponding to an old valley of North-South direction axis.

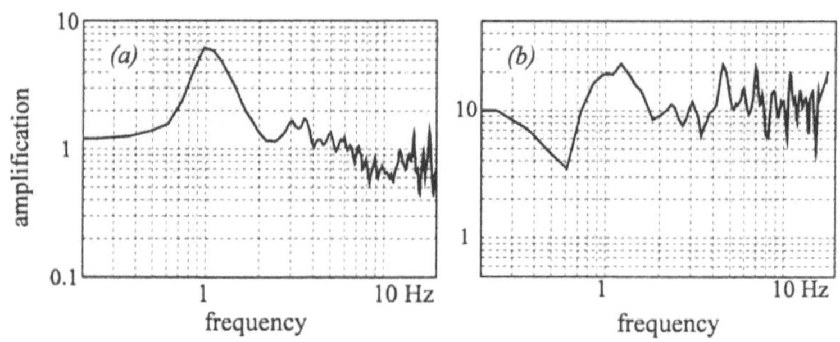

Figure 1. Experimental measurements of the seismic response in the center of Nice (microtremor and real earthquakes).

Earthquake measurements, as well as microtremor analysis (Duval 96a, 96b), clearly indicate that the amplification of seismic motion occurs between 1 and 2 Hz at the center of the alluvial deposit. The curves of figure 1 give horizontal to vertical motion ratios estimated from microtremor measures (left) and transfer functions (right) for real earthquakes measurements (on both alluvial site and reference bedrock site). The values of resonant frequency determined by both methods are very close (figure 1). The amplification factor estimated from microtremor measures is however smaller. Above the thickest alluvial filling, the amplification factor determined by seismic transfer functions reaches a maximum value of 20 around a frequency f=1.5 Hz. Comparing the measurements performed on various sites, the corresponding transfer functions show a strong dependence of resonant frequency on the thickness of the alluvial surface layers.

2 BOUNDARY ELEMENT METHOD

The main advantage of the boundary element method is to avoid artificial truncation of the domain in the case of infinite medium. For dynamic problems, this truncation leads to artificial wave reflections giving a numerical error in the solution. The boundary element method can be divided into two main stages (Dangla 89, Bonnet 95) :

- solution of the boundary integral equation giving displacements and stresses along the border of the domain,
- a posteriori computation for all points inside the domain using an integral representation formula.

The boundary element method arises from the application of Maxwell-Betti reciprocity theorem leading to the expression of the displacement field inside the domain Ω from the displacements and stresses along its border $\partial\Omega$. The following numerical equation has then to be solved (Dangla 88, Dangla 89) :

$$I(y)\, u_j(y) = \int_{\partial\Omega} \left(U_{ij}(x,y)\, t_j(x) - T_{ij}(x,y)\, u_j(x) \right) da(x) \qquad [1]$$

$I(y)$ is 1 if $y \in \Omega$ and 0 in other cases, u_j is the displacement along j, t_j is the j component of the stress vector applied on $\partial\Omega$. $U_{ij}(x,y)$ is the displacement at point x along direction j due a concentrated unit force at point y along direction i. Functions $U_{ij}(x,y)$ are known as Green's functions and $T_{ij}(x,y)$ corresponding stress vectors along $\partial\Omega$. In this article, the model involves the Green's functions of an infinite medium (Dangla 89) or semi-infinite medium (in the case of SH-waves). Numerical solution of equation [1] can be performed by collocation method or by an integral variational approach (Bonnet 95).

3 MODEL FOR A PLANE SH-WAVE

3.1 *Modeling of the geological contour*

An East-West geological section is considered for the numerical simulation with precise description of its contour (figure 2). The alluvial layer is supposed to be homogeneous. The mechanical characteristics of the deposit and of the bedrock are the following :

- alluvial layer : ρ_1=2000 kg/m^3, μ_1=180 MPa giving C_1=300 m/s,
- elastic bedrock : ρ_2=2300 kg/m^3, μ_2=4500 MPa giving C_2=1400 m/s.

The alluvial deposit is deeper on its western part (h=64 m) than on its eastern part (h'=32 m). The loading is a plane SH-wave with vertical incidence (first) and various incidences (afterwards). For a plane SH-wave, the motion is anti-plane with respect to the direction orthogonal to the model plane (only one DOF in each nodes of the mesh).

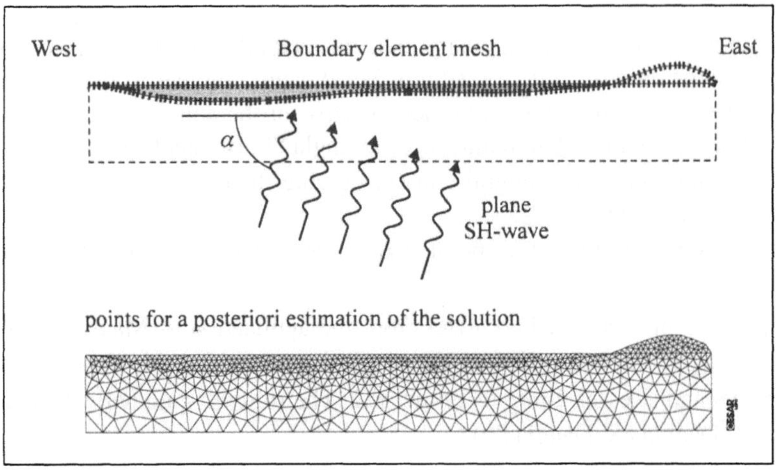

Figure 2. Boundary element mesh for a plane SH wave.

In figure 2, the boundary element mesh is depicted (top) for the solution of the integral equation and a complete mesh (created with a finite element mesh generator, bottom) for the computation of the a posteriori solution inside the domain. This model uses Green's functions of an infinite domain for the alluvial layer and the mount. It involves Green's functions of an semi-infinite domain (easy to determine for SH-waves) to precisely model the bedrock as a subdomain of an infinite half-space. The computation is made using the FEM/BEM code CESAR-LCPC (Humbert 89).

3.2 *Amplification factor for a vertical incidence*

The boundary element method allows for the determination of displacement (and of amplification factor) in all points of the boundary elements. The solution is determined afterwards in all points inside the domain (figure 2).

Figure 3 gives the isovalues of the amplification factor in the alluvial layer and the bedrock for various frequency values (for a plane SH-wave with vertical incidence). Amplification obviously occurs at the surface of the alluvial deposit and reaches a maximum value of 15.0 at frequency 1.6 Hz. The maximum amplification appears in the thickest part of the deposit. However, for a higher frequency (f=2.0 Hz), amplification factor in the thinnest part of the alluvial layer slightly increases.

These results are in good agreement with experimental seismic measurements displayed in figure 1. The first resonant frequency computed numerically is close to the experimental values. The computed amplification factor is not far from the values given by seismic transfer functions. The amplification factor estimated by microtremor H/V ratios is nevertheless smaller.

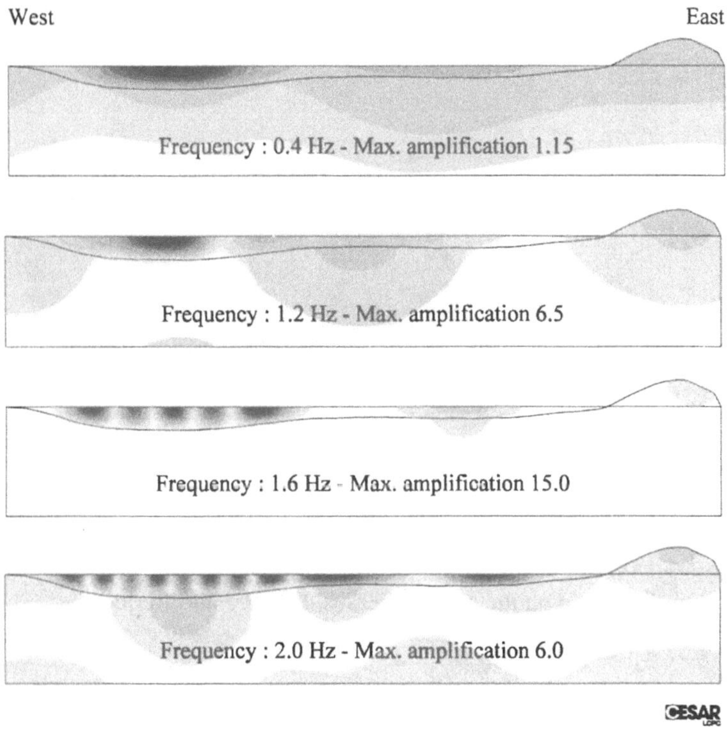

Figure 3. Estimation of amplification factor for a vertical SH-wave at different frequencies.

Three-dimensional graph of figure 4 gives the variations of the amplification factor on the surface of the alluvial layer versus location and frequency. It is then possible to estimate the amplification level at each frequency as well as the location and length of the corresponding maximum amplification area. In figure 4, amplification factor is low for frequencies under 0.8 Hz. Above this frequency value, an area of important amplification appears in the thickest part of the alluvial deposit (West). For higher frequencies, several areas of high amplification factor are detected always in the western part of the deposit. Above 1.5 Hz, amplification factor in the thinnest part of the alluvial filling (East) increases progressively. Between 2.0 and 2.5 Hz, the amplification decreases in the thickest part of the deposit whereas it strongly increases in the thinnest part.

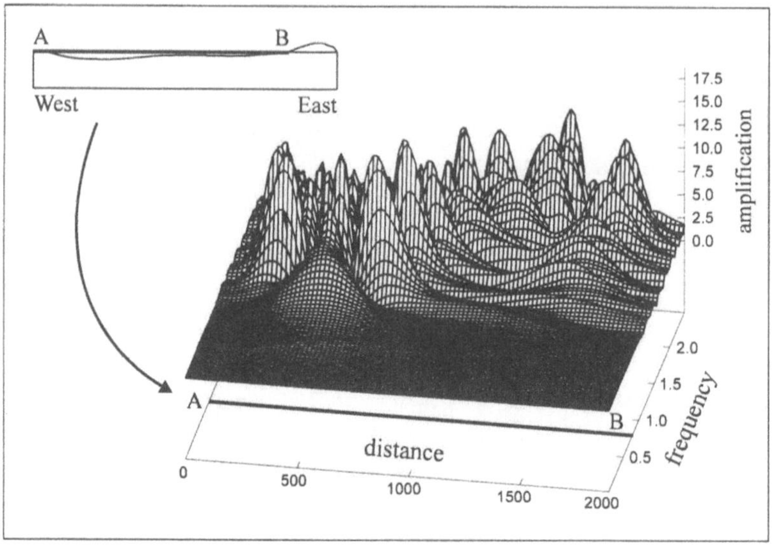

Figure 4. Amplification on the surface versus distance and frequency.

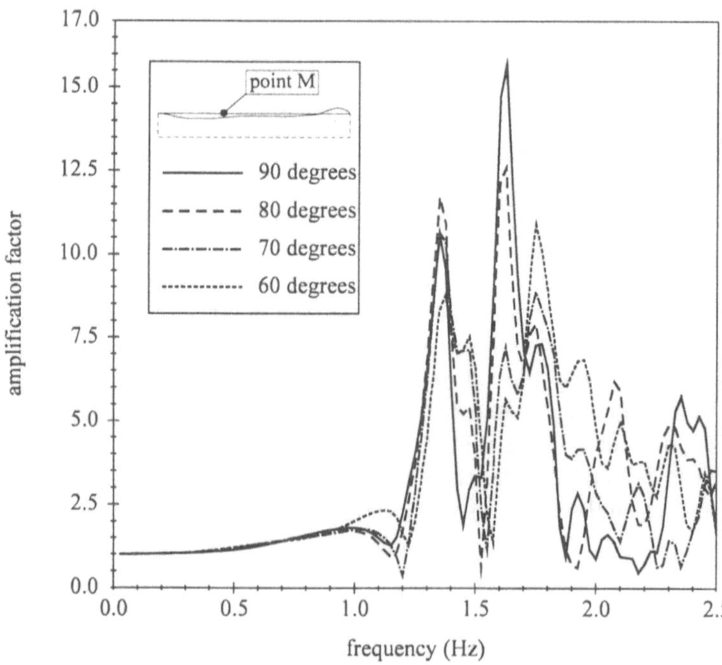

Figure 5. Amplification factor versus frequency at point M (plane SH-wave).

3.3 *Influence of incidence*

As the bedrock is described using Green's functions of an infinite half-space, seismic amplification can be determined, in the case of plane SH-waves, for every values of incidence (no truncation effect). To analyze the effect of incidence, results are firstly given in a specific point versus frequency. Afterwards, amplification factor is studied all along the surface for several values of frequency. Results of figure 5 correspond to the point M for which amplification is maximum at frequency 1.6 Hz (see figures 3 et 5). This point M is located at distance d_M=860 m of the western border of the deposit.

From figure 5, amplification is obviously very low for frequencies under 1.2 Hz. Above this frequency value, amplification factor strongly increases (between 10 and 12 depending on incidence). Around 1.5 Hz, amplification decreases and increases once more up to a maximum value for an approximate frequency of 1.6 Hz. For this specific frequency, amplification factor decreases when incidence is low. Point M corresponds to the maximum amplification for normal incidence but is perhaps not the point of highest amplification for other incidences. The following results will allow the comparison of the locations of site effects for various incidences.

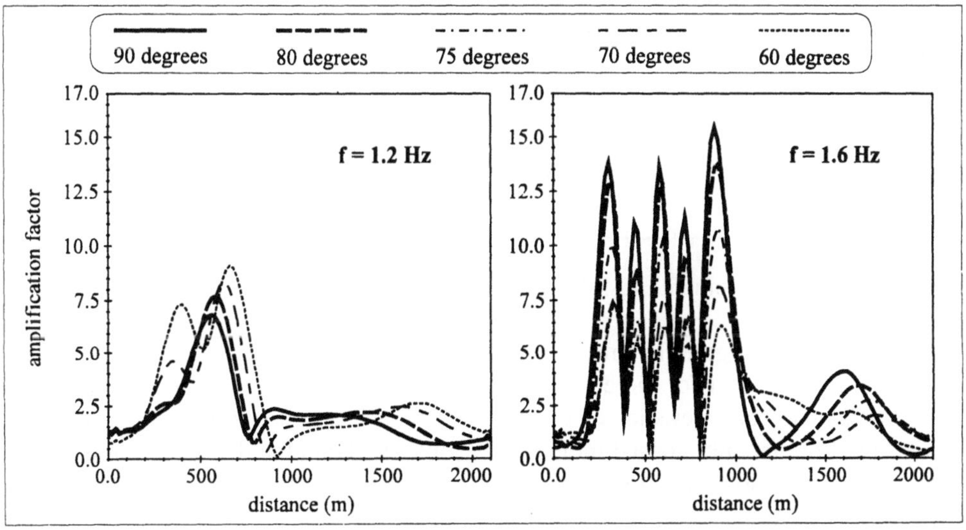

Figure 6. Dependence of amplification factor on incidence for two different frequencies.

Results of figures 6 give the values of the amplification factor in all points of the surface of the deposit for two specific frequency values. It is then possible to investigate the influence of incidence on site effects and their location. From figure 6, it can be seen that the amplification factor at 1.2 Hz increases with decreasing incidence : it has a value of 7.0 for a normal incidence and reaches 9.0 for a 60° incidence. The location of the maximum amplification area has significant changes : it is at a distance of 500 m from the western border of the deposit for a normal incidence and at 700 m for a 60° incidence (figure 6). The numerical modeling gives an interesting estimation of spatial variability of site effects. For a frequency of 1.6 Hz, amplification factor decreases with decreasing incidence. Its value is 15.5 for a normal incidence but is under 7.0 for a 60° incidence. The differences between results corresponding to various incidences are due to the dependence of horizontal and vertical wavenumbers on incidence. The deposit « appears » thicker for decreasing incidence and the amplification factor for low frequencies is then higher for lower incidences.

4 P AND SV-WAVES

For pressure and shear waves (P and SV), the motion is in the propagation plane. It is not very easy to use Green's functions of an infinite half-space (as for SH-waves) and the boundary element mesh has been extended on both sides of the deposit (see figure 7). To avoid numerical errors due to the truncation of the mesh, it has to be sufficiently extended to model correctly the infinite wideness of the bedrock. To minimize these effects, P and SV waves computations are only performed for normal incidence.

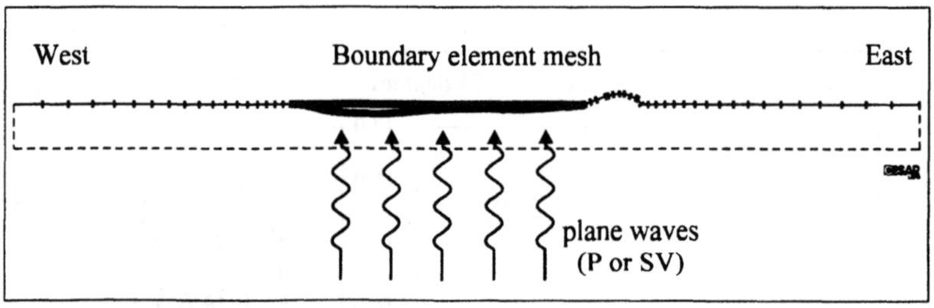

Figure 7. Boundary element mesh for P and SV waves.

Curves of figure 8 give the amplification factor values in every point of the surface for various frequencies and for three different wave types (SH, P and SV). For these simulations, mechanical characteristics of the deposit and of the bedrock are the same as in section 3. Furthermore, only normal incidence is considered and the displayed parameter is the magnitude of the displacement vector.

For low frequencies (1.2 and 1.4 Hz, figures 8*a,b*), P-wave amplification is very low whereas shear-waves amplification (SH and SV) is significant (between 7.0 and 8.0), and even stronger for SH-waves at 1.4 Hz (13.0). The maximum amplification always occurs in the thickest part of the alluvial deposit. For higher frequencies (1.6 and 1.8 Hz, figures 8*c,d*), seismic amplification of SH-waves is maximum (15.5) for 1.6 Hz whereas SV-waves amplification decreases. At 1.8 Hz, site effects are very strong for SV-waves (16.5), increase for P-waves (7.5) and decrease for SH-waves (7.5). The highest amplification factor is always reached in the western (thickest) part of the deposit. Nevertheless, the number and extent of maximum amplification areas depend on the wave type. Above 2.0 Hz (figures 8*e,f*), P-waves have significant amplification (12.5) identical to that of SV-waves, but for SH-waves amplification is under 6.0. At 2.2 Hz, the amplification in the thinnest part of the deposit increases (7.5) for both shear wave types. For all frequencies, maximum amplification factor is then between 12.5 for P-waves and 16.0 for SV-waves. It is always reached in the thickest part of the deposit, whereas the thinnest part has lower amplification at higher frequencies.

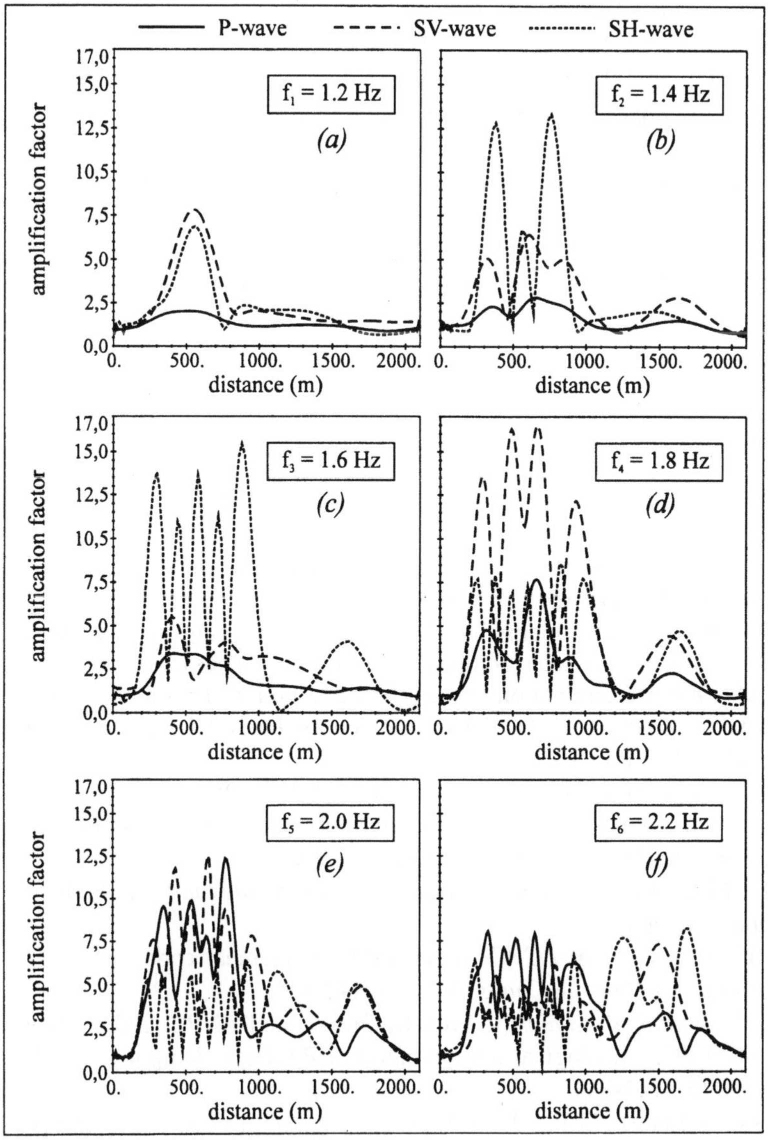

Figure 8. Amplification factor for different wave types at various frequencies.

5 CONCLUSION

Site effects lead to a local amplification of seismic motion. They are significant in the highly builded areas of the center of Nice. In situ experiments performed by the CETE-Méditerranée show that the seismic motion is strongly amplified between 1 and 2 Hz. Numerical simulations based on the boundary element method are in good agreement with these results. For SH-waves, the amplification factor reaches 15.5 at frequency 1.6 Hz. Maximum amplification is located in the thickest part of the alluvial deposit.

Site effects are quantified numerically and appear to be sensitive to incidence. It changes the maximum amplification factor, the frequency at which it occurs and the corresponding affected area. For the various wave types considered (SH, P, SV), seismic motion amplification is very different but always reaches a high level (between 12.5 and 16.0). Shear waves lead to the strongest site effects (figures 8*c,d*). The location of maximum amplification areas is also depending on the features of the seismic loading. The boundary element method seems to be efficient to analyze site effects from a qualitative, as well as quantitative, point of view.

References

Bonnet, M. 1995. Integral equations and boundary elements (in French), CNRS Editions-Eyrolles, 316 p., Paris.

Dangla, P. 1988. A plane strain soil-structure interaction model, *Earthquake Engineering and Structural Dynamics*, 16 : 1115-1128.

Dangla, P. 1989. Finite elements-integral equations coupling in elastodynamics and soil-structure, Ph.D. thesis (Ecole Nationale des Ponts et Chaussées), Paris.

Duval, A.M. 1996. Determination of the seismic response of a site using microtremors. Experimental estimation, Ph.D. thesis (University Paris VI), *Etudes et Recherches des L.P.C*, GT62, L.C.P.C, Paris.

Duval, A.M., Méneroud, J.P., Vidal, S.and Bard, P.Y. 1996. A new method for the estimation of seismic soil response by microtremor measurement (in French), *Bulletin des Laboratoires des Ponts et Chaussées*, 203: 75-90.

Heitz, J.F. 1992. *Waves propagation in non linear medium* (in French). Ph.D. thesis (Grenoble University).

Humbert, P. 1989. CESAR-LCPC : a general finite element code (in French), *Bulletin des Laboratoires des Ponts & Chaussées*, 160: 112-115.

Pecker, A. 1984. *Soils dynamics* (in French), Presses de l'ENPC, 259 p., Paris.

Semblat, J.F. 1997. Rheological interpretation of Rayleigh damping, *Jal of Sound and Vibration* 206(5): 741-744.

Semblat, J.F. & Luong, M.P. 1998a. Wave propagation through soils in centrifuge testing, *Jal of Earthquake Engineering* 2(10): 147-171.

Semblat, J.F. 1998b. Damping and dispersion of waves : physical and numerical points of view (in French), *Revue Française de Génie Civil*, 2(1): 91-111.

ANALYSES OF NATM-TUNNELS WITH 2-D AND 3-D FINITE ELEMENT METHOD

C. Vogt
Stuttgart University, Stuttgart, Germany

P. Bonnier
Plaxis BV, Rhoon, The Netherlands

P.A. Vermeer
Stuttgart University, Stuttgart, Germany

ABSTRACT: To model the NATM shotcrete method of tunnelling we apply both 2D and 3D discretisations. The more or less traditional 2D approach is computationally simple, but it requires certain assumptions about the so-called beta-factor to be made. Therefore we also perform 3D FEM analysis for comparison.

Attention is focused on a tunnel in sedimentary clay rock. This rock is heavily overconsolidated and as a consequence it involves high horizontal stresses. The impact of high K_o-values on tunnel contraction and surface displacements is shown to be significant.

1. INTRODUCTION

In tunnel design the FEM is gaining in importance due to its ability to take into account the interactions between the tunnel lining and the surrounding subsoil. Particularly for the analysis of shallow tunnels the FEM is becoming increasingly important as it also provides predictions of surface settlements.

The stress and strain conditions arround an NATM-tunnel (New Austrian Tunnel Method) are influenced by the parameters of the subsoil and the method of excavation. When a tunnel is excavated in several sections (transverse and longitudinal) there are a large number of different states to be analysed. Every excavation state implies a change to the stresses and strains. The unsupported tunnel stretch causes the ground to arch over the excavated section supported by the lining behind it and by the surrounding ground (Fig. 1).

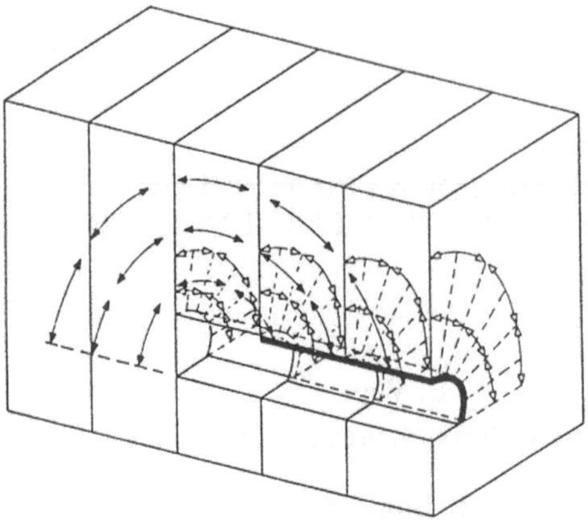

Fig. 1: Stress distribution around face of the heading, Schikora (1988)

2. FEM CALCULATIONS AND ß-METHOD

The stresses and strains due to the excavation of a tunnel can only be modelled realistically by means of 3-dimensional calculations. However, for reasons of time and costs 2-dimensional analyses are usually conducted. All deformations which take place before the lining is introduced are taken into account either through a numerical method of stiffness reduction or a method of load reduction. The magnitude of this reduction must be estimated on the basis of experience or in situ measurements. An example using the ß-method, i. e. a stress reduction method, is given below.

The ß-method consists of unloading by reducing the loads at the nodes along the edge of the tunnel. This means that in the analyses the soil inside the tunnel is first of all replaced by single node loads corresponding to the initial stresses. Then these tunnel support loads are gradually decreased down to a prescribed percentage (ß) of the initial ones. In a second stage the tunnel lining is introduced and the remaining loads are taken by the lining. The stages involved in a multi-staged excavation are shown in Fig. 2. The value of the ß-factor depends principally on the length of the unsupported section, the time schedule of excavation as well as on the geological parameters.

3. RELATIONSHIP BETWEEN ß-FACTOR AND UNSUPPORTED TUNNEL LENGTH

The relationship between the unsupported length and the ß-factor will be explained below with reference to a shallow tunnel in sedimentary rock ($\phi' = 35°$, $c' = 100$ kN/m², $E = 75$ MN/m²). For this purpose 3-dimensional analyses were carried out using the FEM code PLAXIS. The mesh and the geometry of the tunnel are shown in Figure 3. Three different excavation lengths were investigated, resulting in different total mesh lengths.

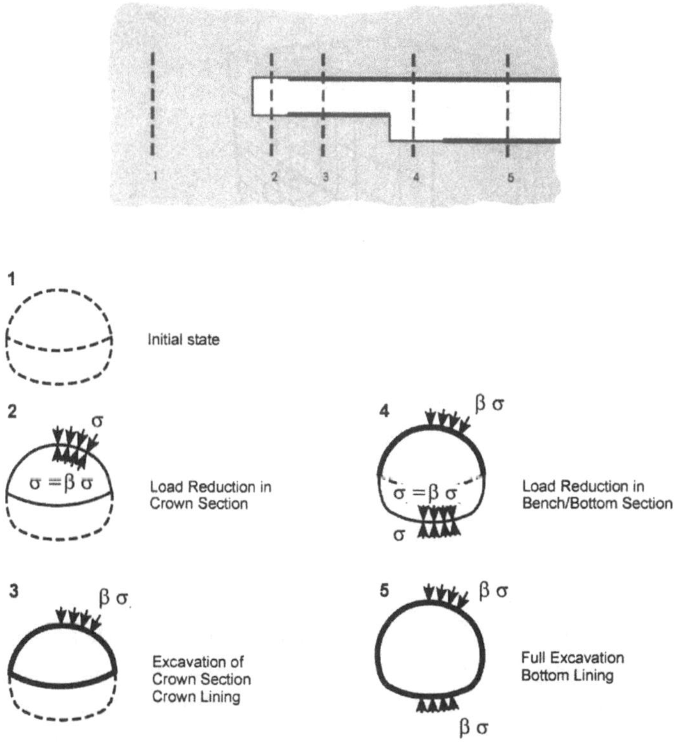

Fig. 2: Load reduction method (ß-method)

In each step of the analysis one section with length L is excavated without support. After calculating the changed stress state as a result of this excavation the support is introduced over this length. The following calculation steps are necessary:

1. Initial stresses (weight of the soil)
2. Excavation and simultaneous introduction of the lining over a starting length
3. Excavation of first section length L (Fig. 3)
4. Introduction of the lining over the first section length L

Step 3 and 4 are repeated again and again.

The results were evaluated with respect to the deformations at the ground surface and at the tunnel roof. Two possible measurement points (Figure 4) are considered. The vertical deformation at these points is shown as a function of the advance of the tunnel heading.

Starting with an initial value resulting from the pre-excavation (Step 2) the deformations at the ground surface and at the tunnel roof increase as the tunnel face approaches (z-direction) towards the points A and B.

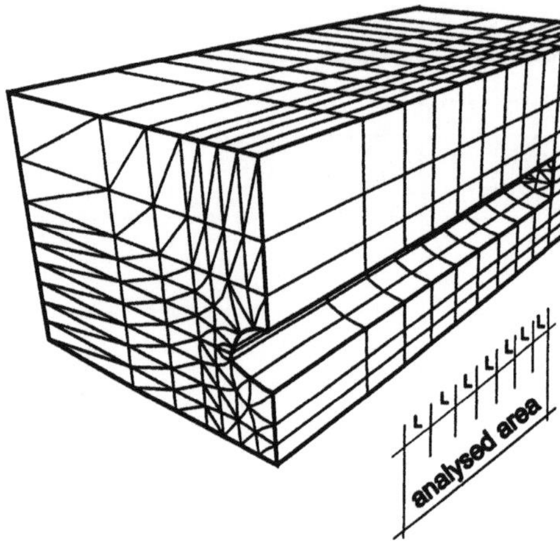

Fig. 3: Geometry of the tunnel and FE-mesh

At the tunnel roof the settlements due to the unloading in front of the face are relatively minor; the main settlement activity occurs as a result of the excavation in the cross section considered (Point B). At the ground surface (Point A) the settlements increase more uniformly and continue after the tunnel face has passed by. In the calculation for a section length of 2.8 m, Fig. 4 shows that the area to be analysed appears to be not long enough and the surface settlement maxima are not reached.

As well as varying the section length, the effect of the typical horizontal prestress found in sedimentary clay rocks is also investigated. In this case the settlements at the surface and the tunnel roof are much smaller. The maximum deformation occurs at the tunnel sides.

In addition to the 3-dimensional analyses, 2D calculations with the same mesh and data sets were carried out. The ß-factor was varied and the settlements at the surface and the tunnel roof were re-evaluated. As expected, the settlements decrease with an increasing ß-factor, i.e. they are smaller the earlier the lining is introduced..

The diagrams of Figure 4 show that in this case the deformations at the tunnel roof with a section length of 5.6 m correspond to a ß-factor of 0.7, i. e. instead of a 3D analysis a 2D analysis with ß = 0.7 will do. In the 3D calculation the settlements at the surface are lower than in the 2D calculation using the ß-method. This could be due to the arching in a longitudinal direction. Indeed, one should never expect full agreement between 2D and 3D analyses.

Generally speaking, the beta-factor only appears to have a minor effect in stable rocks and the deformations do not change appreciably when the beta-factor or the section length are changed. Fig. 4 shows that changing the ß-factor from 0.5 to 1.0 reduce the deformations at the surface from 24.5 mm to 18.5 mm. On the other hand the diagrams show, that the K_0-value influences deformations very much.

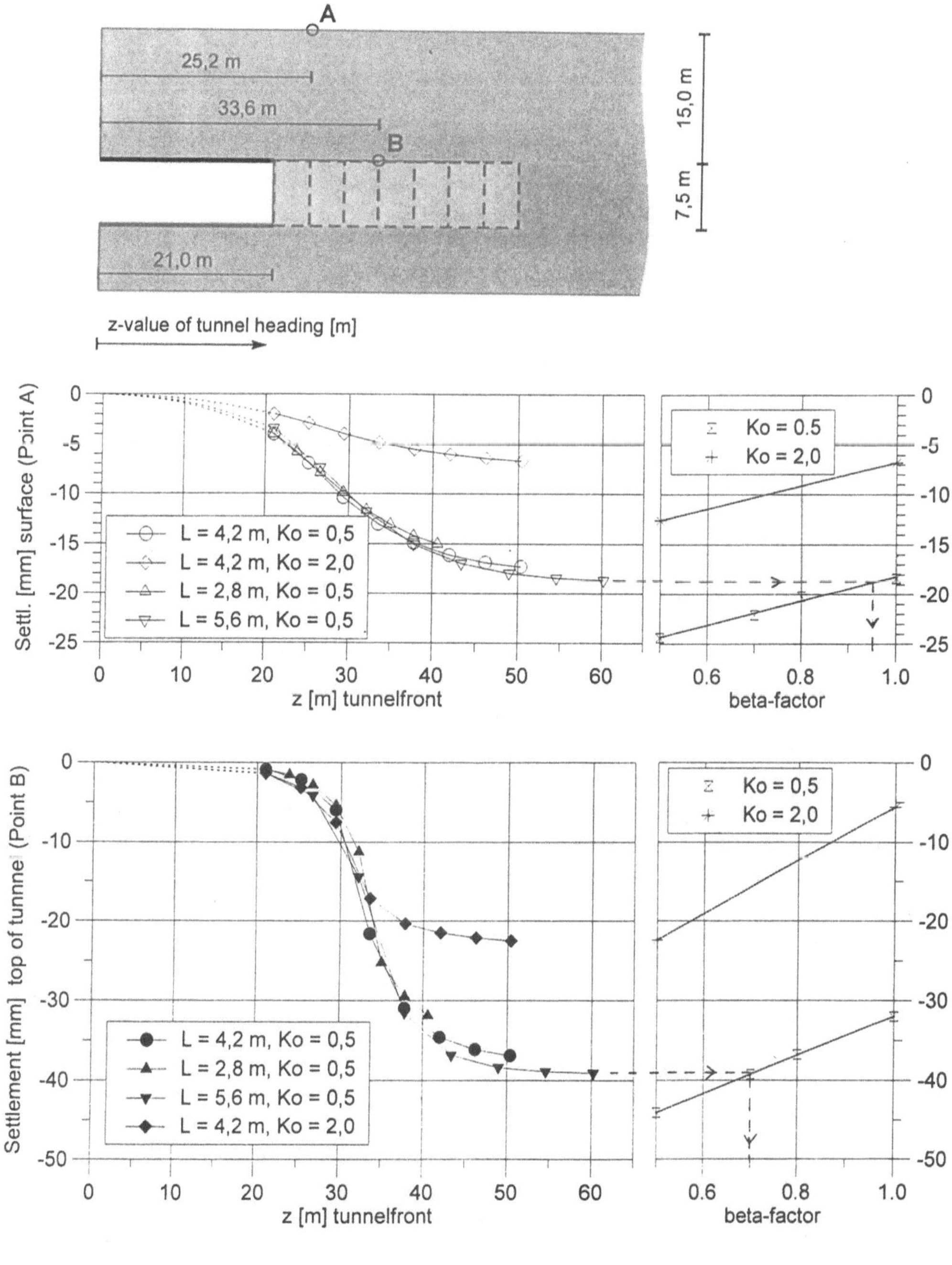

Fig. 4: Vertical deformations of the tunnel roof and the ground surface due to the tunnel advance

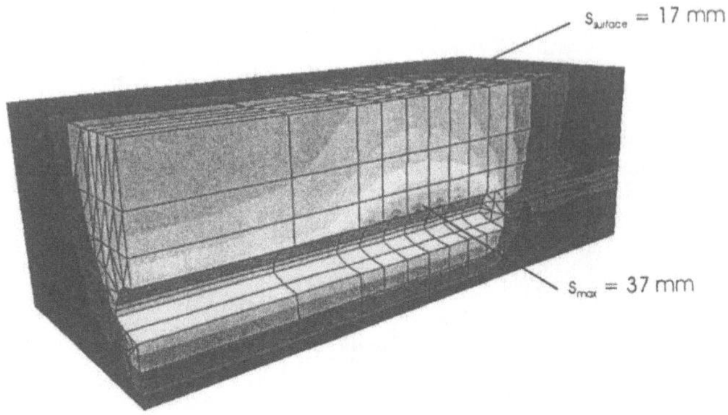

Fig. 5: Deformation contours for $K_o = 0.5$

4. CASE STUDY

For reasons of clarity the relationship between the section length and the ß-factor was described above using an example in which the geometry and the geological aspects were considerably simplified. In practice, however, problems are much more complex, and it is much more difficult to predict the effects of individual factors.

In the following case study 3D analyses will be described for a shallow tunnel in southern Germany where the presence of nearby buildings and railway tracks crossing only 6 m above the tunnel meant that small sections lengths were required to reduce deformations.

The road tunnel, with a typical flattened oval-shaped section, has a clear width of 10.4 m and a clear height of 8.9 m. The excavation of the 98 m² cross section was carried out in two segments, first the crown and then the bottom. To reduce the deformations the crown excavation was lined by a temporary bottom support. The geological situation near the tunnel project comprises 4 layers as shown in Figure 6.

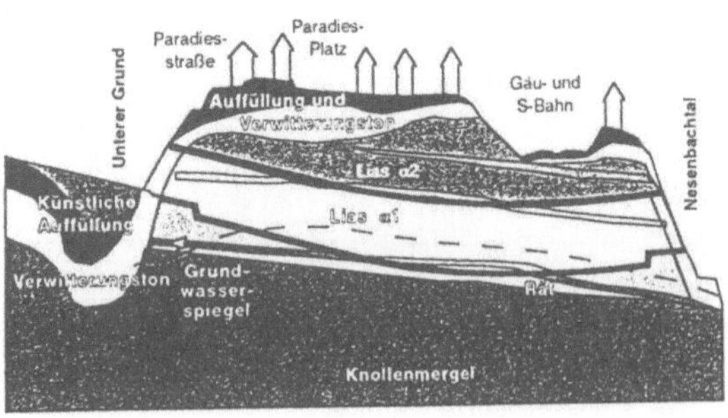

Fig. 6: Longitudinal section of the tunnel ground

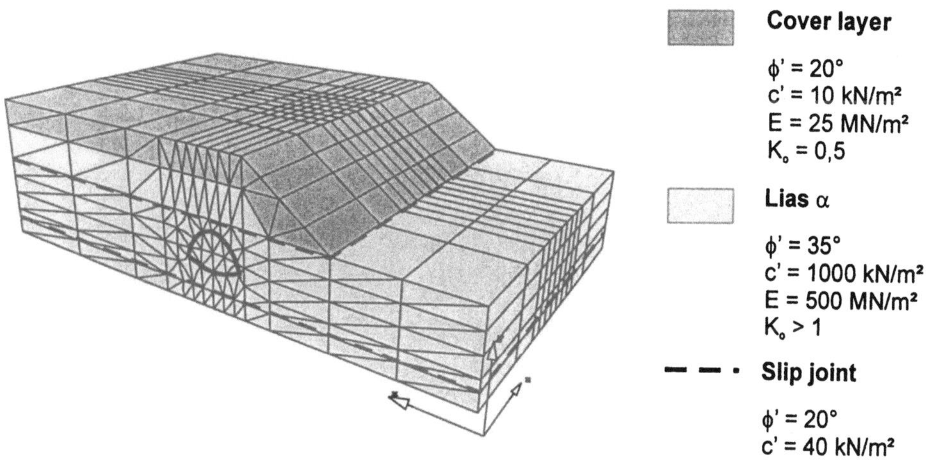

Fig. 7: FEM mesh with soil layers and interfaces

The numerical analyses of the problem described above were conducted using 2046 15-noded prismatic elements. The material is assumed to be homogeneous (per layer) and isotropic and described by the Mohr-Coulomb model. In general, fissures and minor faults could not be taken into account. However two minor slip faults were taken into account and modelled by interface elements as indicated by the dotted lines in Fig. 7.

Although the bilinear elastic plastic Mohr-Coulomb model is often insufficiently accurate for soils, it usually provides acceptable results for stable rocks. The only major simplification with respect to sedimentary rock problems is the assumption of isotropic behaviour. In reality sedimentary rock is characterised by its geological genesis leading to anisotropy. As a result of its former loads the rock type known as lias alpha is horizontally prestressed. This is taken into account in the analyses by selecting K_0-values well beyond unity.

The calculated deformations indicate that the horizontal prestressing results in a considerable convergence of the tunnel sides. The rock beneath the tunnel relaxes and expands into the excavated cross section. At the right hand side of the tunnel (Fig. 8) the horizontal shift involves the entire section up to the surface since the slope offers no resistance to this motion. At the left hand side of the tunnel the direction of the shift changes across the slip joint. At this interface the maximum shear resistance is reached and the upper and lower layers move in different directions.

In Fig. 8 the measured and calculated deformations are compared. The 2D analysis is done with ß-factors between 0 and 1. The diagrams show, that this ß-variation hardly influences the magnitude of deformations. The best conformity with the 3D analysis is obtained with ß = 0.4. On the basis of this ß-value the K_0-values are varied between 2 and 3 in the lower layers and 4 and 5 in the upper layers (Fig. 9). The differentiation of K_0 over the depth should take into account that the effect of prestressing reduces with depth.

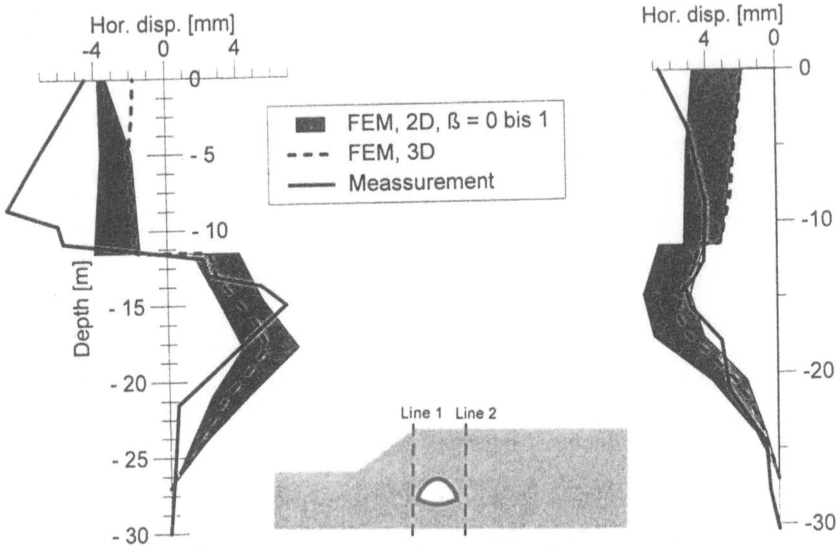

Fig. 8: Comparison between measured and calculated horizontal deformation

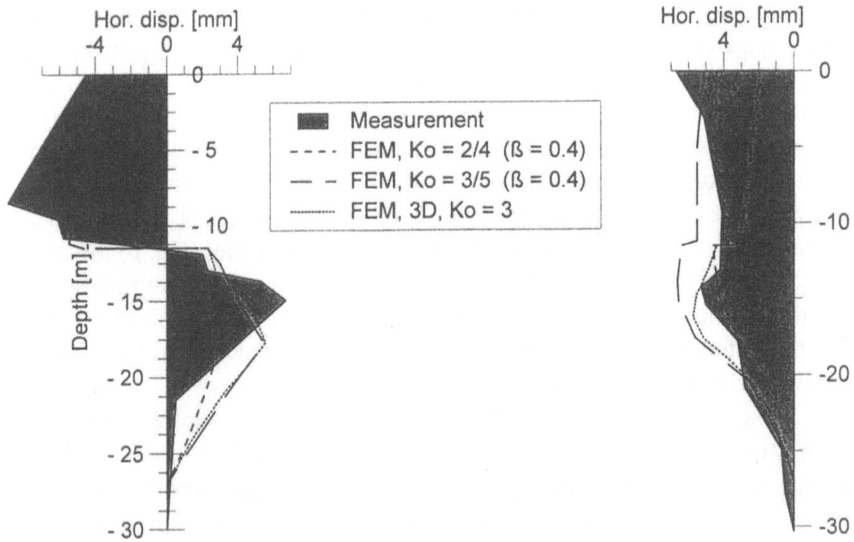

Fig. 9: Horizontal deformations with different K_o-values

5. CONCLUSIONS

For FE-analyses of tunnels 2D calculations are usual. Arching in front of the tunnel head is taken into account with the so-called ß-factor. The determination of the ß-value is based on experience and measurements. The basic drawback of the 2D ß-method is its full empirical character. Moreover one obtains surface settlements and lining loads for the final stage without any information for intermediate construction stages.

Only by means of 3D analyses stress states due to the tunnel excavation can be modelled realistically. The influence of various section lengths, excavation schemes or of horizontal initial stresses can be analysed and in this way the quality of the deformation prediction is enhanced. By an example in sedimantary clay rock the relation between the ß-factor (2D) and the section length (3D) is shown.

Furthermore the results of 3D analyses show the development of the settlements over the excavation process. By early deformation measurements the model can be calibrated and further deformations can be estimated more reliably.

A comparison of the calculated and measured deformations highlights the power of finite element analyses. However, the central prerequisite for FEM is a thorough knowledge of the geological situation and the use of a suitable constitutive model.

The case study points out, that the precise modelling of geological conditions like horizontal stresses and slip joints improves the quality of computational results. However, the convergence of the tunnel due to the excavation is not very much influenced by the section length and the excavation scheme when the stiffness of the rock is high compared to the stiffness of the lining. Hence the present results are not very sensitive to the ß-factor. In this particular case final settlements can well assessed by 2D analyses. However even in this special case the 3D analysis has the advantage that it also yields information on the stability of the tunnel face.

REFERENCES

HAUCK, C.-D. (1997): Die Ostumfahrung von Stuttgart-Vaihingen, 1. Kolloq. "Bauen in Boden und Fels" der TAE, Ostfildern

LANDESHAUPTSTADT STUTTGART (1996): Amtsblatt-Sonderdruck "Ostumfahrung Vaihingen"

MAIDL, B. (1988): Handbuch des Tunnel- und Stollenbaus, Glückaufverlag, Essen.

VERMEER, P. A., BRINKGREVE, R.B.J. (1995): PLAXIS - Finite Element Code for Soil and Rock Analyses. A.A. Balkema/Rotterdam/Brockfield

WITTKE, W. /Beratende Ingenieure für Grundbau und Felsbau GmbH (1996): Geotechnischer Bericht zum Erkundungsprogramm des Österfeldtunnels (in extracts)

2. LABORATORY TEST AND CONSTITUTIVE MODELLING

EFFECT OF STRAIN-SOFTENING IN INTERPRETATION OF LABORATORY COMPRESSION TESTS

L. Andresen and H.P. Jostad
Norwegian Geotechnical Institute, Oslo, Norway

ABSTRACT

The interpretation of post-peak stress-strain relationship from laboratory compression tests performed on strain-softening clay specimens under undrained condition is considered. Since failure of a strain-softening material is accompanied by strain localization, the post-peak stress and strain distributions will be highly inhomogeneous. Standard interpretation techniques will then overestimate the rate of softening significantly, i.e. the steepness of the post-peak stress-strain curve is overestimated. Finite element analyses of a biaxial compression test are used to demonstrate this effect by simulation. The effect of strain-softening is accounted for by an elasto-plastic strain softening material model within the concept of total stresses, using the Tresca yield criterion. Some comments are also made concerning the severe mesh dependency associated with the use of a strain-softening material model.

1. INTRODUCTION

Materials like rock, soil and concrete can exhibit strain-softening during shear deformation. Material strain-softening is defined herein as a constitutive response in which the stress after a peak value, decreases with further increase in strain.

Strain-softening behaviour has been observed in laboratory compression tests under undrained conditions for Norwegian quick or sensitive clays, as reported by Lacasse et. al. (1985), but also for relatively insensitive, normally consolidated clay such as the Boston blue clay (Ladd, 1971) and Chicago glacial clay (Rhee, 1991).

Failure of a body with strain-softening soil is usually accompanied by the forming of shear-bands, i.e. narrow bands in which the shear strain localizes. The forming of shear-bands is a result of material instability. Various patterns of shear-bands, indicating different

EFFECT OF STRAIN-SOFTENING IN INTERPRETATION OF LABORATORY COMPRESSION TESTS

deformation modes, can be seen from "end-of-test" photos of specimens of strain-softening clay which have been subjected to triaxial or biaxial (plane-strain) compression tests.

In biaxial tests, using special testing equipment such as internal strain measurement (Rhee, 1991) or free moving base (Lizcano et. al., 1997), the initiation and growth of shear-bands can also be studied during the test. Such studies have shown that the shear-bands develop rapidly when the peak strength is reached, leading to an inhomogeneous distribution of stresses and strains in the specimen.

It is believed that in some laboratory tests, the rate of softening, i.e. the steepness of the post peak load-displacement curve, can be significantly overestimated due to geometrical softening. Geometrical softening is a consequence of inhomogeneous deformation modes such as localization. Therefore strain-softening, as observed in laboratory tests which post-peak exhibit inhomogeneous deformations, can not be regarded as a material property and caution must be taken when incorporating strain-softening characteristics interpreted from such tests into constitutive models used in numerical analyses.

In this paper a finite element analysis will be presented which demonstrates the effect of strain localization in a saturated clay specimen by simulation of a biaxial test under undrained conditions. The FEM program BIFURC developed at the Norwegian Geotechnical Institute is used for this purpose.

2. MATERIAL MODEL

Strain-softening is accounted for by an elasto-plastic strain softening material model within the concept of total stresses, using the Tresca yield criterion.

The hardening-softening curve in Figure 1 is used by BIFURC. The hardening part of the curve, which originally was proposed by Vermeer and de Borst (1984) is defined through the function

$$\tau = \frac{2 \cdot \tau_p \cdot \sqrt{\gamma_p \cdot \gamma_p^p}}{\gamma_p + \gamma_p^p} \qquad [1]$$

Here τ and τ_p denotes shear stress and peak shear strength respectively, while γ_p and $\gamma_p{}^p$ denotes plastic shear strain and its value at peak strength. The softening part of the curve is defined by a cosine function

$$\tau = \tau_p - 0.5 \cdot [1 - \cos\left(\pi \cdot \frac{\gamma_p - \gamma_p^p}{\gamma_p^r - \gamma_p^p}\right) \cdot (1 - \frac{\tau_r}{\tau_p}) \cdot \tau_p \qquad [2]$$

EFFECT OF STRAIN-SOFTENING IN INTERPRETATION OF LABORATORY COMPRESSION TESTS

in which τ_r/τ_p is the residual shear strength ratio and γ_p^r is the minimum plastic shear strain at residual strength. This curve provides a smooth transition between hardening and softening with a peak value defined by $d\tau/d\gamma = 0$.

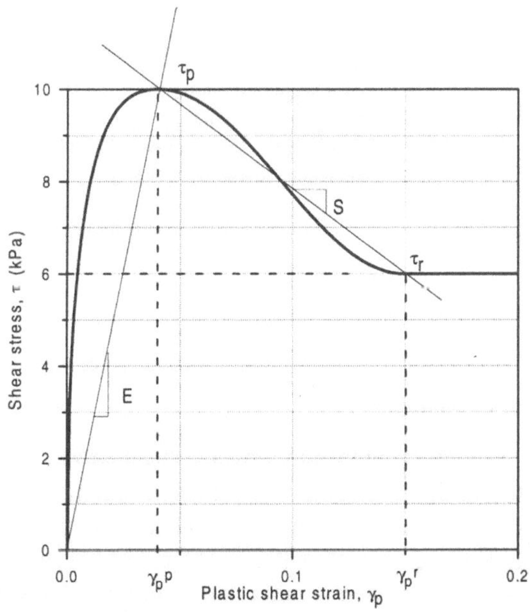

Figure 1 Hardening-softening stress-strain curve used in the analyses

Elastic shear strains are calculated using a constant shear modulus, G, both for loading and unloading. G is generally such large that the elastic shear strains are small compared to the plastic shear strains, i.e. $\gamma_e = \tau/G << \gamma_p$. The softening behaviour can then be characterized by the drop of shear stress from peak to residual, τ_p-τ_r, and the strain-softening ratio S/E.

3. FEM-MODELLING OF BIAXIAL TEST

Figure 2 shows the model of the biaxial test with the material parameters for the strain-softening clay. The clay is modelled as a weightless, isotropic, homogeneous material. Undrained behaviour for the saturated clay is modelled using a bulk modulus $K = 100 \cdot G$, which gives virtually incompressible conditions. Small strain theory is used and inertia effects are not considered. Loading platens at top and bottom are modelled by using very stiff elements. The bottom loading platen is fully fixed while the top platen is only restrained in the horizontal direction. Full friction between the loading platens and the specimen is adopted.

EFFECT OF STRAIN-SOFTENING IN INTERPRETATION OF LABORATORY COMPRESSION TESTS

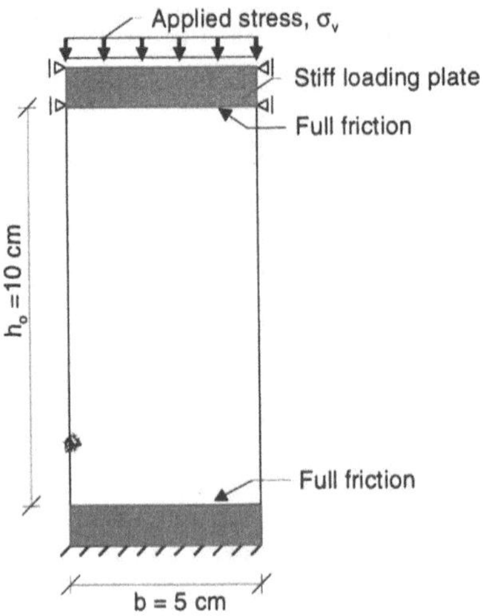

Elastic shear modulus	$G = 1000 \cdot \tau_p$
Bulk modulus	$K = 100 \cdot G$
Peak shear strength	$\tau_p = 10$ kPa
Residual shear strength ratio	$\tau_r/\tau_p = 0.6$
Plastic shear strain at peak strength	$\gamma_p^p = 0.04$ (4 %)
Minimum plastic shear strength at residual strength	$\gamma_p^r = 0.15$ (15 %)

Figure 2 Model of biaxial test

The simulation of the test was done by applying a vertical stress, σ_v, at the top loading platen and then monitoring the calculated vertical displacement of the specimen. An incrementation procedure (Jostad, 1993) based on the arc-length method (Crisfield, 1982) was used to control the loading. Incrementation of the vertical stress was started from a stress- and strain free state. The stress-strain relationship was calculated from the load-displacement curve through the following equations:

$$\tau = \sigma_v/2$$
$$\gamma = 2 \cdot \delta h/h_o \qquad [3]$$

where τ, γ denote, respectively, the maximum shear stress and shear strain, and δh, h_o are the vertical displacement and the initial height of the specimen. This is in accordance with the common practise used for interpretation of stress-strain data from laboratory tests.

The basis analysis was performed with 800 square, eight noded Serendipity elements with reduced integration and material parameters as shown in Figure 2.

It is, however, known that the conventional finite element method is not objective with regard to element size, type and orientation when a strain-softening material model is used. The effect of mesh dependency leads to great differences in post-peak behaviour for different meshes as shown by Pietruszczak and Stolle (1985). The reason for this is that the material models involving strain-softening lacks a geometrical scale factor, i.e. the shear band which develops in the standard finite element solution tends to concentrate in the narrowest shear-bands that can be resolved by the mesh (1-3 elements depending on orientation) whereas experiments show a distinct relationship with a micro-structural length scale. In order to investigate this effect, a second analysis was performed with 200 elements, thus doubling the element size but keeping the material parameters constant.

EFFECT OF STRAIN-SOFTENING IN INTERPRETATION OF LABORATORY COMPRESSION TESTS

Remedial actions can be taken to reduce the problem of mesh dependency. The softening-scaling approach proposed by Pietruszczak and Mróz (1981) is one such method. In this method the rate of softening of the input stress-strain curve is related to the finite elements size in such a way that the ratio between the rate of softening and element size is kept constant. To study this method, a third analysis was performed with 200 elements, but with the minimum plastic shear strain at residual, γ_p^r, changed such that the softening ratio, S/E, was doubled. This keeps the ratio between rate of softening of the input stress-strain curve and the finite element size equal to what it was in the basis analysis.

4. RESULTS AND DISCUSSION

In Figure 3 the stress-strain curve from the basis analysis is compared with the "true" constitutive relation (input stress-strain curve). It can be seen that the imposed stress-strain curve is not reproduced when the stress and strain are calculated from the load-displacement data. The curves match fairly well for pre-peak deformations but in the post-peak regime the backcalculated curve shows a much higher rate of softening. This behaviour is due to the development of shear-bands as can be seen from Figure 4b. The development of shear-bands leads to highly inhomogeneous distributions of stresses and strains in the post-peak regime.

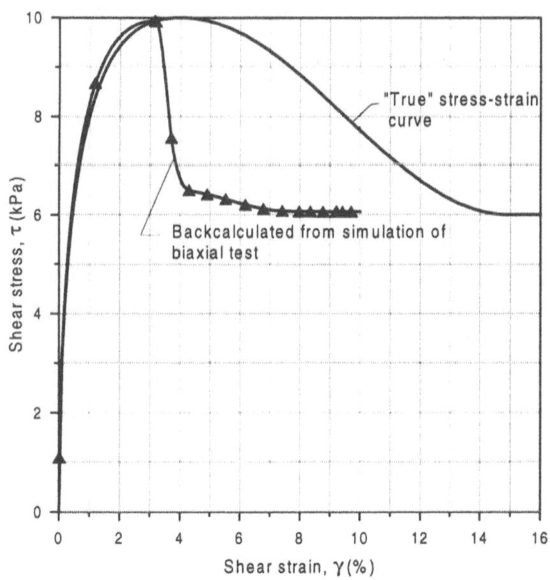

Figure 3 Stress-strain curve interpreted from calculated load-displacement curve shown together with the "true" stress-strain curve

EFFECT OF STRAIN-SOFTENING IN INTERPRETATION OF LABORATORY COMPRESSION TESTS

The distribution of plastic shear-strain in the specimen at peak state, i.e. the state where the specimen carries the maximum load, is shown in Figure 4a with the deformation magnified 8 times. The friction between the end platens and the specimen causes some inhomogeneous distribution of deformation also in the pre-peak regime. At peak state, the plastic-shear strain varies between less than 3 % to slightly greater than 4.7 %, with the highest values in the centre of the specimen and the lowest values concentrated near the end platens. This causes the specimen to react slightly stiffer in the pre-peak regime compared to the "true" stress-strain curve, as can be seen in Figure 3.

At the peak state, parts of the specimen has started to soften, as the plastic-shear strains in the centre of the specimen have slightly passed the limit of 4 % which corresponds to peak strength. In the parts of the eventual rupture surface which are near the free boundaries, with plastic shear strains less than 4 %, the maximum shear stresses have not yet reached peak strength. Thus the average shear resistance of the specimen at failure is slightly less than the peak strength. The specimen is at this point in an unstable equilibrium state, and attempts to increase the load leads to rapid progressive development of two shear bands in which the plastic shear-strain localizes. The shear-stress inside the band then reduces with increasing shear-strain towards the residual shear strength of 6 kPa which is reached at a plastic shear-strain of 15 %. The areas outside of the shear-bands become elastically unloaded and the applied stress must be reduced to maintain equilibrium.

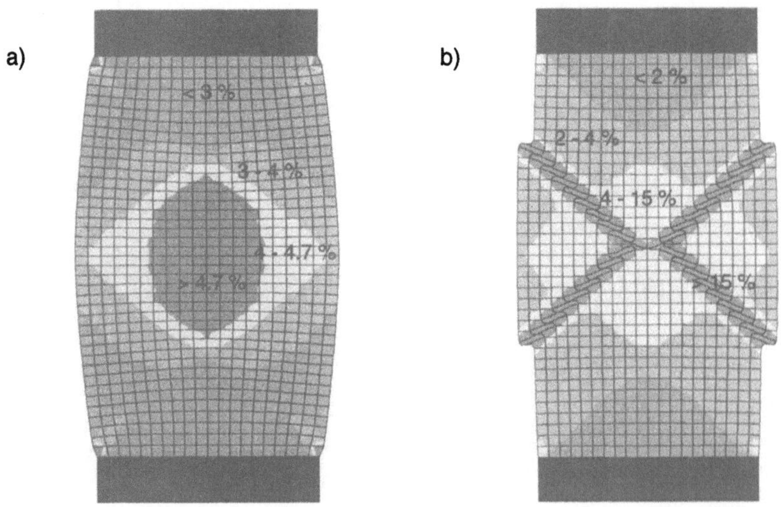

Figure 4 Contours of plastic shear strain plotted on the basis mesh for a) the peak state and b) the residual state

EFFECT OF STRAIN-SOFTENING IN INTERPRETATION OF LABORATORY COMPRESSION TESTS

As mentioned, the simulation slightly underestimates the peak strength, and it can be seen from Figure 3 that the shear strain corresponding to peak strength is 3.0 % compared to 4.0 % for the "true" curve. The "peak strength" obtained in the calculations was 9.93 kPa compared to the peak input strength of 10 kPa. This is due to the progressive failure phenomena as the soil body collapses due to material instability before a complete rupture surface has formed.

It must be pointed out that for a load controlled situation, e.g. construction of an embankment in the field, the descending part of the load-settlement curve is not physical as the soil body experiences a rapid collapse at peak state. The effect of softening will then be balanced by inertia effects, i.e. acceleration of the soil mass which takes part during the failure, or strain rate effects.

Figure 4b shows the distribution of plastic shear-strain at the ultimate state, i.e. the state where the specimen carries the residual load, with the deformed mesh truly magnified. The two shear bands have fully developed and intersect at the centre of the specimen. The same type of deformation mode is often seen from "end-of-test" photos from undrained triaxial or biaxial compression tests performed with rough end platens.

Figure 5 shows the resulting load-displacement curve for all the three analyses performed. The analysis done with doubled element size but unchanged material parameters is first compared with the results from the basis analysis. It can be seen that the post-peak branches of the two calculations are very different. The effect of coarser mesh is a calculated lower rate of softening.

The distribution of plastic shear strains at peak- and residual state are shown on Figure 6. Comparing Figure 6 and 4 it can be seen that in both calculations the shear-band involves three elements. Thus, doubling the element size causes the shear-band thickness also to be doubled. Neither of the two calculations predict the correct shear-band thickness as experiments show that in a clay material such a thickness is invisible and the shear-band is viewed as a slip surface.

The analysis performed with 200 elements and with the minimum plastic shear strain at residual, γ_p^P, changed such that the rate of softening S/E was doubled is now compared with the basis analysis.

It can be seen from Figure 5 that the rate of softening from this calculation is almost equal to what was obtained in the basis calculation. The post peak branch from the new calculation is, however, shifted to the left. This is because the new calculation reaches peak state for a slightly lower load compared to the basis calculation. It is reason to believe that even if it is a linear relationship between the input rate of softening and the element size with respect to predicting the post peak branch of the load-displacement curve, the same may not be true for predicting the peak load.

EFFECT OF STRAIN-SOFTENING IN INTERPRETATION OF LABORATORY COMPRESSION TESTS

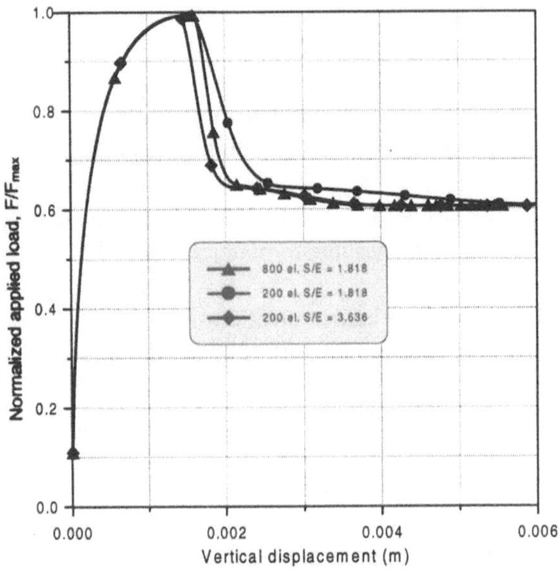

Figure 5 Load-displacement curves from numerical simulation of biaxial test

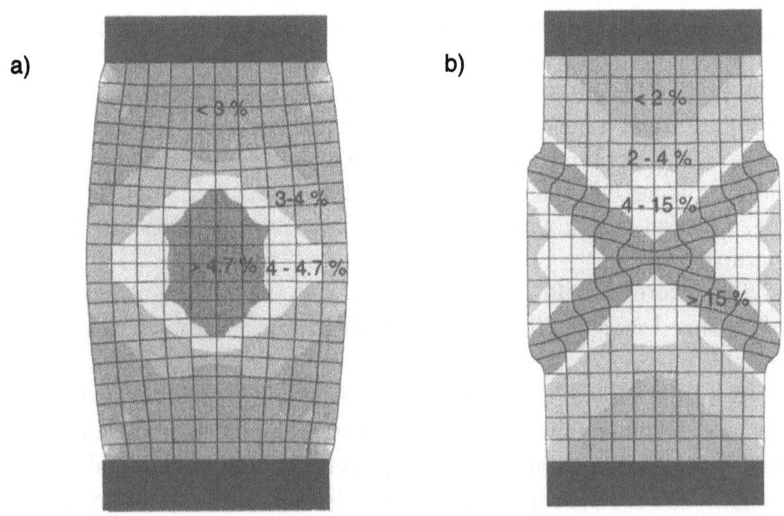

Figure 6 Contours of plastic shear strain plotted on the mesh consisting of 200 elements for a) the peak state and b) the residual state

EFFECT OF STRAIN-SOFTENING IN INTERPRETATION OF LABORATORY COMPRESSION TESTS

5. CONCLUSIONS

Finite element analyses which study the effect of strain localization in an undrained plane strain compression test are presented. The analyses demonstrate that:

- A post-peak deformation mode involving localized shear deformation, as often seen in "end-of-test" photos from laboratory, is successfully calculated by the finite element method.

- The localized deformation is caused by material strain-softening together with friction between end platens and the specimen.

- The highly inhomogeneous distribution of stresses and strains occurring in the post peak regime lead to unreliable stress-strain data if standard interpretation methods are used, e.g. the rate of strain-softening can be significantly overestimated.

- Geometry and boundary conditions in the laboratory compression tests will for moderate strain-softening not significantly influence the interpretation of the peak strength.

- The effect of the softening-scaling method gives good results in form of nearly mesh (or shear-band thickness) independent post-peak load-displacement curve.

- The softening-scaling method gives, however, slightly mesh dependent peak loads. This effect may be more critical for general boundary value problems.

6. REFERENCES

BRINKGREVE, R.B.J. (1994). Geomaterial models and numerical analysis of softening Ph.d. thesis, Delft University of Technology.

CRISFIELD, M.A. (1982). Accelerated solution techniques and concrete cracking. Comp. Meth. in Appl. Mech. and Eng., 33, 585-607

JOSTAD, H.P. (1993). Bifurcation analysis of frictional materials. Ph.d. thesis, University of Trondheim, Norwegian Institute of Technology

LACASSE S., T. BERRE & G. LEFEBVRE (1985). Block Sampling of Sensitive Clays. XI Int. Conf. on Soil Mech. and Found. Eng., San Francisco, Vol. 2: 887-892

LADD, C.C. (1971). Strength Parameters and Stress-Strain Behaviour of Saturated Clays. MIT, Dep. of Civil Eng., Research Report R71-23

EFFECT OF STRAIN-SOFTENING IN INTERPRETATION OF LABORATORY COMPRESSION TESTS

LIZCANO A., I. VARDOULAKIS & M. GOLDSCHEIDER (1997). Biaxial tests on normally, anisotropically consolidated clay. Proc. Def. and Prog. Failure in Geomech., IS-NAGOYA'97: 223-228

PIETRUSZCZAK, ST. & D.E.F. STOLLE (1985). Part I: Objectivity of finite element solutions based on conventional strain softening formulations. Computers and Geotechnics 1, 99-115

PIETRUSZCZAK, ST. & Z. MRÓZ (1981). Finite Element Analysis of Deformation of Strain-Softening Materials. Int. J. Num. Meth. Eng., Vol. 17, 327-334, 1981

RHEE, Y. (1991). Experimental evaluation of strain-softening behaviour of normally consolidated Chicago clays in plane strain compression. Ph.D. thesis, Northwestern University, IL, USA, 1991

VERMEER, P.A. & R. DE BORST (1984)
Non-associated plasticity for soil, concrete and rock. HERON 29, N0. 3, pp. 1-64.

SUPER LOADING YIELD SURFACE CONCEPT FOR THE SATURATED STRUCTURED SOILS

A. Asaoka, M. Nakano and T. Noda
Nagoya University, Nagoya, Japan

ABSTRACT: "Superloading surface concept" is newly developed for the description of elasto-plastic behavior of highly structured soils. The superloading surface, which is similar in shape to the original Cam-clay yield surface, is located above the Cam-clay and the size ratio between two yield surfaces gives the degree of the structure of soil skeleton. The evolution law of the size ratio describes the decay of the structure as plastic deformation proceeds, and the superloading surface falls exactly on the Cam-clay yield surface at the end of the completely remolded state. When the soil is reloaded after elastic unloading, the elasto-plastic behavior of the soil is described by the Hashiguchi's subloading surface, which is also similar in shape to the superloading surface. The size ratio between super and subloading surfaces gives the OCR.

The one dimensional consolidation deformation is numerically solved by the use of super subloading Cam-clay model, in which "secondary consolidation" is naturally observed. The reason for which can be found simply because the delayed collapse of the structure of soil skeleton leads the delayed compression/consolidation.

1 INTRODUCTION

Naturally sedimented alluvial/diluvial clay deposits are mostly in a highly structured nature, and thus they tend to exhibit complicated behavior under applied stresses, that is, the softening under considerably low stress ratio, and secondary consolidation. The first objective of this study is to give a simple description of the "structure" of soil skeleton, based on the unconventional theory of plasticity. The evolution law of the state of the structure is established so that the decay of the structure may proceeds as plastic deformation develops.

The second objective of this study is to clarify the mechanism of "secondary compression/consolidation" on the basis of the non-viscous concept. The mechanism is attributed simply to the delayed collapse of the structure, and the delay is solved under soil-water coupled condition with the use of the equation of continuity and the Darcy's law. In this context, the concept of "inherent time dependent nature of soil skeleton" which is a kernel of the every visco-plastic model is shown to be no need.

2. SUPER-SUB LOADING YIELD SURFACE CAMCLAY MODEL

Naturally sedimented alluvial/diluvial clay deposits are mostly in a highly structured nature, and thus they tend to exhibit complicated behavior under applied stresses. "Softening" under considerably small stress ratio is sometimes attributed to the collapse of "soil structure", and "secondary compression," due to the inherent "time dependent nature" of the saturated soil. Such a structured soil, particularly when close to the normally consolidated state, which in fact differ from the normally consolidated state of remolded clay, (see the new definition of OCR), exists in the so-called "impossible state" in q~p'~e space outside the Roscoe boundary surface (Schofiled and Wroth, 1968) of the fully remolded soil. Thus a necessity has arisen for a constitutive model that cater for the existance of such impossible states, yet retaining the critical state concepts.

In the present study, therefore, the elasto-plastic nature of highly structured soil is newly formulated introducing the concept of "superloading yield surface", which is geometrically similar in shape to the original Cam-clay yield surface and envelopes both the original Cam-clay and the Hashiguchi's subloading yield surfaces(*e.g.* Hashiguchi and Ueno, 1977). Here firstly we present a short review of the yield function of the original Cam-clay model in order to derive the super-subloading surface Cam-clay constitutive relation in the proceeding section.

(1) Yield function of the original Cam-clay model for the remolded soil

Denoting the elastic and plastic components of stretching $\boldsymbol{D}$ by $\boldsymbol{D}^e$ and $\boldsymbol{D}^p$ respectively, it is first assumed that $\boldsymbol{D}$ has the decomposition as, $\boldsymbol{D} = \boldsymbol{D}^e + \boldsymbol{D}^p$. The total volume change from reference time $t=0$ to current time t of the soil skeleton per unit volume is expressed by:

$$\int_0^t J\mathrm{tr}\boldsymbol{D}d\tau = \int_0^t J\mathrm{tr}\boldsymbol{D}^e d\tau + \int_0^t J\mathrm{tr}\boldsymbol{D}^p d\tau, \quad J = \det \boldsymbol{F} = \frac{1+\mathrm{e}}{1+\mathrm{e}_0} \tag{1}$$

in which 1+e and 1+e_0 are the specific volumes at current (time t) and reference ($t=0$) state, respectively. Introducing a material parameter $\tilde{\kappa}$, the first term on the right side of Eq. (1)[1] and material time derivative form of the equation can be expressed in a stress path independent form as,

$$\int_0^t J\mathrm{tr}\boldsymbol{D}^e d\tau = -\frac{\tilde{\kappa}}{1+e_0}\ln\frac{p'}{p'_0}, \qquad \mathrm{tr}\boldsymbol{D}^e = -\frac{\tilde{\kappa}}{1+e}\frac{\dot{p}'}{p'} \tag{2}$$

in which p' and p_0' express the mean effective stress ($-\mathrm{tr}\boldsymbol{T}'/3$) at current and reference state, respectively. $\boldsymbol{T}'$ denotes the Cauchy effective stress tensor in which tensile components are defined positive. The $\tilde{\kappa}$ is the so called swelling index. When isotropic assumption is introduced in addition, the non-linear elastic constitutive relation of the soil skeleton, the following equation is uniquely determined from Eq. (2)[2].

$$\overset{\circ}{\boldsymbol{T}}' = \boldsymbol{E}\boldsymbol{D}^e = \left(\tilde{K} - \frac{2}{3}\tilde{G}\right)(\mathrm{tr}\boldsymbol{D}^e)\boldsymbol{I} + 2\tilde{G}\boldsymbol{D}^e, \quad \tilde{K} = \frac{J(1+e_0)}{\tilde{\kappa}}p', \ \tilde{G} = \frac{3(1-2\nu)}{2(1+\nu)}\tilde{K} \tag{3}$$

In this equation $\overset{\circ}{\boldsymbol{T}}'$ denotes Green-Naghdi (1965) 's objective Cauchy effective stress rate (Dienes, 1979), given by $\overset{\circ}{\boldsymbol{T}}' = \dot{\boldsymbol{T}}' + \boldsymbol{T}'\boldsymbol{\Omega} - \boldsymbol{\Omega}\boldsymbol{T}'$, ($\boldsymbol{\Omega} = \dot{\boldsymbol{R}}\boldsymbol{R}^T$, $\boldsymbol{R}$: rotation tensor), $\boldsymbol{E}$ denotes the elastic modulus tensor and ν denotes Poisson's ratio.

Based on Henkel (1960)'s investigations the total volume change of a soil skeleton should have stress path independent nature and therefore it is a function of only the initial and current effective stresses. This is expressed here as the sum of the isotropic compression term and the term due to dilatancy (Ohta, 1971), *i.e.*

$$\int_0^t J\mathrm{tr}\boldsymbol{D}d\tau = -\frac{\tilde{\lambda}}{1+e_0}\ln\frac{p'}{p'_0} - \frac{\tilde{\lambda}-\tilde{\kappa}}{M(1+e_0)}\frac{q}{p'}, \qquad q = \sqrt{\frac{3}{2}\boldsymbol{S}\cdot\boldsymbol{S}}, \qquad \boldsymbol{S} = \boldsymbol{T}' + p'\boldsymbol{I} \tag{4}$$

where $\tilde{\lambda}$ in Eq. (4)[1] is the compression index and, $D = (\tilde{\lambda}-\tilde{\kappa})/M/(1+e_0)$ is the so-called "coefficient of dilatancy". The term, $\int_0^t J\mathrm{tr}\boldsymbol{D}^e d\tau$ of Eq.(2)[1] is stress path independent. Therefore, by subtracting Eq. (2)[1] from Eq. (4)[1], it can be determined that the second term on the right side of Eq. (1)[1] also becomes stress path independent.

$$F = f(q,p',p'_0) + \int_0^t J\mathrm{tr}\boldsymbol{D}^p d\tau = MD\ln\frac{p'}{p'_0} + D\frac{q}{p'} + \int_0^t J\mathrm{tr}\boldsymbol{D}^p d\tau = 0 \tag{5}$$

Based on the theory of plasticity, F in Eq. (5) should be interpreted as a yield function while $-\int_0^t J\mathrm{tr}\boldsymbol{D}^p d\tau$ is the hardening/softening parameter of the plastic deformation history of the soil skeleton. Eq. (5) is known as the original Cam-clay yield function.

It should be stated here that the term $-\int_0^t J\mathrm{tr}\boldsymbol{D}^p d\tau$ is an approximation of the value of plastic volumetric strain. The Cam-clay model introduced here for the use in finite deformations, therefore, is still an approximation of the original Cam-clay model that was initially developed for infinitesimal strain problems.

(2)Fully remolded soil (non-structured soil) and structured soil

In order to describe the behavior of structured soil which can exist in the impossible state, as mentioned above, the superloading yield surface is newly introduced. This yield surface exists (Fig.1) outside the (original) Cam-clay yield surface. These origin-convex two surfaces are similar in shape with the origin in p' - q space as the center of similarity. When the current stress point (p', q) of the structured soil is on the superloading yield surface, taking the stress point (p'^*, q^*) be on the original Cam-clay yield surface satisfying the relationships $q^*=R^*q$ and $p'^*=R^*p'$ (Fig.1), the following can be derived:

$$F^* = f(q^*,p'^*,p'_0) + \int_0^t J\mathrm{tr}\boldsymbol{D}^p d\tau = 0$$

(Original Cam-clay yield function) (6)

$$F = f(q,p',p'_0) + \int_0^t J\mathrm{tr}\boldsymbol{D}^p d\tau + \mathrm{MD}\ln R^* = 0$$

(Superloading yield function) (7)

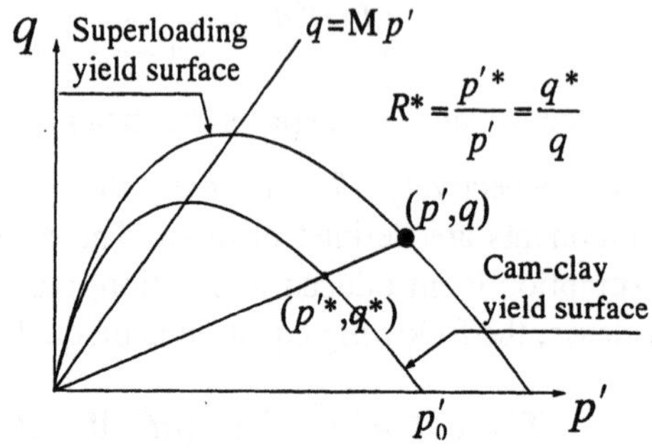

Fig.1 Superloading yield surface for structured soil

In the derivation of these equations, (p', q) of Eq.(5) is taken as (p'^*, q^*). In the process that the soil loses its structure, the superloading yield state approaches the fully remolded state with the plastic deformation. Accordingly the similarity ratio R^* ($0 < R^* \le 1$), which is the index of degree of structure, gradually approaches to 1. When the soil perfectly loses its structure, *i.e.* the two yield surfaces become identical, the soil behaves as the remolded Cam-clay. In the present paper the evolution law of R^*, which is related only to the current plastic deformation rate is, for simplicity, given as:

$$\dot{R}^* = JU_s^* \left\| \boldsymbol{D}_s^p \right\| \ , \ U_s^* = R^*(1-R^{*v_1})/\mathrm{D} \ (>0) \ , \ \boldsymbol{D}_s^p = \boldsymbol{D}^p - (\mathrm{tr}\boldsymbol{D}^p)\boldsymbol{I}/3 \qquad (8)$$

in which v_1 denotes a positive scalar parameter and $\| \ \|$ denotes the Euclidean norm.

(3) Normally consolidated and overconsolidated states

Usually the term “normally consolidated state” of a soil refers to remolded clay, and thus we retain on the same ideology in the proceeding formulation, but with little modification to the term as “remolded normaly consolidated state”, just to distinguish it from the newly defined normally consolidated state of a structured soil. In fact, herein we define the “normally consolidated state” to be the state when stress point lies on the superloading surface and continue to use this term unless otherwise specified. In natural state, however, most of the soils are not in normally consolidated state but in overconsolidated states being due to a result of one or more natural processes such as cementation surcharge removal, water table fluctuations etc. In order to describe the behavior of such overconsolidated states of soils, Hashiguchi’s subloading yield surface concept is applied to the superloading yield surface.

Using the symbols given in Fig.2 and considering the geometric similarity of the superloading yield surface and the subloading yield surface are, the relationships $q = R\bar{q}$, $p' = R\bar{p}'$ and $q^* = R^*\bar{q}$, $p'^* = R^*\bar{p}'$ can be derived. Here (p', q) of Eq.(7) or in Fig.1 is taken as ($\bar{p}'$, $\bar{q}$). R ($0 < \bar{R} \le 1$) is called the "*normal yield subloading surface size ratio, NSR* "(Hashiguchi, 1989) which is the reciprocal of the overconsolidation ratio, OCR. Utilizing the relationships in Eq. (7) the superloading yield function and the subloading yield function are derived respectively:

$$\bar{F} = f(\bar{q}, \bar{p}', p'_0) + \int_0^t J\mathrm{tr}\boldsymbol{D}^p d\tau + \mathrm{MD}\ln R^* = 0$$

(Superloading yield function) (9)

$$F = f(q, p', p'_0) + \int_0^t J\mathrm{tr}\boldsymbol{D}^p d\tau + \mathrm{MD}\ln R^* - \mathrm{MD}\ln R = 0$$

(Subloading yield function) (10)

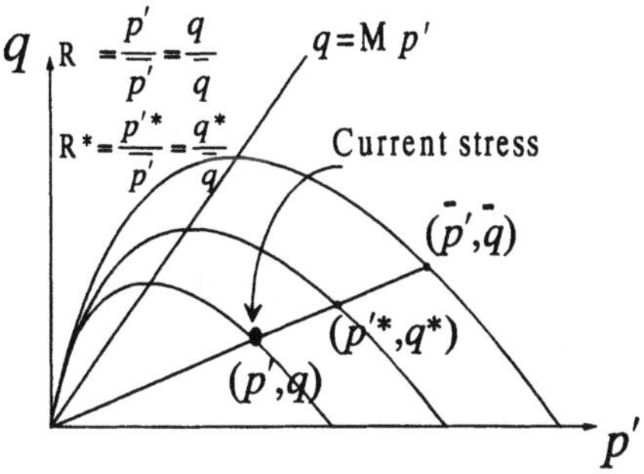

Fig.2 State of "overconsolidated" and structured soil

When the overconsolidated soil on the subloading yield surface is under the loading condition, the overconsolidated state approaches the normally consolidated state, *i.e.* $R \to 1$. On the other hand, if the stress state moves towards inside of the superloading yield surface with elastic response (*i.e.* unloading), the subloading yield surface contracts with current state. Following Hashiguchi's subloading yield surface concept and considering the distinctive evolution law for yield surfaces, in the present formulation R is assumed to satisfy the following simplest form as the evolution law for R :

$$\dot{R} = JU_s\|\boldsymbol{D}_s{}^p\| \ , \ U_s = -\nu_2 \ln R/\mathrm{D} \ (>0) \tag{11}$$

in which ν_2 denotes a positive scalar parameter.

(4) Plastic multiplier and constitutive relation

The current stress point is considered here to be on the subloading yield surface shown in Eq.(10). The associated flow rule for the subloading yield surface Cam-clay, then gives the plastic stretching as:

$$\boldsymbol{D}^p = \lambda \frac{\partial f}{\partial \boldsymbol{T}'} \quad (\lambda > 0) \tag{12}$$

Here, the plastic multiplier λ is determined based on the consistency condition of the yield function, taking the material time derivative of Eq.(10) :

$$\frac{\partial f}{\partial \boldsymbol{T}'} \cdot \dot{\boldsymbol{T}}' + J\mathrm{tr}\boldsymbol{D}^p + \mathrm{MD}\frac{\dot{R}^*}{R^*} - \mathrm{MD}\frac{\dot{R}}{R} = 0 \tag{13}$$

The plastic multiplier λ, which comes from the associated flow rule (Eq. (12)) is determined in terms of stresses using Eq. (13), evolution law of R^*(Eq.(8)[1]), R (Eq.(11)[1]) and $\frac{\partial f}{\partial \boldsymbol{T}'} \cdot \dot{\boldsymbol{T}}' = \frac{\partial f}{\partial \boldsymbol{T}'} \cdot \overset{\circ}{\boldsymbol{T}}'$ and $\boldsymbol{D}_s{}^p = \lambda \frac{\partial f}{\partial q} = \lambda \frac{\mathrm{D}}{p'}$ as:

$$\lambda = \frac{\frac{\partial f}{\partial \boldsymbol{T}'} \cdot \overset{\circ}{\boldsymbol{T}}'}{J\frac{\mathrm{D}}{p'}(\mathrm{M}_s p' - q)} \ , \quad \mathrm{M}_s = \mathrm{M}(1 - \frac{\mathrm{D}U_s{}^*}{R^*} + \frac{\mathrm{D}U_s}{R}) \tag{14}$$

The M_s in Eq.(14) is called here the "apparent critical state constant". When the non-linear elastic response, $\overset{\circ}{\boldsymbol{T}}' = \boldsymbol{E}\boldsymbol{D}^e$ (Eq.(3)[1]) is considered, λ can have another representation in terms of stretching,

$$\Lambda(=\lambda) = \frac{\dfrac{\partial f}{\partial \boldsymbol{T}'}\cdot \boldsymbol{ED}}{\dfrac{\partial f}{\partial \boldsymbol{T}'}\cdot \boldsymbol{E}\dfrac{\partial f}{\partial \boldsymbol{T}'} + J\dfrac{\mathrm{D}}{p'}(\mathrm{M}_s p' - q)} \tag{15}$$

Then the linear rate type elasto-plastic constitutive relation of soil at subloading surface is expressed as,

$$\overset{\circ}{\boldsymbol{T}}' = \boldsymbol{ED} - \Lambda \boldsymbol{E}\frac{\partial f}{\partial \boldsymbol{T}'} \tag{16}$$

(5) Loading criterion

The loading condition, $\Lambda = \lambda > 0$, is first considered using Λ in terms of stretching. Since the subloading yield locus f has the same form as the original Cam-clay, when the evolution law of R^* is adopted in the form of Eq.(8), the denominator of Λ always becomes positive. Therefore, for the Λ being positive, it follows during any loading condition that,

$$\frac{\partial f}{\partial \boldsymbol{T}'}\cdot \boldsymbol{ED} > 0 \tag{17}$$

(6) Elasto-plastic behavior of structured overconsolidated soil

The state of loading is again considered by examining the plastic multiplier $\lambda > 0$ (Eq. (14)[1]) in terms of stresses. When a soil is at remolded normally consolidated state with no structure, *i.e.* $R = 1$ and $R^* = 1$, M_s in Eq.(14) coincides with M since $U_s = U_s^* = 0$. Thus the critical state line, $q = \mathrm{M}p'$ is, what is called the watershed (Asaoka et al.,1994) between hardening ($\frac{\partial f}{\partial \boldsymbol{T}'}\cdot\overset{\circ}{\boldsymbol{T}}' > 0$ when $q < \mathrm{M}p'$) and softening ($\frac{\partial f}{\partial \boldsymbol{T}'}\cdot\overset{\circ}{\boldsymbol{T}}' < 0$ when $q > \mathrm{M}p'$) for the soil (as in original Cam-clay). Similarly, on the other hand, when the soil is structured and overconsolidated, $q = \mathrm{M}_s p'$ gives the watershed between hardening and softening. In fact the magnitude of apparent critical state constant M_s can be M_s >=< M and the magnitude depends on the OCR and the degree of the structure. When the collapse of structure of the soil skeleton proceeds ($R^* \to 1$), M_s increases its value. On the other hand the M_s decreases its value as overconsolidated state approaches to normally consolidated state($R \to 1$). This is schematically illustrated in Fig.3. When the soil completely loses its initial structure, and arrives at the normally consolidated state, the soil then becomes the original "Cam-clay", where $\mathrm{M}_s \to \mathrm{M}$.

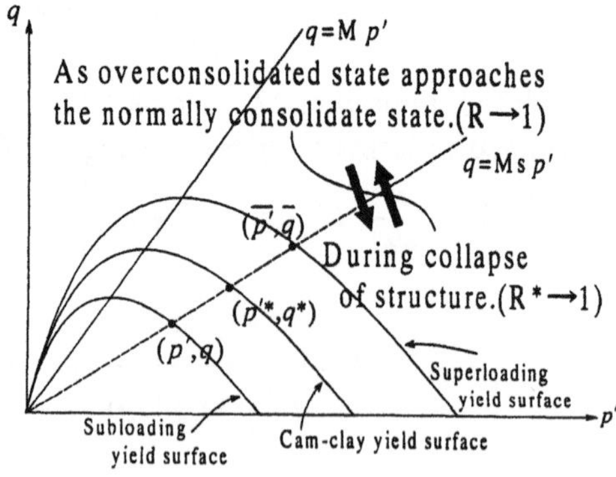

Fig.3 Watershed between hardening and softening for structured soil ($q = \mathrm{M}_s p'$)

Furthermore, the critical state line, $q = \mathrm{M}p'$ is always the watershed between the plastic volume expansion and compression of the soil at any state, since the plastic volume change during loading is given by, $-J\mathrm{tr}\boldsymbol{D}^{p} = -\lambda\frac{\mathrm{D}}{p'^{2}}(q-\mathrm{M}p')$. This suggests that, when $\mathrm{M}_s < \mathrm{M}$, *i.e.* for structured soils, softening behavior under low stress ratios is still found to be possible ($q > \mathrm{M}_s p'$) with plastic volume expansion. When the soil experiences softening under such condition, both the super and the subloading yield surfaces contract in size, while Cam-clay yield surface expands.

3. SOIL-WATER COUPLED COMPUTATION OF ONE-DIMENSIONAL CONSOLIDATION TEST

(1)Constant loading rate consolidation tests

The soil-water coupled one-dimensional consolidation behavior of highly structured and heavily overconsolidated clay specimens were investigated through finite element numerical simulation (Asaoka et al., 1994, 1997). Fig.4 shows the boundary conditions and material parameters of the soil specimen. The specimen was assumed to be initially in isotropic stress condition and uniform and the vertical stress, σ_v was applied at two constant rates: 9.8×10^{-1} and 9.8×10^{-2} kPa/sec.

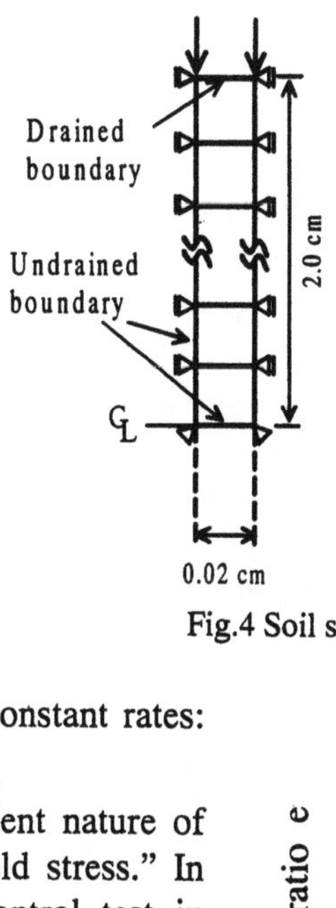

Compression index λ	0.131
Swelling index $\tilde{\kappa}$	0.075
Critical state constant M	1.53
Poisson's ratio ν	0.3
Void ratio on NCL at unit pressure N	1.97
Superloading model constant ν_1	0.5
Subloading model constant ν_2	10
Permeability k (cm/sec)	7.8×10^{-9}
Initial vertical stress σ_{v0} (kPa)	9.8
Initial structure R_0^*	0.05
Initial overconsolidation ratio $1/R_0$	100
Initial void ratio e_0	1.19

Fig.4 Soil specimen for the analysis

Fig.5 shows the rate dependent nature of the so-called "consolidation yield stress." In the figure, the displacement control test in which excess pore pressure did not occur at all are also drawn together. This figure also shows the asymptotic nature of "e~lnσ_v" curve to the completely remolded e~lnσ_v" curve with no time dependency. Figs.6 and 7 illustrate the process of "collapse of structure" and "change to normal consolidated state from overconsolidated state" respectively in

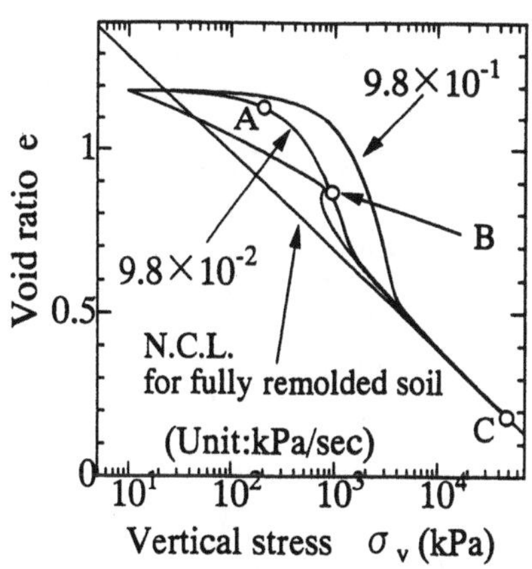

Fig.5 One-dimensional consolidation deformation behavior

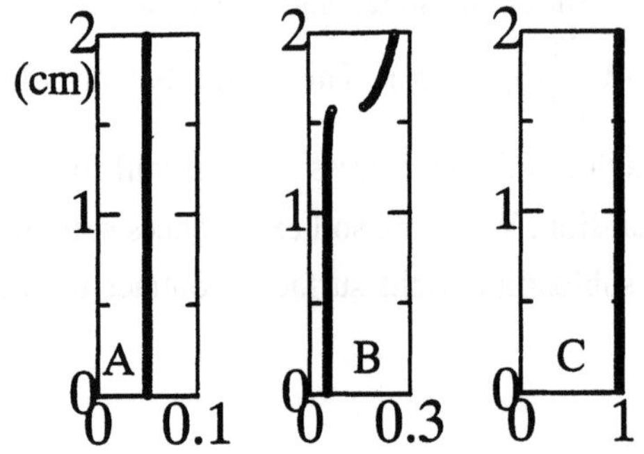

Fig.6 Change of R^* distribution during 1-D loading

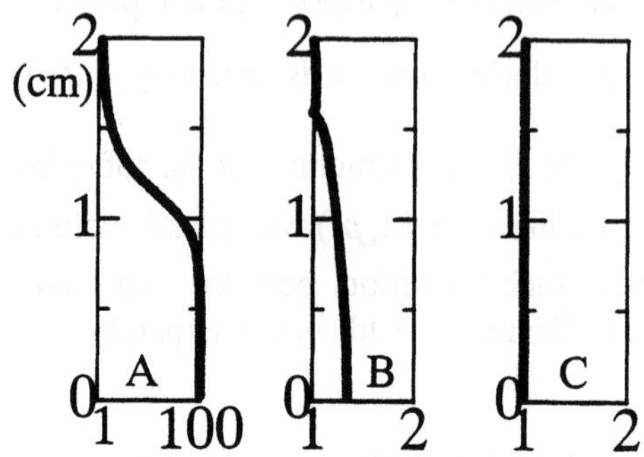

Fig.7 Change of OCR(1/R) distribution during 1-D loading

terms of the vertical distribution of R^* and OCR(1/R) for the loading rate 9.8×10^{-2} kPa/sec. (Plots refer to points A, B and C in Fig.5.)

(2)"Secondary compression/consolidation" of structured soil

Fig.8 shows the one-dimensional consolidation behavior in e~lnσ_v plot under constant load when the applied vertical load was stopped at three stages: *i.e.* at X, Y and Z for the case of 9.8×10^{-1} kPa/sec loading rate. Fig.9 shows the time–settlement behavior from the beginning. As can be seen in Fig.9, a little difference in applied load can cause a significant difference of settlement. In the case of 784.0kPa load, the settlement rate under constant load application is found to be varying, *i.e.* decreasing, increasing and decreasing. The varying settlement rate behavior in this laboratory size specimen is the so-called "secondary compression" as observed for the natural structured soils (*e.g.* see Serge et al., 1985). Figs.10 and 11 show the changing distributions of excess pore pressure and OCR(1/R) with elapse of time (in the 784.0kPa load case). The signs a, b,···, f in these figures correspond to the points a, b,···, f in Fig.9 respectively. Under constant load application, excess pore pressure increased during t = 5000~20000(sec) and then decreased thereafter with the

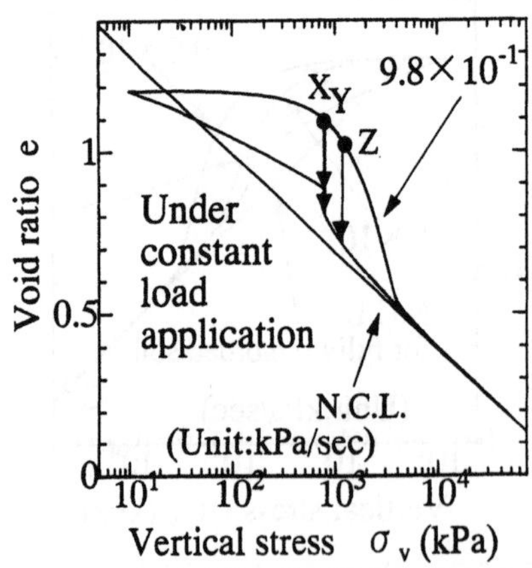

Fig.8 1-D settlement in e~lnσ_v space

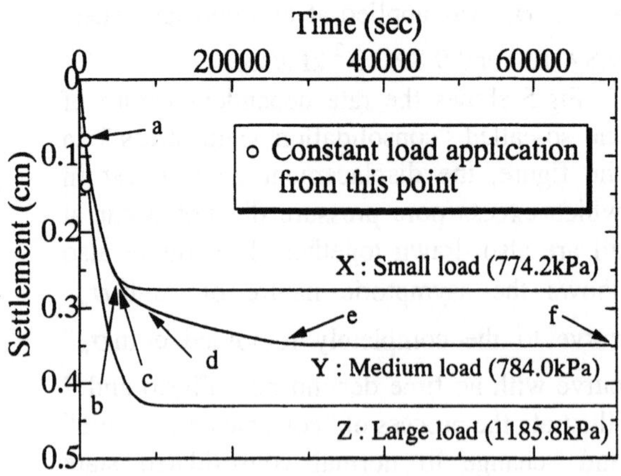

Fig 9 Time-settlement behavior

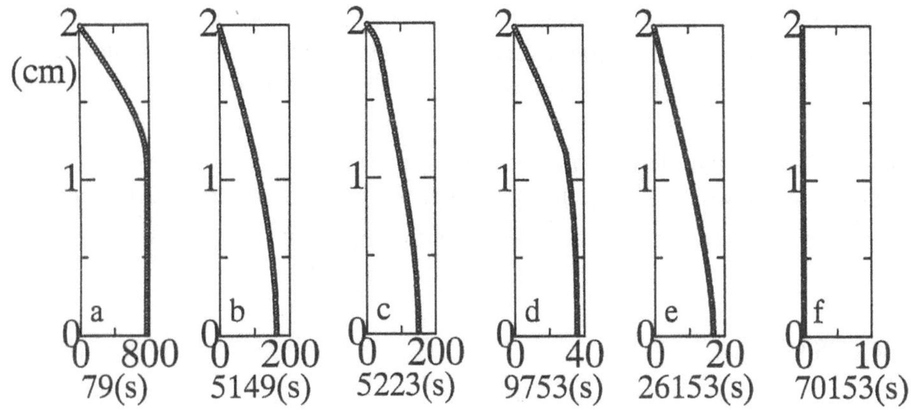

Fig.10 Change of excess pore pressure distribution under constant load application (unit : kPa)

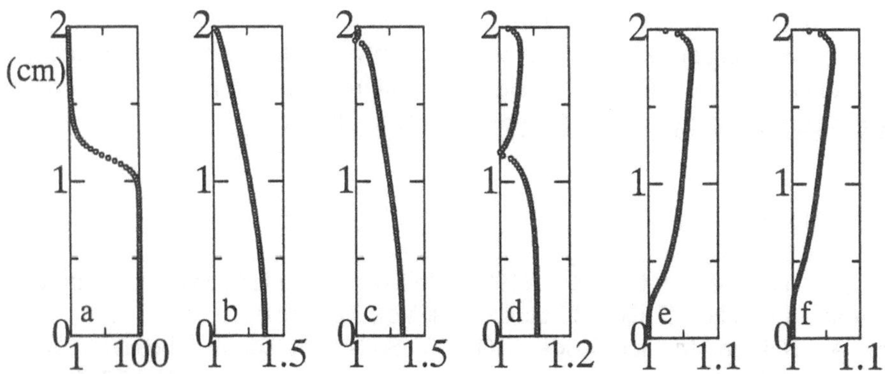

Fig.11 Change of OCR(1/R) distribution under constant load application

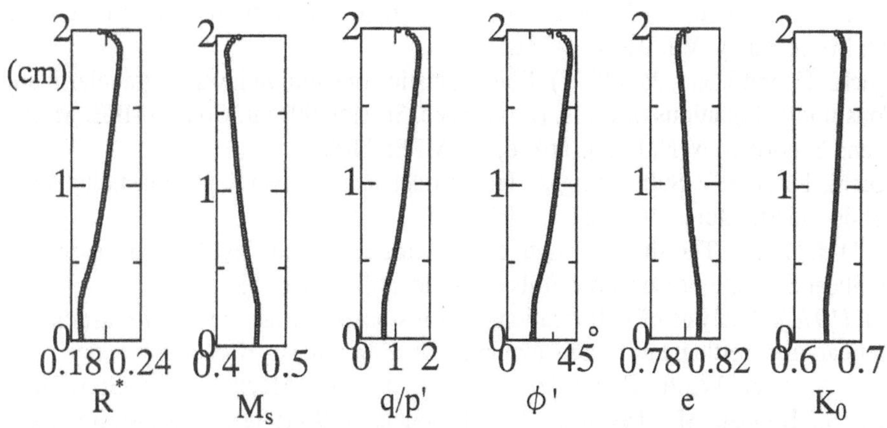

Fig.12 Developed heterogeneity in the soil specimen at the end of 1-D consolidation

OCR value decreasing and increasing repeatedly. This rise of excess pore pressure is attributed to softening due to plastic compression, which is related to the occurrence of the collapse of soil structure, under considerably low stress ratios of highly structured soil (within the region $Mp' > q > M_s p'$). Furthermore this cyclic load (increasing and decreasing OCR) for soil element yielded heterogeneity at the end of consolidation in the soil specimen as shown in Fig.12. In the figure, the distribution of cohesive angle ϕ' is computed from that of M_s and K_0 denotes the earth pressure at rest.

4. CONCLUSIONS

Through this study, the followings are clarified:

(1) When the soil is in a highly structured condition, the plastic softening is still possible even below the critical state line of the remolded soil. The softening is attributed to the collapse of structure of soil skeleton.

(2) Since the softening is possible under a considerably low stress ratio, the softening can be found to occur even under one dimensional compression deformation condition, which triggers the start of "secondary compression/consolidation".

5. REFERENCES

Asaoka, A., Nakano, M. and Noda, T. (1994): Soil-water coupled behavior of saturated clay near/at critical state, Soils and Foundations, Vol. 34: No. 1: 91-105.

Asaoka, A., Nakano, M. and Noda, T.(1997): Soil-water coupled behavior of heavily overconsolidated clay near/at critical state, Soils and Foundations, Vol.37, No.1:13-28.

Dienes, J. K. (1979): On the analysis of rotation and stress rate in deforming bodies, Acta. Mech, Vol. 32: 217-232.

Green, A. E. and Naghdi, P.M. (1965): A General theory of elastic-plastic continuum: Arch. Rat. Mech. Analy, Vol. 18: 251-281.

Hashiguchi, K. and Ueno, M. (1977): Elasto-plastic constitutive laws of granular materials, Constitutive Equations of Soils, (Proc. Spec. Session 9th Int. Conf. SMFE, Murayama, S. and Schofield, A. N. Eds.), Tokyo, JSSMFE: 73-82.

Hashiguchi, K. (1989): Subloading surface model in unconventional plasticity, Int. J. of Solids and Structures, Vol. 25, pp. 917-945.

Henkel, D. J. (1960): The shear strength of saturated remolded clay, Proc. of research Conf. on Shear Strength of Cohesive Soils at Boulder, Colorado: 533-540.

Ohta, H. (1971): Analysis of deformations of soils based on the theory of plasticity and its application to settlement of embankments, Dr. Eng. Thesis, Kyoto University.

Schofield, A. N. and Wroth, C. P. (1968): Critical State Soil Mechanics, McGraw Hill.

Leroueil, S., Kabbaj, M., Tavenas, F. and Bouchard R.(1985): Stress-strain-strain rate relation for the compressibility of sensitive natural clays, Geotechnique, Vol. 35, No.2:159-180.

UNDRAINED CREEP RUPTURE OF NORMALLY CONSOLIDATED CLAY DUE TO BIFURCATION MODE SWITCHING DURING PORE WATER MIGRATION

A. Asaoka, T. Noda, M. Nakano and K. Kaneda
Nagoya University, Nagoya, Japan

ABSTRACT: In the present study, we newly interpret the undrained creep rupture of normally consolidated soil as a result of the change in the initial higher bifurcation mode to lower one during pore water migration within the clay; i.e., "mode switching". The computation is performed not as a single soil element as in the classical soil mechanics, but as a soil-water coupled initial-boundary value problem of a soil specimen, in which only the "inviscid" original Cam-clay model with the subloading surface concept and the Darcy's law are used. As a result, just after the constant applied load, the axial strain is progressing at a very slow rate, however, the strain rate increases immediately when the strain reaches 5.7%. During the time, it is observed that the excess pore pressure tends to be homogeneous and then rises again due to the occurrence of softening in some element.

1 INTRODUCTION

Time dependent behavior is one of the most important characteristic mechanical behaviors of the saturated clays. This behavior has been widely recognized as an inherent mechanical nature of the clay. In order to simulate such time dependent behavior, some elasto *visco-*

plastic constitutive models (e.g., Adachi and Okano, 1974; Sekiguchi, 1977, 1984; Adachi et al., 1987) are developed. However, in these models, the clay is regarded not as a soil specimen with boundary conditions, but as a single soil element retaining the concept of the inherent mechanical nature.

In a previous study (ref. Asaoka et al. 1995), the effect of loading rate on the undrained strength of a remoulded normally consolidated clay has been attributed to the difference in *failure modes* due to pore water migration through triaxial compression test and its numerical simulation. Based on the results, in the present study, the creep rupture, in particular, not of a clay element but of a clay specimen is interpreted by taking notice of *failure modes* due to pore water migration.

2 VARIATION OF UNDRAINED STRNGTH DUE TO THE DIFFERENCE OF FAILURE MODES

2.1 Typical observation in undrained triaxial compression tests

Fig.1 shows the relationship between q_f and $\dot{\varepsilon}_a$ of a series of triaxial compression tests under constant cell pressure on the remoulded normally consolidated Kawasaki clay (Asaoka et al., 1994). The q_f denotes the deviator stress at an axial strain of 17% and $\dot{\varepsilon}_a$, the strain rate of loading. The upper curve in the figure was obtained under partially drained condition with the drainage valve opened, while the lower s-shaped curve was obtained under undrained condition. The shear strengths obtained by both of the tests varied only in the same range of strain rate of $7.0\times10^{-2}\sim1.2$%/min. This result suggests that the difference in the undrained strength between Test A and Test B in the figure should be affected by the migration of pore water. The loading rate is so rapid ($\dot{\varepsilon}_a$ = 4.9%/min) in Test A that almost no migration of pore water should occur during shear, while full migration can be considered in test B($\dot{\varepsilon}_a$ = 9.12×10^{-3} %/min).

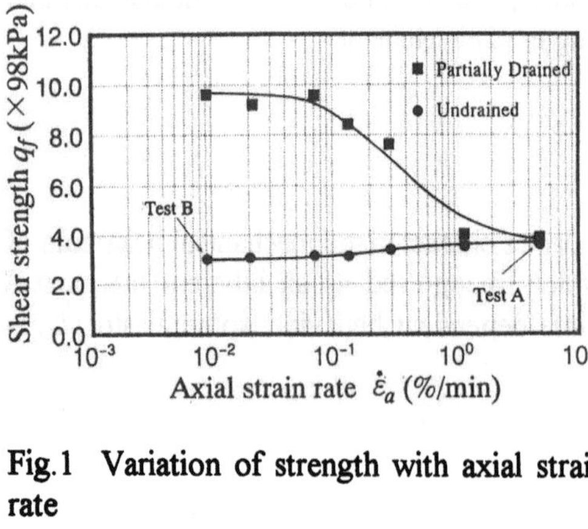

Fig.1 Variation of strength with axial strain rate

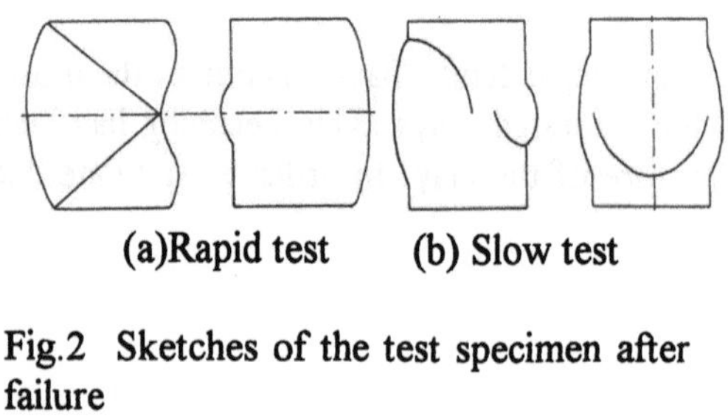

Fig.2 Sketches of the test specimen after failure

However, the effect of pore water migration alone cannot be attributed to the difference in q_f between Test A and Test B. Figs.2 are the key sketches of the test specimen after failure. In the extremely slow loading test, a *single* slip line that crosses diagonally from the side of the specimen is *always* predominant as shown in Fig.2 (a). In the rapid loading test, however, the test specimen *usually* bends and the slip line exhibits *reflectional symmetry* as illustrated in Fig.2 (b). Thus, the difference in q_f between Test A and Test B may also be attributed to the difference in *failure modes* due to pore water migration. Similar findings obtained through numerical simulations will be introduced next.

2.2 Imperfection-sensitive bifurcation of a rectangular Cam-clay specimen

In order to simplify the computation and to avoid holding higher order symmetry during shearing, a plane strain rectangular finite element mesh (B=12cm, H=36cm) was adopted as the soil specimen. Boundary conditions are given in Fig.3. The soil was initially isotropically compressed to 98kPa and then subjected to standard undrained compression test controlled by vertical displacement rate. The axial strain rate on the top of the specimen was $\dot{\varepsilon}_a$ = 4.9%/min for Test A and $\dot{\varepsilon}_a$ = 9.12×10^{-3} %/min for Test B. The behavior of the soil was assumed to follow the original *inviscid* Cam-clay model with the subloading surface concept. The parameters used for numerical analyses are tabulated in Table1.

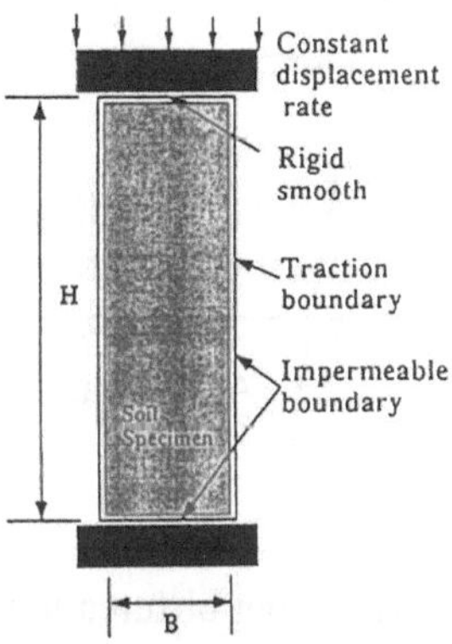

Fig.3 Boundary condition for the plane strain compression test

Table 1 Soil parameters for analysis

Compression index λ	0.108
Swelling index $\tilde{\kappa}$	0.025
Critical state constant M	1.55
Permeability k (cm/sec)	3.7×10^{-8}
Initial consolidation stress p'_0 (kPa)	294.0
Initial void ratio e_0	0.83

When the specimen has no initial imperfection and forms a perfect system, all the elements show perfect undrained path because no porewater migration occur within the specimen. In this case, therefore, the deviator "stress" - the axial "strain" curves as well as the undrained strength are not affect by the axial strain rate of loading.

In actual tests, even though the testing condition is settled almost perfectly, both sides of the specimen can have some initial geometrical imperfection of various order mode. Therefore, in the simulation, the sides are given 2nd + 3rd mode initial imperfections beforehand. In this study, the n-th order imperfection mode is defined as the n-th cosine curve in Eq.(1).

$$(u_1)_{t=0} = A\cos(n\pi X_2/H),$$
$$0 < X_2 < H \quad and \quad n=1, 2, \cdots \qquad (1)$$

Double the amplitude A in Eq.(1) represents the magnitude of the imperfection. As illustrated in Fig.4, the 2nd + 3rd mode is produced by superposition of 2nd + 3rd order cosine curves. The reason why 2nd + 3rd mode was taken as initial imperfection in this numerical calculation is that this mode make the pore water to migrate unsymmetrically within the specimen.

Fig.5 shows the comparison of load-deformation curves between the perfect and the imperfect systems, of which; very rapid loading rate ($\dot{\varepsilon}_a$ = 4.9%/min) and very slow loading rate ($\dot{\varepsilon}_a$ = 9.12×10^{-3} %/min) were employed and, of which the magnitude of the initial imperfection was (2A =) 2/100cm. In the very rapid rate of loading ($\dot{\varepsilon}_a$ = 4.9%/min), the deviator "stress" - the axial "strain" path tends to trace 2nd order bifurcation path, while in the very slow rate of loading ($\dot{\varepsilon}_a$ = 9.12×10^{-3} %/min), full migration of pore water due to initial geometric imperfection makes the path to lead the lower bifurcation path. The migration of pore water naturally yields the development of the spatial variation of the specific volume, excess pore pressure and shear strain of the soil, which naturally disturbs the type of initial geometrical imperfection as illustrated in Figs.6. The failure mode in the slow loading test is found to be the same as 1st order bifurcation mode.

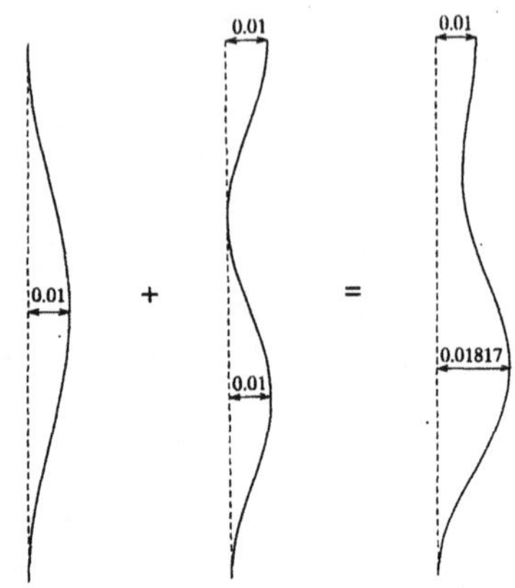

Fig.4 2nd + 3rd order mode initial geometrical imperfection on both sides of specimen

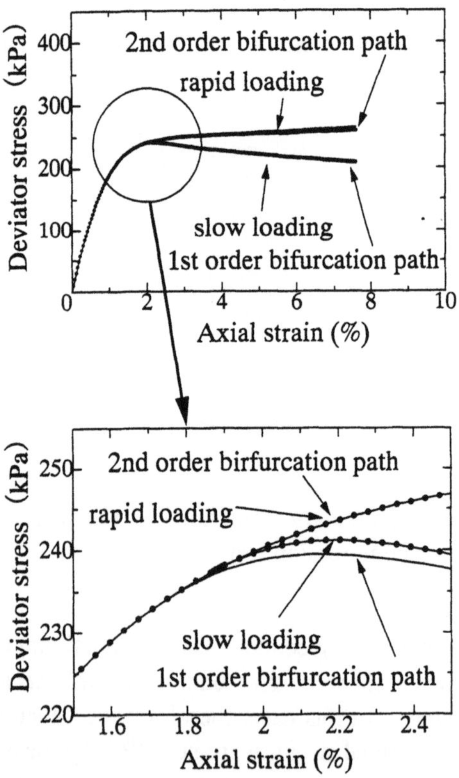

Fig.5 "Stress~strain" curves of specimen with geometrical imperfection

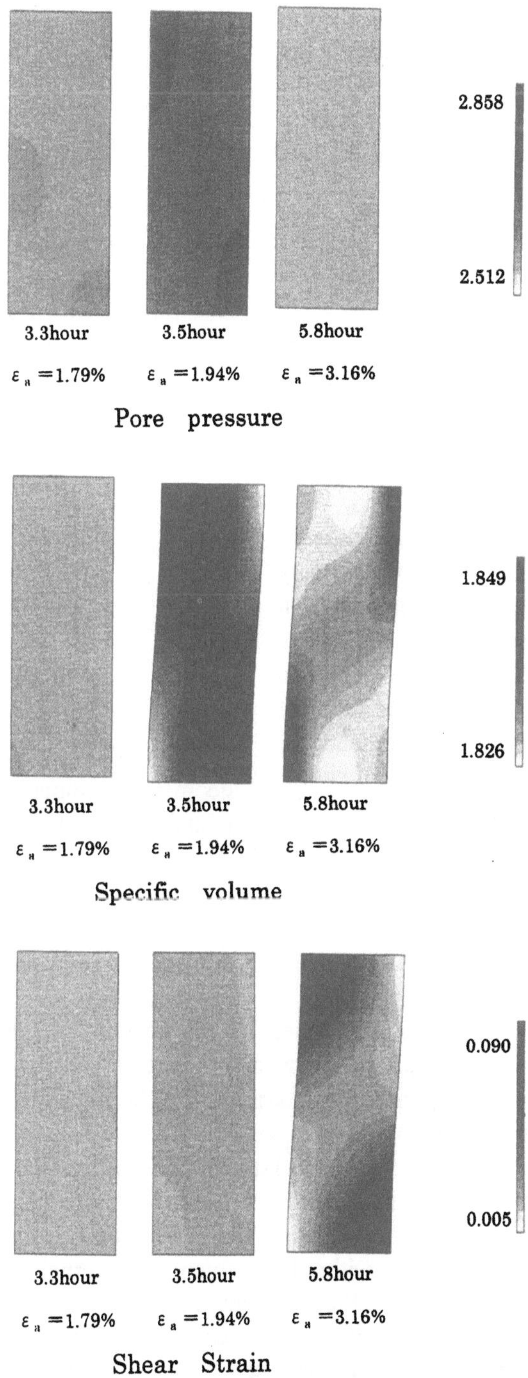

Fig.6 Mode switching procedure

Even the higher mode of initial imperfection can be found only to give an asymptote to the 1st order bifurcation path. This is simply called, in this study, the "mode switching". Since the 1st order bifurcation path gives the lowest strength, a slow loading test *always* yields the lower strength than a rapid loading one.

The important finding obtained here is that in the slow loading test all the loading paths come always asymptote to the 1st order bifurcation path from the upper side. *"An asymptote from above"* can also be observed when the effect of the magnitude of initial imperfection is examined. Since the initial imperfection used was the 2nd + 3rd order one, the effect of the magnitude of initial imperfection of the 2nd mode keeping the 3rd order one constant is examined as an example. The results of the computation are all given in Fig.7. As shown in the figure, all the load - deformation curves approximate the 1st order bifurcation path from above, and the smaller the magnitude of initial imperfection, the higher the degree of the approximation.

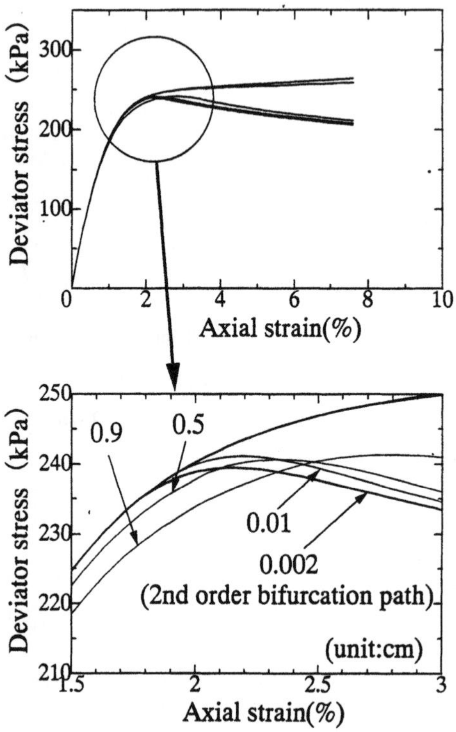

Fig.7 Effect of the magnitude of initial geometrical imperfection on the "stress ~ strain" curves in slow testing

3 UNDRANED CREEP RUPTURE DUE TO "MODE SWITCHING"

Undrained creep rupture can be interpreted by the concept of "mode switching". Fig.8 schematically shows the mechanism of the creep rupture. Initially, the soil specimen is loaded so rapidly without any pore water migration that the specimen tends usually to follow a higher order bifurcation path (A-B), and can sustain a load higher than the 1st bifurcation point, (point B in Fig.8) associated with the 1st order bifurcation path. Next, the specimen is left under constant applied load, and then pore water begins to migrate with time like the behavior in the very slow rate of loading. The migration naturally allows additional deformation to proceed with time without increment of external forces (B-C). When the pore water migration was adequate, the additional deformation might lead the deformation path to the lower 1st order declining bifurcation path from its upper side, as observed before in the slow test, which naturally yields the creep rupture. The "mode switching" should be found, in the creep test, during the period of constant load application.

In the numerical simulation, firstly the specimen was compressed under load control condition with a very rapid loading rate of 3.18×10^3 kPa/sec (point A to B in Fig.8). At the stage of constant load application, the total load at the top of the specimen was held constant with imposing "no bending" condition to the velocities at all the nodal points on the top of specimen, so that the existence of the rigid smooth pedestal can be well simulated (Asaoka et al., 1998).

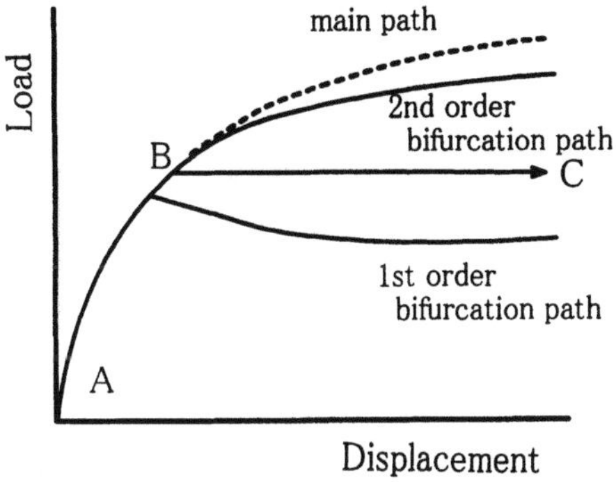

Fig.8 Mechanism of creep rupture

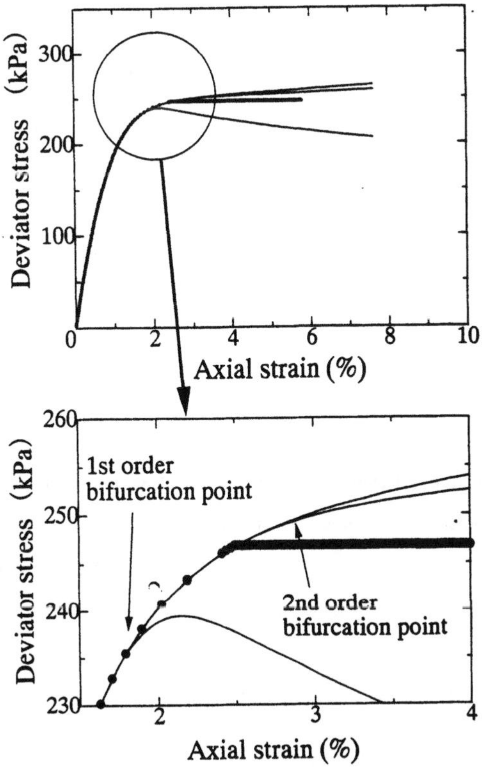

Fig.9 "Stress~strain" relationship of creep rupture test

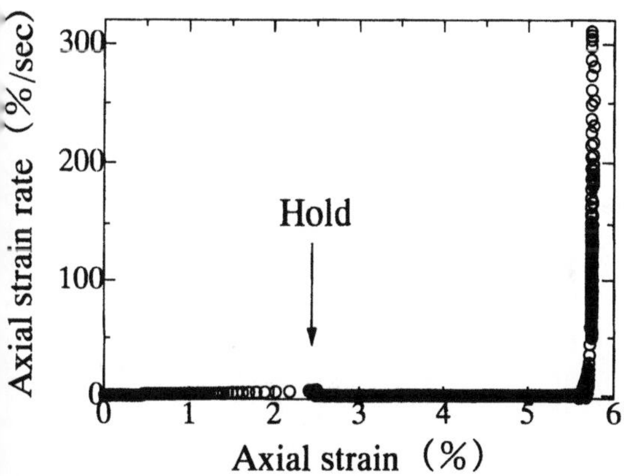

Fig.10 "Strain rate ~ strain" relationship

Fig.9 shows the relationship between the deviator stress and the axial strain of the creep rupture test, including the relationships of the perfect system and 1st and 2nd bifurcation paths. Under the constant applied load of 247kPa, which is between 1st and 2nd bifurcation points, the axial strain is progressing. Fig.10 shows the axial strain rate with the axial strain. Just after the constant load application, the axial strain is progressing at a very slow rate. However, the strain rate increases immediately when the strain reaches at 5.7% level. This indicates the occurrence of creep rupture.

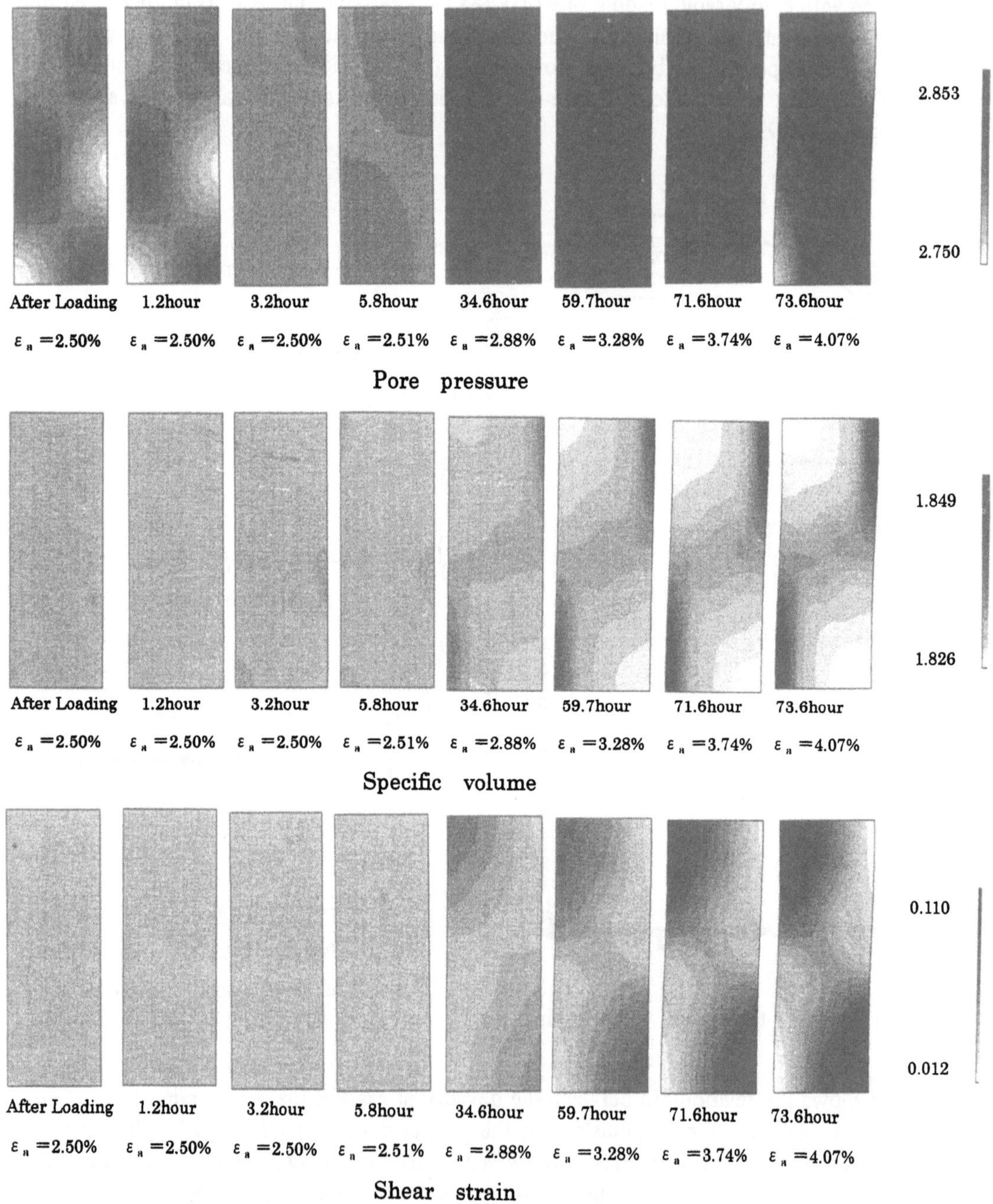

Fig.11 Creep behavior through Mode switching

The spatial variation of the specific volume, excess pore pressure and shear strain of the soil developed due to the pore water migration are illustrated in Fig.11. The "mode switching" from 2nd + 3rd mode to 1st mode can be observed in these figures. However, it should be emphasized that the excess pore pressure tends to be homogeneous at same stage and then rises again due to the occurrence of softening in each element. This variation of softening and unloading behaviors in the specimen is illustrated in Fig.12.

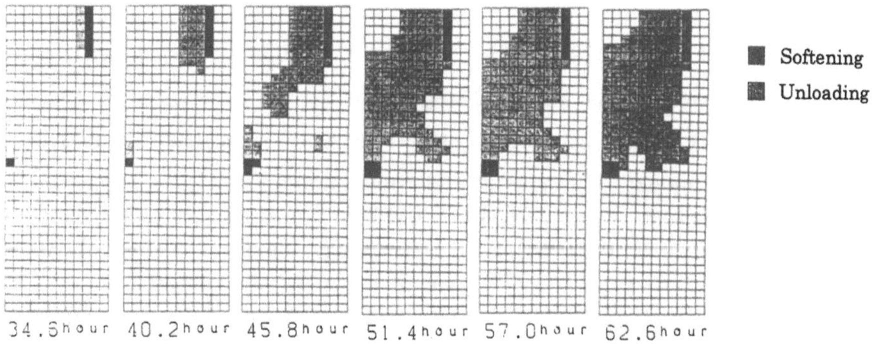

Fig.12 Development of softening and unloading element

4 CONCLUSIONS

In this study, we newly interpret the undrained creep rupture of the normally consolidated soil as due to "mode switching" through numerical analysis by solving a soil-water coupled initial-boundary value problem of a soil specimen. In this numerical analysis, the "inviscid" original Cam-clay model with the subloading surface concept and the Darcy's law are used. The findings of this research are summarized as follows;

(1) The specimens, which are given the 2nd + 3rd order initial geometrical imperfection exhibit failure under rapid strain rate as of 2nd order bifurcation mode. When a very slow strain rate under which full pore water migration can occur, is used, the failure mode is as of 1st order bifurcation mode. The undrained strength of 2nd order bifurcation mode is higher than that of 1st order bifurcation mode.

(2) The undrained creep rupture under constant loading can be interpreted by "mode switching" due to pore water migration. The soil specimen which has initial geometrical imperfection of 2nd + 3rd order mode is loaded so rapidly without any pore water migration thus the specimen tends usually to follow a higher order bifurcation path, and sustains a load higher than the 1st bifurcation point. Under constant load, the pore water migration occur with time, and the deformation path tries to be towards the lower 1st order declining bifurcation path from its upper side, and thus the specimen finally cannot sustain the load higher than the 1st bifurcation point and creep rupture occurs. The "mode switching" should be found, in the creep test, during the period of constant load application.

(3) During constant loading the distribution of excess porewater pressure is initially inhomogeneous, and then tends to dissipate and the axial strain rate is almost zero. However the porewater pressure come into existence again due to the softening of some soil elements and the axial strain rate increases sharply, finally reaching the failure.

5 REFERENCES

Adachi, T. and, Okano, M.(1974) : A constitutive equation for normally consolidated clay, Soils and Foundations, Vol.15, No.4, pp.55-73.

Adachi, T., Oka, F., and Mimura, M.(1987) : Mathematical structure of an overstress elasto-viscoplastic model for clay, Soils and Foundations, Vol.27, No.3, pp.31-42.

Asaoka, A., Nakano, M. and Noda, T. (1994). Soil-water coupled behaviour of saturated clay near/at critical state, Soils and Foundations, Vol.34: No.1, pp.91-105.

Asaoka, A. and Noda, T. (1995): Imperfection sensitive bifurcation of Cam-Clay under plane strain compression with undrained boundaries, Soils and Foundations, Vol.35: No.1, pp.83-100.

Asaoka, A., Nakano, M. and Noda, T. (1997a): Soil-Water coupled behavior of heavily overconsolidated clay near/at critical state, Soils and Foundations, Vol.37, No.1, pp.13-28.

Asaoka, A., Nakano, M., Noda, T. and Kaneda, K. (1997b): Undrained creep rupture of normally consolidated clay triggered by the mode switching during pore water migration, Proc. IS-Nagoya 97 (A. Asaoka, T. Adachi and F. Oka Edts) Nagoya, Japan, pp.21-26.

Asaoka, A., Noda, T. and Kaneda, K. (1998) : Displacement/ Traction Boundary Conditions Represented by Constraint Conditions on Velocity Field of Soils, Soils and Foundations, to appear.

Hashiguchi, K. and Ueno, M. (1977): Elasto-plastic constitutive laws of granular materials, Constitutive Equations of Soils, (Proc. Spec. Session 9th Int. Conf. SMFE, Murayama, S. and Schofield, A. N. Eds.), Tokyo, JSSMFE, pp.73-82.

Hashiguchi, K. (1993): Fundamental requirements and formulation of elasto-plastic constitutive equations with tangential plasticity, Int. J. of Plasticity, Vol. 9, pp.525-549.

Ikeda, K and Goto,S (1993) Imperfection sensitivity for size effect of granular materials, Soils and Foundations, Vol.33, No.2, pp.157-170.

Ikeda, K and Murota, K (1990) Critical initial imperfection of structures, *International Journal of Solid and Structure* Vol.26, No.8, pp.865-886

Sekiguchi, H.(1997) : Rheological characteristics of clays, Proceedings of 9th International Conference of Soil Mech. Found. Eng., Tokyo, Vol.1, pp.289-292.

Sekiguchi, H.(1984) : Theory of undrained creep rupture of normally consolidated clay based on elasto-viscoplasticity, Soils and Foundations, Vol.24, No.1, pp.129-147.

NUMERICAL ANALYSIS OF HIGHER-ORDER CONTINUA IN THE DESCRIPTION OF GRANULAR ASSEMBLIES

K. Bagi and I. Bojtár
Technical University of Budapest, Budapest, Hungary

ABSTRACT: The paper deals with the theoretical and numerical analysis of granular assemblies. Introduction of the re-interpretation of classical continuum-mechanical state variables is followed by a summary of continua with higher degrees of freedom. Results of numerical experiments are shown to point out that classical continua cannot contain enough information for the reliable description of material behaviour and the application of higher-order continua seems to be much more reasonable.

1. INTRODUCTION

This paper concentrates on the analysis of *granular assemblies*. The term 'granular assembly' is used here for a material consisting of randomly packed grains not fixed to each other (in contrast to the other type of granular materials referred to as 'cemented'). The grains in it are able to slip or roll along each other, contacts can be deleted or new contacts created etc., so even the whole internal structure can be rearranged under the external mechanical loading effects. This phenomena makes the behaviour of granular assemblies so interesting both from theoretical and from practical engineering point of view.

The final goal of the mechanics of granular materials is to provide relationships between the external loads acting on the material and the resulting displacements, with special attention on predicting the load-bearing capacity. In classical theories the effect of external loads is expressed by the continuum-mechanical state variable *stress* (relation between loads and the stress field is given by the equilibrium equations, i.e. the Cauchy-equations); deformations are reflected by the other continuum-mechanical state variable *strain* (geometrical equations set the link between translation field and the strain field). Stress and strain are related to each other through the *constitutive equations* (which are expected to contain all the necessary information about the mechanical characteristics of the material). The geometrical and equilibrium equations are clear in continuum-mechanics; but to find the proper constitutive equations for granular assemblies is not as simple at all: for many years, a large number of theoretical and experimental studies have been concerned with the problem, and the results seem to be rather limited. One possible explanation for the unreliable nature of classical theories is the fact that granular assemblies are, by their deepest nature, *not continuous* at all. Instead, they have a very characteristic internal structure that strongly influences their macro-level behaviour and cause them to perform phenomena that cannot be reflected by classical theories. So the aim of our researches is to find *more reliable* methods for the material description.

The work introduced in this paper focus on the quasi-static behaviour of the material; so time effects are not considered at all, and any part of the assembly is assumed to be in equilibrium. The assembly consists of randomly packed separate grains having arbitrary convex shape; contacting grains have discrete contact points transmitting concentrated contact forces. (The loads can be given as external forces that act on the assembly through the contact points with its neighbourhood, as body forces, or as prescribed deformations. For simplicity, body moments are not considered here.)

Chapter 2. deals with the continuum-mechanical fundaments: 2.1. will introduce how to interpret the classical stress and strain tensors in terms of micro-structural characteristics, and 2.2. gives a summary of the most important higher-order continua. Chapter 3. contains the numerical experiments. Finally, Chapter 4. draws the conclusions and summarises the possible directions of further researches.

2. CONTINUUM-MECHANICAL FUNDAMENTS

2.1. Classical continuum

State variables of classical continua are assigned to the points of a continuous domain. The points of the domain have three degrees of freedom, i.e. translations according to the coordinate axes. Gradient of the translation field is the basis of the different versions of *strain tensor*. Its dual variable, *stress tensor*, is symmetric so that moment equilibrium of the elementary volume dV could be satisfied.

In case of materials with discrete internal structure the interpretation of the two state variables is not self-evident at all. Since the limit $dV \rightarrow 0$ has no sense in this case, the assembly has to be replaced by suitably chosen 'equivalent continua' so that the state variables could be defined and expressed in terms of the characteristics of microstructure. In the previous years we invested intensive efforts into the solution of this problem and finally arrived to a discrete defining expression of the two state variables (see Bagi (1995), Bagi (1996) for details).

Since the loads and internal forces are transmitted through the grains while the deformations are carried on by the voids and the internal structure, we defined two geometrical systems, the *material cell system* and the *space cell system* (see Figure 1.a. and Figure 1.b.). They are dual to each other in topological sense (in 3D, cells / faces / edges / nodes in one system correspond to nodes / edges / faces / cells in the other system, respectively; in 2D, cells / edges / nodes in one system correspond to nodes / edges / cells in the other system, respectively).

Fig. 1.a. The material cell system

Fig. 1.b. The space cell system

The material cell system served as the geometrical basis of the microstructural definition of *stress tensor*: volume average of stresses belonging to the space cells could be expressed as

$$\bar{\sigma}_{ij} = \frac{1}{V}\sum_{(L)} V^{L}\bar{\sigma}_{ij}^{\ L} = \frac{1}{V}\sum_{(c)} l_i^c F_j^c$$

where the notations are:

L: index of material cells

c: index of the faces of material cell system (neighbouring pairs of grains)

F_i: contact force between neighbouring grains

l_i: branch vector (the vector connecting the centres of neighbouring grains)

This definition is nearly equivalent to those introduced previously in the literature. However, it is more general since body forces like the weight of the grains can also be considered (this was not possible in the previous versions).

The average displacement gradient tensor of the space cells is, in terms of the complementary area vector (the dual variable of branch vector), and the relative translations of neighbouring grains

$$\bar{e}_{ij} = \frac{1}{V}\sum_{(K)} V^K \bar{e}_{ij}^{\,K} = \frac{1}{V}\sum_{(c)} \Delta u_i^c d_j^c$$

K: index of space cells
c: index of the edges of space cell system
Δu_i : relative translation of neighbouring grains
d_i : complementary area vector
Skew-symmetric part of this tensor reflects the average rigid-body rotation of the system, so not related to deformations. The symmetric part, on the other hand, expresses the distortions of the analysed domain so by our suggestion this is called the microstructural strain tensor of granular assemblies.

2.2. Higher-order continua

In higher-order continua the points of the analysed continuous domain have further degrees of freedom beside those three translations of classical theories. The several different higher-order theories can be distinguished basically according to the applied degrees of freedom. We would like to mention here only those that have or may have the highest importance from the point of application.

According to the classification given by Füzy et al. (1984), the Mindlin-theory is mentioned first. In a Mindlin-continuum beside the 3 translation components the 'point' (more precisely, 'elementary volume') has a micro-deformation-tensor with 9 elements whose skew-symmetric part expresses the rotation and the symmetric part contains the deformation belonging to the analyzed 'point'. Other known theories can be interpreted as special cases of this continuum. The Cosserat-theory, for example, applies 6 degrees of freedom: the 'points' have a translation vector, and a rotation vector independently from that. In pseudo-Cosserat theories the degrees of freedom is reduced to 3 by the constraint that the rotation vector belonging to the 'point' is equal to the rotation of the translation vector (expressed by the skew-symmetric part of the gradient of translation vector). However, the application of rotations is not necessary at all: Füzy et al. (1982) suggest a continuum with 4 degrees of freedom where the 'point' has (beside the 3 translational components) a volume-change possibility, as the isotropic part of Mindlin's micro-deformation-tensor.

The state variables of higher-order theories could probably be interpreted for discrete materials similarly to the classical variables. As far as we know, this is still missing today so to find the microstructural definitions of non-classical state variables is a part of our plans for the future researches.

3. EXPERIMENTAL RESULTS

The experiments were executed with the help of 2D and 3D versions of the dynamic model developed by P. A. Cundall et al. (1982), Cundall et al. (1983). This model considers the material as a set of randomly arranged rigid discs or spheres, with deformable contacts between them. Motion of every grain is followed in time through a step-by-step algorithm, according Newton's force-acceleration law.

The examples will illustrate three of those phenomena that are characteristic of the behaviour of granular assemblies but cannot be reflected by classical continua.

3.1. Grains forming groups

Our first example illustrates that the individual grains in the assembly tend to form *groups* during the macro-level deformation of the material: these groups have displacements relative to each other while they also have their own deformation. Figure 2. introduces the initial arrangement of a 2D assembly. Its geometrical and physical data:

→ size of the domain: 1000*1000
→ number of grains: 100
→ average grain size: 100.4 ; 95.4 < (grain size) < 105.4; uniform distribution
→ porosity: 0.1974; coordination number: 3.725

→ the applied coordinate system:

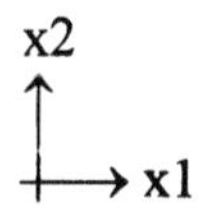

Material characteristics:
grain-grain contacts:
→ normal stiffness (force/disp.): 1.0 e8
→ shear stiffness (force/disp.): 4.0 e7
→ friction coefficient: 0.1
grain-wall contacts:
→ normal stiffness (force/disp.): 1.0 e12
→ shear stiffness (force/disp.): 0.0
→ friction coefficient: 0.0
Initial stress state:
→ average stress:

$$\sigma_{ij} = \begin{bmatrix} -2.251\text{ e5} & 0.1567\text{ e5} \\ 0.1567\text{ e5} & -2.081\text{ e5} \end{bmatrix}$$

→ isotropic component:

$$\sigma_0 = -2.166\text{ e5}$$

→ average contact force: 1.845 e7

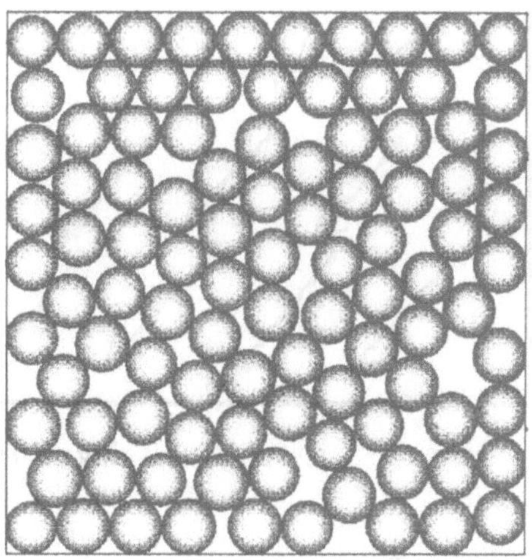

Fig. 2. The initial arrangement

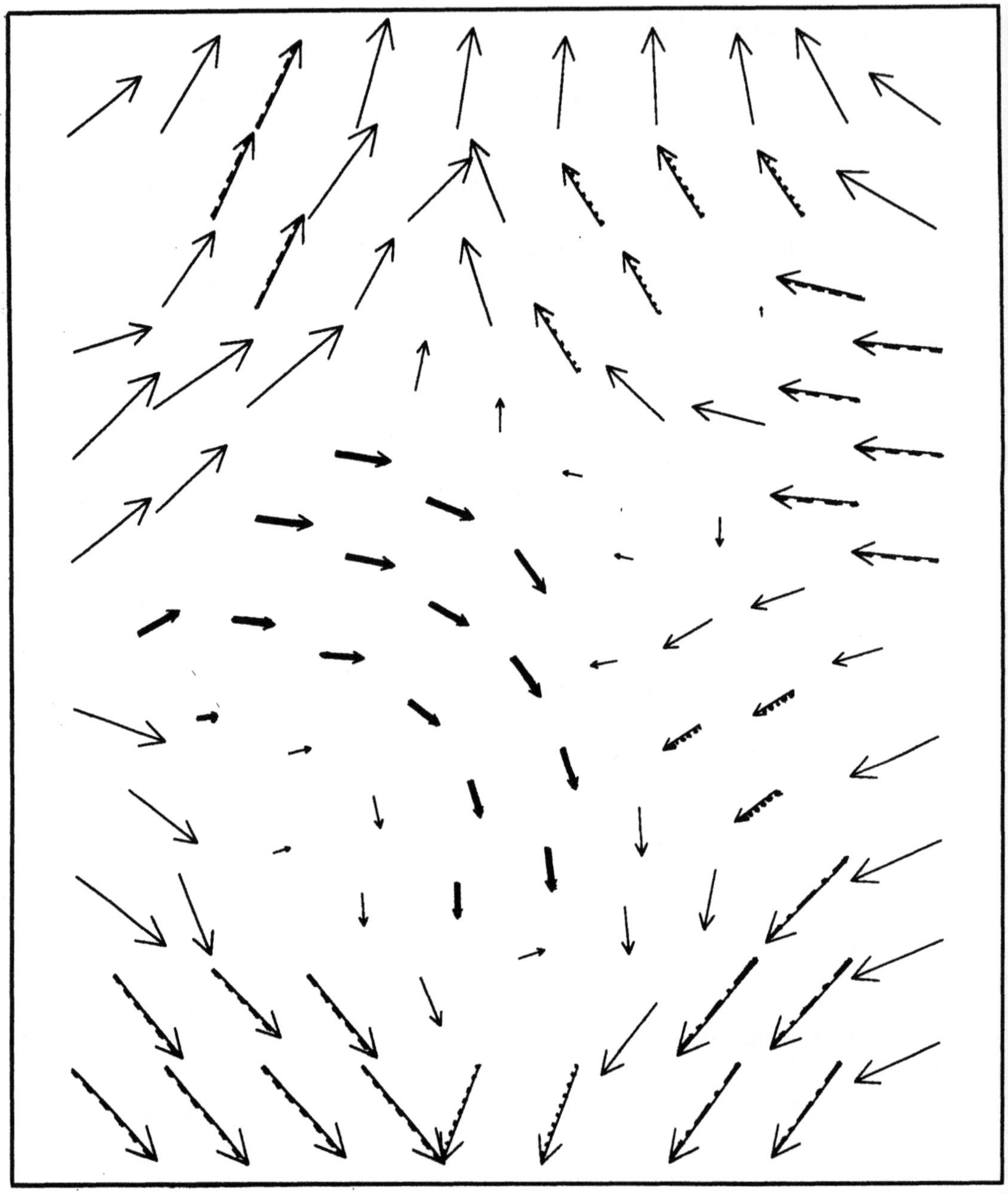

Fig. 3. Translations of the grains

This assembly was loaded by prescribed translations of the boundary walls, resulting in the following strain tensor as prescribed deformation:

$$\bar{e}_{ij} = \begin{bmatrix} -1.00\% & 0 \\ 0 & +1.00\% \end{bmatrix}$$

$$\begin{matrix} & \uparrow & \\ \rightarrow & & \leftarrow \\ & \downarrow & \end{matrix}$$

The arrows in Figure 3. show the translations of individual grains (the lengths of the translation vectors were multiplied by a sufficiently large scalar, to support visuability). Bold arrows indicate grains forming groups (sub-domains) during the deformation. (It is worthwhile to mention that the tendency of grains to collect into groups has been reported by real laboratory tests.) The existence of such sub-domains strongly support those proposals to apply theories with higher degrees of freedom.

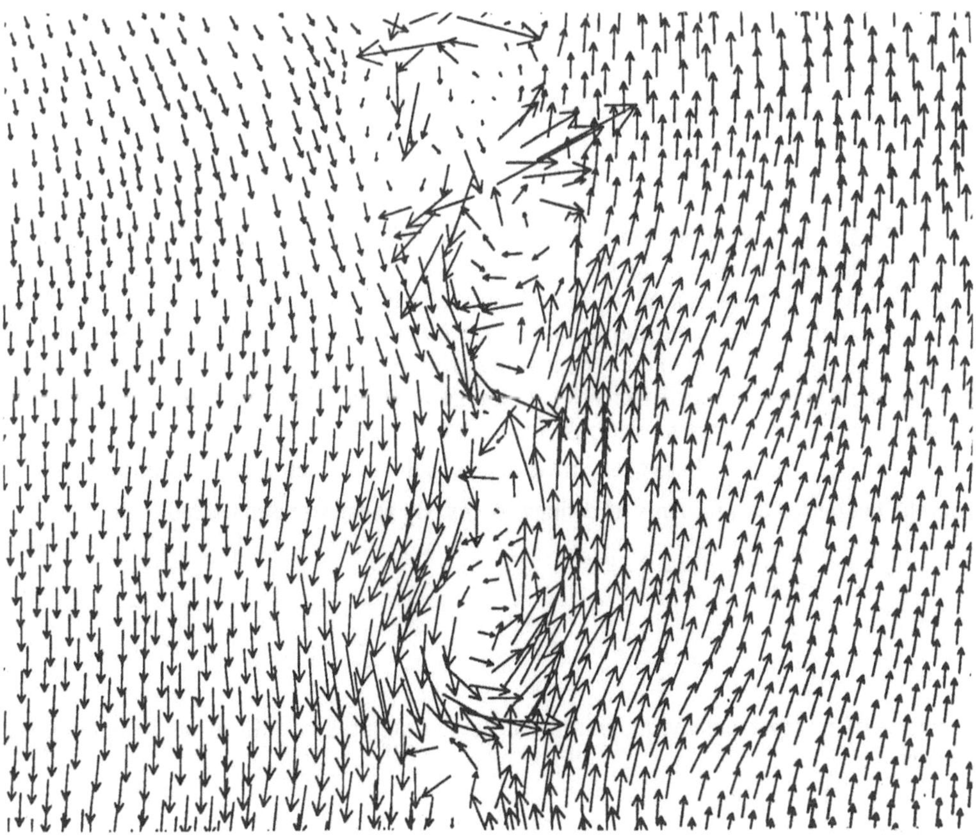

Fig. 4. Grain translations in sheared samples

3.2. Whirls along slip lines

Our next example is a sheared 2D sample. Figure 4. contains the translation vectors of individual grains (left half of the sample was translated downwards and the right half upwards). The figure shows that state when the shear line has already been formed. It is seen that the grains perform a kind of 'curling' displacement system at the shear region.

3.3. Direct analysis of the importance of grain rotations*

The coming measurements were done on a 3D assembly of 432 spheres of nearly the same size. Data of the initial assembly:

Grain size: 216 spheres of 0.256 mm diameter
216 spheres of 0.215 mm diameterl

Material properties of grains: G = 2.9 e10 N/m2 (shear modulus)
$\nu = 0.2$ (Poisson-factor)
f = 0.35 (friction coefficient)
(characteristics of glass beads)

Initial state: $\rho = 0.368$ (porosity)
σ = -1.38 e5 N/m2 (initial hydrostatic compression)
Δ = -13.22 e-4 (initial hydrostatic strain)
nearly isotropic geometrical arrangement and stress state

Two tests were done on the same initial assembly. In the first case ('free') the compressional stress acting on the assembly was increased in the direction of the x3 coordinate axis, and decreased in the x1 and x2 directions, keeping the isotropic stress component to be the same during the whole process. Grains were able to translate as well as rotate. In the second case ('fixed') the loading process was the same but the grains were not able to rotate: they were artificially fixed against rotation. The diagrams in Figure 5. show the shear stress – shear strain relations, according to the following:

Let σ_{ij}^N , and σ_{ij}^T denote the stress tensors calculated separately from the normal and from the tangential components of contact forces. By shear stresses we mean the

$$q^N = -\frac{1}{2}\left(\sigma_{33}^N - \frac{\sigma_{11}^N + \sigma_{22}^N}{2}\right)$$

$$q^T = -\frac{1}{2}\left(\sigma_{33}^T - \frac{\sigma_{11}^T + \sigma_{22}^T}{2}\right)$$

quantities (they represent the magnitude of deviatoric stresses). Shear strain is understood similarly:

$$\gamma = -\frac{1}{2}\left(e_{33} - \frac{e_{11} + e_{22}}{2}\right)$$

* This analysis was done at Cornell University, Dept. of Theoretical and Applied Mechanics, under the supervision of prof. J.T.Jenkins whose help is greatly acknowledged.

Horizontal axes of the diagrams show the shear strain (normalised by the initial hydrostatic deformation) and the vertical axes belong to the shear stresses (again, normalised by the initial hydrostatic stress). As obvious from the diagrams, depriving the grains from their rotational degree of freedom drastically changed the macro-level behaviour of the assembly.

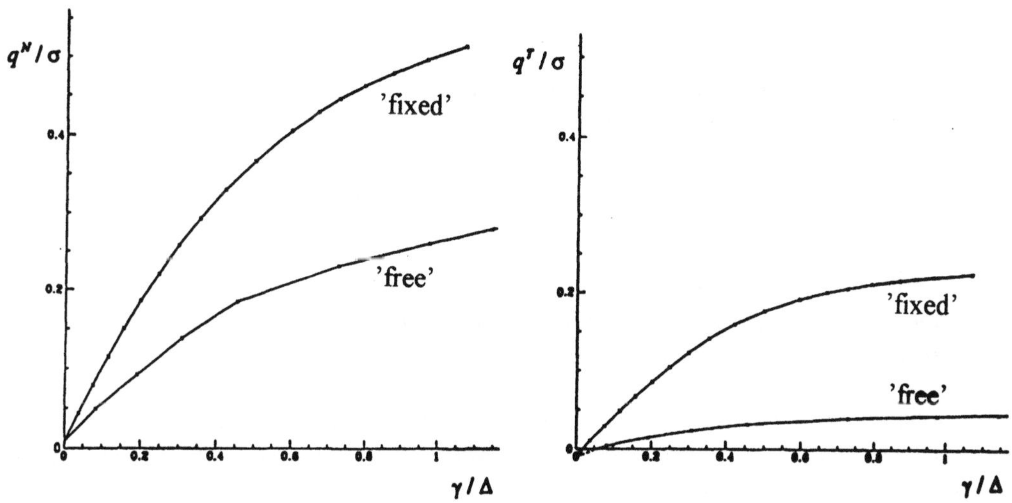

Fig. 5. Shear stress versus shear strain

4. CONCLUSIONS

We think that the above results point out that classical continua are not sufficient for the reliable description of the behaviour of granular assemblies. Beside theoretical problems (confrontation of Hill's lemma; meaningless of $dV \to 0$ limit etc.) they are not able to reflect important characteristics of the real behaviour of the material. Today we see two possible directions for finding more suitable theories:

1. *Higher-order continua* may turn out to be powerful if proper links are established between micro- and macro-level characteristics. The Mindlin-theory and those versions using only some of its degrees of freedom (Cosserat, pseudo-Cosserat, Füzy and Vas etc.) deserve further investigation in our opinion.
2. Pioneering suggestions can be found in the literature to apply a completely *discrete description*, instead of continuum approaches. Graph-theoretical works of Satake (1983), (1994), statistical physical considerations of Gudehus (1997) etc. belong to this line. Though these methods may seem rather unusual for practical engineers, it is possible that they will open the way for more effective theories than those based on continuum approach.

ACKNOWLEDGEMENTS

Our researches were supported by the OTKA program of Hungarian Academy of Sciences, under grants F-22992 and T-25250.

REFERENCES

1. Bagi, K. (1995): Geometrical modelling of granular assemblies. *Acta Technica Acad. Sci. Hung.*, Vol. 107. No. 1-2, pp. 1-16

2. Bagi, K. (1996): Stress and strain in granular assemblies. *Mechanics of Materials*, Vol. 22, pp. 165-177

3. Cundall, P. A. - Drescher, A. - Strack, O. D. L. (1982): Numerical experiments on granular assemblies. *Proc. IUTAM Conference on Deformation and Failure of Granular Materials*, Delft, pp. 355-370

4. Cundall, P. A. - Strack, O. D. L. (1983): Modelling of microscopic mechanisms in granular material. *Mechanics of Granular Materials: New Models and Constitutive Equations*, ed. J. T. Jenkins and M. Satake, Elsevier, pp. 137-149

5. Füzy, J. – Vas, J. (1982): Proposed continuum model for simulating the behaviour of granular materials. *Acta Technica Acad. Sci. Hung.*, Vol. 95. No. 1-4, pp. 49-53

6. Füzy, J. - Vas, J. (1984): Relationships and application possibilities of the theories of micro-elastic continua. *Acta Technica Acad. Sci. Hung.*, Vol. 97. No 1-4, pp. 69-83

7. Gudehus, G. (1997): Attractors, percolation thresholds and phase limits of granular soils. *Powders and Grains 97*, ed. R. Behringer & J.T. Jenkins, Balkema; pp. 169-183

8. Satake, M. (1983): Fundamental quantities in the graph approach to granular materials. In: *Mechanics of Granular Materials: New models and constitutive relations*, ed. J.T.Jenkins and M.Satake, Elsevier, pp. 9-19

9. Satake, M. (1994): Discrete-mechanical formulation of granular materials. Mechanics of Granular Materials (as editor), *report of ISSMFE*, New Delhi, pp. 1-6

A CONSTITUTIVE MODEL COMBINING THE MICROSCOPIC AND MACROSCOPIC BEHAVIOUR OF SANDS IN SHEAR AND VOLUMETRIC DEFORMATION

B. Baharom and S.E. Stallebrass
City University, London, UK

ABSTRACT

The paper describes how dissipation functions and yield surfaces derived by considering the microscopic mechanisms of particle deformation and particle rearrangement of sands (Chandler, 1985) can be combined with the volumetric constraints of the framework of Critical State Soil Mechanics (Schofield and Wroth, 1968) to give a model that provides a consistent link between the features of the shear deformation of sands at large strain and the volumetric state of the sand. The difficulties of relating microscopic parameters used to formulate yield surfaces and flow rules to conventional macroscopic critical state parameters are discussed. The potential of the approach is demonstrated by comparison with laboratory test results.

1. INTRODUCTION

For coarse grained materials such as sands it is now generally accepted that at microscopic scale there are two possible mechanisms of plastic deformation; particle rearrangement and particle breakage (e.g. Coop and Lee, 1993, Pestana and Whittle, 1995). Particle breakage is the main mechanism of volumetric deformation of sands. The other interesting feature of the behaviour of sands is the effect of the process by which they are deposited. Unlike clays which achieve a given state by preconsolidation, sands can also be mechanically compacted to reach a required state. Coop (1990) highlighted the difference in behaviour between sands that had been compacted and those preconsolidated to the same state, demonstrating the importance of distinguishing between these processes when analysing and modelling soil data.

Very few recently developed constitutive models for sand explicitly account for particle breakage or particle deformation in their formulation. The main exceptions are McDowell et al. (1996), who used a model for particle breakage to derive a unique normal compression line, and Chandler (1985) who developed a continuum model which allowed for particles that deform as well as rearrange. In addition, most current models for sand include the effect of the volumetric state of the soil using the state parameter, defined with respect to the critical state line (e.g. Jefferies, 1993; Muir Wood et al., 1994; Manzari and Dafalias, 1997), rather than overconsolidation ratio, defined with respect to a normal compression line. None of these models distinguish clearly between soil that has been preconsolidated to its current state or soil that has been compacted.

This paper describes a model for the large strain deformation of sands which seeks to combine recent understanding of the effect of particle breakage on both volumetric and shear behaviour with the overall framework of Critical State Soil Mechanics which was found to be valid for sands by Coop and Lee (1993). The complete model consists of a formulation that allows for a systematic distinction between behaviour of compacted and preconsolidated sands, but this paper will concentrate on large strain deformation of sands at normally compressed states and overconsolidated states reached by preconsolidation.

2. EXPERIMENTAL OBSERVATIONS OF SAND BEHAVIOUR

Over the past five to ten years there have been many advances in laboratory testing techniques. Although in general these have been aimed at the measurement of small strain stiffness, the most significant advances in laboratory testing for sands have been those that have allowed tests to be carried out at elevated pressures. For example (Coop, 1990) tested sands at pressures in the range 50 kPa to 50 MPa and Yamamuro et al. (1996) tested sands at pressures of up to 100 MPa. All the tests identified particle breakage as the main mechanism of plastic volumetric compression at high stresses.

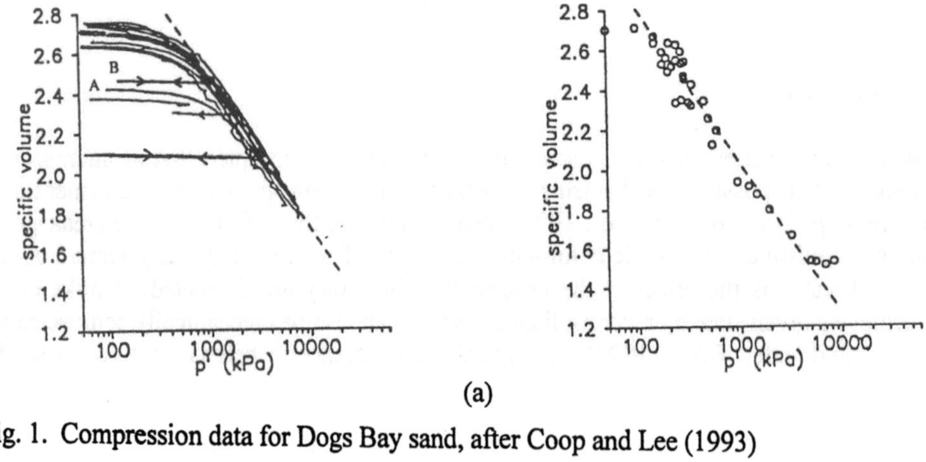

(a) (b)

Fig. 1. Compression data for Dogs Bay sand, after Coop and Lee (1993)

Figure 1a shows data from isotropic compression tests on Dog's Bay sand, which is a carbonate sand, (Coop, 1990; Coop and Lee, 1993) and Figure 1b shows end of test states for shear loading of the same sand. These data identify an isotropic normal compression line and a critical state line which are parallel in v:lnp′ space and unique for a given sand. Compression data presented by Pestana and Whittle (1995) confirmed the presence of a limiting line at high stresses for a wide range of sands, although they called this a "limiting compression line". The fact that this line has been clearly defined for many different sands makes it particularly useful for the purposes of normalisation. In contrast, the position of the critical state line which is used to calculate the state parameter (Been and Jefferies, 1985) may be uncertain. This is usually because of the difficulty of shearing samples to large enough strains. Some of the data shown in Figure 1b are for tests that did not reach critical state at the end of shearing. This uncertainty means that the critical state line may well be less suitable as a basis for the development of a constitutive model.

Figure 1a also illustrates two different responses for samples compressed from states inside the normal compression line. As described by Coop (1990) samples that have not previously been compressed to high stresses, such as A in Figure 1a, undergo significant deformation before reaching the normal compression line, whereas those that are loaded following unloading from the normal compression line, such as B, only deform elastically by very small strains. Coop (1990) concluded that some particle breakage was occurring before the normal compression line was reached in compacted samples that had not previously undergone a high stress, whereas no particle breakage would occur in a sample which had already been preconsolidated to a high stress. A standard critical state model would not distinguish between these two states

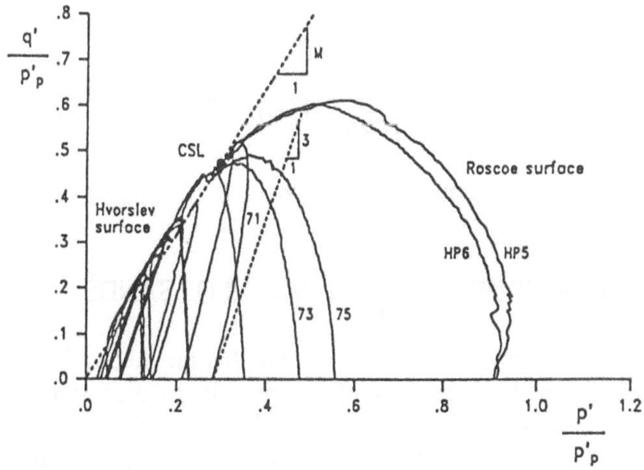

Fig. 2. Shearing data for decomposed granite (after Lee, 1991).

Figure 2 shows data from drained tests on a decomposed granite from Korea (Coop and Lee, 1993). These data are normalised with respect to volume by dividing the stress invariants by the current preconsolidation pressure p'_p, and they define a single state boundary surface.

Coop and Lee (1993) observed that the critical state point reached in the tests is not at the apex of the state boundary surface, in common with the other sands that they tested. Samples sheared from a state wet of critical reach a maximum deviatoric stress greater than the deviator stress at critical state. If the state boundary surface is a yield surface, this implies that flow will not be associated. This conclusion is supported by observations of strain increment vectors during shear tests (Coop and Lee, 1993).

As noted above a number of authors have identified particle breakage as the main mechanism of volumetric deformation. Data from Coop and Lee (1993) shown in Figure 3 demonstrate that particle breakage occurs during shearing as well as compression and furthermore that particle breakage during shearing is caused by both deviatoric and mean effective stresses. The figure shows the relative breakage B_r (Hardin, 1985) at the end of shearing and compression tests plotted against mean effective stress, p′, where p′ is the value at critical state for those samples that were sheared. Two separate lines can be identified. One represents the relationship between B_r and p′ for samples compressed isotropically and the other samples sheared to critical state irrespective of loading path. The presence of two separate lines indicates that particle breakage occurs at constant p′ due to changes in shear stress.

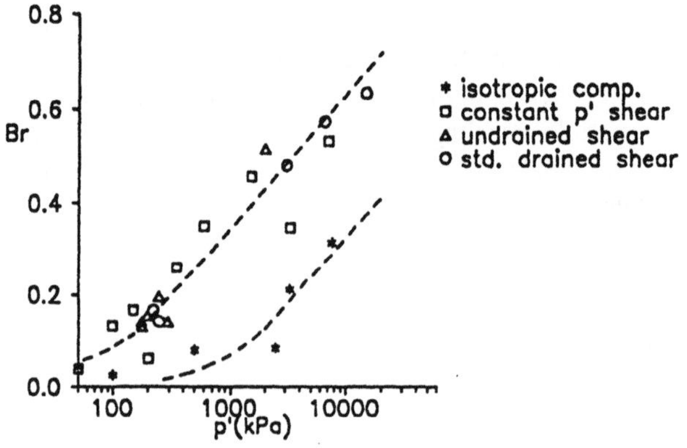

Fig. 3. Particle breakage during tests on Dog's Bay sand (Coop and Lee, 1993)

3. MODELLING THE MICROSCOPIC BEHAVIOUR OF SANDS

The laboratory test data described above demonstrate that the mechanism of particle breakage has a significant effect on the behaviour of a sand. It therefore seems reasonable that a constitutive model for sand should be derived from flow rules and yield functions obtained by considering energy loss due to both particle breakage and particle rearrangement, not just the latter. A model for a particulate material in which particle deformation and particle rearrangement occur was proposed by Chandler (1985). The mechanisms of particle deformation and particle breakage are analogous because they both involve a process which uses energy to increase the contact area between particles. Hence this model is a useful basis for the new constitutive model proposed in this paper. However there are some important differences between breakage and deformation which will be noted later. The model described by Chandler (1985) uses the theory of plasticity and an energy based equation with a dissipation function which allows for the material to strain plastically by both particle rearrangement and deformation. The yield surfaces and flow rules that define the model are then obtained by assuming conservation of energy and using the theory of envelopes. The basic equations defining the model will be presented in terms of conventional triaxial stress and strain invariants, although Chandler (1985) used generalised stress and strain quantities.

The model assumes a rigid-plastic response and shear and volumetric plastic strains are subdivided into components caused by the two mechanisms; rearrangement, $\delta\varepsilon^R$, and deformation, $\delta\varepsilon^D$, see equation 1.

$$\delta\varepsilon_v = \delta\varepsilon_v^D + \delta\varepsilon_v^R \quad \text{and} \quad \delta\varepsilon_s = \delta\varepsilon_s^D + \delta\varepsilon_s^R \tag{1}$$

An energy dissipation function is then formulated which accounts for energy loss due to friction during rearrangement and an increase in contact area during deformation of particles, equation 2.

$$\text{energy dissipated} = \left[k^2\left(\delta\varepsilon_s^D\right)^2 + l^2\left(\delta\varepsilon_v^D\right)^2 + \left(\mu p'\right)^2\left(\delta\varepsilon_s^R\right)^2\right]^{1/2} \tag{2}$$

The first two terms in this equation represent energy dissipated due to deformation of the particle, where l and k have dimensions of stress. These parameters describe the resistance to normal and shear deformation of the contact area between particles, which is dependent on the size of this contact area and the strength of the particles (Chandler, 1990a). This approach contrasts with the model for dissipation of energy due to an increase in surface area by fracture, which was proposed by McDowell et al. (1996). However, by simulating particle breakage as particle deformation in this way, the increase in contact area is clearly caused by changes in both shear and normal stress. This is equivalent to the important experimental observation by Coop and Lee (1993) that relative breakage occurs because of changes in both shear and mean effective stresses. The third term in equation 2 is the energy dissipated due to friction during particle rearrangement, where μ is a form of coefficient of friction. Chandler (1985) linked the shear and volumetric components of plastic deformation due to particle rearrangement by a rearrangemental dilation constant, ν, equation 3.

$$\nu = -\frac{\delta\varepsilon_v^R}{\delta\varepsilon_s^R} \tag{3}$$

Chandler (1990a) describes how this rearrangemental dilatancy can be split into two components, dilatancy due to rearrangement of particles and dilatancy due to the non-circular shape of a particle . Hence, increasing confining stress was said to lead to a denser packing of particles and deformation of particles to a more non-uniform shape, both of which increase rearrangemental dilatancy. However, if particle breakage is occurring rather than particle rearrangement the effect of increasing confining stress may be to reduce the angularity and size of particle and hence reduce rearrangemental dilatancy.

The new constitutive model adopts the dissipation function given in equation 2, but assumes that the two factors that might affect the rearrangemental dilation cancel each other out and hence that the angle of dilation, ν, is a constant. Flow rules and yield surfaces can then be derived following the procedures described in Chandler (1990a). The two yield surfaces that are obtained are illustrated in Figure 4 in stress space. The surfaces are defined so as to form a continuous yield criterion, with associated flow at low stress ratios when behaviour is governed purely by particle breakage (yield surface 1). At stress ratios exceeding the rearrangemental dilation constant both particle breakage and particle rearrangement occur and the flow rule is non-associated (yield surface 2). It is possible to derive an expression for a psuedo-plastic potential (Chandler, 1990b), which plots as a series of lines defining the plastic strain increment ratio as shown in Figure 4 and leads to a critical state point which is not at the apex of the surface.

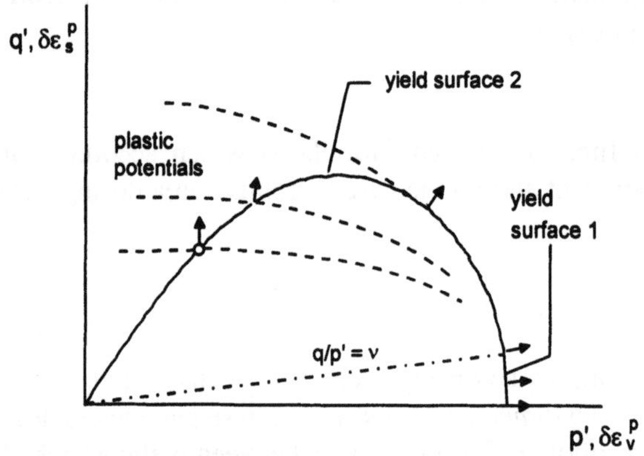

Fig. 4 Yield surfaces and plastic potentials

Modelling energy loss due to particle breakage as well as particle rearrangement has generated a yield surface defining yield stresses whose magnitude has been firmly linked to the strength of the sand particles. In addition the interaction between breakage and rearrangement has automatically generated a yield criterion and a stress ratio corresponding to failure at critical state which corresponds to observations from test data. These yield surfaces and flow rules form the first component of the new model.

4. MODELLING THE MACROSCOPIC BEHAVIOUR OF SANDS

The second component of the model comes from experimental evidence that for sands that are normally compressed or preconsolidated a unique state boundary surface in p′, q′ and v space can be defined, which is also a yield surface describing stress states at which significant plastic deformation occurs for soils with these histories of deposition. Compacted sands deform significantly prior to reaching this state boundary surface and will be discussed separately. The yield surface defined by considering the microscopic mechanisms of deformation represents the projection of this state boundary surface along an elastic wall, such that the size of this surface increases with decreasing specific volume according to a volumetric hardening law identical to that used in standard critical state models (Schofield and Wroth, 1968). The assumption of volumetric hardening is valid because a unique state boundary surface was obtained from the experimental data solely by normalising with respect to volume. This implies that the shape of the yield surface is constant with increasing stress, supporting the decision to keep rearrangemental dilatancy constant, and ensures that the locus of critical states in stress space is a straight line. To ensure the model is realistic the deformation of soil preconsolidated to states inside the state boundary surface is elastic and not negligible.

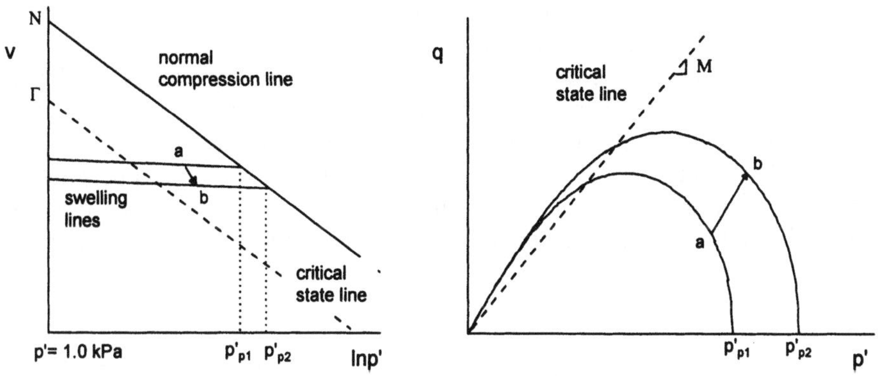

Fig. 5. New constitutive model at normally compressed and preconsolidated states

Figures 5a and 5b illustrate the resulting model in stress space and volumetric space respectively. Use of the yield surface and flow rules proposed by Chandler (1985, 1990a) with a hardening law and definition of critical state taken from standard critical state soil mechanics results in links between the soil properties characterising the microscopic behaviour used by Chandler, the standard critical state parameters, κ, λ, G′, M, N and Γ and state parameters such as p_p' the preconsolidation pressure and p_c' the mean effective stress at critical state. The following features of the new model compared to both the original model proposed by Chandler and critical state models such as Cam Clay should be noted:

1. The parameters k and l are related by a fixed ratio α, which determines the peak deviatoric stress that can be reached wet of critical
2. Although the shape of the yield surface is constant for a given sand it is not constant for the model. Consequently, the ratio between p_p' and p_c' (the spacing ratio, r) is also only unique for a given soil. The spacing ratio calculated from the critical state parameters, Γ, N, λ and κ and resulting from the choice of microscopic parameters, μ, ν, and α must be consistent.
3. The microscopic parameter l is equivalent to p_p', such that the normal compression line describes the increase in the strength of the contact area between particles, as specific volume decreases.
4. There is a relationship between coefficient of friction, M, the spacing ratio, r, and the microscopic parameters μ and ν.

$$r^2[\nu(M-\nu)] = \mu^2 \tag{4}$$

5. The assumption that ν is a constant implies that at critical state continuous breakage of particles is cancelling out the effect of the constant rearrangemental dilation.

At present it is not possible to determine values for the microscopic parameters μ,ν and α, directly from triaxial tests and they have to be obtained from parametric studies. The parameters μ and ν are not independent as they are related to the measurable quantities M and the spacing ratio, as shown in equation 4. To use the model it is effectively necessary to define nine parameters. These comprise six standard critical state parameters, as the spacing ratio is not fixed, and the three microscopic parameters μ,ν and α, two of which are related to the critical state parameters as noted above.

One important aspect of the behaviour observed in laboratory tests which has not been included in the model defined above is the difference between compacted and preconsolidated soils. Having accepted that this model is a basic framework for sands at normally compressed and preconsolidated states, the different behaviour of compacted soils can be simulated using bounding surface plasticity (Dafalias and Herrmann, 1982) and an inner loading surface representing the previous maximum stress applied to the sand. The effect of this addition will be discussed elsewhere.

5. COMPARISON OF PREDICTED BEHAVIOUR WITH EXPERIMENTAL DATA

To enable predictions to be compared to test data it was necessary to derive a set of model parameters for Dog's Bay sand. The critical state parameters were obtained conventionally from data reported by Coop (1990). The microscopic parameters were computed using an iterative process that looked for the best fit to the shape of state boundary surface (Coop, 1990), whilst ensuring that the values of the parameters were consistent with points 2 and 4 noted above.

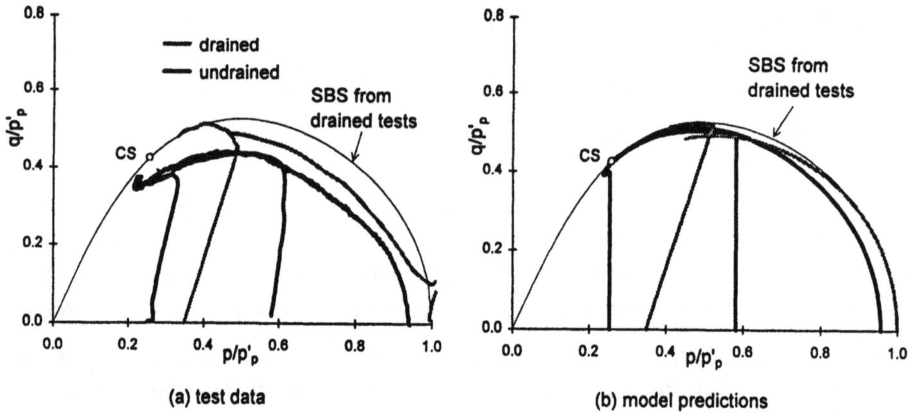

Fig. 6. Normalised stress paths for undrained and drained tests on Dog's Bay sand.

A programme of tests has been carried out looking specifically at the behaviour of sands sheared from normally compressed states or states reached by preconsolidation. All the samples were initially compressed isotropically to either 14 MPa or 13 MPa. Normalised stress paths from these tests are shown in Figure 6a and predicted normalised stress paths in Figure 6b. Normalised paths for both drained and undrained tests are essentially straight until they meet a limiting surface which they move along to reach critical state. The constant slope of the undrained paths indicates that deformations are elastic but slightly anisotropic, whereas the model predicts isotropic elastic deformation. However there is no evidence of significant plastic deformation until very near the limiting surface justifying the assumption that, for sands with this depositional history, the state boundary surface is a yield surface. The observed anisotropy may result from orientation of the elongated particles during sample preparation. The test data shown do not appear to define a unique surface which coincides with the surface identified by Coop (1990), whereas as might be expected the predicted stress paths all reach this surface, which has been defined by the choice of microscopic parameters. It is not clear at present why there is a difference in the position of the observed normalised surfaces, but it may be caused by creep during the test which is significant for this soil.

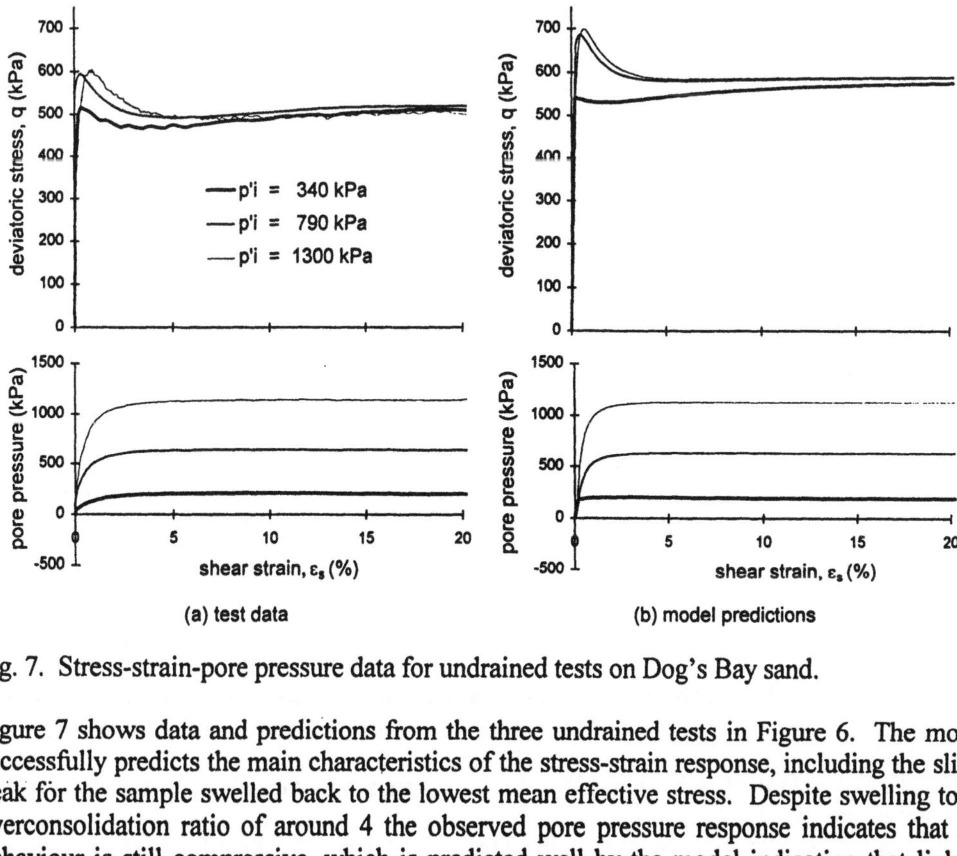

Fig. 7. Stress-strain-pore pressure data for undrained tests on Dog's Bay sand.

Figure 7 shows data and predictions from the three undrained tests in Figure 6. The model successfully predicts the main characteristics of the stress-strain response, including the slight peak for the sample swelled back to the lowest mean effective stress. Despite swelling to an overconsolidation ratio of around 4 the observed pore pressure response indicates that the behaviour is still compressive, which is predicted well by the model indicating that linking volumetric state and the yield surfaces and flow rules derived by considering the microscopic response gives a reasonable framework for soils with this depositional history.

6. SUMMARY AND CONCLUSIONS

Including particle breakage in the derivation of the yield surface has ensured that the model simulates important features of sand behaviour such as non-associated flow and a critical state point which is not at the apex of the yield surface. Particle breakage is also defined as the main mechanism of purely volumetric compression. The introduction of the framework of Critical State Soil Mechanics associates this yield surface with an experimentally determined state boundary surface. This leads to the assumption of a volumetric hardening rule which is defined by a parameter representing the strength of sand particles and a volumetric framework governing the response of soils to shear loading. Yield surfaces defined in constitutive models can be derived directly from experimentally observed state boundary surfaces only if the state boundary surface represents first yield. For preconsolidated sands the comparisons shown above indicate that this assumption is reasonable and that the resulting model generates the correct pattern of shear response for sands with different volumetric states. Thus a valid framework has been set up for preconsolidated sands which can easily be extended, using simple bounding surface plasticity (Dafalias and Hermann, 1982), to accommodate the plastic deformation due to particle breakage inside the state boundary surface that occurs with compacted sands

7. REFERENCES

BEEN K & JEFFERIES M.G. (1985). A state parameter for sands. Geotechnique, 35:99-112.

CHANDLER H.W. (1985). A plasticity theory without Drucker's postulate, suitable for granular materials. J.Mech. Phys. Solids, 33:215-216.

CHANDLER H.W. (1990a). A model for the deformation and flow of granular materials undergoing monotonic shear loading. Geotechnique, 40: 379-388.

CHANDLER H.W. (1990b). Homogeneous and localised deformation in granular materials : A mechanistic model. Int. J. Engng. Sci., 28: 719-734

COOP M.R. (1990). The mechanics of uncemented carbonate sands. Geotechnique, 40: 607-626

COOP M.R. & LEE I.K. (1993). The behaviour of granular soils at elevated stresses. Predictive Soil Mechanics - Proc. Wroth Memorial Symposium, Oxford. Thomas Telford, London, 186-198.

DAFALIAS Y.H. & HERMANN L.R. (1982). Bounding surface formulation of soil plasticity. Soil mechanics - Transient and cyclic loads (eds. G.N. Pande & O.C. Zienkiewicz). John Wiley and Sons Inc., London, 253-282.

HARDIN B.O. (1985). Crushing of soil particles. J. Geotech. Engng. Div. Am. Soc. Civ. Engrs., 111: 1177-1192.

JEFFERIES M.G. (1993). Nor-Sand: A simple critical state model for sand. Geotechnique, 43: 91-103

LEE I.K. (1991) Mechanical behaviour of compacted decomposed granite soil. PhD Thesis, City University

MANZARI M.T. &DAFALIAS Y.F. (1997). A critical state two-surface plasticity model for sands. Geotechnique, 47: 255-272.

MCDOWELL G.R., BOLTON M.D. & ROBERTSON D. (1996). The fractal crushing of granular materials. J. Mech. Phys. Solids, 44:2079-2102.

MUIR WOOD D., BELKHEIR K. & LIU D.F. (1994). Strain softening and state parameter for sand modelling. Geotechnique, 44: 335-339.

PESTENA J.M. & WHITTLE A.J. (1995). Compression model for cohesionless soils. Geotechnique, 45: 611-631.

SCHOFIELD A.N. & WROTH C.P. (1968). Critical State Soil Mechnaics. McGraw-Hill, London.

YAMAMURO J.A., BOPP P.A. & LADE P.V. (1996). One-dimensional compression of sand at high pressures. J. Geotech. Engng. Div. Am. Soc. Civ. Engrs., 122: 147-154.

NUMERICAL SIMULATION OF FLOWS IN POROUS MEDIA USING A MICROSTRUCTURAL MODEL

M. El Yazidi, M. Hellou and J. Martinez
Institut National des Sciences Appliquées, Rennes, France

ABSTRACT : This work presents numerical results of flow simulations through granular media modelled by a regular assembly of monosized cylinders. The adopted numerical approach consists in determining the unknown coefficients of the analytical solution by using least square method in order to satisfy the boundary conditions. We examine the influence of the geometry of the medium on the microscopic and macroscopic quantities which are characteristic of the flow. At the microscopic scale of the elementary cell, the results show some regions of fluid recirculations between the obstacles. The presence of these regions may favour the retention of the substance carried by the fluid. Furthermore by means of a homogenization technique we have determined the macroscopic permeability and analysed its evolution for several assembly configurations.

1. INTRODUCTION

Several microstructural models of porous media can be found in the literature which are differentiated either by the shape of obstacles or cavities. The principal families are formed by capillary channels or by separated fibers assemblies. Several shapes of fibers (cylindrical, elliptical...) which simulate the solid matrix are used. Recently a model constituted by an array of rectangular fibers has been introduced by Wang (1996). Other authors have simulated porous media by one or two rows of cylinders (Moreau and al., 1998).

Bruschke and Advani (1993) have selected two-dimensional media consisting of periodic lattices of cylindrical fibers in order to examine the effect of lattice geometries (square or hexagonal) on the permeability. The theoretical study of Caltagirone and Arquis (1986) concerning the Stokes flow through a network of identical cylinders has shown that for some values of the geometrical parameters some recirculating flows between two cylinders have been observed.

In the present paper, our goal is to simulate the flow in porous media by slow flows around circular obstacles. The porous media model chosen is composed of a regular array of cylinders whose centers form isoceles or equilateral triangles. First, our efforts are devoted to the microscopic analysis of the flow structure in order to detect the dead regions and to estimate their extent. Furthermore, by means of a homogenization method we determine the permeability of the model. The calculations are carried out for different porosities values.

2. FORMULATION

2.1. Models and boundary conditions

This work concerns the study of plane flow through a porous media model composed of a regular assembly of circular cylinders which allows to approximate a real porous media with high tortuosities. This model is defined by two geometrical parameters $2x_o$ and $2y_o$ which correspond to the distance between the centers of cylinders along the x-direction and the y-direction respectively (Fig. 1a).

The fluid is viscous and incompressible with constant physical properties ; its dynamic viscosity is denoted by μ. The plane flow perpendicular to the cylinder axes is created by imposing macroscopic flow rate (per depth), Q, at the entry of the porous media. The Reynolds number of this flow is supposed to be sufficiently small to ensure the validity of Stokes regime hypothesis. Under these conditions, the equation of motion in the flow domain is given by :

$$\nabla^2 (\nabla^2 \psi) = 0 \qquad (1)$$

where ψ is the stream function and ∇^2 a second-order differential operator.

Due to the symmetry of the problem we consider only the streaked domains in Figure 1a. Figure 1b shows an elementary cell of this domain including cartesian and polar axes. This cell is delimited by ABCDEFA boundary and the fluid flow enters through the face EF and exits through the face BC.

The boundary conditions, normalized by taking the radius R and the average velocity V_o on the BC line as reference length and reference velocity respectively, are written in dimensionless form below :

Boundary	*Conditions*			
AB	$\zeta = 0,$	$\psi = 0$		(2a)
BC	$V_y = 0,$	$\dfrac{\partial P}{\partial y} = 0$		(2b)
CD	$V_x = 0,$	$V_y = 0,$	$\psi = Y_o - 1$	(2c)
DE	$V_y = 0,$	$\zeta = 0$		(2d)
EF	$V_y = 0,$	$\dfrac{\partial P}{\partial y} = 0$		(2e)
FA	$V_x = 0,$	$V_y = 0$		(2f)

V_x and V_y are the velocity components in the cartesian coordinates, P and ζ represent pressure and vorticity. Moreover, $X_o = x_o/R$ and $Y_o = y_o/R$ define the cell adimensional characteristics and $V_o = \dfrac{Q}{y_o - R}$ the adimensional average velocity.

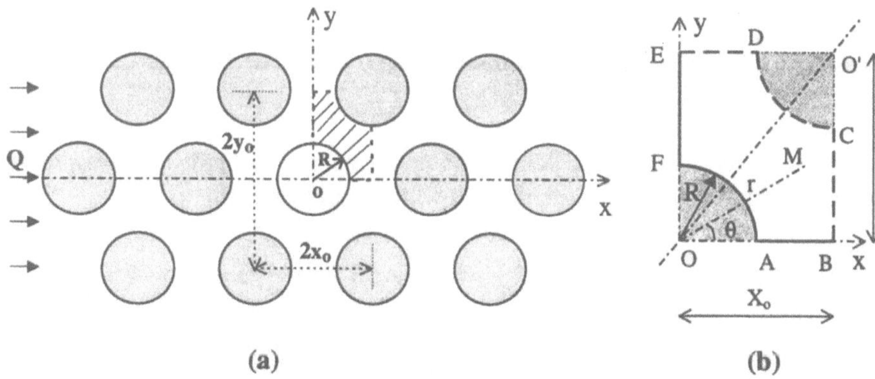

Fig. 1 : Regular assembly of cylinders **(a)** ; elementary cell for calculations **(b)** ; Boundary conditions satisfied : analytically ———— , numerically - - - - - -

2.2. Analytical solution and problem unknowns

With above boundary conditions along EFAB, the solution of (1) given in the polar coordinates is expressed in the following form :

$$\psi(r,\theta) = [b_1(\frac{1}{2r} + r\log r - \frac{1}{2}r) + d_1(r^{-1} - 2r + r^3)]\sin\theta + \sum_{n=3,5,\dots}^{\infty}[b_n(\frac{1-n}{n}r^{-n} + r^{2-n} - \frac{1}{n}r^n) + d_n(\frac{1}{n}r^{-n} + r^{n+2} - \frac{1+n}{n}r^n)]\sin n\theta \tag{3}$$

where the b_n and d_n coefficients are the problem unknowns, calculated numerically by using the remaining boundary conditions (2b), (2c) and (2d) along BCDE.

3. NUMERICAL SOLUTION

3.1. Used method

The quadratic minimization method has been used in order to satisfy the boundary conditions. The efficiency of this method has often been proved, for example in Moreau and al. (1998). This method consists in minimizing the quadratic difference between the imposed conditions on the boundaries and the values calculated by the series (3). In our case the minimum is given by the following expression :

$$\frac{\partial}{\partial A_i}\left[\int_{DE}(V_y^2+\varsigma^2)\,ds+\int_{CD}(V_x^2+V_y^2+(\psi-Y_o+1)^2)\,ds+\int_{BC}(V_y^2+(\frac{\partial P}{\partial y})^2\,ds\right]=0 \quad (4)$$

with i=1, 2, ..., 2N and where A_i represents the arbitrary coefficients b_n and d_n, N the number of terms retained in the series (3).

We can write the expressions of the velocity components, the vorticity, the pressure, and the stream function, extracted from (3), in the following condensed form :

$$V_x=\sum_{j=1}^{2N}A_j X_j,\quad V_y=\sum_{j=1}^{2N}A_j Y_j,\quad \varsigma=\sum_{j=1}^{2N}A_j \varsigma_j,\quad P=\sum_{j=1}^{2N}A_j P_j,\quad \psi=\sum_{j=1}^{2N}A_j \psi_j \quad (5)$$

where X_j, Y_j, ζ_j, P_j and ψ_j are elementary functions of the problem coordinates derived from (3). Discretization of (4) leads to the following linear system of 2N equations :

$$\sum_{j=1}^{2N}A_j\Big[\sum_{m=1}^{Nx}[Y_iY_j+\varsigma_i\varsigma_j] + \sum_{m=1}^{Nq}[X_iX_j+Y_iY_j+\psi_i\psi_j] + \sum_{m=1}^{Ny}[Y_iY_j+(\frac{\partial P}{\partial y})_i(\frac{\partial P}{\partial y})_j]\Big]=\sum_{m=1}^{Nq}\psi_i(Y_o-1) \quad (6)$$

where i = 1,...2N

This discretization is made by dividing up the segment BC on N_y points, the curve CD on N_q points and the DE segment on N_x points ; the sum of N_q, N_x and N_y forms the total number of points NT.

3.2. Accuracy of calculations

The calculations concern the flow structure and several physical data for a large range of porosities. Some data values may be very small particularly in the recirculating domains. That is why a high accuracy of calculations is necessary to detect clearly the stagnant regions between two neighbour cylinders. This accuracy is given by the average residual values of minimization (denoted ε) calculated on the dashed line of Fig. 1b whose expression is :

$$\varepsilon = \overline{V_y} + \overline{\varsigma} + \overline{V} \tag{7}$$

with :

$$V = \sqrt{Vx^2 + Vy^2}$$

where $\overline{V_y}$, $\overline{V}$ and $\overline{\varsigma}$ are the average adimensional values of the velocities on BC, CD and the vorticity on DE respectively. In our calculations, the adequate parameters are NT = 200 and 2N = 40. In the most considered cases, the calculations have been made with an excellent accuracy (better than 10^{-3}) as shown in Fig. 2.

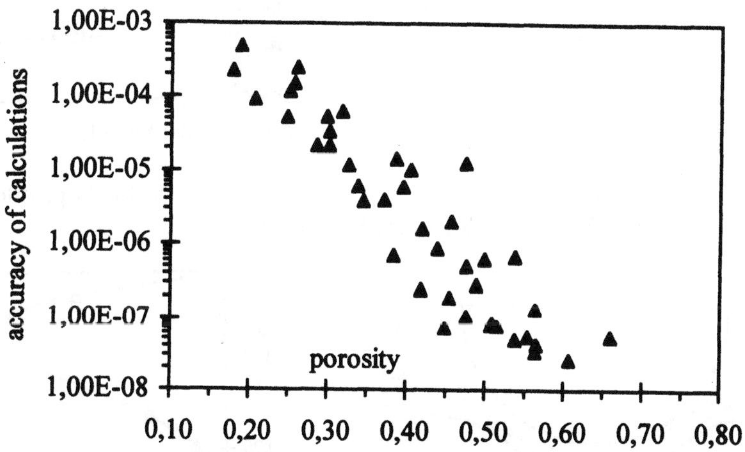

Fig. 2 : Accuracy of calculations for different geometric configurations

4. MICROSCOPIC ANALYSIS : FLOW STRUCTURE

The flow microscopic structure is analysed by means of the streamlines within the elementary cell representative of the pore space. Two kinds of flow structure can be found according to the shape of the streamlines. Figure 3a shows a flow structure containing only open streamlines. In this case streamlines become parallel to the axis of flow in the region between the cylinders 1 and 2. The velocity profile along the vertical axis of symmetry

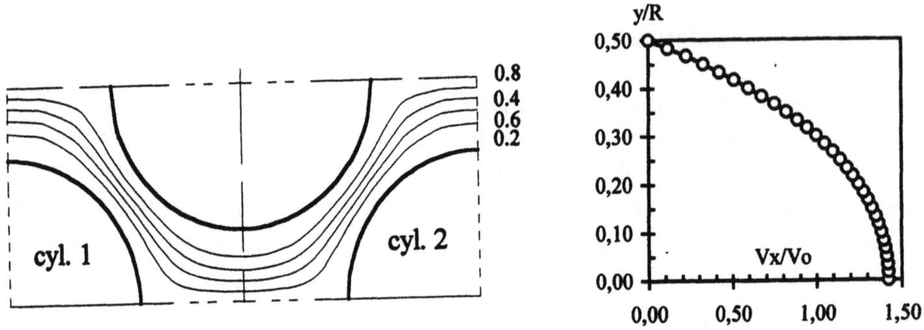

Streamlines Velocity profile

(a) : X_o =1.80 - Y_o = 1.50 - n = 0.42

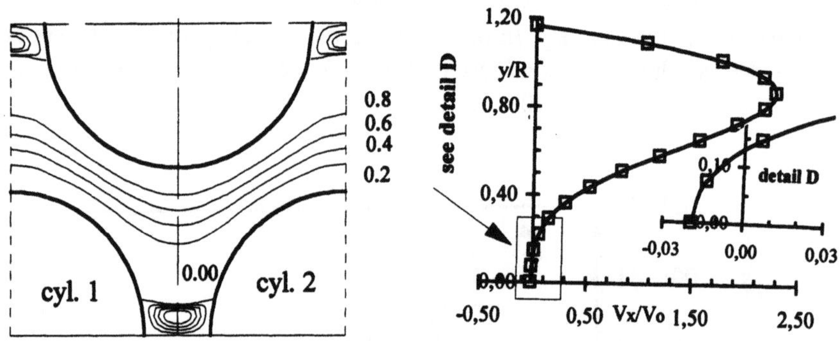

Streamlines Velocity profile

(b) : X_o =1.25 - Y_o = 2.165 - n = 0.42

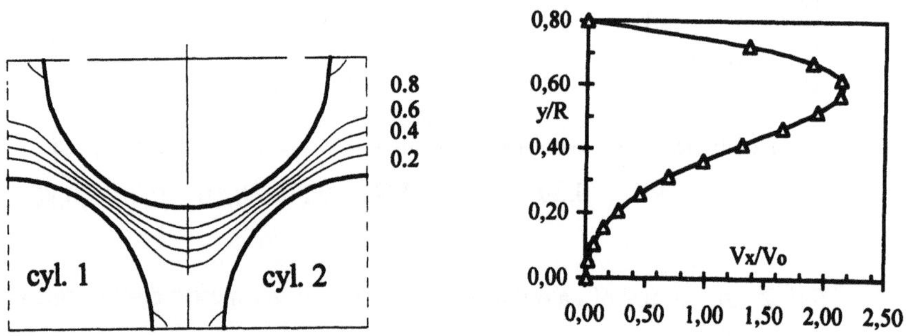

Streamlines Velocity profile

(c) : X_o =1.25 - Y_o = 1.80 - n = 0.30

Fig. 3 : Flow structure without recirculation domain **(a)** ; the appearance of a full eddy between the cylinders 1 and 2 **(b)** ; the appearance of corner eddies near the cylinders 1 and 2 **(c)** ; 0.00, 0.2, 0.4, 0.6 and 0.8 are values of stream function related to the value of stream function on DE line.

shows that the sign of the horizontal velocity is kept constant and that the maximum is reached near the axis of the flow. Figure 3b presents an other kind of flow structure in which both open streamlines and closed ones forming a cellular domain or stagnant region are found. In the cellular domain containing here only one eddy, the average normalized velocity is very small (approximately 10^{-3}) and a change of the sign of the velocity is observed near the horizontal axis. The cellular domain can also include more than one eddy (El Yazidi, 1997) or just corner eddies as shown in Figure 3c. This last case corresponds to transition between the above structures.

The extent of the cellular region is an important parameter regarding contamination problems as it contributes to pollutant retention due to the low velocity. This stagnant zone is quantified here by the ratio of the area of the stagnant domain to the area of the pore space. Figure 4 shows that the stagnant domain can reach up to 20 % for low values of the global porosity but does still exist for higher values of this last (n = 50 %).

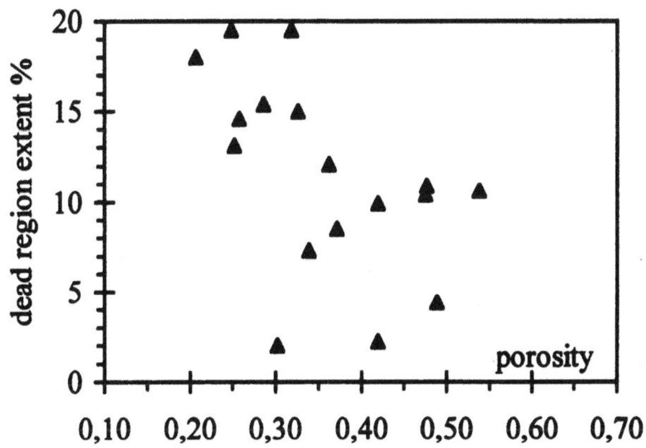

Fig. 4 : Evolution of the extent of the stagnant zone with porosity

5. MACROSCOPIC RESULTS : PERMEABILITY

5.1. Expression of porosity and permeability

porosity

The porosity (n), defined by the ratio of the area of pore space to the total area of the elementary cell, is expressed by :

$$n = 1 - \frac{\pi}{2\,X_o\,Y_o} \tag{8}$$

macroscopic permeability

The equivalent permeability of our porous media model is calculated by using Darcy's law hypothesis justified by the Stokes regime (small Reynolds number). Given a known rate flow (Q) through the array of cylinders, we calculate the pressure drop (P_o-P_1) over the elementary cell. Under this consideration, the application of the Darcy's law gives a linear relation between the macroscopic gradient of pressure and the flow rate. This law is expressed as following :

$$\frac{Q}{(y_o - R)} = \frac{k}{\rho\,g}\,\frac{P_o - P_1}{x_o} \quad \text{or} \quad \frac{Q}{(y_o - R)} = \frac{K}{\mu}\,\frac{P_o - P_1}{x_o} \tag{9}$$

with :

k :	Darcy permeability depending on solid matrix and fluid,
P_o, P_1 :	pressure at cell entrance and exit with P_o is taked arbitrary equal to zero,
μ, ρ, g :	dynamic viscosity of the fluid, fluid density and gravity respectively,
K :	geometric permeability depending on solid matrix only.

The normalized equivalent permeability K^* is : $$K^* = \frac{K}{R^2} = \frac{X_o}{\Delta p^*} \tag{10}$$

with : $\Delta p^* = \dfrac{P_o - P_1}{\mu \dfrac{Q}{(y_o - R)R}}$: the normalized pressure drop.

5.2. Results analysis

Figure 5 illustrates the variation of the normalized permeability of the porous media versus the porosity. We observe that the normalized permeability values are located within a large range and increase with the porosity as in real porous media.

More precisely, we consider the Kozeny-Karman relationship between the permeability K' and the global porosity n :

$$K' = \frac{n^3}{t_k\,\Sigma^2} \tag{11}$$

where Σ is the specific surface and t_k a Kozeny constant.

For a porous media formed by an assembly of spheres t_k has found equal to 5. Compared with the Kozeny formula with t_k=5, the calculated permeability of our model seems in good agreement essentially for porosity between 0.30 and 0.50.

With our model, the specific surface is related with the porosity by :

$$\Sigma = 2(1-n) \qquad (12)$$

The permeability can thus be written as a function of the porosity only. Such an approximate relationship is given in Figure 5.

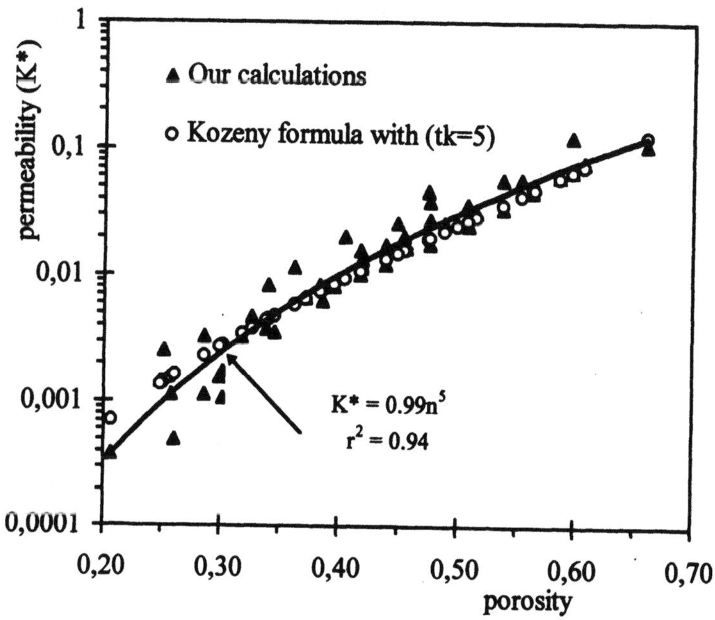

Fig. 5 : Permeability evolution and comparison with Kozeny formula.

6. CONCLUSION

This numerical study has allowed us to show the presence of some recirculation phenomena which may appear between the solid matrix in the real porous media. We have demonstrated that the extent of the stagnant domains can reach about 20 % of the global space pore and its evolution is closely linked to the geometric parameters of the porous media. The macroscopic permeability is well correlated with porosity. The present microstructural model is thus representative of observed fluid motion phenomena in soils.

REFERENCES

BRUSCHKE M.V. & S.G. ADVANI (1993). Flow of generalized Newtonian fluids across a periodic array of cylinders. J. Rheol., Vol. 37, No. 3, pp. 479-498.

CALTAGIRONE J.P. & E. ARQUIS (1986). Recirculations en milieux poreux. C. R. Acad. Sc. Paris, t. 302, série II, n° 14, pp. 843-846.

EL YAZIDI M. (1997). Contribution à l'étude microstructurelle des écoulements de Stokes dans les milieux poreux. Thèse de Docteur, INSA de Rennes, France.

MOREAU F., M. HELLOU & M. EL YAZIDI (1998). Ecoulements cellulaires de Stokes dans un canal plan obstrué par une file de cylindres. Z. angew. Math. Phys. Vol. 49, pp. 31-45.

WANG C.Y. (1996). Stokes flow through an array of rectangular fibers. Int. J. Multiphase Flow, Vol. 22, pp. 185-194.

NUMERICAL ANALYSIS OF DEEP EXCAVATIONS

INFLUENCE OF THE CONSTITUTIVE MODEL, DIFFERENT STRUT LEVELS AND WALL STIFFNESS

M.G. Freiseder and H.F. Schweiger
Technical University Graz, Graz, Austria

ABSTRACT: Finite element calculations for predicting deformations due to deep excavations are presented. As it is a much debated issue how complex the constitutive model has to be in order to obtain the results required from the practical point of view a numerical study is made for a deep excavation. The problem has been chosen such it can be regarded as a simplified analysis of a real construction site. The influence of the constitutive model is shown by comparing results obtained from a Mohr-Coulomb and a Hardening Soil model. In addition, the influence of different strut levels and the wall stiffness on deformation behaviour and bending moments is investigated.

1 INTRODUCTION

Construction and design of deep excavations is a challenging task for both geotechnical and structural engineers. Because of the site specific properties of the surrounding soils and structural system and the complexity of the construction sequence it is impossible to develop close-form solutions. This is the main reason why numerical methods have gained acceptance as a tool for designing deep excavations. The purpose of an analysis of a deep excavation is to design an optimum system as well as the construction sequence that will ensure that the deformation of the system and the surrounding strpctures will be within predecided tolerable limits. In order to assess the merits of various construction methods and excavation sequences with respect to ground deformations numerical analyses prove to be of considerable help. A number of researchers have addressed therefore the problem of analysing deep excavations and numerical studies have been presented ranging from fundamental earth pressure studies (e.g. Potts & Fourie 1986) to more practical oriented analyses (e.g. Powrie & Li 1991, Whittle et al. 1993).

Although significant progress has been made in numerical modelling of deep excavations a number of problems still exist at least as far as the application in practice is concerned. It is a much discussed issue, how complex the constitutive law has to be to obtain results to the accuracy required from the practical point of view. It is frequently argued that even with very sophisticated models results will be reliable only to a certain degree because of uncertainties in input parameters which are inevitable in geomechanics. Besides the important question of constitutive modelling which may have a significant influence on the results in terms of the overall deformation behaviour (e.g. Simpson 1992, Whittle & Hashash 1994) one has to be clear that it is almost impossible to include the actual construction sequence into the model in great detail. The latter aspect is especially true if 2-D analyses are performed, an approach which is commonly adopted although occasionally results from 3-D analyses are published (e.g. Ou et al. 1996).

2 GEOMETRY AND SOIL PROPERTIES

This study is concentrated on a symmetric plane strain excavation geometry. Fig. 1 shows a cross section of an excavation which has been chosen for a comparison of the influence of the constitutive law. According to this Figure the excavation is 30 m wide and 12 m deep. The construction steps, prop levels and a simplified soil profile are also indicated. The finite element code PLAXIS Version 7 (Brinkgreve et al. 1998) is used for the finite element simulation. Different types of elements are used in an attempt to reflect the actual behaviour of each components of the system: the soil is modelled utilizing 6-noded triangle elements. Diaphragm walls and struts simulated by beam elements and bar elements respectively. Interface elements are used to describe the relative slip between soil and wall. The wall is assumed to be in place i.e. the construction of the wall has not been modelled. The diaphragm wall was considered as linear elastic with parameters given in Table 1. The strut stiffness was chosen with 1.0E+05 kN/m.

diaphragm wall	E (MPa)	ν (-)
concrete	22 000	0.15

Table 1. Stiffness parameters for diaphragm wall

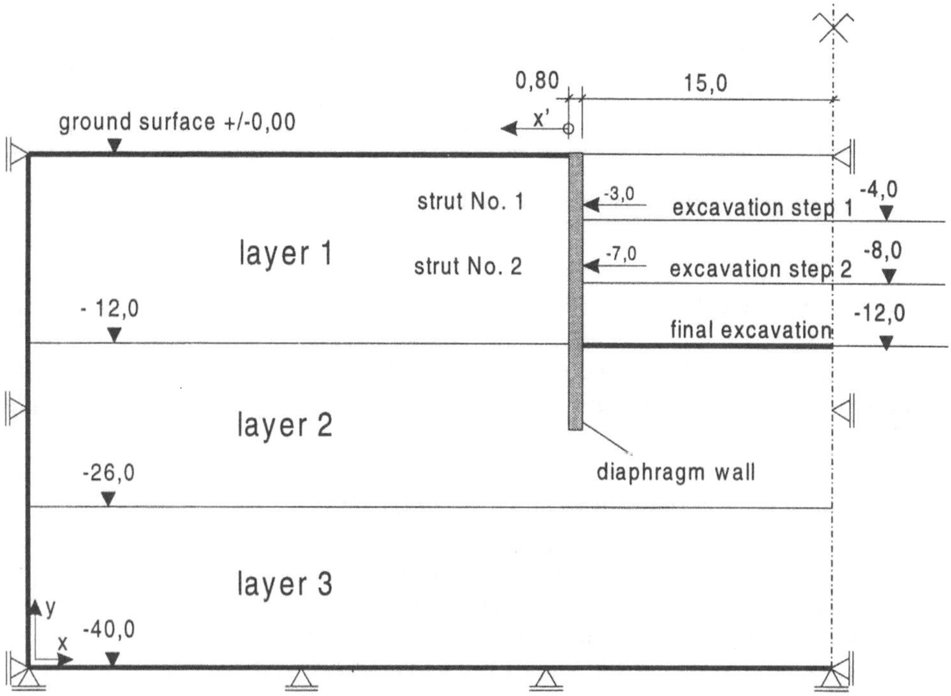

Fig. 1. Cross section of excavation

The finite element analyses followed the construction sequence (except for the installation stage of the diaphragm wall itself) as described in the following steps:

step 1: initial conditions
(layer 1: $K_o = 0.5$, layer 2+3: $K_o = 0.65$) and self weight of diaphragm wall
step 2: excavation to a depth of 4.0m
step 3: excavation to a depth of 8.0m and prop in -3.0m in place
step 4: excavation to a depth of 12.0m and props in -3.0 and -7.0m in place.

Two different soil models, namely a Mohr-Coulomb and a Hardening Soil model have been investigated by comparing results obtained from these models.

2.1 Mohr-Coulomb model

The strength and stiffness parameters for the soil layers used in the drained, elastic perfectly plastic analysis using a Mohr-Coulomb failure criterion, which serves as a reference calculation are given in Table 2.

soil layer	frict. angle (°)	dilat. angle (°)	cohesion (kPa)	E (MPa)	ν (-)	K_o (-)
layer 1	35.0	5.0	2.0	20.0	0.30	0.50
layer 2	26.0	0.0	10.0	12.0	0.40	0.65
layer 3	26.0	0.0	10.0	80.0	0.40	0.65

Table 2. Strength and stiffness parameters for Mohr-Coulomb model

2.2 Hardening Soil model

The Hardening Soil model is intended to model the behaviour of gravel, sand and heavily overconsolidated cohesive soils but may be also used successfully for excavation problems in soft soils.

Some basic characteristics of the model (Schanz 1997) are:

- Stress dependent stiffness according to a power law.
- Hyperbolic relationship between strain and deviatoric stress.
- Distinction between primary deviatoric loading and unloading/reloading.
- Failure behaviour according to the Mohr-Coulomb model.

The assumed hyperbolic relationship between vertical strain and deviatoric stress is described in (1):

$$\varepsilon_1 = \frac{1}{E_i}\frac{q}{1-q/q_a} \tag{1}$$

E_i is the initial stress dependent Young's modulus and is given by the following equation:

$$E_i = E_i^{ref}\left(\frac{c\cot\varphi - \sigma_3'}{p^{ref}}\right)^m \tag{2}$$

For unloading and reloading stress path, another stress dependent stiffness modulus is used:

$$E_{ur} = E_{ur}^{ref}\left(\frac{c\cot\varphi - \sigma_3'}{p^{ref}}\right)^m \tag{3}$$

Additional parameters for the soil layers used in the drained, elastoplastic analysis using Hardening Soil model are given in Table 3. The unloading / reloading Poisson ratios may appear somewhat unusual. However, they are chosen to allow comparison to results obtained by the Mohr-Coulomb model.

soil layer	E_{50}^{ref} (MPa)	E_{ur}^{ref} (MPa)	ν_{ur} (-)	m (-)	p^{ref} (kPa)	R_f (-)
layer 1	7.0	20.0	0.3	0.50	100	0.9
layer 2	4.0	12.0	0.4	0.50	100	0.9
layer 3	27.0	80.0	0.4	0.50	100	0.9

Table 3. Parameters for Hardening Soil model

E_{50}^{ref} ..... Primary loading stiffness (secant modulus at 50% strength)
E_{ur}^{ref} ..... Unloading / reloading stiffness
ν_{ur} ..... Unloading / reloading Poisson ratio
m ..... Constant in stiffness law which controls stress depending stiffness
R_f ...... Failure ratio

3 COMPARISON OF RESULTS

Fig. 2 and Fig. 3 illustrate the results obtained from the Mohr-Coulomb model and the Hardening Soil model. The heave of the wall is shown in Figure 2. The Hardening Soil model results in 50% smaller vertical displacement than the Mohr-Coulomb model. The heave in the middle of the bottom of the excavation (Fig. 3) obtained from the Hardening Soil model is also less pronounced if compared to the Mohr-Coulomb model. The Hardening Soil model results in larger horizontal deflections of the top of the wall (Fig. 2) which are directed towards the excavation in all loadcases (+ ve in Fig. 2). In general, the deformations obtained by the Hardening Soil model are considered as more realistic than results obtained from the Mohr-Coulomb model.

4 EFFECT OF WALL AND STRUT STIFFNESS

The following studies are performed using a Mohr-Coulomb model and the parameters previously described. Diaphragm walls with various thicknesses d and various strut stiffnesses D at the final stage of excavation are compared with respect to wall bending moments (Fig. 4) and horizontal wall deflection (Fig. 5). Nine cases were considered, with

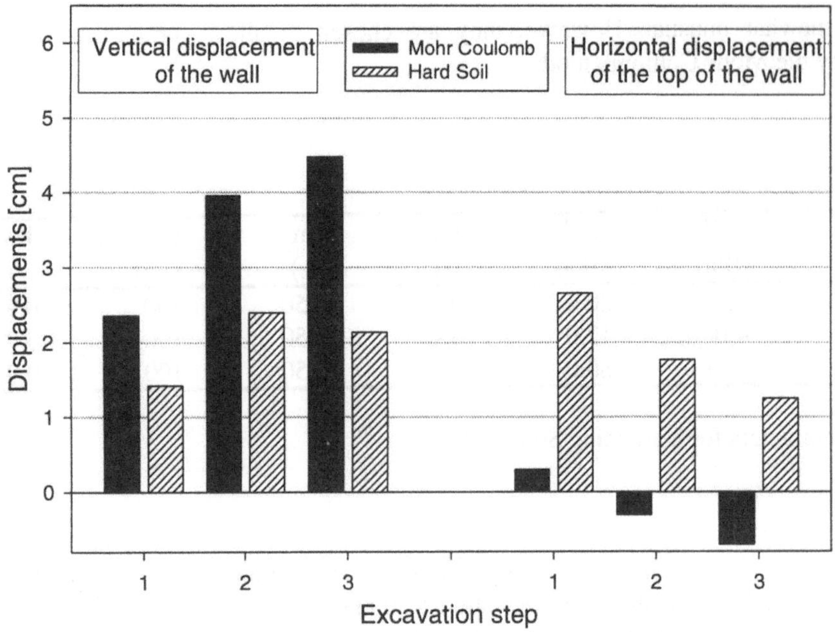

Fig. 2. Heave and horizontal displacement of the top of the wall

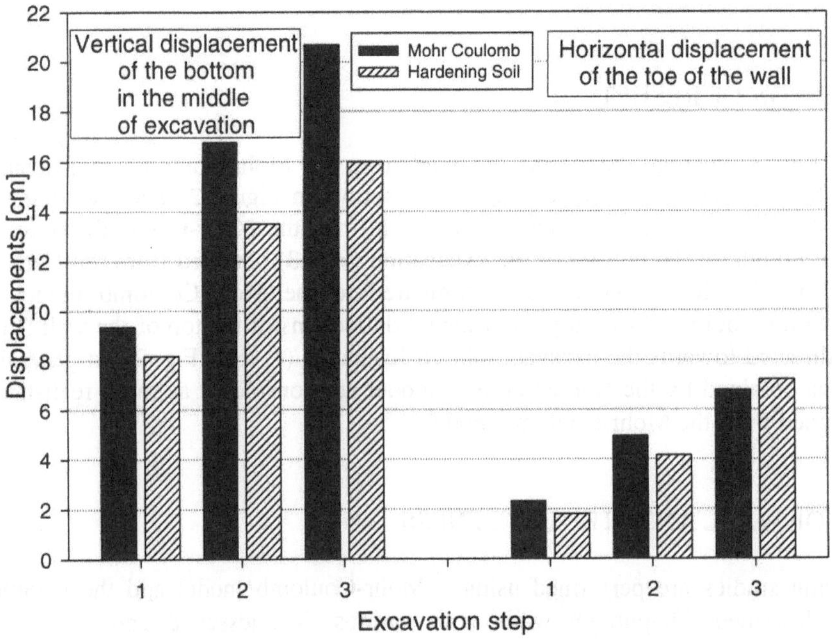

Fig. 3. Horizontal displacement of the toe of wall and heave in excavation

wall thickness d of 0.6 m, 0.9m and 1.2m combined with strut stiffness D of 1E+04 kN/m, 5E+04 kN/m and 1E+10 kN/m respectively. A strut stiffness of 1E+10 kN/m can be considered as extremely rigid whereas strut stiffnesses of 1E+04 kN/m and 5E+04 kN/m are in the range often used in practice. As expected the more rigid the struts and the wall become the smaller the deformation of the wall will be. On the other hand this will result in large bending moments in the wall.

Increasing the thickness of the wall from 60 cm to 90 cm reduces displacements by about 5mm only. Further increasing the thickness from 60 cm to 120 cm results in about 10 mm less displacements but leads to an increase of the bending moments of up to 240%. In general, for reducing the wall displacements it is more efficient to install rigid struts than increasing the wall thickness. By optimizing both bending moments and wall displacements best results are obtained by the combination of rigid struts and thin walls.

The differences of the maximum bending moment in the chosen examples exceeds 300%. Therefore Fig. 4 and 5 clearly indicate the importance of correctly assuming wall- and strut stiffnesses, if finite element analysis are used to design support systems.

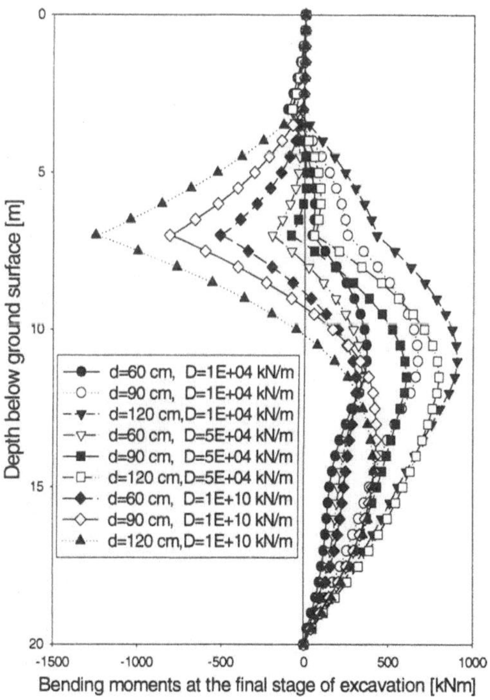

Fig.4. Bending moments

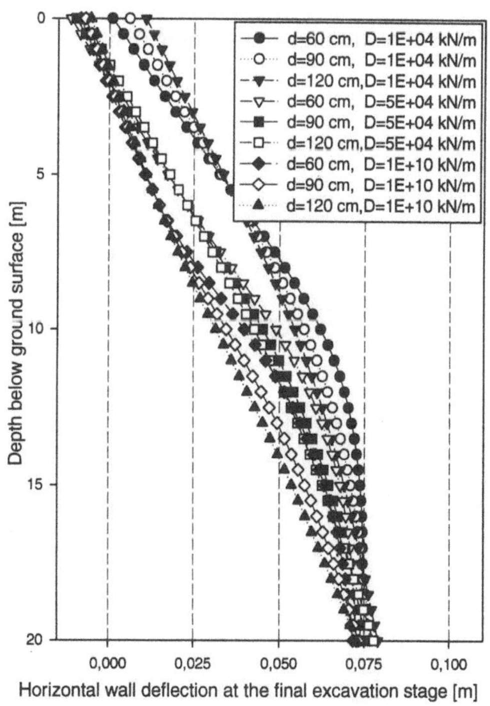

Fig. 5. Horizontal wall displacements

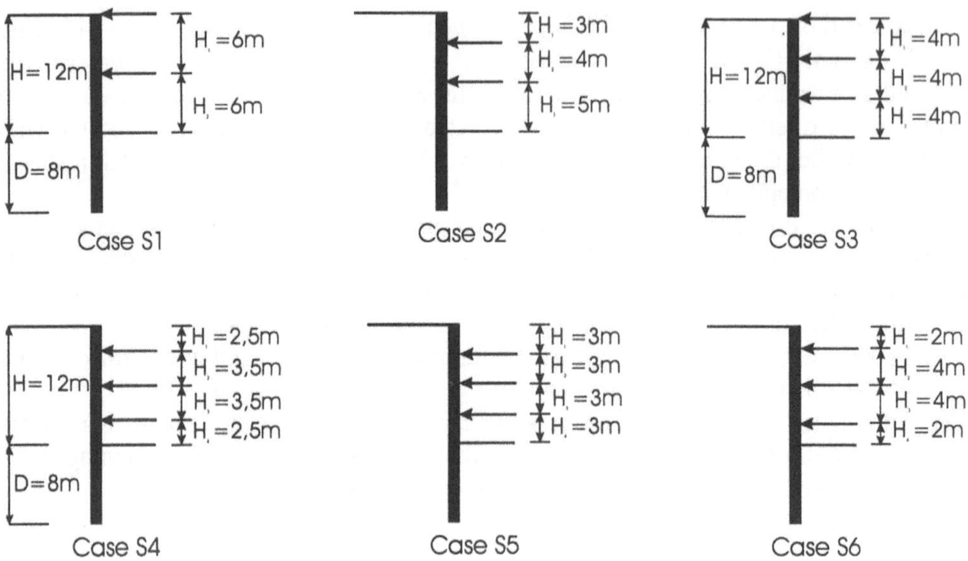

Fig. 6. Positioning and spacing of struts

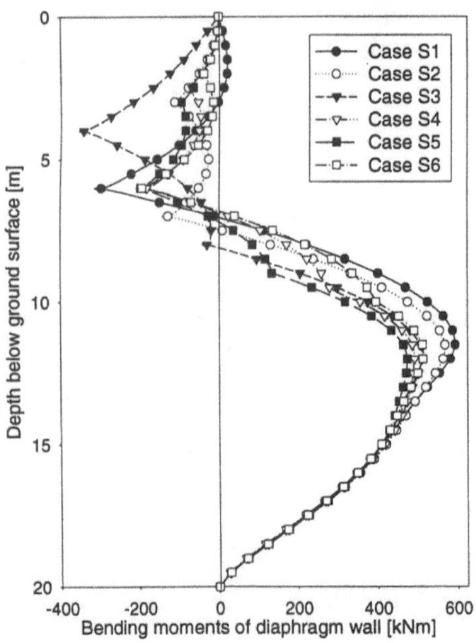

Fig. 7. Bending moments

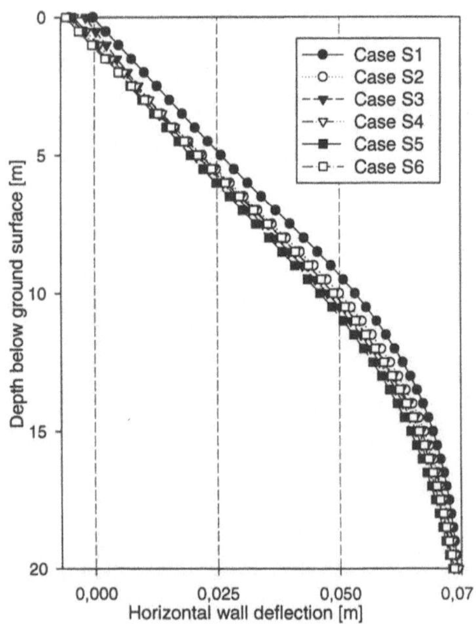

Fig. 8. Deflections of the wall

5 EFFECT OF POSITION AND SPACING OF STRUTS

The following studies are also performed using a Mohr-Coulomb model and the parameters previously described. The strut stiffness was assumed to be 5.0E+05 kN/m. In this study six cases are compared as shown in Fig. 6. The results of this part of the numerical study are illustrated in Fig. 7 and Fig. 8. Case 5 results in the lowest horizontal displacements and bending moments of the wall. Case 1 leads to highest wall deflections and bending moments of the wall, which are 25% higher if compared to Case 5. The position and the number of struts have a significant influence on the bending moments but less effect on the horizontal displacements of the wall. As a consequence, proper arrangement of the struts may result in favourable technical and economic factors.

6 CONCLUSION

A numerical study of a deep excavation in soft soil has been presented. The influence of the constitutive law on the deformation behaviour has been shown by comparing results obtained by a Mohr-Coulomb and a Hardening Soil model. The deformations obtained by the Hardening Soil model are in general considered as more realistic than results obtained from the Mohr-Coulomb model.

The parametric studies indicate that the thickness of the wall has significant effect on the deformation and bending moment. The assumed range of the parameters in the chosen examples leaded in differences in the maximum bending moments more than 300%. Therefore the results of the performed numerical study clearly indicate the importance of correctly assuming wall- and strut stiffnesses, if finite element analysis are used to design support systems.

The effect of number and position of the struts on deformation and bending moment, as presented in this paper, has proven to be an important factor, as well. For the considered examples the differences have been about 25 % for the maximum bending moment. One can conclude, that is worth to carry out numerical studies in order to optimize support systems of deep excavations.

Due to the fact of uncertainties in input parameters which are inevitable in geomechanics one should emphasize the interaction between in-situ measurements and numerical computation. It is important for the designer to obtain field measurements of crucial indicators of performance during early stages of construction. These should include heave within the excavation, settlement behind the wall, pore water pressure and the measurements of wall deflection at a few key locations. If these measurements disagree with the predicted behaviour in essential points, analysis and design should be reviewed and revised as necessary.

REFERENCES

BRINKGREVE R.B.J. & P.A. VERMEER. PLAXIS – Finite Element Code for Soil and Rock Analyses. Version 7. A.A.Balkema, Rotterdam, Brookfild (1998).

OU C.-Y., D.-C. CHIUO & T.-S. WU (1996). Three - Dimensional Finite Element Analysis of Deep Excavations. J. of Geotech. Engineering 122: 337-345.

POTTS D.M. & A.B. FOURIE (1986). A numerical study of the effects of wall deformation on earth pressures. Int. J. Num. Anal. Meth. Geom. 10: 382-405.

POWRIE W. & E.S.F. LI (1991). Finite element analyses of an in situ wall propped at formation level. Geotechnique 41: 499-514.

WHITTLE A.J., Y.M.A. HASHASH & R.V. WHITMAN (1993). Analysis of deep excavation in Boston. J. of Geotech. Engineering 119: 69-90.

WHITTLE A.J. & Y.M.A. HASHASH (1994). Soil modeling and prediction of deep excavation behavior. In (Shibuya et al. eds.), Proc. Int. Symp. Prefailure Deformation of Geomaterials: 589-594. Rotterdam, Balkema.

SCHANZ T. (1997) Introduction to Hard Soil model. Proc. Course for experienced PLAXIS users. Stichting Postacademisch Onderwijs. Netherlands.

SIMPSON B. (1992) Retaining Structures: displacement and design. Geotechnique 41: 541-576.

ON THE EFFECTS OF PARAMETERS IN A NONLINEAR THERMO-HYDRO-MECHANICAL SOIL MODEL

B. Gatmiri and B. Jenab
CERMES, École Nationale des Ponts et Chaussées, Paris, France

ABSTRACT : Investigation on the thermo-hydro-mechanical behaviour of two-phase and multi-phase geomaterials is of a great importance in the problems of geotechnical and environmental engineering and more particularly in industrial, urban and nuclear waste disposals. The set of field equations of fully coupled thermo-hydro-mechanical behaviour of a saturated clayey engineering barrier based on the concept of the state surface of void ratio with a non-linear (hyperbolic) elastic response of the solid skeleton has been presented by (Gatmiri 1995). Study of the influence of different parameters on the thermo-hydro-mechanical response of soils in two dimensional case is the subject of this paper. Permeability, thermal conductivity and heat capacity of the medium seem to have a significant effect on the behaviour of the medium. It is concluded that the coefficient of permeability and the thermal conductivity have identical effects on the rate of consolidation, in such a way that the increase in their values causes an increase in the rate of consolidation. The contours of temperature, water pressure distribution and settlement at certain time steps after the application of loads with different values of heat capacity, thermal and hydraulic conductivity are presented.

1. INTRODUCTION

A great deal of attention is focused on coupled thermohydromechanical behaviour of fluid-saturated porous media in diverse areas such as : extraction of oil or geothermal energy, road subgrade and pavement subjected to heating-cooling cycles, frictional heating in fault zones in rocks and soils and specially high-level radioactive waste disposal.

From review of models treating the thermo-hydro-mechanical behaviour of two phase porous media, it appears that an important number of models is based on extension of isothermal consolidation theory to account for the effect of thermal expansion of elastic

matrix (e.g., Schiffman 1972). In these studies the effect of convection has been disregarded and the effect of thermal expansion and compressibility of pore fluid were neglected (indirect coupling between the energy and fluid flow, Noorishad et al. 1984).
Based on their previous works in which, the solid is assumed to be elastic linear and saturated by an incompressible fluid, Aboustit et al. (1985) have presented a finite element approach for thermo-elastic consolidation. They have ignored the convection. The coupling terms between temperature and pressure does not appear in their formulation. This results in a symmetric general matrix.
Booker and Savvidou (1985) provided analytical solutions for a thermoelastic consolidation which takes into account the differential thermal expansion of the pore water and soil skeleton. The temperature field is uncoupled from pressure and displacements by ignoring the mechanical contributions to energy balance and the convective terms.
An extension of the isothermal theory of Rice and Cleary (1976), considering the compressibility of fluid and solid constituents, as well as thermal expansion of both phases is given by McTigue (1986). This theory is a fully linearized theory, and in addition to many limitations, the convective heat transfer is also neglected. Lewis et al. (1986) have presented a coupled finite element model for the analysis of the deformation of elastoplastic porous media under the effect of heat and fluid flow. The convection terms are neglected in this formulation.
In this paper, based on the model of THM behaviour of a non linear elastic saturated soil presented by Gatmiri (1995), Gatmiri and Delage (1997), the influence of different parameters on the soil behaviour is investigated. In this model, the influence of temperature on Young elastic modulus, bulk modulus and hydraulic and thermal properties of the porous medium, has been considered. Effects of changes in soil characteristics such as, permeability, heat capacity and thermal conductivity are also envisaged with the aid of results of a two dimensional example of the presented THM model.

2. BRIEF PRESENTATION OF NONLINEAR ELASTIC MODEL

In this part of the paper, set of governing equations for thermo-hydro-mechanical behviour of a nonlinear elastic saturated soil based on the concept of "thermal state surface of void ratio" is presented (Gatmiri, 1995).
Governing equations for the behaviour of solid skeleton are the followings;
- equilibrium equation ; $\sigma_{ij},_j + \rho g_i = 0.$ (eq. 1)

- constitutive law under small strain assumption ; $\sigma''_{ij} = D_{ijkl}(\varepsilon_{kl} - \varepsilon^p_{kl} - \varepsilon^t_{kl})$. (eq. 2)

and so for mechanical parameters we have ; σ_{ij}'' : effective stress,

$$\sigma_{ij}'' = \sigma_{ij} - \alpha\, p\, \delta_{ij} - \beta\, T\, \delta_{ij}, \quad \text{(eq. 3)}$$

with ; α : Biot hydromechanic coupling coefficient, β : thermomechanical coupling coefficient, $\beta = (2G + 3\lambda)\alpha_s = 3K'\alpha_s$, (eq. 4)

in which α_s is solid thermal expansion coefficient. K' is effective bulk modulus ;

$K' = (2G+3\lambda)/3$, (eq. 5)
where: G, λ : Lame's coefficients, ε_{kl} : strain tensor, D_{ijkl} : stress-strain relationship tensor, ε^{p}_{kl} : thermal volumetric deformation due to isotropic compression produced by pore water pressure, ε^{t}_{kl} : thermal deformation.
The equations concerning pore fluid motion are as follows;

- fluid mass conservation ; $\dot{\zeta} = -\dot{w}_{i,i} = \alpha\dot{\varepsilon}_{ii} + C_a\dot{P} + C_h\dot{T}$. (eq. 6)
- Darcy's law ; $\dot{w}_i = -\Lambda_{ij}(p_{,i} + \rho_w g_i)$. (eq. 7)

hence, hydraulic parameters appearing in the above-mentioned equations are as follows; C_a: mixed fluid-solid compressibility, $C_a = (\alpha - n)/K_s + n/K_w$, (eq. 8)

with, K_s : solid bulk modulus, K_w : fluid bulk modulus, C_h : coefficient of undrained thermal expansion of mixture, $C_h = (\alpha - n)\alpha_s + n\,\alpha_w$, (eq. 9)

where
α_s : coefficient of thermal expansion of soil, α_w : coefficient of thermal expansion of water, ζ: volumetric fluid content, Λ_{ij} : permeability which depends both on temperature and mechanical loading by the following equation ;

$$\Lambda_{ij} = \Lambda_{wij}\left(\frac{\nu_i}{\nu}\right)\left(\frac{e}{e_i}\right)^3 \Rightarrow \Lambda_{ij} = f(T,e). \quad \text{(eq. 10)}$$

Energy transport equations and Fourier law are the followings;
- energy conservation equation ; $\rho_w C_w T_{,i}\dot{w}_{i,i} - h_{i,i} = \beta T_0\,\dot{\varepsilon}_{ii} + C_b\dot{T} - C_T T_0\dot{p}$. (eq. 11)
- Fourier law as an equation governing the heat flow is given by ; $h_i = -k_{ij}\,T_{,j}$. (eq. 12)

where thermal parameters are given as follows;
h_i : heat flow vector, k_{ij} : solid-fluid mixture thermal conductivity tensor ;

$$k_{ij} = nk^{w}_{ij} + (1-n)k^{s}_{ij}, \quad \text{(eq. 13)}$$

with
k^{w}_{ij} : heat conductivity tensor of fluid, k^{s}_{ij} :, heat conductivity tensor of solid, T_0 : initial absolute temperature, $C_b = (\rho C)_M - \{(1-n)C_s\rho_s\alpha_s + nC_w\rho_w\alpha_w\}T_0$, (eq. 14)

with
$(\rho C)_M$: solid-fluid mixture heat capacity ;
$(\rho C)_M = (1-n)\rho_s C_s + n\rho_w C_w$, (eq. 15)

C_s : specific heat capacity of solid, C_w : specific heat capacity of fluid,

$$C_T = (1-n)C_s\rho_s/K_s + nC_w\rho_w/K_w\,. \quad \text{(eq. 16)}$$

Application of classical finite element principles to the above-mentioned equations gives the final matrix form as follows :

$$\begin{bmatrix} [R] & [L] & [C] \\ [B_T] & \theta\Delta t[K_T]+[C_{TT}] & [C_{TW}]+\theta\Delta t[G_T] \\ [C]^T & [C_{WT}] & \theta\Delta t[K_W]+[C_{WP}] \end{bmatrix} \begin{Bmatrix} \Delta U \\ \Delta T \\ \Delta P_w \end{Bmatrix} = \Delta t \begin{bmatrix} 0 \\ F_{q_0} - [K_T]T_0 - [G_T]p_{W_0} \\ F_{w_0} - [K_W]p_{w_0} \end{bmatrix} + \begin{Bmatrix} \Delta F_\sigma \\ \theta\Delta t\Delta F_q \\ \theta\Delta t\Delta F_w \end{Bmatrix},$$

where ;

$$[R]=\int_\Omega B^T D B d\Omega \qquad [L]^T=\int_\Omega N^T \beta m^T B d\Omega$$

$$[C]=\int_\Omega B^T m\alpha N d\Omega \qquad \{F_\sigma\}=\int_\Gamma \bar{N}^T \bar{\sigma} d\Gamma$$

$$[C]^T=\int_\Omega N^T.\alpha.m^T.B.d\Omega \qquad [C_{wp}]=-\int_\Omega N_W^T.C_a.N_w.d\Omega$$

$$[C_{wt}]=-\int_\Omega N_T^T.C_h.Nd\Omega \qquad [K_W]=\int_\Omega (\nabla N_W)^T.\Lambda_W.\nabla N_W d\Omega$$

$$\{F_W\}=\int_\Omega (\nabla N_W)^T \Lambda_W \gamma_W \nabla z d\Omega + \int_\Gamma \bar{N}_W^T.q_W.d_\Gamma \qquad [K_T]=\int_\Omega (\nabla N_t)^T.k_t.\nabla N_t.d\Omega$$

$$[B_T]=-\int_\Omega N^T.\beta T_o.m^T[B]d\Omega \qquad [C_{TT}]=-\int_\Omega N_t^T.C_b.Nd\Omega$$

$$[C_{TW}]=-\int_\Omega N_t^T.C_T T_0.N_t.d\Omega \qquad \{F_q\}=\int \bar{N}^T.q_t.d\Gamma+\int \nabla N.k_t.\nabla z.d\Omega$$

3. PARAMETRIC STUDY OF A BIDIMENSIONAL CASE OF A SATURATED SOIL

In this part of the paper, thermo-elastic consolidation analysis in two dimensions for a case of gasification is studied. Fig. 1 illustrates the stratification. The material properties are listed in Table 1.

For obtaining better and more precise results, we have refined the finite element mesh near the heat source where temperature changes are important.

In Fig. 2, surface settlement is presented at different times after thermal loading for elastic linear and elastic non linear cases. As it can be observed, settlements in non linear case are greater than linear ones which can be explained by nonlinear effects in mechanical behaviour and a degradation of soil mechanical properties in high temperatures. The combined effects of temperature and and pore pressure and their rates of distribution in the medium and deformations produced in the soil in each step of loading have been used for determining the deformation moduli. These deformation moduli govern the obtained settlement values.

It should be mentioned that in all the results presented hereafter, the parameters that are not tabulated remain unchanged.

In Fig.3, the temperature distribution for two case studies in which the parameters are presented in Table 2, are shown. It should be mentioned that layers are numbered from bottom to top. As can be observed, by increasing the heat conductivity of soil layers, an

increase in rate of heat distribution in the medium occurs. Indeed,the greater the value of heat conductivity, the greater are the temperature values of soil for a given time.
The settlement contours are shown in Fig. 4 for 13 and 21 months after thermal loading for two cases tabulated in Table 3. From the results, it can be concluded that, the values of soil settlements increase with increasing coefficients of permeability. In other words,

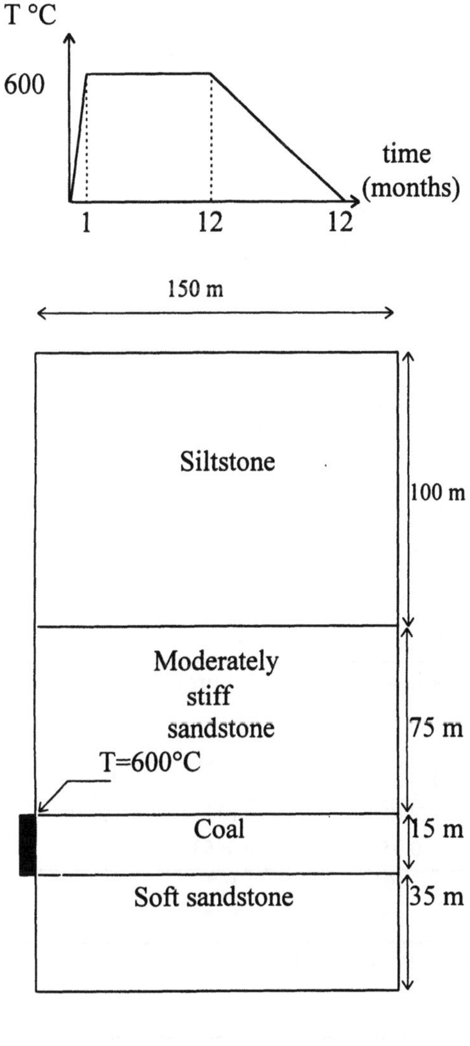

Fig. 1 : Thermal loading and strata description of the problem.

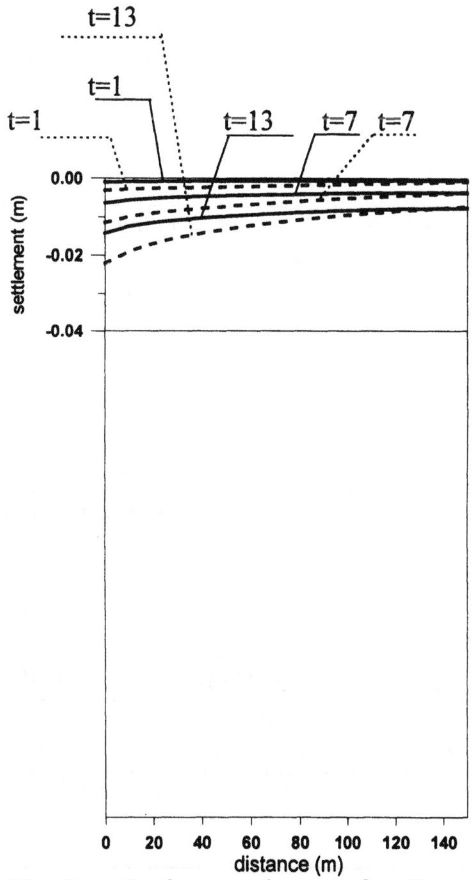

Fig 2 : Surface settlement for linear elastic (——) and non linear elastic (········) models at different times after thermal loading (time in months).

	Soft sandstone	Coal	Moderately stiff sandstone	Siltstone	Units
ρ	2.2	2.2	2.2	2.2	t/m^3
ϕ	30	30	30	30	degree
C	1.5	1.5	1.5	1.5	t/m^2
α_s	$0.8*10^{-5}$	$0.5*10^{-5}$	$0.8*10^{-5}$	$0.8*10^{-5}$	1/ °C
k_s	0.0463	0.05	0.0463	0.05	cal/ °C ms
C_s	0.32	0.19	0.32	0.32	Kcal/m³ °C
K_w	10^{-7}	$2*10^{-7}$	10^{-7}	$2*10^{-7}$	m/s
Mechanical parameters, linear elastic model :					
E	$0.2*10^8$	$0.2*10^8$	$0.6*10^8$	$0.1*10^8$	t/m^2
ν	0.2	0.2	0.2	0.2	-
B	$1.1*10^7$	$1.1*10^7$	$0.33*10^7$	$0.55*10^7$	t/m^2
Mechanical parameters, non linear(hyperbolic) elastic model :					
K_l	$0.2*10^7$	$0.2*10^7$	$0.6*10^7$	$0.1*10^7$	-
K_u	$0.2*10^7$	$0.2*10^7$	$0.6*10^7$	$0.1*10^7$	-
n	0.5	0.5	0.5	0.5	-
R_f	0.75	0.75	0.75	0.75	-
K_b	$1.1*10^6$	$1.1*10^6$	$0.33*10^6$	$0.55*10^6$	-
m	0.3	0.3	0.3	0.3	-

Table 1 : Material properties for the treated two dimensional problem.

		layer 1	layer 2	layer 3	layer 4
case (a)	K_w (m/s)	10^{-8}	$2*10^{-8}$	10^{-8}	$2*10^{-8}$
	k_s (cal/°C ms)	0.0463	0.05	0.0463	0.05
case (b)	K_w (m/s)	10^{-8}	$2*10^{-8}$	10^{-8}	$2*10^{-8}$
	k_s (cal/°C ms)	0.5	0.12	0.67	0.214

Table 2 : Values used for parametric study concerning Figs. 3, 5.

		layer 1	layer 2	layer 3	layer 4
case (a)	K_w (m/s)	10^{-7}	$2*10^{-7}$	10^{-7}	$2*10^{-7}$
	k_s (cal/°C ms)	0.5	0.12	0.67	0.214
case (b)	K_w (m/s)	10^{-8}	$2*10^{-8}$	10^{-8}	$2*10^{-8}$
	k_s (cal/°C ms)	0.5	0.12	0.67	0.214

Table 3 : Values used for parametric study concerning Figs. 4, 6.

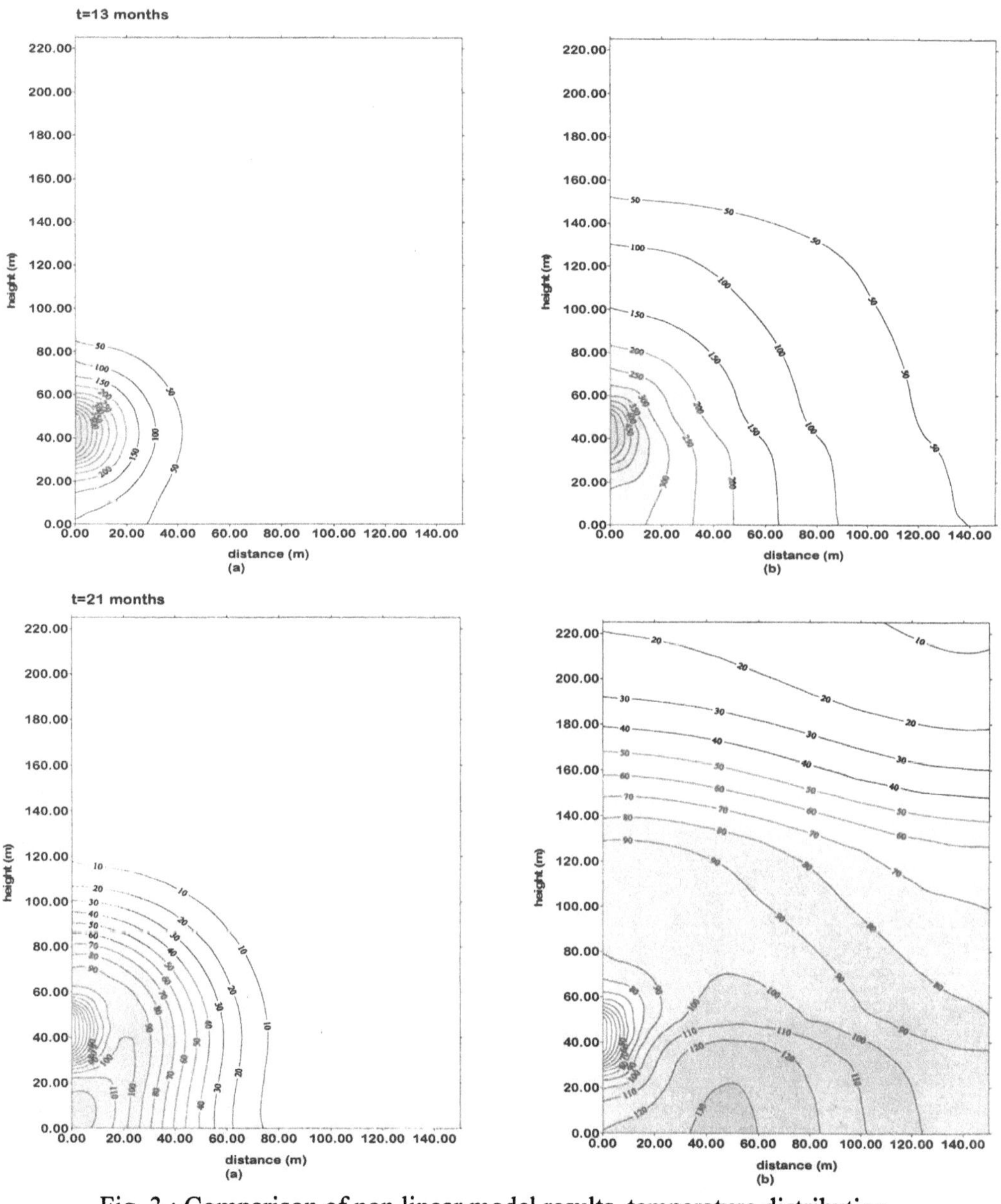

Fig. 3 : Comparison of non linear model results, temperature distribution.

consolidation phenomenon is more rapid in a more permeable soil which gives greater amounts of soil deformation for a given time.

Experimentally, it is shown that pore water pressure increases with temperature increase (Campanella & Mitchell, 1968). Hence, in a soil with higher thermal conductivity, one expects to have more important pore pressure values in a given time. Predictions for

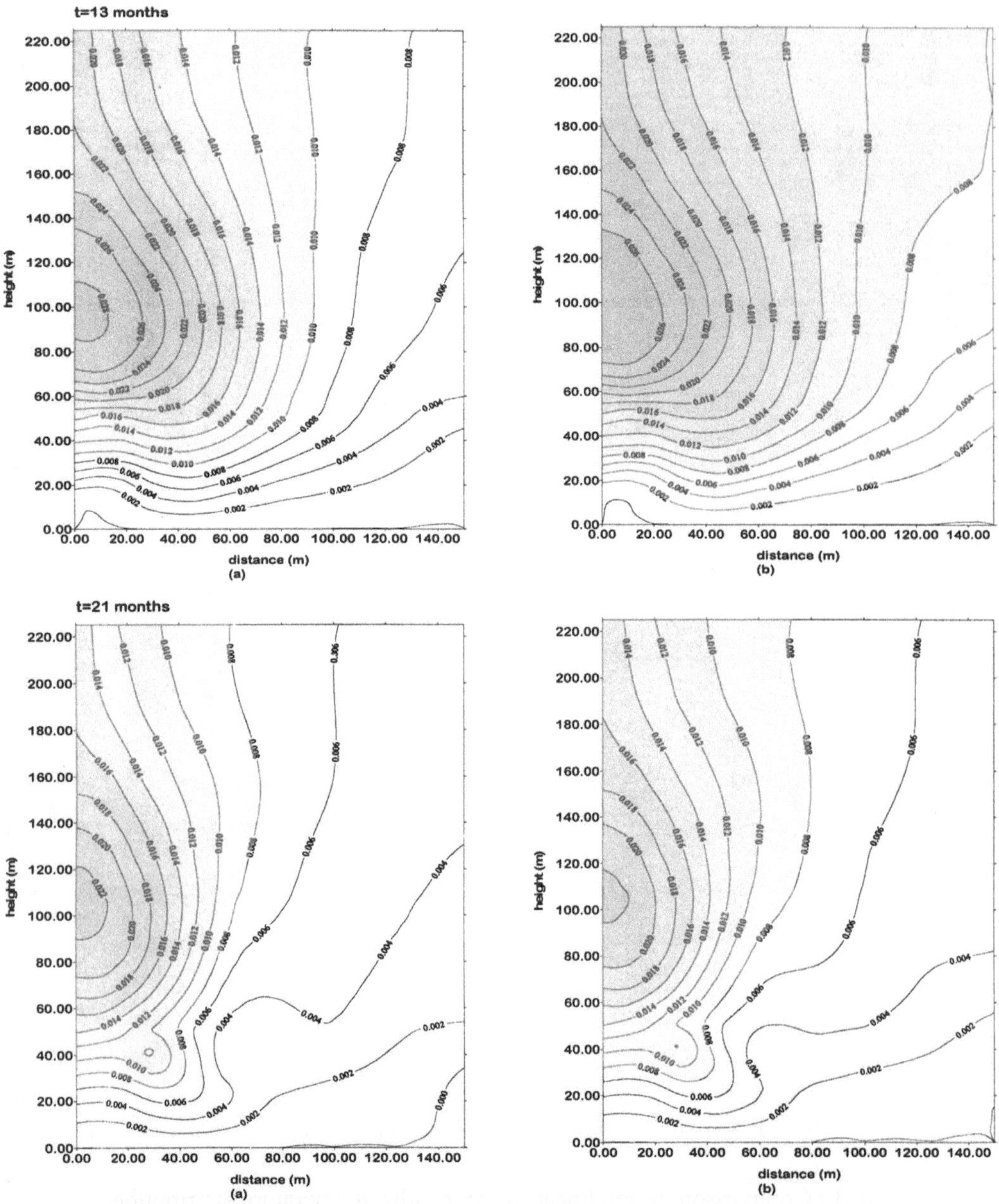

Fig 4 : Comparison of non linear model results, settlement cotours.

this case are presented in Fig. 5 for values given in Table 2. In more permeable soils, based on the above-mentioned experimental phenomenon, pore water pressure values are lesser because of the relatively higher dissipation rates of thermally generated pore pressures in these soils. Pore pressure predictions for permeabilities shown in Table 3, 13 months after thermal loading are presented in Fig. 6.

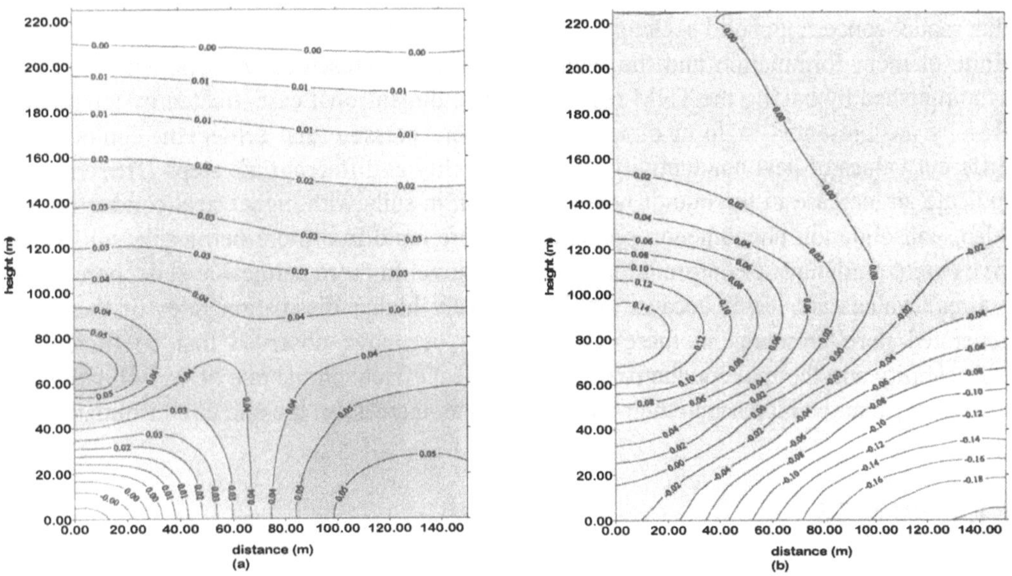

Fig. 5 : Comparison of the non linear model results, pore pressure distribution after 13 months.

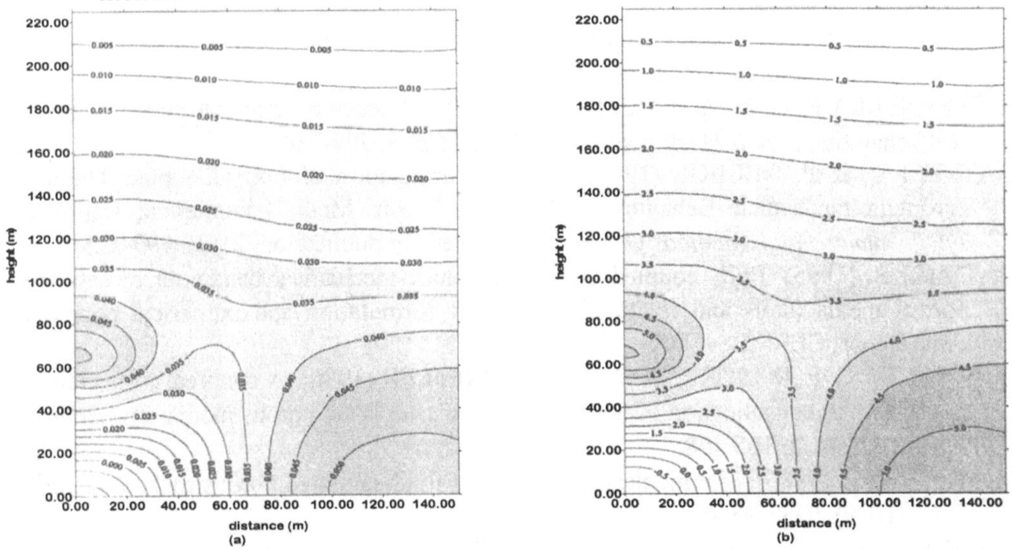

Fig. 6: Comparison of the non linear model results, pore pressure distribution after 13 months.

4. CONCLUSIONS

The main objective of this paper was to perform a parametric study of an elastic non linear (hyperbolic) thermo-hydro-mechanical model for a saturated soil. The basic equations of this model concerning solid skeleton, fluid phase and thermal transfer, as well as, the final finite element formulation and the related matrices are presented. A parametric study is accomplished by having the THM results of a two dimensional case, treated by the model. Results are presented in form of temperature, pore pressure and settlement contours for different values of heat conductivity and permeability at different time steps. These results indicate an increase in the rate of heat distribution in soils with higher heat conductivities. Also, consolidation phenomenon seems to be more rapid in a more permeable soil which gives greater amounts of deformation for a given time. In more permeable soils, pore water pressure values are lesser because of the relatively higher dissipation rates of thermally generated pore pressures in these soils. Finally, we have observed that coefficient of permeability and thermal conductivity have identical effects on the rate of consolidation, in such a way that the increase in their values causes an increase in the rate of consolidation.

5. REFERENCES

ABOUSTIT B. L., S. H. ADVANI & J. K. LEE (1985). Variational principles and finite element simulations for thermoelastic consolidation, *Int. J. Numer. Anal. Methods Geomech.*, **9**, *49-69.*

BOOKER J. R. & SAVVIDOU C. (1985). Consolidation around a point heat source, *Int. J. Numer. Anal. Methods Geomech.*, **9,** 173-184.

CAMPANELLA R. G. & MITCHELL J. K. (1968). Influence of temperature variations on soil behaviour, *J. Soil Mech. and Foun. Div. ASCE,* **3**, 709-734.

GATMIRI B. & P. DELAGE (1997). A New Formulation of Fully Coupled Thermal-Hydraulic-mechanical Behaviour of Saturated Porous Media - Numerical Approach, *Int. J. Numer. Anal. Methods Geomech*, accepted for publication. **21**(3), 199-225.

GATMIRI B. (1995). Fully coupled thermal-hydraulic-mechanical behaviour of saturated porous media (soils and fractured rocks),new formulation and numerical approach, Final Report CERMES-EDF.

LEWIS R. W. C. E. MAJORANA & B. A. SCHREFLER (1986). A coupled finite element model for the consolidation of non-isothermal elastoplastic porous media, *Transport In Porous Media*, **1**, 155-178.

McTIGUE D. E. (1986). Thermoelastic response of fluid-saturated porous rock, *J. Geophy. Res.*, **91**(B9), 9533-9542.

NOORISHAD J., C.F. TSANG & P. A. WITHERSPOON (1984). Coupled thermal-hydraulic-mechanical phenomena in saturated fractured porous rocks : Numerical approach, *J. Geophy. Res.*, **89**(B12), 365- 373.

RICE J. R. & M. P. CLEARY (1976). Some basic stress-diffusion solution for fluid-saturated elastic porous media with compressible constituents, *Rev. Geophys. Space Phys.*, **14**,.227-241.

SCHIFFMAN R. L. (1972). A thermoelastic theory of consolidation, *Environmental and geophysical Heat Transfer*, **4**, *Heat Transfer Div., ASME.*

MODELLING PENETROMETER INSTALLATION USING A MODIFIED STRAIN PATH METHOD APPROACH

D.R. Gill and B.M. Lehane
Trinity College, University of Dublin, Dublin, Ireland

ABSTRACT

The Strain Path Method (SPM) is presently the most realistic theoretical model available for predicting the disturbance to a clay soil caused by the installation of a pile or penetrometer. The method is based on the analogy that a penetrometer pushed into the ground can be modelled as fixed body located within an ideal inviscid fluid flow. The model assumes that the soil deformation is independent of its shearing characteristics and consequently a single flow solution, that of an ideal inviscid flow, is deemed satisfactory to predict soil movements. This paper examines the effect of assuming a flow that is not ideal and investigates the influence of viscosity and boundary conditions on predicted soil strain fields. A variety of solutions are presented for a penetrometer with a 60° conical tip. These facilitate an assessment of the implications of the SPM solution and, following appropriate experimental observations, will allow deduction of more realistic strain fields around penetrometers.

1. INTRODUCTION

In geotechnical problems involving retaining walls, slopes or shallow foundations, the stresses in the ground can be related directly to the depth of overburden and the applied loads. However, the kinematic constraints present at depth during the advance of a penetrometer or driven pile are such that the strains induced in the ground are relatively insensitive to the precise shearing characteristics of the soil. Such insensitivity was observed in a series of laboratory simulations of penetrometer installation by Baligh (1975) who subsequently developed what is now known as the Strain Path Method (SPM). The SPM assumes that, during undrained penetration, the relative position of soil particles to a penetrometer are the same as the streamlines followed by an inviscid incompressible fluid as it flows around a fixed penetrometer. Baligh (1985) predicted the path of such streamlines using potential fluid flow theory and modelled penetrometer geometries through a combination of sources and sinks in a uniform flow. The simplest case, that of a single source in a uniform flow, generates what has been termed the 'simple pile' solution; this solution elucidated the strongly two-dimensional nature of undrained penetrometer installation in a very succinct manner and succeeded in changing the course of research on driven piles and on the interpretation of Cone Penetration Test (CPT) data.

Advances to the SPM have been made by a number of workers, most notably at MIT and at Oxford University. These advances have been aimed primarily at improving predictions of stresses from strains and have assumed that the strain paths predicted using potential theory for an inviscid fluid flow are valid. Teh (1987) did, however, use a finite difference fluid model to predict strain paths, but assumed that, as in the potential theory, the fluid had no viscosity and that full slip took place at the surface of the penetrometer, i.e., no shear stresses develop on the penetrometer during installation.

This paper compares the strain paths adopted by the SPM with those predicted using a finite element fluid dynamics package which enabled the effects of fluid viscosity (or soil shear strength) and shear stresses on the penetrometer to be considered; these facilitate an assessment of the implications of the SPM solution. Gill (1998) and Gill & Lehane (1998) extend this work by using experimental observations of displacement fields around penetrometers to derive a means for estimating more realistic strain fields around penetrometers.

2. STRAIN PATHS USED BY THE SPM

Typical strain paths predicted using potential theory for Baligh's 'simple pile' (Gill 1998) and predicted for a 60° cone by Teh (1987) using a finite difference fluid model are presented in Figs. 1 and 2. Both sets of predictions assume that the penetrometer is perfectly smooth and that the soil can be modelled as an ideal, incompressible, inviscid and irrotational fluid.

The strain paths shown on Figs. 1 and 2 correspond to those followed by an element that is initially located at a radial distance of one penetrometer radius (r_0 = R) from the centreline. The paths are fully described by the three strain components E_1, E_2 and E_3, where:

$$E_1 = \varepsilon_{zz} \qquad E_2 = \left(\varepsilon_{rr} - \varepsilon_{\theta\theta}\right)/\sqrt{3} \qquad E_3 = 2\varepsilon_{rz}/\sqrt{3} \tag{1}$$

and ε_{zz}, ε_{rr} & $\varepsilon_{\theta\theta}$ are the direct vertical, radial & circumferential strains; ε_{rz} is the complimentary shear strain in the *r-z* plane. E_1, E_2 and E_3 correspond to strains induced under triaxial, pressuremeter (cylindrical cavity expansion) and direct simple shear loading.

The predictions illustrate the relatively complex series of strain changes that soil elements are likely to experience adjacent to a penetrometer installation. Each component of strain has a maximum when the soil element is at a different location relative to the

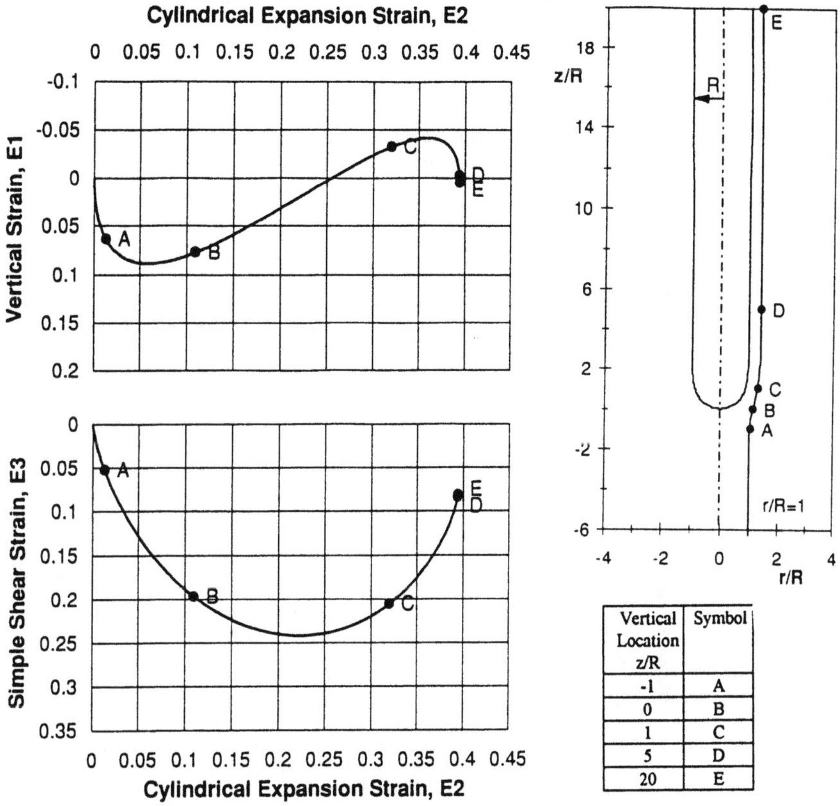

Vertical Location z/R	Symbol
-1	A
0	B
1	C
5	D
20	E

Fig. 1. Spreadsheet prediction of strain paths during simple pile penetration

penetrometer tip; vertical stresses are a maximum below the tip, whilst cavity expansion strains reach a maximum near the shoulder of the penetrometer. The differences evident in Figs. 1 and 2 illustrate the strong influence of the penetrometer geometry on the strain paths caused by penetration. E_1 and E_3 show a similar trend in both figures but have greater peaks for the cone shape. E_2 however, shows a reversal after the soil passes the cone shoulder, while E_2 increases monotonically for the simple pile.

3. FLOW MODELLING

The first step in modelling the penetration process is to solve the fluid flow problem of an axisymmetric fixed body located within a uniform flow field. The ANSYS (1997) finite element fluid dynamics package was used for this purpose. The use of finite elements allows the modelling of different flow types and penetrometer shapes to be achieved far more simply than with potential theory, and is more suited to solving problems with complex boundaries than finite difference models (Huebner 1995).

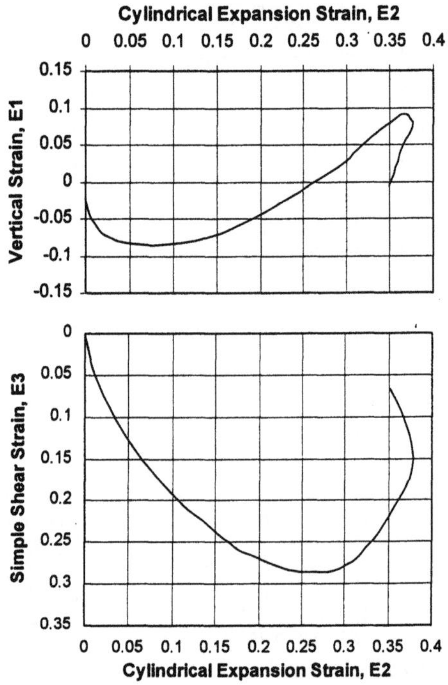

Fig. 2. Strain paths due to 60° cone penetration (Teh 1987)

The fluid flow solutions presented in this paper are steady-state incompressible Newtonian flow analyses produced by the FLOTRAN computational fluid dynamics solver within ANSYS. A two-dimensional axisymmetric co-ordinates system is used with the penetrometer located on the axis of symmetry. The fluid flow problem is defined by the laws of conservation of mass, momentum and energy. The sequential solution of all the governing equations constitutes a global iteration. The stream function values (ψ) are derived from the calculated velocities and stored on a nodal basis at the end of each iteration. The number of iterations required to achieve a converged solution depends on the grid meshing and stability of the problem. All the flow problems were meshed with elements comprising four nodes (quadrilateral shaped); see Fig. 3. The analyses performed for fluids with high viscosities produced smooth, ordered velocity fields and converged easily. Following the recommendations of the ANSYS authors solutions for flows with very low viscosities had to be solved first employing a higher viscosity than required; the problem would then be re-started from the previous solution with a new lower viscosity.

The boundary conditions adopted for modelling the flow distribution around a cone ended penetrometer are shown on Fig. 3. For all solutions, a uniform velocity (*U*) profile is applied at the inlet *AF* (upstream) and far field boundary (*FE*); this includes the specification of a zero velocity condition normal to the direction of flow. On the axis of symmetry (r = 0 axis), the velocity normal to the flow (the r-direction) is also set to zero and but unconstrained in the z-direction. On the penetrometer face, a zero normal velocity is again applied and the tangential velocity (v_{pent}) is assigned a value between zero and the flow field velocity (*U*), giving conditions referred to as slip, non-slip or partial slip. v_{pent}/U varies from zero in non-slip flow to unity for full slip conditions (as assumed in the SPM). A zero relative pressure is specified at the outlet (*DE*) for all solutions.

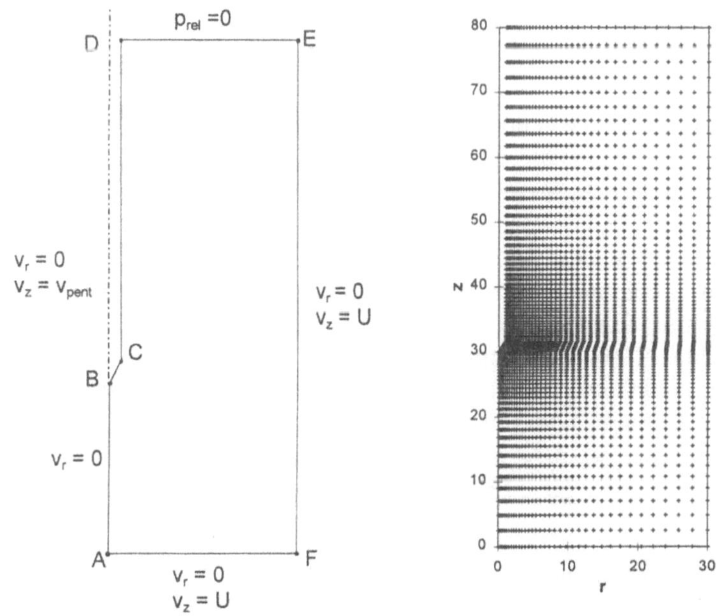

Fig. 3. Flow Model: (a) boundary conditions; (b) nodal grid

4. INTERPOLATING PARTICLE PATHS

The SPM derives strains in soil elements by integration of strain rates along the streamlines (or flow paths) followed by the fluid flow. Each streamline is defined by a constant value of ψ, termed the stream function, and is given directly by potential theory. ANSYS computations do not, however, predict the location of streamlines but instead output the ψ value at each node.

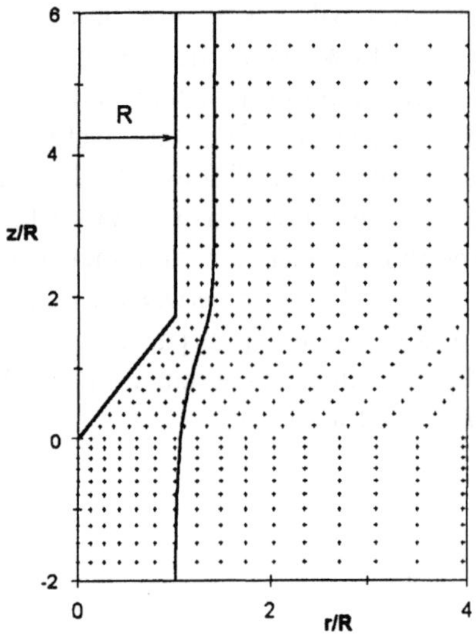

Fig. 4. Interpolated streamline in nodal grid

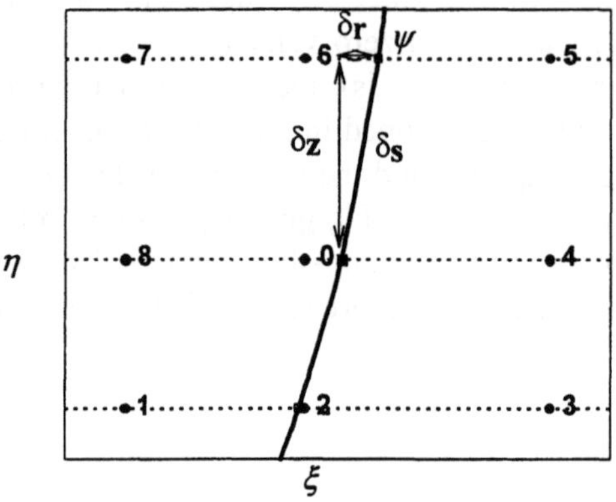

Fig. 5. Streamline interpolation through the nine-node local mesh

The location of streamlines in the ANSYS analyses were therefore found, as shown on Fig. 4, by interpolating for a constant stream function through the nodal grid. For this purpose a polynomial interpolation function that can be integrated or differentiated without difficulty is required (see Huebner 1995 for complete description of interpolation functions). In two dimensions, a complete *n*th-order polynomial in the *r-z* plane may be written as:

$$P_n(r,z) = \sum_{k=1}^{T_n^{(2)}} \alpha_k r^i z^j , \qquad i + j \leq n \tag{2}$$

where the number of terms in the polynomial is

$$T_n^{(2)} = (n+1)(n+2)/2$$

and the coefficients (α_k), termed the generalised co-ordinates of the element, control the magnitude of the distribution of the two-dimensional field variable. The field variable for the fluid flow solutions is the stream function (ψ) and the nine-node local mesh system used by Teh (1987) was chosen for its interpolation grid. The shape of the short streamline segment through this grid is governed by the local variation of the field variable given by the nine discrete nodal ψ values. The polynomial in (3) which is expressed in terms of the local co-ordinates (ξ,η) was chosen to allow a matrix solution at each node whilst also maintaining the symmetry of the equation. The local mesh shape and numbering system are shown in Fig. 5. Each node in the local mesh is expressed in terms of (3) to form a square

(9x9) matrix. This matrix is first used to solve for $\{\alpha\}$ and later to derive the position of a particular streamline.

$$\psi(\xi,\eta)=\alpha_0+\alpha_1\xi+\alpha_2\eta+\alpha_3\xi^2+\alpha_4\xi\eta+\alpha_5\eta^2+\alpha_6\xi^2\eta+\alpha_7\xi\eta^2+\alpha_8\xi^2\eta^2 \tag{3}$$

The corresponding matrix formulation for (3) is:

$$\{\psi\}=[M]\{\alpha\} \tag{4}$$

and $\{\alpha\}$ is found from:

$$\{\alpha\}=[M]^{-1}\{\psi\} \tag{5}$$

This set of simultaneous equations gives the generalised co-ordinates for the central node in the local mesh and is applied to every node in the grid (except the boundary nodes).

It is now possible to interpolate streamline paths (of constant stream function, ψ) through the nodal grid. The position of the streamline is found in terms of the local co-ordinates (ξ,η) relative to the central node. This is achieved by substituting the interpolation constants $\{\alpha\}$ into (3) and solving for the local co-ordinates (ξ,η). The position of the streamline is found where it crosses each horizontal grid row, and therefore the local y-direction co-ordinate, η is zero. Equation (3), for this case, reduces to a quadratic equation which can easily be solved:

$$\psi=\alpha_0+\alpha_1\xi+\alpha_3\xi^2 \tag{6}$$

5. CALCULATION OF STRAINS AND DISPLACEMENTS

The calculation of strains from the ANSYS output files is an adaptation of the method presented by Teh (1987) for finite difference solutions. Following derivation of the location of a streamline, as outlined, this involves calculation of the velocities and the strain rates at each interpolated point along the flow path. The accumulated strain can then be calculated along the streamline by integration of the strain rates at each interpolated point.

The velocities and strain rates are obtained by differentiating the polynomial in (3); expressions for these functions are given in Appendix I. The strain accumulated along a streamline is then calculated by dividing the streamline along its length into linear segments determined by the streamline points found on the horizontal grid lines. As shown on Fig. 5, these linear sections are approximations of the path length (δs) between the established points. Each strain increment is calculated at the mid-point between the streamline points and the integration moves from the bottom grid row upwards in the direction of flow. The cumulative strain is given by:

$$\varepsilon_{ij} = \sum_{k=1}^{n} \dot{\varepsilon}_{ij} \frac{\delta s}{\bar{v}_s} \tag{7}$$

where $\bar{v}_s$ is the average velocity in the direction of the streamline between points i and j.

The radial displacement at node n (δr_n) may be easily determined since the streamline position has been found and is the path assumed to be taken by any given soil particle. The vertical displacement can be found from the relative velocity of the fluid (or soil) to the penetrometer. In this solution, the observer is travelling with the penetrometer and the uniform flow field velocity (U) is the steady state condition. Therefore, the increment of vertical displacement (δz_n) is given by the relative vertical velocity multiplied by the time for which this velocity acts; the total vertical displacement at each streamline point is the sum of these increments.

$$\delta r_n = r_n - r_0 \tag{8}$$

$$\delta z_n = \sum_{k=1}^{n} \left(v_{rel} . t\right)_k = \sum_{k=1}^{n} \left(\left(v_z - U\right)\frac{dz}{U}\right)_k \tag{9}$$

6. STRAIN PATH PREDICTIONS

The analyses presented in this paper sought to determine the influence of (a) shear stresses on the penetrometer boundary and (b) fluid viscosity during penetration of a 60° cone. In all, seven ANSYS analyses were performed and the relevant details of each are summarised in Table 1. To facilitate comparison between the analyses, predictions of deviatoric strains

Run	Viscosity, μ (N s m^{-2})	Velocity on Penetrometer Boundary, v_{pent}/U	Reynolds Number, Re†	Flow Type Description
A*	1×10^{-60}	1	1.2×10^{62}	Very low viscosity full slip flow
B	1.2	0	2000	Viscous non-slip flow
C	1.2	1	2000	Viscous full slip flow
D	100	1	24	high viscosity full slip flow
E	1×10^{4}	1	0.24	high viscosity full slip flow
F	1.2	0.975	2000	viscous partial-slip flow
G	1.2	0.75	2000	viscous partial slip flow

* Fluid density, ρ = 1 kg/m^3 and uniform flow field velocity, U = 2 m/s. For all other solutions ρ = 2000kg/m^3 and U = 0.02 m/s

† For internal duct flows; $\mathrm{Re} = \rho V D_h / \mu$, where D_h is the hydraulic diameter and V is the fluid velocity.

Table 1. ANSYS Flow Solutions

(E_1, E_2 & E_3), shown on Figs. 6 and 7, correspond to those on same streamline, which is located at an initial radial distance of one penetrometer radius from the centreline i.e. $r_0/R = 1$. These figures also provide information on how strains change along the streamline by annotating conditions at five relative vertical positions (z/R = -1, 0, 1, 5 & 20) and may be compared directly with Fig. 2 which shows the corresponding strains for 'full slip' conditions in an inviscid fluid.

A comparison of Fig. 2 with the solution presented on Fig. 6 for flow A (with a 'full slip' condition and a very low viscosity) indicates good agreement and validates the procedure outlined. There is however, a difference between the paths followed by E_3 at large distances above the cone tip. The ANSYS prediction for the final reversal of shear for flow A is a consequence of the unconstrained boundary condition at the outlet (boundary *DE* on Fig. 3) and would be eliminated if a condition of uniform velocity was specified on this boundary.

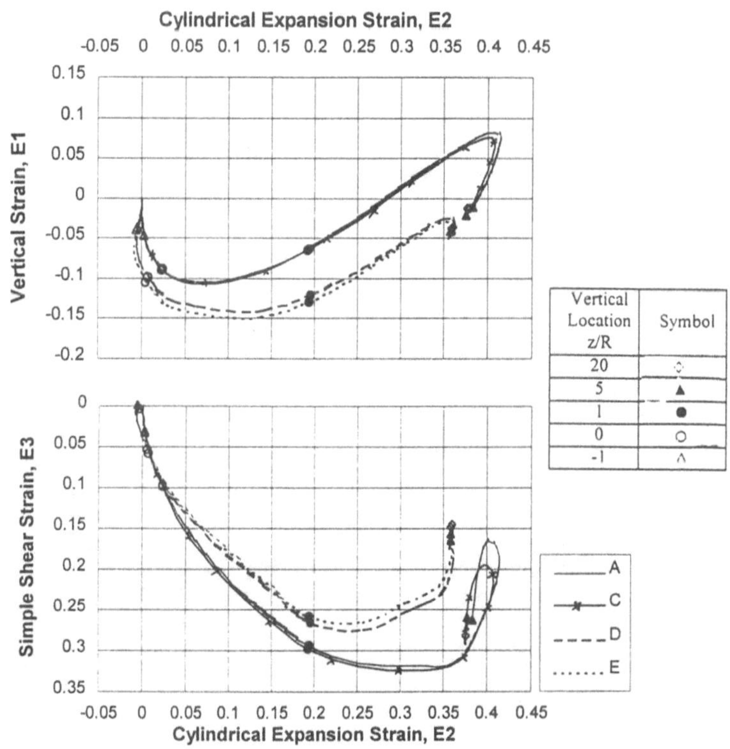

Fig. 6. Deviatoric strain paths during penetration of a 60° cone for 'full slip' flow.

6.1 *Viscous effects*

The predictions plotted on Fig. 6 illustrate the effect of fluid viscosity on 'full slip' flow. It is evident that high viscosity flows with low Re values (flows D & E) produce distinctly different strain paths to flows with lower viscosity (A & C):

- High viscosity results in greater compressive strains (E_1, which is negative when compressive) as the penetrometer tip approaches. The elements for flows D & E are seen in Fig. 6 to have a net vertical compression throughout the penetrative process.
- For 'full slip' flow, E_2 and E_3 are lower for viscous flows (D & E) than for low viscosity flows (A & C). However, as seen below, friction on the penetrometer surface can cause the opposite effect on the magnitudes of E_2 and E_3.
- The greatest changes in strain occur in the small band of viscosity ranging from 1 to 100 Ns/m^2 (compare flows C and D). The strain paths produced are therefore most sensitive to changes in viscosity over this range.

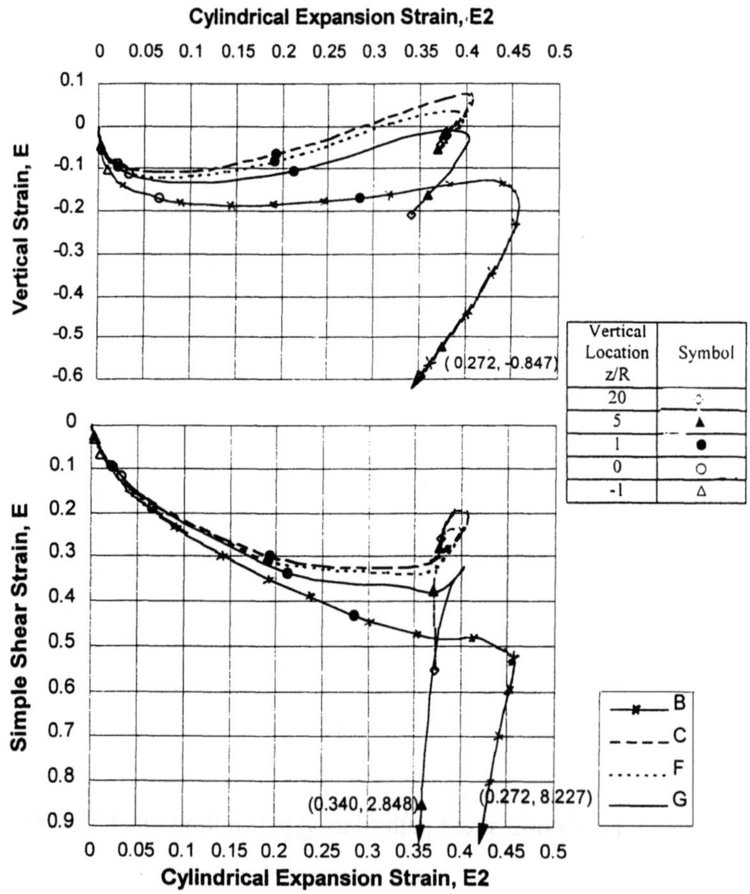

Fig.7. Deviatoric strain paths during penetration of a 60° cone for constant viscosity.

In general, all analyses performed (including those involving partial slippage on the penetrometer surface) indicated that the overall effect of increasing the fluid viscosity is to increase the range of influence of the penetrometer and its specified boundary conditions on the surrounding elements.

6.2 Boundary effects

The predictions plotted on Fig. 7 examine the influence of the slip condition on the boundary of the penetrometer for a fluid with a viscosity approximately 1000 times that of water. The strain paths for flows with 'full slip' (C), 'non-slip' (B), and two 'partial slip' conditions (F & G) are presented. Reducing the velocity of the penetrometer increases the friction on its surface and this results in:

- large increases in the shear strain (E_3) once the cone shoulder has past (particularly evident in flows B & G)
- increased vertical compressive (negative) strain E_1 (flows B & G remain in compression throughout the penetrative process)
- greater peaks in E_2 (the cylindrical expansion strain) at the cone shoulder, but also greater reversals in E_2 as the distance from the tip (z/R) increases.

7. CONCLUSIONS

Baligh's Strain Path Method (SPM) postulates that strain fields induced in the ground during penetrometer installation may be approximated by assuming that the installation process is analogous to the flow of an inviscid fluid around the penetrometer, with full slip of the fluid on the penetrometer's surface. This paper examines the sensitivity of the predicted strain fields to the fluid viscosity and slip condition. It became evident that increasing fluid viscosity increases the penetrometer's range of influence in the soil and accentuates the influence the boundary conditions. The effect of partial slip on the penetrometer boundary results in relatively significant changes at large distances from the penetrometer tip; partial slip also leads to lower vertical extension strains but higher cavity expansion and shear strains. These findings may be used in conjunction with experimental measurements of displacement fields around penetrometers to deduce more realistic strain fields than those currently adopted by the SPM.

APPENDIX I

The velocities and strain rates are expressed below; initially using potential theory in terms of the stream function in a cylindrical global co-ordinate system, and then in terms of the local co-ordinates and interpolation constants. (Note the local and global co-ordinates can simply be interchanged in the differentiation process as $\delta r = \delta\xi$ and $\delta z = \delta\eta$.).

$$v_r = -\frac{1}{r}\frac{\partial\psi}{\partial z} \qquad v_z = \frac{1}{r}\frac{\partial\psi}{\partial r} \tag{10}$$

Where v_r is the radial velocity and v_z the vertical velocity. The radial, vertical, circumferential and shear strain are expressed as

$$\dot{\varepsilon}_{rr} = \frac{\partial v_r}{\partial r} \qquad \dot{\varepsilon}_{zz} = \frac{\partial v_z}{\partial z} \qquad \dot{\varepsilon}_{\theta\theta} = \frac{v_r}{r} \qquad \dot{\varepsilon}_{rz} = \frac{1}{2}\left(\frac{\partial v_r}{\partial z} + \frac{\partial v_z}{\partial r}\right) \tag{11}$$

Then from (3)

$$v_r = -\frac{1}{r}\left(\alpha_2 + \alpha_4\xi + 2\alpha_5\eta + \alpha_6\xi^2 + 2\alpha_7\xi\eta + 2\alpha_8\xi^2\eta\right) \tag{12}$$

$$v_z = \frac{1}{r}\left(\alpha_1 + 2\alpha_3\xi + \alpha_4\eta + 2\alpha_6\xi\eta + \alpha_7\eta^2 + 2\alpha_8\xi\eta^2\right) \tag{13}$$

$$\dot{\varepsilon}_{rr} = \frac{1}{r^2}\left\{\alpha_2 + \alpha_4(\xi - r) + 2\alpha_5\eta + \alpha_6\xi(\xi - 2r) + 2\alpha_7\eta(\xi - r) + 2\alpha_8\xi\eta(\xi - 2r)\right\} \tag{14}$$

$$\dot{\varepsilon}_{zz} = \frac{1}{r}\left(\alpha_4 + 2\alpha_6\xi + 2\alpha_7\eta + 4\alpha_8\xi\eta\right) \tag{15}$$

$$\dot{\varepsilon}_{\theta\theta} = -\frac{1}{r^2}\left(\alpha_2 + \alpha_4\varepsilon + 2\alpha_5\eta + \alpha_6\varepsilon^2 + 2\alpha_7\xi\eta + 2\alpha_8\xi^2\eta\right) \tag{16}$$

$$\begin{aligned}\dot{\varepsilon}_{rz} = \frac{1}{2r^2}\big[&-\alpha_1 + 2\alpha_3(r - \xi) - \alpha_4\eta - 2\alpha_5 r + 2\alpha_6\eta(r - \xi) \\ &- \alpha_7(2\xi r + \eta^2) - 2\alpha_8\xi(\xi r + \eta^2)\big]\end{aligned} \tag{17}$$

REFERENCES

ANSYS (1997), Help Manual for Revision 5.3, SAS IP, Inc.

BALIGH M.M. (1985). Strain path method, J. Geotech. Eng. ASCE, 111, 1108-1136.

BALIGH M.M. & R.F.SCOTT (1975), Quasi-static deep penetration in clays, , J. Geotech. Eng. ASCE, 101, 1119-1133.

GILL D.R. (1998). Experimental and theoretical investigations of undrained penetrometer installation, Ph.D. Thesis, Trinity College, Univ. of Dublin.

GILL D.R. & B.M.LEHANE (1998). Extending the Strain Path Method analogy for modelling penetrometer installation, (submitted to Int. J. Numer. Anal. Meth. Geomech).

HUEBNER K.H., E.A.THORNTON & T.G.BYROM (1995). The Finite Element Method For Engineers, John Wiley & Sons, New York.

TEH C.I. (1987). An analytical investigation of the cone penetration test, Ph.D. Thesis, Dept. Eng. Sci., University of Oxford, U.K..

A NOTE ON FINITE ELEMENT MODELLING OF UNSATURATED CONSOLIDATION PROBLEMS

C. Jommi
Polytechnic of Milan, Milan, Italy

J. Vaunat
Polytechnic University of Catalunya, Barcelona, Spain

ABSTRACT: A first comparison between the performance of different models which may be adopted in the finite element analysis of unsaturated consolidation problems is presented. The fully coupled governing equations are recalled and the numerical formulation is presented. The numerical examples, performed assuming an elastic behaviour for the soil skeleton, show how models written in terms of different stress variables sets may be equally adopted in the analysis. Special attention is payed to the numerical implementation of the fully coupled problem on the onset of saturation and desaturation.

1. INTRODUCTION

The thermo-hydro-mechanical behaviour of unsaturated soils is receiving considerable attention nowadays, due to the rising importance of many enginnering problems, mainly related to environmental control, involving soils undergoing changes in water content and temperature.

Due to the inherent coupling between thermal, hydraulical and mechanical effects, and to the resulting, highly non-linear, soil behaviour, numerical methods are nearly always mandatory for the study and the analysis of problems concerning this class of problems.

Although much research on constitutive and numerical modelling of unsaturated soils has been performed in the last years, still a lack of a generally accepted framework may be observed. This lack is primarly due to the *neverending* question on which stress variables should be adopted for a correct modelling of the mechanical behaviour of unsaturated soils.

Secondarily, even restricting the interest to quasi-static, isothermal processes and thus concentrating on hydro-mechanical coupling, the numerical implementation of the governing equations still presents open problems, if a general and robust solution procedure is sought.

In this context, focusing the attention on the numerical implementation of the fully coupled unsaturated consolidation problem, the governing equations, assuming soil skeleton displacements, water pressure and air pressure as primary variables, are recalled. Models written in terms of different combinations of total stress tensor, pore water pressure and pore air pressure may be equally adopted, without significant difficulties. In the implementation, special attention is payed to the compatibility of the unsaturated formulation with the saturated one, when the gas phase ceases to be continous, and the effective stress tensor must be recovered in the description of the mechanical behaviour of the soil skeleton.

The numerical formulation, which should be efficient both in unsaturated and saturated conditions if reliable analyses have to be performed, is presented and discussed.

The results of one-dimensional numerical tests, performed assuming an elastic behaviour for the soil skeleton, allow for a preliminary comparison between some of the consequences of the different choices in constitutive modelling of unsaturated soils.

In the following, boldface letters, when adopted, will denote vectors and tensors.

2. GOVERNING EQUATIONS FOR FLUIDS FLOW

Restricting the attention to quasi-static, isothermal processes, the mass conservation equations for unsaturated flow are generally written under the assumption that both fluid phases (a = air, w = water) are continuos. In this case, omitting source or sinks terms, they read, for air and liquid water respectively:

$$\frac{\partial(\rho^w n S_r)}{\partial t} + \nabla(\rho^w \mathbf{v}^w) = 0 \tag{1}$$

$$\frac{\partial(\rho^a n(1 - S_r))}{\partial t} + \nabla(\rho^a \mathbf{v}^a) = 0 \tag{2}$$

where ρ^α is the mass density of fluid phase α; $\mathbf{v}^\alpha$ is the discharge velocity of fluid phase α; n is the porosity; and S_r is the water saturation (water volume/void volume).

Generalized Darcy's law is assumed to govern the fluid phases flow [Bear, 1972]:

$$\mathbf{v}^\alpha = -\mathbf{k}_\alpha \nabla(p^\alpha + \rho^\alpha g z) \tag{3}$$

$$\mathbf{k}_\alpha = \mathbf{K}\frac{k_{r\alpha}}{\mu_\alpha} \tag{4}$$

in which **K** is the absolute permeability tensor, μ_α is viscosity of phase α, and $k_{r\alpha}$ is the relative permeability of phase α, function of the saturation degree S_r.

The phase pressures are related by the capillary pressure p^c

$$p^a = p^w + p^c(S_r) \tag{5}$$

which is a generally non-linear and hysteretic function of saturation degree. For the sake of simplicity, a non-hysteretic function will be assumed in the sequel. The mass density of water is generally assumed constant, while, assuming for air a perfect gas behaviour, its mass density is linearly related to its pressure

$$\rho^a = (M^a / RT)p^a = \beta p^a \tag{6}$$

with M^a is the molecular mass of water, R the gas constant, T the absolute temperature, β constant for isothermal processes.

In this widely adopted form, the mass conservation equations may be identified with continuity equations, and hold, by definition, when the relevant phase is continuous. Equation (2) thus loose its validity on the onset of saturation, when continuity of the gas phase is lost, and the same happens to equation (1) as saturation degree approaches its residual value, when the continuity of the liquid phase is lost.

A physically based approach, leading to the so-called compositional formulation [Panday & Corapcioglu, 1989; Olivella & al., 1994], which overcomes this difficulty, is briefly recalled here. A more detailed presentation of the formulation may be found in Vaunat & al. [1997]. Starting from the consideration that besides free air, dissolved air is present in the mixture, the latter may be taken into account in the mass balance eqaution. The volume of dissolved air in the pore water is governed by Henry's law, and is equal to $H\,n\,S_r$, with H representing Henry's constant.

The mass conservation equation can be then rewritten taking into account both free air and dissolved air, transported by water, in the following form:

$$\frac{\partial(\rho^a n(1 - S_r + HS_r))}{\partial t} + \nabla(\rho^a(v^a + Hv^w)) = 0 \tag{7}$$

If in saturated conditions air pressure is expressed as a function of dissolved air density, and not of gas pressure, the mass balance equation (7) holds too. Water pressure and air pressure evolve independently, and desaturation may occur naturally, as soon as the water pressure reaches the air pressure.

It's worth noting that, following a similar path of reasoning, the mass balance equation for water may be written taking into account both liquid water and water vapour, transported in the gas phase.Vapour moisture transport would become significant only at very low saturation degrees, or when the effects of temperature gradients must be taken into account. None of these situations will be considered in the sequel. Continuity for the liquid phase, expressed through equation (1), will be assumed.

3. EQUILIBRIUM EQUATIONS, DISPLACEMENTS COMPATIBILITY AND CONSTITUTIVE MODELS FOR THE SOIL SKELETON

Adopting soil mechanics conventions (tractions negative), the equilibrium equations for the whole mixture reads

$$\frac{\partial \sigma_{ij}}{\partial x_j} - b_i = 0 \qquad (i = 1, 2, 3) \tag{8}$$

where σ_{ij} is the total stress tensor and b_i bulk forces. Summation over index j is assumed.

In the standard finite element formulation displacements of the soil skeleton are adopted as primary variables. Thus the stress-strain relationship must be substituted in (8) and strains written as a function of displacement derivatives.

Although many constitutive models have been already proposed for soils in the unsaturated state, still the discussion is open on which stress variables should be adopted for a correct modelling of their mechanical behaviour [Gens, 1996; Wheeler & Karube, 1996].

Here, the relevant advantages and shortcomings of the adoption of different stress variables will not be discussed, neither the capabilities of different models in reproducing the observed experimental behaviour will be examined. The attention will be focused only on the numerical implementation of two classes of these models, in order to show how they can equally be adopted in a finite element analysis. Following this point of view, we assume that the stress-strain relationship may be written in finite form, for the sake of simplicity, through a general expression of the type:

$$\hat{\sigma}_{ij} = D_{ijhk}(\varepsilon_{hk} - \varepsilon^s_{hk}) \tag{9}$$

where $\hat{\sigma}_{ij}$ may be any combination of total stress, air pressure and water pressure, and ε^s_{hk} are strains due to suction. For $\hat{\sigma}_{ij}$ the name *constitutive stress* will be adopted. The attention will be focused here only on two of the possible choices, namely the adoption of net stress

$$\hat{\sigma}_{ij} = \sigma_{ij} - p^a\delta_{ij} \tag{10}$$

and the adoption of the difference between total stress and an average of the fluid pressures, weighted by saturation degree,

$$\hat{\sigma}_{ij} = \sigma_{ij} - S_r p^w \delta_{ij} - (1 - S_r)p^a\delta_{ij} \tag{11}$$

Although for (11) the name *effective stress* has sometimes been used, in order to avoid confusions with the saturated effective stress, here *intergranular stress* will be used, in the sense that it is thought to be a macroscopic measure of the average soil skeleton stress.

Rewriting the equilibrium equations in the following form, respectively:

$$\frac{\partial \hat{\sigma}_{ij}}{\partial x_j} + \frac{\partial p^a}{\partial x_j}\delta_{ij} - b_i = 0 \qquad (i = 1, 2, 3) \tag{12a}$$

$$\frac{\partial \hat{\sigma}_{ij}}{\partial x_j} + \frac{\partial (S_r p^w)}{\partial x_j}\delta_{ij} + \frac{\partial ((1-S_r)p^a)}{\partial x_j}\delta_{ij} - b_i = 0 \qquad (i = 1, 2, 3) \qquad (12b)$$

the stress-strain relationship and the compatibility equations may easily be inserted, thus recovering the usual displacements formulation.

4. DISCRETIZATION AND NUMERICAL SOLUTION

A standard Galerkin finite element discretization in space is adopted. Besides soil skeleton displacements, fluid pressures p^a and p^w are adopted as primary variables. Water content has been used in numerical analyses of unsaturated flow too [see f.e. Forsyth & al., 1995], as well as saturation degree [Dangla & Coussy, 1997]. But while fluid pressures are continuos regardless eventual domain inhomogeneities, water content and saturation degree may be discontinuous. The former approach should thus be preferred.

Finite element discretization in space leads to a discretized equation system of the type:

$$\mathbf{M}\frac{\partial \mathbf{q}}{\partial t} + \mathbf{K}\mathbf{q} + \mathbf{F} = \mathbf{0} \qquad (13)$$

in which $\mathbf{q}$ collects all primary variables, $\mathbf{M}$ is the so called mass matrix, and $\mathbf{K}$ is the stiffness matrix.

Time discretization may be performed through a finite difference θ-scheme, leading to a final system of the type

$$(\mathbf{M} + \theta\Delta t\mathbf{K})\mathbf{q}^{n+1} = (\mathbf{M} + (\theta - 1)\Delta t\mathbf{K})\mathbf{q}^{n} - \Delta t\mathbf{F} \qquad (14)$$

where $\mathbf{q}^n$ is the solution at time t^n.

The choice of a fully implicit scheme is widely accepted, and adopted herein. The choice is mainly due to stability considerations, in view of the highly non-linear behaviour of the governing equations. In fact, although any θ-scheme with $\theta \geq 0.5$ is proved to be unconditionally stable, non-phisical oscillations may occur, and have been observed, in unsaturated flow analysis for any value of θ. It has been pointed out that satisfaction of the stability criterion $\theta \geq 0.5$ does not ensure a non-oscillating solution. For linear elements, it has been proved [Ouayang & Xiao, 1994] that a non-oscillating solution can be obtained for time steps Δt satisfying the following conditions:

$$-\min_{ij}\left[\frac{M_{ij}}{\theta K_{ij}}\right]_{(i \neq j)} \leq \Delta t \leq \min_{i}\left[\frac{M_{ii}}{(1-\theta)K_{ii}}\right] \qquad (15)$$

If a backward scheme ($\theta = 1$) is adopted, the upper limit disappears, but Δt is still subjected to the lower bound. In order to obtain a non-oscillating solution, the mass matrix $\mathbf{M}$ should be lumped. In this case, in fact, with $M_{ij} = 0$ (for $i \neq j$), the minimum time step becomes zero, regardless of the element size.

It is worth noting, anyway, that mass lumping is a sufficient condition to eliminate oscillations in the solution, only provided that the coefficients of the stiffness matrix **K** are different from zero. If any K_{ii} tends to zero, as for example in the case of air permeability coefficients on the onset of saturation, mass lumping cannot guarantee any more the absence of spurious oscillations in the (convergent) solution. A small residual value for permeability coefficients is sufficient to eliminate these oscillations, but the behaviour of the solution may posses different characteristics from the true solution. This aspect will be further discussed with reference to the second numerical example presented in Section 5.

A Newton-Raphson iterative scheme is adopted for the non linear solution at time t^{n+1} of the discretized system (14). It has been pointed out that, in spite of all the already mentioned cautions, this numerical formulation leads to mass balance errors, which increase with time step and element size. This is due to the approximation introduced in the iterative solution procedure, when the partial derivatives of saturation degree with respect to the fluid pressures (which may be used for the evaluation of the storage term, which appears in the mass matrix **M**) are numerically computed during the time step. In order to eliminate the error resulting from this approximation, fluid mass changes may be calculated explicitly during the time step, following the approach described for example in Celia & al. [1990]. The latter approach is proved to be perfectly mass conservative regardless of the time step, leading to a substantial improvement in the solution quality.

Many different analytical expressions have been proposed for the description of the relationship between saturation degree and suction. If the adopted function presents discontinuous derivatives with respect to fluid pressures for some value of suction, a smoothening in the derivatives discontinuity is adopted, in order to avoid convergence and oscillation problems.

A final note concerns the behaviour of the two constitutive stress variables on the onset of saturation. While the intergranular stress naturally degenerates in the saturated effective stress for $S_r = 1$, the same effective stress may be recovered from net stress if the following definition is adopted:

$$\sigma_{ij}^{net} = \sigma_{ij} - p^w \delta_{ij} - s\delta_{ij} \tag{16}$$

where suction s is defined as:

$$s = (p^a - p^w) \text{ when } S_r < 1 \;;\; s = 0 \text{ when } S_r = 1 \tag{17}$$

A discontinuity in the derivatives of suction occurs for $S_r = 1$. Again, in order to avoid further convergence and oscillation problems, a smoothening in the derivatives is performed.

5. NUMERICAL EXAMPLES

One dimensional numerical examples are presented in order to highlight the capabilities of the formulation. The first example shows how different stress variables may be equally adopted in the analysis of the fully coupled unsaturated-saturated consolidation problems. In the analysis, in order to highlight the features of the stress variables choice alone, an isotropic linear elastic model is assumed for the soil skeleton. The second example allows for a discussion of some implications of the numerical implementation.

5.1 Stresss controlled compression tests

Assuming oedometric conditions, one dimensional compression tests were performed, both with an elastic model written in terms of net stress and suction, and a model written in terms of the intergranular stress. The numerical results are shown in Fig.1.

Starting from an unsaturated state, an increasing vertical load is applied at the top of the sample (fig.1a), imposing a constant water content throughout the test. One test was performed at constant air pressure, and in a second test the air content was kept constant.

In the first case, due to compression and constant water content assumption, the sample progressively saturates, reaching at some stage a fully saturated condition. In fig.1b) the fully saturated state is reached for a total vertical stress equal to 800 kPa. The water pressure on the onset of saturation equals the atmospheric pressure.

If the load is further increased, due to fully saturation and oedometric conditions, only an increase in water pressure must be observed, as constant water content implies undrained conditons. In fact, starting from time $t = 5.e^5$ sec, only an increase in water pressure, equal to the increase in total vertical stress is observed. The constitutive stress (fig. 1c)), which in saturated conditions coincides with the effective stress, remains constant during further loading, as one would expect. If afterwards the total vertical stress is decreased, an equal decrease in water pressure is observed, until desaturation begins to occur, and the costitutive stress begins to decrease too. This correct behaviour is shown whichever stress variables set is adopted in the analysis.

The comparison between the evolution of net stress and intergranular stress, shown in fig. 1c) for the constant air pressure case, shows that although the two models exhibit the same behaviour pattern, the constitutive stresses assume different values. This aspect may be easily explained on the basis of the stress-strain relationships adopted in the analysis. For the model which adopts net stress, the constitutive law may be written in the form:

$$\varepsilon_{ij} = C^1_{ijhk}(\sigma_{hk} - p^a\delta_{hk}) + C^2_{ijhk}s\delta_{hk} \tag{18}$$

where C^1_{ijhk} and C^2_{ijhk} are constant, generally independent coefficients. Conversely, the elastic part of models written in terms of intergranular stress is usually expressed in the form:

$$\varepsilon_{ij} = C_{ijhk}(\sigma_{hk} - p^a\delta_{hk}) + C_{ijhk}S_r s\delta_{hk} \tag{19}$$

and this expression, with constant C_{ijhk}, was assumed here.

The comparison between equations (18) and (19) show that if constant C^1_{ijhk}, C^2_{ijhk} and C_{ijhk} coefficients are assumed, the two models will predict different strains states for identical total stress and fluid pressures states. Conversely, if the same strain state and fluid pressures are imposed, the predicted stress state will be different. Moreover, for constant soil skeleton compliances, the first model predicts a linear dependency of strains on suction, while in the second one this dependency is non linear as the product $S_r s$ appears.

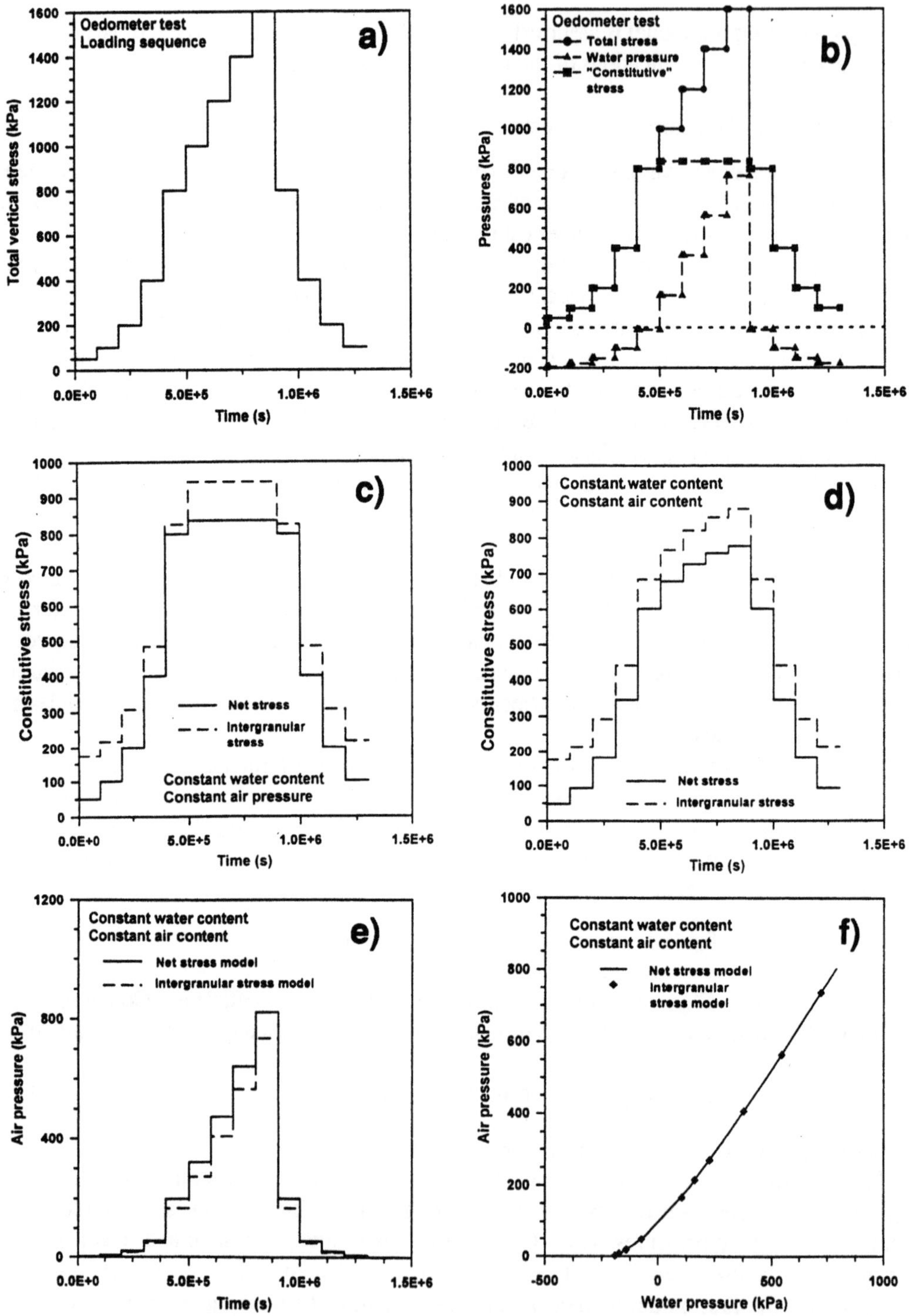

Fig. 1. Stress controlled compression tests: numerical results.

In the case of constant air content, saturation degree increases with the vertical load, but full saturation could be reached only if the total amount of air initially present in the sample could be dissolved in water. During the numerical loading programme full saturation was never reached, and a continuous evolution of the constitutive stresses may be observed from the results presented in fig. 1d).

The evolution of air pressure is shown in fig. 1e), and it can be observed that the different stiffness of the two models influences the values of air pressure too, although the same overall pattern is observed in any case. The ratio between air and water pressure is governed by the initial volume fractions of the two phases, and it is thus the same irrespective of the constitutive model adopted (fig. 1f).

5.2 Vertical drainage from a sand column

Some results of the numerical simulation of the drainage of water from a vertical 1m high sand column are presented. The sand column is initially fully saturated and under hydrostatic water pressure distribution. At $t = 0$ vertical drainage from the bottom begins to be allowed, keeping water pressure at the lower boundary equal to the atmospheric pressure. The top of the column is impermeable to water, and no lateral displacements are allowed. As soon as the gravity-driven water flow begins, a desaturation process occurs. The problem is thus numerically analysed through two-phase unsaturated flow formulation. The air pressure is kept equal to the atmospheric pressure both at the top and at the bottom of the column. A linear elastic behaviour is assumed for the soil skeleton.

Results of many different numerical analyses of this problem have already been presented [Jommi & al., 1997]. Only some selected results are shown here, in order to highlight the problems related to the numerical implementation of the fully coupled unsaturated flow problem. Dangla & Coussy [1997] demonstrated analitically that, due to the hydraulical properties of the sand column, a desaturation front must be expected, and that around the desaturation front, a discontinuty in the air pressure gradient appears. On the unsaturated upper side of the front, an infinite air pressure gradient should be observed.

In fig. 2a) and 2c) the results of the numerical analyses for air pressure evolution, performed respectively with a 20 and a 200 equal-sized elements discretization of the column, are shown. Obviously, a finer discretization allows for a more accurate solution, and the expected infinite gradient around the desaturation front is better modelled. But the overall evolution pattern is already well kept in the analysis performed with only 20 elements. Due to the high non linearity of the problem, this should be considered a good proof of the robustness of the formulation.

It's worth noting that the numerical solution for air pressure oscillates around the desaturation front. This may be better observed probabily with reference to fig. 2b) and 2d), where suction evolution is depicted. With mesh refinement oscillations persist, although less evident, even if mass lumping is performed. The oscillating behaviour is intrinsically related to the null air permeability around the front, as in this case the lower bound criterion (15) cannot be satisfied regardless of the elements size.

A way to smoothen the solution, often adopted, is to keep a small residual permabilty for air in saturated conditions too. The results of an analysis performed with the introduction of

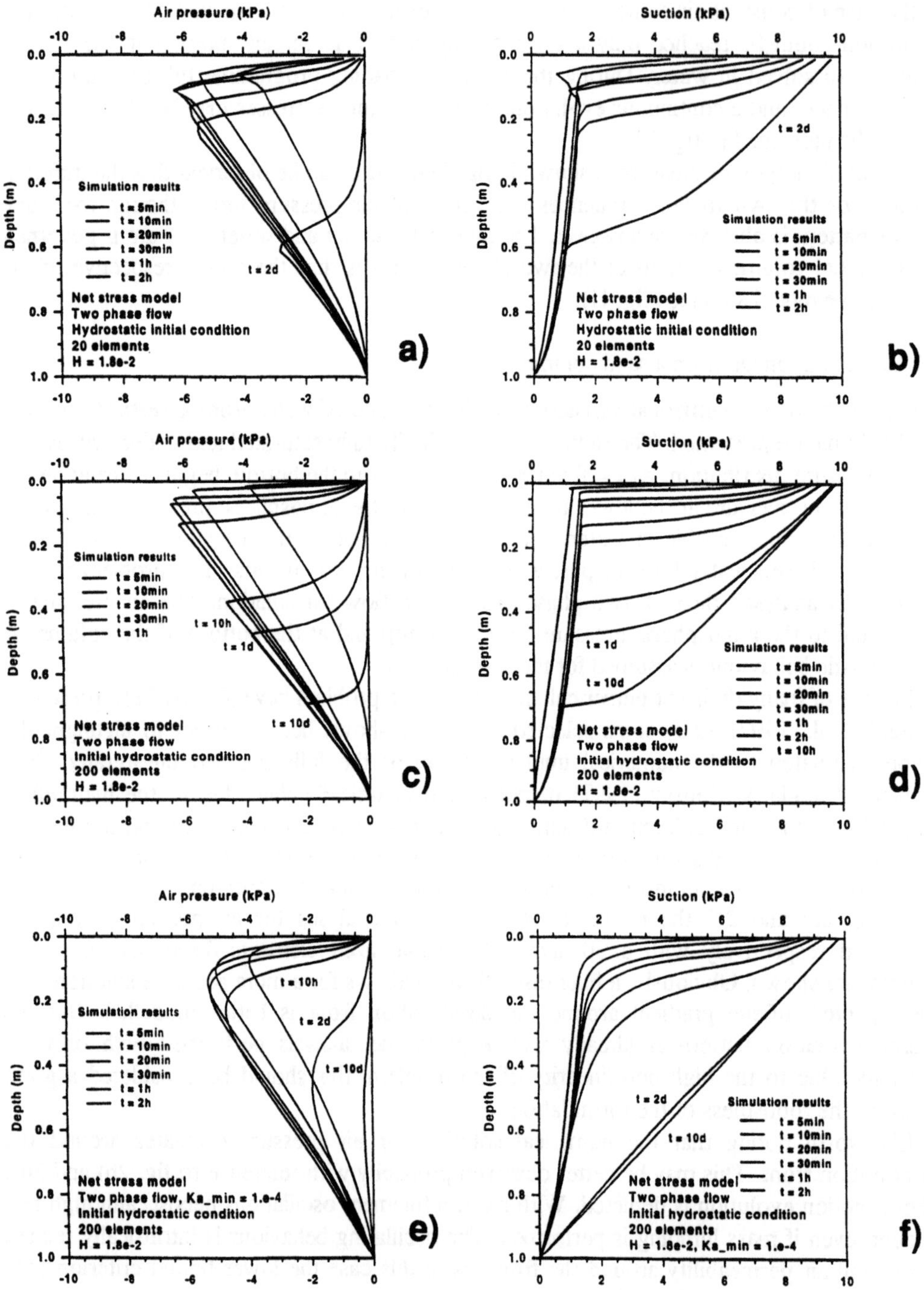

Fig. 2. Vertical drainage from a sand column: numerical results.

a residual permeability value of $1.e^{-4}$ are shown in fig. 2e) and 2f). Although very low, the introduced residual value allows for a complete smoothening of the solution, irregardless of the elements size if mass lumping is performed. Nevertheless, the consequences of the numerical *trick* on the numerical solution are evident in fig. 2e) and 2f). The overall patterns of both air pressure and suction evolution change noticeably, comparing the results with those depicted in fig. 2c) and 2d), especially for larger times. With reference, for example, to suction profile at time t = 10 days, in fig. 2d) still a desaturation front may be observed, below which suction is still very low. The smoothened solution, on the contrary, shows that steady state conditions are almost reached, thus giving sensitively lower values for suction (and saturation degree) with respect to the true ones in all the lower part of the domain. This feature may appear dangerous, when the influence of suction on the mechanical behaviour of the soil skeleton has to be taken into account properly.

6. CONCLUDING REMARKS

Finite element analysis of isothermal unstaurated-saturated consolidation problems may be performed taking into account dissolved air in water (and eventually water vapour in the gas phase), thus allowing for a compositional formulation which holds both in saturated and usaturated (and eventually dry) states.

Fluid pressures and soil skeleton displacements appear to be the most favourable choice for primary variables, at least in the analysis of quasi-static processes, in which full saturation may occur at least in a part of a generally non-homogeneous medium.

Attention must be payed in the numerical implementation due to the higly non-linear resulting discretization. A fully implicit time integration scheme, mass matrix lumping, and a fully mass conservative formulation are recommended.

The simple examples presented show how both constitutive models written in terms of two separate stress variables (net stress and suction) or in terms of a single stress variable (intergranular stress) may be equally implemented.

With the adoption of net stress as constitutive stress, two compliance tensors are necessary for the complete definition of the elastic behaviour of the soil skeleton. Conversely, the adoption of intergranular stress allows for suction induced strains without the need for any other parameter except the saturated compliance. On the other hand, it is obvious that a less flexible model will result in the latter case, as the compliance which accomplishes for suction is inherently linked to the saturated compliance, while in the first class of models the two compliance tensors, accounting for net stress and suction, may be completely independent.

As a final remark, it's worth noting that the adoption of elasto-plastic constitutive laws is necessary to model most of the experimentally observed aspects of the mechanical behaviour in a comprehensive framework. All the elasto-plastic models proposed until now, even those written in terms of intergranular stress, recover in the end two stress variables, at least for the modelling of plastic volumetric strains which may occur during wetting [Gens, 1996].

Nevertheless, the basic characteristics highlighted here, with reference to the linear elastic behaviour of a model written in terms of intergranular stress, namely a natural transition to saturated states, the advantage (payed in terms of less flexibility) of the need for a fewer number of new parameters besides the saturated ones, the necessity of the knowledge of the retention curve, a more difficult representation of suction controlled tests (where the saturation degree history, which is often not available experimentally, need to be defined) are retained in elasto-plastic modelling too.

ACKNOWLEDGMENTS

The financial support of ALERT-Geomaterials is gratefully acknowledged.

REFERENCES

BEAR J. (1972). Dynamics of Fluids in Porous Media. American Elsevier Pub., New York.

CELIA M.A., E.T.BOULOUTAS & R.L.ZARBA (1990). A general mass-conservative numerical solution for the unsaturated flow equation. Water Resour. Res., Vol.26, No.7, pp. 1483-1496.

DANGLA P. & O.COUSSY (1997). Consequences of a vanishing diffusion process in the drainage from a one dimensional sand column. Proc. NAFEMS World Congress, Vol.2, Stuttgart, pp.1326-1337.

FORSYTH P.A., Y.S.WU & K.PRUESS (1995). Robust numerical methods for saturated-unsaturated flow with dry initial conditions in heterogeneous media. Adv. Water Res., Vol.18, pp. 25-38.

GENS A. (1996). Constitutive modelling: application to compacted soils. Keynote Lecture, 1st Int. Conf. Unsaturated Soils, E.E.Alonso & P.Delage Edts., Vol.3, pp.1179-1200.

JOMMI C., J.VAUNAT, A.GENS, D.GAWIN & B.A.SCHREFLER (1997). Multiphase flow in porous media: a numerical benchmark. Proc. NAFEMS World Congress, Vol.2, Stuttgart, pp.1338-1349.

OLIVELLA S., J.CARRERA, A.GENS & E.E.ALONSO (1994). Nonisothermal multiphase flow of brine and gas through saline media. Transp. Porous Media, Vol.15, pp.271-293.

OUYANG H. & D.XIAO (1994). Criteria for eliminating oscillation in analysis of heat conduction equation by finite-element method. Comm. in Num. Meth. Engng., Vol.10, pp.453-460.

PANDAY S. & M.Y. CORAPCIOGLU (1989). Reservoir transport by compositional approach. Transp. Porous Media, Vol.4, pp.369-393.

VAUNAT J., C.JOMMI & A.GENS (1997). A strategy for numerical analysis of the transition between saturated and unsaturated flow conditions. NUMOG VI, S.Pietruszczak & G.N.Pande Edts., Balkema, Rotterdam, pp.297-392.

WHEELER S.J. & D.KARUBE (1996). Constitutive modelling. State of Art Report, 1st Int. Conf. Unsaturated Soils, E.E.Alonso & P.Delage Edts., Vol.3, pp.1323-1356.

MODELLING OF NON-LINEAR BEHAVIOUR OF A ROAD EMBANKMENT

P. Koskela and O. Ravaska
University of Oulu, Oulu, Finland

ABSTRACT: A large number of methods applicable to the calculation of traffic load-induced stresses and strains are available today, but pavements are commonly designed using conventional, simplified methods. This paper deals with non-linear stresses and strains of a road embankment according to the Kenlayer program, which is used to analyse an elastic multilayer system of a flexible pavement under a circular loaded area. The program solves the stresses, strains and non-linear elastic moduli of the layers. FEM analyses are then carried out using the Plaxis program with a Mohr-Coulomb's yield criterion. The influence of lateral stresses and Poisson's ratio, which depend on the stress state, on the behaviour of the road embankment are analysed in particular. Finally, the calculation results are compared with results obtained by conventional means.

1. INTRODUCTION

The conventional Finnish pavement design procedure is based on the achievable bearing capacity of the road, the bearing capacity of the subgrade and the quality of construction materials. The achievable bearing capacity may be defined for each course layer and each construction phase, and it depends on traffic load, defined as the number of loading cycles of a standard axle. Its magnitude is given for each road class as an E-modulus value with a dimension of MPa, and in a geotechnical sense it should be regarded as a "bearing value" rather than bearing capacity. Dimensioning of the structural layers can be determined by

Odemark's formula (Odemark, 1949). The bearing value depends on the deformation modulus (Young's modulus), regarded constant for each layer and the thickness of the layer. The deformation modulus and bearing values correspond to the rebound modulus E_2, obtained from a standard plate bearing test. This test is used to control the as-built bearing value of each structural course in the field.

Dimensioning of the structural courses described above is successful only if well known, conventional materials are used. If new materials are used, such as recycled or other potential materials of today, dimensioning may not be successful because of the lack of empirical data on the proper deformation moduli and their suitability for the design. Problems of the same kind exist in the design of road structures reinforced with geogrids or steel grids. Another basic problem is that dimensioning cannot be made according to "true" parameters such as tensile stresses in the pavement or settlement of the structural courses and the subgrade. The usability of an aged road depends mainly on factors such as rutting, uneven settlement and cracking, but only seldom on the bearing capacity of the road embankment.

The design of road structural courses using "true" parameters for soil materials needs to be based on modelling of stresses and strains. For this purpose, dynamic cyclic tests on course materials have been and are being carried out worldwide. Numerous research records indicate that the stress-strain behaviour of these materials is not linearly elastic even at low strain levels, but the stiffness of the material depends non-linearly on the prevailing stress state. A change in Poisson's ratio has also been clearly observed in the tests performed at the Geotechnical Laboratory of the University of Oulu. These test results will be published later, but they gave the writers an initiative to report on the effect changes in these parameters has on the behaviour of the road embankment in relation to the conventional design of the road structural courses.

2. MODELLING

2.1 Kenlayer program

A Kenlayer computer program (Huang, 1993) was used for the first modelling of the multilayer system present in the road embankment so the non-linear stress-strain behaviour of road course materials could be taken into account. Dynamic triaxial test results obtained at the Geotechnical Laboratory of the University of Oulu and by Kolisoja (1997) were used as input values for these calculations.

The Kenlayer program can be applied only to flexible pavements with no joints or rigid layers. The backbone of Kenlayer is the solution for an elastic multilayer system under a circular loaded area. As a result, Kenlayer can be applied to layered systems under single or various sets of wheels, with each layer behaving differently, being either linearly elastic, non-linearly elastic or viscoelastic. The program output includes final stress-strain states and deformation moduli for each layer iteratively obtained using a non-linear relationship between the stress state and the elastic modulus E, according to the equation (1) presented by Brown & Pell (1967) for the resilient modulus M_r of a granular material:

$$M_r = K_1(\theta)^{K_2} \tag{1}$$

where M_r is the resilient modulus, θ is the sum of the principal stresses and K_1 and K_2 are material parameters. The following model is used for cohesive soils:

$$E = K_1 + K_3(K_2 - q),\ for\ q < K_2 \tag{2}$$

$$E = K_1 + K_4(q - K_2),\ for\ q > K_2 \tag{3}$$

where q is deviatoric stress and K_1, K_2, K_3, K_4 are material parameters.

According to dynamic triaxial tests performed at the Geotechnical Laboratory of the University of Oulu, neither of the functions above was able to adequately represent the subgrade behaviour, because deformation properties depend on both hydrostatic and deviatoric stresses. In order to have a better relationship between Eqs. (2), (3) and the laboratory tests, the modulus values were calculated at a stress level corresponding to that of the course layer in the embankment.

A better model would have been the equation by Uzan (1985):

$$M_r = k_1\theta_0(\theta/\theta_0)^{k_2}(q/\theta_0)^{k_3} \tag{4}$$

where θ_0 is a reference stress of 100 kPa and k_1, k_2, k_3 are material parameters. This relationship could not be used, however, because the Kenlayer program uses only the ones presented by Eqs. (1), (2) and (3).

2.2 Model of the road

According to the Finnish practice, a road embankment is composed of a wearing course, a base course, a subbase course and a filter course. The road embankment was modelled using two thicknesses for the wearing course and the filter course. The effect of lateral stress, varied by the coefficient of lateral pressure K_x and Poisson's ratio ν, on the stress and deformation state of the road embankment was studied in the preliminary calculations. A cross-section of the model is presented in Fig. 1 and the values used in the calculations are given in Table 1. The shaded boxes in the table include the variable values.

The Kenlayer program was only used to calculate the E moduli of the layers, because it did not include any proper failure criteria for the soil material, and the presentation of the stress state was in tabular form and thus difficult to figure out. Therefore, the Plaxis program was used for the stress-strain calculation, and the E moduli obtained from the Kenlayer program were used as initial values for these calculations.

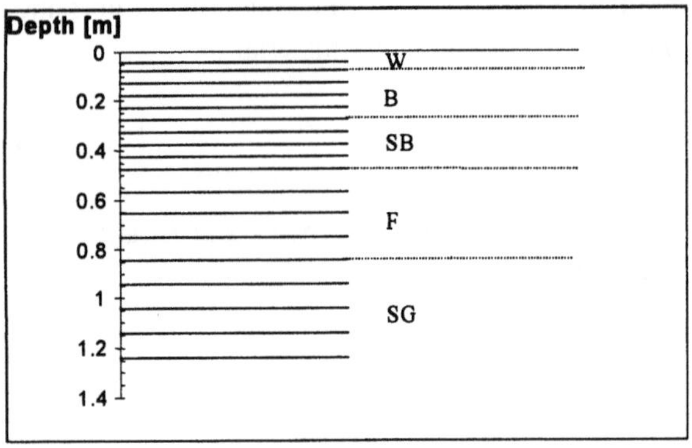

Fig. 1. Principal figure of the cross-section in the Kenlayer calculations.

Course	Thickness [mm]	Poisson's ratio ν	K_x
Asphalt concrete	80,160	0.01, 0.18, 0.35	0.6, 1.3, 2
Base	200	"	"
Subbase	200	"	"
Filter	360, 560	"	"
Subgrade		"	"

Table 1. Course thicknesses and calculation parameters of the road model.

In the Plaxis calculations, the layer subdivision presented in Fig. 1 was followed as far as possible. However, the limitations of the program made it necessary to reduce the number of layers, and a finite element mesh shown in Fig. 2 was used. The Plaxis calculations were carried out in an axi-symmetric manner using 12 noded triangle elements.

3. CALCULATION RESULTS

3.1 Preliminary calculations

At first, preliminary calculations were performed to figure out the effect of lateral stress caused by compaction, for instance, and Poisson's ratio ν on the stress state. Horizontal stress was varied using different values for the coefficient of lateral pressure K_x. A large number of calculations was included in the program, and the results can be summarized as follows: changes in Poisson's ratio and the coefficient of lateral pressure affected stresses and strains in the vertical and horizontal directions differently. The impact of the changes was the same in general but the structure also had a certain effect on the stresses and strains. As an example, the effect of ν and K_x is presented for an AB80/F360 (asphalt concrete 80

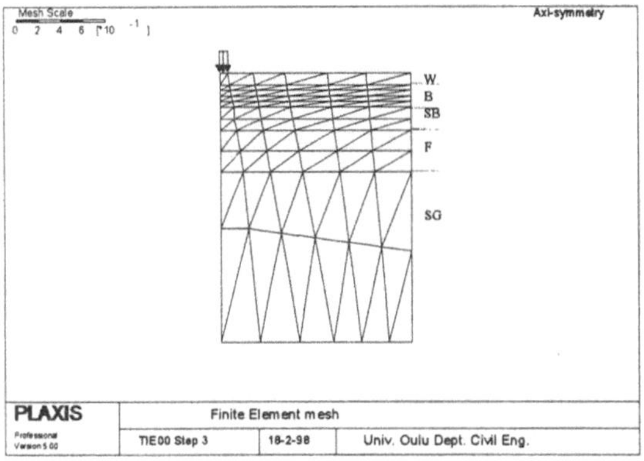

Fig. 2. Cross-section of the road model used in the Plaxis calculations.

mm, filter course 360 mm) structure in Fig. 3. The figures on the left show that a decrease in ν increases vertical stresses and considerably affects radial (horizontal) stresses in the base course. Vertical strains increase in the upper parts of the base course and the subgrade, while they decrease in the subbase and filter courses. Radial strains decrease in all courses with decreasing ν, which is a natural consequence. In general, the effect of Poisson's ratio is linear, i.e., changes in stresses and strains are directly proportional to ν.

The effect of the change in K_x on the stress-strain state depends on the course structure. Examples of the calculation results for one structure are presented on the right in Fig. 3. As K_x increases, vertical stresses remain almost unchanged while horizontal stresses decrease markedly in the lower part of the base course and in the subbase. Vertical strains also decrease in the filter course and the subgrade, and horizontal strains decrease in all the courses. The changes in the stresses and strains are no longer linear, as was the case for Poisson's ratio. Most of the changes take place as K_x increases from 0.6 to 1.3, but a further increase to 2.0 causes only small changes.

3.2 Improved calculations

The basic aim of the Plaxis calculations was to follow the development of stresses and strains in the course layers using deformation moduli given by the Kenlayer program as input values for the calculations. In elastic calculations, minor principal stresses are small in the vicinity of the loaded areas, and for friction soils, distinct plastic zones are generated in the base and subbase course, Fig. 4a. This means that the structure would fail, resulting in high deformations in the road embankment. Extremely high strength parameters corresponding to a very dense structure only reduce the extent of the plastic zone, Fig. 4b. This unrealistic behaviour in a granular material can be avoided by assuming that high lateral pressures will

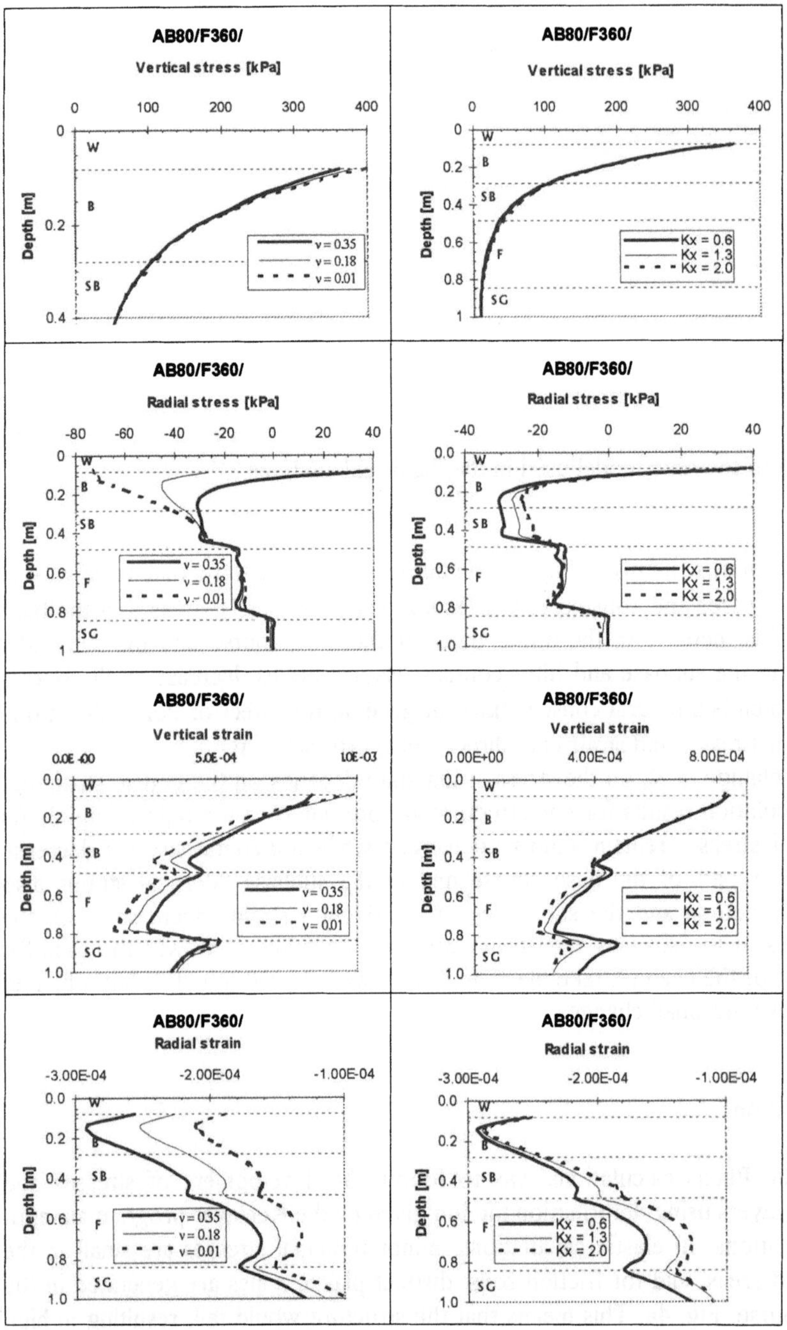

Fig. 3. An example of the preliminary calculations: Effects of ν and K_x on stresses and deformations.

be generated in the base and subbase courses during compaction, and after that a residual stress higher than the earth pressure at rest will remain. Duncan & Seed (1986), for example, have presented measurement data and a calculation method for evaluating compaction-induced lateral stresses resulting from the placement and compaction of fill layers, either in the free field or adjacent to vertical, non-deflecting soil-structure interfaces. The measurements gave lateral pressures in the range of 10-20 kPa, compacted by a 1.3 ton vibratory roller adjacent to the vertical, non-deflecting wall. A stress state caused by the weight of the soil and compaction-induced lateral stresses is called a K_c-state in the following.

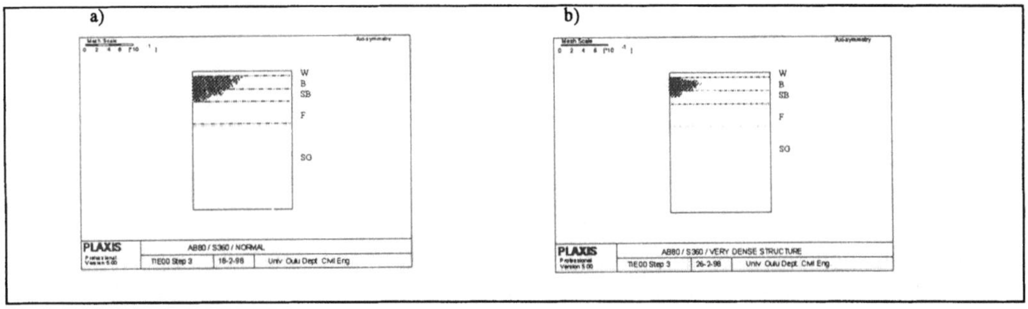

Fig 4. Plastic zone: a) average strength parameters; b) high strength parameters.

According to the Finnish design guide for building excavations (Rakennuskaivanto-ohje, 1989), horizontal stress of about 25 kPa corresponds to the earth pressure developed during compaction in layers by a 6 ton vibratory roller, which is a normal compaction device for road base and subbase courses.

After a K_c-state had been obtained, the stresses and strains of the courses were recalculated using Kenlayer and Plaxis programs. In addition, the stress-dependence of Poisson's ratio was taken into consideration by assigning a value for each layer obtained from dynamic triaxial tests under similar conditions in the laboratory. For comparison, a linear calculation was also performed using average modulus values for each course layer calculated by the Kenlayer program. An example of the calculation results is shown in Fig. 5, including a linear calculation, a calculation with a stress-dependent elastic modulus E in a K_0-state ($K_x = K_0 = 0.6$ and $\nu = 0.35$) and with stress-dependent E and ν in a K_c-state.

Fig. 5 shows that stresses and strains calculated using average deformation moduli differ considerably from those obtained from the non-linear calculations, and according to the laboratory tests, they cannot be considered reliable. Changes in lateral stress and Poisson's ratio affect both the vertical strain and horizontal stress-strain states and thus their influence should also be taken into account.

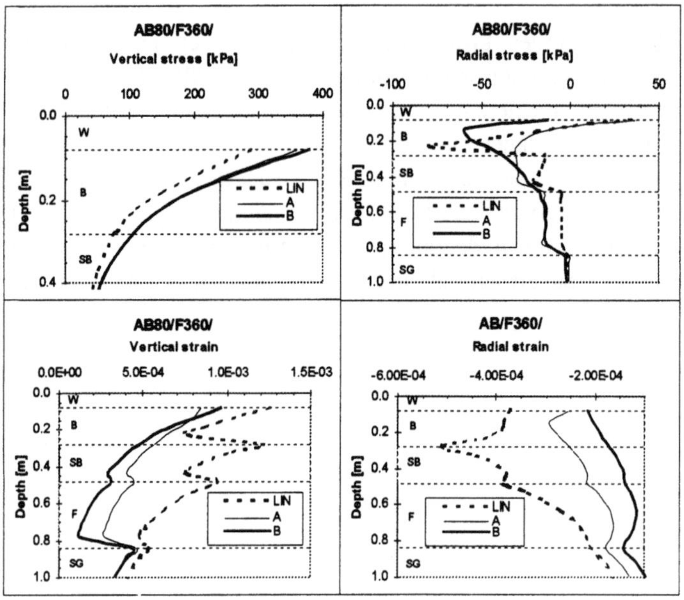

Fig. 5. Calculation result of a road structure using linear and non-linear models. LIN = linearly elastic, A = stress-dependent E ($K_x = K_0 = 0.6$, $\nu = 0.35$), B = stress-dependent E and ν in a K_c-state

3.3 Deformation durability

The effect of the deformation modulus and Poisson's ratio is finally discussed in the light of duration criteria presented in the literature for road structures. In these criteria, the deformation durability of the unbound layers and the subgrade is estimated by means of vertical stresses or deformations at the top of the layer. According to Ullidtz (1987), most of the empirical relations for permissible stresses and strains being used today are of the type:

$$\sigma \quad or \quad \varepsilon = K \cdot (N/10^6)^{(-1/a)} \cdot (E/E')^b \tag{5}$$

where N is the number of loads to cause a certain deterioration, E' is the reference modulus and K, a, b are constants or functions of materials and environmental conditions. The permissible number of loading cycles can be determined from Eq. (5) in the form:

$$N = \sigma^{-a} \quad or \quad \varepsilon^{-a} \cdot K^a \cdot (E/E')^{ab} \cdot 10^6 \tag{6}$$

Let us consider stress on top of the base course and settlement on top of the subgrade. According to Fig. 3, a change of 0.35 to 0.01 in Poisson's ratio increases vertical stress in the base course 40 kPa and vertical strain in the subgrade $4*10^{-5}$. Correspondingly, a change of 0.6 to 1.3 in the coefficient of earth pressure changes the vertical strain by 10^{-4}. The effect of the changes on the number of loading cycles is presented in Table 2.

If we take vertical stress to be a criterion, Eq. (6), a change in ν or K_x has only a negligible effect on the allowable number of loadings, as presented in Table 2. The reason for this is the change of vertical stress in non-linear calculation due to the change in the E modulus, causing practically no change in N, whereas for vertical strain, the effect is clear: 28-66 %. Table 2 also includes the allowable number of loading cycles in the base course and the subgrade, calculated from the values presented in Fig. 5. In the linear case (LIN) the difference compared with stress-dependent cases (A and B) was about half.

Layer	Criterion	Change	N	Difference
Base course (AB 80)	$N(\sigma_z)$	Poisson's ratio 0.35 → 0.01 (Fig 3.)	$2.36*10^5$ → $2.52*10^5$	6 %
Base course (AB 80)	$N(\sigma_z)$	Coefficient of lateral earth press. 0.6 → 1.3 (Fig 3.)	No effect	0 %
Subgrade	$N(\varepsilon_z)$	Poisson's ratio 0.35 → 0.01 (Fig 3.)	$5.67*10^9$ → $7.84*10^9$	28 %
Subgrade	$N(\varepsilon_z)$	Coefficient of lateral earth press. 0.6 → 1.3 (Fig 3.)	$7.84*10^9$ → $2.31*10^{10}$	66 %
Base course (AB 80)	$N(\sigma_z)$	Elastic → non-linear (Fig 5.)	$1.41*10^5$ → $2.52*10^5$	44 %
Subgrade	$N(\varepsilon_z)$	Elastic → non-linear (Fig 5.)	$4.45*10^9$ → $7.84*10^9$	43 %

Table 2. Effect of Poisson's ratio ν and the coefficient of lateral earth pressure K_x, given by the constant and variable parameters, on the allowable number of loading cycles N.

4. CONCLUSIONS

If the design of pavements includes the determination of course thicknesses and the estimation of duration time using "true" parameters, the non-linearity of materials evidently has to be taken into account. The non-linearity of the deformation modulus clearly affects the vertical stress and strain states and thus, also the allowable number of loading cycles, which increases to double the value.

FEM calculations using K_0-conditions in the road embankment led to unrealistic deformations because the failure state was exceeded even though extremely high strength parameters were used. Taking the compaction-induced residual stresses into account caused the course layers to behave realistically.

According to laboratory tests, values of Poisson's ratio vary in the range 0.01-0.1 for an elastic state and in the range 0.35-0.4 for a plastic state having a considerable effect on the stress-strain state of the road structure.

In the detailed design of road structures, e.g., in dimensioning reinforced structures, attention has to be paid to the stress-strain state taking into account stress-dependence of the deformation modulus and Poisson's ratio and also compaction induced residual stresses. In the design of deformation durability by means of the number of loading cycles, stress-dependence of Poisson's ratio and compaction induced residual stresses were not, however, of importance, as vertical stress was taken to be a criterion. For vertical strain the change in the allowable number of loading cycles was clear.

Non-linearity was taken into account in these calculations only in the vertical direction. In more accurate calculations the structure should be regarded as being non-linear also in horizontal direction, i.e., the deformation modulus should be stress-dependent also in that direction.

REFERENCES

Brown, S.F., Pell, P.S. (1967). An experimental investigation on the stresses, strains and deflections in a layered pavement structure subjected to dynamic loads. Proc. of the 2nd Int. Conf. on Structural Design of Asphalt Pavements. Ann Arbor, USA, pp. 478-504.

Duncan, J.M., Seed, R.B., (1986)Compaction-Induced Earth Pressures under K_0-Conditions. Journal of Geotechnical Engineering, ASCE, Vol. 112, No. 1, pp. 1-22.

Huang, Y.H. (1993). Pavement Analysis and Design. University of Kentucky. Prentice Hall, New Jersey, USA.

Kolisoja, P. (1997). Resilient Deformation Characteristics of Granular Materials. Tampere University of Technology, Publications 223. Tampere, Finland.

Odemark, N. (1949). Undersökningar av elasticitetsegenskaperna hos olika jordarter samt teori för beräkning av beläggningar enligt elasticitetsteorin. Statens Väginstitut. Meddelande 77, Stockholm, Sweden, 1949.

Rakennuskaivanto-ohje (1989). Finnish design guide for building excavations (in Finnish). RIL 181-1989. Vammala, Finland.

Ullidtz, P. (1987). Pavement Analysis. Developments in Civil Engineering, 19. The Technical University of Denmark, Lyngby. Elsevier Publishers Company Inc., Amsterdam, The Netherlands.

Uzan, J. (1985). Characterization of granular material. Transportation Research Record 1022, pp. 52-59.

DIMENSIONAL REDUCTION APPLIED TO SPECIFIC PROBLEMS OF CONSOLIDATION

P. Kuklík, M. Sejnoha and J. Mares
Czech Technical University, Prague, Czech Republic

Abstract

This paper explores various aspects of consolidation of clay type soils. Formulation is provided for two specific problems. First, the isotropic consolidation with reference to the CAM clay model is investigated to reconcile the mutual dependency of individual state variables and provide certain modification to the existing models in order to obtain better agreement with experiments. Next, the improved model is implemented in the study of consolidation of the soil body drained by a system of concentric cylindrical drains. Several numerical examples are presented to support theoretical formulation.

1 Introduction

When solving the long-term deformation of soils, we are faced with the problem of finding the distribution of two functions p (the pore pressure) and $\boldsymbol{u}$ (the displacement field) in general three-dimensional environment at an arbitrary time instant t, which is not an easy task. Fortunately, in certain applications when employing a suitable method of attack the problem can be significantly simplified by reducing the dimension of the original problem, so that only a one variable function is to be determined.

Herein, two such problems in which the dimensional reductions follows directly from the formulation accompanied by introducing several modifications are discussed. The numerical solutions are tackle using the collocation method furnished by cubic spline for discretization of space variables. The numerical integration is realized by employing the generalized mid-point rule.

2 Isotropic consolidation

This section is devoted to the analysis of isotropic consolidation with emphasis on the constitutive modeling. To this end, the popular CAM clay model is selected to represent the behavior of variety of soils. Our prime objective is to provide several modifications to the existing formulation aiming to improve the modeling of the behavior of partially saturated soils.

Owing to the significant troubles in measuring the actual amount of gas and liquid pressure in the pores we limit our considerations to the two phase medium only and incorporate the actual level of saturation indirectly through the modified coefficient of filtration. Consequently, the pore pressure is regarded herein as an average quantity of both the gas and water pressures. This approach has been first introduced in [3]. Here, the formulation is reexamined and provided in rather appealing form accompanied by several results to support the theoretical expectation.

2.1 Governing equations

We begin by an overview of the basic relations governing the behavior of a multiphase medium with reference to the isotropic consolidation.

The starting point is the equation of continuity which, when combined with Darcy's law, attains the usual form

$$-\frac{k}{\gamma_w}\Delta p + \dot{\epsilon}_v = 0. \tag{1}$$

where k and γ_w are the coefficient of filtration and the specific weight of water, respectively; p represents the pressure found in the pores and $\dot{\epsilon_v}$ is the velocity of the volumetric strain; Δ stands for the Laplace operator. When using the CAM clay model, the volumetric strain is provided by

$$\epsilon_v = -\frac{\lambda}{(1+e_0)} ln\left(-\sigma_m^{'}\right). \tag{2}$$

Then, differentiating above equation with respect to time gives

$$\dot{\epsilon}_v = -\frac{\lambda}{(1+e_0)}\frac{\dot{\sigma}_m^{'}}{\sigma_m^{'}}. \tag{3}$$

Note that Eq. (3) represents not only the elastic, but also the inelastic deformations. This fact can be easily verified by multiplying above equation thru by dt, so that Eq. (3) becomes

$$\mathrm{d}\epsilon_v = -\frac{\lambda}{(1+e_0)}\frac{\mathrm{d}\sigma_m^{'}}{\sigma_m^{'}}. \tag{4}$$

The plastic deformation then follows from Eq. (4) by eliminating the elastic strain, which in the case of the CAM clay model attains the familiar form

$$d\epsilon_v^e = -\frac{\kappa}{1+e_0-(\lambda-\kappa)ln\,(\sigma_m^{'})}\frac{d\sigma_m^{'}}{\sigma_m^{'}}. \tag{5}$$

Clearly, Eq. (5) would enter the analysis provided an unloading during the consolidation process was considered. To conclude our introductory survey of equations we introduce Eq. (3) into Eq. (1) to get

$$\dot{\sigma}_m^{'} + \frac{k\,(1+e_0)\,\sigma_m^{'}}{\gamma_w\lambda}\Delta p = 0. \tag{6}$$

2.2 Description of isotropic consolidation in triaxial apparatus

We now proceed and simplify the above formulation by considering the isotropic consolidation carried out in the triaxial apparatus. This appears to be advantageous, particularly when identifying significance of individual parameters in the model, since a direct comparison with the experimental measurements exists.

Now, consider a triaxial apparatus with a drain placed at the upper base of the cylindrical specimen. In such a case the total mean stress is constant and is given by

$$\sigma_m = \sigma_m^{'} - p = const. \tag{7}$$

Differentiating Eq. (7) with respect to time leads us to the equation of equilibrium in the form

$$\dot{\sigma}_m^{'} - \dot{p} = 0. \tag{8}$$

Then, when introducing Eq. (8) in Eq. (6), we find that

$$\dot{p} = -\frac{k(1+e_0)\sigma_m^{'}}{\gamma_w\lambda}\Delta p, \tag{9}$$

Note that above expression is the well known Terzaghi's parabolic equation. In this case, however, the coefficient of consolidation is not constant but rather depends on the actual amount of pressure taken by pores. The resulting reduction of Eq. (9) to a one dimensional problem follows not only from the assumed axial symmetry of the problem, but a detailed analysis would reveal that also no changes in the radial direction are permitted. Hence, the above equation can be recast as follows

$$\frac{\partial p}{\partial t} = -\frac{k(1+e_0)\sigma_m^{'}}{\gamma_w\lambda}\frac{\partial^2 p}{\partial z^2}. \tag{10}$$

In addition, the formulation must be supplemented by the boundary conditions. In this case, the corresponding boundary conditions are given in terms of the zero pore pressure prescribed at the locations of drains together with the initial condition which

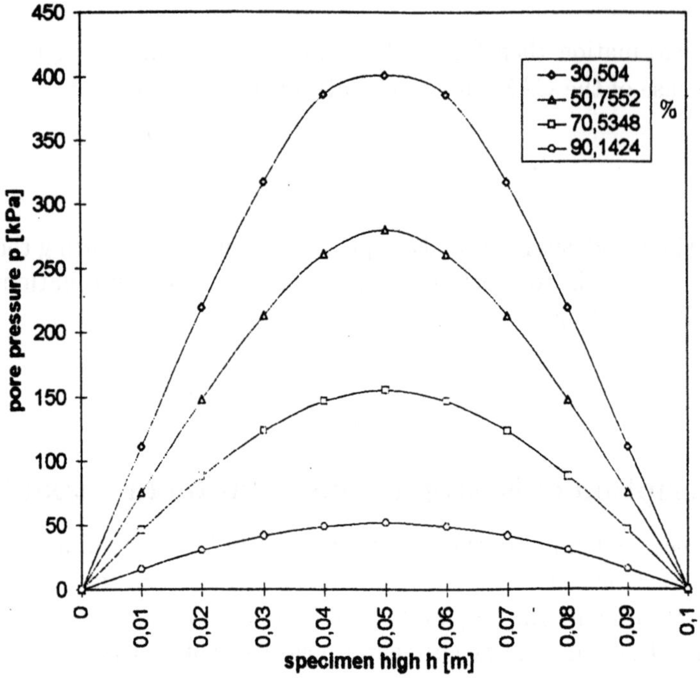

Figure 1: Variation in pore pressure at various stages of consolidation.

states that in the case of the fully saturated specimen the applied load is equal to the pore pressure only.

For more details regarding the numerical implementation we refer the reader to [3], in which the aforementioned modification of the coefficient of filtration k has been carefully examined. The recommended form of k reads

$$k = k_0 \left(\frac{e}{e_0}\right)^{m_1+m_2} \left(\frac{1}{Sr_0}\right)^{m_2} \tag{11}$$

2.3 Results 1

The nonlinear solution of Eq. (10) is illustrated in Fig. (1) indicating an evolution of the pore pressure during consolidation. In this particular example the reconstituted clay-stone from the mine activity located in the northwest Bohemia was considered. The material respective parameters characterizing the CAM clay model are: the coefficient of filtration $k = 2.4\ 10^{-11}$ (m/s), the specific weight $\gamma = 20$ (kN/m^2), the slope of NCL $\lambda = 0.06$, the initial value of the effective mean stress $\sigma'_{m,in} = -4$ (kPa), and the initial void ratio $e_0 = 1$. The selected specimen height h was equal to 0.1 m and the prescribed chamber pressure σ_m was set equal to -400 (kPa). The percentage of consolidation was computed from the relation

$$100 \frac{\int_V \sigma'_m \mathrm{d}V}{\int_V \sigma_m \mathrm{d}V}.$$

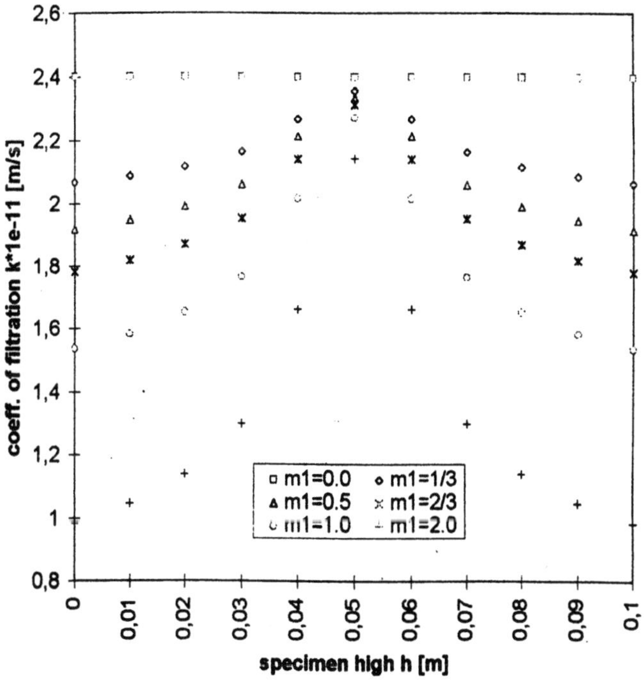

Figure 2: Distribution of coefficient of filtration corresponding to 30% of consolidation for different values of m_1

It is evident from Fig. 1 that at the state corresponding to the 30 % of consolodation there exists an appreciable region in the specimen with no change in the pore pressure. This is attributed to the fact that when applying the typical initial condition, representing equality of the pore pressure and the prescribed chamber pressure, results in zero value of the coefficient of consolidation at all points in the specimen. The remedy is partially provided by prescribing a small initial value of the effective mean stress $\sigma'_{m,in}$. This modification starts up the consolidation at once without changing its character. Moreover, Eq. (6) shows that the only parameter which may alter a rather slow initial process of consolidation is the coefficient of filtration.

To obtain a better agreement with the experimental measurements the coefficient of filtration must decrease as the effective mean stress σ'_m increases. This statement is in correspondence with the proposed modification of k, Eq. (11). Applicability of such a modification is illustrated in Fig. 2 assuming the 30% of consolidation. The model parameters are taken from the first example.

Dot diagrams used for individual distributions indicate the piecewise constant distribution of k thru the height of the specimen to comply with the selected computational algorithm. Should the more rigorous variation in k to be considered, the odd derivatives would have had to be included as indicated by following equation

$$p_{,t} = \frac{k(1+e_0)}{\gamma_w \lambda}(-\sigma_m - p - \sigma'_{m,in}) \left[p_{,zz} + \frac{k_{,z} p_{,z}}{k} \right] . \tag{12}$$

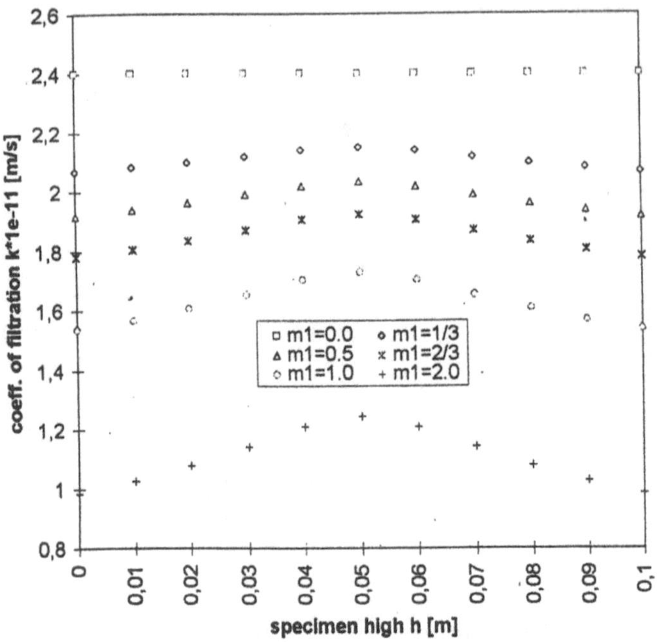

Figure 3: Distribution of coefficient of filtration corresponding to 50% of consolidation for different values of m_1

Solution to Eq. (12), however, may prove quite laborious. Furthermore, when using the finite element technique, such an "improvement" appears rather impractical.

Numerical experiments further revealed that with increasing parameter m_1 the difference between the coefficient of filtration inside the specimen also increases. This change, however, is not very dramatic.

Similar results as for the 30% of consolidation are presented for the stage corresponding to the 50% of consolidation, see Fig. 3. Evidently, when the process of consolidation progresses the difference between the coefficients of filtration at various locations becomes less pronounced and approaches the estimated lower bound.

3 Consolidation of earth body using a system of drains

Our objective now is to incorporate essential features of the proceeding formulation into more practical applications.

We know that the presence of water in the soil significantly reduces its shear resistance, which may eventually lead to a loss of stability. From this point of view, an appropriate drainage still appears as the most efficient mean preventing such an undesirable results.

Here, we limit our attention to a periodic distribution of the system of circular drains.

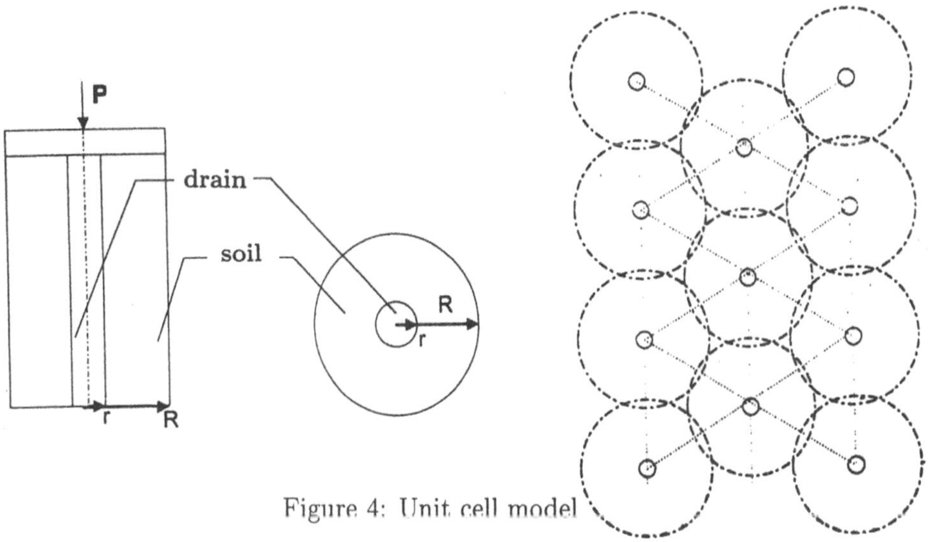

Figure 4: Unit cell model

In such a case, the solution to this problem is conveniently provided by a unit cell model. Selected periodicity of drains together with characteristic loading conditions then offer another simplification to the general three-dimensional formulation by assuming an axial symmetry, thus reducing the space considerations by one. A typical unit cell is shown in Fig. 4.

We assume that at a distance R, which corresponds to the periodic distribution of drains, no water flow is observed. The water itself is carried away by drain of radius r_1. Moreover, we assume that the loading is applied thru a rigid plate as shown in Fig. 4. Owing to variation of the total mean stress, a simple formulation such as the one presented in the previous sections is no longer available, so that one of the classical methods in mechanics must be called. In the present study, the more favorable deformation method is used to serve this purpose.

3.1 Formulation

We begin by writing the displacements in the form

$$u = u(r,t), \qquad v = 0, \qquad w = a(t)z, \tag{13}$$

where u, v and w are the components of the displacement field given in the cylindrical r, φ, z coordinate system. Then, using the strain-displacement relations together with the constitutive equations provides the components of the total stresses. Note that due to absence of the shearing stress components, the present state of stress is the principal one. Thus,

$$\begin{aligned} \sigma_r &= M\frac{\partial u}{\partial r} + 2G\frac{\nu}{1-2\nu}\left(\frac{u}{r} + a(t)\right) - p, \\ \sigma_\varphi &= M\frac{u}{r} + 2G\frac{\nu}{1-2\nu}\left(\frac{\partial u}{\partial r} + a(t)\right) - p, \end{aligned} \tag{14}$$

$$\sigma_z = Ma(t) + 2G\frac{\nu}{1-2\nu}\left(\frac{\partial u}{\partial r} + \frac{u}{r}\right) - p,$$

where M is the oedometric modulus and G is the shear modulus of elasticity. $\sigma_r, \sigma_\varphi, \sigma_z$ are the normal stresses in directions of r, φ and z, respectively. In the case of the CAM clay model, the material parameters M, G can be continuously modified depending on the actual value of the effective mean stress.

The governing equations of the problem are provided by the equilibrium condition in the radial direction written as

$$M\left(\frac{\partial^2 u}{\partial r^2} + \frac{1}{r}\frac{\partial u}{\partial r} - \frac{u}{r^2}\right) = \frac{\partial p}{\partial r}. \tag{15}$$

and by Darcy's law together with the equation of continuity given by

$$\frac{k}{\gamma_w}\left(\frac{\partial^2 p}{\partial r^2} + \frac{1}{r}\frac{\partial p}{\partial r}\right) = \frac{\partial}{\partial t}\left(\frac{\partial u}{\partial r} + \frac{u}{r} + a(t)\right). \tag{16}$$

Eqs. (15) and (16) are used to provide for the unknown functions $u(r,t)$ and $p(r,t)$. The unknown function $a(t)$ follows from the equilibrium condition written in the z direction.

To proceed, we let us now introduce the following notation

$$g(r,t) = \frac{k}{\gamma_w}\frac{\partial p}{\partial r} - \frac{\partial u}{\partial t}. \tag{17}$$

Hence, the equation (16) becomes

$$\frac{\partial}{\partial t}g(r,t) + \frac{1}{r}g(r,t) = \dot{a}(t) \tag{18}$$

Solution to above equation with respect to r assumes the form

$$g(r,t) = \frac{C_{10}(t)}{r} + \frac{1}{2}\dot{a}(t)r. \tag{19}$$

Employing the boundary condition at points $r = R$ in the form

$$\frac{\partial p}{\partial r} = 0, \qquad \frac{\partial u}{\partial t} = 0, \tag{20}$$

gives

$$C_{10}(t) = -\frac{1}{2}\frac{\partial a(t)}{\partial t}R^2. \tag{21}$$

Introducing Eqs. (16), (18) and (20) into Eq. (14) provides the final form of the equation of equilibrium such that

$$\frac{\partial^2 u}{\partial r^2} + \frac{1}{r}\frac{\partial u}{\partial r} - \frac{u}{r^2} = -\frac{\dot{a}(t)}{2M}\frac{\gamma_w}{k}\left(\frac{R^2}{r} - r\right) + \frac{\frac{\partial u}{\partial t}}{M}\frac{\gamma_w}{k}. \tag{22}$$

3.2 Numerical implementation and results

In the present approach the collocation method furnished with cubic splines to approximate the distribution of individual variables is used to carry out the numerical solution of Eqs. (15) and (16). The generalized mid-point rule is selected to provide a reliable scheme for numerical integration. As an example of this technique we present a discretized form of Eq. (22) as

$$M_j^* + \frac{1}{r_j}\left[\frac{u_{j+1}^* - u_j^*}{h} - \frac{M_j^*}{3}h - \frac{M_j^*}{6}h\right] - \frac{u_j^*}{r_j^2} = \frac{a^{i+1} - a^i}{\Delta t 2M\frac{k}{\gamma_w}}\left[\frac{R^2}{r_j} - r_j\right] + \frac{u_j^{i+1} - u_j^i}{\Delta t M\frac{k}{\gamma_w}}, \quad (23)$$

$$\begin{aligned} M^* &= \frac{\partial^2 u^*}{\partial r^2} \\ M_j^* &= M_j^i(1-\tau) + M_j^{i+1}\tau \\ u_j^* &= u_j^i(1-\tau) + u_j^{i+1}\tau \end{aligned}$$

Similarly we could arrived at the discretized form of Eq. (16) and the continuity condition imposed on radial stresses at the drain-soil interface. The reader may familiarize himself with the detailed numerical algorithm in [4].

It is important to point out that the closed form solution describing the distribution of displacements within the drain is available, provided the pore pressure in the drain is assumed to be equal to zero. Consequently, the zero pore pressure and the continuity of the radial stresses and displacements is enforced along the drain-soil interface. The initial condition is identical to the one used when solving the problem of isotropic consolidation.

Solution of the problem at a given time $t = 10 hours$ is illustrated in Fig. 5. Since at this time the consolidation was completed, we can compare the numerical result with the known elastic solution, and indeed both results are as expected in a very good agreement. A rather high coefficient of filtration was assumed resulting in the fast consolidation process.

4 Conclusion

Two specific problems of consolidation employing the CAM clay model for constitutive modeling were addressed in this paper. First, the case of isotropic consolidation suitable for model calibration was investigated. It has been shown that a proper modeling of clays requires a modification of the existing model. Two such modifications, particularly an application of the initial value of the effective mean stress together with improved version of the coefficient of filtration were proposed. Comparing with the available experimental measurements both modifications improve the prediction of the actual behavior of the clay type soils.

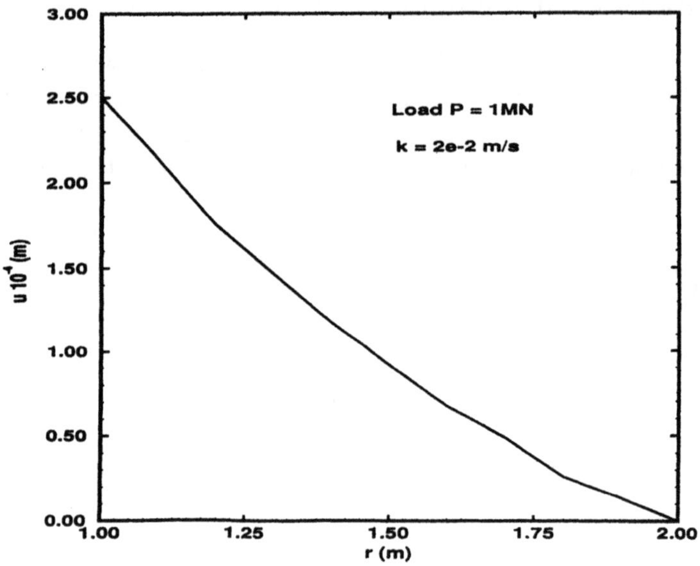

Figure 5: Distribution of the displacement u at time $t = 10hours$.

The modified CAM clay model was then used to analyze a consolidation of the earth body using a system of concentric drains. Although various experimental tests to verify present constitutive model and proposed numerical procedure are still being conducted, we see that the current formulation provides a number of useful informations which are not commonly available. At the present time, an applicability of the present approach is supported by the fact that with progress in consolidation the solution approaches the known elastic solution.

Acknowledgment

Financial support for this project was provided by GAČR 103/98/0276.

References

[1] KUKLÍK, P., et. al., (1994) Nonlinear analysis of the coefficient of consolidation, pp. 81-83, Balkema, Rotterdam.

[2] KUKLÍK, P., ZÁLESKÝ, J., (1995) Analysis of some experiments carried on the CAM clay model, pp. 357-358, WORKSHOP 95, Prague.

[3] KUKLÍK, P., et. al., (1998) Verification of CAM clay model in description of consolidation, to be presented at SSCC98, Durban, South Africa.

[4] MAREŠ, J., (1997) Solution of consolidation using the CAM clay model, Master thesis, Czech Technical University in Prague, (in Czech).

A SOIL MODEL FOR THE OVERCONSOLIDATED REGION OF CLAYS

T. Länsivaara and S. Nordal
Norwegian University of Science and Technology, Trondheim, Norway

ABSTARCT: The paper presents an elastoplastic model developed for the overconsolidated region of clays. In the model, the stiffness of the soil is related to its previous history introducing memory points. The stiffness is thus depending on the size and the direction of the new stress/strain increment compared to previous stress/strain states. The plastic strains are constructed from three components. The first plasticity mode is for pure volumetric behavior, the second one for pure shear behavior, and the third one is a Coulombian plasticity term which includes coupling effects. The performance of the model is discussed in relation to laboratory tests. The model has been found to comprise several essential characteristics of low strain level soil behavior, like stress history and direction dependency of stiffness and strain accumulation in cyclic loading within a relatively simple framework.

1 INTRODUCTION

The significance of both the nonlinearity and the high stiffness at low strain levels has been recognized for quite a long time. The conventional soil models can not handle these aspects properly and problems like overprediction of deformations in the far field areas occur, (Kirkebø et al. 1994). Several models with a variety of complexity and overall applicability have been presented to overcome this problem. In a PhD-project at our institute, a comprehensive experimental study on the deformation properties of undisturbed clays is made. This is done in combination with theoretical modeling for improving our NTNU soil models. The experimental results include well known stiffness reduction curves as given by figure 1. The theoretical model combines aspects of the Brick model by Simpson (Simpson 1992) and the membership function by Klisinski (Klisinski 1988), with the Mobilized Friction Model developed at NTNU (Nordal et al. 1989). This paper presents some preliminary results from the development of the theoretical model, concentrating on the behavior in the overconsolidated region of clays. For the normally consolidated region additional elements, like an anisotropic yieldsurface will be added.

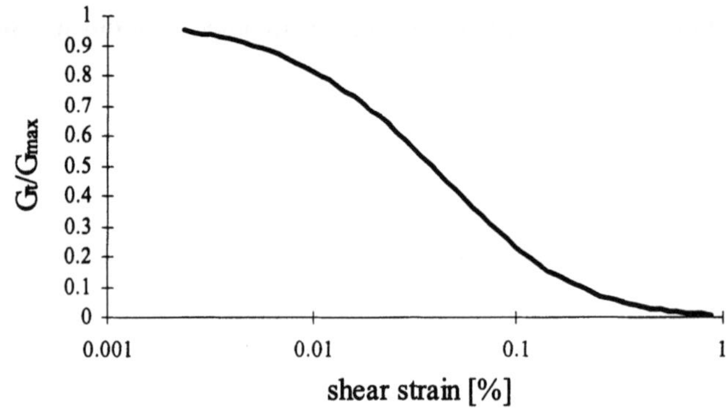

Fig.1 A typical modulus reduction curve for the shear modulus observed in laboratory.

2 A MEMORY MODEL FOR STRESS HISTORY

Soil memory is often related to a preconsolidation pressure or a more generalized yield surface. However soil remembers previous loading also within the OC region. Evidently, the stiffness varies with the direction of loading relative to past stress history. One approach to model this is to use a finite number, n, of memory points $\overline{\sigma}_i^j$, $j = 1...n$, see figure 2. The memory model has borrowed its main principles from (Simpson 1992). According to the model the stress is pulling the memorypoints behind as brick in a rope. The directions from these memory points to present stress state, $\overline{\sigma}_i$, is given by a memoryvector $\overline{m}_i^j = \overline{\sigma}_i - \overline{\sigma}_i^j$. All of the memorypoints $\overline{\sigma}_i^j$ correspond to a different distance $|m^j|$ back in stress history having thus memories of different length. If the distance from the memorypoint $\overline{\sigma}_i^j$ to the new stress state $\overline{\sigma}_{i+1}$ is less than the length of the memory $|m^j|$ (a slack rope), then the memorypoint will remain in its position and be inactive causing no stiffness reduction. If however the distance is greater than $|m^j|$, the memorypoint $\overline{\sigma}_i^j$ has to be updated, and the stiffness reduces depending on the angle between stress increment vector $d\overline{\sigma}$ and the memoryvector $\overline{m}_i^j$. Softest response is obtained in the direction of the memoryvector, that is in the direction which according to that particular memorypoint represents stress history. The influence of direction is illustrated in figure 2.

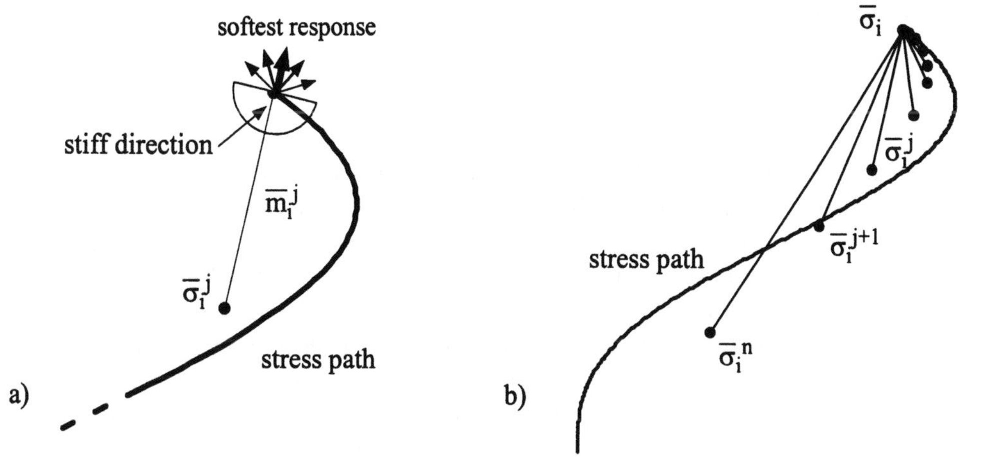

Fig.2 The effect of the direction of stress increment to the stiffness contribution of one memorypoint. a), and a set of memorypoints pulled behind a changing stress state b).

The plasticity model to be presented involves uncoupled volumetric and deviatoric plasticity, with different plastic modulus for volume and shear. It is therefore convenient, that the memory model for stiffness gives only one stiffness number $\Gamma \in (0,1)$. It is a parameter similar to the membershipfunction by Klisinski (Klisinski 1988). The stiffness number Γ is calculated from:

$$\Gamma = 1 - \sum_{j=1}^{n} \Delta\Gamma^j = 1 - \sum_{j=1}^{n} \cos\theta^j \cdot w^j \tag{1}$$

were n = number of memorypoints

θ^j = the angle between $d\overline{\sigma}$ and $\overline{m}_i^j$

w^j = weight of memorypoint $\overline{\sigma}_i^j$, $w^j \in (0,1)$

3 AN ELASTO-PLASTIC MODEL FOR THE OC-REGION

As in classical elasto-plasticity theory, the strains are divided to an elastic and a plastic contribution. Elasticity is described with a maximum shear modulus G_{max} and a maximum bulkmodulus K_{max}, which both describe the initial stiffness of the soil. Both G_{max} and K_{max} are made dependent on the preconsolidation pressure using Janbu stiffness formulation (Janbu 1963).

The plastic strains are decomposed into three parts, i.e;

$$d\bar{\varepsilon}^p = d\bar{\varepsilon}_p^p + d\bar{\varepsilon}_q^p + d\bar{\varepsilon}_f^p \tag{2}$$

The subscripts p, q and f refer to volumetric, deviatoric and mobilization plasticity. The different plasticity modes can be characterized with three fully kinematic yield surfaces, which define the direction of the plastifying stress increment independently for each mode.The volumetric plasticity mode provides plasticity in the hydrostatic axis direction, the deviatoric mode in the Π-plane and the mobilization plasticity mode according to a Coulombian strength mobilization criteria.The volume and shear modes are made dominant for low and moderate degrees of mobilization, while the mobilization mode is introduced to provide both a smooth transition to failure and dilatancy/contractancy effects for higher degrees of mobilization.Since the plastic modulus is determined by the stiffness number only the directions of the surface gradient is of interest. The plastic multipliers for the different plasticity modes are defined as:

$$d\lambda_i = \frac{1}{H_i}\frac{\partial F_i}{\partial \bar{\sigma}}d\bar{\sigma} \tag{3}$$

The subscript i stands for the different plasticity modes p,q and f, H_i is the plastic modulus. In figure 3 principle sketches and equations for the plastic moduli are given. The reduction of Γ is restricted to a value $\Gamma_{min} > 0$ corresponding to *large strain stiffness* in the OC region.

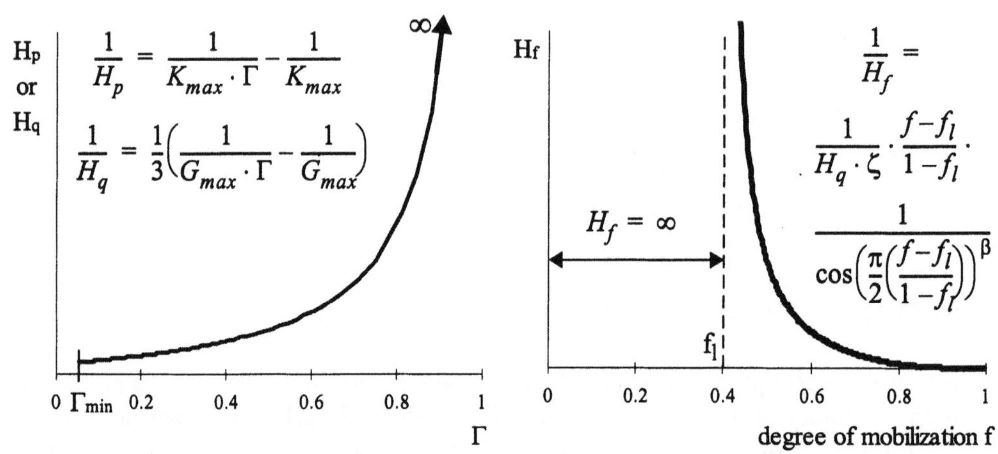

Fig.3 Principle sketches and equations for the plastic moduli, f_l is a lower limit for mobilization plasticity, ζ and β are positive numbers.

For the mobilization plasticity a Mohr Coulombian mobilization was chosen, but it could as well be replaced by any other. The non-associated flow rule used utilizes a dilatancy angle ψ to give the relation between volumetric and deviatoric strain for the mobilization plasticity mode. The value of ψ is given as a function of the degree of mobilization and effective mean stress p'.

4 PERFORMANCE OF THE MODEL

In the following some examples will be given on the performance of the model. The objectivity is to put focus on the ability of the model rather than on quantitative measures of single examples.

Example 1: Effect of stress history

Richardson (1988) carried out a set of triaxial tests on reconstituted London clay, concentrating on the accurate measurements of small strains (Simpson 1992). In figure 4 one of his test series is presented as replotted by Simpson. Richardson consolidated al specimens first to point A, were after he took them by various routes, such as AOX, AOBOX, AOCOX and AODOX to OX. In figure 4 b) the tangential stiffness of the final path OX is given. As can be seen from the figure, highest stiffness was found for path AODOX which included a full stress reversal at the beginning of path OX. For stress paths AOX and AOCOX with 90° change in stress direction a bit softer response was found, while clearly softest response was obtained for path AOBOX which continues in the same stress direction also after passing point O.

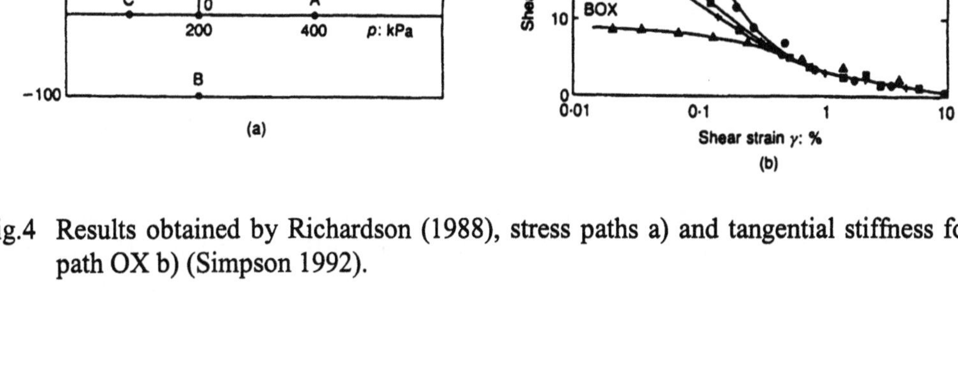

Fig.4 Results obtained by Richardson (1988), stress paths a) and tangential stiffness for path OX b) (Simpson 1992).

In figure 5 a similar kind of test is numerically simulated by the proposed model. The stiffness-strain curves for path OX are given in figure 5 b) for the different stress paths AOX, AOBOX, AOCOX and AODOC. As in Richardsons tests the highest stiffness is found for path AODOX and the lowest stiffness for path AOBOX. However, Richardson got a slightly higher stiffness for path AOX than for AOCOX, while the model gives the opposite. The reason for the behavior of the model is that in case AOX all of the memorypoints lay between O and A at the start of path OX, while for AOCOX most of the memory points lay between O and C but some are between O and A. Therefore for path AOCOX the memory-points are more concentrated at the starting point O and will thus give a bit higher stiffness for path OX. With shorter memory lengths or with higher stress, as in Richardsons tests, paths AOX and AOCOX would give similar stiffness reduction.

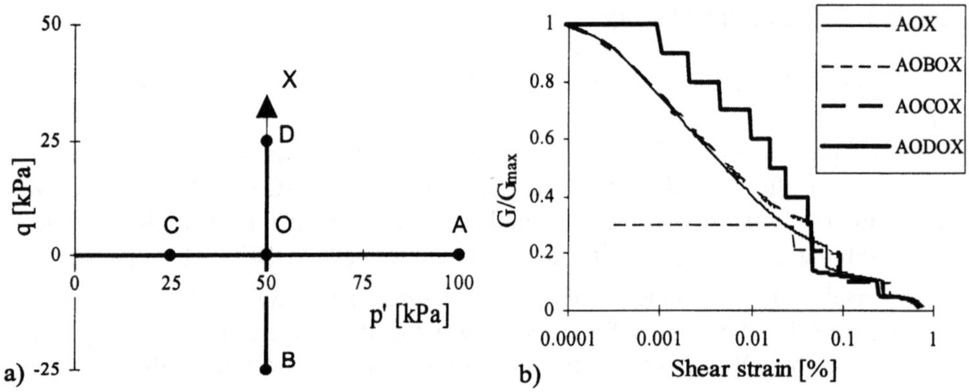

Fig.5 Stress paths modeled a), and corresponding shear stiffness b).

Example 2: Effect of stress path

Smith, Jardine and Hight (1992) presented yielding characteristics of Bothkennar clay using stress-path triaxial tests. In their main test series they consolidated samples along the path ABCD, were point D represents in situ stresses, figure 6 a). From point D the samples which will here be discussed, were subjected to continuous drained probing tests according to the figure, were the number stands for the angle of the stress-path.The results are given in figure 6 b) for the bulk modulus and in figure 7 for the shear modulus.

Corresponding test is simulated numerically by the model as shown in figure 8. The stresses are taken from point A to B and then unloaded to C. From point C seven different stress-paths are followed to points D, E, F, G, H, I and J. The results given by the model are presented in figure 9.As expected paths ABCG and ABCH give the softest response, while stress-paths with stress reversal like ABCD, ABCE and ABCJ give the stiffest response. However all of the results are not so self-explanatory and will be discussed in more detail.

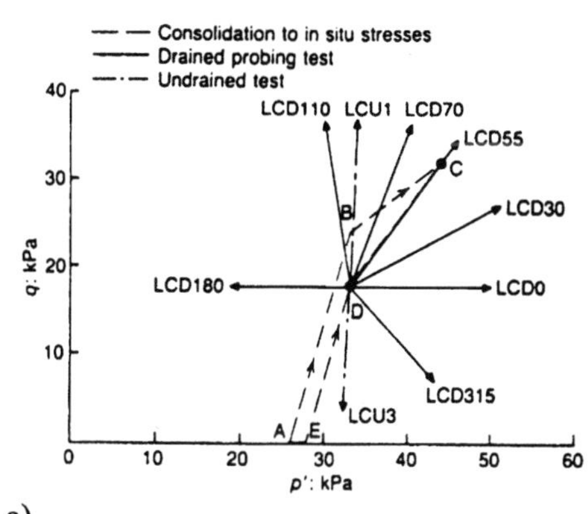

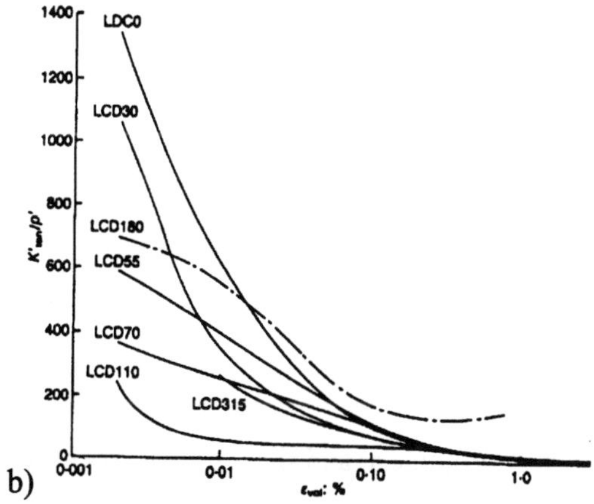

Fig.6 a) Stress-paths used by Smith, Jardine and Hight (1992) and b) observed tangential bulk moduli.

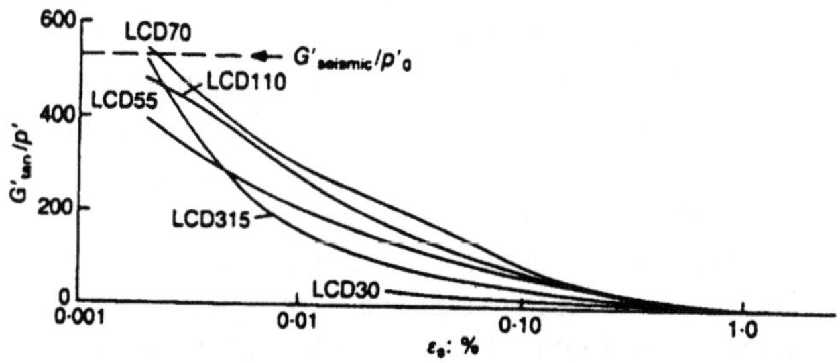

Fig.7 Tangential shearmodulus for the stress path given in figure 6 (Smith, Jardine and Hight 1992).

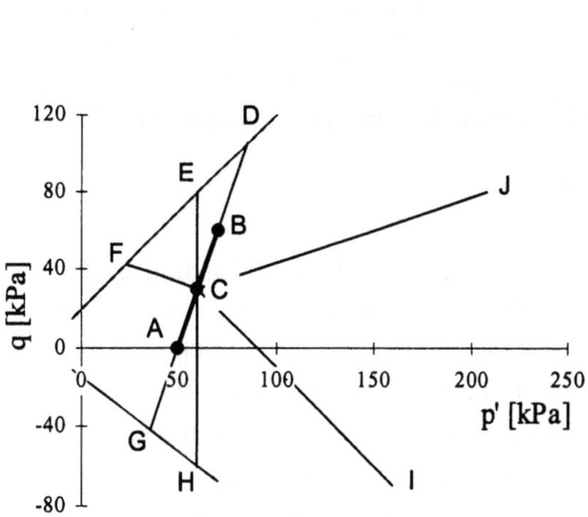

Fig.8 Stress-paths used in the modeling example.

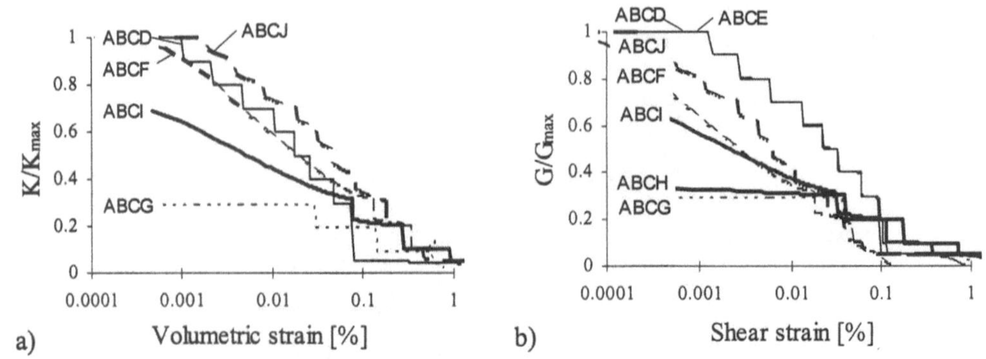

Fig.9 Volume a) and shear b) stiffness curves for the different stress-paths given in figure 8.

For shear, the stiffest response is obtained along paths ABCD and ABCE which show almost similar stiffness reduction curves. Path ABCJ show a clearly softer response. However for volumetric stiffness path ABCJ gives the stiffest response, while path ABCD shows a slightly softer behavior. This is due the stress direction dependency of the stiffness number given by the memory model. For example path CJ could be decomposed into a change in effective mean stress and a change in deviatoric stress. Assuming that the path lies under the limit for mobilization plasticity only the change of effective mean stress causes volume change and only the change of deviatoric stress causes shear strain. Still both the change of effective mean stress and the change of deviatoric stress contributes to the reduction of the stiffness number. Therefore the stiffness for shear reduces more than the distance travelled in deviatoric stress and result in a softer response in comparison to paths CD and CE. Similarly for path CD the main contribution to the reduction of the stiffness number comes from the change of deviatoric stress and the volumetric stiffness reduces more rapidly than for path CJ. In the test results by Smith, Jardine and Hight a similar behavior can be observed. They obtained stiffer response in shear for paths LCD70 and LCD110 than for path LCD55 even though path LCD55 has a full stress reversal to previous direction, figure 7. The same thing can be found in figure 6 b) for the bulk modulus, were the stiffest response is obtained for the path LCD0 while the path LCD55 shows softer behavior.

Example 3: Cyclic loading

In figure 10 results from a simulation of cyclic loading is illustrated. Accumulated shear strain predicted by the model is given as a function of cyclic shear stress and cyclic mobilization level. The abbreviation f_{cyc} is used for the degree of mobilization at the top of the loops. In the example f_l = 0,6 is used, which causes that no strain accumulation occurs for loops done with f_{cyc} lower than this. As can bee seen the model gives increasing strain accumulation with both increasing cyclic shear stress and degree of mobilization. This is a common feature obtained in cyclic loading on clays, see for example Andersen (Andersen 1988). To separate the effects of cyclic shear stress and cyclic mobilization level from pore pressure build up drained conditions were considered. For a contractive soil in undrained loading the strain accumulation would also increase with number of cycles as the mobilization would rise with increasing pore pressure.

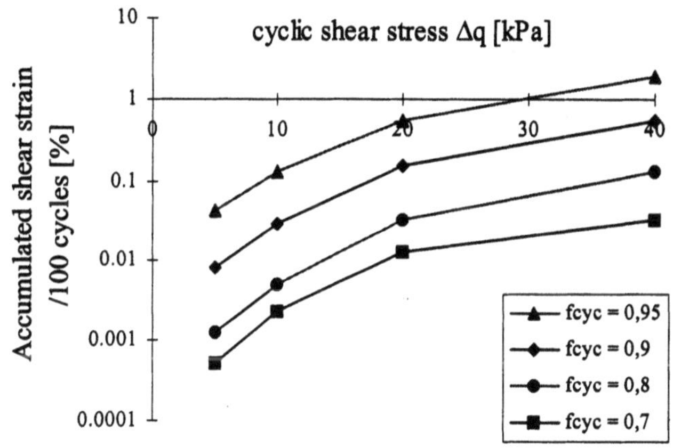

Fig.10 Example of strain accumulation given by the model as a function of cyclic shear stress and cyclic mobilization level.

5 CONCLUSIONS

In the paper preliminary results from a model under development for the overconsolidated region of clays are presented. The model utilizes memorypoints to describe the stress history and direction dependency of stiffness in the overconsolidated region. The preliminary results of comparing the behavior predicted by the model to laboratory results have been promising. The model seems to comprise several essential characteristics of low strain level soil behavior within a relatively simple framework and with consistent parameters.Although it was developed for the overconsolidated region of clays, a similar concept can also be used for sands.

6 ACKNOWLEDGEMENTS

The founding of the project by the Research Council of Norway is greatly acknowledged.

7 REFERENCES

ANDERSEN K. H. (1988). Properties of soft clay under static and cyclic loading. NGI Publication No. 176, Oslo.
JANBU N. (1963). Soil compressibility as determined by oedometer and triaxial test.

Proc.3rd ECSMFE, Wiesbaden, 1: 19-25.

KIRKEBO S., NORDAL S. & SVANØ G. (1994). A numerical study of delayed failure of excavations. Proc. ECONMIG 94, Manchester, 355 - 361.

KLISINSKI M. (1988). Plasticity theory based on fuzzy sets. Journal of Engineering Mechanics, Vol. 114, No.4:563-582.

NORDAL S., JOSTAD H. P., KAVLI A. & GRANDE L. (1989). A coulombian soil model applied to an offshore platform. Proc. 12th Int. Conf. Soil Mechanics and Foundation Engineering, Rio de Janeiro, 4: 471- 474

RICHARDSON D. (1988). Investigations of treshold effects in soil deformations. PhD thesis, City University, London.

SIMPSON B. (1992). Retaining structures: displacement and design. Géotechnique, 42: 541-576.

A TWO DIMENSIONAL FINITE ELEMENT CONSOLIDATION MODEL FOR UNSATURATED SOIL

K. Nesnas and I.C. Pyrah
Napier University, Edinburgh, UK

ABSTRACT:

A fully coupled model involving the transient flow of air and water in a deformable unsaturated soil has been developed and implemented in a finite element program. The elastoplastic constitutive model representing the behaviour of the unsaturated soil uses net stress and suction as stress parameters enabling wetting induced swell and collapse to be modelled. In this paper the numerical model is used to study the collapse and swell mechanisms occurring in a column of soil subjected to infiltration.

1. INTRODUCTION

The most commonly used approaches for studying the behaviour of geotechnical structures are the undrained and drained analyses. These two approaches, which are of the uncoupled type, are usually more convenient than the fully-coupled flow-deformation approach which is demanding both in terms of computer time and computer power. Numerical modelling has already been carried out for these uncoupled problems (Nesnas, Pyrah and Wheeler 1998, Nesnas and Pyrah 1998) involving the drained and undrained analysis of unsaturated soils. This paper is an extension of this work to simulate the coupled flow-deformation of unsaturated soils; this is analogous to a fully-coupled consolidation analysis for a fully saturated soil.

The mathematical modelling of the coupled flow-deformation problem of unsaturated soils is based on a simplification that the solid phase (soil skeleton) and the fluid phases (air and water) are continua, and the fluid phases occupy the entire flow domain. The general procedure followed in deriving the mathematical model is based on establishing the conservation and equilibrium equations in terms of the appropriate variables and representing the behaviour of the materials (i.e. solid and fluids) with appropriate constitutive laws. The resulting mathematical model describes the time and space evolution of the variables which are displacement, pore water pressure and pore air pressure.

Unsaturated soils, with a potential for shrinkage or collapse when subjected to a source of water, are one of the principal causes of damage to civil engineering structures. Matyas and Radhakrishna (1968) and Fredlund and Morgenstern (1977) showed that the use of a single stress variable (effective stress) is not sufficient to predict this collapse caused by a significant reduction of volume due to plastic volumetric strain; a pair of variables, net stress ($\bar{\sigma}$, total stress minus air pressure) and suction (s, defined as the amount of air pressure above water pressure), must be introduced in defining a relationship for the volume changes of unsaturated soils. Alonso, Gens and Josa (1990) developed an elastoplastic model (Barcelona model) using these two stress variables. They modified the Cam Clay model by redefining the stresses in terms of net stress and by adding the variable suction as a new stress parameter. In addition, a Loading-Collapse (LC) yield curve is introduced in the $\bar{p}-s$ plane which produces an Elliptical Loading-Collapse (ELC) yield surface in the $\bar{p}-q-s$ space. In this way they were able to model features of behaviour characteristic of unsaturated soils such as swell and collapse and/or shear failure occurring under wetting.

2. CONSTITUTIVE RELATIONS

2.1 STRESS STRAIN RELATIONS

Using the classical theory of plasticity and the concept of initial strain, the relationship linking increment of stress to increment of strain for the Barcelona model is:

$$\{\delta\bar{\sigma}\} = \{D_{ep}\}(\{\delta\varepsilon\} - \{\delta\varepsilon_o\}) \qquad (1)$$

The bar symbol is used to denote net stress and $\delta\bar{\sigma}$ is the vector of incremental net stress; $\lfloor D_{ep} \rfloor$ is the elastoplastic stiffness matrix dependent on the net stress level (or effective stress level if the soil is saturated) and suction, $\{\delta\varepsilon\}$ represents the total strain vector for the soil skeleton and $\{\delta\varepsilon_o\}$ is an initial strain vector caused by suction. $\{D_{ep}\}$ and $\{\delta\varepsilon_o\}$ are defined as follows:

$$[D_{ep}] = [D_e] - \frac{[D_e]\left\{\frac{\partial Q}{\partial \bar{\sigma}}\right\}\left\{\frac{\partial F}{\partial \bar{\sigma}}\right\}^T [D_e]}{-\frac{\partial F}{\partial \varepsilon_v^p}\frac{\partial Q}{\partial \bar{p}} + \left\{\frac{\partial F}{\partial \bar{\sigma}}\right\}^T [D_e]\left\{\frac{\partial Q}{\partial \bar{\sigma}}\right\}} \tag{2}$$

and $\{\delta\varepsilon_o\}$ as:

$$d\bar{\sigma} = [D_{ep}]\left\{\{d\varepsilon\} - \left(\{m\}\frac{\partial \varepsilon_v}{\partial s} + \frac{[D_e]\left\{\frac{\partial Q}{\partial \bar{\sigma}}\right\}\frac{\partial F}{\partial s}}{-\frac{\partial F}{\partial \varepsilon_v^p}\frac{\partial Q}{\partial \bar{p}}}\right) ds\right\} \tag{3}$$

where $[D_e]$ is the elastic stiffness matrix, Q is the plastic potential functional, F is the yield surface functional, $\{\sigma\}$ vector of net stress, ε_v total volumetric strain, ε_v^p plastic volumetric strain and $\bar{p}$ is the mean net stress.

2.3 AIR AND WATER VELOCITIES

The flow of air and water is assumed to be laminar and thus both follow Darcy's law expressed as:

$$v_\alpha = \frac{k_\alpha k_{\alpha r}}{\mu_\alpha}\nabla(u_\alpha + \rho_\alpha g y) \tag{4}$$

where the subscript α indicates the phase (air or water); k permeability; μ dynamic viscosity; u is the pressure; ρ density; g gravity acceleration and y elevation above an arbitrary datum.

The permeability for air is based on the relationship proposed by Yoshimi and Osterberg (1963):

$$k_a = d(e(1 - S_r))^c \gamma_a / \mu_a \tag{5}$$

where d and c are constant parameters, e is the void ratio and S_r is degree of saturation.

and the permeability of water is given by the relationship proposed by Gardner(1958):

$$k_w = \left(\frac{k_{wo}}{1 + as^b}\right) \tag{6}$$

where k_{wo} is the permeability at saturation, a and b are constant parameters and s is the suction.

The degree of saturation S_r is a function of suction and is given by the relation proposed by Gens , Vaunat & Ledesma, (1995) for clay:

$$S_r = 0.995 - p(1 - EXP(-fs)) \tag{7}$$

where the parameters p and f are constant.

3 GOVERNING EQUATIONS

3.1 MASS CONSERVATION AND EQUILIBRIUM EQUATIONS

The mass conservation for phase α filling a volume V_α is obtained as:

$$\int_{V_\alpha}\left(\frac{\partial\rho_\alpha}{\partial t} + \nabla.\left(\rho_\alpha\{v_\alpha\}\right)\right)dV_\alpha = 0 \tag{8}$$

Where $\alpha = a$ or w, ρ_α is the density of phase α, v_α is the velocity of phase α.

Considering an arbitrary volume for each phase equation (8) is rewritten for the water phase and the air phase as:

$$\frac{\partial}{\partial t}\left[\rho_a n(1 - S_r + HS_r)\right] + \nabla.\left[\rho_a\left(\{v_a\} + H\{v_w\}\right)\right] = 0 \tag{9}$$

$$\frac{\partial}{\partial t}(\rho_w n S_r) + \nabla.\left(\rho_w\{v_w\}\right) = 0 \tag{10}$$

where n is the soil porosity and H is Henry's constant.

The equilibrium of an infinitesimal volume of soil subjected to the stress tensor $[\sigma]$ and the body force vector $\{b\}$ is expressed as:

$$\nabla.[\sigma] + \rho\{b\} = 0 \tag{11}$$

where ρ is the mass bulk density of the soil skeleton.

3.2 SPATIAL AND TIME DESCRETISATION

The spatial discretisation of equations (9), (10) and (11) is obtained by approximating the variable of the problem, namely displacements $\{d\}$, water pressure $\{u_w\}$ and air pressure $\{u_a\}$ in terms of their respective nodal values $\{a\}$, $\{u_w^n\}$ and $\{u_a^n\}$ through the respective shape functions $[N]$, $[N_w]$ and $[N_a]$

$$\{d\} = [N]\{a\} \tag{12}$$

$$\{u_w\} = [N_w]\{u_w^n\} \tag{13}$$

$$\{u_a\} = [N_a]\{u_a^n\} \tag{14}$$

$$\begin{bmatrix} 0 & 0 & 0 \\ 0 & H_{ww} & 0 \\ 0 & H_{aw} & H_{aa} \end{bmatrix} \begin{Bmatrix} a \\ u_w^n \\ u_a^n \end{Bmatrix} + \begin{bmatrix} K_{dd} & K_{dw} & K_{da} \\ K_{wd} & K_{ww} & K_{wa} \\ K_{ad} & K_{aw} & K_{aa} \end{bmatrix} \frac{\partial}{\partial t} \begin{Bmatrix} a \\ u_w^n \\ u_a^n \end{Bmatrix} = \begin{Bmatrix} F_d \\ F_w \\ F_a \end{Bmatrix} \tag{15}$$

where,

$$K_{dd} = \int_V [B]^T [D_{ep}][B] dV$$

$$K_{dw} = \int_V [B]^T [D_{ep}]\{C\} N_w dV$$

$$K_{da} = \int_V [B]^T \{m\}[N_a] dV - \int_V [B]^T [D_{ep}]\{C\}[N_a] dV$$

$$F_d = \int_A [N]^T dA \frac{\partial\{\tau\}}{\partial t} + \int_V [N]^T dV \frac{\partial\{b\}}{\partial t}$$

$$H_{ww} = \int_V \nabla[N_w]^T [k_w] \rho_w \nabla[N_w] dV$$

$$K_{wd} = \int_V [N_w]^T \rho_w S_r a_1 [D_{ep}][B] dV$$

$$K_{ww} = \int_V [N_w]^T \{\rho_w S_r a_1 [D_{ep}]\{C\} - \rho_w (na_2 + S_r a_3) + nS_r C_w\}[N_w] dV$$

$$K_{wa} = \int_V [N_w]^T [-\rho_w S_r a_1 [D_{ep}]\{C\} + \rho_w (na_2 + S_r a_3)][N_a] dV$$

$$F_w = -\int_V \rho_w^2 g \nabla[N_w]^T [k_w] \nabla y dV + \int_{A'_1} [N_w]^T q_w dA$$

$$H_{aw} = \int_V \nabla[N_a]^T H[k_w] \rho_a \nabla[N_w] dV$$

$$H_{aa} = \int_V \nabla[N_a]^T [k_a] \rho_a \nabla[N_a] dV$$

$$K_{ad} = \int_V [N_a]^T \rho_a (1 - (1 - H)S_r) a_1 [D_{ep}][B] dV$$

$$K_{aa} = \int_V [N_a]^T \{-\rho_a (1 - (1 - H)S_r) a_1 [D_{ep}]\{C\} + \rho_a (-n(1 - H)a_2 + (1 - (1 - H)S_r) a_3) b_2 + (n(1 - S_r) + HnS_r) C_a\}[N_a] dV$$

$$K_{aw} = \int_V [N_a]^T [\rho_a B_1 [D_{ep}]\{C\} - \rho_a B_2][N_w] dV$$

$$F_a = -\int_V \rho_a^2 g \nabla[N_a]^T [k_a] \nabla y dV - \int_V \nabla[N_a]^T H[k_w] \rho_a \rho_w g \nabla y dV + \int_{A'_2} [N_a]^T q_a dA$$

and the coefficient $\{C\}$ is defined as

$$\{C\} = \{m\} \frac{\partial \varepsilon_v}{\partial s} + [D_e] \left\{ \frac{\partial Q}{\partial \bar{\sigma}} \right\} \frac{\partial F}{\partial s} \Big/ -\frac{\partial F}{\partial \varepsilon_v^p} \frac{\partial Q}{\partial \bar{p}}$$

The coefficients a_1, a_2, a_3 are defined as follows:

$$a_1 = \frac{\partial n}{\partial v}\left[\left(\frac{\partial v}{\partial \bar{p}} + \frac{\partial v}{\partial \bar{p}_o}\frac{\partial \bar{p}_o}{\partial \bar{p}}\right)\left\{\frac{\partial \bar{p}}{\partial \bar{\sigma}}\right\} + \left(\frac{\partial v}{\partial q} + \frac{\partial v}{\partial \bar{p}_o}\frac{\partial \bar{p}_o}{\partial q}\right)\left\{\frac{\partial q}{\partial \bar{\sigma}}\right\}\right];\ a_2 = \frac{\partial S_r}{\partial s};\ a_3 = \frac{\partial n}{\partial v}\left[\frac{\partial n}{\partial s} + \frac{\partial n}{\partial \bar{p}_o}\frac{\partial \bar{p}_o}{\partial s}\right]$$

Equation (15) is rewritten in a condensed form as:

$$[K]\frac{\partial}{\partial t}\{X\} + [H]\{X\} = \{F\} \tag{16}$$

The time discretisation of equation(16) results in:

$$\left(\left[H\left(X_{t+\theta}^{k-1}\right)\right]\theta\Delta t + \left[K\left(X_{t+\theta}^{k-1}\right)\right]\right)\Delta X^k = -\left[H\left(X_{t+\theta}^{k-1}\right)\right]\Delta t X_t + (1-\theta)\Delta t F_t + \theta\Delta t F_{t+\Delta t}^k \tag{17}$$

where k is the current iteration number, Δt is the time step, θ parameter defined as $0 \le \theta \le 1$ and $X_{t+\theta}^{k-1}$ is defined as:

$$X_{t+\theta}^{k-1} = (1-\theta)X_t + \theta X_{t+\Delta t}^{k-1} \tag{18}$$

4. APPLICATION

This section describes an application of the above model to a cylindrical vertical column of soil subjected to infiltration at its surface. Axial symmetry is assumed and the column is modelled using 10 quadrilateral elements of equal size. Each element is cubic in terms of displacements and linear in terms of water pressure and air pressure. The vertex nodes of the element have 4 degrees of freedom (2 for displacements, 1 for water pressure and 1 for air pressure) whereas the midside nodes have 2 degrees of freedom (2 displacements). Each element is assigned the soil parameters for Boom clay; these are obtained from the paper by Gens et al (1995), see Table 1.

The vertical sides of the column are restrained horizontally and assumed to be impermeable to water; the air pressure is specified to be zero. The bottom of the column is restrained vertically and horizontally and both air pressure and water pressure are maintained at their initial values (when simulating the drained analysis, only the air pressure is maintained at its initial value); along the top the air pressure is zero. At t=0 the displacements are zero, the water pressure is equal to -100kPa and the air pressure is equal to zero throughout the soil column. The in situ stress are assumed to be isotropic $\bar{\sigma}_x = \bar{\sigma}_y = \bar{\sigma}_z = 250\text{kPa}$, with the stress history of the soil defined by a preconsolidation pressure at saturation $\bar{p}_o(0) = 150\text{kPa}$. At t>0 the water pressure at the top of the soil column is decreased incrementally by 1kPa to reach zero under partially drained conditions.

Prior to examining the evolution of suction and displacements with time, and as part of the verification of the coupled flow-deformation program, the program was first run to simulate a drained analysis using a very large value of permeability (of order 10^5). The vertical displacement of the top of the soil column for different values of suction are plotted in figure 1 and compared with those obtained from the uncoupled drained analysis, i.e. a deformation analysis where the suction is specified externally by the user.

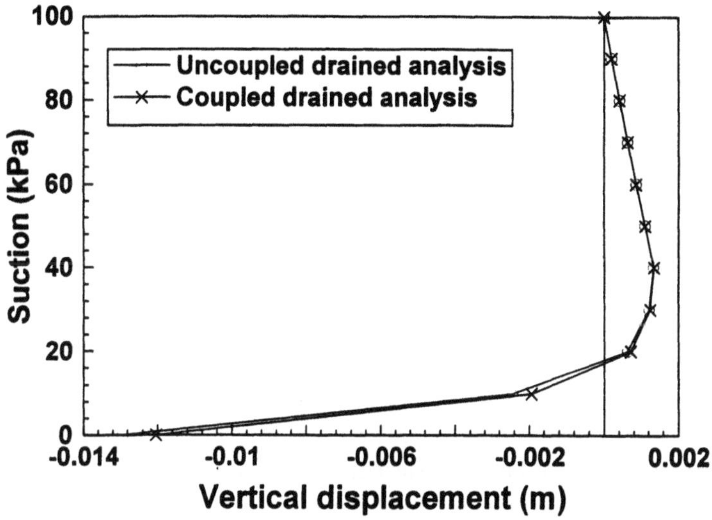

Fig. 1 Comparison of vertical displacements at the top side of the soil column from uncoupled drained analysis and coupled drained analysis

It is seen that both analyses predict the same results. The top of the soil column initially starts to heave (positive displacement) as the suction decreases from its initial value of 100kPa to about 40kPa. As the suction decreases further the heave decreases progressively and a net settlement (due to the occurrence of plastic volumetric strain) occurs when the suction reduces below 20kPa; as suction approaches zero both analyses predict large settlements.

In the above drained analyses, the suction is uniform with depth for each stage of the analysis. The evolution of suction and deformation with time can be modelled by using the coupled "consolidation" program and a more realistic value of permeability as given in table 1.

Figure 2 shows the variation of suction with elevation for five values of elapsed time. At the base (zero elevation), the suction remains at its initial value of 100kPa because at the bottom of the soil column, the air pressure and water pressure are maintained at zero and -100kPa respectively. Above the base the suction decreases progressively.

Compressibility under mean stress changes	$\lambda(0) = 0.026$, $\kappa = 0.00265$, $\beta = 0.0544 kPa^{-1}$ $r = 0.564$, $p^c = 60. kPa$
Compressibility under suction changes	$\kappa_s = 0.00322$
Shear modulus	$\upsilon = 0.333$
Shear strength	$M = 1$, $k = 0.00732$
Permeability of air	$c=1.31\ 10^{-19} m/s$, $d=2.47$
Permeability of water	$k_{wo} = 1.3\ 10^{-14}$ m/s, $a= 1.35\ 10^{-10}$ $b=1.692$
Degree of saturation	$p=0.918$, $f=8.61\ 10^{-5}$

Table 1 Material parameters for natural Boom clay

The surface and mid-height vertical displacement with time are shown in figure 3. It is observed that heave occurs at both the surface and at the mid-height of the soil column for times less than 144mins. Beyond this time the middle of the column experiences further heave, but this is not the case at the top of the column where the vertical displacement decreases i.e. settlement occurs.

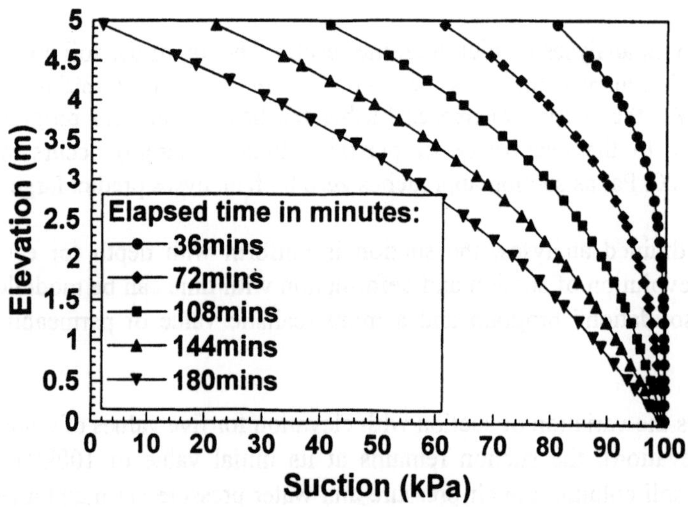

Fig. 2 Profile of suction

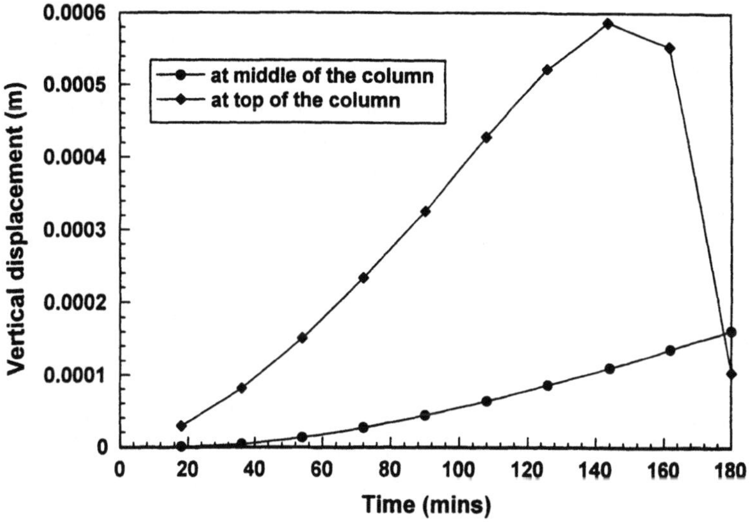

Fig. 3 Variation of the vertical displacement with time

To obtain a clearer picture of what is happening locally in the column, the variation of the vertical strain with depth was examined (figure 4). Up to an elapsed time of 108 minutes the whole column experiences swell (negative strain) resulting in the surface heave indicated in figure 3.

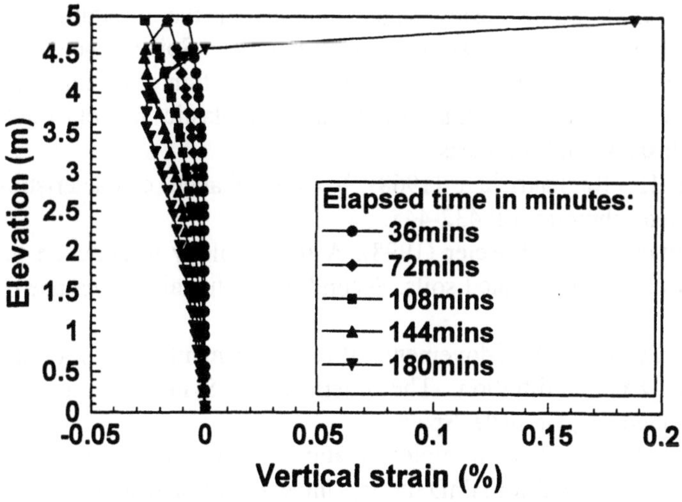

Fig. 4 Profile of vertical strain

Between 108 and 144 minutes the small collapse strains in the top 0.5m (i.e. elevation 4.5 to 5.0m) are insufficient to cancel out the effect of the swelling occurring in the bottom 4.5m of the soil column and the resulting surface displacement is continuing heave. Between 144 and 180 minutes, however significant collapse strains occur in the top metre of the column whilst below an elevation of 3.5m the soil continues to swell. These swell strains are relatively small compared to the collapse strains in the upper part of the column resulting in surface settlement rather than heave (figure 3). At the mid-height of the column continuous swell occurs as no collapse strains occur below this level.

5. CONCLUSIONS

Equations for fully-coupled flow-deformation behaviour of an elastoplastic unsaturated soil have been developed and implemented in a finite element program. The application to a column of soil subjected to infiltration at the surface shows that the numerical model is able to predict the occurrence of swell and collapse which may occur at different locations within the soil column. Examination of the variation of the vertical strains in the column helps to explain the complex pattern of surface settlement caused by surface infiltration.

6. REFERENCES

Alonso E.E., A.Gens & A.Josa (1990). A constitutive model for partially saturated soils, Geotechnique, 40: 405-430.

Fredlund, D.G. and N.R.Morgenstern (1977). Stress state variables for saturated soils. Proc. ASCE, 103, No.GT5: 313-312.

Gardner W.R. (1958). Some steady state solutions of the unsaturated moisture flow equation with application to evaporation from a water table. Soil Science, 85: 228-232.

Gens A., J.Vaunat & A.Ledesma (1995). Analysis of hydration of an engineered barrier in a radioactive waste repository scheme using an elastoplastic model. First International Conference on Unsaturated Soil, Paris.

Matyas E.L. & H.S.Radhakrishna (1968) Volume change characteristics of partially saturated soils, Geotechnique, 18: 432-448.

Nesnas K., I.C.Pyrah & S.J.Wheeler (1998). A methodology to simulate the drained and undrained behaviour of unsaturated soils. Second International Conference on Unsaturated Soils, Beijing.

Nesnas K. & I.C.Pyrah (1998). Movements of road pavements due to a rise in the ground water table and water infiltration. The International Symposium on Subdrainage in Roadways and Subgrades, Granada, Spain.

Nesnas K. (1995). A finite element implementation of a critical state model for unsaturated soil to simulate drained condition, PhD thesis, University of Sheffield.

Yoshimi, Y. and J.O. Osterberg (1963). Compression of partially saturated cohesive soils. ASCE J. Soil Mech. Found. Div., 89: 1-24.

NUMERICAL SIMULATION OF
THE THERMOVISCOPLASTIC BEHAVIOUR
OF BITUMINOUS CONCRETE

P. Royis and J.F. Seignol
École Nationale des Travaux Publics de l'Etat, Vaulx-en-Velin, France

ABSTRACT: The general framework of the paper deals with the finite element modelling of thermomechanical problems involving viscous materials. The study focuses on the statement of constitutive equations describing the thermoviscoplastic behaviour of bituminous concrete, as well as on their implementation in a finite element program. They are discretized thanks to two different schemes. Eventually, two sets of numerical results are presented. The first one is used to compare those schemes, whereas the second one consists of numerical simulations of real-size experiments performed on a road structure subjected to thermal and mechanical loadings.

1 Introduction

The general framework of this paper deals with the finite element modelling of thermomechanical problems involving viscous materials such as bituminous concrete. For this material, a realistic model must take into account the thermoviscoplastic component of its behaviour, if one wants to lay out the corresponding structures correctly (Di Benedetto & Yan, 1994). The use of viscoplastic constitutive equations for describing the non-linear and rate-sensitive behaviour of other geomaterials, such as clays, is now well established and many rheological models have been issued from this formalism (Perzina, 1963). The study described in this paper focuses on the statement of constitutive equations describing the thermoviscoplastic behaviour of bituminous concrete, as well as on their implementation in the finite element program ELFIMTH developed in previous works for the numerical resolution of boundary value problems in thermomechanics (Royis & Seignol, 1998). The paper consists of three main sections. In each of them the sign conventions of the continuum mechanics will be used for stresses and strains.

The first main section is devoted to the statement of the thermomechanical problem. The second section focuses on the time discretization of the viscoplastic component of the constitutive equations considered. For that purpose two different schemes are used: the classical θ-scheme and an original method based on a direct integration of the constitutive equations over finite time steps. Eventually, the third section presents two sets of numerical examples. The first one is used to compare the different time-discrete schemes. The second one is the simulation of real-size experiments performed on a road structure. By comparing the numerical results with experimental ones, it allows us to test the finite element program on a more realistic problem.

2 The thermomechanical problem

2.1 General considerations

This paper deals with determining both histories $T(t,\cdot)$ and $\mathbf{u}(t,\cdot)$ of the temperature and displacement fields of a continuum $\Omega \subset \mathbb{R}^3$ over the time interval $[0, \mathrm{T}]$. The initial conditions are assumed to be known and both thermal and mechanical boundary conditions are given. In the first case, they can be of three sorts (Dirichlet, Neuman and Fourier conditions), and in the second case they are prescribed displacements or else prescribed stresses on every point of the boundary Γ. Eventually, we restrain our study to the following asumptions: we assume small strains and small displacements, and we consider only quasistatic problems for which the influence of thermodynamical coupling effects on the temperature field are very weak.

Let $N \in \mathbb{N}^*$ and let $t_0, t_1, \ldots, t_N$ be an increasing sequence of time values such that $t_0 = 0$ and $t_N = \mathrm{T}$. In the following we are interested in the displacement fields $\mathbf{u}(t_n, .)$ and in the temperature fields $T(t_n, .)$ relating to the time values t_n, $n \in \{1, \ldots, N\}$. We put, $\forall n \in \{0, \ldots, N\}$ and $\forall x \in \Omega$, $\mathbf{u}_n(x) = \mathbf{u}(t_n, x)$, as well as analogous notations for t, T, and so on. Eventually, we denote as Δt_n the time-increment $t_n - t_{n-1}$, $n \in \{1, \ldots, N\}$. Thanks to the last asumption, we can solve the thermal problem for each time t_n, then solve the mechanical problem for the same time, using the field $T_n(\cdot)$ lastly computed.

For the sake of conciseness, we shall not present the resolution of the thermal problem. For further details, one may refer to (Royis, Seignol, 1998).

Let $\mathbf{b}(t, .)$ be the vectorial field at time t of the body forces acting per unit volume on Ω. We shall denote as $\Gamma_1(t)$ the part of the boundary Γ of Ω on which we have, at time t, the values of the stress vector prescribed and equal to $\mathbf{g}(t, .)$ and as $\Gamma_2(t) = \Gamma - \Gamma_1(t)$ the part of the boundary with prescribed displacements.

Let now $\varepsilon(t, .)$ be the linearized tensorial field of the small strains in Ω at time t, and let $\sigma(t, .)$ be the tensorial field of the Cauchy stresses in Ω at the same time. This paper focuses on the finite element modelling of geomaterials, such as bituminous concrete, the behaviour of which is described by a thermoviscoplastic constitutive law. For that we assume an additive decomposition of the total strain rate tensor $\dot{\varepsilon}$ into linear elastic, viscoplastic and thermal components, as follows:

$$\dot{\varepsilon} = \mathbf{C} : \dot{\sigma} + \mathbf{F}(\sigma, T) + \alpha(T)\dot{T}\mathbf{I}_2 \tag{1}$$

where $\mathbf{C}$ is the classical fourth-order tensor of elastic compliances and T is the absolute temperature. The expression of the tensorial function $\mathbf{F}$, which generalizes the one-dimensional relations obtained by (Di Benedetto & Yan, 1994) from experimental results on axisymmetric triaxial tests, will be described in the following subsection 2.2. Eventually, $\dot{T}$ is the rate of temperature, $\alpha(T)$ the coefficient of thermal expansion, which is a function of the absolute temperature T, and $\mathbf{I}_2$ denotes the second-order unit tensor.

Then the problem which consits in determining the history $\mathbf{u}(t,.)$ of the displacements of Ω over the time interval $[0,\mathrm{T}]$ is written on the basis of the balance principle of linear and angular momentum, in the absence of body and surface couples. The developments made in subsection 3 devoted to the time discretization of the constitutive equation (1) lead, for given $n \in \{1,\ldots,N\}$ and $x \in \Omega$, to the following formal relation

$$\varepsilon_n(x) = \mathbf{H}_{nx}\left(\sigma_n(x), \mathcal{H}_{nx}\right) \tag{2}$$

where $\mathcal{H}_{nx}$ denotes the set of memory parameters different from $\sigma_n(x)$ at point x and at time t_n. The tensorial function $\mathbf{H}_{nx}$ is assumed to be one-to-one, according to the principle of determinism. So we can write, formally

$$\sigma_n(x) = \mathbf{H}_{nx}^{-1}\left(\varepsilon_n(x), \mathcal{H}_{nx}\right) \tag{3}$$

This assumption allows us to discretize the mechanical problem. Then we transform it into a classical one-field weak-formulation. Since the above relation brings into play strong rheological non-linearities, we shall use the robust Newton method in order to linearize the variational problem.

Let then $V = \left(H^1(\Omega)\right)^3$. For any given $n \in \{1,\ldots,N\}$ we denote as V_n the closed subspace of V of the functions $\mathbf{v} \in V$ such that $\mathbf{v}\,|\Gamma_2(t_n) = \mathbf{0}$. Let now $V_h \subset V$ be a finite element space and let us denote as $\mathbf{u}_n^{(r)}$ the solution of the linearized problem $(\mathbf{P}_{nh}^{(r)})$ relating to iteration (r). Then the problem $\left(\mathbf{P}_{nh}^{(r+1)}\right)$ takes the following form:

$$(\mathbf{P}_{nh}^{(r+1)})\left\{\begin{array}{l} \text{Find } \mathbf{u}_n^{(r+1)} \in V_h \text{ such that } \forall \mathbf{v} \in V_h \cap V_n \\ \displaystyle\int_\Omega \varepsilon(\mathbf{v}) : \mathbf{G}_{nx}^{-1}\left(\sigma_n^{(r)}, \mathcal{H}_{nx}\right) : \varepsilon(\mathbf{u}_n^{(r+1)})\,\mathrm{d}\Omega = \int_{\Gamma_1(t_n)} \mathbf{g}_n.\mathbf{v}\,\mathrm{d}\Gamma \\ \qquad \displaystyle + \int_\Omega \varepsilon(\mathbf{v}) : \left(\mathbf{G}_{nx}^{-1}\left(\sigma_n^{(r)}, \mathcal{H}_{nx}\right) : \mathbf{H}_{nx}\left(\sigma_n^{(r)}, \mathcal{H}_{nx}\right) - \sigma_n^{(r)}\right)\mathrm{d}\Omega \\ \qquad \displaystyle + \int_\Omega \mathbf{b}_n.\mathbf{v}\,\mathrm{d}\Omega \\ \mathbf{u}_n = \mathbf{u}_{in} \text{ on } \Gamma_2(t_n) \end{array}\right. \tag{4}$$

where $\mathbf{G}_{nx} = \mathbf{grad}_\sigma(\mathbf{H}_{nx})$ is the gradient tensor of $\mathbf{H}_{nx}$.

2.2 Description of the viscoplastic component of the strain rate tensor

The viscoplastic criterion proposed by (Di Benedetto & Yan, 1994) from experimental data obtained by Yan on cylindrical samples of bituminous concrete is defined, in the orthonormal space $\mathbb{R}^3$ of principal stresses, by two cone-shaped parts with apex on the trisector $\sigma_1 = \sigma_2 = \sigma_3$. The intersections of both of these conic parts with the deviatoric planes are equilateral triangles, the vertices of which are located on the projections (onto these planes) of the negative half-axes of $\mathbb{R}^3$. In addition we have also the following expression (Di Benedetto & Yan, 1994) of the breaking axial stresses $\sigma_1^{(c)}$, for axisymmetric triaxial tests in compression on cylindrical samples of bituminous concrete with constant lateral stress $\sigma_2 = \sigma_3$ and constant axial strain rate $\dot{\varepsilon}_1 < 0$.

$$\left\{ \begin{array}{l} \dfrac{-\sigma_1^{(c)}}{\sigma_u} = \alpha \dfrac{-\sigma_3}{\sigma_u} + \beta(T, c_p) \ln \left(\dfrac{-\dot{\varepsilon}_1 + \delta(T)}{\dot{\varepsilon}_u} \right) + \gamma(T, c_p) \\ \text{with } \alpha = \alpha_c \text{ if } \sigma_3 \leq 0 \text{ and } \alpha = \alpha_t \text{ if not} \end{array} \right. \tag{5}$$

where β, δ and γ are three given functions, σ_u and $\dot{\varepsilon}_u$ are units of stress and strain rate, respectively, whereas α_c and α_t are two given constants.

We can then generalize the one-dimensional relation (5) in order to obtain the tensorial function $\mathbf{F}$ of equation (1). Let first $\mathbf{s}$ be the deviatoric part of the stress tensor σ, Then, if $\mathbf{d_s}$ denotes the direction of $\mathbf{s}$ defined by $\mathbf{d_s} = \mathbf{s}/\|\mathbf{s}\|$ where $\|\mathbf{s}\| = \sqrt{s^{ij}s^{ij}}$ is the Euclidian norm of $\mathbf{s}$, we obtain:

$$\dot{\varepsilon}^{(vp)} = \mathbf{F}(\sigma, T) = \phi_1(\sigma, T) \left[C_1 - C_2 \mathrm{tr}(\mathbf{d_s}^3) \right] [\phi_2(\sigma, T) - \delta(T)] [-C_3 \mathbf{I}_2 + C_4 \mathbf{d_s}] \tag{6}$$

with:

$$\left\{ \begin{array}{rcl} \phi_1(\sigma, T) & = & Y\left(\sigma_0^{(c)} - \sigma_0^{(cr)}\right) Y \left(-\mathrm{tr}\sigma + \dfrac{3\sigma_0^{(cr)}}{\alpha_t - 1} \right) \\ \phi_2(\sigma, T) & = & \dot{\varepsilon}_u \exp \left(\dfrac{1}{\beta(T, c_p)} \left(\dfrac{\sigma_0^{(c)}}{\sigma_u} - \gamma(T, c_p) \right) \right) \\ \sigma_0^{(cr)} & = & \sigma_u \left(\beta(T, c_p) \ln \dfrac{\delta(T)}{\dot{\varepsilon}_u} + \gamma(T, c_p) \right) \\ \sigma_0^{(c)} & = & -\mathrm{tr}\sigma + (2 + \alpha) \max\{\sigma_1, \sigma_2, \sigma_3\} \\ \multicolumn{3}{l}{C_1 = 2\dfrac{1 + \nu_\infty}{1 + 4\nu_\infty} \quad C_2 = \sqrt{6}\dfrac{2\nu_\infty - 1}{1 + 4\nu_\infty} \quad C_3 = \dfrac{1 - 2\nu_\infty}{3} \quad C_4 = \sqrt{\dfrac{2}{3}}(1 + \nu_\infty)} \\ \multicolumn{3}{l}{\text{with } \alpha = \alpha_c \text{ if } \max\{\sigma_1, \sigma_2, \sigma_3\} \leq 0 \text{ and } \alpha = \alpha_t \text{ if not}} \end{array} \right. \tag{7}$$

where Y denotes the Heaviside function and ν_∞ is the tangent Poisson ratio at viscoplastic flow for axisymmetric triaxial tests in compression. For details, the reader can refer to (Royis & Seignol, 1998).

3 Time discretization of the constitutive equations

The present section is devoted to the time discretization of the constitutive equation (1). For that purpose two different time-discrete schemes have been implemented in the finite element program ELFIMTH developed in previous works for the numerical resolution of boundary value problems in thermomechanics (Royis & Seignol, 1998). Let us first make some general considerations.

To begin with note that the time-integration of the elastic component of the strain rate tensor is immediate. And indeed, independently of the choice of any time-discrete scheme and taking into account the initial stress field σ_0, we have

$$\varepsilon^{(e)} = \mathbf{C} : (\sigma - \sigma_0) \tag{8}$$

On the other hand the thermal component of the strain rate tensor will be approximated, with $\mathcal{O}(\Delta t_n^2)$ accuracy, by using the well-known Crank-Nicolson scheme, which gives

$$\varepsilon_n^{(th)} - \varepsilon_{n-1}^{(th)} = \frac{\alpha(T_{n-1}) + \alpha(T_n)}{2} (T_n - T_{n-1}) \, \mathbf{I}_2 \tag{9}$$

3.1 Use of the θ-scheme for the time dicretization of $\dot{\varepsilon}^{(vp)}$

For the sake of conciseness we first consider the following ordinary differential equation $\dot{y}(t) = f(y,t)$. The use of the θ-scheme for its time-dicretization over the interval $[t_{n-1}, t_n]$ gives

$$y_n - y_{n-1} = \Delta t_n \left((1-\theta) f(y_{n-1}, t_{n-1}) + \theta f(y_n, t_n) \right) \tag{10}$$

where $\theta \in [0,1]$. If $\theta = 0$ we get the fully explicit Euler scheme, if $\theta = 1$ the fully implicit one, and if $\theta = 1/2$ the Crank-Nicolson scheme.

So the use of this classical scheme for the time discretization of the viscoplastic component $\dot{\varepsilon}^{(vp)}$ coming from relations (6) and (7) gives

$$\varepsilon_n^{(vp)} - \varepsilon_{n-1}^{(vp)} = \Delta t_n \left((1-\theta) \mathbf{F}(\sigma_{n-1}, T_{n-1}) + \theta \mathbf{F}(\sigma_n, T_n) \right) \tag{11}$$

By omitting the space variable x and taking into account relations (8) and (9), this leads to the following expression of the formal relation (2)

$$\left\{ \begin{array}{rcl} \varepsilon_n & = & \mathbf{H}_n(\sigma_n, \mathcal{H}_n) \\ & = & \mathbf{C} : (\sigma_n - \sigma_{n-1}) + \Delta t_n \left((1-\theta)\mathbf{F}(\sigma_{n-1}, T_{n-1}) + \theta \mathbf{F}(\sigma_n, T_n) \right) \\ & + & \dfrac{\alpha(T_{n-1}) + \alpha(T_n)}{2} (T_n - T_{n-1}) \, \mathbf{I}_2 + \varepsilon_{n-1} \end{array} \right. \tag{12}$$

From the expressions (6) and (7), one can determine (Royis & Seignol, 1998) the fourth-order tensor $\mathbf{G}_n = \mathbf{grad}_\sigma(\mathbf{H_n})$ involved in (4) and relating to the iteration (r):

$$\left\{ \begin{array}{rcl} \mathbf{G}_n\left(\sigma_n^{(r)}, \mathcal{H}_n\right) & = & \mathbf{C} + \theta \Delta t_n \phi_1\left(\sigma_n^{(r)}, T_n\right) \left[\mathbf{G}_1\left(\sigma_n^{(r)}, \mathcal{H}_n\right) \right. \\ & & \left. + \mathbf{G}_2\left(\sigma_n^{(r)}, \mathcal{H}_n\right) + \mathbf{G}_3\left(\sigma_n^{(r)}, \mathcal{H}_n\right) \right] \end{array} \right. \tag{13}$$

where the scalar function ϕ_1 is given by the first of relations (7). As to the fourth-order tensorial functions $\mathbf{G}_1$, $\mathbf{G}_2$ and $\mathbf{G}_3$ of $\sigma_n^{(r)}$ and of the memory parameters $\mathcal{H}_n$, they are given by the following expressions (14), in which $\sigma_n^{(r)}$ has been replaced by σ for the sake of simplicity.

$$\left\{\begin{array}{l}\mathbf{G}_1(\sigma,\mathcal{H}_n)=\dfrac{C_4}{\|s\|}\left[C_1-C_2\mathrm{tr}(\mathbf{d}_s^3)\right][\phi_2(\sigma,T_n)-\delta(T_n)]\left[\mathbf{I}_4-\dfrac{1}{3}\mathbf{I}_2\otimes\mathbf{I}_2-\mathbf{d}_s\otimes\mathbf{d}_s\right]\\ \mathbf{G}_2(\sigma,\mathcal{H}_n)=\dfrac{3C_2}{\|s\|}[\phi_2(\sigma,T_n)-\delta(T_n)][C_3\mathbf{I}_2-C_4\mathbf{d}_s]\otimes\left[\mathbf{d}_s.\mathbf{d}_s-\dfrac{1}{3}\mathbf{I}_2-\mathrm{tr}(\mathbf{d}_s^3)\mathbf{d}_s\right]\\ \mathbf{G}_3(\sigma,\mathcal{H}_n)=\dfrac{\phi_2(\sigma,T_n)}{\beta(T_n,c_p)\sigma_u}\left[C_1-C_2\mathrm{tr}(\mathbf{d}_s^3)\right][C_3\mathbf{I}_2-C_4\mathbf{d}_s]\otimes\left[\mathbf{I}_2-(2+\alpha)\dfrac{\partial m}{\partial\sigma}\right]\end{array}\right. \tag{14}$$

where $m = \max\{\sigma_1, \sigma_2, \sigma_3\}$, with σ_1, σ_2 and σ_3 the principal stresses.
The expressions of C_1, C_2, C_3, C_4, α and ϕ_2 are detailed in (7).

3.2 Direct integration of $\dot{\varepsilon}^{(vp)}$

This scheme is based on the following $\mathcal{O}(\Delta t_n^2)$-accurate approximation of $\sigma(t)$

$$\forall t \in [t_{n-1}, t_n],\ \sigma(t) = \sigma_{n-1} + \frac{\sigma_n - \sigma_{n-1}}{t_n - t_{n-1}}(t - t_{n-1}) + \mathcal{O}(\Delta t_n^2) \tag{15}$$

For the sake of simplicity, we shall consider that the value of the temperature used to compute the parameters β, γ and δ in equation (6) is constant on each interval $]t_{n-1}, t_n[$ and equal to $T_{n-\frac{1}{2}} = \frac{T_{n-1}+T_n}{2}$. We will first describe the direct integration scheme by considering the following ordinary differential equation, where u and v play parts analogous to those of σ and ε, respectively.

$$\dot{v}(t) = f(u(t)) \tag{16}$$

Then the previous approximation (15) gives:

$$\dot{v}(t) = f\left(u_{n-1} + \frac{u_n - u_{n-1}}{t_n - t_{n-1}}(t - t_{n-1})\right) + \mathcal{O}(\Delta t_n^2) \tag{17}$$

and the direct integration of equation (17) leads to :

$$v_n - v_{n-1} = \int_{t_{n-1}}^{t_n} f\left(u_{n-1} + \frac{u_n - u_{n-1}}{t_n - t_{n-1}}(t - t_{n-1})\right) dt \tag{18}$$

which is a $\mathcal{O}(\Delta t_n^3)$-accurate relation between u and v. In the case of the constitutive equation (1), this scheme leads to the following expression of the formal relation (2):

$$\left\{\begin{array}{rcl}\varepsilon_n &=& \mathbf{H}_n(\sigma_n,\mathcal{H}_n)\\ &=& \mathbf{C}:(\sigma_n-\sigma_{n-1})+\displaystyle\int_{t_{n-1}}^{t_n}\mathbf{F}\left(\sigma_{n-1}+\frac{\sigma_n-\sigma_{n-1}}{t_n-t_{n-1}}(t-t_{n-1}),T_{n-\frac{1}{2}}\right)dt\\ &+& \dfrac{\alpha(T_{n-1})+\alpha(T_n)}{2}(T_n-T_{n-1})\mathbf{I}_2+\varepsilon_{n-1}\end{array}\right. \tag{19}$$

and the fourth-order tensor $\mathbf{G}_n = \mathbf{grad}_\sigma(\mathbf{H}_n)$ relating to iteration (r) has the following form :

$$\mathbf{G}_n\left(\sigma_n^{(r)}, \mathcal{H}_n\right) = \mathbf{C} + \int_{t_{n-1}}^{t_n} \frac{t-t_{n-1}}{t_n-t_{n-1}} \mathbf{grad}_\sigma \mathbf{F}\left(\sigma_{n-1} + \frac{\sigma_n^{(r)}-\sigma_{n-1}}{t_n-t_{n-1}}(t-t_{n-1}), T_{n-\frac{1}{2}}\right) dt \quad (20)$$

in which we have, by setting

$$\sigma^{(r)}(t) = \sigma_{n-1} + \frac{\sigma_n^{(r)} - \sigma_{n-1}}{t_n - t_{n-1}}(t - t_{n-1}) \quad (21)$$

$$\left\{\begin{array}{rcl} \mathbf{grad}_\sigma \mathbf{F}\left(\sigma^{(r)}(t), T_{n-\frac{1}{2}}\right) & = & \phi_1\left(\sigma^{(r)}(t), T_{n-\frac{1}{2}}\right)\left[\mathbf{G}_1\left(\sigma^{(r)}(t), T_{n-\frac{1}{2}}\right)\right. \\ & + & \left.\mathbf{G}_2\left(\sigma^{(r)}(t), T_{n-\frac{1}{2}}\right) + \mathbf{G}_3\left(\sigma^{(r)}(t), T_{n-\frac{1}{2}}\right)\right] \end{array}\right. \quad (22)$$

where $\mathbf{G}_1$, $\mathbf{G}_2$ and $\mathbf{G}_3$ are computed from (14) after replacing σ by $\sigma^{(r)}(t)$ and T_n by $T_{n-\frac{1}{2}}$. Both integrals defined by equations (19) and (20) are evaluated by using the Gauss-Legendre quadrature method.

4 Numerical simulations

In order to compare the different schemes presented in section (3), two problems have been simulated. The first one is aimed to validate the schemes and to compare their accuracy and stability. The second one is a thermomechanical loading of a road, under the same conditions as a real-size test made by the LAVOC (LAboratoire des VOies de Circulation) of the EPFL (Ecole Polytechnique Fédérale de Lausanne). It allows us to test the finite-element program on a more realistic problem.

4.1 Axisymmetric triaxial test

The first series of computations are aimed to simulate an homogenous axisymmetric triaxial test (compression and relaxation) on a cylindrical sample of bituminous concrete.

In this subsection, we are interested in two properties of the different schemes: their stability and their accuracy. And indeed, it is well known that for viscous constitutive equations, implicit schemes show a better stability than explicit ones (Hughes & Taylor, 1978). We also focused on second-order accurate schemes for which we are expecting a good accuracy.

4.1.1 Stability

In order to compare the stability of the different schemes, computations are made with increasing values of the time step Δt. Four schemes are used in these simulations: for the sake of simplicity, they will be referred to by the following acronyms: explicit Euler scheme EE (θ-scheme with $\theta = 0$), implicit Euler scheme IE ($\theta = 1$), Crank-Nicolson scheme CN ($\theta = 0.5$) and direct integration scheme DI. We obtain good results for IE, CN and DI for big values of Δt (up to $\Delta t = 48\,\mathrm{s}$), whereas EE begins to diverge as soon as Δt is closed to 6 s. Moreover, IE gives the best results, whereas DI starts oscillating for $\Delta t = 48$s. The performances of implicit Euler scheme, Crank-Nicolson scheme and direct-integration scheme are similar to those observed by (Royis, 1995) for linear and non-linear viscoelastic constitutive equations.

4.1.2 Accuracy

The second step to compare the schemes is to evaluate their precision. Since an analytical solution cannot be found out, the reference solution is given by the application of the classical fourth-order Runge-Kutta method to the one-dimensional constitutive equation binding the axial stress σ_{22} and the axial strain ε_{22}.

Axial strain rate	0.25 %min^{-1}	1.0 %min^{-1}	4.0 %min^{-1}
CN	$9.92\,10^{-5}$	$4.22\,10^{-4}$	$4.50\,10^{-3}$
EE	$4.70\,10^{-3}$	$1.21\,10^{-2}$	$1.17\,10^{-1}$
IE	$4.60\,10^{-3}$	$1.12\,10^{-2}$	$2.75\,10^{-2}$
DI	$9.92\,10^{-5}$	$2.28\,10^{-4}$	$2.20\,10^{-3}$

Table 1: Precision of the schemes

We first note that the error increases for higher strain-rate, whatever the scheme is. The theory predicted that second-order accurate schemes would give better results than first-order accurate ones. It can be seen in table 1: implicit and explicit Euler schemes give errors 50 times bigger than the other ones. Moreover, direct integration scheme seems to be slightly better than Crank-Nicolson one.

4.1.3 Computation speed

The above properties of the different schemes allow us to hope a faster computation by using non-explicit schemes. Indeed, thanks to the stability, one can use bigger time-steps, which means a lower number of computations. Moreover, the good precision of the non-explicit schemes leads to obtain so good results with these schemes using big time-steps as EE using little time-steps. Hence, one can compare the minimal number of iterations needed by the different schemes to obtain an *a priori* given precision (we first checked that the computation time of an iteration was constant whatever the scheme we used was). For instance, a maximal relative error of $e_{\mathrm{Max}} = 3.8\,10^{-3}$ is obtained 5.6 times faster with CN or DI than with EE.

4.2 Thermomechanical loading of a road structure

Bituminous concrete is widely used in road construction. So it seems natural to use our finite-element program to simulate the behaviour of a road structure. This simulation is based on a real-size experiment performed at the LAVOC in 1992 (Recordon, Bovet, Dumont, Salhi & Gueye, 1992). The structure consists of three rectangular layers. The first one is bituminous concrete and is 8 cm thick. We model it with the previous constitutive equations (1), (6) and (7). The second layer and third layer are made of sand and are 80 and 120 cm thick, respectively. Both are governed by the same incremental law involving interpolations together with a Drucker-Prager criterion. All the numerical values for the material parameters can be found in (Royis & Seignol, 1998).

The experiment consists of three phases: a thermal one, a mechanic one with quasistatic loadings and a dynamic one. In this paper, we only focus on the second one. It can be modelized as the sollicitation of a cylinder, with prescribed normal displacement on the basis and all around, whereas the upper side is free except near its centre, where a force is applied through a 30 cm diameter disc. It is modelled like an uniform stress under the surface of the whole disc. The mechanical loading (global force F) is then applied as follows: four phases of time-linear load and unload are applied with growing maximal force. Eventually, a force is progressively applied and maintained. This problem is modelled by using a 429-dof-mesh. The Crank-Nicolson scheme has been used for the discretization of the constitutive equations of the bituminous concrete. As an instance, the figure 1 shows the radial strains obtained from the program (continued lines, referred to by numbers in brackets) compared with experimental data provided by the LAVOC (dotted lines, referred to by numbers in square brackets).

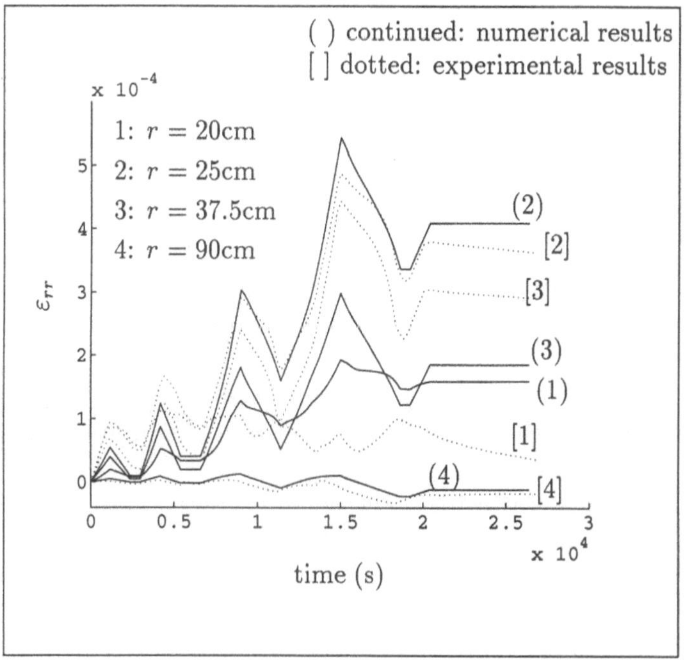

Fig. 1: Radial strains ε_{rr} on the surface of the road

Globally, this simulation gives good results. The general evolution (in both space and time) of the strains and the stresses are well simulated by the program. One can although notice some discrepancies in the numerical values. They can be partially explained by experimental reasons, but there remains some investigations to be led, more particularly for explaining the program's underestimation of strain rate during creep phases.

5 Conclusion

In order to model the thermoviscoplastic behaviour of bituminous concrete, a tensorial constitutive equation has been stated on the basis of one-dimensional relations coming from experimental results obtained at the LGM of the ENTPE, and then implemented in the finite element program ELFIMTH developed in previous works for the numerical resolution of boundary value problems in thermomechanics. That implementation was carried out by using two different time-discrete schemes for the numerical integration of the constitutive equations considered over finite time steps. A set of numerical simulations of homogeneous triaxial tests has shown the good properties of these schemes when compared with the classical fully explicit Euler one. The resulting finite element program was also used to simulate a real size experiment performed at the LAVOC of the EPFL. The comparison between numerical results and experimental data shows a fair agreement for that more complex and realistic problem, even if some discrepancies can be observed.

Acknowledgements

The work presented in this paper was partly supported by the "Geomaterials EURO-GRECO".

References

DI BENEDETTO H. & X.YAN (1994). Comportement mécanique des enrobés bitumineux et modélisation de la contrainte maximale. Materials and Structures, 27: 539-547.

HUGHES T.J.R. & R.L.TAYLOR (1978). Unconditionally stable algorithms for quasi-static elastovisco-plastic finite element analysis. Computer and Structures, 8: 168-173.

PERZINA P. (1963). The constitutive equations for rate sensitive plastic materials. Quarterly of Applied Mathematics, 20(4): 321-332.

RECORDON E., M.BOVET, M.DUMONT, A.SALHI & I.GUEYE (1992). Modélisation numérique des enrobés bitumineux. Rapport scientifique de l'Euro-Greco Géomateriaux, 1:20-34.

ROYIS P. (1995). Integration by FEM of viscous models: a comparison between three time-discrete schemes. Proc. of the 5th Int. Conf. on Numerical Models in Geomechanics: 423-428, A.A Balkema, Davos, Switzerland.

ROYIS P. & J-F.SEIGNOL (1998). Finite element simulation of the thermoviscoplastic behaviour of bituminous concrete. To be published in Int. J. for Numerical and Analytical Methods in Geomechanics.

A NEW CREEP LAW TO DESCRIBE THE TRANSIENT AND SECONDARY CREEP PHASE

K. Salzer
Institut für Gebirgsmechanik, Leipzig, Germany

H. Konietzky
ITASCA Consultants GmbH, Gelsenkirchen, Germany

R.-M. Günther
Institut für Gebirgsmechanik, Leipzig, Germany

ABSTRACT

For stability analyses as well as for predictions of the future geomechanical behaviour of underground openings in salt rock strata, proper calculation tools are needed. Numerical calculation methods and material laws are required to describe the mechanical behaviour of salt. A proposed new material salt creep law describes the deformation-hardening behaviour of the transient creep phase, which gives a good description for a relatively short time after the excavation phase and can, by considering the recovery of the hardening, also be extended to the stationary creep phase, which is significant for long times after the creation of openings. The developed material law also includes inverse transient creep, which takes place during load reductions. This material law is used for the back-analysis of all common types of lab tests (compression, creep and relaxation) as well as for some field measurements. The material creep law was implemented into the ITASCA codes *FLAC* (ITASCA 1998) and *FLAC3D* (ITASCA 1997).

1. INTRODUCTION

Since time immemorial, salt deposits have been used for the production of salt minerals. Within the last decade, the use of leaching caverns in salt rock formations for the storage of gas has become increasingly important. Also, the use of salt deposits for storing chemical-toxic and radioactive wastes have become a topic of great interest. The visco-plastic behaviour (creep) of salt rock formations leads to practical impermeability and, in the long term, to a complete sealing of openings. Therefore, the salt rock formations fulfil the

demands that were made for host rock formations for underground storage caverns and repositories. For stability analyses as well as for predictions of the future geomechanical behaviour of underground openings in salt rock formations, proper calculation tools are needed. Therefore, efficient numerical calculation methods and material laws have been developed to describe the mechanical behaviour of salt.
In principle, the creep curve (i.e. the deformation-time curve (Fig. 1) under constant load), can be characterised by three phases:

(1) a primary or transient creep phase (Hardening dominates during that part of the creep curve.);
(2) a secondary, or stationary, creep phase during which hardening and recovery are in dynamic equilibrium; and
(3) a tertiary, or accelerating, creep phase that involves the initiation of a creep fracture. Damage inside the rock grows more quickly than it can be repaired by relaxation and heal-up processes.

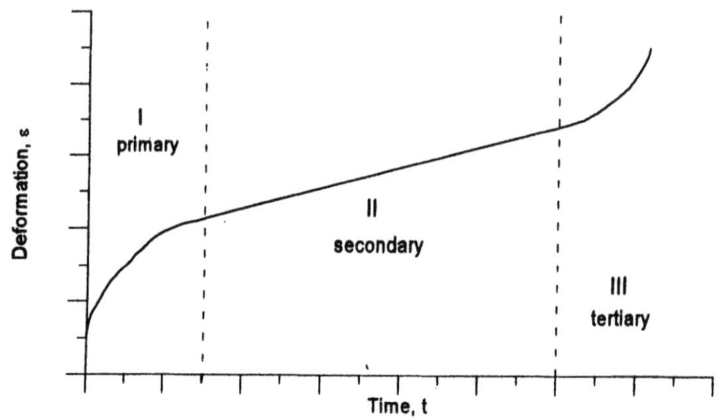

Figure 1: The three phases of a creep curve

The observed and measured convergence underground, especially during the phase immediately after the creation of openings, cannot be described with the secondary creep approach alone. Figure 2 shows a typical example of the measured vertical compression of a Sylvenite pillar compared to the calculated values according to the secondary creep approach (IfG 1992). The dashed bold line shows the calculated pillar compression using the secondary creep approach when the excavation time point corresponds to real time and the secondary creep parameters are fitted to the displacement velocities after 4 years. The deformation parameters for the deformation-hardening approach, which were deduced from laboratory tests in advance of pillar creation, lead to a good agreement between the in-situ measured deformations and the calculated values (see the thin line in Figure 2). For longer times, the measured values lie above the calculated ones because transition into the stationary phase has already taken place. The deformation-hardening approach with power functions for stress and deformation has been used by several authors (e.g. LEMAIRTE (1970) and MENZEL (1977)). This approach has had good success in describing the creep behaviour of salt. However, it does not include the secondary creep phase.

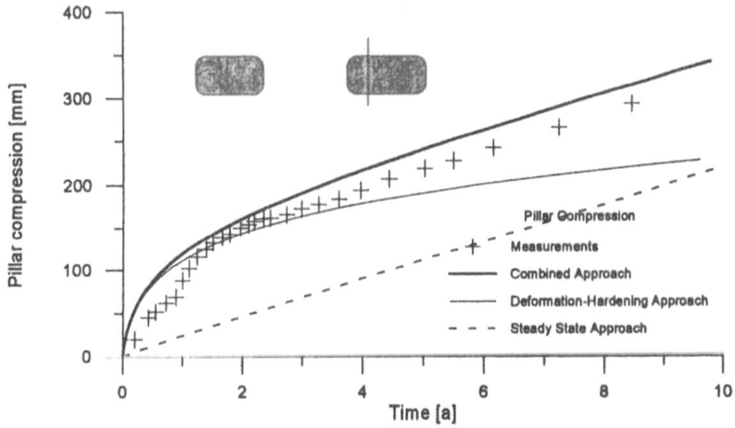

Figure 2: Comparison between in-situ measured pillar compression and several calculated pillar compression values using different creep approaches

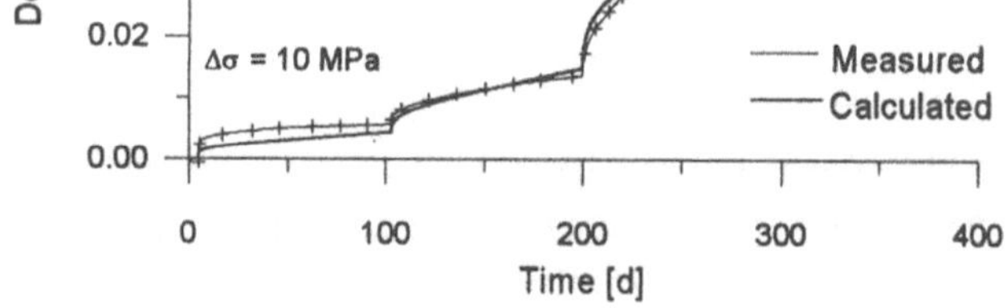

Figure 3: The inverse transient creep demonstrated for deformation of a halite sample during a Karman test (HUNSCHE 1994)

This paper presents a new combined creep approach that contains both creep phases by recovery of the hardening, whereby the hardening-part of the creep deformation, ε^{V}, is used as an internal state variable. The inverse transient creep is also described with this creep approach. An example of the inverse transient creep that appears during the reduction of the

deviatoric stresses is shown in Figure 3 (HUNSCHE 1995), which shows the deformation time curve of a triaxial test on a cylindrical salt sample in the so-called "Karman box" with 5 stress levels.
Consideration of these effects is necessary for all geomechanical processes in which the deviatoric stress level decreases (e.g. load changes in a salt storage cavern or the interaction between the salt rock and the sealing construction).

2. MATHEMATICAL DESCRIPTION OF THE NEW MATRIAL LAW

The strain rate tensor, $\dot{\varepsilon}_{ij}$, is given by the following equation:

$$\dot{\varepsilon}_{ij} = \dot{\varepsilon}_{ij}^{el} + \dot{\varepsilon}_{ij}^{cr} \quad (1)$$

where ε_{ij}^{el} and ε_{ij}^{cr} are the elastic and creep portions of the strain tensor, respectively. The corresponding strain rate components are:

$$\dot{\varepsilon}_{ij}^{el} = -\frac{\nu}{E} \bullet \dot{\sigma}_{kk}\delta_{ij} + \frac{(1+\nu)}{E} \bullet \dot{\sigma}_{ij} \quad (2)$$

$$\dot{\varepsilon}_{ij}^{cr} = \frac{3}{2}\dot{\varepsilon}_{eff}^{cr}\frac{S_{ij}}{\sigma_{eff}} \quad (3)$$

In the deformation-hardening approach, which results in a good description of the primary creep phase with non-inverse transient creep, ε^{cr}_{eff} is given by the following relation:

$$\dot{\varepsilon}_{eff}^{cr} = A^1 \frac{\sigma^{\beta^1}}{(\varepsilon_{eff}^{cr})^{\mu}} \quad (4)$$

The secondary creep can be considered as a special case of this approach, with $\mu = 0$.
In the proposed combined creep approach, the total creep deformation, ε^{cr}_{eff}, is divided into a hardening part, ε^{V}_{eff} (which represents the internal state variable), and the remainder, ε^{R}_{eff}:

$$\varepsilon_{eff}^{cr} = \varepsilon_{eff}^{V} + \varepsilon_{eff}^{R} \quad (5)$$

The following equations apply in the combined approach:

$$\dot{\varepsilon}_{eff}^{cr} = A^1 \frac{\sigma^{\beta^1}}{(\varepsilon_{eff}^{V})^{\mu}} \quad (6a)$$

$$\dot{\varepsilon}^{V}_{eff} = A^{1} \frac{\sigma^{\beta^{1}}_{eff}}{(\varepsilon^{V}_{eff})^{\mu}} - \frac{\varepsilon^{V}_{eff}}{t_0} \tag{6b}$$

The second term in equation (6b) describes the recovery of that part of the deformation that acts as hardening. If the first term of equation (6b), which describes the hardening and which is equal to the growth of the total creep deformation [compare to equation (6a)] is ignored ($\sigma = 0$, total unloading of the sample), the integration of equation (6b) leads to an exponential decay of ε^{V}_{eff} with time, due to the recovering. The approach used to describe the recovery of the hardening is based on the thermal nature of this process.
For short times after the creation of the openings, and with corresponding small values for $\varepsilon^{\alpha}_{eff}$ and ε^{V}_{eff}, the second term of equation (6b) is negligible; hence, the creep behaviour can be described by the deformation-hardening approach alone. For larger values of ε^{V}_{eff}, the importance of the second term increases. The state of stationary creep is reached if ε^{V}_{eff} does not grow further—i.e. $\dot{\varepsilon}^{V}_{eff} = 0$. Recovery and hardening are then in dynamic equilibrium. For that creep state, the value of ε^{V}_{eff} is given by:

$$(\varepsilon^{V}_{eff})^{II} = (A^{1}\, \sigma^{\beta^{1}}_{eff}\, t_0)^{\frac{1}{1+\mu}} \tag{7a}$$

and the following relations apply:

$$(\dot{\varepsilon}^{cr}_{eff})^{II} = A^{II}\, \sigma^{\beta^{II}}_{eff} \tag{7b}$$

where

$$A^{II} = A^{I} \left(\frac{1}{A^{1} t_0} \right)^{\frac{\mu}{1+\mu}} \tag{7c}$$

$$\beta^{II} = \frac{\beta^{1}}{1+\mu} \tag{7d}$$

As will be shown later, the adjustment of t_0 to the in-situ measured deformation values gives recovery times between 10 and 20 years.
During the inverse transient creep, ε^{V}_{eff} lies above the equilibrium hardening, $(\varepsilon^{V}_{eff})^{II}$, for the new reduced deviatoric stress level, so that ε^{V}_{eff} reduces to a new equilibrium combined with a corresponding increase of $\varepsilon^{\alpha}_{eff}$ as a result of the recovery process.

Due to the thermal nature of the recovery, it is reasonable to use the ARRHENIUS approach for the temperature dependence on the recovery time, t_0:

$$t_0 \approx e^{\frac{Q_r}{RT}} \tag{8a}$$

Based on equation (7c), one can deduce the known temperature dependence of the secondary creep—i.e. factor A^{II}:

$$A^{II} \approx e^{-\frac{Q_r \mu}{RT(1+\mu)}} \tag{8b}$$

The implementation of the new creep law into *FLAC* and *FLAC3D* is based on the direct numerical evaluation of the equations (6a) and (6b) for each zone and each timestep. Especially during the initial creep phase ($\varepsilon^{V}_{eff} = 0$), extremely small timesteps are necessary. The assumption of a small primary hardening of the order of 10^{-5} accelerates the calculation process during the initial phase and shows only a negligible influence on the calculation results.

3. VALIDATION OF THE NEW COMBINED MATERIAL LAW

3.1 LABORATORY TESTS

As shown in Figure 3, the new creep law provides a good description of creep tests (load tests including multi-level creep and inverse transient creep).
The following three figures demonstrate that, with the same parameter set describing a specific salt type, the stress-deformation behaviour for three different load regimes can be reproduced. These three load regimes are: a standard creep test (σ_{eff} = constant; Figure 4), a triaxial test with constant deformation rate ($\dot{\varepsilon}$ = constant; Figure 5) and a triaxial test with relaxation phases ($\dot{\varepsilon} = 0$; Figure 6).
For the halite investigated, the following material parameters were determined based on creep tests (see, for example, Figure 4):

Young's modulus = 30 GPa; Poisson's ratio = 0,23
$A^I = 1{,}0 \cdot 10^{-58}$; $\beta^I = 28$; $\mu = 7$ $t_0 = 10$ years
The system units are: σ [MPa]; t [h]

The parallel shifting of the creep curve occurs because primary compression loading was applied in several steps in the laboratory test, and in only one step in the numerical simulation. Figure 5 shows the stress-deformation curve for a triaxial compression test under constant deformation velocity with relaxation phases at 1% and 5% of the sample compression. Figure 6 shows the corresponding stress-time diagram. Figure 5 also documents the influence of increased deformation velocity on the stress-deformation behaviour. After the second relaxation, at a compression of 5%, the deformation velocity was increased from $5 \cdot 10^{-6} s^{-1}$ to $2.5 \cdot 10^{-5} s^{-1}$. An increase of the deformation velocity is connected with a stress increase at the same deformation. This is also reproduced

quantitatively by the computer model.
Figure 6 shows the result of a back-analysis of both relaxation phases. The relaxation behaviour is also reproduced by the computer model with sufficient accuracy. Therefore, it can be assumed that complicated load regimes can also be modelled properly with the material model as long as the deformation has not reached the tertiary creep phase.

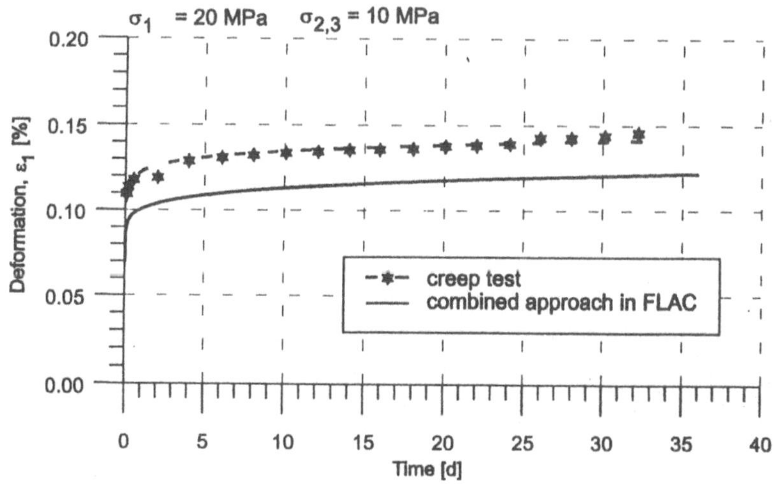

Figure 4: Comparison between measured and calculated values for a creep test of 'Leinesteinsalz' (halite)

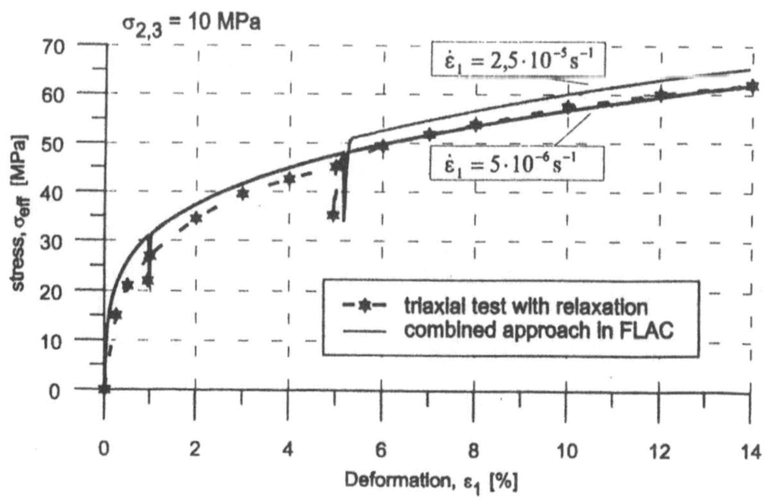

Figure 5: Comparison between measured and calculated values for a triaxial test on 'Leinesteinsalz' (halite) with two interposed relaxation phases

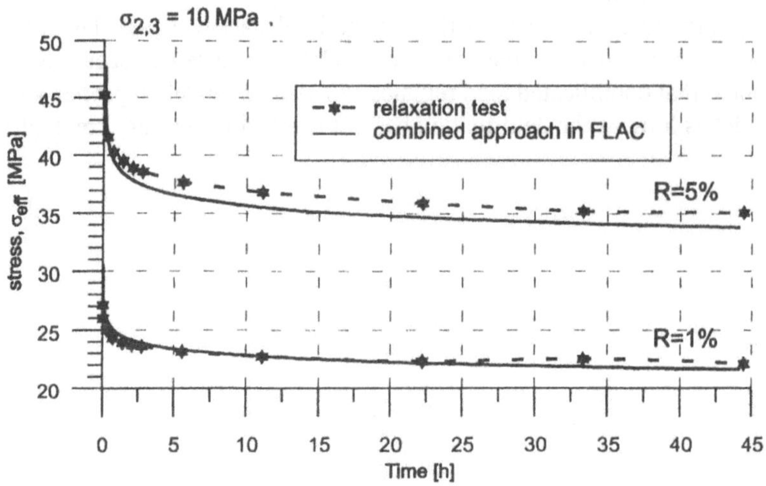

Figure 6: Comparison between measured and calculated displacements during the relaxation phase for 'Leinesteinsalz' (halite)

3.2 FIELD MEASUREMENTS

With the help of the two following examples, the assessment of the recovery time, t_0, is demonstrated. The assesment is made on the basis of the adjustment of the calculated deformation values to the in-situ measured values.

3.2.1 SYLVENITE PILLAR

The measured in-situ pillar compression of a sylvenite pillar was compared with the calculated values (Figure 2). In this example, a quadratic pillar was transformed into a column shape that can be modelled with the axisymmetric calculation option. The zero measurement (reference) was conducted 25 days after creation of the pillar and was taken into consideration in Figure 2. With creep parameters for the deformation-hardening approach determined during the early 1970s (IFG 1992): $A^I = 1{,}15 \bullet 10^{-56}$; $\beta^I = 25$ and $\mu = 7{,}3$; the later (1978 - 1987) observed pillar compression could be described properly. Only in the last 5 years of the total measuring period did an underestimation of the pillar compression became apparent. Before investigations with the new creep approach were begun, the factor A^{II} was fitted in such a way that the pillar compression rates within the stationary phase were reproduced properly (dashed bold line). On the basis of the so-determined A^{II}, the parameter t_0 can be estimated according to equation (7c). For this example, t_0 amounts to 20,9 years. With these parameters for the combined creep behaviour, the bold line was calculated; it shows good agreement between the measured and calculated pillar compression for all considered times.

3.2.2 SALT PILLAR

The second example (SALZER 1996) compares calculated and measured deformation in a relatively thick salt pillar situated within a room-and-pillar system under work. The pillar width is 35 m, the room width is 20 m, and the mean pillar height is 35 m. After the excavation of a roof chamber 20-m wide and 5-m tall, the total headroom is mined by underhand stoping and haulaged through a system of excavations in the floor (Figure 7).

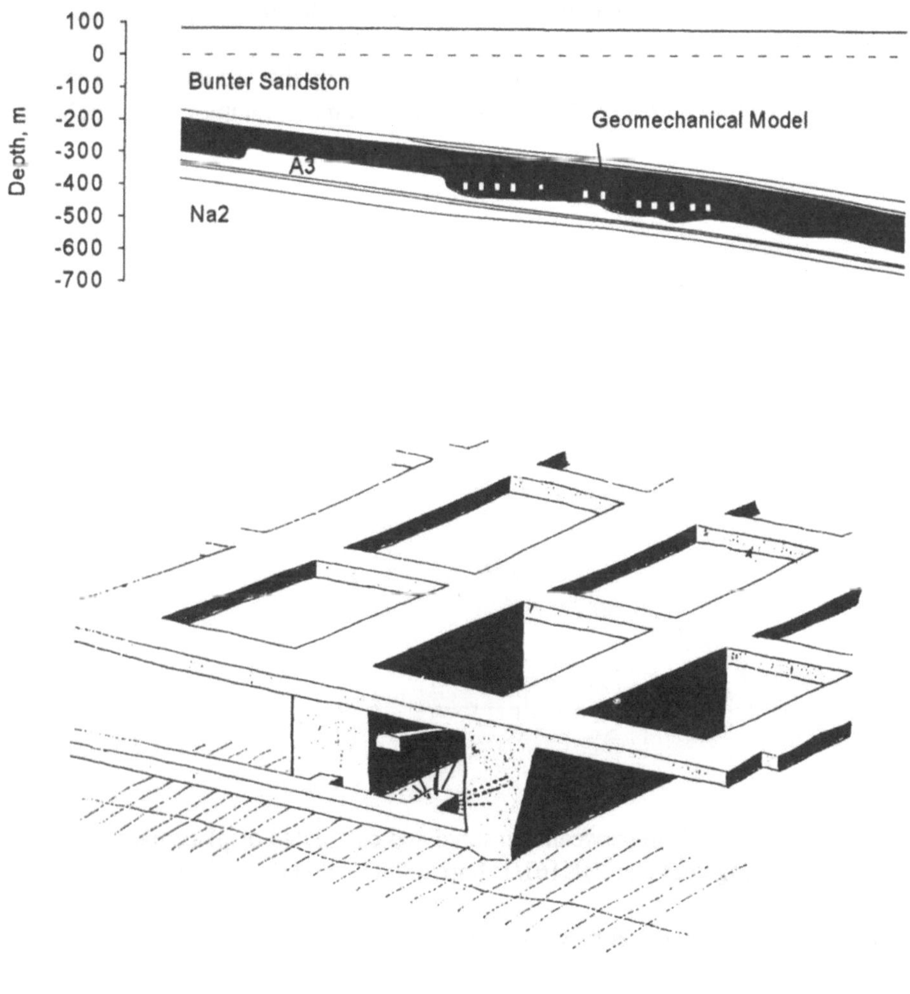

Figure 7: Immediate surroundings of the measuring chamber (plane and perspective view)

A measuring system (MK2) was installed with the objective of monitoring the stress-deformation behaviour during the room-creation phase. The second excavation phase corresponds to the excavation of the room located Southeast 1612; the third excavation phase corresponds to the room located Northwest 1613; and the fourth phase includes the drivage of a conveyer road in the pillar itself (Figure 7). The vertical compression of the pillar and its lateral deformation is measured with 3 extensometers in boreholes; the extensometers are oriented vertical, and horizontal into the Northeast and Northwest directions. Figure 7 shows a cross-section through the measuring plane, which serves also as the basis for the rock mechanics model. A test calculation showed that the stationary creep approach results in an unsatisfactory agreement between the in situ measured and calculated values. This is because the measurements include the creation process and, therefore, the transient creep part can not be neglected. Subsequently, the new creep approach was used, with recovery time t_0 chosen in such a way that the measured surface settlements above the mining area could be reproduced in a more complex rock mechanics model.

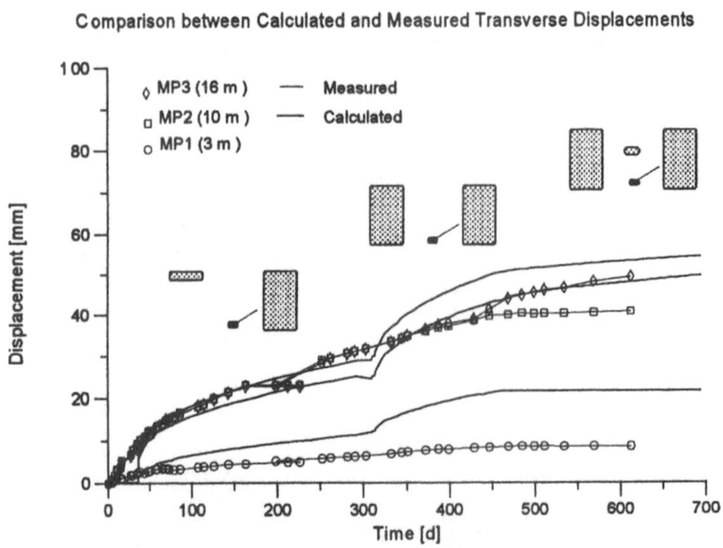

Figure 8: Comparison between measured and calculated displacements (measuring chamber MK161/1; horizontal to slightly inclined)

The following deformation parameters were chosen:

Young's modulus= 25 Gpa Poisson's ratio = 0,25 μ = 1,6

$A^I = 1{,}7 \bullet 10^{-22}$; $\beta^I = 10$; t_0 = 10 years

Figure 8 shows a comparison between the calculated and measured displacements which were obtained from the measuring chamber. It also shows a comparison for the displacements between the control points inside the measuring chamber and points at distances of 3, 10 and 16 m from the chamber periphery. The calculated values for the 10-m and 16-m measuring lines show good agreement with the measured values; for the 3-m measuring line, the 3-dimensional effect of the chamber face becomes important. The calculated minor principal stresses correspond quite well with the shut-in pressure values obtained from hydraulic fracturing measurements.

4. CONCLUSIONS

The proposed new combined creep law, which is implemented into the codes *FLAC* and *FLAC3D*, allows a proper description of the secondary, transient and inverse transient creep phases. The proposed combined creep law corresponds to the addition of a new component in the deformation-hardening model formulation to account for a better description of the transient creep phase by consideration of hardening recovery and allow modelling of secondary creep phase and inverse creep. The examples presented show good agreement between the measured and calculated deformations and for the new calculated values according to the new combined creep approach, although the deduced creep parameters, A^I, β^I and μ were already determined on the basis of load tests before the excavation had taken place and only recovery time t_0 was adjusted to the in-situ measurements. A first assessment of t_0 under mining temperature conditions (30° C) gives values between 10 and 20 years

5. ACKNOWLEDGEMENT

The presented work was granted by the Federal Ministry of Research and Technology under the contract 02-C0062-8.

6. LITERATURE

HUNSCHE U., SCHULZE, O. (1995). Personnel communication

IfG (1992). Überprüfung vorhandener Modellvorstellungen zum Konvergenzverhalten und zur Standsicherheit komplexer Grubengebäude im Salzgestein. Forschungsvorhaben BMFT-FB/02 E 8161 6, IfG GmbH Leipzig, Januar 1992

ITASCA Consulting Group, Inc. (1997). FLAC^{3D}, Version 2.0, Minneapolis, Minnesota: ICG

ITASCA Consulting Group, Inc. (1998). FLAC, Version 3.4, Minneapolis, Minnesota: ICG

LEMAITRE J. (1970). Sur la determination des lois de compartement des materiaux. ela viscoplastiques, thesis, Public., O.N.E.R.A N° 135 81970), Paris

MENZEL W., SCHREINER W. (1977). Zum geomechanischen Verhalten von Steinsalz verschiedener Lagerstätten der DDR. Teil II: Das Verformungsverhalten, Neue Bergbautechnik 5, 8: 565 - 571

SALZER K., SCHREINER W. (1996): Long-term safety of salt mines in flat beddings. 4th Conference on the Mechanical Behavior of Salt, Montreal

APPLICATION OF DAMAGE MECHANICS TO DEEP TUNNELS

G. Swoboda
University of Innsbruck, Innsbruck, Austria

M. Zaki
Minia University, Minia, Egypt

ABSTRACT

The paper utilizes an anisotropic nonlinear damage model in strain space to describe the behavior of jointed rock and applies it to mechanical analysis of tunneling. This work focuses on rate-independent and small-deformation behavior during static isothermal processes. The prime results include: (1) the properties of damage-dependent elasticity tensors based on geological information of the jointed rock mass; (2) the damage evolution law presented on the basis of thermodynamics and combined with the endochronic theory; (3) the symmetric form of the effective stress concept which held between the elasticity tensor of the damaged material and that of the fictitious undamaged material; and (4) the global damage tensor based on the work-equivalence principle and the local geological data of multi-joint sets. Finally the numerical results of a tunnel intersection in jointed rock is presented to illustrate the mechanical behavior of this model.

1. INTRODUCTION

Jointed rock has complex mechanical behavior, such as anisotropy, hysteresis, dilatancy, irreversible strain and strongly path-dependent stress-strain relationships, which is generally associated with the existence of a great deal of cracks and their propagation. In recent years, much work has been performed on the description and calculation of anisotropic mechanical behavior caused by cracks in solids.

Micromechanical approaches attempted to predict the macroscale thermomechanical response of heterogeneous materials based on mesostructural models of a representative volume element (RVE) within the material (Krajcinovic 1983, Krajcinovic *et al.* 1991, Stumvoll & Swoboda 1993, Halm & Dragon 1996). Micromechanical models have the distinct advantage of being able to capture structure details at the microscale and mesoscale, and of permitting formulation of the kinetic equation for damage evolution based on the physical process involved. But these models can be computationally inefficient in many practical applications and can therefore only be applied to limited cases.

Continuum Damage Mechanics (CDM), which employs some continuum variables to describe cracks and joints, provides another way of modeling jointed rock mass (Kachanov 1958, Murakami & Ohno 1980, Kawamoto *et al.* 1988, Yazdani & Karnawat 1996, Swoboda *et al.* 1995). CDM is based on the thermodynamics of irreversible process, internal state variable theory and relevant physical considerations. Formal CDM modelling of the cracked rock was first suggested by (Kawamoto *et al.* 1988). Thereafter additional advances were reported in some simple cases of anisotropic damage analysis (Chan *et al.* 1997). In spite of all the efforts described above, it has not been possible until now to directly use the anisotropic damage theory for jointed rock mass in the mechanical analysis of relatively large engineering problems.

In this paper, the phenomenological second-order tensor which is directly related to the geological data of the jointed rock mass is chosen as the damage tensor of CDM. The concept of the active damage tensor is introduced in order to reflect the phenomenon that the cracks in the rock mass may close in compression, which means the damage tensor may be changed by compression or tension of strain state but not its propagation. A one-parameter damage-dependent elasticity tensor is formulated by tensorial algebra and thermodynamic requirements. The damage evolution law is formulated in the conjugate force space based on the irreversible thermodynamics and the endochronic theory. The engineering application to tunneling is presented at the end of the paper.

2. DAMAGE ELASTICITY

During the development of the damage theory, the definition of the damage tensor was also changed from a scalar formulation (Kachanov 1958, Lemaitre 1990) to a higher-order tensor form (Murakami & Ohno 1980, Kawamoto *et al.* 1988, Halm & Dragon 1996, Ju 1989). However, all these damage models can not be conveniently used in engineering calculations of jointed rock mass without any further improvement. The difficulties come either from calculation of the local damage tensor for one set of cracks, or from the global tensor, which should not be based on a simple summation over the entire local damage tensor. The damage model adopted here is a second-order tensor whose definition for one set of parallel joints is given by:

$$\Omega_{ij} = \omega n_i n_j, \quad i,j = 1,3 \tag{1}$$

where $\boldsymbol{n} = [l, m, n]^T$ is the direction vector of damage tensor Ω of the crack set; and ω is the separation factor of one set of cracks, which is taken as the crack density. The separation factor is part of the geological information of the jointed rock mass. For a multiple set of joints, the global tensor should be calculated according to the work-equivalence principle. The details of its calculation will be presented in a subsequent section.

2.1 Modelling of damage-dependent elasticity tensor

This subsection presents the development of an explicit expression of a damage-dependent elasticity tensor $\mathbf{D} = \mathbf{D}(\boldsymbol{\Omega})$. A one-parameter damage elasticity tensor is introduced as:

$$\mathbf{D} = \phi \cdot \mathbf{D}^0 \cdot \phi \tag{2}$$

$$\tilde{D}_{ijkl} = \lambda \phi_{ij} \phi_{kl} + \mu(\phi_{ik}\phi_{jl} + \phi_{il}\phi_{jk}) \tag{3}$$

$$\phi_{ij} = \delta_{ij} - m\tilde{\Omega}_{ij} - (1-m)\tilde{\Omega}_{im}\tilde{\Omega}_{mj}, \qquad 0 \leq m \leq 1 \tag{4}$$

where m is assumed to be the only damage-related material constant and λ and μ are the Lame's constants of the undisturbed rock mass. For jointed rock m is related to the crack system and is a function of the crack layer thickness. The condition $0 \leq m \leq 1$ is needed to satisfy the positive definite requirement. Evidently, the elasticity tensor satisfies all the general principles of CDM (Malvern 1969). To describe the closing and opening of the joint system in rock, an active damage tensor is introduced in the following subsection.

2.2 Active damage tensor

In rock-like material, the damage appears in the form of planar cracks. The cracks may close in compression and open in tension. In order to reflect this character of rock damage, the concept of active damage tensor is introduced here. By decomposing one original crack vector in the principal strain coordinate system, it is assumed that each equivalent crack is composed of only these components in tensile principal strain directions. Thus, the active damage tensor can be calculated by the transformation

$$\tilde{\boldsymbol{\Omega}} = \mathbf{P}^+ : \boldsymbol{\Omega} \tag{5}$$

where $\mathbf{P}^+$ is just the so-called "positive projection tensor" (Ortiz 1985),

$$P^+_{ijkl} = Q^+_{ik} Q^+_{jl}, \qquad \text{where} \qquad \mathbf{Q}^+ = \sum_{i=1}^{3} \hat{H}(\varepsilon_i)\mathbf{p}_i\mathbf{p}_j \tag{6}$$

where Q^+_{ij} is the positive (tensile) spectral projection tensor, $\mathbf{p}_i$ are the i-th principal direction vectors of strain tensor $\boldsymbol{\varepsilon}$ and ε_i are the principal strain values. $\hat{H}(\cdot)$ is the Heaviside function:

$$\hat{H}(\varepsilon_i) = \begin{cases} 1 & \text{if} \quad \varepsilon_i > 0 \\ 0 & \text{otherwise} \end{cases} \tag{7}$$

The active damage tensor is symmetric, ensured by eqns. (5) and (6). In some geomaterials, the Heaviside function $\hat{H}$ will overestimate the difference between compression and tension. For example, in jointed rock, the joints are not perfectly in contact but contain some filling materials or roughness. In order to take these factors into account, the Heaviside function $\hat{H}$ in eqn. (6) can be replaced with

$$H(\varepsilon_i) = (1-h)\hat{H} + h = \begin{cases} 1 & \text{if} \quad \varepsilon_i > 0 \\ h & \text{otherwise} \end{cases} \tag{8}$$

where h $(0 \leq h < 1)$ is a material constant to reflect the properties of crack contact.

3. DETERMINATION OF DAMAGE TENSOR USING THE GEOLOGICAL DATA

The orientation of the joint set is usually identified in the field by sampling along a line, as in a borehole, or over an area, as on an outcropping surface. Statistical descriptions of the orientation data are always expressed as the orientation of the normal line of the joint plane. Spherical coordinates (α, β, r) or the direction cosines (l, m, n) are generally used for this purpose. These are related to the dip angle, which is defined as the angle between the horizontal and the line of maximum dip of the joint plane, β and the dip direction, which is defined as the angle between north and the horizontal projection of the line of maximum dip of the joint plane measured in clockwise direction, α, together with the direction angle of the tunnel axis γ, the angle between the tunnel axis and north, as follows:

$$\begin{cases} l = sin\beta cos[2\pi - (\alpha + \gamma)] \\ m = sin\beta sin[2\pi - (\alpha + \gamma)] \\ n = cos\beta \end{cases} \tag{9}$$

The direction vector of the damage tensor Ω according to eqn. (1) is defined on the basis of the direction vector $\boldsymbol{n} = [l, m, n]^T$. Here an example for calculating the damage tensor from the geological data is given. The dip direction $\alpha = 45^0$, dip angle $\beta = 45^0$, tunnel axis direction angle $\gamma = 0$, damage parameter $\omega = 0.3$ are based on the geological investigation. Using the above mentioned calculation process, the initial damage tensor would be:

$$\boldsymbol{\Omega}_0 = \begin{bmatrix} 0.075 & -0.075 & 0.106 \\ -0.075 & 0.075 & -0.106 \\ 0.106 & -0.106 & 0.15 \end{bmatrix}$$

3.1 Global damage tensor for multiple sets of joints

For jointed rock mass, the most common damage state is the presence of multiple sets of joints which intersect each other. Presented here are the techniques for determining the global damage tensor which can represent the damage effect of the multiple joint sets system. The damage tensor related to each set of joints can be obtained with the techniques introduced above.

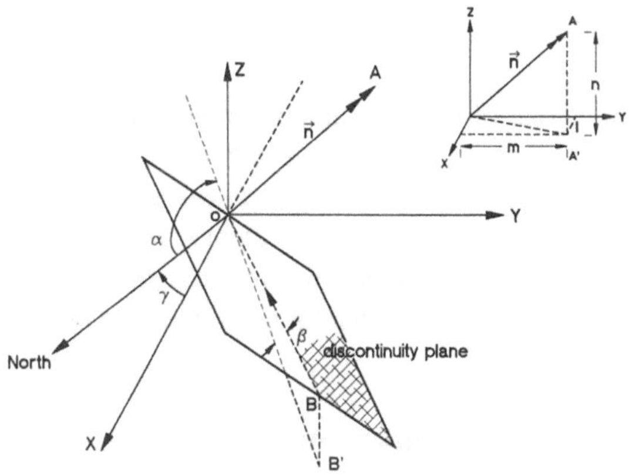

Fig. 1: Relationship between the orientation and the spherical coordinate system.

Calculation of the global damage tensor is based on the principle of energy equivalence: The energy dissipated by the global damage tensor is equivalent to the sum of the energy dissipated by the damage tensor related to each set of joints.

The equation for calculating the global damage tensor is given in the following:

$$\boldsymbol{\Omega}_g = \mathbf{I} - \left[\sum_{n=1}^{N} (\mathbf{I} - \boldsymbol{\Omega}_n)^{-1} - (N-1)\mathbf{I} \right]^{-1} \tag{10}$$

where $\boldsymbol{\Omega}_g$ is the global damage tensor; $\boldsymbol{\Omega}_n$ is the damage tensor for the nth joint; N is the total number of joint sets; $\mathbf{I}$ is a second-order unit tensor.

Thed calculation of the global damage tensor is given for three sets of joints whose geological data are:

$$\begin{array}{lll} \gamma = 0, & & \\ \alpha_1 = 0, & \beta_1 = \frac{\pi}{2}, & \omega_1 = 0.2 \\ \alpha_2 = -\frac{\pi}{4}, & \beta_2 = \frac{\pi}{2}, & \omega_2 = 0.6 \\ \alpha_3 = \frac{\pi}{4}, & \beta_3 = \frac{\pi}{4}, & \omega_3 = 0.3 \end{array}$$

Thus, the damage tensor for each set of joints is:

$$\boldsymbol{\Omega}_1 = \begin{bmatrix} 0.2 & 0.0 & 0.0 \\ 0.0 & 0.0 & 0.0 \\ 0.0 & 0.0 & 0.0 \end{bmatrix} \boldsymbol{\Omega}_2 = \begin{bmatrix} 0.3 & 0.3 & 0.0 \\ 0.3 & 0.3 & 0.0 \\ 0.0 & 0.0 & 0.0 \end{bmatrix} \boldsymbol{\Omega}_3 = \begin{bmatrix} 0.075 & -0.075 & 0.106 \\ -0.075 & 0.075 & -0.106 \\ 0.106 & -0.106 & 0.15 \end{bmatrix} \tag{11}$$

The global damage tensor for the three joint sets, according to eqn. (10), is:

$$\boldsymbol{\Omega} = \begin{bmatrix} 0.45946 & 0.19459 & 0.01 \\ 0.19459 & 0.38595 & -0.1 \\ 0.01 & -0.1 & 0.15243 \end{bmatrix}$$

4. DAMAGE EVOLUTION

Damage evolution law is one of the most important aspects of the constitutive relationship of rock. In this section it is assumed that the damage evolution is completely determined by its conjugate force, which is also known as *the strain energy density release rate* (Krajcinovic *et al.* 1991). Using the principle of endochronic theory (Valanis 1990, Chow & Chen 1992), this evolution law for the isothermal static damage problems is established in the framework of thermodynamics. No softening or hardening related to damage process is considered.

4.1 Description of the damage evolution law

Firstly, following the form given by (Chow & Chen 1992), the potential function ϕ for damage evolution can be expressed as follows:

$$\phi^2 = \frac{1}{2} R_{ij} B_{ijkl} R_{kl}, \qquad B_{ijkl} = B I_{ijkl} \tag{12}$$

where R_{ij} is the conjugate force tensor, I_{ijkl} is the fourth-order unit tensor and B is a material constant governing the behavior of damage propagation.

According to the theory of thermodynamics the damage propagation $\dot{\Omega}_{ij}$ is obtained from the relation:

$$\dot{\Omega}_{ij} = \dot{\lambda} \frac{\partial \phi}{\partial R_{ij}} = \dot{\lambda} \frac{B_{ijkl}}{2\phi} R_{kl} \tag{13}$$

where $\dot{\lambda}$ is a multiplier.

Then, the endochronic theory is introduced into the calculation of damage evolution. Endochronic theory assumes that the ratio of an internal variable is proportional to its conjugate force. Based on this principle the damage evolution in the intrinsic time space is obtained by:

$$\frac{d\Omega}{dz} = C_{ijkl} R_{kl} \tag{14}$$

where C_{ijkl} is a constant coefficient tensor; and dz is the infinitesimal increment of intrinsic time scale defined in elastic strain ε^e_{ij} space:

$$dz = \sqrt{de_{ij} de_{ij}}, \qquad e_{ij} = \varepsilon^e_{ij} - \frac{1}{3} \delta_{ij} \varepsilon^e_{kk} \tag{15}$$

The following relation is obtained by comparing eqn. (13) and eqn. (14):

$$C_{ijkl} = \dot{\lambda} \frac{B_{ijkl}}{2\phi} \tag{16}$$

Replacing Newton time with the intrinsic time dz, the damage evolution equation can be written as follows:

$$\frac{d\Omega}{dz} = \dot{\lambda} \frac{B_{ijkl}}{2\phi} R_{kl} \tag{17}$$

Damage propagation is controlled through the threshold value of the damage potential ϕ^0, which is the second parameter for damage evolution. The meaning of $\dot{\lambda}$ is defined as follows:

$$\dot{\lambda} = \begin{cases} 1 & \text{if } \phi > \phi^0; \quad \text{propagation.} \\ 0 & \text{if } \phi \leq \phi^0; \quad \text{elastic.} \end{cases} \tag{18}$$

Therefore, the equation for an infinitesimal increment of the damage tensor is obtained by:

$$d\Omega = \begin{cases} \frac{B_{ijkl}}{2\phi} R_{kl} dz & \text{if } \phi > \phi^0 \\ 0 & \text{if } \phi \leq \phi^0 \end{cases} \tag{19}$$

For a given loading process, the calculation of the conjugate force R_{ij} is:

$$R_{ij} = P^{+}_{ijkl} \tilde{R}_{kl} = P^{+}_{ijkl} (\frac{1}{2} \varepsilon^{e}_{ij} \frac{\partial D_{ijkl}}{\partial \Omega_{mn}} \varepsilon^{e}_{kl}) \tag{20}$$

Making integration over eqn. (19) can provide the increment in the damage tensor $\Delta\Omega_{ij}$ corresponding to a load increment.

4.2 Calculation of symmetric effecitve stress

To account for further calculations for the elastoplastic damage, the effective stress concept and the energy equivalence principle should be held between elasticity tensor of the damaged material and that of the fictitious undamaged material. The calculation of the symmetric effective stress of fictitious undamaged state is given as (Zheng & Betten 1996):

$$\boldsymbol{\sigma}^{effective} = \frac{\boldsymbol{\sigma}^{Cauch}}{(\mathbf{I} - \boldsymbol{\Omega})} \tag{21}$$

Then, in order to get a symmetric stress tensor, the following formulation is used:

$$\tilde{\sigma_{ij}} = \sqrt{\sigma^{effective}_{ij} \sigma^{effective}_{ji}} \tag{22}$$

where no summation for i and j.

Therefore we can make the operation of square root of damage tensor in the principal space of damage tensor and get:

$$\tilde{\boldsymbol{\sigma}} = (\mathbf{I} - \boldsymbol{\Omega})^{-\frac{1}{2}} \boldsymbol{\sigma}^{Cauchy} (\mathbf{I} - \boldsymbol{\Omega})^{-\frac{1}{2}} \tag{23}$$

For the stress $\boldsymbol{\sigma}^{Cauchy}$ which is calculated in the global coordinate system, it is necessary to make a go and back transfer with the eigen-matrix $\mathbf{T}$ of the damage tensor Ω. Then we get the effective stress tensor $\tilde{\boldsymbol{\sigma}}$ in global coordinate system:

$$\tilde{\boldsymbol{\sigma}} = \mathbf{T}(\mathbf{I} - \boldsymbol{\Omega})^{-\frac{1}{2}} (\mathbf{T}^T \boldsymbol{\sigma}^{Cauchy} \mathbf{T}) (\mathbf{I} - \boldsymbol{\Omega})^{-\frac{1}{2}} \mathbf{T}^T \tag{24}$$

5. APPLICATION TO A TUNNEL INTERSECTION IN JOINTED ROCK

The 3-km-long Schönberg Tunnel was newly built for the by-pass around the city of Schwarzach (Salzburg). The tunnel had to be constructed in heavily jointed phyllite rock. The construction was a combination of a TBM pilot tunnel and NATM for final construction. For the intersection between the main tunnel and one of the escape tunnels a three-dimensional model was used to simulate the impact of the jointed rock on the construction. The parameters of the rock mass are:

$$E_1 = 20.0MPa, \quad \mu_1 = 0.25 \tag{25}$$

and that of the shotcrete lining are:

$$E_2 = 20.0MPa, \quad \mu_2 = 0.2 \tag{26}$$

In the numerical model it was very important to very carefully simulate all construction steps from the pilot tunnel, to the crown excavation, to the escape tunnel and the final lining. The intersection was analyzed with two systems of jointed rock. The first joint system was a set of joints with the following orientation and initial damage:

$$\alpha = 90^o, \quad \beta = 45^o, \quad \gamma = 180^o, \quad \omega = 0.5 \tag{27}$$

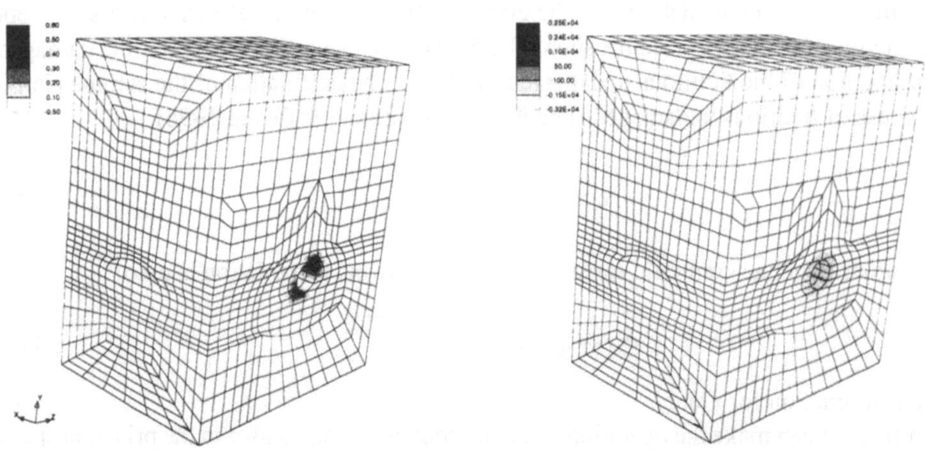

Fig. 2: Distribution of maximum principal active damage and effective stress: loading 1

Due to the initial ground stress all cracks are closed, which results in an active damage tensor, namely zero. Already after excavation of the pilot tunnel, as shown in Fig. 2, the cracks start to open, presented by the active damage tensor and the effective stress. The cracks open mainly in this part of the tunnel, where they are parallel to the tunnel surface. During construction of the main tunnel the damage region increases in the same direction, as shown in Fig. 3. The result of excavation of both tunnels is shown in Fig. 4, which shows that rock loosening appears

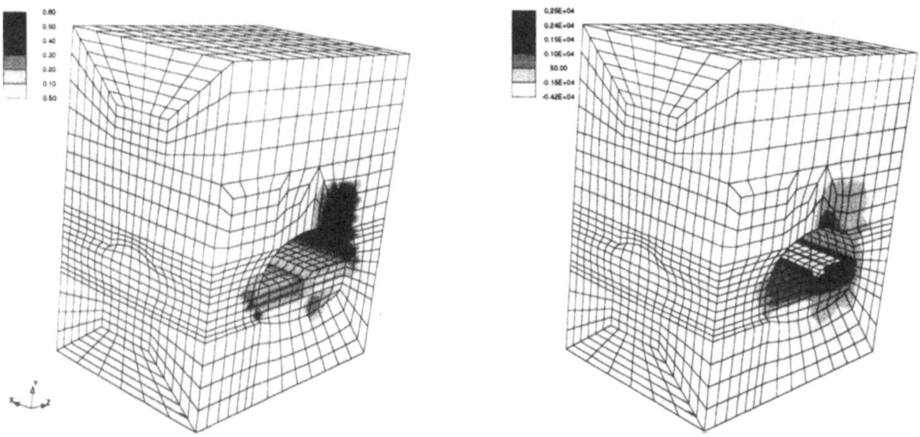

Fig. 3: Distribution of maximum principal active damage and effective stress: loading 4

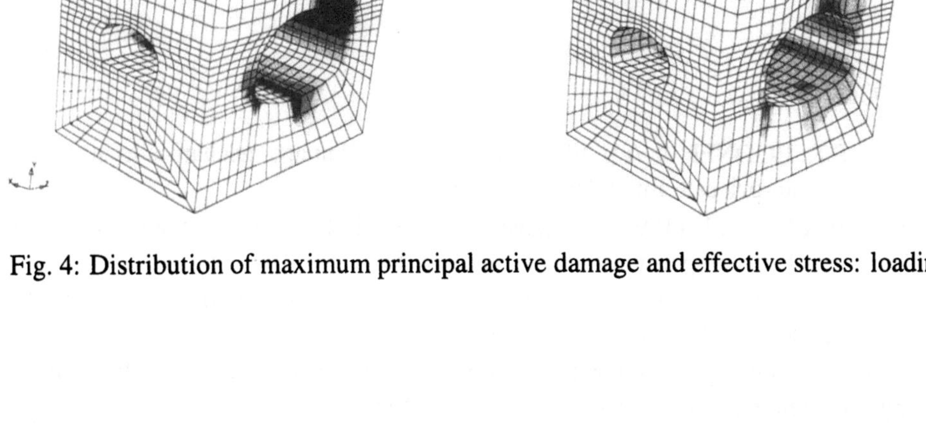

Fig. 4: Distribution of maximum principal active damage and effective stress: loading 10

in a great area around the tunnel. Damage propagation was not very serious; the maximum principal damage value increased from 0.5 to a value of 0.58. It is important to mention that a conventional elastoplastic calculation with the material parameters used above and the Mohr-Coulomb yielding criteria did not show any plastic region at all. The maximum damage and the maximum tensile effective stress in the main tunnel are not only close to the lining; but they are found at some distance from the excavation. This is well known by tunnel engineers; it is the so-called support ring found in the rock around the tunnel. The active damage zone is also very important for the tunnel engineer, namely as a support for laying out the anchoring system.

6. CONCLUSION

Determination of the initial damage tensor of rock mass is one of the principal difficulties in the application of CDM to geotechnical engineering. In this paper the initial damage tensor of rock mass is determined on the basis of the geological data of joint sets obtained by *in situ* measurements. The influence of closure of cracks under compression is considered by adopting the concept of *active damage tensor*. Degradation of the elasticity tensor resulting from damage evolution was investigated with the Finite Element Method. The damage evolution law was presented on the basis of thermodynamics and the endochronic theory. The calculation of the effective stresses ensure the results of damage propagation. The damage theory presented in this paper was applied in the mechanical analysis of the Schönberg Tunnel in Austria. The numerical results of damage distribution are reasonable and coincide well with *in situ* engineering experience.

ACKNOWLEDGMENTS

The work reported here was supported by the Austrian National Science Foundation, Fonds zur Förderung der wissenschaftlichen Forschung, project S08004-TEC.

REFERENCES

CHAN K.S., S.R. BODNER & A.F. FOSSUM (1997). A damage mechanics treatment of creep failure in rock salt, Int. J. Dama. Mech., 6: 121-152.

CHOW C.L. & X.F. CHEN (1992). An anisotropic model of damage mechanics based on endochronic theory of plasticity. Int. J. Fracture, 55: 115-130.

HALM D. & A. DRAGON (1996). A model of anisotropic damage by macrocrack growth; unilateral effect. Int. J. Dama. Mech., 5:384-402.

JU J. W (1989). On energy-based coupled elastoplastic damage theories: constitutive modelling and computational aspects. Int. J. Solids Structure, 25: 803-833.

KACHANOV L.M. (1958). Time of the rupture process under creep condition, IVZ Akad Nauk, S.S.R., Otd Tech Nauk, 8: 26-31.

KAWAMOTO T., Y. ICHIKAWA & T. KYOYA (1988). Deformation and fracturing behavior of discontinuous rock mass and damage mechanics theory. Int. J. Numer. Anal. Meth. Geomech., 12: 1-30.

KRAJCINOVIC D. (1983). Constitutive equations for damaging materials. J. Appl. Mech., 50: 355-360.

KRAJCINOVIC D., M. BASISTA & D. SUMARAC (1991). Micromechanically inspired phenomenological damage model. J. Appl. Mech., 58: 305-316.

LEMAITRE J. (1990). A Course on Damage Mechanics. 2nd Edition, Springer, Berlin.

MALVERN L.E. (1969). Introduction to the mechanics of a continuous medium. Prentice-Hall, Inc., Englewood Cliffs, N.J.

MURAKAMI S. & N. Ohno (1980). A continuum theory of creep and creep damage. Proc. 3rd IUTAM Symp. on Creep in Structures, Ponter, A. R. S. and Hayhust, D. R. Ed., Springer-Verlag, Berlin, 422-443.

ORTIZ M. (1985). A constitutive theory for the inelastic behaviour of concrete. Mech. Mater., 4: 67-93.

STUMVOLL M. & G. SWOBODA (1993). Deformation behaviour of ductile solids containing anisotropic damage. J. Engng. Mech., ASCE, 19: 1321-1332.

SWOBODA G., Q. YANG & F. ITO (1995). Damage propagation model and its application to rock engineering. Proc. 8th Int. Congress on Rock Engineering, Tokyo, 159-163.

VALANIS K.C. (1990). A theory of damage in brittle materials. Engng. Fracture Mech., 36: 403-416.

YAZDANI S. & S. KARNAWAT (1996). A constitutive theory for brittle solids with application to concrete. Int. J. Dama. Mech., 5: 93-110.

ZHENG Q.-S. & J. BETTEN (1996). On damage effective stress and equivalence hypothesis. Int. J. Dama. Mech., 5: 219-239.

A COMPARATIVE STUDY ON THE ASSESSMENT OF MECHANICAL PROPERTIES OF POROUS AND HETEROGENEOUS ROCKS BY VARIOUS AVERAGING METHODS

N. Tokashiki
Ryukyu University, Okinawa, Japan

Ö. Aydan
Tokai University, Shimizu, Japan

T. Kyoya
Tohoku University, Sendai, Japan

K. Sugawara
Kumamoto University, Kumamoto, Japan

ABSTRACT: There are various approaches available in literature to asses mechanical properties of porous and/or heterogeneous rocks. However, engineers are always confused which one to use for designing structures in rock masses since these approaches do not always provide unique values. In this article, we first review the fundamental concepts of various approaches for modelling porous and/or heterogeneous rocks as equivalent continua. Then, the applications of these approaches to rocks having various types of pores or inclusions are compared and their validity is discussed.

1 INTRODUCTION

Depending upon the formation process of rocks in nature, the geometry of pores or inclusions in rocks may vary from a simple spherical shape to very complicated slit-like shapes. Figure 1 shows some pictures of typical porous or heterogeneous rocks. Timur et al. (1971) investigated the geometrical shape of pores in sedimentary rocks and reported the existence of various shapes. Tokashiki et al. (1995) investigated the shape of pores in Ryukyu limestone on rock slices using the stereology technique. They concluded that the shape of pores in Ryukyu limestone is elipsoidal rather than circular or spherical. Nevertheless, this conclusion is only valid for Ryukyu limestone, similar investigations on other rocks are necessary before, making a generalization.

Clastic rocks and fracture zones, which consist of hard inclusions with soft matrix, are good examples for heterogenous media. Clastic rocks are generally found in volcanic regions while fracture zones, which are found in tectonically disturbed regions of the earth's crust, consist of soft clayey matrices with hard fragments from adjacent parent rocks.

The increase of constructions of rock engineering structures in/on this type of rocks makes necessary to understand the thermo-hydro mechanical behaviour of such rocks. Therefore] modelling porous and/or heterogeneous rocks has been recently receiving a particular interest among rock mechanicians and rock engineers. There have been mainly three approaches for modelling such rocks, namely, *Experimental approach*, *Discrete modelling approach*, and *Equivalent continuum modelling approach.*

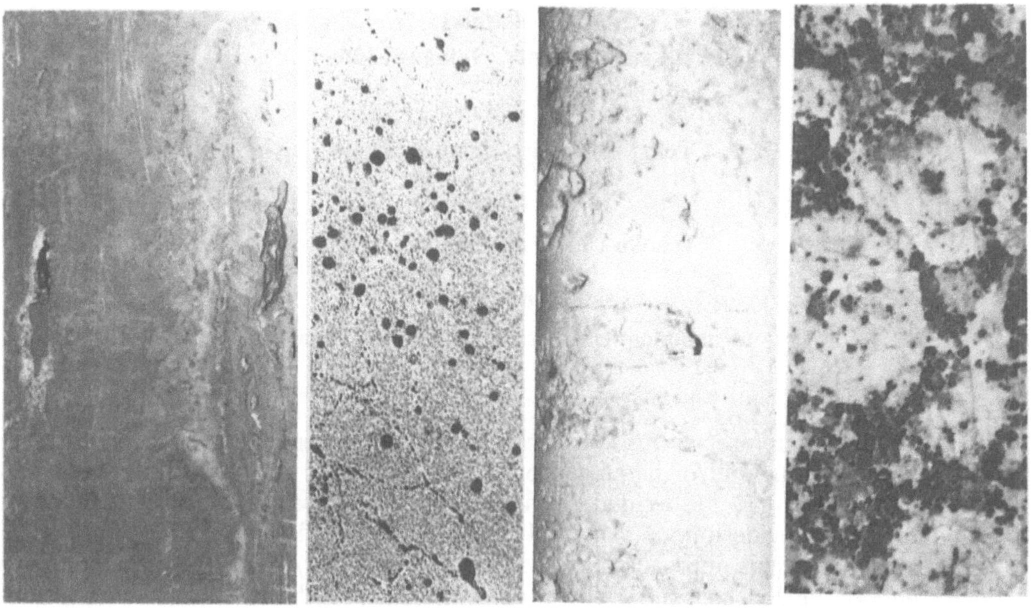

Figure 1 Pictures of some porous and heterogeneous rocks

Experimental approach is probably the simplest and easiest approach to assess mechanical properties of porous and/or heterogeneous rocks. As long as specimen sizes are appropriate in comparison with pore sizes, this approach provide the most reliable values for mechanical properties. Since such rocks are considered to be equivalent continua in macroscopic scale, it is difficult to assess the mechanical behaviour in a microscopic scale, which may be very important in long-term behaviour. Furthermore, the variation of pore or inclusion geometry and quantity may require a very large number of tests, which may increase the cost of testing.

Discrete modelling approach, although, is not common, pores or inclusions are modelled explicitlyas inclusions. For simple pore or inclusion geometries and loading conditions, one may obtain closed form solutions (Aydan 1997). However, when the geometry and loading conditions are complicated, numerical methods are generally used (Tokashiki et al. 1996). This approach is generally used to check the validity of equivalent continuum approaches to be explained in detail in the next section.

Equivalent continuum modelling approach starts with the first attempts by Voigt (1910) and Reuss (1929) for multiphase materials. Since then many models are proposed and the models applicable to porous and/or heterogeneous rocks in rock mechanics are ***Mixture theory, Micro-mechanics models, Micro-structure models, and Homogenization technique.***

In this article, we first review the fundamental concepts of the models in the category of ***equivalent continuum modelling approaches*** and discuss their formulations. Then, the estimations of equivalent mechanical properties of porous rocks by these approaches are compared with each other as well as with those from experiments and discrete modelling approaches by considering some fundamental examples and their validity is discussed.

2 MAIN CHARACTERISTICS OF VARIOUS EQUIVALENT CONTINUUM MODELS

The main characteristics of the equivalent continuum modelling approaches can be summarised as follows:

1-) Pore or inclusion geometry is known and pores are distributed in rock matrix in a periodic manner so that it is possible to define a representative elementary volume (REV).

2-) Pores are mechanically modelled as inclusions having zero stiffness when they are filled with air. If there is fluid type inclusions in pores, the volumetric response of inclusion material is taken into account only.

3-) The constitutive law of rock matrix is generally assumed to be linear elastic.

4-) The stress tensors acting on pores or inclusions and rock matrix are associated with the tractions and displacement along the boundary of the REV through some tensorial influence operators.

5-) The tensorial influence operators associate the macroscopic constitutive equation with those of rock matrix and pores or inclusions.

2.1 Mixture theory

This model is the earliest equivalent model and is the most widely used approach in multi-phase modelling. The theory has its rootes in the models of Voigt (1910) and Reus (1929). It is quite surprising that the actual properties of porous and/or heterogeneous materials are bounded by those obtained from these two models. Therefore, they are also known as ***bounding solutions.*** Tensorial influence operators are isotropic tensors associated with the volumetric fraction of inclusions. In the case of porous rocks, the volumetric fraction corresponds to porosity. In deriving specific forms of tensorial influence operators, REV is assumed to be subjected to either isostrain or isostress (Figure 2). The former is known as Voigt's model (Voigt 1910) and the latter is known as Reus' model (Reus 1929). When pores are empty, Reus' model could not be used for porous rocks.

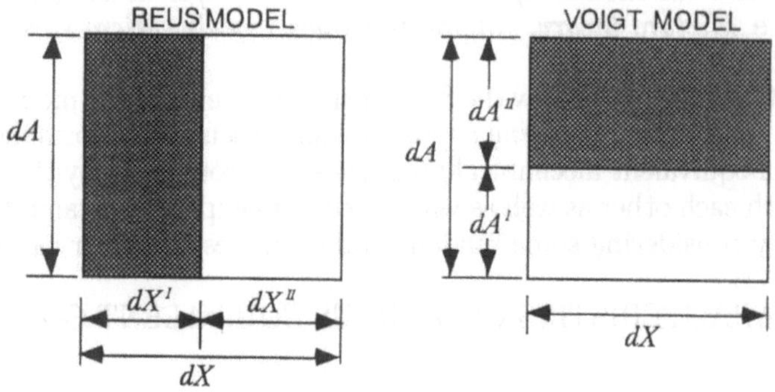

Figure 2 One dimensional illustration of mechanical concept of mixture models

2.2 Micro-mechanics Model (MMM)

Following the mixture theory, Eshelby (1957) introduced S-transformation tensor to represent the mechanical effect of elipsoidal inclusions. The size of inclusions was assumed to be small in comparison with the size of REV. This model is known as ***dilute solutions*** and they utilize elastic solutions for elipsoidal inclusions embedded in infinite elastic domains. As the interactions between inclusions are assumed to be negligible, the estimations given by this model become unacceptable as the volume fraction of pores increases.

Following ***dilute solutions***, solutions for non-dilute concentrations are proposed. They are mainly classified into two classes:

1-) ***Mori-Tanaka Method:*** This method is based on Eshelby's S-tensor and the shape of pores or inclusions is assumed to be spheroidal (Mori-Tanaka 1973).

2-) ***Self-consistent Method:*** This method is based on the method of Hershey (1954) and Kröner (1958), originally proposed for aggregates of crystals. It is employed to porous and cracked materials (i.e. Hill 1963, Budiansky & O'Connel 1976).

As the interactions between inclusions are taken into account, these solutions can handle with 100 % volumetric fractions of pores (inclusions). Therefore estimations given by these models are acceptable as compared with those given by ***dilute solutions.***

2.4 Micro Structure Models

Aydan et al. (1992,1996) proposed two models for discontinuous rock masses based on the micro-structure theory of mechanics (Jones 1975). Stress and strain fields of each constitutient are related to each other using two concepts, namely, ***Globally Series and Locally Parallel Model: GSLPM*** and ***Globally Parallel and Locally Series Model: GPLSM.*** One dimensional illustrations of these models are shown in Figure 3. The constitutive law is not restricted to elasticity and it can be any kind which can describe the mechanical response of inclusions and rock matrix. This model can also be easily extended to estimate permeability, heat conductivity and diffusion coefficients of porous and/or heterogeneous rocks.

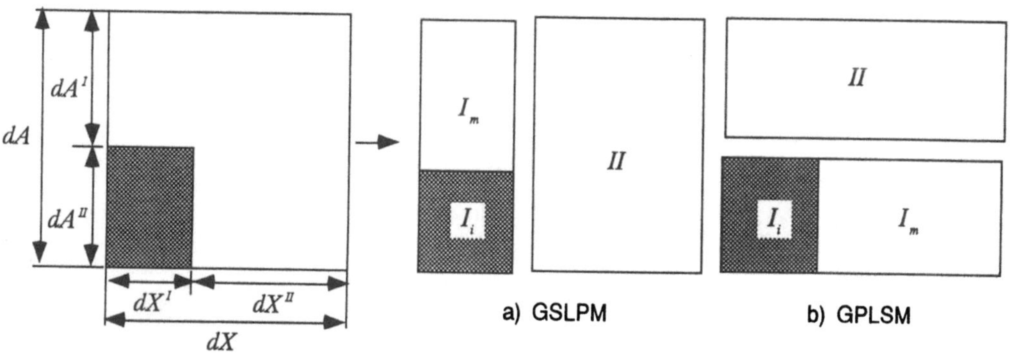

Figure 3 One dimensional illustration of mechanical concept of micro-structure models

2.5 Homogenization Technique

The homogenization technique was mainly used to obtain the equivalent characteristics of composites (Bakhvalov & Panasenko 1984, Sanchez-Palencia 1980). It has been recently applied to soils (Auriault 1983) and rocks (Kawamoto and Kyoya 1993). Stress and strain fields of constitutients are obtained from a perturbation of displacement field. An influence tensor, which is a gradient of 6 vectorial functions called ***characteristics deformation functions*** for a given representative elementary volume (unit cell) is used to establish relations between the homogenised elasticity tensor and those of constitutients. Except very simple cases, the equivalent parameters are obtained using a numerical method such as FEM.

3 COMPARISONS AND DISCUSSIONS

3.1 Elastic Modulus

In the first example, a body with a periodic cylindrical cavities is analysed. Figure 4 shows the calculated results by various methods including an exact solution (Aydan 1997) together with experimental results. As it is noted from the figure, the mixture model(MM) gives the upper bound while the GPLSM yields the lower bound. It is interesting to note that the estimations given by homogenization technique are inbetween those from the GPLSM and GSLPM. The results by the FEM analysis, on the other hand, are close to those by the GPLSM. While the results obtained by the homogenization technique and the FEM analysis lie in the region bound by the solutions provided by the both model, the experimental results are even less than those obtained from the GPLSM. Estimations by dilute and non-dilute solutions of micro-mechanics theory are not close to experimental results. Particularly, the estimations from dilute solutions are unacceptable when the volume fraction exceeds 0.1.

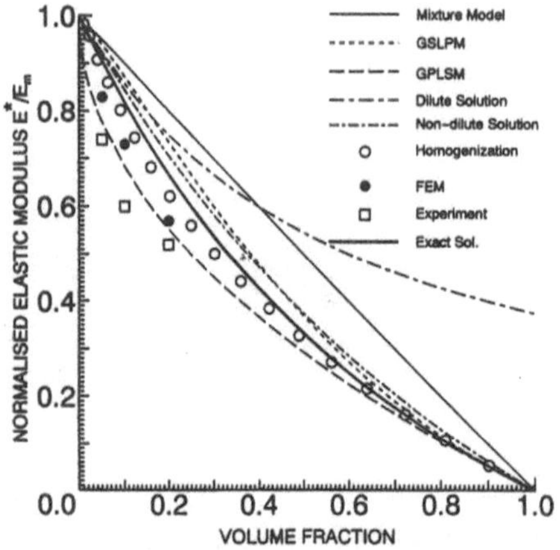

Figure 4 Comparison of estimations of normalised elastic modulus by various models with experiments (cylindrical pores)

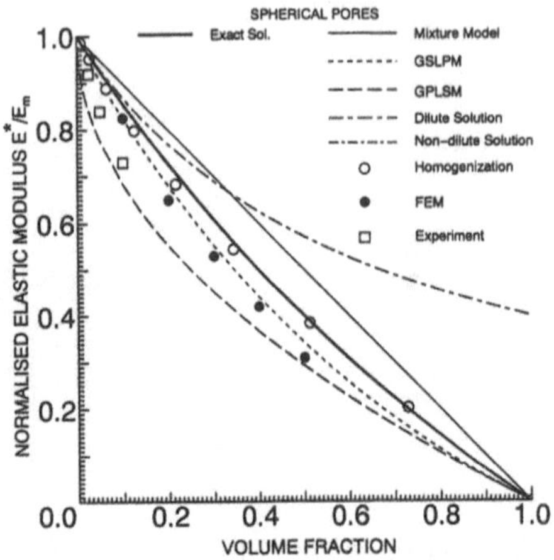

Figure 5 Comparison of estimations of normalised elastic modulus by various models with experiments (spherical pores)

Figure 5 compares the results obtained by various averaging techniques including an exact solution for a body with spherical pores (cavities). In this particular example, it is again noted that the mixture model yields the upper bound while the GPLSM does the lower bound. The GSLPM and the homogenization technique give almost the same results. Estimations from homogenization technique and non-dilute solutions are very close to each other. Experimental results plotted in the figure are obtained by Ito et al. (1990), who carried out tests on mortar samples having spherical cavities created by ver soft spherical rubber inclusions. The results by the exact solution coincides with those by the non-dilute solution of micromechanics. In this particular case, the averaged elastic moduli obtained from experiments are close to estimations by the GPLSM. The other approaches underestimates the effect of pores.

5appropriate to be used with pores which have an anisotropic geometry.

An interesting series of tests were carried out by Sakurai et al. (1991). In the tests, they considered two cases:

Case 1: Matrix is hard and inclusions are soft

Case 2: Matrix is soft and inclusions are hard

The ratios of moduli were approximately 20/1 and 1/20 for each respective case. Tests results together with predictions by various approaches are shown in Figures 6 and 7. It is of great interest that the test results are very close to predictions. Nevertheless, the predictions by the mixture approach, the homogenization technique and the GSLPM model are greater than those measured in experiments.

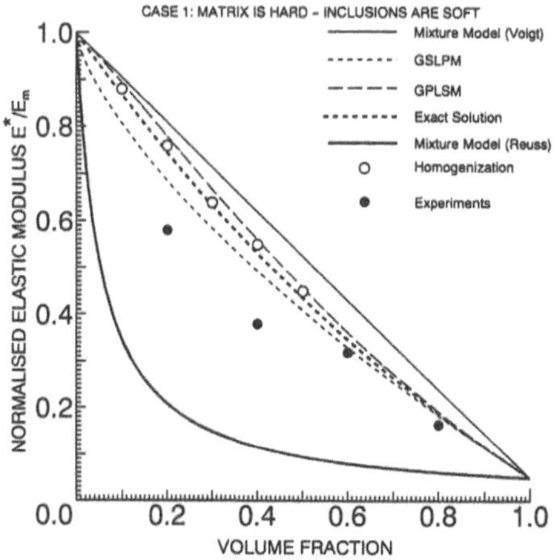

Figure 6 Comparison of estimations of normalised elastic modulus by various models with experiments (Case 1)

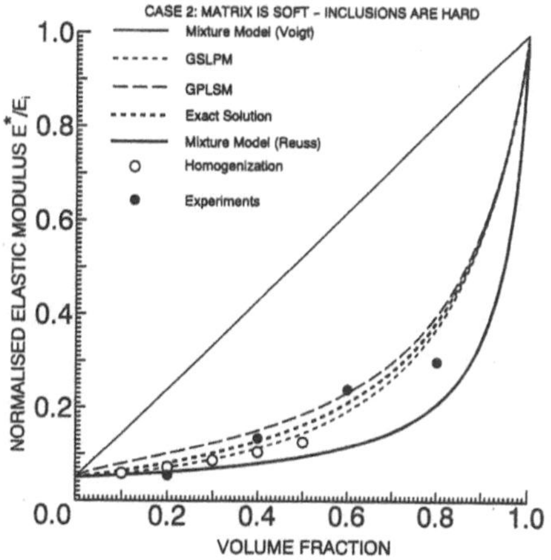

Figure 7 Comparison of estimations of normalised elastic modulus by various models with experiments (Case 2)

3.2 Poisson's Ratio

An example of application of the homogenization method and the micro-structure theory is made to evaluate the effect of volume fraction of pores in Ryukyu limestone on its Poisson's ratio. Figure 8 show the effect of volumetric fraction of pores on the equivalent Poisson's ratio of the limestone. As seen from Figure 8, experimental results scattered on both sides of the predicted curves for various Poisson's ratios. The scatterness may be attributed to the non-periodic character of the natural materials.

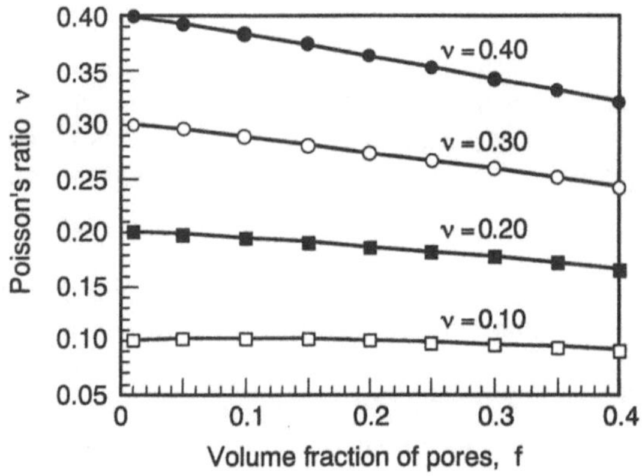

Figure 8 Variation of equivalent Poisson's ratio estimated from homogenization technique

3.3 Uniaxial Compressive Strength

Figure 9 shows the calculated results for average uniaxial strength of mortar samples containing cylindrical pores by various methods together with experimental results by assuming that the failure of porous rocks is brittle. The reduction of uniaxial strength of mortar computed from various methods are the same as that of equivalent elastic modulus. In this particular example, the experimental results are somewhat close to estimations given by the mixture model, the GSLPM of micro-structure theory and non-dilute solutions of micro-mechanics theory.

Figure 10 compares the results obtained by various averaging techniques for a body with spherical pores (cavities). In contrast to the experimental results in the previous case, the reduction of compressive strength of mortar samples is very drastic. It seems that the models mentioned in the figure is incapable of estimating the experimental results. Such a discrepancy may be resulting from non-uniform distribution of spherical cavities in experiments and stress concentrations in samples which are not taken into account in the averaging methods.

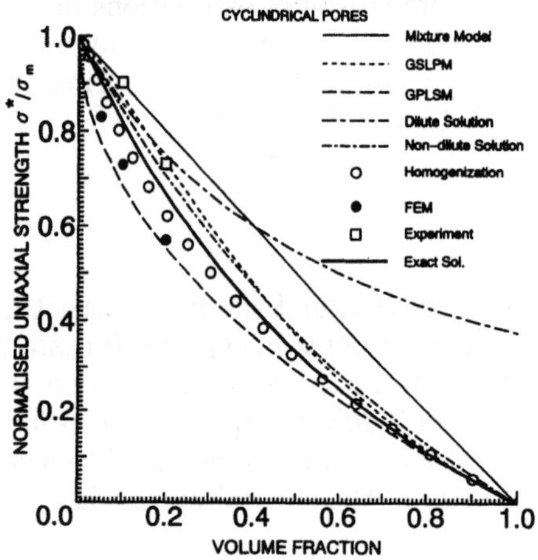

Figure 9 Comparison of estimations of normalised uniaxial strength by various models with experiments (cylindrical pores)

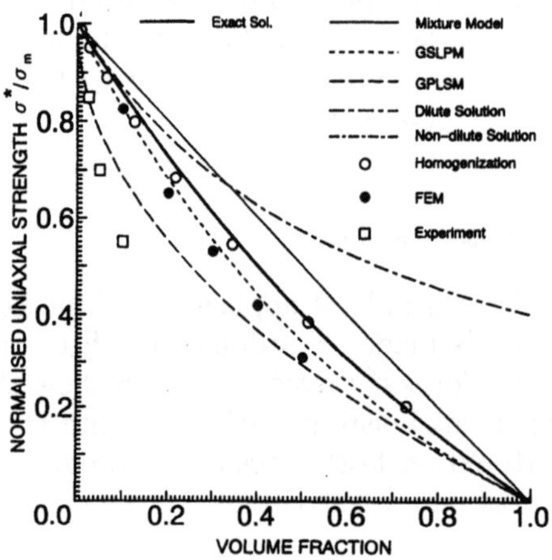

Figure 10 Comparison of estimations of normalised uniaxial strength by various models with experiments (spherical inclusions)

Tests results reported by Sakurai et al. (1991) and quoted in Figures 6 and 7 are shown together with predictions by various approaches are shown in Figures 11 and 12. It is quite interesting that the test results are very close to predictions. Nevertheless, the predictions by the mixture approach, the homogenization technique and the GSLPM model are slightly greater than those measured in experiments.

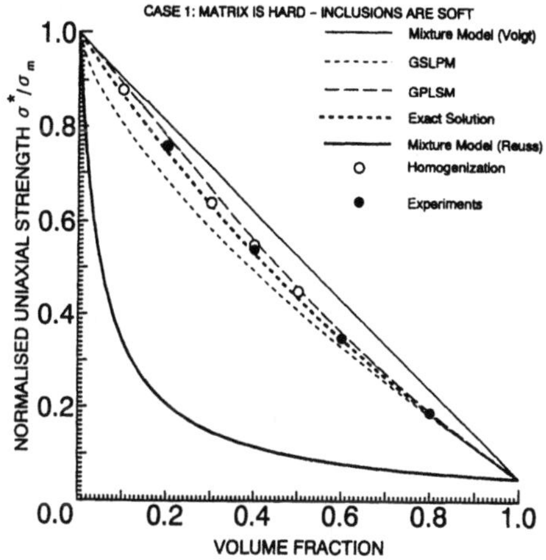

Figure 11 Comparison of estimations of normalised uniaxial strength by various models with experiments (Case 1)

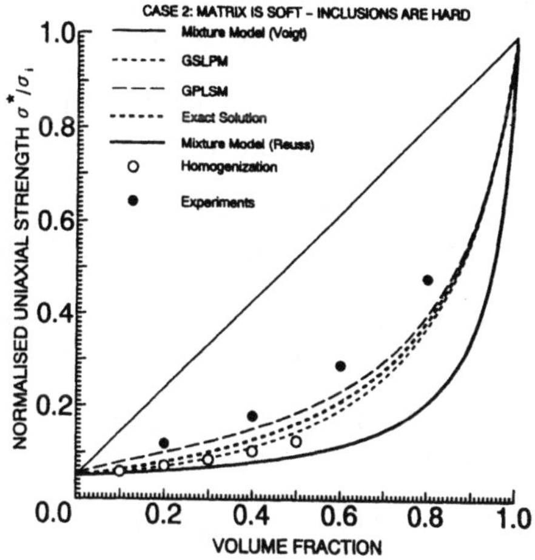

Figure 12 Comparison of estimations of normalised uniaxial strength by various models with experiments (Case 2)

4 CONCLUSIONS

Available averaging techniques have been applied to assess the elastic coefficients and strength of porous and heterogenous rocks. From the comparisons presented in the previous section, it may be said that none of equivalent models can exactly predict the mechanical characteristics of such rocks. Particularly, dilute solutions of micromechanics models is not recommended to be used when the volume fraction of pores or inclusions exceed 0.1. The equivalent modelling approaches generally provide either upper or lower bound solutions to actual characteristics of porous and/or heterogeneous rocks. Furthermore, the equivalent models may be cumbersome to use when the mechanical behaviour of rock masses deviates from the linear elasticity.

REFERENCES

AURIAULT, J.L. (1983). Homogenisation: application to porous saturated media. Summer School on Two-Phase Medium Mechanics, Gdansk.

AYDAN, Ö. (1997). Pressure dependency of porosity and permeability of rocks. Bulletin of Rock Mechanics, Ankara, TSRM, **13**, 37-52.

AYDAN, Ö., N. TOKASHIKI, T. SEIKI (1996). Micro-structure models for porous rocks to jointed rock mass *APCOM'96*, 3, 2235-2242.

AYDAN, Ö., T. AKAGI, Y. SHIMIZU, T. KAWAMOTO (1997). Test for mechanical properties of fracture zones, *ARMS'97*, Seoul.

BAKHVALOV, N. & G. PANASENKO (1984). Homogenization: Averaging processes in periodic media, Kluwer Academic Pub.

BUDIANSKY, B., R.J. O'CONNEL (1976). Elastic moduli of a cracked solid. Int. J. Solids & Structures, 12, 81-97.

HERSHEY, A.V. (1954). The elasticity of an isotropic aggregate of anisotropic cubic crystals. *J. Appl. Mech.*, **21**, 236-240.

HILL, R. (1963). Elastic properties of reinforced solids: Some theoretical principles. J. Mech. Phys. Solids, Vol. 11, pp. 357-372.

JONES, R.M. (1975). Mechanics of composite materials, Hemisphere Pub. Co., New York.

KAWAMOTO, T., T. KYOYA (1993). Some applications of homogenization method in rock mechanics. Seminar on Impacts of Computational Mechanics to Engineering Problems, Sydney.

KRÖNER, E. (1958). Berechnung der elastischen Konstanten des Vielkristalls aus den Kanstanten des Einkristalls. **Z. Phys.**, **151**, 402-410.

LITEWKA, A., A. SAWCZUK (1984). Experimental evaluation of the overall anisotropic material response on continuous damage. Mechanics of Material Behaviour, ***Studies in Applied Mechanics***, Vol. 6, 239-252.

REUSS, A. (1929). Berechnung der Fliessgrense von Mischkristallen auf Grund der Plastizitätsbedingung für Feinkristalle. Z. Agnew. Math. u. Mech., Vol. 9, 49-58.

SANCHEZ-PALENCIA, E. (1980). Non-homogenous media and vibration theory. Lecture Note in Physics, No. 127, Springer, Berlin.

TİMUR, A., W.B. HEMPKINS, R.M. WEINBRANDT (1971). Scanning electron miroscope study of pore systems in rocks. *J. Geophy. Res.* **76**, 4932-4948.

TOKASHIKI, N., T. KYOYA and Ö. AYDAN (1995). A research on modelling porous rocks. *Symp. on Num. Methods in the design and development of Agricultural Engineering Machinery*, Agricultural Eng. Machinery Soc. of Japan, Okinawa, 65-79.

VOIGT, W. (1910). Lehrbuch der Kristallphysik, Teubner, Berlin.

3. NUMERICAL ALGORITHMS AND THEORETICAL ASPECTS

CREEP-LIKE DELAYED FAILURE OF CLAYEY GROUND AFTER THE END OF EMBANKMENT CONSTRUCTION

A. Asaoka, T. Noda, G.S.K. Fernando and E. Yamada
Nagoya University, Nagoya, Japan

ABSTRACT: Even in normally or lightly overconsolidated clay foundations, failures have been observed some days/weeks after the end of the embankment construction. The clay foundation stays apparently stable just after the completion of construction, but with the passage of time instability develops gradually instead of expected consolidation and it may undergo catastrophic failure in an unexpected moment. In this paper such a creep-like delayed failure of a homogeneous normally consolidated clay foundation is investigated through the soil-water coupled finite deformation computation with the use of the invicid subloading surface Cam-clay model.

When the embankment loading is stopped at a point near the limit load, the foundation may experience failure with the elapse of time. The occurrence of delayed failure is due to the continued softening of soil that causes the increase of excess pore pressure. With time excess pore pressure isochrones expand until failure. When the end of construction load level is comparatively low, only the consolidation process proceeds.

1.0 INTRODUCTION

In the classical description of the behavior of normally consolidated or lightly overconsolidated clay foundation to embanking, a foundation is said to be safe if it was stable at the end of load application. After the end of embankment construction the foundation is expected to undergo consolidation that would result a higher factor of safety. By contrast there have been observed some failures of soft clay foundations (e.g. Changi airport 1995, Crawford et al, 1995, Eide and Holmberg, 1972) some days/weeks after the end of construction. The foundation seems apparently stable just after the completion of construction, but with the passage of time instability develops gradually instead of expected consolidation process and it may undergo catastrophic failure in an unexpected moment. In heavily overconsolidated clays this type of failures are somewhat common (Skempton, 1970) and known as the creep failures. As reported by Crawford et al (1995), Eide and Holmberg (1972), in soft clays foundations, measured excess pore pressures in some regions in the foundation tend to increase even after the loading was finished and eventually reached failure. Also complete failure of the structure had occurred within a very short period after a first crack appeared. The usual stability calculations for those failed embankments had confirmed safety with a considerable margin (Eide and Holmberg (1972)).

Naturally such delayed failures are attributed to creep behavior of the clay which is commonly determined through creep tests that consider the soil as a single soil element. Therefore, these problems are mostly approached using elasto-visco-plastic constitutive models. However, the elasto-plastic soil-water coupled finite deformation approach (Asaoka et al, 1994, 1995, 1997), in which the problem is considered as an initial boundary value problem, has shown success in describing some naturally observed phenomena relating to saturated clays such as time rate effects, hardening/softening etc. As a matter of fact these rational descriptions are based on the critical state soil mechanics framework. In this paper we will discuss the occurrence of delayed failure of an embankment in soft clay foundation only using the elasto-plastic soil-water coupled finite deformation analysis.

2.0 SOIL-WATER COUPLED FINITE DEFORMATION COMPUTATION

The analysis in this research was based on the soil-water coupled finite deformation theory which considers both the material non-linearity and the geometric non-linearity. (The comp--lete mathematical formulation can be found in our previous publications; Asaoka et al, 1994, 1995, and, 1997), The subloading surface Cam-clay model (for details ref. Asaoka et al. 1997), which can adequately represent the behavior of all the states of soil (from normally consolidated state to any overconsolidated state), was taken as the soil model.

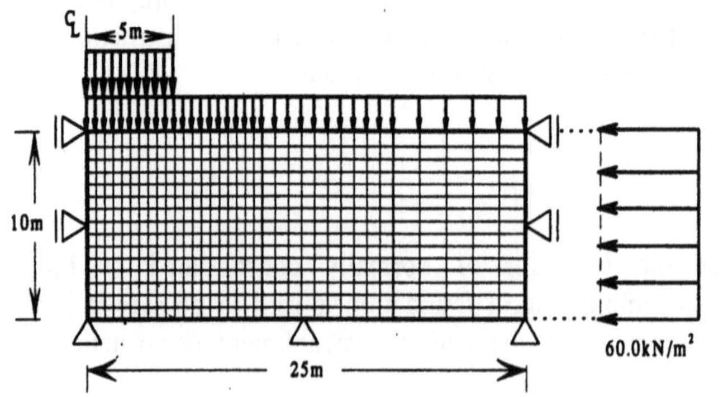

Fig. 1 Domain for the analysis and initial conditions

Compression index λ	0.131
Swelling index κ	0.016
Critical state parameter M	1.53
Poisson's ratio ν	0.30
Initial specific volume v_0	2.038
Initial mean pressure p'_0 (kN/m^2)	60.0
Permeability k (m/day)	3.2x10^{-5}

Table 1 Material Parameters

The clay foundation was taken to be 25m wide and 10m deep and was assumed to be at normally consolidated state with the isotropic preconsolidation pressure of 60kN/m^2. For simplicity, the clay was taken as weightless and thus having the rectangular shaped initial vertical effective stress distribution as illustrated in Fig. 1. Also the half loaded area considered in this plane strain model was 5m in width similar to a long embankment with no slope. The top and bottom boundary of the domain was set to be permeable and the load was applied at different rates firstly to determine failure loads. This is in fact the "build up to failure". As for the occurrence of delayed failure the coupled analysis was performed again by stopping the load application at different load levels which were less than the limit load. Table 1 shows the Kawasaki clay parameters used in the analysis.

3.0 LIMIT LOAD OF THE FOUNDATION

The response of a saturated clay foundation to an external load application largely depends on the drainage conditions within the clay mass which in fact related to many factors. In a normally consolidated clay foundation, when loading is continued under impeded drainage the foundation reaches failure at low loading magnitudes. When partial drainage within the soil mass is possible at least to a certain degree during load application, then the load bearing capacity of the foundation should naturally reach somewhat higher value.

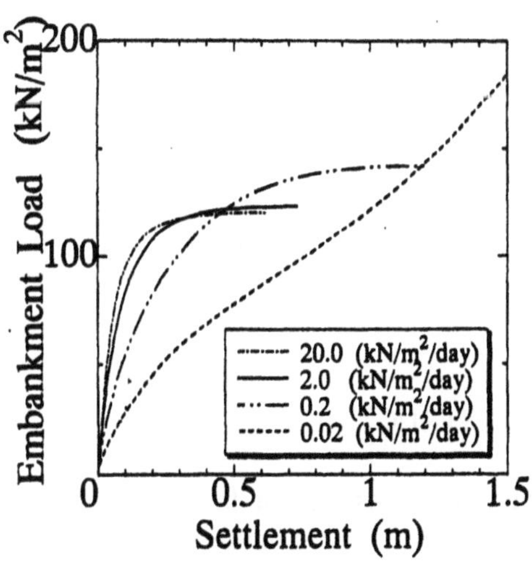

Fig. 2 Load~settlement behavior under different loading rates

Fig. 2 shows the computed load-settlement behavior for different loading rates in which the top and bottom boundaries were set to be permeable. The settlement axis represents the surface settlement at the center of the loaded area. When the load application is done at slower rates, *e.g.* 0.2kN/m^2/day, the foundation can sustain higher loads. At a very slow rate, *e.g.*0.02kN/m^2/day, the bearing capacity ever increases; this is in fact the numerical result showing ever hardening soil elements. The loading rate dependency on load bearing capacity diminishes at higher loading rates; in this example it is somewhere near 20kN/m^2/day, which is equivalent to 1m high soil layer-

-placement per day.

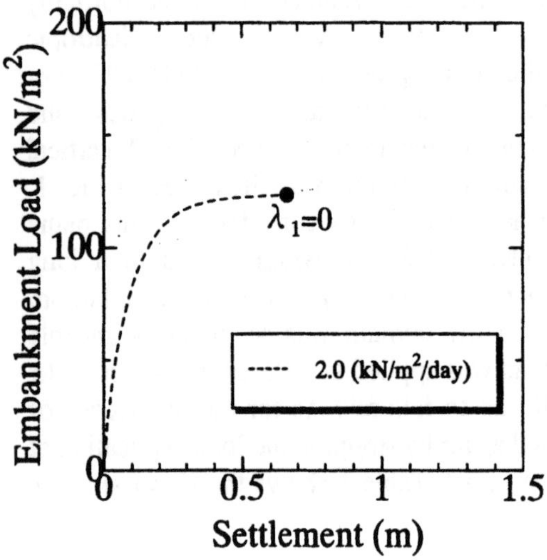

Fig. 3 Zero eigen position

The eigen value analysis of the incremental tangent stiffness matrix that was performed for those cases, (by writing the incremental equilibrium equation in the form $K\Delta\boldsymbol{u} = \Delta\boldsymbol{f}'$ (Asaoka et al, 1995) in which K is the non-symmetric tangential stiffness matrix obtained by disregarding the corresponding terms for known displacement boundary and $\Delta\boldsymbol{f}'$ the equivalent load increment vector effective to incremental displacement vector $\Delta\boldsymbol{u}$) showed that when the load-settlement curve became flat one eigen value was reaching zero. Also at the same time the components of $\Delta\boldsymbol{f}'$ are approaching zero. This indicates that the clay foundation is reaching the stationary condition or the limit load under the particular applied loading rate.

The zero eigen position is indicated in the load-settlement curve in Fig. 3. For comparison the velocity vector field and eigen vector field at the limit load point are plotted in Figs. 4 and 5 respectively. The similarity of these two figures confirm that deformation mode at the limit load is as of the eigen mode.

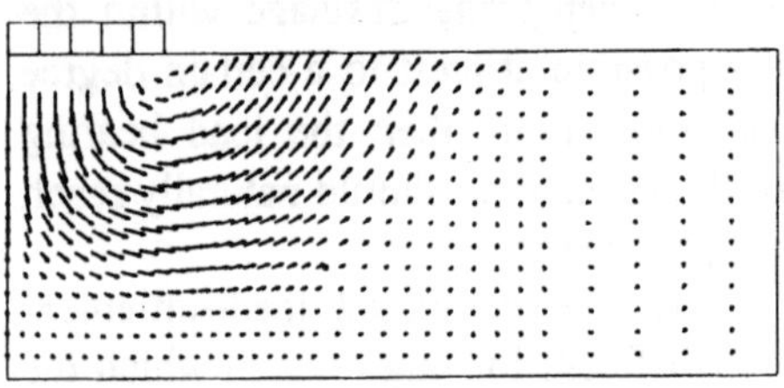

Fig. 4 Velocity vectors at limit load

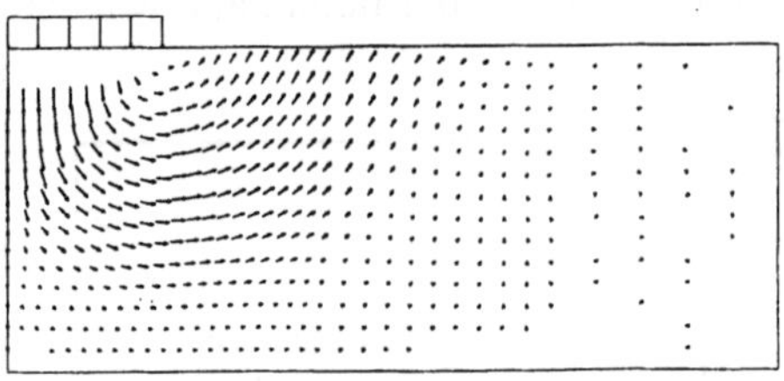

Fig. 5 Eigen mode at limit load

4.0 DELAYED FAILURE AFTER THE END OF CONSTRUCTION

Loading up to failure, as described above is the most familiar case for the failure problems. However, as observed in-situ, failure of a foundation may occur some days/weeks after the end of load application. Excavated slopes or natural slopes in heavily overconsolidated clays show delayed failures quite often and such occurrences are usually attributed to creep failures in which the clay is said to be reached fully softened shear strength. As per the classical interpretation, such creep like failures have mainly been solved considering the viscosity as inherent in the soil skeleton . However, in this paper we will interpret the occurrence of such delayed failure, specifically in a normally consolidated clay foundation as due to pore water migration of softening/hardening soil by solving a soil-water coupled initial boundary value problem.

4.1 Creep-like delayed failure in a normally consolidated clay foundation

In this case we proceeded the coupled deformation analysis by keeping the applied total load constant at different magnitudes. Figs. 6 and 7 illustrate the result for the case of 2.0kN/m^2/day loading rate when load application was held at 123.16kN/m^2 and 120kN/m^2. It is to be noted that the 123.16kN/m^2 load level was very much close to the limit load (for 2.0kN/m^2/day loading rate) as in Fig. 3. Naturally one may expect consolidation to proceed; *i.e.* dissipation of the excess pore pressure, due to top and bottom drained boundaries once the loading is stopped at these magnitudes. By contrast, when the loading was stopped at 123.16kN/m^2, the foundation experienced sudden failure at 8.56th day after the end of load application (Fig. 7). However, if loading was stopped at 120kN/m^2 level, in this example, foundation followed consolidation. The velocity vectors at same instances for these two

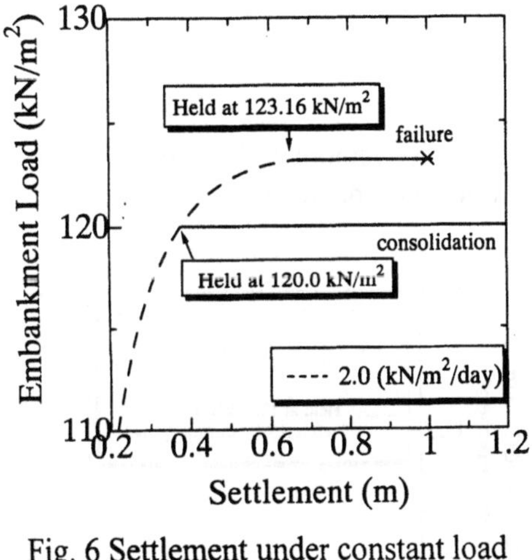

Fig. 6 Settlement under constant load (Loading rate=2.0kN/m^2/day)

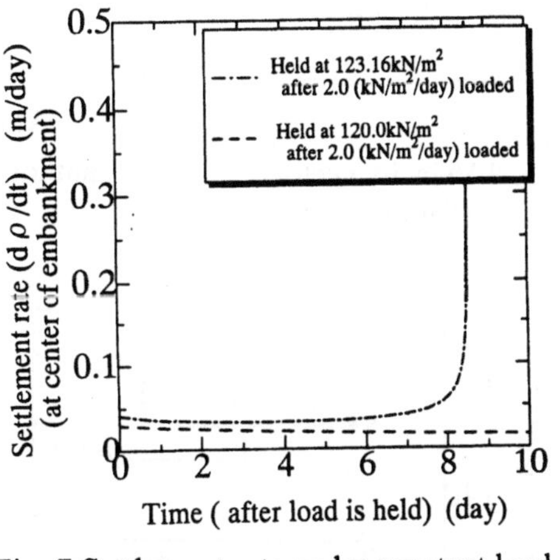

Fig. 7 Settlement rate under constant load (Loading rate=2.0kN/m^2/day)

load levels are shown in Figs 8 and 9. As can be seen in Fig. 8 foundation was apparently stable until the catastrophic failure at 8.56th day after the end of construction. (In fact foundation undergoes settlement at a constant rate until failure). Similar delayed failure was observed even for slower loading rates (Figs. 10 and 11) such as 0.2kN/m^2/day, when loading was stopped close to the limit load. In this case the total load at the end of construction was higher compared with the higher loading rate (2.0kN/m^2/day). For a

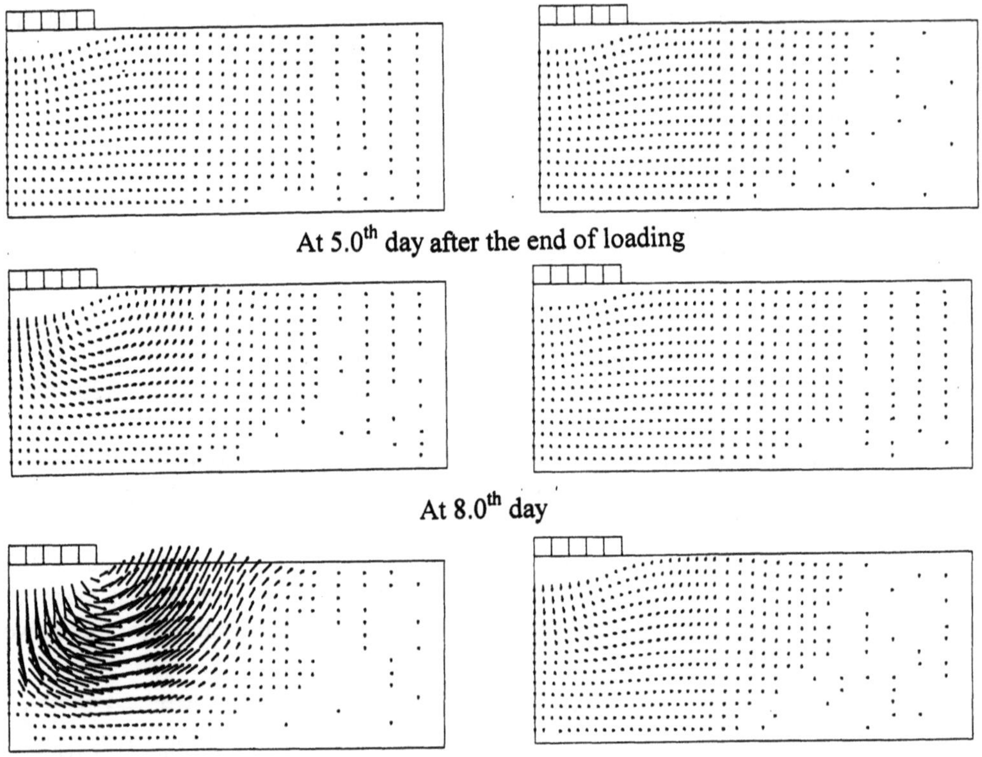

Fig. 8 Velocity vectors under constant load (Failure case)

Fig. 9 Velocity vectors under constant load (Non-failure case)

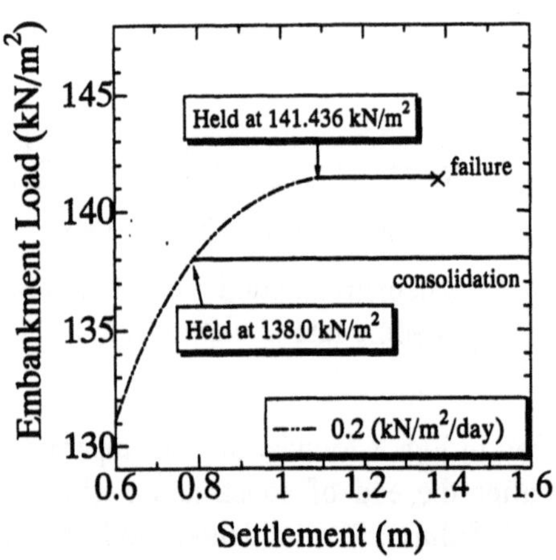

Fig. 10 Settlement under constant load (Loading rate=0.2kN/m²/day)

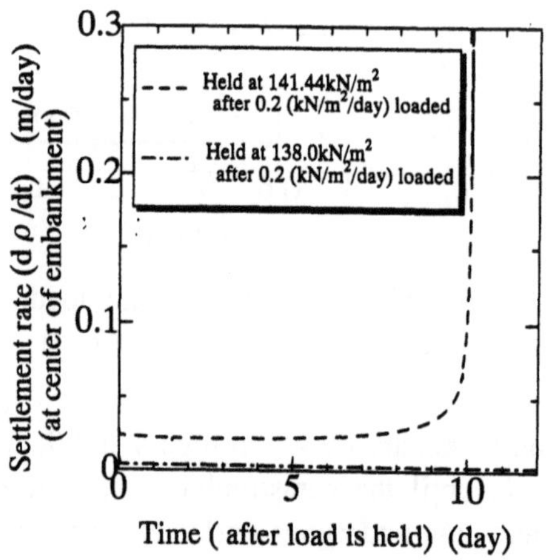

Fig. 11 Settlement rate under constant load (Loading rate=0.2kN/m²/day)

slower loading rate case foundation stays apparently stable longer period (compared to higher loading rates) until the occurrence of sudden failure. In this example, as in Fig. 11, failure occurred at 10.2nd day after the end of construction. These numerical results are in similarity as reported by some researchers (*e.g.* Crawford et al, 1995, Eide, O. and Holmberg, S., 1972) on embankment failures that occurred some days after the end of construction. In fact, in these cases failure had occurred within a very short time lag after observing the first signs of the initiation of failure. The question now remained to be answered is how such a behavior could happen in saturated normally consolidated clays.

Even under constant external load after the completion of construction, soil elements in some local regions in the foundation, specially centerline region in these examples, can be at critical state under shear and due to the surrounding pore pressure field those elements may undergo softening. This should in fact increase the excess pore pressure of those elements. (As per the Terzaghi's one dimensional consolidation theory analogue, softening should result a negative stiffness; minus C_v, which should result positive rate of change of excess pore pressure due to concaveness of isochrones). If nearby elements are also already near/at critical state, it is possible the softening process continue to those elements too depending on the current seepage gradient. The examination of stress states of elements revealed that most of the elements in the centerline region of the foundation were either at

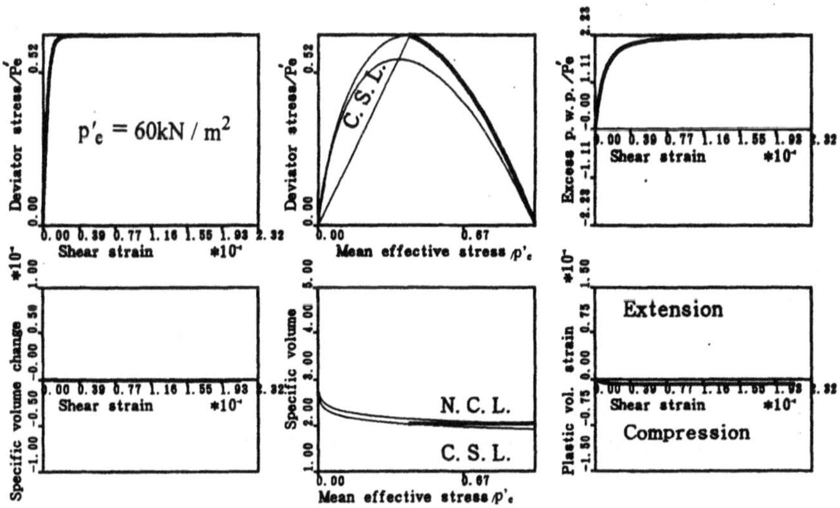

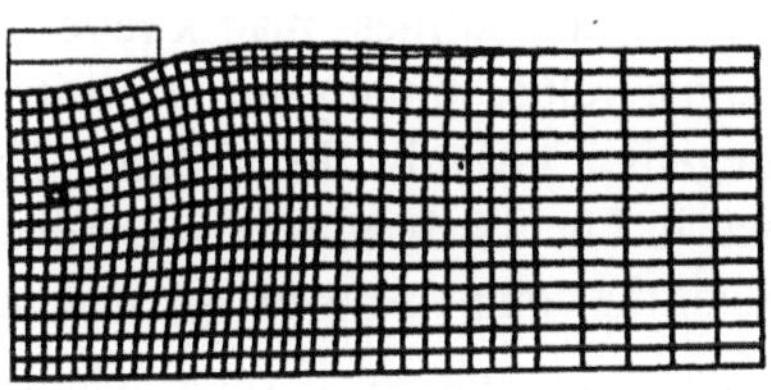

Fig. 12 Element-wise soil behavior (Loading rate=2.0kN/m^2/day)

the start of softening or continue softening. A typical element-wise behavior is shown in Fig. 12. In fact the softening is not easily distinguishable in this figure due to the normalization of deviator stress by p'_e (*i.e.* initial consolidation pressure). In such an event the excess pore pressure isochrones in the softening region should expand eventually reaching the outright failure of the foundation.

Fig. 13 shows the excess pore pressure distributions (for 2.0kN/m^2/day loading rate) at four instances after the end of loading until failure, taking at the end of loading pressure as the base for clarity. As shown here, with the elapse of time the excess pore pressure contours (initially having higher value at the centerline portion) expand from centerline region outwards indicating the occurrence of softening under this constant load. This process in the centerline region continue with time until the 8.56th day at which plastic flow occurs under the external load. If piezometric head in the affected region is measured, it should indicate a gradual rise until failure as observed in actual failures (*e.g.* Crawford et al, 1995, Eide, O. and Holmberg, S., 1972). Also as indicated in the Fig. 13 the excess pore pressure distribution becomes highly non-homogeneous towards the failure. In the case of non-failure (when loading was stopped at 120kN/m^2; Fig. 6) the excess pore pressure decreased gradually resulting consolidation settlements.

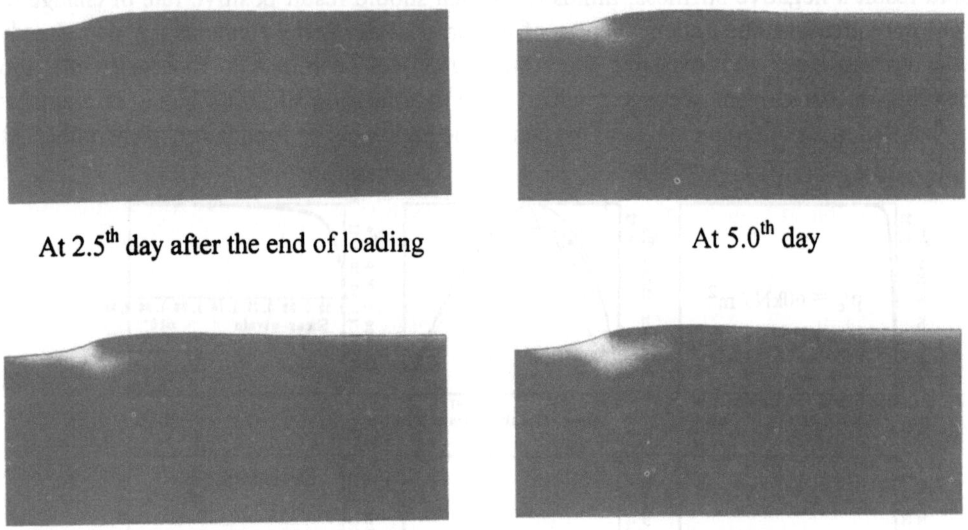

At 2.5th day after the end of loading

At 5.0th day

At 8.0th day

At 8.56th day

Fig. 13 Excess pore water pressure variation under constant load (Loading rate=2.0kN/m^2/day)

(in kN/m^2)

3.24

-23.13

The eigen value investigation of the tangential stiffness matrix revealed that in the failure case, the minimum eigen value, initially being positive when loading was ended, became negative towards the failure. The variation of this eigen value is illustrated in Fig. 14. It is to be mentioned here that at the moment failure occurred all the components of $\Delta f'$ reached zero. However, in the non-failure case, the minimum eigen value retained positive with increasing value with consolidation (Fig. 15).

If the overall load-settlement response of the foundation until failure under the constant load is considered (see Figs. 6 and 10) one may interpret this as the creep like behavior.

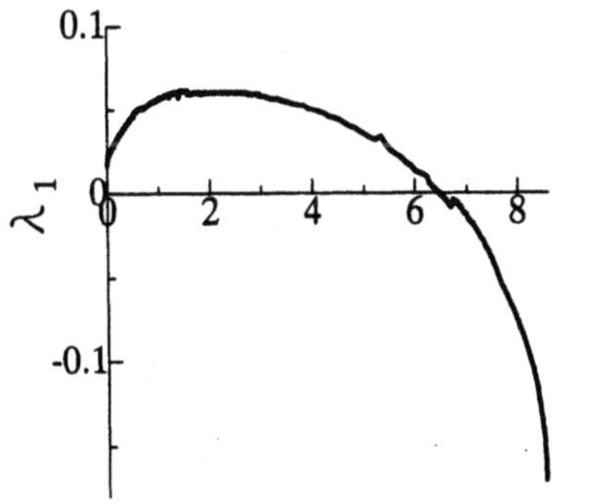

Fig. 14 Variation of minimum eigen (Failure case)

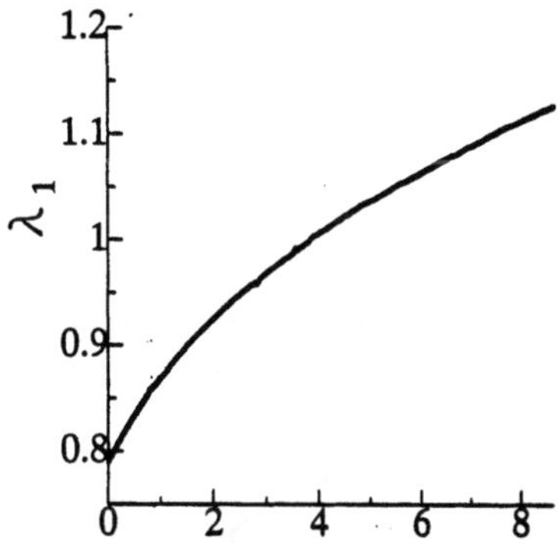

Fig. 15 Variation of minimum eigen (Non-failure case)

Of course, as explained above, this has been the result of pore water migration within hardening/softening saturated clays. However, when the magnitude of load at the end of construction is not high enough to make stress levels of soil elements near/at critical state, that may eventually lead to softening, the soil should undergo consolidation resulting a stable foundation under that load.

5.0 CONCLUSIONS

The occurrence of creep-like delayed failure of a normally consolidated clay foundation has been investigated using the soil-water coupled finite deformation analysis employing the subloading surface Cam-clay model. The findings of this research are summarized as below;

With drained boundaries a clay foundation can experience failure, instead of consolidation, with the elapse of time after the end of load application. The foundation stays apparently stable until it experiences a sudden failure. Such an occurrence of failure is due to the continued softening of soil causing the increase of excess pore water pressure under this constant load. With time, excess pore pressure isochrones expand outwards until failure. Such pore pressure increase under constant load is in similarity with observed actual cases that registered increase piezometric heads. The occurrence of delayed failure is of course depends on the total load level at the end of construction.

6.0 REFERENCES

Asaoka, A., Nakano, M. and Noda, T. (1994): Soil-water coupled behavior of saturated clay near/at critical state, Soils and Foundations, Vol. 34: No. 1: 91-105.

Asaoka, A. and Noda, T. (1995): Imperfection sensitive bifurcation of Cam-clay under plane strain compression with undrained boundaries, Soils and Foundations, Vol. 35: No. 1: 91-105.

Asaoka, A., Nakano, M. and Noda, T. (1997a): Soil-water coupled behavior of heavily overconsolidated clay near/at critical state, Soils and Foundations, Vol. 37, No. 1: 13-28.

Asaoka, A., Noda, T. and Fernando, G.S.K. (1997b): Effects of changes in geometry on the linear elastic consolidation deformation, Soils and Foundations, Vol. 37, No. 1: 29-39.

Asaoka, A., Nakano, M., Noda, T. and Kaneda, K. (1997c): Undrained creep rupture of normally consolidated clay triggered by the mode switching during pore water migration, Proc. IS-Nagoya 97 (A. Asaoka, T. Adachi and F. Oka Eds.) Nagoya, Japan: 21-26.

Asaoka, A., Noda, T. and Fernando, G.S.K. (1998): Consolidation deformation behavior of lightly and heavily overconsolidated clay foundations, Soils and Foundations, Vol. 38, No. 2: 75-91.

Changi Airport (1995) Private Communication with Prof. H. Sekiguchi, Dept. of Civil Engineering, Kyoto University, Kyoto, Japan.

Crawford, C.B., Fannin, R.J. and Kern, C.B. (1995): Embankment failures at Vernon British Columbia, Canadian Geotechnical Journal, Vol. 32: 271-284

Eide, O. and Holmberg, S. (1972): Test fills to failure on the soft Bangkok clay, ASCE Proc. Of the Specialty Conference on Performance of Earth and Earth Supported Structures, Purdue University, Lafayette, Ind., June 11-14, Vol. 1: 151-180.

Fernando, G.S.K. (1996): Non-linear consolidation deformation analysis of soils under embankment loading, D. Eng. Thesis, Dept. of Civil Engineering, Nagoya University.

Hashiguchi, K. and Ueno, M. (1977): Elasto-plastic constitutive laws of granular materials, Constitutive Equations of Soils, (Proc. Spec. Session 9th Int. Conf. SMFE, Murayama, S. and Schofield, A. N. Eds.), Tokyo, JSSMFE: 73-82.

Hashiguchi, K. (1993): Fundamental requirements and formulation of elasto-plastic constitutive equations with tangential plasticity, Int. J. of Plasticity, Vol. 9: 525-549.

Henkel, D. J. (1960): The shear strength of saturated remolded clay, Proc. of research Conf. on Shear Strength of Cohesive Soils at Boulder, Colorado: 533-540.

Proceedings of International Symposium on Trial Embankments on Malaysian Marine Clays (1989). Vols. 1 and 2, Malaysian Highway Authority, Kuala Lumpur, Malaysia.

Schofield, A. N. and Wroth, C. P. (1968): Critical State Soil Mechanics, McGraw Hill.

Tavenas, F. and Leroueil, S. (1980): The behavior of embankments on clay foundations, Can. Geotech. J., Vol. 17:.236-260.

Serge Leroueil, Jean-Pierre Magnan and Francois Tavenas. (1990): Embankments on Soft Clays (Translator David Muir Wood), Ellis Horwood, England.

Skempton, A.W. (1970): First-time slides in overconsolidated clays, Geotechnique, 20: 320-325.

A HYBRID FINITE/DISCRETE ELEMENT METHOD FOR ANALYSIS OF UNPAVED ROADS REINFORCED BY GEOTEXTILES

C. Athanasiu and N. Bergersen
NOTEBY AS Consulting Engineers, Oslo, Norway

ABSTRACT:

A 3-D discrete/finite element method for analysis of unpaved roads reinforced with geotextiles is presented. The foundation soil is soft, normally consolidated soil and is modelled by 3-D finite elements. The geotextile, fill and the contacts geotextile – soft soil and geotextile – fill are modelled by discrete elements. The foundation soft soil is modelled using a modified Tresca failure criterion and associated flow rule to account for stiffness and strength degradation due to cyclic traffic loading. A PC program based on the method is developed. The program consists of a pre-processor module helping the user to prepare input data and to preliminarily design the geotextile and the thickness of the fill, analysis module and a post-processor module displaying the results. The use of the method is illustrated by examples.

1 INTRODUCTION

The design of unpaved roads reinforced by geotextiles requires that the mechanism of load transfer between coarse subgrade material and the foundation soft soil by means of geotextiles should be accounted for. It is widely recognized that the function of the reinforcement is a combination of tensioned membrane effect and the effect of carrying the shear stresses between subgrade material and the underlaying soft soil. The paper presents a hybrid method using both finite and discrete elements that aimes to model the interaction coarse fill-geotextile-soft soil. The discrete elements simulate the normal and shear stress transfer at contacts coarse fill / geotextile and geotextile / soft soil while the eight-noded, 3-D, finite elements model the granular subgrade and the soft soil mass.

2 SIMULATION OF NORMAL AND SHEAR STRESS TRANSFER AT INTERFACES

The mechanism of stress transfer by geotextile at the contact coarse subgrade/soft soil is idealized as shown in Fig. 1. The geotextile carries the normal stress, p_z and the shear stresses $\tau_{xz\,u}$ and $\tau_{xz\,u}$ on the upper surface in contact with the fill. Similarly, normal stress, σ_z, and shear stresses $\tau_{xz\,l}$ and $\tau_{xz\,l}$, develop on the lower face towards the soft soil. These "face" stresses are in equilibrium with the in-plane stresses, σ_x and σ_y.

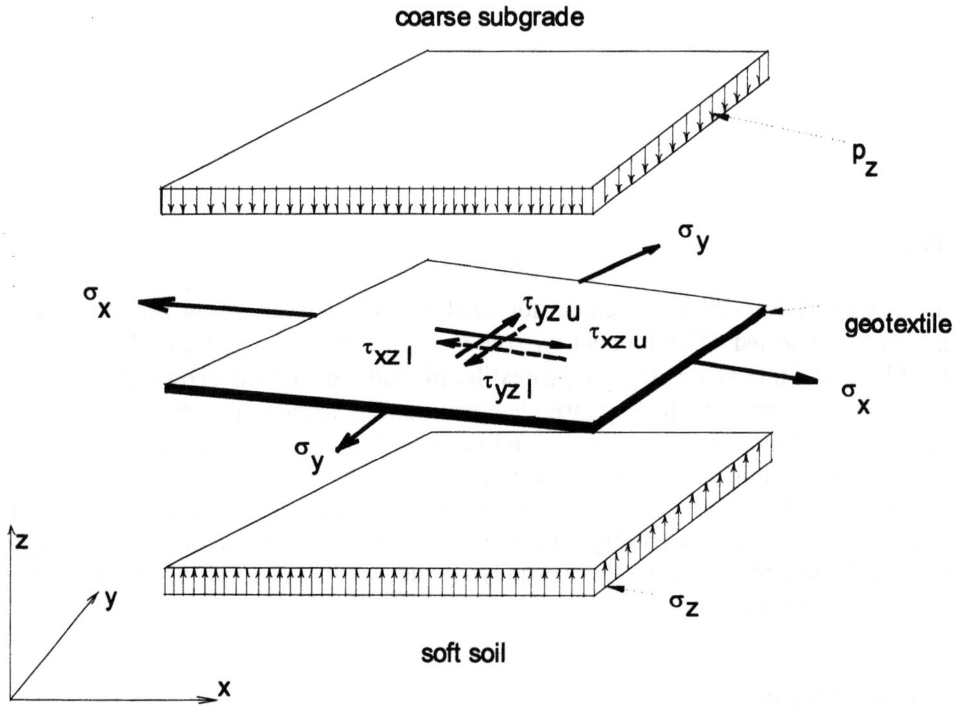

Fig.1 Stress transfer between coarse fill material and soft soil by geotextile.

The geotextile is modelled by a thin, 3-D non-linear elastic element while the contact fill/geotextile and geotextile/soft soil are modelled by interface elements carrying the "face" stresses.

A variable secant modulus technique is used to model the non-linear behaviour of the interface discrete elements. The friction and cohesion mobilisation with strain is specified as mobilisation curves shown in Fig. 2. If the shear strain, γ, obtained from one iteration is

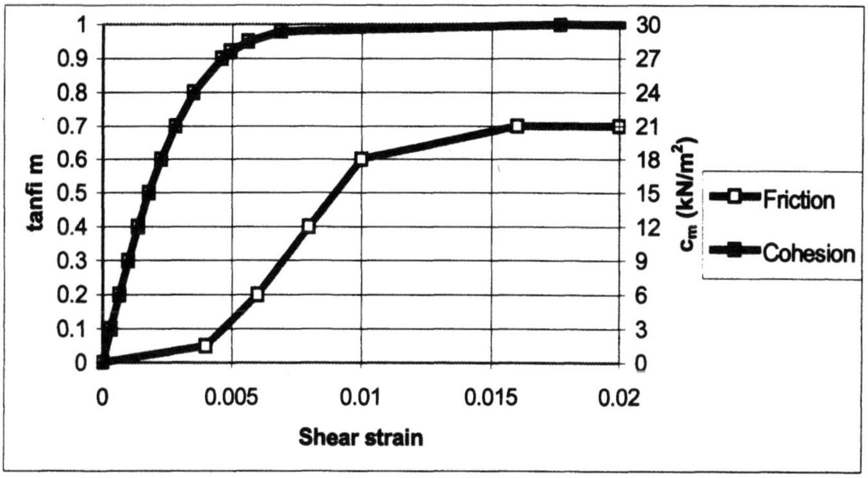

Fig. 2 Mobilised friction and cohesion (c_m) as a function of relative shear strain fill-geotextile or geotextile-soft soil (shear strain of the interface discrete element).

used to determine the mobilised friction, $\tan\varphi_m$ and mobilised cohesion, c_m, from the curves shown in Fig. 2, then the mobilised shear stress, τ_m, is obtained by Mohr-Coulomb equation $\tau_m = \sigma_z \cdot \tan\varphi_m + c_m$, where σ_z is the vertical stress acting on the surface of interface element. If the difference between τ_m and the shear stress obtained in previous iteration, τ, is within the specified tolerance, convergence is achieved for the interface element. If however the difference is greater than specified tolerance then a new secant shear modulus, $G_{sec} = \tau_m / \gamma$, is calculated and the iteration process is repeated until convergence is obtained in all interface elements.

3 SOFT SOIL BEHAVIOUR UNDER CYCLIC LOADING

The stress-strain relationship of soft soil (Fig.3) with undrained shear strength, s_u, and initial, in situ, shear stress, τ_o, is modelled as follows. The static (first cycle) undrained behaviour of soft soil is defined by the secant shear modulus ratio as a function of mobilisation degree:

$$\frac{G_{sec}}{G_o} = \frac{G_{ult}}{G_o} + (1 - \frac{G_{ult}}{G_o}) \cdot (1 - D_m)^{\alpha} \tag{1}$$

where: D_m is the mobilisation degree, $D_m = (\tau - \tau_o)/(s_u - \tau_o)$; G_{ult} is the ultimate secant modulus, $G_{ult} = (s_u - \tau_o)/\gamma_{ult}$; G_{sec} is the secant shear modulus, $G_{sec} = (\tau - \tau_o)/\gamma$; G_o is the initial shear modulus, $G_o = (s_u - \tau_o)/\gamma_r$; γ_r is the reference shear strain; α is an empirical exponent. The strain ratio can thus be expressed as:

$$\left(\frac{\gamma}{\gamma_r}\right)_{st} = \frac{D_m}{G_{sec}/G_o} \tag{2}$$

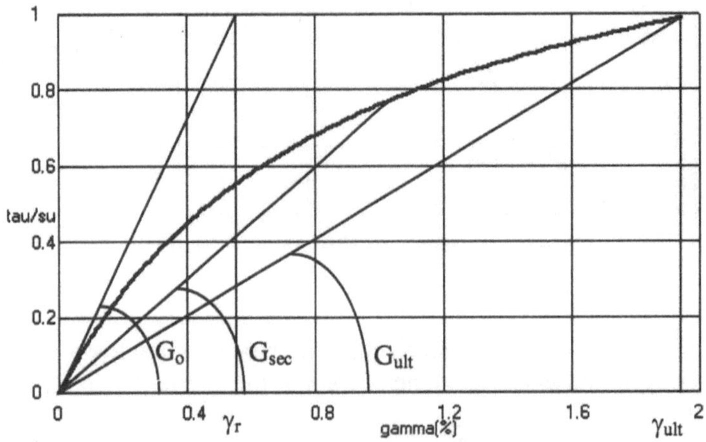

Fig. 3. Undrained stress-strain

Cyclic triaxial tests on plastic clays (see for instance Andersen et al., 1988) have shown that cyclic shear strain increases with the number of cycles and that for each cyclic shear stress level, D_m, there is critical number of cycles, N_{cr}, at which large shear strains (failure) occur (Fig. 4).

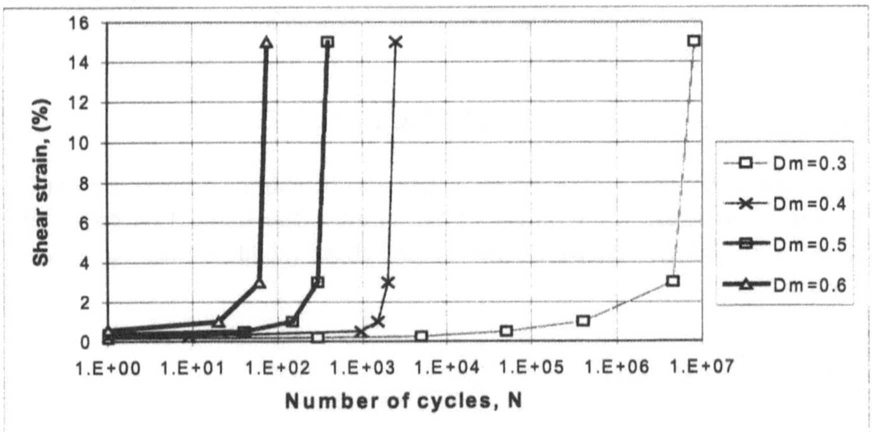

Fig. 4 Cyclic shear strain as a function of applied cyclic shear stress level, Dm, and number of cycles (Plastic Drammen Clay)

The variation of critical number of cycles with the applied cyclic shear stress ratio, D_m, can be expressed as:

$$\log(N_{cr}) = K_{cy} \cdot \left(\frac{1-D_m}{D_m}\right)^{\beta_3} \tag{3}$$

where: K_{cy} and β_3 are constants determined experimentally.

The relation between the applied number of cycles, N, and the corresponding cyclic shear strain ratio, Fig.5, can be expressed as:

$$\left(\frac{\gamma}{\gamma_r}\right)_{cy} = \left(\frac{\gamma}{\gamma_r}\right)_{st} + \frac{\log(N)}{R} \tag{4}$$

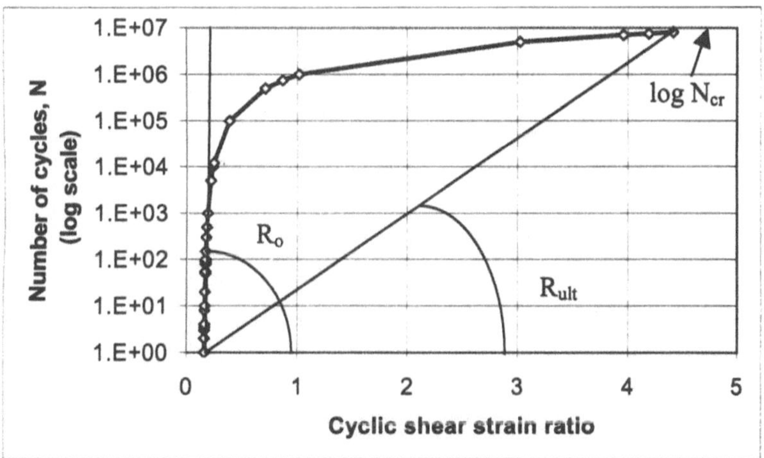

Fig. 5 The relation between applied number of cycles, N, and the corresponding cyclic shear strain ratio, γ/γ_r

where: $(\gamma/\gamma_r)_{cy}$ is the cyclic shear strain ratio (corresponding to N shear cycles); R is the cyclic strain accumulation secant modul. The cyclic shear strain accumulation modul has the same type of variation as a function of cyclic ratio, log(N)/log(Ncr), as the shear secant modulus has as a function of mobilization degree:

$$R = R_o \cdot \left(\frac{R_{ult}}{R_o} + \left(1 - \frac{R_{ult}}{R_o}\right) \cdot \left(1 - \frac{\log(N)}{\log(N_{cr})}\right)^{\beta_2}\right) \tag{5}$$

where: R_o is the initial cyclic shear strain accumulation modul. As Fig. 4 illustrates, the higher mobilization degree is used in a cyclic test the lower initial cyclic shear strain accumulation module is obtain (i.e. cyclic shear strains accumulates more rapidly):

$$R_o = r_o \cdot \left(\frac{1 - D_m}{D_m}\right)^{\beta_1} \tag{6}$$

The ultimate cyclic modulus, R_{ult}, can be expressed as:

$$R_{ult} = \frac{\log(N_{cr})}{\gamma_{ult} / \gamma_r} \tag{7}$$

4 SHORT PROGRAM DESCRIPTION

A Fortran program, **armjrd**, is developed to analyse the mechanism of the load tranfer between coarse fill of unpaved roads and the soft foundation soil.

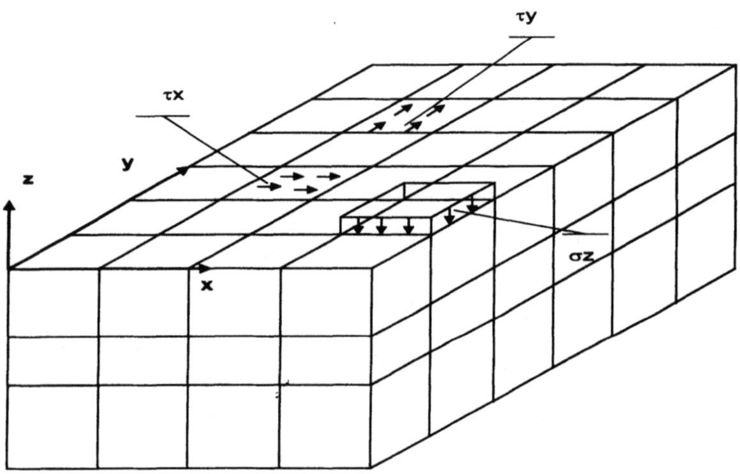

Fig. 6 Element mesh and specified loads

The element mesh is specified by three sets of planes normal to the three coordinate axes, x, y and z, (Fig.). The geometry is specified by the x-coordinates of the vertical planes parallel to the yOz plane, the y-coordinates of the vertical planes paralell to the xOz plane and by the z-coordinates of the horisontal planes paralell to the xOy plane. Boundary conditions are automatically set to restrained displacements in x- direction for the nodes in the outer vertical planes paralell to yOz plane and restrained displacements in y-direction for the nodes in the vertical planes parallel to xOz plane as shown in Fig. 6. The nodes on the lower boundary have restrained displacements in z-direction.

The loads are specified on the surface of the last (upper) element layer as either even distributed normal stress, σ_z, or even distributed shear stress, τ_x, or τ_y or any combination of these (Fig.). Concentrated loads can also be specified at any nodal point. It is assumed that all the loads are applied applied cyclically with a specified number of cycles, N.

The analysis is performed in an iterative procedure. The secant moduli of interface discrete elements for a new iteration are adjusted to correspond to the mobilized friction and cohesion during the previous iteration. A modified Tresca criterion is used for soil elements in elasto-plastic analysis where the plastic modulus required to form the elasto-plastic element stiffness matrix is derived from stress-strain relationships (eqs. 3 to 7).

The program runs in Windows environment and input data and results are displayed at the run time in separate windows for quick inspection. The results are also written on output file. A pre-processor module, **armjrd-pre**, is developed to help the user to prepare input data and to preliminary design the geotextile and the thickness of the fill. A post-processor module, **armjrd-post**, is also developed to display the results. The program can, however, be operated without pre- and post-processors.

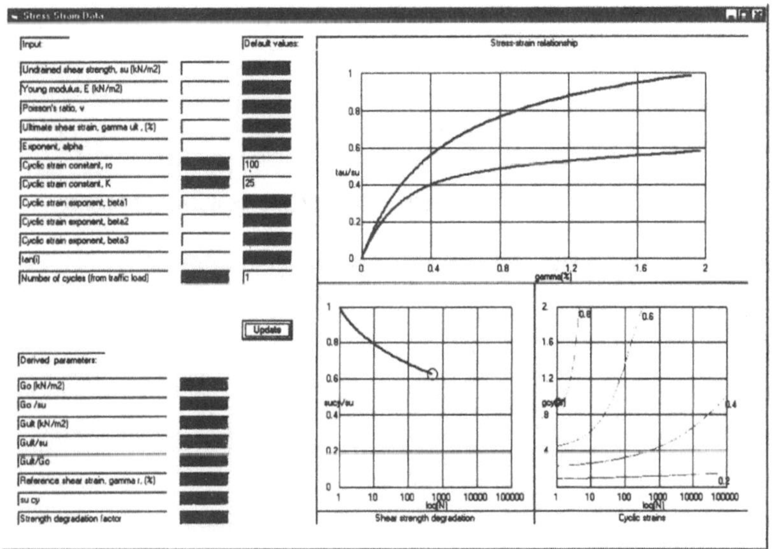

Fig.7. Selecting the soil parameters

5 EXAMPLES

A parametric study including back-calculation of full scale field tests is in progress and will be published elsewhere. The use of the program and the display of the results are only illustrated here by simple examples. Selection of stress-strain parameters using the preprocessor **armjrd-pre** is illustrated in Fig.7. It displays the static and cyclic stress-strain relationships (upper right corner) for the selected parameters. It also illustrates the strength degradation for the selected number of cycles (lower left corner) and the general pattern of cyclic strains as a function of cyclic stress level and number of cycles (lower right corner).

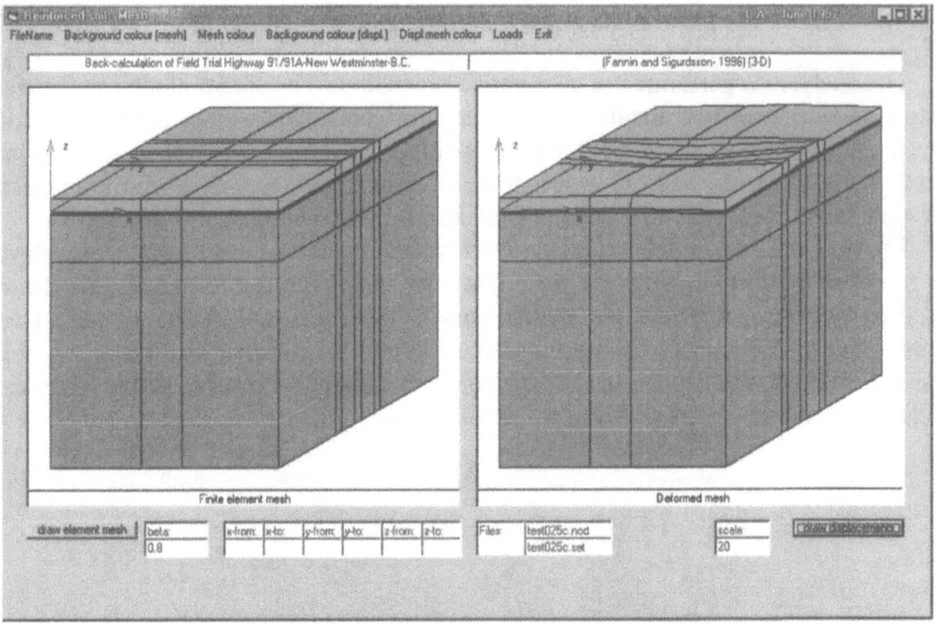

Fig.8. Analysis of a double wheel prints as a 3-D problem

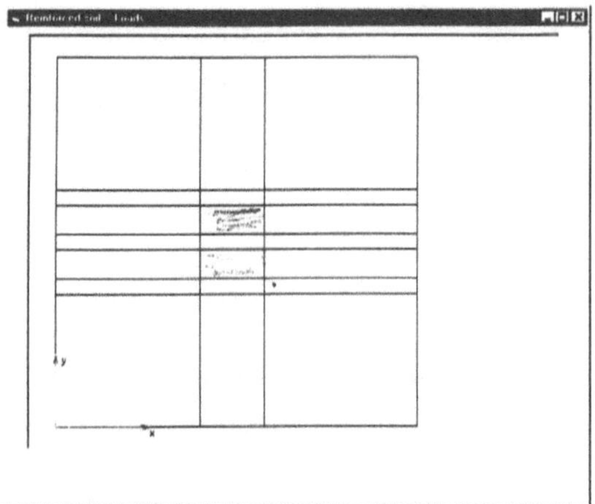

Fig.9. Double wheel print

Simple 3-D analysis is illustrated in Figs.8 and 9 showing the element mesh, the deformed mesh and the applied load.

If an unpaved road is rutted then the problem may be treated like a "plane strain" problem as illustrated in Figs. 10 and 11.

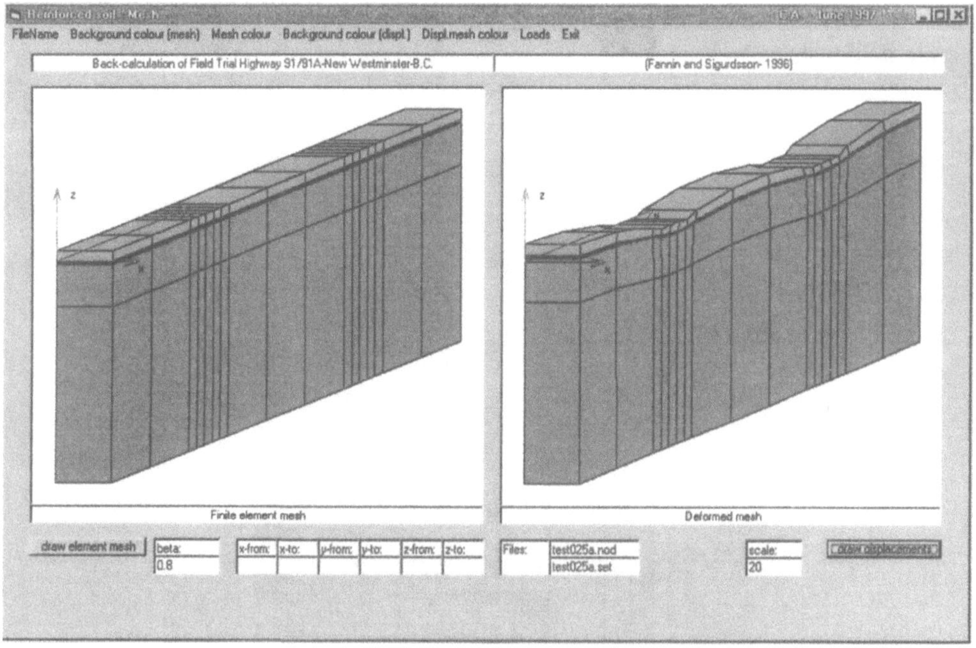

Fig.10. Element mesh and deformations for a rutted road.

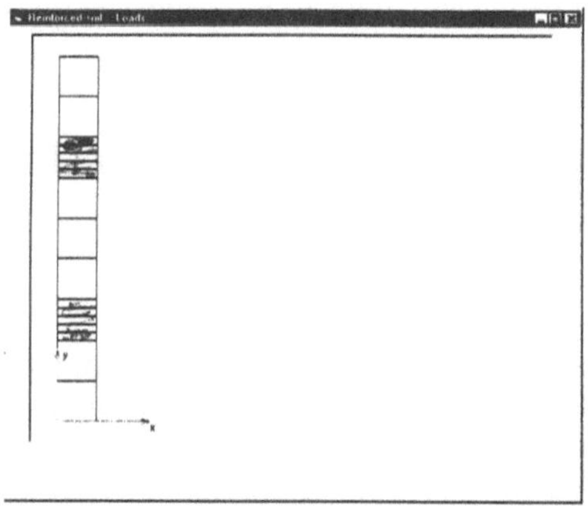

Fig.11. Axel load

The preliminary design of the geotextile and of the fill thickness is illustrated in Fig.12.

6 CONCLUSIONS

A hybrid finite discrete element method for analysis of the unpaved roads reinforced by geotextiles is presented. The method is able to qualitatively simulate the load transfer between the coarse subgrade and the foundation soft soil by meas of geotextiles. A parametric study is in progress to "calibrate" the input parameters from back-calculation of full scale field tests.

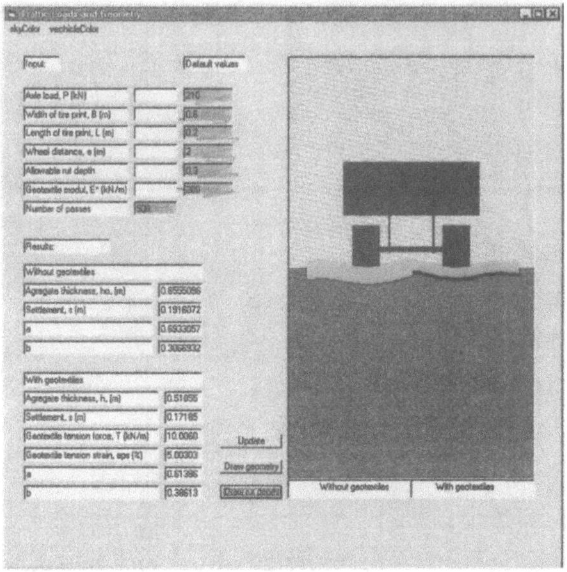

Fig.12. Preliminary design of geotextile and fill thickness.

7 AKNOWLEDGEMENTS

The financial supports to this project from Multiconsult Holding AS and from Norwegian Geotechnical Society are greatefully aknowledged

8 REFERENCES

Andersen K.H., A.Kleven and D.Heien – (1988) – Cyclic Data for Design of Gravity Structures. ASCE Journal of Geotechnical Engineering, vol. 114,No.5.

Fannin, R.J. & O. Sigurdsson – (1996) – Field observations on stabilization of unpaved roads with geotextiles. ASCE Journal of Geotechnical Engineering, Vol.122, No.7.

Giroud, J-P. and L. Noiray – (1981)- Geotextile-Reinforced unpaved road design. ASCE Journal of the Geotechnical Engineering Division, Vol.107, No.GT9.

Houlsby, G.T. & R.A. Jewell – (1990) – Design of reinforced unpaved roads for small rut depths. Geotextiles, Geomembranes and Related Products, Den Hoedt (ed.) Balkema, Rotterdam.

A NUMERICAL PROCEDURE FOR THE ANALYSIS OF SOIL CONSOLIDATION

E. Ausilio and E. Conte
University of Calabria, Cosenza, Italy

A.O. Ficchì
Civil Engineer, Catanzaro, Italy

ABSTRACT: A numerical procedure is presented to analyse consolidation of soils with compressible solid particles and pore fluid. The governing differential equations are reduced to ordinary differential equations by applying a Fourier transform along with a Laplace transform, and the solution is found in the finite element fashion without great computational efforts. A number of comparisons with existing solutions are shown in order to assess the accuracy of the employed procedure.

1. INTRODUCTION

The equations governing the consolidation of saturated soils were first developed by Terzaghi (1943) for one-dimensional conditions. Generalisation of Terzaghi's theory to three dimensions has given rise to two different approaches which are known as diffusion theory (Rendulic, 1936) and Biot's theory (1935). The latter is preferable from a theoretical point of view because it provides a coupling between the magnitude and the progress of settlement (Schiffman et al., 1969). However, the resulting equations are more complicated and closed form solutions have been derived only for simple geometrical and physical conditions (McNamee and Gibson, 1960; Gibson et al., 1970). The constitutive assumptions usually made are of linear elasticity, homogeneity and isotropy of the medium.

More general situations involving anisotropy, heterogeneity and non-linearity of soil can be fully analysed making use of numerical methods. The finite element method has found extensive application for the analysis of soil consolidation (Sandhu and Wilson, 1969; Christian and Boehmer, 1970; Hwang et al., 1971; Yokoo et al., 1971; Ghaboussi and Wilson, 1973; Cividini and Zavelani Rossi, 1983). This is because the method is in principle general with respect to the geometry of the problem, boundary conditions, and variation in material properties and stress-strain relationship. In practice, however, the solution can be very costly in computer time and often requires a large amount of data preparation. Recently, the boundary element method has also been employed (Cheng and Liggett, 1984; Dargush and Banerjee, 1989); this method partly reduces the computational costs in comparison with FEM.

Anyway, in many circumstances it is desirable to have a method available that allows rapid and efficient calculation of the time-settlement relationship due to consolidation. In this regard, Booker and Small (1982) have proposed an attractive method which uses a Fourier transform to represent the field variables in the horizontal plane and a marching type scheme as a time integration process. Afterwards, the same authors (Booker and Small, 1987) have employed the Laplace transforms to solve the time-dependent problem. This eliminates the need to march the solution through a number of time steps.

In this paper, a numerical procedure is presented that is based on the same principles as that developed by Booker and Small (1982). However, in order to treat the time-dependent problem more efficiently we have used the Laplace approach. Moreover, a more general set of field equations has been used where the assumption of incompressibility of pore fluid and solid phases has been relaxed. As a consequence, the present procedure could be employed in several actual situations to analyse consolidation of partially saturated soils. This is the case in which the degree of saturation of soil is sufficiently high (greater than 90 per cent) so that air contained in the pores is occluded and cannot flow as a separate continuous fluid. In these circumstances, the air bubbles and pore water behave as a homogeneous compressible fluid flowing under water pressure gradients (Barden, 1965).

In the following the procedure is first presented, then a number of comparisons with existing solutions are shown, in order to demonstrate the applicability of the method and to assess its accuracy.

2. THE FIELD EQUATIONS

The essential assumptions on which the method is based are: (1) the soil skeleton is linearly elastic; (2) small strains occur; (3) fluid flow is governed by Darcy's law; (4) dynamic and secondary effects are ignored; (5) the soil deposit has infinite lateral extension and is homogeneous or horizontally layered. All soil parameters in the problem are treated as given material constants, although in general they are functions of stresses or strains. These assumptions determine some restrictions in the use of the method.

Following Dargush and Banarjee (1989), the field equations of consolidation are:

- *the equation of equilibrium*

$$\sigma'_{ij,j} + \delta_{ij}\,\chi\; p_{,j} = 0 \qquad (1)$$

where σ'_{ij} denote the increase in the components of the effective stress tensor due to applied load, p indicates the excess pore pressure, and χ is a dimensionless material parameter that can be related to the bulk modulus of the soil B, and to that of the solid constituents B_S, from χ=1-B/B_S.

- *the stress-strain relationship*

$$\sigma'_{ij} = C_{ijkl}\ \varepsilon_{kl} \tag{2}$$

in which C_{ijkl} is the matrix of the elastic constants of soil, and ε_{kl} represents the components of the strain tensor.

- *the mass conservation law*

$$\chi\dot{\varepsilon}_{ii} = v_{i,i} + \beta\ \dot{p} \tag{3}$$

where $\dot{}$ indicates partial differentiation with respect to time t, β is a parameter that accounts for compressibility of pore fluid (Verruijt, 1969; Dargush and Banerjee, 1989), and v_i indicates the components of the superficial velocity of the pore fluid that are given by Darcy's law.

The above equations provide a set that has to be solved with the appropriate boundary conditions. If we suppose that strains are restricted tó the x-z plane (plane strain conditions), the set of equations becomes

$$\begin{cases} A\dfrac{\partial^2 u_x}{\partial x^2} + C\dfrac{\partial^2 u_z}{\partial x \partial z} + \dfrac{F}{2}\left(\dfrac{\partial^2 u_x}{\partial z^2} + \dfrac{\partial^2 u_z}{\partial x \partial z}\right) + \chi\dfrac{\partial p}{\partial x} = 0 \\ \dfrac{F}{2}\left(\dfrac{\partial^2 u_x}{\partial x \partial z} + \dfrac{\partial^2 u_z}{\partial x^2}\right) + C\dfrac{\partial^2 u_x}{\partial x \partial z} + D\dfrac{\partial^2 u_z}{\partial z^2} + \chi\dfrac{\partial p}{\partial z} = 0 \\ \chi\left(\dfrac{\partial u_x}{\partial x} + \dfrac{\partial u_z}{\partial z}\right) = -\dfrac{1}{\gamma_w}\displaystyle\int_0^t\left(K_H\dfrac{\partial^2 p}{\partial x^2} + K_V\dfrac{\partial^2 p}{\partial z^2}\right)dt + \beta\ p \end{cases} \tag{4}$$

in which u_x and u_z are the components of soil displacement; A, C, D and F denote the elastic constants of soil. For an isotropic material A=D=λ+$2G$, F=$2G$ and C=λ, where λ is Lamè constant and G is the shear modulus. Moreover, γ_w is the unit weight of water; K_H and K_V denote the horizontal and vertical coefficients of permeability, respectively.

In order to formally eliminate the time dependence from Eqs. (4), a Laplace transform is applied. In addition, for a soil system of infinite lateral extension, the dependence on the spatial co-ordinate x can be also removed by the application of a Fourier transform. By these transformations, Eqs. (4) are reduced to the ordinary differential equations:

$$\begin{cases} -\alpha^2 A\overline{U}_x + \alpha C\dfrac{\partial \overline{U}_z}{\partial z} + \dfrac{F}{2}\left[\dfrac{\partial^2 \overline{U}_x}{\partial z^2} + \alpha\dfrac{\partial \overline{U}_z}{\partial z}\right] + \alpha\chi\overline{P} = 0 \\ \alpha\dfrac{F}{2}\left[\dfrac{\partial \overline{U}_x}{\partial z} + \alpha\overline{U}_z\right] + \alpha C\dfrac{\partial \overline{U}_x}{\partial z} - D\dfrac{\partial^2 \overline{U}_z}{\partial z^2} - \chi\dfrac{\partial \overline{P}}{\partial z} = 0 \\ \chi\left(\alpha\overline{U}_x - \dfrac{\partial \overline{U}_z}{\partial z}\right) - \dfrac{K_V}{s\gamma_w}\dfrac{\partial^2 \overline{P}}{\partial z^2} + \left[\alpha^2\dfrac{K_H}{s\gamma_w} + \beta\right]\overline{P} = 0 \end{cases} \tag{5}$$

where $\overline{U}_x, \overline{U}_z$ and $\overline{P}$ denote the transformed displacements and pore pressure, s is the Laplace transform parameter, and α is the wave number to describe the variation of the field variables in x direction. The bar over a variable indicates that a Laplace transform has been applied to that variable. The advantage of Eqs. (5) over (4) is that, for a given value of α and s, the unknown transformed variables only depend on the depth z. From a computational point of view, this considerably simplifies the solution as the original problem, which is essentially two-dimensional and time-dependent, is formally reduced to a "static" one-dimensional problem, in that only one independent variable is involved.

4. THE NUMERICAL SOLUTION

In order to obtain a numerical solution of Eqs. (5), the soil deposit is discretized by a finite number of one-dimensional elements. Within each element, the field quantities can be approximately represented by their values at the nodes of the element and assumed shape functions. The discretization of the system is carried out along depth only, and most types of element can be employed for this purpose; the high-continuity quadratic elements (*HC* elements) proposed by Aristodemo (1985) included. These interpolation functions enforce continuity of the field variables and their first derivative at the interfaces of each element. This provides a significant reduction of the total number of unknowns, if compared with the usual quadratic shape functions. In fact, when the material is homogeneous one node per element is only required, except for the end elements (Fig.1). The use of *HC* elements presents no particular difficulty when the soil deposit consists of layers with different material properties. In this case, it is necessary to discretize each layer by means of elements of equal thickness (however the thickness may be different from layer to layer) and to add a node at the junction of different materials.

The finite element formulation of the governing equations leads to a set of equations that can be written in the matrix form as

$$\mathbf{CF} = \mathbf{R} \tag{6}$$

where **F** denotes the vector of the nodal unknown variables, **R** is the vector of the known terms containing the prescribed external loads, **C** is a matrix whose terms depend on the soil properties, the thickness of the elements, and the transform parameters α and s.

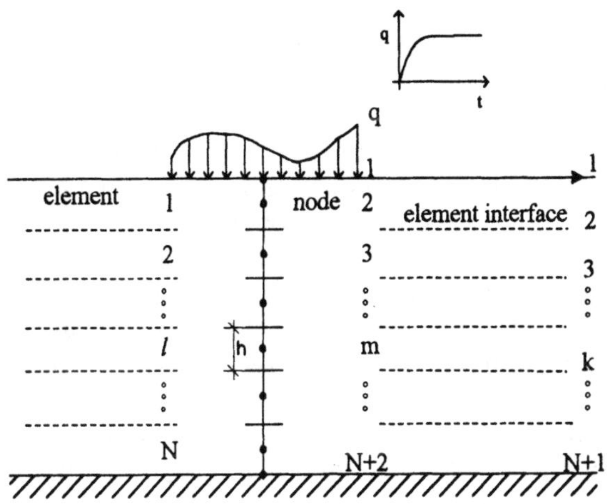

Fig. 1 A pattern of the discretization scheme by using *HC* elements.

Solution of (6) is a function of α and s, inversion is thus required to recover the original physical variables. The main steps of the procedure can be summarised as follows. Given a time-depending load applied over a soil deposit discretized as in Fig. 1, the solution technique requires first that the load be expanded by a Fourier transform and a Laplace transform, and then that Eq. (6) be solved for each value of s and α, in order to obtain the transformed displacements and pore pressure as discrete functions of s and α. Finally, once these quantities are found, the actual displacements and pore pressure at the nodes can be calculated by numerical inversion of the Laplace transform and the Fourier transform. The proposed procedure can be found in more detail in Conte (1998).

In order to simplify the calculation, it is also possible to treat the field variables as periodically spaced functions in x direction by using a Fourier series. In this case, however, the spatial period must be set wide enough (greater than 5 times the foundation width) to avoid any interaction among the loaded areas and to minimise the unfavourable effects on the solution due to the presence of rigid and impervious vertical walls fictitiously introduced whenever the load periodically repeats itself. As regards the Laplace transform, a number of techniques are available to perform numerical inversion, each of those is affected by a different degree of approximation. In the present work, an algorithm based on Durbin's formula has been used that is very accurate to invert numerically a Laplace transform (Manolis and Beskos, 1988).

5. COMPARISONS

In this section, the results of a series of examples are shown. The purpose is to compare the feasibility and the accuracy of the presented procedure against other existing methods.

Also, it is an opportunity to investigate the influence of some discretization parameters on the solution. All of the results shown in this study have been obtained discretizing the soil system by *HC* elements.

The first problem considered refers to the consolidation of a half-space subjected to a uniform load, q, over a flexible strip foundation. Solid particles and pore fluid are assumed to be incompressible. The solution to this problem was obtained analytically by Schiffmann et al. (1969). Figure 2 shows the distribution of excess pore pressure for the vertical line through the center of the loaded area, at a given time. From the figure, it is evident that the agreement between the results is very good. The slight lack of accuracy occurring at some depths could be due to the fact that in this study the half space has been approximated by a finite layer. The thickness of this layer has been set equal to $14a$, being a=semi-width of the foundation.

A second comparative study concerns the one-dimensional consolidation of a soil layer containing a compressible pore fluid. This problem was previously examined by Ghaboussi and Wilson (1973) via FEM. The time-settlement curves for several values of $\eta=1/\beta C_v$ are shown in Fig.3. As can be seen, the agreement between the results is noteworthy. It should be noted that in order to analyse one-dimension consolidation, we have used a Fourier cosine series to synthesise a uniform load that is applied over a finite area and spaced by a period equal to the foundation width.

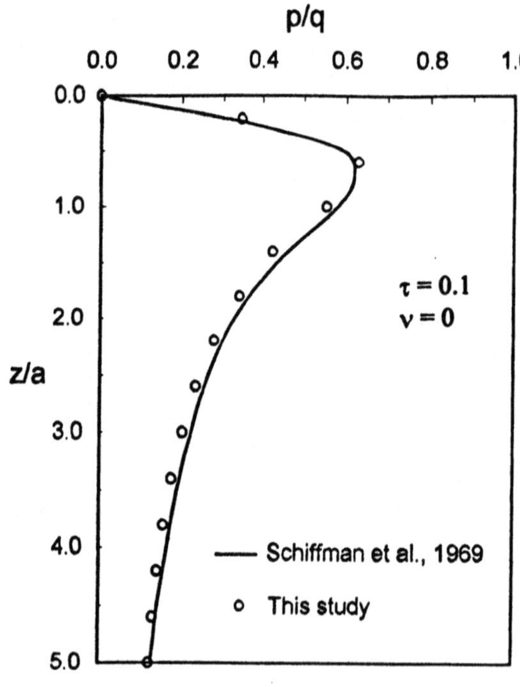

Fig. 2 Excess pore pressure distribution at a given time (adapted from Schiffman et al, 1969).

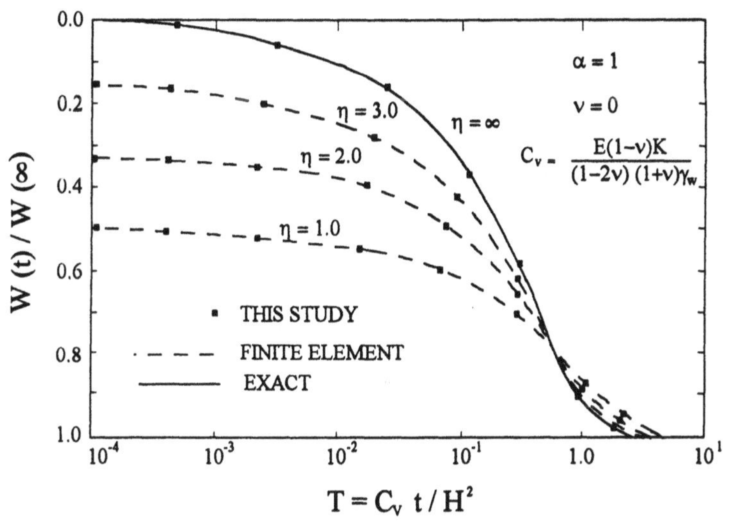

Fig. 3 Surface displacement in one-dimensional consolidation of a soil layer with compressible pore fluid (adapted from Ghaboussi & Wilson, 1973).

Another example analysed is that of a strip load of width $2B$ applied to the surface of a homogeneous isotropic layer with thickness $H=B$. Solid particles and pore fluid are incompressible. Fig.4 shows the distribution of excess pore pressure beneath the centre of the loaded area, at various time factors $\tau = \dfrac{2GKt}{\gamma_w H^2}$, where K=coefficient of permeability.

The results obtained are in good agreement with those presented by Booker and Small (1982) that are also indicated in the figure. However, due to the different discretization techniques employed in the two procedures, differences between the results can be observed especially when the time factor is large. In the present study, the solution of the time-dependent problem is achieved by applying a Laplace transformation to the field variables. Consequently, the solution is obtained directly at any time and does not depend on the results at previous times. Booker and Small (1982), on the contrary, used a marching scheme as a time integration process where the solution is found from one at the previous time. Therefore, although the solutions agree well at the smaller time, differences between the results may arise at large time values (Booker and Small, 1987).

The results of Fig.5 are relative to the same example, but they has been obtained by varying the number of discretization elements, N. Vertical profiles of displacement are also shown in Fig.6, where B=foundation width. As can be seen, the solution may be significantly affected by the number of elements used, especially at small values of τ. By increasing N, however, convergence of the solution is achieved. Furthermore, the figures show that the calculation of p generally requires a finer mesh than u_z.

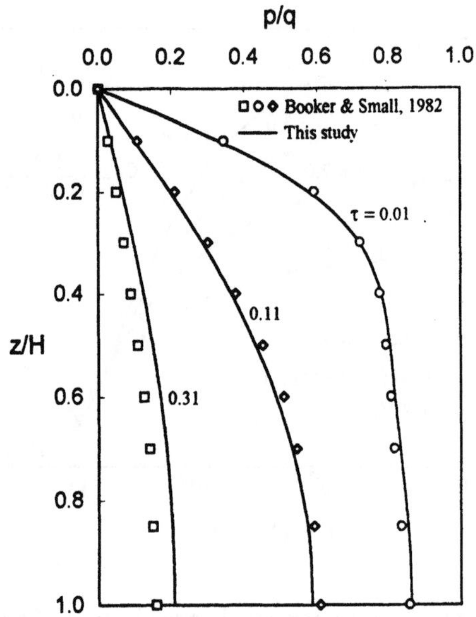

Fig. 4 Excess pore pressure profile at various time factors (adapted from Booker and Small, 1982).

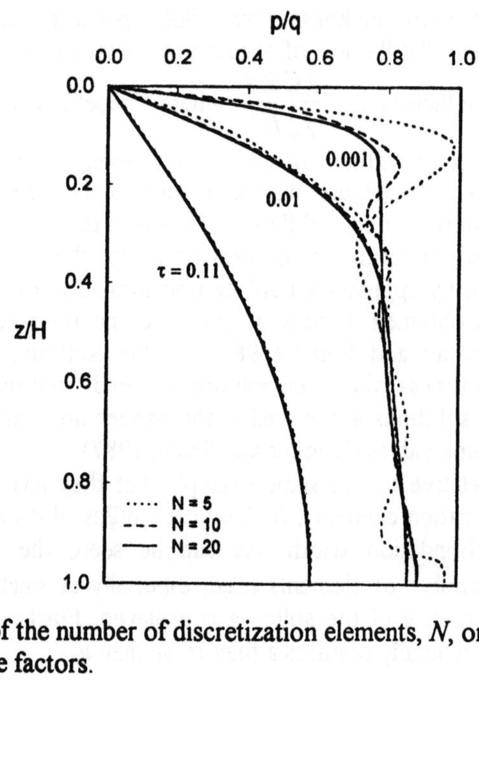

Fig. 5. Influence of the number of discretization elements, *N*, on excess pore pressure at various time factors.

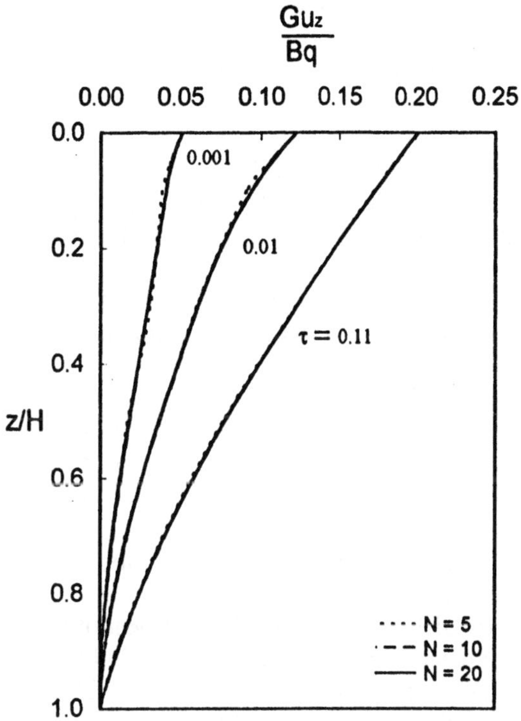

Fig. 6 Influence of the number of discretization elements, *N*, on vertical displacement at various time factors.

6. CONCLUSIONS

A numerical procedure has been presented for the analysis of the coupled consolidation involving soils with compressible pore fluid and solid particles. The method requires the use of a Fourier transform (or Fourier series) along with a Laplace transform. Solution to the governing equations in the transformed domain is obtained in the finite element fashion by subdividing the thickness of the soil deposit into a finite number of one-dimensional elements.

From a series of comparisons carried out, close agreement is found between the results obtained by the present procedure and those of other existing solutions.

REFERENCES

ARISTODEMO M. (1985). A high-continuity finite element model for two-dimensional elastic problems, Comp. & Struct., 21(5): 987-993.

BARDEN L. (1965). Consolidation of compacted and unsaturated clays, Geotechnique, 15(3): 267-286.
BIOT M. A. (1935). Le problem de la consolidation des matieres argileuses sous une charge, Annaies de la Societe Scientifique de Bruxelles, Series B.55: 110-113.
BOOKER J. R. & SMALL J. C. (1982). Finite layer analysis of consolidation. I, Int. J. Numer. Anal. Methods Geomech., 6: 151-171.
BOOKER J. R. & SMALL J. C. (1987). A method of computing the consolidation behaviour of layered soils using direct numerical inversion of Laplace transforms, Int. J. Numer. Anal. Methods Geomech., 11: 363-380.
CHENG A. H-D & LIGGETT J. A. (1984). Boundary integral equation method for linear porous elasticity with applications to soil consolidation, Int. J. Numer. Methods Eng., 20: 255-278.
CHRISTIAN J. T. & BOEHMER J. W. (1970). Plane strain consolidation by finite elements, J. Soil Mech. Found. Div., ASCE, 96(SM4): 1435-1457.
CIVIDINI A. & ZAVELANI ROSSI A. (1983). The consolidation problem treated by a consistent (static) finite element approach, Int. J. Numer. Anal. Methods Geomech., 7: 435-455.
CONTE E. (1998). Consolidation of anisotropic soil deposits, Soils and Foundations (accepted for publication).
DARGUSH G. F. & BANERJEE P. K. (1989). Development of BEM for transient poroelasticity, Int. J. Numer. Methods Eng., 28: 2423-2449.
GHABOUSSI J. & WILSON E. L. (1973). Flow of compressible fluid in porous elastic solids, Int. J. Numer. Methods Eng., 5: 419-442.
GIBSON R. E., SCHIFFMAN R. L. & PU S. L. (1970). Plane strain and axially symmetric consolidation of a clay layer on a smooth impervious base, Q. J. Mech. Appl. Math., 23: 505-520.
HWANG C. T., MORGENSTERN & MURRAY D. W. (1971). On solutions of plane strain consolidation problems by finite element methods, Canad. Geotech. J., 8: 109-118.
MANOLIS G. D. & BESKOS D. E. (1988). Boundary element methods in elastodynamics, Unwin Hyman Ldt., London.
MCNAMEE J. & GIBSON R. E. (1960). Displacement functions and linear transforms applied to diffusion through porous elastic media, Q. J. Mech. Appl. Math., 13: 98-111.
RENDULIC L. (1936) "Porenziffer und porenwasserdruck in tonen", Der Bauingenieur, 17: 559-564.
SANDHU R. S. & WILSON E. L. (1969). Finite element analysis of seepage in elastic media, J. Eng. Mech. Div., ASCE, 95(EM3): 641-652.
SCHIFFMAN R. L., CHEN A. T-F. & JORDAN J. C. (1969). An analysis of consolidation theories, J. Soil Mech. Found. Div., ASCE, 95(SM1): 285-312.
TERZAGHI K. (1943). Theoretical soil mechanics. John Wiley and Sons, New York.
VERRUIJT A. (1969). Elastic storage of aquifers, in R. J. M. de Wiest (ed.), Flow Through Porous Media, Academic Press, N.Y., 331-376.
YOKOO Y., YAMAGATA K. & NAGAOKA, P. (1971). Finite element method applied to Biot's consolidation theory, Soils and Foundations, 11(1): 29-46.

INTERACTION ANALYSIS OF RIGID FOUNDATION-BUILDING SYSTEM

Z. Czap
Technical University of Budapest, Budapest, Hungary

ABSTRACT

Numerical procedures – related to reinforcement concrete plates placed on Winkler bedding – are widespread in the civil engineering practice. Unfortunately these can not be used for the estimation of the effect of nearby buildings.

As semi-analytical simplified FE procedure is suggested for minicomputers taking in consideration-layered subsoil and several rigid plate-building structures.

The rigid plate-building structures are modelled by polygon elements. The material properties of the soil layers can be obtained from in-situ and/or laboratory tests. The flexibility matrix of the layered soil is calculated on the basis of the Boussinesq theory.

Although the applied method is much quicker than a 3D or elastic plate analysis, correct results are obtained. The paper shows the details of the analysis and the results of checking calculations.

1 INTRODUCTION

An important part of the design of structures is the control over settlements. The usual method used in Civil Engineering practice consists in calculating settlements of the foundation and comparing calculated values with set settlement criteria. Criteria are usually expressed in terms of absolute and relative settlements, differential settlements, tilt, sag and deflection (Fig. 1). In such a calculation the interaction between footings is generally not considered.

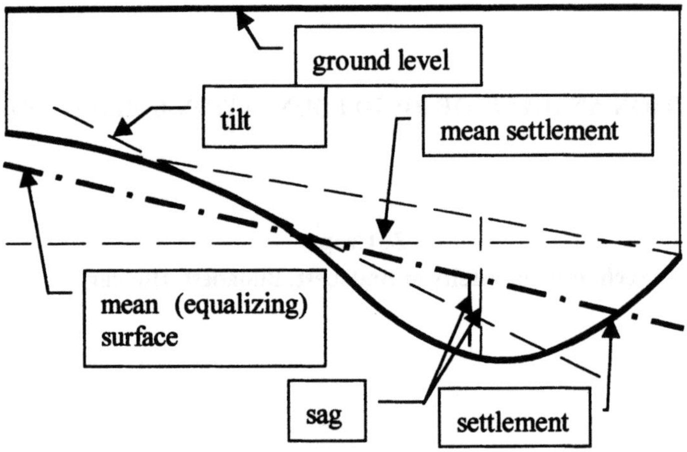

Fig. 1

Foundation slabs are mostly calculated by some FEM slab programme where the soil model commonly applied is the Winklerian bedding. Of the structure, the foundation slab as a separate member is only considered while the stiffening effect of the rest of the structure is not taken into account.

For the design of flexible foundation slabs the Author together with Co-author Tornyos has developed a method which is essentially more accurate than methods based on the Winkler model. Based on the procedure described [1], a programme module was produced for use in conjunction with the static code named FEM Design, developed jointly by Skanska Software (Sweden-Malmö) and PiHun Ltd. (Hungary-Budapest).

It is a common case that the structure as a whole can be considered as rigid in respect of settlements that is deflections and sagging in the structure itself are negligibly small as compared to settlements. Then the fact whether the foundation structures is flexible e.g. a thin slab is indifferent as far as soil response is concerned.

It is also a common case that a major building is divided into parts by expansion joints or there are adjacent but independent buildings, which may not even, have been constructed at the same time. In such cases, individual buildings or major parts of them may also be considered as rigid (with respect to the subsoil) as long as deflections in the structure or in structural elements are again negligibly small relative to the equalising plane of the settlement surface.

Author has worked out a calculation method applicable to the above cases, based on literature [2]. His programme, specifically suited to the calculation of settlements of rigid buildings, or major rigid parts of buildings, by taking into account interaction between them was first used in practice in 1990 at the design of a condominium development project near Budapest. The programme furnishes contact pressures under the base as well as non-uniformly distributed bedding coefficients which values can in turn be used as input data for a FEM foundation calculation programme.

2 THE STRUCTURAL MODEL

A building structure, or a part of it, is considered as rigid by itself. Its settlement, s_i, at a point defined by the co-ordinates $[x_i, y_i]$ is given by

$$s_i = a \cdot x_i + b \cdot y_i + c. \tag{1}$$

Such points are not nodal points as those used in FEM analyses but raster points at centre of elements (Fig. 2). By combining results from n points tested of the structure, (2) gives the cumulative settlement for individual points of the structure.

$$\mathbf{s} = \mathbf{G} \cdot \mathbf{m}, \quad \mathbf{s} = \begin{bmatrix} s_1 \\ s_2 \\ \vdots \\ s_n \end{bmatrix}, \quad \mathbf{G} = \begin{bmatrix} x_1 & y_1 & 1 \\ x_2 & y_2 & 1 \\ \vdots & \vdots & \vdots \\ x_n & y_n & 1 \end{bmatrix}, \quad \mathbf{m} = \begin{bmatrix} a \\ b \\ c \end{bmatrix}. \tag{2}$$

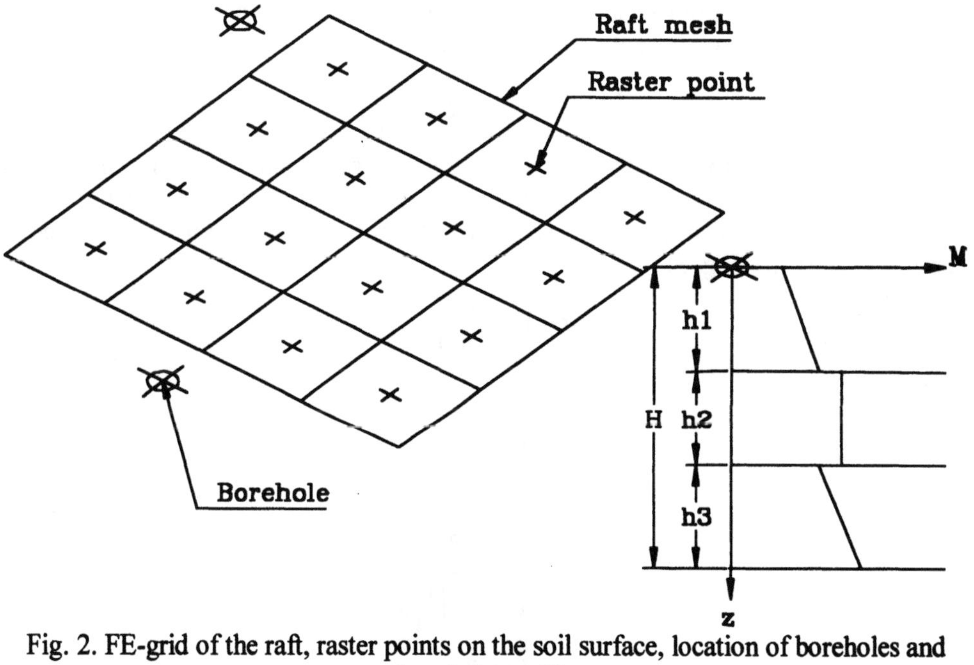

Fig. 2. FE-grid of the raft, raster points on the soil surface, location of boreholes and borehole profile

When the number of building parts considered is k, the equation becomes

$$\mathbf{s}_m = \mathbf{G}_m \cdot \mathbf{m}_m, \quad \mathbf{s}_m = \begin{bmatrix} \mathbf{s}_1 \\ \mathbf{s}_2 \\ \vdots \\ \mathbf{s}_k \end{bmatrix}, \quad \mathbf{G}_m = \begin{bmatrix} \mathbf{G}_1 & & & \\ & \mathbf{G}_2 & & \\ & & \ddots & \\ & & & \mathbf{G}_k \end{bmatrix}, \quad \mathbf{m}_m = \begin{bmatrix} \mathbf{m}_1 \\ \mathbf{m}_2 \\ \vdots \\ \mathbf{m}_k \end{bmatrix}. \tag{3}$$

3 THE SOIL MODEL

3.1 Basic assumptions

The soil mass considered in this method is pseudo-elastic, generally layered halfspace. In any layer, the physical constant – the M(z) compression modulus – can vary linearly with depth (Fig. 2). In the special case of a single layer with constant modulus, the classical Boussinesq halfspace applies. The compression modulus of the k-layer is calculated from the following formula:

$$M_k(z) = M_{k0} + m_k \cdot (z - \sum_{i=1}^{k-1} h_i), \quad 0 \le z \le H \tag{4}$$

where M_{k0} is the compression modulus at the top of the k-layer, m_k is a factor of modulus change in the k-layer, z is the depth under the foundation plate, h_i is the thickness of the i-layer, H is the limit depth (the soil is considered only down to this limit depth H). The elastic properties of the soil have the following relation:

$$M = E \cdot \frac{1-\mu}{(1+\mu)\cdot(1-2\cdot\mu)} \tag{5}$$

where E is the elasticity modulus, μ is the Poisson ratio and M is the compression (oedometric) modulus. For each raster point on the raft base, the data of the nearest borehole profile is applied (Fig. 2). The vertical stresses in the soil are obtained by Boussinesq theory [3] (Fig. 3)

$$\sigma = P \cdot \frac{3 \cdot z^3}{2 \cdot \pi \cdot R^5}. \tag{6}$$

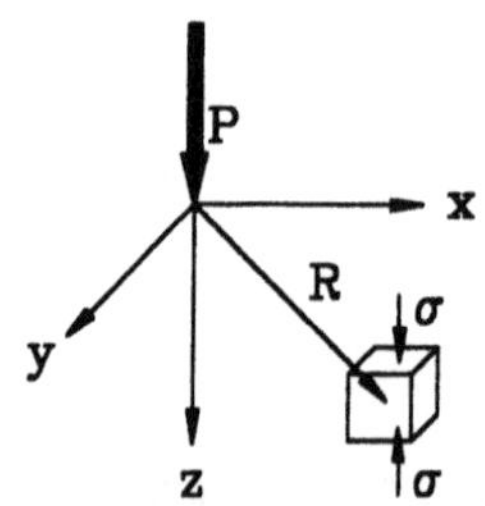

Fig. 3.

Although this theory strictly applies to the homogeneous, elastic infinite half-space, it can be reliably used in practice for settlement calculations. In the calculation of settlements, the surface load is assumed to the soil via the plate elements resting on the soil with no tensile stresses between slab and soil. In order to obtain the stress (σ) at a point at depth z, the cumulative effect of all plate elements is considered. The settlement is calculated in the usual way from the stresses and from the compression modulus, as:

$$s = \int_H \frac{\sigma(z)}{M(z)} dz \tag{7}$$

where s is the settlement. The value of H is normally determined by national standards. In Hungary, it is taken equal to the depth where the excess soil stress σ due to loading reduces to 20 % of the effective overburden pressure, σ_0:

$$\sigma(H) = 0.2 \cdot \sigma_0(H) \tag{8}$$

Alternatively, H can be obtained from the dimensions of the raft, for example, with large plates the limit depth

$$H = 0.5 \cdot B \tag{9}$$

where B is the smaller dimension (width) of the slab.

3.2 The flexibility matrix of the soil

A FE-grid will be considered at the surface of the soil, which exactly matches the FE-grid for the foundation slab (Fig. 1). The centre of each slab element is taken as a raster point. In order to obtain the flexibility matrix, each raster point is considered to be acted upon by a unit concentrated (point) vertical force. The elements of the soil flexibility matrix $[s_{ij}]$ are the deflections pertaining to the unit load, as computed from (7)

$$\mathbf{S} = [s_{ij}] = \left[\sum_{k=1}^{L} \int_{h_k} \frac{\sigma_{ij}(z)}{M_{ki}(z)} dz \right] \quad 0 \le z \le H \tag{10}$$

where subscript i denotes the location, i.e. the raster point for which the settlement is calculated from the unit loads; subscript j denotes the raster point which is acted by the unit load; L is the number of layers at the borehole profile relevant to i^{th} raster point; z is the depth below the foundation level; h_k is the thickness of the k-layer; $\sigma_{ij}(z)$ is the stress function.

When i=j, then in order to avoid singularities the stress function needs to be adjusted by dividing the area of the slab element pertaining to the raster point considered into triangles and calculating the stress by integrating the partial effects obtained from (6), (Fig. 4).

In a polar co-ordinate system, the partial stress obtained by integration with respect to r over the area of a triangular sub-element is

$$\sigma_\Delta(z) = \frac{1}{2\pi A_\Delta} \int_{-\vartheta_1}^{\vartheta_2} 1 - \left(\frac{z \cdot \cos\vartheta}{\sqrt{m^2 + (z \cdot \cos\vartheta)^2}} \right)^3 d\vartheta$$

$$x_p = t \cdot x_2 + (1-t) \cdot x_1 \ ; \ y_p = t \cdot y_2 + (1-t) \cdot y_1 \, . \tag{11}$$

$$t = \frac{(y_c - y_1)\cdot(y_2 - y_1) + (x_c - x_1)\cdot(x_2 - x_1)}{(y_2 - y_1)^2 + (x_2 - x_1)^2}$$

The angles ϑ_1 and ϑ_2 are to be interpreted according to the sign convention assumed (i.e. if $t<0$, then $\vartheta_1<0$, and for $t>1$, $\vartheta_2<0$); A_Δ is the area of a triangular. The integration with respect to ϑ is carried out numerically, and the overall value of the stress σ_{ii} is calculated by summing up the partial values obtained for the triangles.

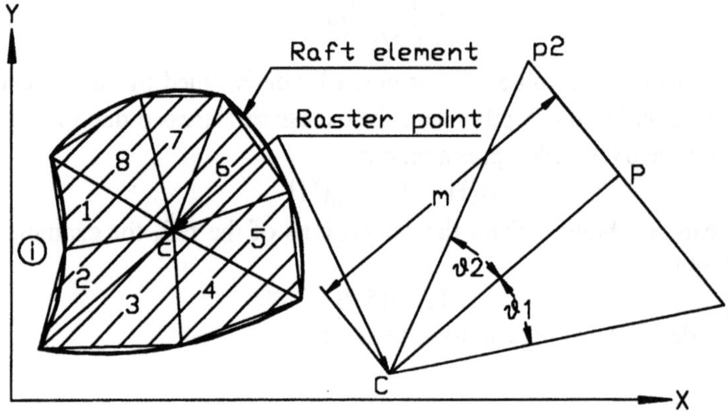

Fig. 4.

When i≠j then the stress function is

$$\sigma(z)_{ij} = \frac{3 \cdot z^3}{2 \cdot \pi \cdot R_{ij}^5} \tag{12}$$

where $R_{ij} = \sqrt{(x_i - x_j)^2 + (y_i - y_j)^2 + z^2}$ and x_i, x_j, y_i, y_j: are the co-ordinates of the ith and j[th] raster points, respectively.

The stiffness matrix of the subsoil is obtained as the inverse of the flexibility matrix:

$$\mathbf{K} = \mathbf{S}^{-1} \tag{13}$$

4 RELATION BETWEEN SOIL AND STRUCTURE

The relationship between the settlement of a point on the surface of the soil (i.e. the mean settlement of the slab element above that point) and the force acting on it (i.e. the resultant of contact pressures acting on the slab element) is given by (14),

$$\mathbf{K} \cdot \mathbf{s}_m = \mathbf{q}, \quad \mathbf{s}_m = \begin{bmatrix} s_1 \\ s_2 \\ \\ s_T \end{bmatrix}, \quad \mathbf{q} = \begin{bmatrix} F_1 \\ F_2 \\ \\ F_T \end{bmatrix}, \quad T = \sum_{i=1}^{k} n_i \tag{14}$$

where s_m and k are as defined by (3); n_i is the number of points under the ith building (part of building); F_j is the resultant force acting on the jth element.

By substituting s_m according to (3) and by multiplying by G_m^T from the left we obtain

$$\mathbf{G}_m^T \cdot \mathbf{K} \cdot \mathbf{G}_m \cdot \mathbf{m}_m = \mathbf{G}_m^T \cdot \mathbf{q} \tag{15}$$

$$\mathbf{K}_m \cdot \mathbf{m}_m = \mathbf{q}_m, \quad \mathbf{K}_m = \mathbf{G}_m^T \cdot \mathbf{K} \cdot \mathbf{G}_m, \quad \mathbf{q}_m = \mathbf{G}_m^T \cdot \mathbf{q} \tag{16}$$

In (16)

$$\mathbf{G}_m^T \cdot \mathbf{q} = \begin{bmatrix} \mathbf{G}_1^T \cdot \mathbf{q}_1 \\ \mathbf{G}_2^T \cdot \mathbf{q}_2 \\ \vdots \\ \mathbf{G}_k^T \cdot \mathbf{q}_k \end{bmatrix}, \quad \mathbf{G}_j^T \cdot \mathbf{q}_j = \begin{bmatrix} \sum\limits_{(j)} x_i \cdot F_i \\ \sum\limits_{(j)} y_i \cdot F_i \\ \sum\limits_{(j)} F_i \end{bmatrix} \tag{17}$$

give for each group of buildings the moments about axis y and x, respectively, and the resultant force.

With displacement characteristics a_i, b_i and c_i for the ith building part known, settlement of individual points can be obtained by using (2), whereas (14) gives contact pressures (by dividing by the base area of the element).

Resulting contact pressures as well as bedding coefficient values obtained as the quotient of contact pressure and settlement can then be introduced into the calculation of the foundation slab.

5 CHECKING OF RESULTS

As an example of verification tests, the calculation of a rigid circular plate is given in the following (Fig. 5).

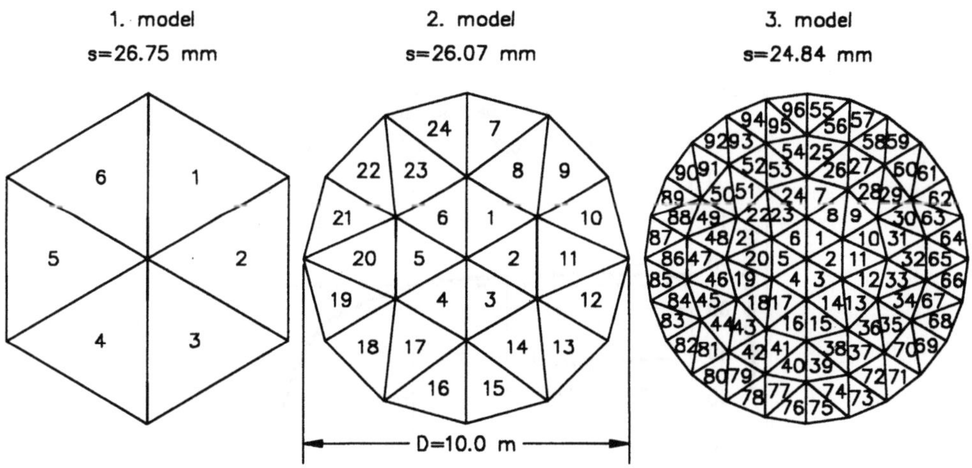

Fig. 5. FE grid of test models

Initial data:

- D=10 m;
- H=10 m;
- M=20 MPa;
- p=100 kPa (uniform load).

The grid covering the circular base area is made up of equilateral triangles.
Results:

- Relationship between contact pressures and subdivision of elements: Fig. 6.
- Comparison of settlements calculated by numerical and analytical methods: Fig. 7. Note that analytical calculation requires values of modulus of compressibility and of Poisson's number as given in (5).
- Relationship between settlement and subdivision of elements: Fig. 8.

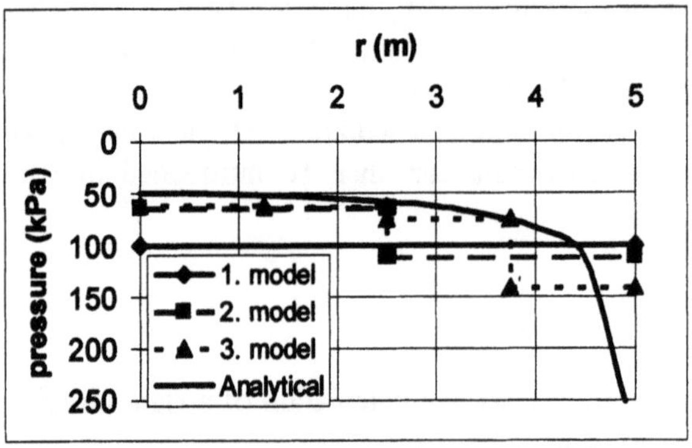

Fig. 6. Soil pressures under test models

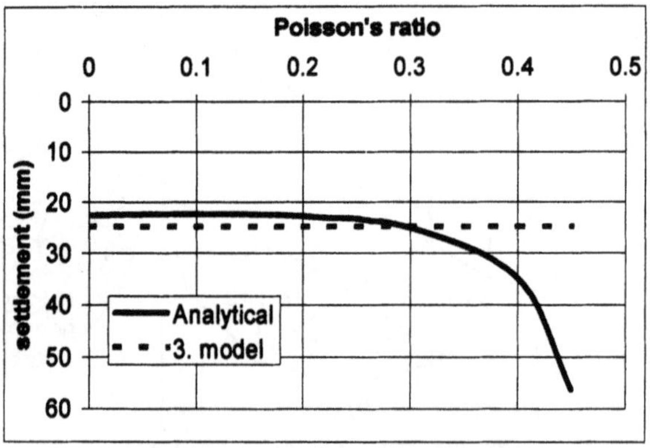

Fig. 7. Settlement of test and analytical model

The method described was first applied to the re-calculation of settlements of a hospital complex built in South-Hungary a few years ago. Following the completion of construction, serious damages to buildings occurred due to significantly different loads acting on the central high-rise section and on other lower parts of the complex. The Department of Geotechnics in TUB was involved as a consultant in the rehabilitation study of the case.

Initial data were as follows.

- Plan area of building 90 m by 72 m;
- H=10.8 m as per (8);
- M=5 MPa (silty to clayey soils);
- p=32.6-261 kPa;
- number of elements: 51.

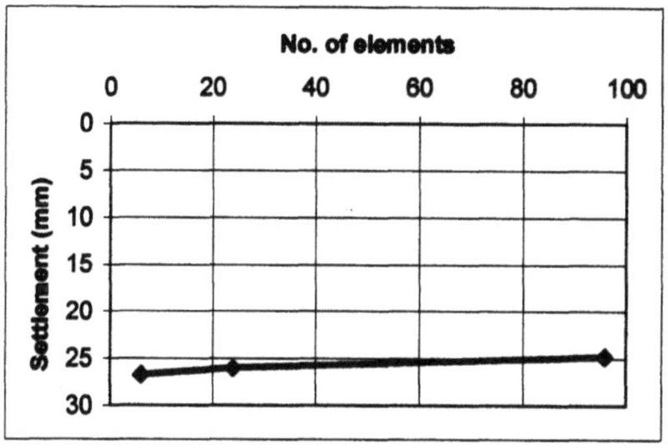

Fig. 8. Settlement of test models

PhD student Mohammed Roxy carried out calculations.

Settlements for the complex divided by expansion joints are shown in Fig. 9. The max. calculated settlement of 367 mm differs only by a few cm from the measured value.

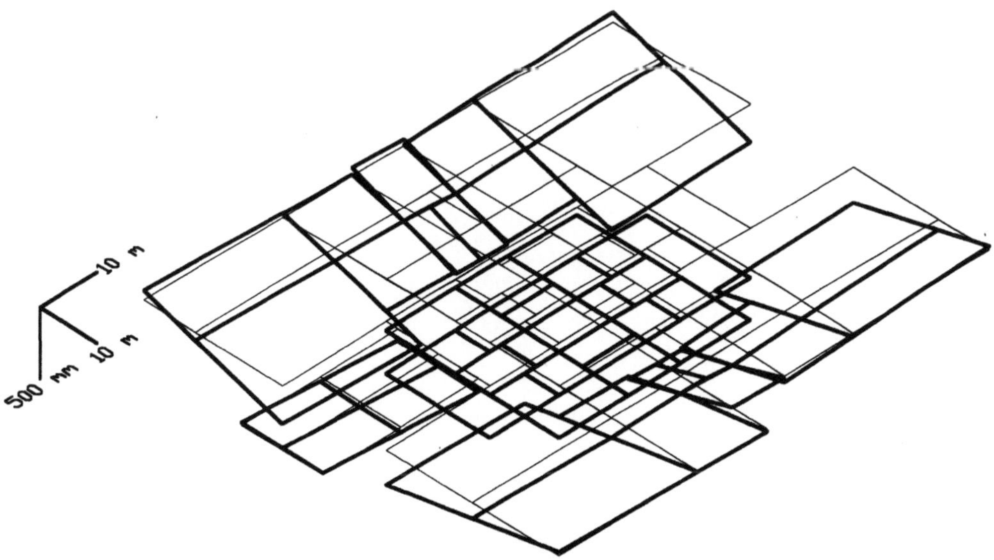

Fig. 9. Settlement of a hospital building

CONCLUSION

The Author developed, on the basis of conventional settlement analysis, a novel algorithm which is applicable to the calculation of settlements under rigid buildings or building groups with due consideration to interaction effects. The program based on the algorithm can handle cases involving multiple soil exploration profiles, stratified soil, different moduli of compressibility changing linearly with depth in each layer. Concentrated loads, line loads or distributed surcharge can equally be handled. The results are settlements and contact pressures under the building (group of buildings) and derived supplementary bedding coefficients which all can be introduced as input data in current computer programmes used for foundation calculations.

ACKNOWLEDGEMENTS

The present study was supported by the Hungarian National Scientific and Research Foundation (OTKA # T 022138).

REFERENCES

[1] Z. CZAP - Á. TORNYOS. Foundation interaction analysis with uncoupled iterative method. Computer Methods and Advances in Geomechanics, Yuan (ed.). Balkema, Rotterdam, 1997.
[2] E. DULÁCSKA (ed.). Soil Settlement Effects on Buildings. Elsevier, Amsterdam, 1992.
[3] K. SZÉCHY. Foundation I-II. Műszaki Könyvkiadó, Budapest, 1971.

A MODEL FOR THE BEHAVIOUR OF SOIL-STRUCTURE INTERFACE

V. De Gennaro
CERMES, École Nationale des Ponts et Chaussées-LCPC, Paris, France

G.N. Pande
University of Wales, Swansea, UK

ABSTRACT: The study of the mechanical behaviour of the soil-structure interfaces in the context of the theoretical framework of strain localization is presented in this paper. Classical conditions for localization of deformation are compared with other existing criteria and applied, by means of deviatoric hardening constitutive model, to the analysis of the stability in interface tests performed in a purpose-built apparatus. The post-peak behaviour is described applying an homogenisation technique. Results are compared with the observed behaviour.

1 INTRODUCTION

In many geotechnical problems failure or collapse is often preceded by the formation of narrow shear bands or cracks where localization of strains takes place. This localized mode of deformation is encountered in many geotechnical engineering problems such as retaining walls, soil nailed structures, piles, earth-dams and in some cases may arises from the fact that geotechnical structures often involve interaction between soil and structural elements. In the latter case, the shear band virtually corresponds to the interface layer between the soil and the structural element and represents a zone over which transition of stresses and displacements (slip) takes place.

Tests on initially homogeneous specimens often lead to strain localization phenomena in all of the most widely used apparatus for soil testing (among others: Vardoulakis and Sulem, 1994; Desrues, 1990; Arris et al., 1995; Lerat, 1996). Mathematically bifurcation corresponds to the loss of uniqueness of the solution of the governing equilibrium equations. In the context of the classical continuum theories, early works of Drucker (1950) and later on Hill (1958), Mandel (1966) and Rudnicki and Rice (1975) give criteria for possible bifurcations (localized or diffused) in elastoplastic materials with both associated or non-associated flow rules. Loss of uniqueness, loss of material stability and any possible bifurcation correspond to the loss of positive definiteness of the rate of the second order work. This condition is also referred as loss of ellipticity of the equilibrium equations and corresponds to the zero value of the determinant of the acoustic tensor. If only conditions on the kinematics of the mode of deformation are involved necessary condition corresponds to the loss of strong ellepticity (Bigoni and Hueckel, 1991). This occurs when the symmetric part of the acoustic tensor is positive semidefinite. In the theoretical framework mentioned above a study of the instability phenomena associated to the mobilisation of shear stress at the interface between the soil and the structure is reported based on the experiments of Lerat (1996) on a ring simple shear test (ACSA). An elastoplastic strain hardening model has been used, testing its capability to predict the inception of localization of deformations. Based on the original technique adopted by Pietruszczak and Niu (1992), the post-localized behaviour has been described by means of a model based on a homogenisation technique. Comparison between experimental observations and numerical predictions are then presented. The results of the numerical study show the capability of the proposed approach and its suitability for modelling problems involving the description of soil-structure interfaces.

2 INTERFACE ANALYSIS AND MATERIAL INSTABILITY

The ring simple shear apparatus simulates the shearing of a thick walled cylinder of granular material in plane strain by a cylindrical inclusion subjected to torsion. Figure 1a shows a schematic view of the ring simple shear apparatus. Problem is axisymmetric with respect to the z axis. Analysis has been developed admitting that each surface parallel to the middle plane of the specimen and perpendicular to the axis of geometrical symmetry is under plane strain. We will focus our attention to the conditions corresponding to the constant volume (undrained conditions) tests. The geometry of the problem is shown in Fig. 1b, soil sample is dry, so that analysis can be performed in terms of total as well as effective stresses. Moving from the particular geometry of the problem a possible failure mechanism could be the one shown in Fig. 1b. Experimental results seems to corroborate this possible scheme, in fact they show a strong localization of deformation at the interface between soil and structural element (Lerat, 1996). The surrounding material is disturbed up to a small distance from the inner cylinder, denoted a in Fig. 1b. We will suppose that the overall mechanical behaviour is guided by the stress and strain fields inside the deformed region A'BCD'. The latter, as it will be shown, will be treated as a composite material consisting of a shear band of thickness t surrounded by the remaining intact material. Irrespective of the particular strain rate vector considered, a material is stable (Drucker, 1950; Hill, 1958) if the second order rate work is always positive. Necessary condition for loss of material stability is then:

$$d\varepsilon_{ij}\, d\sigma_{ij} = 0 \tag{1}$$

This type of bifurcation criterion can also be rewritten considering the elastoplastic stiffness matrix **D** of the materials. In this case the condition for any type of bifurcation to occur is:

$$\det \mathbf{D}^{s} = 0 \tag{2}$$

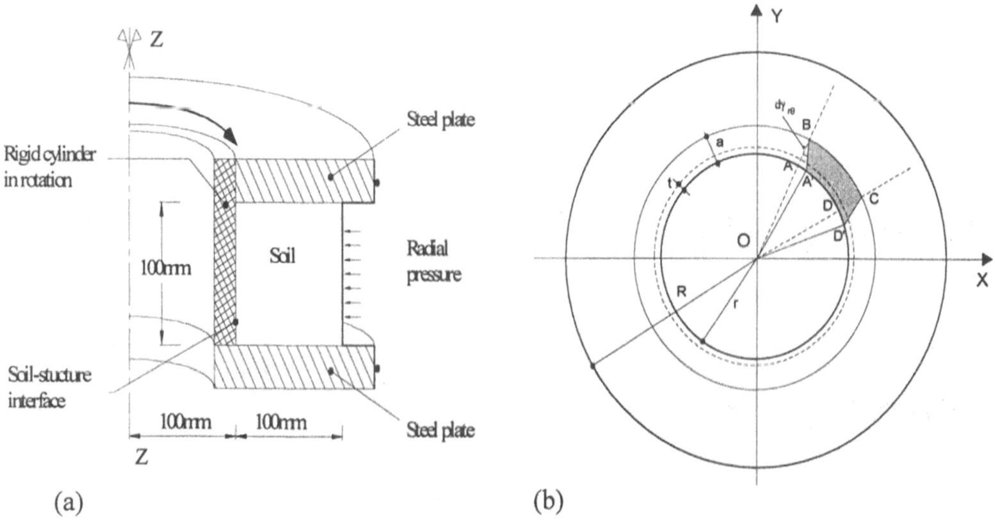

Fig. 1: (a) Schematic view of the ring simple shear device; (b) possible failure mechanism

where $\mathbf{D}^{s}$ is the symmetric part of the stiffness matrix, $D^{s}_{ij} = 1/2(D_{ij} + D_{ji})$. Equation (2) can also be reduced to a lower bound condition for the hardening modulus (Maier and Hueckel, 1979). If an homogeneous body is susceptible to localize, the strain field inside the discontinuity must obeys the conditions corresponding to the kinematics compatibility of the discontinuous mode of deformation with the strain field of the remaining intact-body. Assuming that in the deformed body stress and strain fields are uniform, imposing equilibrium across the plane of discontinuity and assuming intuitively at the inception of localization $\mathbf{D} = \mathbf{D}^{L}$, where $\mathbf{D}^{L}$ is the stiffness matrix of the localized material, the necessary condition for the formation of the shear band reduce to the condition of the loss of ellipticity of the general problem of equilibrium (Rudnicki and Rice, 1975):

$$\det \mathbf{Q} = 0 \quad ; \qquad \mathbf{Q} = (\mathbf{D}\, \underline{n})\, \underline{n}^{t} \tag{3}$$

where $\underline{n}$ is the normal to the plane of discontinuity, **Q** is the acoustic tensor. If the hypothesis adopted for the stiffness matrices of the two materials is now relaxed, the necessary condition for the inception of localized deformation kinematically compatible corresponds to the so-called loss of strong ellipticity (Bigoni and Huekel, 1991). In this case bifurcation modes in the form of shear band occur when:

$$\det \mathbf{Q}^{s} = 0 \tag{4}$$

where $\mathbf{Q}^{s}$ is the symmetric part of the acoustic tensor. This criterion can be satisfied prior to or at the same time as the criterion in eq. (3). In materials with non-associated flow rule criterion in eq. (4) is fulfilled before the condition given in eq. (3), while identical predictions are obtained if material obey an associated flow rule. Indeed bifurcation modes can exist only in the case of strain softening if an associated flow rule is observed.

3 NUMERICAL MODELLING OF INTERFACE BEHAVIOUR

Interface behaviour has been widely studied in recent years. Many issues have been raised in the literature concerning both experimental and theoretical aspects (among others: Desai, 1981; Boulon and Nova, 1990). Constitutive relationships similar to those used for rock joints and interfaces can also be applied to the shear band behaviour. The non-linear behaviour of the material before strain localization has been described using an extension of the Mohr-Coulomb criterion for cohesionless soils incorporating deviatoric strain hardening (Pietruszczak and Stolle, 1987; Pietruszczak and Niu, 1992). In axisymmetric conditions and along with the hypothesis of plane strain as obtained in ACSA apparatus the yield function (assuming compression positive) can be written as:

$$F = \sqrt{\alpha} - \eta\left(\sigma_{\theta\theta} + \sigma_{rr} + \sigma_0\right) = 0 \qquad \alpha = \left(\sigma_{\theta\theta} - \sigma_{rr}\right)^2 + 4\sigma_{r\theta}^2 \tag{5}$$

$$\eta = \eta_0 + (\eta_i - \eta_0)\frac{\varepsilon^p}{A + \varepsilon^p} \qquad \varepsilon^p = \frac{1}{2}\sqrt{\left(\varepsilon^p_{\theta\theta} - \varepsilon^p_{rr}\right)^2 + \left(\gamma^p_{r\theta}\right)^2} \tag{6}$$

η is the mobilised friction angle, η_0 defines the elastic region, η_i is the peak friction coefficient of the intact material, σ_0 is a constant of the model, A is a parameter controlling the rate of the deviatoric hardening and, finally, ε^p is the plastic shear strain.
For the plastic potential function the following form is assumed:

$$Q = \sqrt{\alpha} + \beta\left(\sigma_{\theta\theta} + \sigma_{rr}\right)\ln\left(\frac{\sigma_{\theta\theta} + \sigma_{rr}}{\sigma_0}\right) = 0 \tag{7}$$

In the above equation β is the slope of the phase transformation line that defines the threshold from compaction ($\eta < \beta$) to dilatancy ($\eta > \beta$) and σ_0 is the initial value of the effective normal stress on the plane of the shear band (i.e. radial direction). Experimental findings (Lerat, 1996) show that, in undrained conditions, shear tests on dense or medium dense sand specimens are characterised by an increase of the radial pressure at the interface as a consequence of the dilatant characteristics of the material. The initial phase of restrained dilatancy decreases during shear and equals zero for the larger tangential displacement, this phenomenon is more evident at higher initial external radial pressures imposed. For that reason, to avoid excessive dilatancy, a special form has been selected for the β coefficient, based on the expression of the dilatancy ratio $d\varepsilon_v / d\gamma_{r\theta}$ obtained in the context of the strain hardening elastoplasticity theory. Thus a possible expression for β is:

$$\beta = \beta_0 + (\eta - \beta_0)\left[1 - \mathrm{sech}\,(\mathrm{B}\,\varepsilon^p)\right] \tag{8}$$

value of β_0 corresponds to the initial slope of the phase transformation line while the parameter B controls the rate of dilatancy. For the deviatoric hardening model a path-independent criterion has also been proposed (Pietruszczak and Stolle, 1987). In this study only the classical bifurcation conditions are investigated, given by conditions in eqs. (2), (3) and (4), the latter is the set of relations that are considered in order to make a comparison between the different bifurcation criteria. When the homogeneous sample undergoes a unstable response during the incremental loading process, corresponding to the formation of a localized deformation pattern at the soil-structure interface, response of the specimen in terms of shear resistance shows a softening branch in the stress-strain graph. Several approaches have been addressed to the possibility to give a satisfactory description of the non-homogeneous localized deformation. In this paper the problem of the description of the localized behaviour is solved based on an homogenisation technique (Pietruszczak and Niu, 1992). The material can then be treated as a composite consisting of the weak material inside the interface bounded by the remaining intact material that usually unloads and behaves in an elastic manner. A modified Mohr-Coulomb model including deviatoric strain softening (Schweiger et al., 1997) has been assumed for the interface. The yield function and the plastic potential are, therefore, of the type:

$$\mathrm{F} = \sigma_{r\theta} - \mathrm{M}\,\sigma_{rr} = 0 \qquad ; \qquad \mathrm{Q} = \sigma_{r\theta} - \mathrm{M}_c\sigma_{rr} = 0 \tag{9}$$

$$\mathrm{M} = \mathrm{M}_r - (\mathrm{M}_r - \mathrm{M}_l)\exp\left[-\mathrm{A}_1\,(g^p_{r\theta})^2\right] \tag{10}$$

$$\mathrm{M}_c = \mathrm{M}_{cr} - (\mathrm{M}_{cr} - \mathrm{M}_{cl})\exp\left[-\mathrm{B}_1\,(g^p_{r\theta})^2\right] \tag{11}$$

where M is the mobilised friction, M_r is the residual friction, M_l is the value of friction at bifurcation (same convention for the indices of the dilatancy M_c). Again, the parameters A_1 and B_1 control the rate of deviatoric softening and $g^p_{r\theta}$ is the strain softening parameter.

Assuming perfect bonding at the interface of the two materials (intact material and material inside the shear band). The stress and strain in the homogenised medium can be defined via the mixture:

$$\underline{d\sigma} = \mu_l \underline{d\sigma}^l + \mu_n \underline{d\sigma}^n \quad ; \quad \underline{d\varepsilon} = \mu_l \underline{d\varepsilon}^l + \mu_n \underline{d\varepsilon}^n \quad ; \quad \mu_l + \mu_n = 1 \tag{12}$$

In the following indices "n" and "l" denote non-localized and localized material, μ_l and μ_n are the volume fractions. If "t" is the thickness of the shear band, according with the geometry of the ACSA apparatus (Fig. 2b), we can write:

$$\mu_l = \frac{t}{a} \quad \text{and} \quad \mu_n = 1 - \mu_l \tag{13}$$

The kinematics constraint and the equilibrium conditions, in agreement with the hypothesis of perfect bonding, can be established so that the continuity of the shear and normal stress between the two materials (localized and non-localized) is verified. For the sake of brevity details of calculation are omitted, similar developments can be found in Pietruszczak and Niu (1992). Invoking for both intact material and interface an elastoplastic behaviour, the functional form of the constitutive matrices has been derived from the two models discussed previously, namely: the deviatoric hardening model for the intact material and the deviatoric softening model for the material of the shear band. The constitutive matrix for the homogenised material can then be derived from equation (12) by employing the constitutive matrices defined for the two materials. It will depends on the thickness of the interface (*t*) that plays the role of an internal length allowing to scale the problem to the dimension of the localized zone. This is slightly different in comparison with the previous studies above mentioned, where the thickness of the shear band is assumed to be negligible.

4 NUMERICAL STUDY OF INTERFACE TESTS

The analysis proposed has been implemented in a computer program with a view to match experimental results obtained from tests at constant volume performed on the ACSA apparatus on dense Hostun RF sand (Lerat, 1996).
Tests analysed has been performed at an initial confining pressure of 100 kPa and 400 kPa. At first the numerical study has been addressed to the determination of the instability conditions. The parameters adopted for the model have been directly derived from the test, elastic parameters (E, ν) are those derived by Dupla (1995) from cavity expansion tests on Hostun RF sand. For the test *s1905dl* numerical results in terms of evolution of the determinant of the acoustic tensor and of its symmetric part are shown in Fig. 2 (a and b). Results are traced for four different increments of the tangential displacement and varying the orientation of the shear band. In both cases the last increment corresponds to the inception of instability according to condition (3) and (4) respectively.
Curves are closer to the condition of zero determinant for increasing increments, which means increasing tangential displacement, bifurcation is reached for condition (3) and (4)

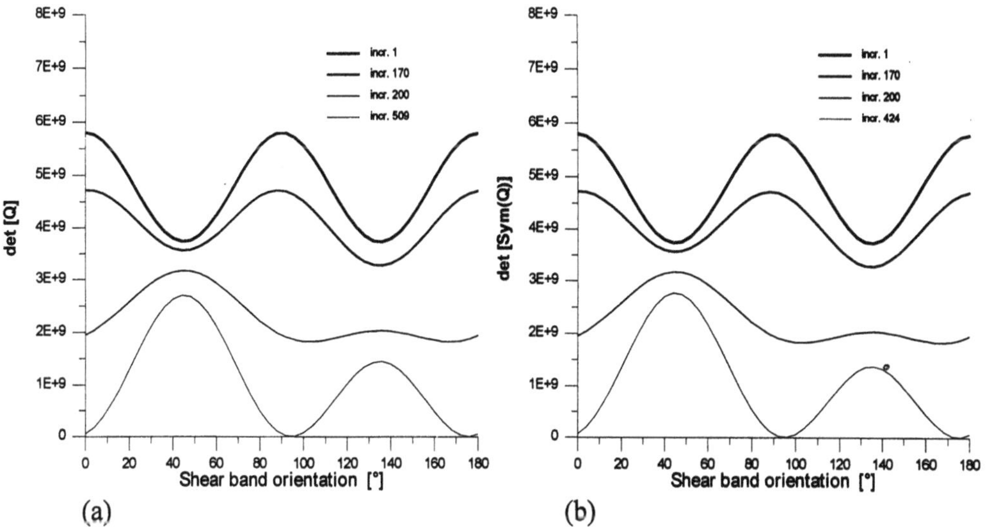

Fig. 2: Evolution of the determinant of the acoustic tensor (a) and of its symmetric part (b) during the incremental process varying the shear band orientation (test *s1905dl*)

when the normal at the discontinuity plane is oriented at 94°, corresponding to θ_{sb} = 4° ("sb" means shear band). These values are in a good agreement with the observed localization at the interface in a thin layer parallel to the circular surface of the inner cylinder.

In Fig. 3, for the current test, comparison between measured and predicted values of shear stress and radial pressure at the interface versus tangential displacement are plotted assuming an homogeneous state of deformation, points corresponding to the different localization criteria are also indicated. For this test a satisfactory prediction of the inception of bifurcated solutions is provided from both Hill's criterion and the condition corresponding to the loss of strong ellipticity. Moreover the instability in both cases is detected at the same increment (incr. 424). Loss of ellipticity, eq. (3), is activated later on and is always in the hardening regime.

As it has been already reported the general bifurcation criterion given by Hill (eq. (2)) is a necessary condition for any type of bifurcation. Kinematic compatibility between non localized and localized material as well as identical stiffness matrices for the materials inside and outside the shear band at the inception of localization are not required. The latter assumptions are necessary if the usual analysis via the acoustic tensor is undertaken. On the other hand loss of strong ellipticity requires only a kinematically compatible mode. As a consequence the results presented in Fig. 3 seems to give the possibility to conclude that in this particular case the proposed failure mechanism (Fig. 1b) well corresponds to the real deformation mode at the inception of bifurcation.

Starting from the point of detection of the loss of uniqueness of the solution, modelling of

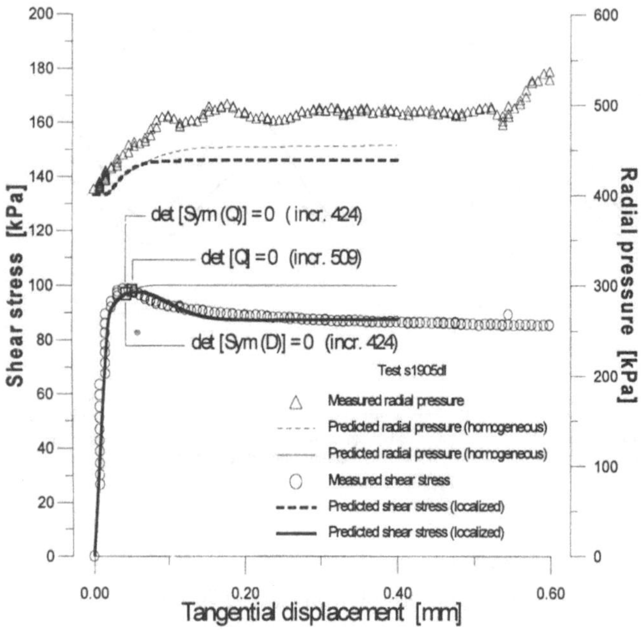

Dev. hardening	values
Young's Mod. E	15 MPa
Poisson ratio, ν	0.3
Peak friction, η_i	0.18
Init. friction, η_o	0
Param. β_o	0.017
Param. σ_o	1 kPa
Param. A	$7\ 10^{-3}$
Param. B	650
Dev. softening	
Norm. Stiffness	6.7 GN/m
Tang. Stiffness	1.9 GN/m
Res. friction, M_r	0.10
Dilatancy , M_{cl}	0.08 β
Res. dilat. M_{cr}	$5\ 10^{-4}$
Param. A_1	5000
Param. B_1	8000
Interf. thick., t	0.003 m

Fig. 3: Predicted instabilities and results of the numerical simulations of the experiments

the post-peak behaviour is described with the above mentioned homogenisation technique, assuming that the material within the shear band has a deviatoric softening mode of deformation. Initial conditions in terms of mobilised friction and state of stress are those obtained from deviatoric hardening model when bifurcation occurs. The remaining material is always treated as a deviatoric hardening material. Nevertheless during the incremental response it has been observed that after localization, intact materials unloads (dF < 0) and consequently it behaves in an elastic manner.

Results of simulation involving the post-peak behaviour are also shown in Fig. 3. In Fig. 4 results of the simulation of the test *s1205dl* performed at 100 kPa of confining pressure are presented. For the sake of brevity similar analysis to those shown in Fig. 2 for the test *s1905dl* are not presented. Again satisfactory prediction of the inception of bifurcated solutions is provided from both Hill's criterion and also the condition corresponding to the loss of strong ellipticity. The latter in this case is fulfilled for a value $\theta_{sb} = 4°$. The comparison between experiments and simulations seems to give satisfactory results.

As it has been already mentioned the description of the post-peak behaviour has been obtained applying an homogenisation technique to the composite material inside the region A'BCD' in Fig. 1b. The value of *t* in this study has been chosen as 10 D_{50} (among others: Mühlhaus and Vardoulakis, 1987).

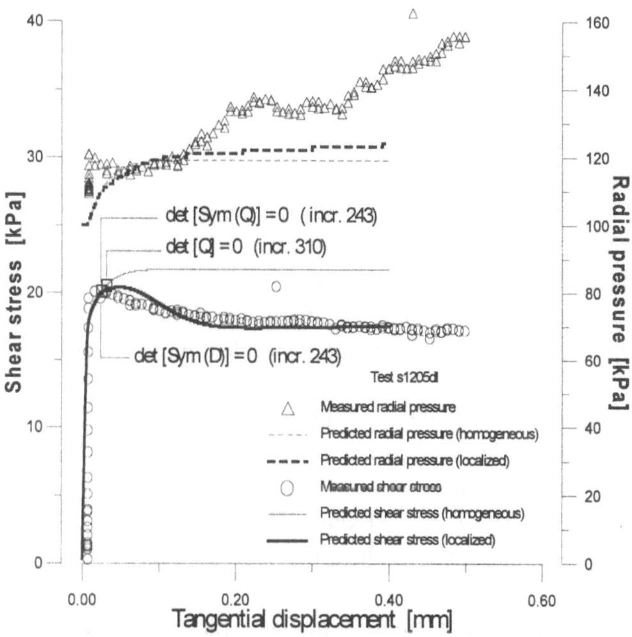

Dev. hardening	values
Young's Mod. E	15 Mpa
Poisson ratio, ν	0.3
Peak friction, η_i	0.18
Init. friction, η_o	0
Param. β_o	0.017
Param. σ_o	1 kPa
Param. A	$7\ 10^{-3}$
Param. B	650
Dev. softening	
Norm. Stiffness	6.7 GN/m
Tang. Stiffness	1.9 GN/m
Res. friction, M_r	0.10
Dilatancy , M_{cl}	0.08 β
Res. dilat. M_{cr}	$5\ 10^{-4}$
Param. A_1	5000
Param. B_1	8000
Interf. thick., t	0.003

Fig. 4: Predicted instabilities and results of the numerical simulations of the experiments

5 CONCLUSION

The problem of the stability of the mechanical response of the soil-structure interfaces has been discussed in this paper, based on some simple mechanical and geometrical assumptions. Results of a numerical study investigating the behaviour of the interfaces in a purpose-built apparatus has been presented. Moving from a possible failure mechanism at the interface layer, in the context of the theoretical framework of strain localization, a detailed study of the conditions that lead to an unstable behaviour have been reviewed. The classical necessary condition for localization corresponding to the Hill's stability criterion has been verified to give satisfactory results for the analysis performed. Possible bifurcations with a compatible mode of deformation (loss of strong ellipticity), furnish same conditions as that given by Hill's criterion. The estimated value of shear band orientation seems to be in agreement with the observed experimental behaviour. The description of the post-peak behaviour has been developed using a homogenisation technique, treating the material involved in the deformation process as a composite material consisting of a shear band and the remaining intact material. The response predicted by the model seems to fit well with the basic features of the phenomena that develop at the soil-structure interface.

6 REFERENCES

BIGONI D. & T. HUECKEL (1991). Uniqueness of localization - associated and non-associated elastoplasticity. Int. J. Solids Structures 28: 197 - 213.

BOULON M. & R. NOVA (1990). Modelling of soil-structure interface behaviour a comparison between elastoplastic and rate type laws. Comp. and Geoth., 9: 21 - 46.

DESAI C. S. (1981). Behaviour of interfaces between structural and geological media. Proc. Int. Conf. Rec. Adv. in Geot. Earthq. Eng. & Soil Dyn. Univ. of Rolla, Missouri.

DESRUES J. (1990). Shear band initiation in granular materials: experimentation and theory. Geomaterials: Constituve Equations and Modelling, Elsevier, Rotterdam.

DRUCKER D. C. (1950). Some applications of work hardening and ideal plasticity. Q. Appl. Math., 7: 411-418.

DUPLA J. C. (1995). Application de la sollicitation d'expansion de cavité cylindrique à l'evaluation des caracteristques de liquefaction d'un sable. Thése de Doctorat de l'Ecole Nationale des Ponts et Chaussees, Paris, France.

HARRIS W. W., G. VIGGIANI, M. A. MOONEY & J. FINNO (1995). Use of stereophotogrammetry to analyze the development of shear bands in sand. Geot. Testing J., GTJODJ, vol. 18(4): 405-420.

HILL R. (1958). A general theory of uniqueness and stability in elasto-plastic solids. J. of Mech. and Physics of Solids, 6: 236-249.

LERAT P. (1996). Etude de l'interface sol-structure dans les milieux granulaires a l'aide d'un nouvel appareil de cisaillement annulaire. Thése de Doctorat de l'Ecole Nationale des Ponts et Chaussees, Paris, France.

MAIER G. & T. HUECKEL (1979). Non-associated and coupled flow rules of elastoplasticity for rock-like materials. Int. J. Rock Mech. Min. Sci., 16: 77-92.

MANDEL J. (1966). Condition de stabilité et postulat de Drucker. Rheologie et Mecanique des Sols, Kravtchenko and Sirreys Eds., Springer, Berlin.

MAÜHLHAUS H. B..& I. VARDOULAKIS (1987). The thickness of the shear bands in granular materials. Géotechnique, 37(3): 271-283.

PIETRUSZCZAK S. & STOLLE D. F. E. (1987). Deformation of strain softening materials. Part II. Comp. and Geoth., 4: 99-115.

PIETRUSZCZAK S. & X. NIU (1992). Numerical evaluation of bearing capacity of a foundation in strain softening soil. Comp. and Geoth., 13: 187-198.

RUDNICKI J.W. & J. R. RICE (1975). Conditions for the localization of deformation in pressure-sensitive dilatant materials. J. of Mech. and Ph. of Solids, 23, pp. 371 - 394.

SCHWEIGER H. F., M. KARSTUNEN & G. N. PANDE (1997). Modelling strain localisation in anisotropically consolidated samples. Proc. Int. Symp. Num. Models in Geomech., Montreal, Canada: 181-186.

VARDOULAKIS I. & J. SULEM (1994). Bifurcation analysis in Geomechanics. Blackie Ed.

PLANE-STRAIN MODELLING FOR UNSATURATED SOIL

D. Gallipoli, M. Karstunen and S.J. Wheeler
University of Glasgow, Glasgow, UK

ABSTRACT: The general analytical formulation for the plane-strain deformation regime of an elasto-plastic unsaturated soil is presented and then applied to two types of stress path: wetting at constant in-plane net stress state and in-plane loading at constant suction. Two different implementations of the plane-strain deformation regime are compared. The first one is correct whereas the second one assumes improperly that each single component of the total out-of-plane strain has to be null. This incorrect formulation is believed to be implemented in some finite element codes due to historical reasons. Modelling the soil behaviour by the constitutive model proposed by Alonso *et al.*(1990), substantially different load-displacement relationships are obtained with the two different formulations for the wetting paths. In the case of loading at constant suction the results from the two different formulations coincide in the elastic field and small discrepancies are observed when plastic deformations occur.

1 INTRODUCTION

In geotechnical engineering design the assumption of a plane-strain deformation mode is often made in order to simplify the analysis of three-dimensional problems. In many cases the geometry of the problem and the boundary conditions applied are sensibly constant along one dimension and, for this reason, a plane strain assumption is reasonable. It is therefore crucial to model correctly the plane-strain deformation in design tools, such as finite element codes, which are widely used by geotechnical engineers.

This paper addresses the plane-strain implementation for an unsaturated soil with elasto-plastic mechanical behaviour. It shows the consequences arising from an incorrect modelling for some typical stress paths involving elasto-plastic deformation. For each of these stress paths the load-displacement behaviour has been predicted using two different approaches in modelling the plane-strain deformation regime: one is correct while the other introduces some further unnecessary constraints on the deformation mode of the soil. It is thought that due to historical reasons this incorrect implementation is present in some codes.

A complete load-displacement relationship has been achieved by modelling the stress paths by means of a finite element code, "Compass" (Thomas and He in press 1998), which is able to perform fully coupled flow-deformation analyses in unsaturated soil. This code has been developed by Prof HR Thomas and colleagues at the University of Wales, Cardiff. Although the constitutive model of Alonso *et al.* (1990) has been assumed, the modelling of the plane-strain deformation regime discussed in the next section has general relevance for any unsaturated soil with an elasto-plastic mechanical response.

2 PLANE-STRAIN FORMULATION FOR UNSATURATED SOIL

The stress state variables needed for modelling the mechanical behaviour of unsaturated soil cannot be reduced to a single effective stress tensor (Jennings and Burland 1962, Wheeler and Karube 1995). The stress variables used most frequently are the net stress tensor, σ', and the scalar suction, s, which for bi-dimensional problems (i.e. plane strain, plane stress and axi-symmetric) are defined in vectorial form as:

$$\underline{\sigma}' = \underline{\sigma} - \underline{m}^T \cdot u_a \qquad (1)$$

$$s = u_a - u_w \qquad (2)$$

where $\underline{\sigma} = [\sigma_x \;\; \sigma_y \;\; \tau_{xy} \;\; \sigma_z]^T$, $\underline{m} = [1 \;\; 1 \;\; 0 \;\; 1]$, u_a is the pore air pressure and u_w the pore water pressure.

For an elasto-plastic unsaturated soil the vector of increments of total strain, $\dot{\underline{\varepsilon}}$, is made up of three components:

$$\dot{\underline{\varepsilon}} = \dot{\underline{\varepsilon}}^{\sigma e} + \dot{\underline{\varepsilon}}^{se} + \dot{\underline{\varepsilon}}^{p} \qquad (3)$$

where $\dot{\underline{\varepsilon}}^{\sigma e}$ is the vector of increments of elastic strains due to a change in the net stresses, $\dot{\underline{\varepsilon}}^{se}$ is the vector of increments of elastic strain due to a change in suction and $\dot{\underline{\varepsilon}}^{p}$ is the vector of increments of plastic strain.

The incremental stress-strain relationship linking net stress increments to $\underline{\dot{\varepsilon}}^{\sigma e}$ is given by:

$$\underline{\dot{\sigma}}' = \mathbf{D}\,\underline{\dot{\varepsilon}}^{\sigma e} \tag{4}$$

where **D** is the elastic tangent stiffness matrix.

The other two strain increment component vectors are given by:

$$\underline{\dot{\varepsilon}}^{se} = \frac{1}{3}\cdot A_s \cdot \underline{m}^T \cdot \dot{s} \tag{5}$$

$$\underline{\dot{\varepsilon}}^{p} = \dot{\lambda}\frac{\partial Q}{\partial \underline{\sigma}'} \tag{6}$$

where A_s is the elastic volumetric compressibility of the soil due to a change in suction, $\dot{\lambda}$ is the plastic multiplier and Q is the plastic potential.

The equations above, namely Eqs.(3), (4), (5) and (6), have to be simultaneously satisfied at each material point together with the consistency condition which imposes that, during plastic flow, the stress state must lie on the yield locus. The consistency condition is defined analytically through the equation:

$$dF\,(\,\underline{\sigma}', s, h\,) \;=\; 0 \tag{7}$$

where h is the hardening parameter.

This set of equations represents the general formulation of the elasto-plastic constitutive model for unsaturated soil in bi-dimensional problems. It introduces a non-linear relationship between the increment of net stresses and the increments of strains and suction.

When the general formulation given above is applied to the particular case of plane-strain deformation, the condition of nullity has to be imposed on the out-of-plane component of the vector of increments of total strain, $\underline{\dot{\varepsilon}}$. Therefore for the plane-strain case the particular definition for the vector of increments of total strain, $\underline{\dot{\varepsilon}}$, is given by:

$$\underline{\dot{\varepsilon}} = \left[\dot{\varepsilon}_x \;\; \dot{\varepsilon}_y \;\; \dot{\gamma}_{xy} \;\; \dot{\varepsilon}_z\right]^T = \left[\frac{\partial \dot{u}_x}{\partial x} \quad \frac{\partial \dot{u}_y}{\partial y} \quad \frac{\partial \dot{u}_x}{\partial y}+\frac{\partial \dot{u}_y}{\partial x} \quad 0\right]^T \tag{8}$$

where $\dot{u}_x$ and $\dot{u}_y$ are the x and y components of the increment of displacement.

In some finite element codes for unsaturated soils the plane-strain assumption is implemented incorrectly by imposing the condition of nullity not only on the out-of-plane component of the total strain vector but on the out-of-plane components of each strain contribution represented in Eq.(3). A possible explanation for this inconsistency is due to the history of the development of finite element programs: they have often been improved by adding new features to existing versions. As a result, parts of the code which were correct in the previous versions have often been imported in the improved one without carrying out all the changes which were required. In particular, the first versions of finite element programs included only a linear-elastic constitutive model. In this case the plane strain condition can be independently imposed on the total strain vector or on the elastic strain vector, because these two coincide. When plastic behaviour was introduced, the constitutive model used most frequently for geomaterials was the elastic-perfectly plastic Mohr-Coulomb model with an associated flow rule. In this case the out-of-plane component of the plastic strain vector is always null, if it is assumed that the out-of-plane stress remains the intermediate principal stress throughout the analysis, because the plastic potential is independent of the intermediate principal stress. When, subsequently, more sophisticated constitutive models and plastic flow rules were introduced, it was possible that the assumption of nullity of the out-of-plane component of the plastic strain vector was erroneously retained.

In the case of unsaturated soil there is a third contribution to the total strain vector due to the elastic deformation caused by a change in suction. When the pre-existing codes were extended to include constitutive models for unsaturated soil, the elastic contribution of strain due to change in suction had to be introduced and, in some implementations the out-of-plane component of this contribution was assumed to be null for the plane-strain deformation.

In this paper two different implementations for plane strain deformation are compared. The first one (version A) is the correct one and has been described previously in this section, the second one (version B) is based on the assumption that the out-of-plane component of each single contribution of the total strain vector in Eq.(3) is null. Version B of the plane-strain formulation therefore assumes that the total strain vector, $\underline{\dot{\varepsilon}}$, is still defined according to Eq.(3) but the two contributions given by $\underline{\dot{\varepsilon}}^{se}$ and $\underline{\dot{\varepsilon}}^{p}$ are now defined as:

$$\underline{\dot{\varepsilon}}^{se} = \frac{1}{2} \cdot A_s \cdot \underline{n}^T \cdot \dot{s} \tag{9}$$

$$\underline{\dot{\varepsilon}}^{p} = \left[\dot{\varepsilon}^p_x \;\; \dot{\varepsilon}^p_y \;\; \dot{\gamma}^p_{xy} \;\; \dot{\varepsilon}^p_z\right]^T = \left[\dot{\lambda}\frac{\partial Q}{\partial \sigma'_x} \;\; \dot{\lambda}\frac{\partial Q}{\partial \sigma'_y} \;\; \dot{\lambda}\frac{\partial Q}{\partial \tau_{xy}} \;\; 0\right]^T \tag{10}$$

where $\dot{\lambda}$, A_s and Q have been previously defined and $\underline{n} = [1\ 1\ 0\ 0]$. Eq.(10) violates the plastic flow rule.

These further unnecessary constraints on the out-of-plane components of the two strain vectors, $\underline{\dot{\varepsilon}}^{se}$ and $\underline{\dot{\varepsilon}}^{p}$ cause the out-of plane component of the increment of elastic strain due to a change in net stresses, $\underline{\dot{\varepsilon}}^{\sigma e}$, to be equal to zero and, as a result, the increment of the out-of-plane net stress is related to the other two in-plane increments through the relationship:

$$\dot{\sigma}'_z = \nu' \cdot (\dot{\sigma}'_x + \dot{\sigma}'_y) \tag{11}$$

where ν' is the Poisson's ratio of the soil skeleton. This well-known relationship is valid when the plane-strain condition is applied to elastic deformation for saturated soils. It is in fact possible that the above relationship, Eq.(11), has sometimes been confused with the definition of the plane-strain condition itself. For plane strain behaviour of an unsaturated soil, Eq.(11) is, however, only true for the specific case of elastic behaviour at a constant value of suction.

3 CHOICE OF MECHANICAL PARAMETERS

The constitutive model adopted to represent the soil behaviour in the numerical simulations is the one proposed by Alonso *et al.* (1990). The set of model parameters have been determined using the results from tests on compacted unsaturated Speswhite Kaolin (Sivakumar 1993, Zakaria 1995). These mechanical parameters are summarised in Table 1.

ELASTIC PARAMETERS	
Shear Modulus (G)	**10000 kPa**
Swelling index for changes in suction (κ_s)	**0.020**
Swelling index for changes in mean net stress (κ)	**0.025**
STRENGTH PARAMETERS	
Slope of critical state line in constant suction planes (M)	**0.9**
Parameter controlling cohesion increase with suction (k)	**0.5**
NORMAL COMPRESSION LINES AND LC YIELD CURVE	
Reference pressure (p^c)	**2000 kPa**
Slope of normal compression line (NCL) at zero suction ($\lambda(0)$)	**0.13**
Specific volume on NCL at zero suction for $p'=p^c$ ($N(0)$)	**1.66**
Parameter controlling ratio of NCL slopes at $s\rightarrow\infty$ and $s=0$ (r)	**1.5**
Parameter controlling variation of NCL slope with suction (β)	**0.01 kPa-1**
INITIAL STATE	
Initial location of LC yield curve (p_o^*)	**200 kPa**

Table 1. Values of mechanical parameters

The choice of a value greater than 1 for the parameter r in Table 1 may seem surprising since in the model of Alonso *et al.* (1990) values smaller than 1 are suggested, on the assumption that the slope of the normal compression line at constant suction, $\lambda(s)$, decreases with increasing suction. However, the tests of Sivakumar (1993) show an opposite response for compacted Speswhite Kaolin, with $\lambda(s)$ generally increasing when suction increases. In order to match the experimental results, a value of r greater than 1 has been chosen.

The choice of a value for the parameter r has to be combined with an appropriate choice of value for the reference pressure, p^c, in order to predict realistic shapes for the LC yield curves over an appropriate stress range. In general, whenever a value of r greater than 1 is selected, the value of the reference pressure, p^c, has to be considerably higher than the range of values of interest for the hardening parameter p_o^*. In Figs.1 and 2 the LC yield curves corresponding to different values of the hardening parameter are plotted in a linear and logarithmic scale, for the choice of material parameters shown in Table 1.

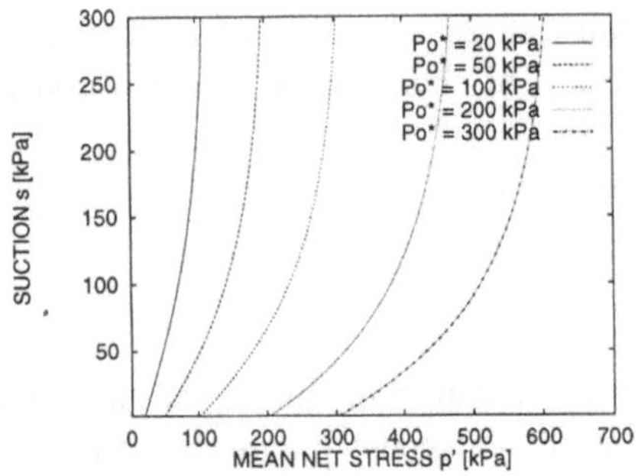

Fig.1 LC curves (linear scale)

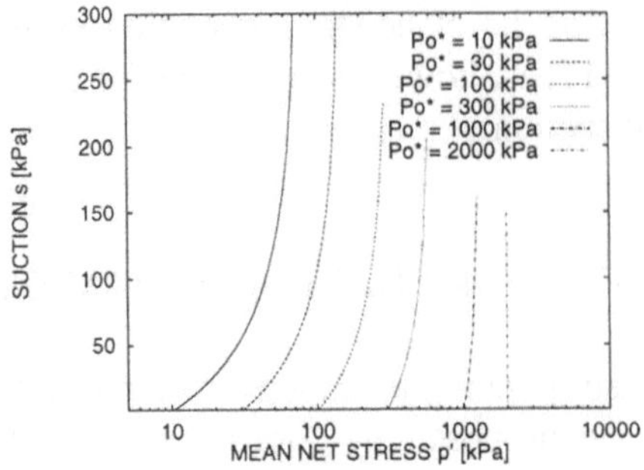

Fig.2 LC curves (logarithmic scale)

As can be observed in Fig.2, the shape of the LC curve tends to a vertical straight line when the value of the hardening parameter, p_o^*, tends to the reference pressure, p^c. However, when the LC curves are plotted in a linear scale (Fig.1), it can be seen that, for low values of the hardening parameter, the inclination of the LC curve increases as the yield curve expands, as has been observed in laboratory tests on Speswhite Kaolin (Wheeler and Sivakumar, 1995). The inclination of the LC curve is a maximum at a particular value of p_o^*, and then decreases with the curve becoming a vertical straight line for $p_o^*=p^c$.

Another difference with the model of Alonso *et al.* (1990) is the introduction of the parameter *N(0)*, which defines the specific volume on the NCL at zero suction and for $p'=p^c$. The original model of Alonso *et al.* (1990) is able to predict the changes of specific volume due to a change in the stress state but is unable to calculate the absolute value of the specific volume corresponding to a given stress state. The introduction of the *N(0)* parameter changes this relationship from incremental to absolute, hence establishing a unique correspondence between the stress state and the specific volume.

4 NUMERICAL SIMULATIONS

The numerical analyses carried out simulate two types of bi-axial tests for an unsaturated soil. These are: wetting of the soil at constant in-plane net stresses and in-plane loading at constant suction. Both versions A and B of the plane-strain formulation described in Section 3, have been implemented in the finite element code "Compass" in order to make a comparison. Although the code is able to carry out coupled analyses, in the present work each simulation is modelled as fully drained. This means that the pore air pressure and the pore water pressure are imposed at nodes and they are constant for the case of the in-plane loading whereas their variation is controlled for the case of wetting.

The mesh is very simple since the study is performed at material point level and the stress-strain state is homogeneous at each Gauss point. The mesh with boundary conditions is presented in Fig.3. Quadrilateral eight noded isoparametric elements are used with same shape functions for all the nodal unknowns (displacement, pore air pressure and pore water pressure).

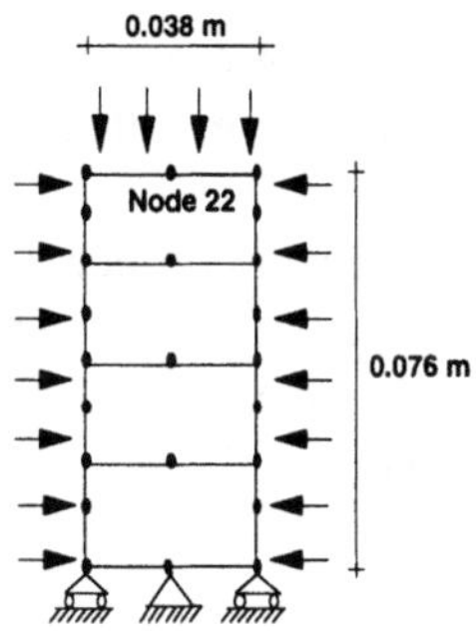

Fig.3 Mesh and boundary conditions

Wetting case - In Fig.4 the projections of the three wetting paths on the (s,p') plane are presented for the correct version of the plane-strain formulation (version A) and for the incorrect one (version B). The initial net stress state is assumed to be isotropic and equal to 100 kPa, 250 kPa or 400 kPa. The starting value of suction is taken equal to 200 kPa in all three simulations. In both correct and incorrect formulations the in-plane net stresses are held constant during wetting. With the correct formulation (version A) the mean net stress, p', varies during wetting, because of changes induced in the out-of-plane net stress which arise in order to maintain the plane-strain condition. In contrast, version B incorrectly predicts no change of out-of plane net stress, and hence no change of mean net stress.

In Figs.5, 6 and 7 the suction-displacement curves of Node 22 (Fig.3) are plotted for the wetting paths chosen. In each figure a comparison is made between the correct and incorrect plane-strain formulations. Fig.5 shows that even when the wetting stress path remains in the elastic region the two versions give substantially different results. In particular the incorrect version (version B) tends to overestimate the wetting-induced swelling of the soil. Inspection of Eq.(9) shows that in version B the increment of elastic axial strain due to a change in suction, ε_y^{se} is incorrectly predicted as:

$$\dot{\varepsilon}_y^{se} = \frac{1}{2} \cdot A_s \cdot \dot{s} \tag{12}$$

In contrast, for this type of elastic wetting path under constant in-plane net stresses the value of ε_y^{se} is correctly given by version A as:

$$\varepsilon_y^{se} = (1+\nu') \cdot \frac{1}{3} \cdot A_s \cdot \dot{s} \tag{13}$$

Eqs.(12) and (13) coincide only if the Poisson's ratio, ν' is equal to 0.5. It is also evident from Fig.6 that when the wetting path intersects the yield surface the two versions predict different values of suction. This is a consequence of the fact that the stress paths in the elastic region are different. In particular the correct formulation predicts an increase in mean net stress whereas in the incorrect one the mean net stress remains constant (Fig.4). Moreover in the correct version a deviatoric component of net stress arises due to the change in the out-of-plane net stress: therefore in this case the yield point is not situated on the LC yield curve. When plastic deformations occur (Figs.6 and 7) the incorrect version tends to overestimate the collapse of the soil produced by a decrement of suction. For large wetting-induced plastic deformation (Fig.7) the two curves seem to become parallel and the percentage error due to the use of the incorrect version becomes less significant as the magnitude of the collapse strains increases.

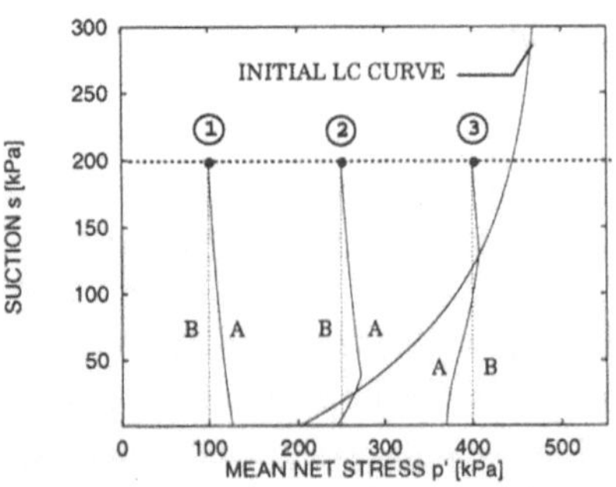

Fig.4 Simulations 1, 2, 3 (wetting)

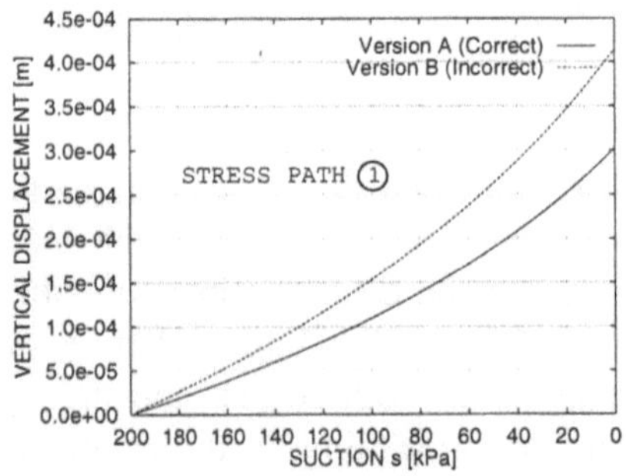

Fig.5 Suction-displacement (wetting path 1)

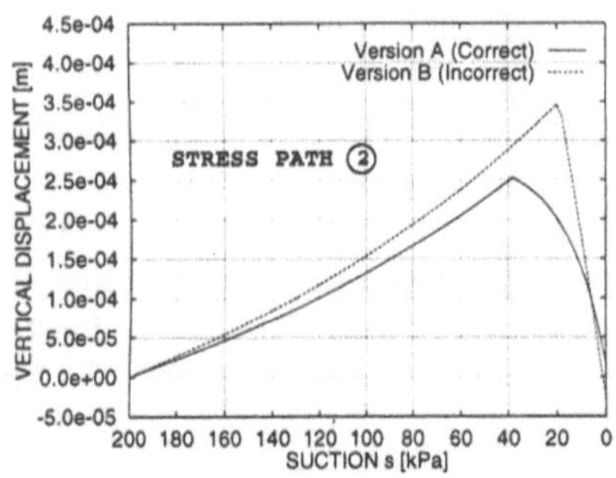

Fig.6 Suction-displacement (wetting path 2)

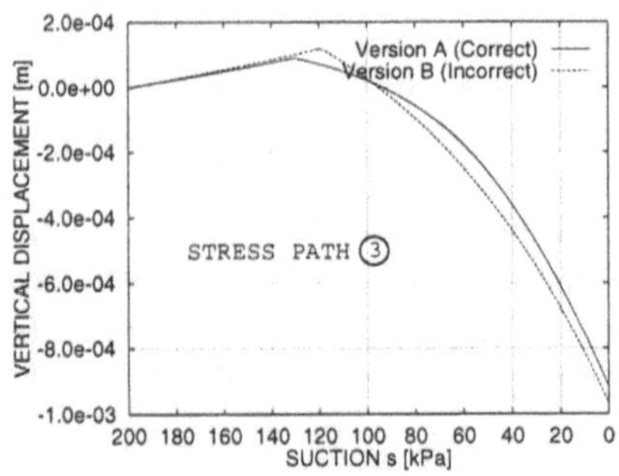

Fig.7 Suction-displacement (wetting path 3)

Constant suction loading case - In Fig.8 the projections of the two constant suction in-plane loading stress paths on the (s,p') plane are presented for both plane-strain formulations. In this case the projections of the stress paths for the two formulations overlap each other and, for clarity, they have been plotted slightly separated in Fig.8. Each simulation is carried out at a constant value of suction of either 100 kPa or 0 kPa (fully saturated). The initial stress state is isotropic and always equal to 100 kPa. Loading is then performed applying the same normal stress in the two in-plane directions. In Figs.9 and 10 the load-displacement curves of Node 22 (Fig.3) are plotted for the two constant suction loading paths. Each figure presents a comparison between the two versions of the plane-strain formulation. It is clear that the two versions give exactly the same results when the soil behaves elastically. This is because the component of the elastic strain due to change in suction, $\dot{\underline{\varepsilon}}^{se}$, which was present in the wetting paths, is absent and the total strain vector, $\dot{\underline{\varepsilon}}$, coincides with the vector of increments of elastic strain due to a change in the net stress state, $\dot{\underline{\varepsilon}}^{\sigma e}$. In the plastic field the incorrect version tends to overestimate the compression of the soil produced by an increment of the in-plane stress. However this tendency is less evident when large plastic strain occur and, again, the two curves become almost parallel. Therefore, also in this case the percentage error produced by the use of the incorrect formulation (version B) becomes less important for large values of plastic strains.

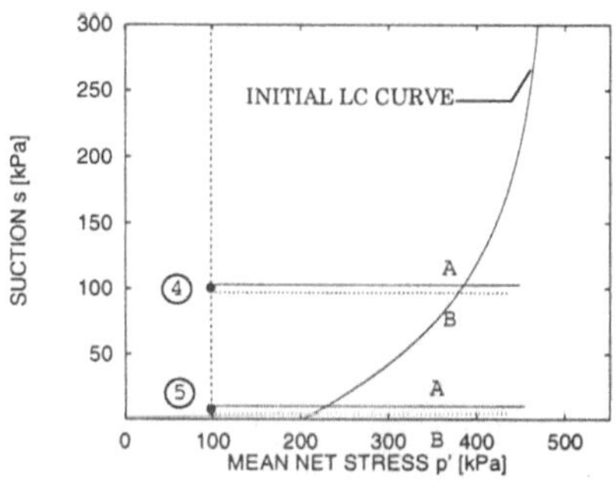

Fig.8 Simulations 4,5 (constant suction loading)

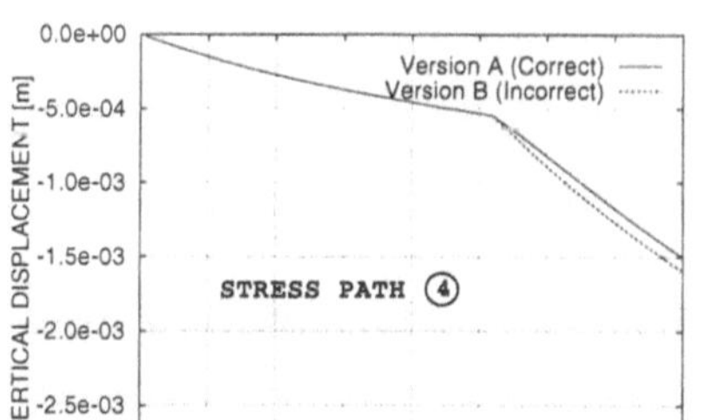

Fig.9 Load-displacement (constant suction path 4)

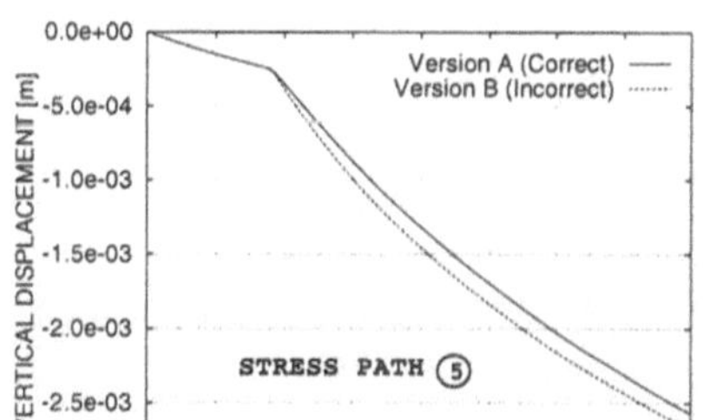

Fig.10 Load-displacement (constant suction path 5 - fully saturated)

5 CONCLUSIONS

Two different formulations for plane-strain deformation mode of an unsaturated soil have been presented: the first one is correct whereas the second one is based on the incorrect assumption that individual components of the out-of-plane strain must all be null. The formulations have been compared for two typical stress paths (wetting and in-plane loading at constant suction) by means of a finite element program. For wetting, the incorrect version tends to overestimate substantially the swelling of the soil in the elastic phase. It tends also to overestimate the collapse due to an increment of suction in the plastic phase. This effect, however, is more evident for small values of plastic strains whereas for large values of plastic strains the predicted suction-displacement curves seem to become parallel. For constant suction loading, the two versions give identical results in the elastic field. When plastic strains occur the incorrect version predicts greater deformations for the same increment of in-plane stress. This tendency is, however, less evident than in the wetting case and, for large values of plastic strains, the predicted load-displacement curves seem again to become parallel. This also suggests that the incorrect formulation could be difficult to detect in numerical codes implementing the traditional models for saturated soil, as can be deduced from the results of the simulation of loading test in saturated conditions (Fig.10). For both types of stress paths the percentage error due to the use of the incorrect version becomes less important for large values of plastic strains.

6 ACKNOWLEDGEMENTS

The present work is part of a PhD project at the Department of Civil Engineering of the University of Glasgow funded by the Faculty of Engineering of the University of Glasgow. The Authors are grateful to Professor HR Thomas and his colleagues from the University of Wales, Cardiff for providing the finite element code "Compass" used in this work.

7 REFERENCES

ALONSO E.E., A. GENS & A. JOSA (1990). A constitutive model for partially saturated soils. Géotechnique, 40: 405-430.

JENNINGS J.B.E. & J.B. BURLAND (1962). Limitations to the use of effective stresses in partially saturated soils. Géotchnique, 12: 125-144.

SIVAKUMAR V. (1993). A critical state framework for unsaturated soil. PhD thesis, University of Sheffield, UK.

THOMAS H.R. & Y. HE (in press 1998). Modelling the behaviour of unsaturated soil using an elasto-plastic constitutive model. Géotechnique.

WHEELER S.J. & D. KARUBE (1995). State of the art report - Constitutive modelling. Proc. 1st Int. Conf. on Unsaturated soils, Paris, 3.

WHEELER S.J. & V. SIVAKUMAR (1995). An elasto-plastic critical state framework for unsaturated soil. Géotchnique, 45: 35-53.

ZAKARIA I. (1995). Yielding of unsaturated soil. PhD thesis, University of Sheffield, UK.

LARGE STRAIN CONSOLIDATION FOR ELASTO-PLASTIC SOILS

J.M. Dluzewski
Warsaw University of Technology, Warsaw, Poland

ABSTRACT: Large stain formulation for two phase medium i.e. soil skeleton and ground water, is studied in the paper. The updated Lagrange procedure is developed for the fully coupled consolidation process. The pseudo-visco-plastic iteration procedure is applied to solve the large strain elasto-plastic consolidation problems. The equations are developed in the finite element code. The numerical tests for elastic and elasto-plastic soil skeleton are performed. The obtained results for plain stains problem are reported and studied in details. As the engineering example, the consolidation of the mud layer below the earth dam is studied in the paper.

1. INTRODUCTION

An analysis of the behaviour of the soft soils such as peat, gytia, mud etc. combines several non-linear effects which should be taken into account. The effects are superimposed and not easy to model. The proper prediction of the bearing capacities of the subsoil and final displacements of the soil makes the problem coupled and complicated. On the other hand, it is very important from the practical point of view, to predict settlements of the structures (as road embankments, dikes, earth dams, etc.) risen on the soft saturated soils.

In the paper a two phase medium is considered (soil skeleton and pore water). The fully coupled consolidation process is described in the framework of finite element method, Lewis & Schrefler (1987). The theory of viscoplasticity, after Perzyna (1966), is used to model the elasto-plastic behaviour of soils. The pseudo visco-plastic iterative procedure is applied in the numerical algorithms, Zienkiewicz & Cormoau (1974). To described the large stains, the updated Lagrange procedure is adopted, Bathe (1982). The current permeability is defined for orthotropic materials in relation for the current void ratio. The explicit time stepping algorithms are used to model consolidation with non-linear effects.

2. TWO PHASE MEDIUM - SOIL SKELETON AND WATER

The behaviour of two phase medium (soil skeleton and pore water), under loading is described by means of the coupled set of differential equations. The following assumptions are taken.

1) The total stresses are the sum

$$\sigma = \sigma'' + \alpha\, \mathbf{m}\, p_{sum} \tag{1}$$

where $\sigma = [\sigma_x, \sigma_y, \sigma_z, \tau_{xy}, \tau_{xz}, \tau_{zxy},]^T$ is the Cauchy stress tensor, σ'' is effective Cauchy stress tensor, α is the Biot's coefficient, $\mathbf{m} = [1, 1, 1, 0, 0, 0]^T$, p_{sum} is the pore pressure,

2) Strains are large, the velocity of strains is the sum of elastic and visco-plastic parts

$$\dot{\varepsilon} = \dot{\varepsilon}^e + \dot{\varepsilon}^{vp} \tag{2}$$

3) The viscous-plastic strain rate is defined after Perzyna (1966)

$$\dot{\varepsilon}^{vp} = \gamma < \phi(F) > \frac{\partial Q}{\partial \sigma} \tag{3}$$

where Q is the potential, F is the yield function, γ is a viscosity coefficient and $<\phi(F)>$ is a special function. Non-associated flow rule is used in the paper. The brackets $< >$ stand for an on/off switch. If $F > 0$ then the switch is on, if $F < 0$ the switch is off. We assume that the function takes the following form

$$\phi(F) = \delta F^{\beta} \tag{4}$$

We will apply the visco-plastic formulation to model the elasto-plastic behaviour of soil. The time which appeares in equation (5) is found as a parameter of the process and called pseudo-time. Applying the Euler scheme for pseudo time integrating, we have

$$\Delta\varepsilon^{vp} = \dot{\varepsilon}^{vp}\, \Delta t \tag{5}$$

where Δt is here understood as a time parameter, not a real time. The stress increment takes the form

$$\Delta\sigma'' = \mathbf{D}_e\, (\Delta\varepsilon - \Delta\varepsilon^{vp}) \tag{6}$$

4) The continuity equation takes the form

$$\nabla^T \mathbf{v}\rho_w + \frac{\partial(\rho_w\, n)}{\partial t} = 0 \tag{7}$$

where $\nabla^T = \left[\frac{\partial}{\partial x}, \frac{\partial}{\partial y}, \frac{\partial}{\partial z}\right]$ is the differential operator, $\mathbf{v}^T = [v_x, v_y, v_z]$ is the seepage velocity, n is the porosity. No internal sources are taken into consideration.

5) The velocity of the pore fluid is defined by Darcy's flow rule

$$\mathbf{v} = \frac{\overline{\mathbf{k}}}{\mu} \nabla p_{sum} \tag{8}$$

where $\overline{\mathbf{k}}$ is the array of absolute permeability coefficients of porous medium to fluid, μ is the dynamic viscosity of the fluid.

$$\overline{\mathbf{k}} = \begin{bmatrix} \overline{k}_{xx} & 0 & 0 \\ 0 & \overline{k}_{yy} & 0 \\ 0 & 0 & \overline{k}_{zz} \end{bmatrix} \tag{9}$$

6) The variation of the permeability for orthotropic properties of the soil is defined by the following rule (Dłużewski 1997), confirmed by the exprimental results for the peat of orthotropical properties (Chaciński & Dłużewski, 1995)

$$\log_{10}\left(\frac{k_{ij}}{k_{ij}^0}\right) = \frac{e - e_0}{A} \tag{10}$$

where e_0 and e are the initial and current void ratios respectively, A is a material parameter. For isotropic materials the similar relation has been proposed by Monte & Kritzen (1976)

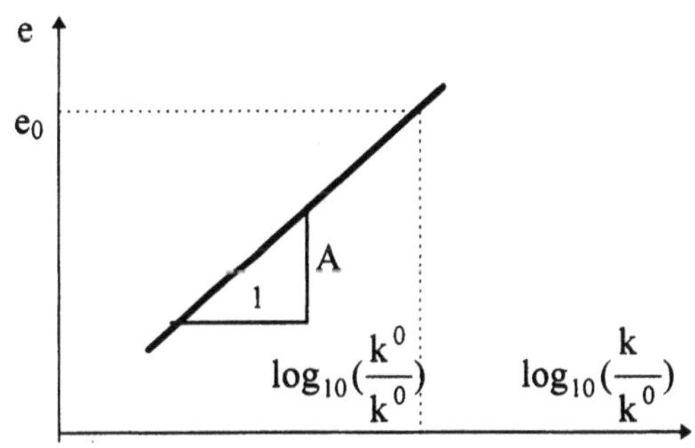

Fig.1 The graphical interpretation of the equation (10)

The updated Lagrange procedure (Bathe, 1982) is adopted for large strains during consolidation process. According to the procedure, at the beginning of the increment, the virtual work principle is defined in the initial local configuration of the body. The II Piola-Kirchhoff stress tensor and Green strain tensor are the conjugate variables in the energy balance. At the end of increment, the virtual work principle can be defined in the final configuration of the body, where the Cauchy stress tensor and linear part of the Almansi strain tensor are the conjugate variables in the virtual work principle. The rules of transformations between the initial and current configuration of the body can be found in book by Crisfield (1997).

3. NUMERICAL PROCEDURE

The virtual work principle, continuity equation with boundary conditions are the starting points for numerical formulation. The finite element method is applied to solve initial boundary value problems. Several procedures stemming from elasto-plastic modelling can be coupled with the time stepping algorithm during the consolidation process. The elasto-plastic soil behaviour is modelled by means of visco-plastic theory (Perzyna, 1966). The finite element formulation for the elasto-plastic consolidation and for large strains combines overlapping numerical processes.

The elasto pseudo-viscoplastic algorithm for numerical modelling of elasto-plastic behaviour is used after Zienkiewicz and Cormeau (1974). The stability of the time marching scheme was proved by Cormeau (1975). The pseudo-viscous algorithm developed in finite element computer code HYDRO-GEO is successfully applied to solve a number of boundary value problems, Dłużewski (1993). The visco-plastic procedure was extended to cover the geometrically non-linear problems by Kanchi et al (1978) and also developed for large strains in consolidation, Dłużewski (1997). The pseudo-viscous procedure is adopted herein for modelling elasto-plastic behaviour in consolidation. In the procedure two times appear, the first (*t*) is the real time of consolidation and the second time (t) is only a parameter of the pseudo-relaxation process.

The global set of equations for the consolidation process is derived as follows

$$\begin{bmatrix} \mathbf{K}_T & \mathbf{L} \\ \mathbf{L}^T & -(\mathbf{S}+\theta\Delta t\ \mathbf{H}^i) \end{bmatrix} \begin{bmatrix} \Delta\mathbf{u}^i \\ \Delta\mathbf{p}^i \end{bmatrix} = \begin{bmatrix} \mathbf{0} & \mathbf{0} \\ \mathbf{0} & -\Delta t\mathbf{H}^i \end{bmatrix} \begin{bmatrix} \mathbf{u}^i \\ \mathbf{p}^i \end{bmatrix} + \begin{bmatrix} \Delta\mathbf{F}^i \\ \Delta\mathbf{q} \end{bmatrix} \tag{11}$$

where $\mathbf{K}_T$ is the tangent stiffness array, considering large strains effects, $\mathbf{L}$ is the coupling array, $\mathbf{S}$ is the array responsible for the compressibility of the fluid, $\mathbf{H}$ is the flow array, $\mathbf{u}$ are the nodal displacements, $\mathbf{p}$ are the nodal excesses of the pore pressure, $\Delta\mathbf{F}^i$ is the load nodal vector defined below

$$\Delta\mathbf{F}^i = \Delta\mathbf{F}_L + \Delta\mathbf{R}_I{}^i + \Delta\mathbf{R}_{II}{}^i \tag{12}$$

$\Delta\mathbf{F}_L$ is the load increment, $\Delta\mathbf{R}_I{}^i$ is the vector of nodal forces due to pseudo-visco iteration, $\Delta\mathbf{R}_{II}{}^i$ is the unbalanced nodal vector due to geometrical nonlinearity. $\Delta\mathbf{R}_I{}^i$ takes the following form

$$\Delta\mathbf{R}_I^i = \int_{{}^{t+\Delta t}_{(i-1)}V} \mathbf{B}_{(i-1)}^T \mathbf{D}\left({}^{t+\Delta t}_{t+\Delta t}\Delta\varepsilon_i^{vp}\right)^{t+\Delta t}_{(i-1)} dv \tag{13}$$

and is defined in the current configuration of the body. The subscripts indicate the configuration of the body, and superscripts indicate time when the value is defined (notation after Bathe (1982)). $\Delta\mathbf{R}_I^i$ stands for the nodal vector which results from the relaxation of the stresses. For each time step the iterative procedure is engaged to solve the material non-linear problem. The i-th indicates steps of iterations. Both local and global criterion for terminating the iterative process are used. The iterations are continued until the calculated stresses are acceptable close to the yield surface, F < Tolerance at all checked points, where F is the value of the yield function. At the same time the global criterion for this procedure is defined at the final configuration of the body. The global criterion takes its roots from the conjugated variables in the virtual work principle, where the Cauchy stress tensor is coupled with the linear part of the Almansi strain tensor. For two phase medium, the unbalanced nodal vector $\Delta\mathbf{R}_{II}^{i}$ is calculated every iterative pseudo-time step.

$$\Delta\mathbf{R}_{II}^{i} = \mathbf{P} - \int_{{}^{t+\Delta t}_{(i-1)}V} \mathbf{B}_{(i-1)}^{T} \left({}^{t+\Delta t}_{t+\Delta t}\sigma'^{(i-1)} + \mathbf{m}\, {}^{t+\Delta t}_{t+\Delta t}p^{(i-1)} \right) {}^{t+\Delta t}_{(i-1)}dv \tag{14}$$

The square norm on the unbalanced nodal forces is used as the global criterion of equilibrium. The iterative process is continued until both criterions are fulfilled.

4. TEST RESULTS

Two dimensional consolidation problem is studied. The plane strain conditions are assumed. The problem has been solved analytically (for small strains and elastic soil) by Gibson et al (1970). The plane layer located on a rigid smooth and impervious foundation is loaded by uniformly distributed traction as shown in the Fig. 2. The following material parameters are taken for the layer: G = 100 kPa, $\nu = 0.1$, $k_x = k_y = 5\bullet10^{-7}$ m/s, $\gamma_w = 10$ kN/m^3 . For the cases of large strains, to describe the permeability changes, two more parameters are needed, Fig.1., void ratio calculated from porosity n = 0.75 and A = 1. The layer is assumed to be 10m thick. To compare the numerical results with analytical solutions the area of 250m long is taken for numerical modelling. The length of the layer influences the results especially at the beginning of the consolidation process. Since the analytical solution has been obtained for the semi-infinite medium, long stratum is taken for numerical calculations. The time stepping process is started from very short time steps and continue until the decay of the excess pore pressure is obtained. The loading (100kPa=5x20 kPa) is applied in five equal increments at the beginning. The results of calculations are reported for six cases described in the Table 1. For the elasto-plastic analysis the Coulomb-Mohr yield surface is taken. The nonassociated flow rule is assumed to confirm the incompressibility of the soil skeleton by taking the dilatancy angle equal zero for all cases.

-	Strains	Model (cohesion c [kPa], friction angle [°])
A	large	elastic
B	small	elastic
C	large	elasto-plastic (c=30, φ=45)
D	small	elasto-plastic (c=30, φ=45)
E	large	elasto-plastic (c=15, φ=30)
F	small	elasto-plastic (c=15, φ=30)

Table 1. Assumptions for the analysed cases

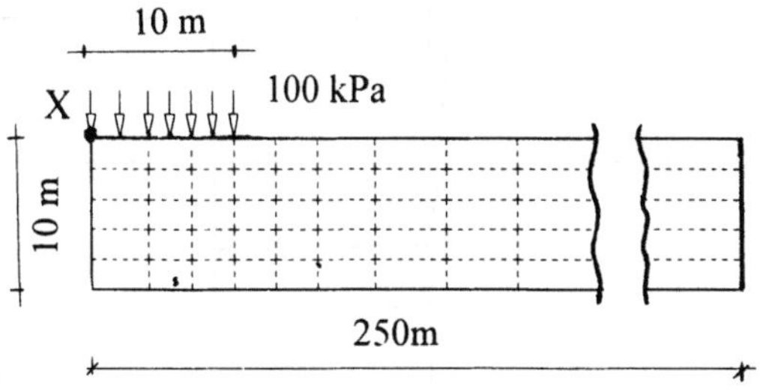

Fig.2. Plain strain problem

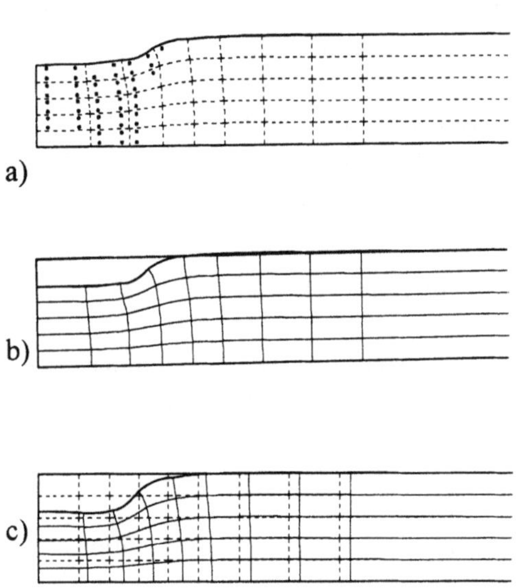

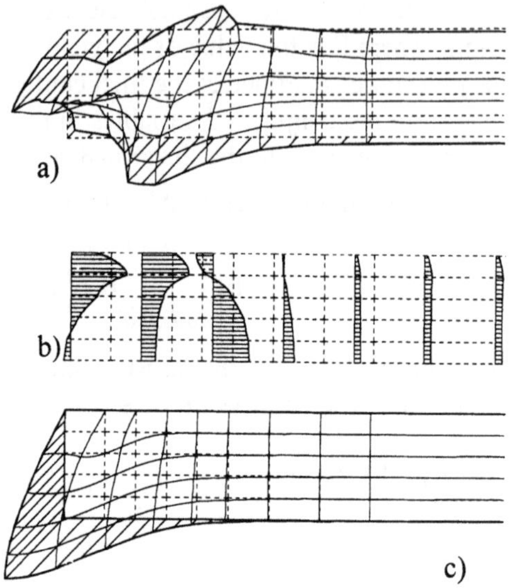

Fig.4. Case C, the deformed mesh, a) 5th time step, plastic points, b) 10th time step c) 32nd time step

Fig. 5. Excess pore pressure a) , b) case F 10th time step c) case E 32nd time step

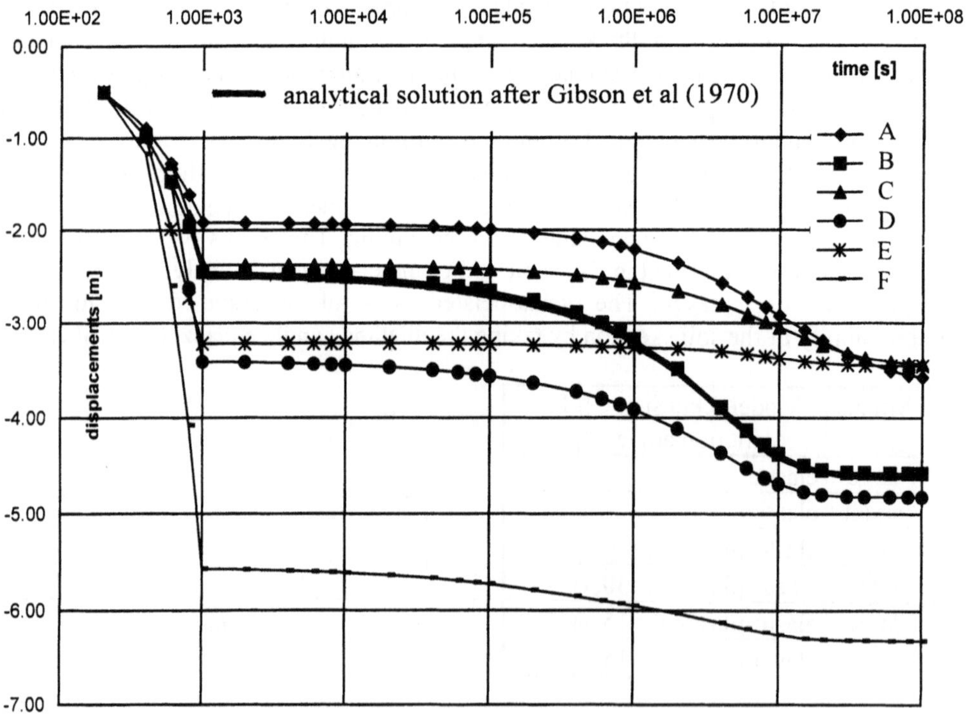

Fig. 6. Settlement for the central point (point X in the Fig.2)

The deformed mesh at the beginning, in between and at the end of time stepping process is shown in real scale, in the Fig. 4 a, b, c. The plastic points are shown at the background of deformed mesh in the Fig. 4 a. The plastic behaviour is mostly observed at the beginning of the calculation during five increments of loading. Next the decay of the excess pore pressure cause the elastic unloading of the soil skeleton. The distribution of the excess pore pressure for the elasto-plastic model for the soil is shown in the Fig.5. The comparison of the settlements for the central point X (Fig.2) at the ground level is shown in the Fig. 6. The analytical solution obtain by Gibson et al (1970) for the small strains and elastic material is drown with the solid line. It is case B in numerical analysis. The obtained deformation for large strains are much smaller comparing the small strains. Considering the elasto-plastic behaviour of the soil skeleton, we observe much greater deformation of the subsoil at the beginning of the process because of plastic behaviour. On the other hand it is not possible to generate high excess pore pressure in the subsoil what influence the final settlement. The obtained settlements for the elastic and elasto-plastic consolidation are composed of two main contributors, elasto-plastic deformation during short period when the loading is applied, and nearly elastic settlement during consolidation process. The settlement for large strain analysis are always smaller comparing the small strain results.

5. MUD LAYER BELLOW THE EARTH DAM

The Mietków dam has been risen on the multilayer subsoil. Below the ground level the mud layer ranging from 0.5m to 2.5m is found at the old river beds. Next, the gravel with stones form the layer of 5m thickness. Below the gravel the clay formation is located, Fig. 7. The material parameters are listed in the Table 2.

No.	soil	E [kPa]	v	c [kPa]	φ [°]	ψ [°]	γ [kN/m^3]	k_x=k_y [m/day]	n	A
1	clay	37400	0.35	80	17	0	21	0.000001	-	-
2	gravel	37100	0.3	20	30	0	22	100.	-	-
3	mud	2350	0.4	12	5	0	16	0.000001	0.75	0.3
4	sand-gravel	21700	0.3	12	42	0	19.5	100.	-	-

Table 2. Material parameters

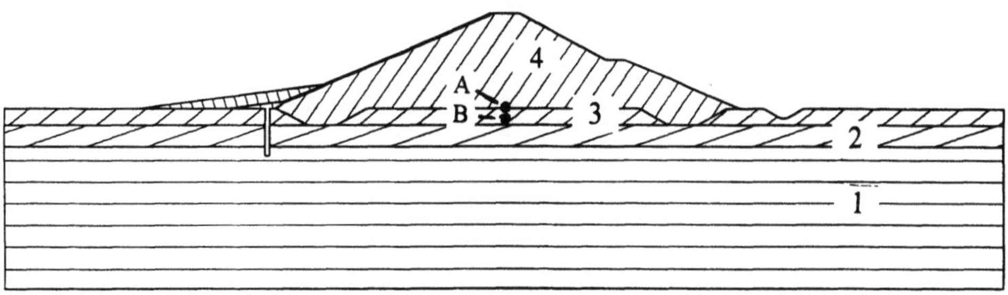

Fig. 7. The cross section of the dam

The dam is built from sand-gravel mix. The concrete plates are performed from the upper water side. The clay-concrete screen of 0.8 m thickness is performed. The high of the screen ranges from 15 m to 28 m. The length of the dam is more than 3 km. The dam is partly located on the old river beds where the mud layer is formed. Before rising the dam, two trenches have been perform to make teeth for better interaction between the dam and subsoil, Fig.8. The installed benchmarks show continuos process of the settlement. The consolidation process of the mud layer is observed. There is also consolidation of the clay but, due to high stiffness of the clay, the increase in displacements is relatively small.

Stage	State	Year	Time increment	Load
1	introduction of initial stresses	-	-	-
2	rising of dam and state before filling	1983 1984 1985	1 1 1	twelve increments 0
3	reservoir filling, no activity	1986 from 1987 . to 2002	1 fifteen time steps	five increments 0 . 0

Table 3. Schedule of the dam rising and filling of the reservoir

The calculations are performed in three stages, Table 3. In the first stage the initial stresses are introduced into the subsoil. In the second one, the soil dam is risen by adding elements. In the third stage, the loading caused by filling of the reservoir is applied. The calculation are performed for both small and large strains. For large strains the variation of the permeability is taken into account during consolidation process. The elasto-plastic soil models based on the Coulomb-Mohr yield surface and nonassociated flow rule are used. The final state of the deformation is the sum of the elasto-plastic consolidation process due to loading caused by mass weight of the dam and forces from filling of the reservoir. The soft mud layer causes wide range of plastic zone located at the central part of the earth dam, Fig. 8. The calculated excess pore pressure at the mud layer both for small and large strains analysis are depicted in the Fig.9.

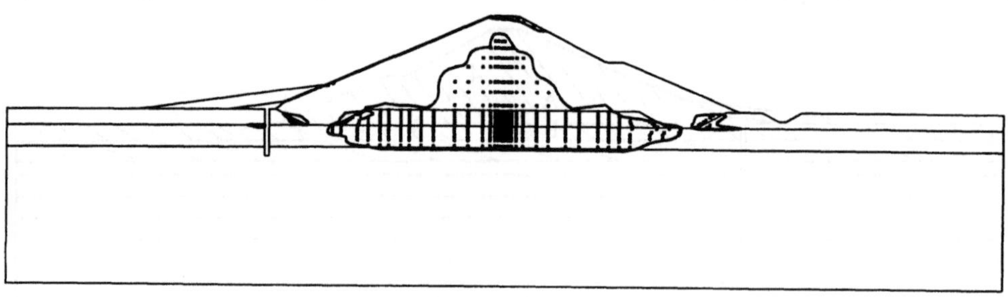

Fig. 8. Plastic zones, after dam rising

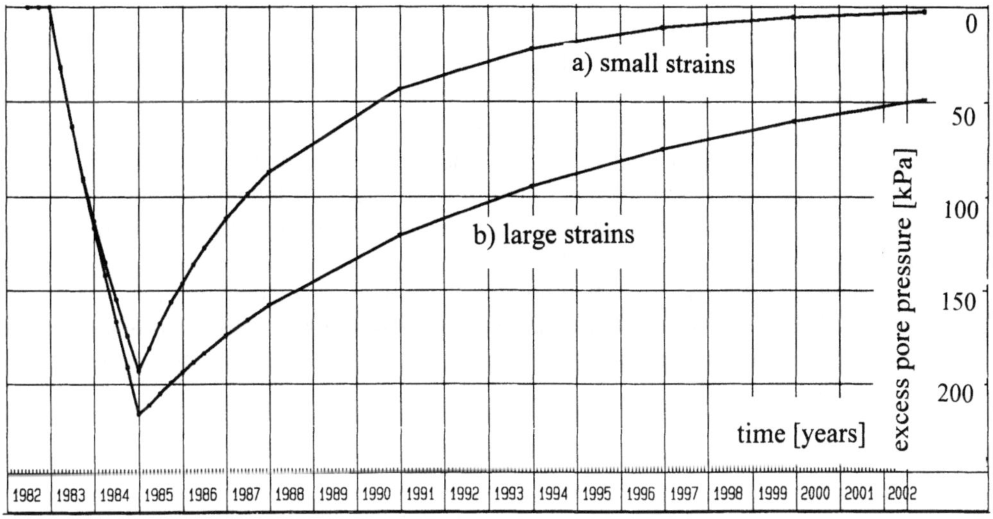

Fig. 9. Excess of pore pressure at mud layer (point B Fig.7): a) small strains b) large strains.

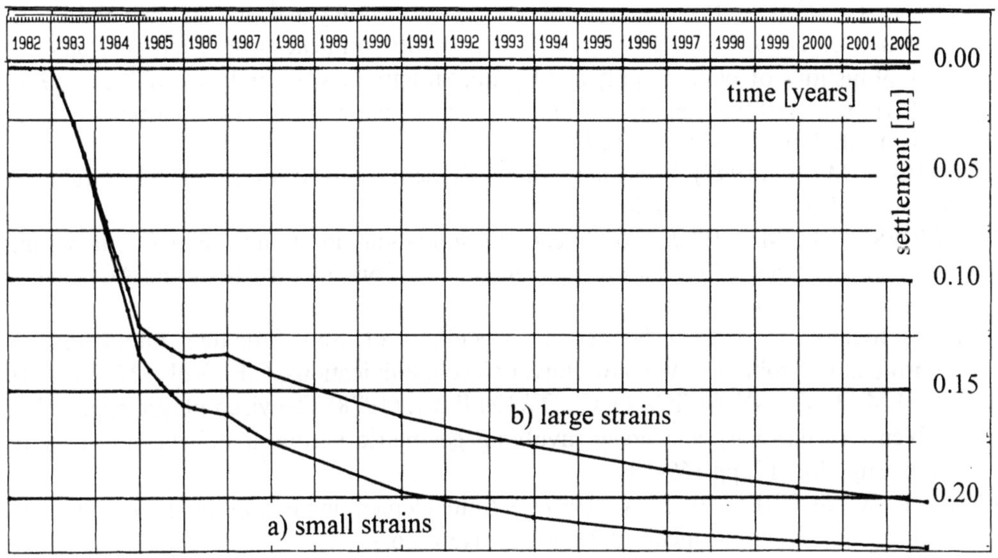

Fig. 10. Displacements for point A, Fig.7 a) small strains, b) large strains

Closing of the pores influence time of the excess pore pressure redistribution, Fig.9. The displacements for the chosen point (initially located at the ground level in the middle of the dam, point A, Fig.7) are drown for both cases of analysis in the Fig.10. The obtained displacements are smaller for the cases of the large strain analysis.

6. CONCLUSIONS

The consolidation process for elasto-plastic materials allowing for large strains can be successfully mixed with the updated Lagrange procedure. The elasto-plastic behaviour of the soil skeleton can be efficiently modelled by means of the pseudo-visco plastic iterative procedure in such coupled situations. The procedure is sensitive on the amount of load increment. The iterative process is found to be divergent if the loading increase is too large especially for the axially symmetric cases. The variation of the permeability, during deformation process is important and influence the time of the excess pore pressure redistribution. The differences in the obtained results for small and large strains are important and should be taken into account especially where soft organic soils are located in the foundations. The calculated settlements for the large strains analysis is found to be smaller comparing the small strain approach.

7. ACKNOWLEDGEMENT

The paper is supported by KBN Project No 765/T07/97/13 what is gratefully appreciated.

8. REFERENCES

BATHE K. J. (1982) Finite element procedures in engineering analysis, Prentice-Hall,

CHACIŃSKI Z., DŁUŻEWSKI J.M. (1995) Permeability of peat - test results (in Polish), II Sem. of Institute of Water Supply and Hydro. Structures, Warsaw Univ. of Tech., 69-75

CORMEAU I. C. (1975) Numerical stability in quasi static elasto-visco-plasticity" Int. J. Num. Meth. Eng., Vol. 9, 109-127,

CRISFIELD M. A. (1997) Non-linear finite element analysis of solids and structures, Advanced Topics, Vol. 2, Wiley

DŁUŻEWSKI J. M. (1997) Nonlinear consolidation in finite element modelling, Proceedings of the IX International Conference on Computer Methods and Advances in Geomechanics, Wuhan, China

DŁUŻEWSKI J. M. (1993) Numerical Modelling of Soil Structure Interactions in Consolidation problems, Warsaw Univ. of Tech. Publications, C.E., Vol. 123, pp 1-116

KANCHI M. B., ZIENKIEWICZ O. C., OWEN P.R.J. (1978) The visco-plastic approach to problems of plasticity and creep involving geometric nonlinear effects, Int. J. Num. Meth. Eng., Vol.12, pp 169-181

LEWIS R. W., SCHREFLER B. A. (1987) A finite element method in the deformation and consolidation of porous media, John Wiley, New York

MEROI E.A. SCHREFLER B. ZIENKIEWICZ O.C (1995) Large strain static and dynamic semisaturated soil behaviour, Int. J. Num. Anal. Meth. Geomech., Vol.19 81-106

MONTE J. L., KRITZEN R. J. (1976) One dimensional mathematical model for large strain consolidation, Geotechnique 26(3)

PERZYNA P. (1966) Fundamental problems in viscoplasticity Advan. in Appl Mech.Vol. 9

ZIENKIEWICZ O.C. CORMEAU I. (1974) Visco-plasticity-plasticity and creep in elastic solids - a unified numerical solution approach, Int. J. Num. Meth. Eng., Vol. 8, 821-845

IMPLICIT INTEGRATION OF NON-STANDARD PLASTICITY MODELS

O.M. Heeres and R. De Borst
Delft University of Technology, Delft, The Netherlands

Abstract

The observation that cyclic behaviour of granular materials cannot be described appropriately with standard plasticity models has led to the introduction of numerous non-standard plasticity models. Currently, most of these models are still integrated using an explicit formulation, thus potentially leading to inaccuracy and unstable behaviour. In this paper we give a general method for the fully implicit integration of the rate equations resulting from non-standard plasticity theories. The method is illustrated showing the implicit integration of two of such models.

1 Introduction

In geotechnical engineering, robust and accurate finite element prediction of the inelastic behaviour of soils subjected to cyclic loading has become more and more important, resulting in the introduction of a host of non-standard plasticity models. In these models plasticity does not only occur when the stress point is situated on the yield surface, but also *within* this yield surface. As a consequence hysteresis loops and a smooth transition from an elastic state into a fully plastic state can be described.

As the first in the class of cyclic plasticity theories with multiple *translating yield surfaces*, the multi-surface model was introduced by Mróz *et al.* (1981), assuming a large number of surfaces. Dafalias and Popov (1975) simplified this model to the two-surface model by assuming a single loading surface within the yield surface. Dafalias and Popov (1977) further simplified the two-surface model by letting the loading surface contract to a point. A model related to this has been presented by Pastor *et. al.* (1990). In the remainder of this paper we refer to this model as the zero surface model.

In the class of cyclic plasticity models with multiple *expanding/contracting yield surfaces*, Hashiguchi (1980) presented the initial subloading model, which is rather identical to the bounding surface model by Dafalias and Herrmann (1980). Later, the initial subloading model was modified to the extended subloading model by adding kinematic hardening. An example is the extended subloading model with rotational hardening by Hashiguchi and Chen (1998).

It is not easy to find detailed information on the way these models have been implemented. It seems that most of the algorithms for the integration of the equations resulting from cyclic plasticity theories are still based on explicit formulations. Consequently, problems can arise in locations where a stable behaviour of the stress integration algorithm is of capital importance.

On the other hand, attemps to apply implicit integration to static geomechanical problems have been made. Examples are Borja (1991), Jeremić and Sture (1997), Macari *et al.* (1997), Groen (1997) and Suiker (1998). In this study we will apply implicit integration to two cyclic plasticity models. The algorithm which we use seems also applicable to the integration of viscoplastic models, for instance Wang (1997) and hypoplastic models, Kolymbas *et al.* (1995).

In Section 2 an outline of the integration method for the rate equations is given. Section 3 deals with the implicit integration of the zero surface model, while Section 4 demonstrates application of implicit integration to the extended subloading model with rotational hardening. In Section 5 some concluding remarks are presented.

2 Numerical implementation of nonstandard plasticity theory

2.1 The rate equations

In classical plasticity theory, yielding occurs when the current stresses satisfy the yield criterion

$$f(\boldsymbol{\sigma}, \kappa) = 0, \tag{1}$$

where $\boldsymbol{\sigma}$ contains the stresses and the hardening parameter κ determines the amount of hardening or softening in the material. Loading/unloading is established by means of the Kuhn-Tucker relations

$$\dot{\lambda} \geq 0,\ f \leq 0,\ \dot{\lambda} f = 0, \tag{2}$$

where a superimposed dot denotes a time derivative, and $\dot{\lambda}$ is the rate of the consistency parameter.

For granular materials, a more appropiate description of the material behaviour is given by plasticity theories in which the current stress point is assumed to be always on the loading surface. It has been shown in different studies, Mróz and Zienkiewicz (1984), Hashiguchi (1993), that in this case a sufficient condition for loading and unloading is

$$\begin{aligned} \mathbf{nD}\dot{\boldsymbol{\varepsilon}} > 0 &\rightarrow \text{loading} \\ \mathbf{nD}\dot{\boldsymbol{\varepsilon}} = 0 &\rightarrow \text{neutral loading} \\ \mathbf{nD}\dot{\boldsymbol{\varepsilon}} < 0 &\rightarrow \text{unloading,} \end{aligned} \tag{3}$$

with $\boldsymbol{\varepsilon}$ the strains, $\mathbf{n}$ the gradient of the plastic potential function with respect to the stresses and $\mathbf{D}$ the elastic constitutive matrix.

Plastic behaviour is characterized by the presence of irreversible strains on load removal, which is obtained by the usual additive decomposition of the strains into an elastic and a plastic part,

$$\boldsymbol{\varepsilon} = \boldsymbol{\varepsilon}^e + \boldsymbol{\varepsilon}^p. \tag{4}$$

The stresses are related to the elastic strains by

$$\boldsymbol{\sigma} = \mathbf{D}(\boldsymbol{\varepsilon} - \boldsymbol{\varepsilon}^p). \tag{5}$$

The flow rule expresses the plastic strain rate $\dot{\boldsymbol{\varepsilon}}^p$ as

$$\dot{\boldsymbol{\varepsilon}}^p = \dot{\lambda}\frac{\partial g(\boldsymbol{\sigma},\ \kappa)}{\partial \boldsymbol{\sigma}}, \tag{6}$$

in which $g(\boldsymbol{\sigma},\ \kappa)$ is the plastic potential function. In this study we assume strain hardening, so that the rate of the hardening parameter κ is a positive, monotonically increasing function of the plastic strain rate,

$$\dot{\kappa} = \dot{\kappa}(\dot{\boldsymbol{\varepsilon}}^p). \tag{7}$$

2.2 Integration of the rate equations

The integration algorithm for the rate equations described in the previous subsection essentially consists of solution of a system of equations which are assembled in a residual vector

$$\mathbf{r} = \mathbf{r}(\boldsymbol{\sigma}(\mathbf{a}(\boldsymbol{\varepsilon})), \mathbf{a}(\boldsymbol{\varepsilon}), \boldsymbol{\varepsilon}) = \mathbf{0}. \tag{8}$$

In (8), the *primary* variables are collected in the vector $\mathbf{a}$, which generally can be divided into three categories; a *stress category*, a *consistency* category and the category of the *hardening laws.* Note that as the problem is fully *strain driven*, $\mathbf{r}$ is fully determined by $\boldsymbol{\varepsilon}$. Now, in contrast to the usual and often complicated elimination of all stress components and hardening variables, all primary variables including the consistency parameter, are *solved simultaneously* on integration point level using a regular Newton-Raphson procedure

$$\mathbf{r}_{n+1} = \mathbf{r}_n + \frac{\partial \mathbf{r}_n}{\partial \mathbf{a}_n} \Delta \mathbf{a}_n = \mathbf{0} \tag{9}$$

where n stands for the local iteration number.

2.3 Evaluation of the consistent tangent operator

As has been shown by Simo and Taylor (1985), the tangent operator $\mathbf{D}^{ep}$ must be obtained by *consistent linearization* of the stress update at the end of the local iteration $n+1$, in order to obtain quadratic convergence of the equilibrium iterations:

$$\mathbf{D}^{ep} = \left. \frac{\mathrm{d}\boldsymbol{\sigma}(\boldsymbol{\varepsilon},\, \mathbf{a})}{\mathrm{d}\boldsymbol{\varepsilon}} \right|_{n+1} . \tag{10}$$

Elaboration of this differentiation gives the total differential

$$\frac{\mathrm{d}\boldsymbol{\sigma}(\boldsymbol{\varepsilon},\, \mathbf{a})}{\mathrm{d}\boldsymbol{\varepsilon}} = \frac{\partial \boldsymbol{\sigma}}{\partial \boldsymbol{\varepsilon}} + \frac{\partial \boldsymbol{\sigma}}{\partial \mathbf{a}} \frac{\partial \mathbf{a}}{\partial \boldsymbol{\varepsilon}} . \tag{11}$$

Because the strains are constant during integration we have

$$\frac{\mathrm{d}\mathbf{r}}{\mathrm{d}\boldsymbol{\varepsilon}} = \frac{\partial \mathbf{r}}{\partial \boldsymbol{\varepsilon}} + \frac{\partial \mathbf{r}}{\partial \mathbf{a}} \frac{\partial \mathbf{a}}{\partial \boldsymbol{\varepsilon}} = \mathbf{0} \tag{12}$$

and obtain $\partial \mathbf{a}/\partial \boldsymbol{\varepsilon}$ as

$$\frac{\partial \mathbf{a}}{\partial \boldsymbol{\varepsilon}} = - \left[\frac{\partial \mathbf{r}}{\partial \mathbf{a}} \right]^{-1} \frac{\partial \mathbf{r}}{\partial \boldsymbol{\varepsilon}} . \tag{13}$$

Substitution of eqn. (13) into eqn. (11) gives the consistent tangent operator

$$\frac{\mathrm{d}\boldsymbol{\sigma}(\boldsymbol{\varepsilon},\, \mathbf{a})}{\mathrm{d}\boldsymbol{\varepsilon}} = \frac{\partial \boldsymbol{\sigma}}{\partial \boldsymbol{\varepsilon}} - \frac{\partial \boldsymbol{\sigma}}{\partial \mathbf{a}} \left[\frac{\partial \mathbf{r}}{\partial \mathbf{a}} \right]^{-1} \frac{\partial \mathbf{r}}{\partial \boldsymbol{\varepsilon}} . \tag{14}$$

Generally, $\partial \mathbf{r}/\partial \mathbf{a}$ is non-symmetric, while $\partial \boldsymbol{\sigma}/\partial \mathbf{a}$ and $\partial \mathbf{r}/\partial \boldsymbol{\varepsilon}$ do not commute, resulting in a non-symmetric tangent operator.

3 The "zero surface" model

In 1990, Pastor, Zienkiewicz and Chan presented a plasticity model for the description of the behaviour of sand and clay. This model has the remarkable feature that the *gradients* to the yield and plastic potential functions are defined explicitly, instead of these functions themselves.

In this section we shall first give a review of the model, after which we show that a fully implicit integration of the model can be done in a simple way.

3.1 Model definition

In modelling triaxial tests, where the stress state is in principle axialsymmetric, it is convenient to work with the invariants of the *effective* stresses defined as $p = \frac{1}{3}(\sigma_{xx} + \sigma_{yy} + \sigma_{zz})$ and $q = \sqrt{\frac{3}{2}\boldsymbol{\xi}^{\mathrm{T}}\mathbf{R}\boldsymbol{\xi}}$, where the deviatoric stresses $\boldsymbol{\xi}$ are $\boldsymbol{\xi} = \boldsymbol{\sigma} - \boldsymbol{\pi}p$, $\mathbf{R} = diag(1,1,1,2,2,2)$ and $\boldsymbol{\pi} = [1,1,1,0,0,0]^T$. The volumetric strain ε_v and the second invariant of the deviatoric strains ε_s are $\varepsilon_v = \varepsilon_{xx} + \varepsilon_{yy} + \varepsilon_{zz}$ and $\varepsilon_s = \sqrt{\frac{2}{3}\boldsymbol{\gamma}^{\mathrm{T}}\mathbf{R}^{-1}\boldsymbol{\gamma}}$, respectively, with the deviatoric strains $\boldsymbol{\gamma} = \boldsymbol{\varepsilon} - \boldsymbol{\pi}\varepsilon_v$.

The gradients to the yield and the plastic potential functions are

$$\mathbf{n} = \left[\begin{array}{cc} \partial f/\partial p & \partial f/\partial q \end{array}\right]^T = \frac{1}{\sqrt{1+d^2}}\left[\begin{array}{cc} d & 1 \end{array}\right]^{\mathrm{T}} \tag{15}$$

and

$$\mathbf{m}_L = \left[\begin{array}{cc} \partial g/\partial p & \partial g/\partial q \end{array}\right]^T = \frac{1}{\sqrt{1+d_g^2}}\left[\begin{array}{cc} d_g & 1 \end{array}\right]^{\mathrm{T}} \tag{16}$$

for loading, while for unloading we have

$$\mathbf{m}_U = \left[\begin{array}{cc} -\mathrm{abs}\; m(1) & m(2) \end{array}\right]^{\mathrm{T}}, \tag{17}$$

In eqns. (15), (16) and (17), d and d_g stand for the *dilatancy* which, following constant q/p tests on sand, are given by

$$\begin{aligned} d &= (1+\alpha)(M + \frac{q}{p}) \\ d_g &= (1+\alpha)(M_g + \frac{q}{p}), \end{aligned} \tag{18}$$

with α, M and M_g model parameters. Formally, the yield and plastic potential functions can be obtained by integration of $\mathbf{n}$ and $\mathbf{m}$.

In expression (18), M is a function of Lode's angle θ:

$$M = \frac{18M_{crit}}{18 + 3(1 - \sin 3\theta)}, \tag{19}$$

with M_{crit} the slope of the critical state line. The hardening function for loading H_L allows stress states beyond the critical state line and is a function of the effective stresses and the isotropic hardening parameter

$$H_L = H_L(p, q, \kappa(\boldsymbol{\varepsilon}^p)). \tag{20}$$

Similarly, for unloading we have

$$H_U = H_U(p, q, \kappa(\boldsymbol{\varepsilon}^p)), \tag{21}$$

where eqn. (3) is used to distinguish between loading and unloading.

3.2 Numerical integration of the rate equations

The deviatoric stresses at stage t_{n+1} are determined by

$$\boldsymbol{\xi}_{n+1} = \boldsymbol{\xi}_n + 2G\mathbf{R}^{-1}(\Delta\boldsymbol{\gamma}_{n+1} - \Delta\boldsymbol{\gamma}^p_{n+1}), \tag{22}$$

where G is the shear modulus. Substituting the flow rule, eqn. (6), in discrete form and the flow direction, eqn. (16) or (17) into eqn. (22), we obtain after rewriting an explicit expression for the deviatoric stress:

$$\boldsymbol{\xi}_{n+1} = \frac{q_{n+1}}{q_{n+1} + 3G\Delta\lambda_{n+1} m_{L/U}(2)_{n+1}} \boldsymbol{\xi}^{trial}. \tag{23}$$

Using the definition for the second invariant of the deviatoric strains, q_{n+1} is then found as

$$q_{n+1} = q^{trial} - 3\Delta\lambda_{n+1} G m_{L/U}(2)_{n+1}. \tag{24}$$

The volumetric stresses are updated as

$$p_{n+1} = p^{trial} - \Delta\lambda_{n+1} K m_{L/U}(1)_{n+1}, \tag{25}$$

where K stands for the bulk modulus. In eqns. (24) and (25), p^{trial} and q^{trial} are the elastic stress predictors.

As a consequence of the model definition, the consistency requirement cannot be expressed in the form of eqn. (1). Therefore, we require that

$$\Delta\lambda_{n+1} = \left.\frac{\mathbf{n}^T\mathbf{D}\Delta\boldsymbol{\varepsilon}}{H_{L/U} + \mathbf{n}^T\mathbf{D}\mathbf{m}_{L/U}}\right|_{n+1} \tag{26}$$

is satisfied.

With the update equations (25, 24 and 26), a vector consisting of residuals, eqn. (8), is assembled. The category of residuals in which the *stress updates* are collected reads

$$\mathbf{r}_1 = \left\{ \begin{array}{l} p_{n+1} - p^{trial} - \Delta\lambda_{n+1} K m_{L/U}(1)_{n+1} \\ q_{n+1} - q^{trial} - 3\Delta\lambda_{n+1} G m_{L/U}(2)_{n+1} \end{array} \right\}, \tag{27}$$

while the *consistency* category reads

$$r_2 = \Delta\lambda_{n+1} - \left.\frac{\mathbf{n}^T\mathbf{D}\Delta\boldsymbol{\varepsilon}}{H_{L/U} + \mathbf{n}^T\mathbf{D}\mathbf{m}_{L/U}}\right|_{n+1}. \tag{28}$$

Finally, the *hardening* category consists of

$$r_3 = H_{n+1} - H_{L/U}(p_{n+1}, q_{n+1}, \kappa(\varepsilon^p_{n+1})) \tag{29}$$

and the complete integration algorithm consists of solving

$$\mathbf{r}_{n+1} = \mathbf{0}. \tag{30}$$

In (30), the variables are also divided into *three categories*: The *stress category* which contains the variables p_{n+1} and q_{n+1}, the *consistency category* which contains $\Delta\lambda_{n+1}$ and finally the *hardening category* with H_{n+1}. The consistent tangent operator is determined as outlined in Section 2.3.

4 The extended subloading model with rotational hardening

In this section we shall apply implicit integration to the extended subloading model with rotational hardening, reported by Hashiguchi and Chen (1998). This model is an extension of the initial subloading model, with translational and rotational kinematic hardening.

The extended subloading model features a subloading surface which expands, contracts, translates and rotates within a yield surface. This surface passes through the current stress point and keeps a similarity with respect to the yield surface, while the center of similarity can translate. Next, the plastic modulus is a monotonically decreasing function of the ratio of the size of the subloading surface and that of the yield surface.

As has been shown by Hashiguchi and Chen (1998), this model satisfies continuity between loading and (partial) unloading. Furthermore, the *Masing effect*, which means that the curvature of the unloading-reloading load-displacement curves decreases with respect to the initial curve, can be controlled, Masing (1926).

4.1 Model definition

The yield surface is defined by

$$f(\hat{\chi},\ \hat{p}) - F(\hat{\boldsymbol{\sigma}}, \kappa(\boldsymbol{\varepsilon}^p)) = 0. \tag{31}$$

In eqn. (31), $\hat{\chi}$ is a measure for the magnitude of the deviatoric stresses with respect to the rotated hydrostatic axis of the yield surface,

$$\begin{aligned} \hat{\boldsymbol{\eta}} &= \frac{\hat{\boldsymbol{\xi}}}{\hat{p}} - \boldsymbol{\beta} \\ \hat{\chi} &= \frac{1}{\hat{m}}\|\hat{\boldsymbol{\eta}}\|. \end{aligned} \tag{32}$$

Here, $\hat{p}$ and $\hat{\boldsymbol{\xi}}$ are the hydrostatic and deviatoric stresses on the yield surface, respectively. The rotational hardening variable $\boldsymbol{\beta}$ is used to describe anisotropy, $\hat{m}$ is a model parameter and F a function of the stresses $\hat{\boldsymbol{\sigma}}$ on the yield surface and the isotropic hardening parameter κ. The subloading surface inside the yield surface is described by

$$f(\bar{\chi},\ \bar{p}) - RF(\hat{\boldsymbol{\sigma}}, \kappa(\boldsymbol{\varepsilon}^p)) = 0, \tag{33}$$

where $\bar{\chi}$ and $\bar{p}$ are computed as in eqn. (32), but with the stresses with respect to the (translated) origin of the subloading surface, $\bar{\boldsymbol{\sigma}}$. In eqn. (33) R stands for the ratio of the sizes of the subloading and the yield surface, $0 \leq R \leq 1$. Hence, $R = 1$ means that the subloading and yield surfaces coincide, while $R = 0$ means that the subloading surface is a point.

The rotational hardening rate $\dot{\boldsymbol{\beta}}$ is assumed to be proportional to the magnitude of the deviatoric strain,

$$\dot{\boldsymbol{\beta}} = C_\beta \|\boldsymbol{\gamma}^p\| \, \|\bar{\boldsymbol{\eta}}\| \bar{\boldsymbol{\eta}}_b, \\ \bar{\boldsymbol{\eta}}_b = \bar{m}_b \frac{\bar{\boldsymbol{\eta}}}{\|\bar{\boldsymbol{\eta}}\|} - \boldsymbol{\beta}, \tag{34}$$

with C_β and $\bar{m}_b$ model parameters, and $\|\boldsymbol{\gamma}^p\|$ the magnitude of the deviatoric plastic strain. The stresses $\bar{\boldsymbol{\sigma}}$ are expressed in the current stress $\boldsymbol{\sigma}$, R and $\mathbf{S}$ as

$$\bar{\boldsymbol{\sigma}} = \boldsymbol{\sigma} - (1 - R)\mathbf{S}, \tag{35}$$

while

$$\bar{\boldsymbol{\sigma}} = R\hat{\boldsymbol{\sigma}}. \tag{36}$$

In eqn (35), R evolves according to

$$\dot{R} = C_R \left(\frac{1}{R} - 1 \right) \|\boldsymbol{\varepsilon}^p\|, \tag{37}$$

with C_R a model parameter. In this way the the stress point approaches the yield surface in an asymptotic, smooth manner. The evolution of the similarity center $\mathbf{S}$ is defined such that $\mathbf{S}$ always stays inside the yield surface. It has be shown by Hashiguchi and Chen (1998) that

$$\dot{\mathbf{S}} = C_S \|\boldsymbol{\varepsilon}^p\| (\boldsymbol{\sigma} - \mathbf{S}) + \frac{1}{F} \left\{ \dot{F} - \frac{\partial f(\chi_S, p_S)}{\partial \boldsymbol{\beta}} \dot{\boldsymbol{\beta}} \right\} \mathbf{S} \tag{38}$$

satisfies this requirement, with χ_S computed with $\mathbf{S}$ as in eqn. (32), $p_S = 1/3(S_{xx} + S_{yy} + S_{zz})$ and C_S a model parameter.

4.2 Numerical integration of the rate equations

Using eqns. (5) and (6) a residual with respect to the *stresses* at stage t_{n+1} is formed as

$$\mathbf{r}_1 = \boldsymbol{\sigma}_{n+1} - \boldsymbol{\sigma}^{trial} + \Delta\lambda_{n+1} \mathbf{D} \mathbf{m}_{n+1}, \tag{39}$$

where $\boldsymbol{\sigma}^{trial}$ is an elastic stress predictor. Next, with help of eqns. (34), (37) and (38), residuals regarding the *kinematic hardening* variables are defined as

$$\begin{aligned} \mathbf{r}_2 &= \boldsymbol{\beta}_{n+1} - \boldsymbol{\beta}^{trial} - C_\beta \|\boldsymbol{\gamma}^p_{n+1}\| \, \|\hat{\boldsymbol{\eta}}_{n+1}\| \hat{\boldsymbol{\eta}}_{b,n+1} \\ r_3 &= R_{n+1} - R^{trial} - C_R \left(\frac{1}{R_{n+1}} - 1 \right) \|\boldsymbol{\varepsilon}^p_{n+1}\| \\ \mathbf{r}_4 &= \mathbf{S}_{n+1} - \mathbf{S}^{trial} - C_S \|\boldsymbol{\varepsilon}^p_{n+1}\| (\boldsymbol{\sigma}_{n+1} - \mathbf{S}_{n+1}) \\ &+ \frac{1}{F_{n+1}} \left\{ F_{n+1} - F^{trial} - \frac{\partial f_{n+1}(\chi_S, p_S)}{\partial \boldsymbol{\beta}_{n+1}} (\boldsymbol{\beta}_{n+1} - \boldsymbol{\beta}_n) \right\} \mathbf{S}_{n+1} \end{aligned} \tag{40}$$

in which F^{trial}, $\boldsymbol{\beta}^{trial}$, R^{trial} and $\mathbf{S}^{trial}$ are the converged values at the end of the previous time step. The residual for the evolution of the *isotropic hardening* is defined as

$$r_5 = F_{n+1} - F(\hat{\boldsymbol{\sigma}}_{n+1}, \kappa(\boldsymbol{\varepsilon}^p_{n+1})) \tag{41}$$

Because the subloading surface eqn. (33) always passes through the current stress point, the equation involving *consistency* is

$$r_6 = f(\bar{\chi}_{n+1}, \bar{p}_{n+1}) - RF(\hat{\boldsymbol{\sigma}}_{n+1}, \kappa(\boldsymbol{\varepsilon}^p_{n+1})) \tag{42}$$

Now, the algorithm is satisfied solving the system of equations

$$\mathbf{r}_{n+1} = \mathbf{0}, \tag{43}$$

for $\boldsymbol{\sigma}_{n+1}$, $\boldsymbol{\beta}_{n+1}$, R_{n+1}, S_{n+1}, F_{n+1} and $\Delta\lambda_{n+1}$. The consistent tangent matrix is evaluated as described in Section 2.3.

5 Concluding remarks

The algorithm addressed in this study is based upon satisfaction of a system of equations consisting of residuals. In this system we have three categories: a *stress category*, a *consistency category* and a *hardening category*.

When the yield functions are not defined explicitly, the equations involving consistency have to be reformulated. In this paper we have shown that after doing this, implicit integration of these class of models can be done in a simple way.

Next, we have derived an algorithm for the implicit integration of the extended subloading model with kinematic hardening. Here, the formation of residuals resulted in a simple algorithmic treatment of the rate equations.

References

BORJA, R. (1991). Cam-Clay plasticity, Part II, Implicit integration of constitutive equations based on a nonlinear stress predictor. *Comp.Meth.Appl.Mech.Engng.*, **88:** 225–240.

DAFALIAS, Y. F. & HERRMANN, L. R. (1980). A bounding surface soil plasticity model, In Pande, G. & Zienkiewicz, O., editors, *Proc. Int. Symp. Soils ander Cyclic Loading*, pages 335–345, Swansea, Balkema, Rotterdam.

DAFALIAS, Y. F. & POPOV, E. P. (1975). A model of nonlinearly hardening materials for complex loading. *Acta Mechanica*, **21:** 173–192.

DAFALIAS, Y. F. & POPOV, E. P. (1977). Cyclic loading for materials with a vanishing elastic region. *Nucl.Eng.Design*, **41:** 293–302.

GROEN, A. E. (1997). *Three-Dimensional Elasto-Plastic Analysis of Soils.* Dissertation, Delft University of Technology.

HASHIGUCHI, K. (1980). Consitutive equations of elastoplastic materials with elastic-plastic transition. *J.Appl.Mech.*, **102:** 266–272.

HASHIGUCHI, K. (1993). Mechanical requirements and structures of cyclic plasticity models. *Int.J.Plast.*, **9:** 721–748.

HASHIGUCHI, K. & CHEN, Z.-P. (1998). Elastoplastic constitutive equation of soils with the subloading surface and the rotational hardening. *Int.J.Num.Anal.Meth.Geom.*, **22:** 197–228.

JEREMIĆ, B. & STURE, S. (1997). Implicit integration in elastoplastic geotechnics. *Mech.Coh.Frict.Mat.*, **2:** 165–183.

KOLYMBAS, D., HERLE, I., & WOLFFERSDORFF, P. (1995). Hypoplastic constitutive equation with internal variables. *Int.J.Num.Anal.Meth.Geom.*, **19:** 415–436.

MACARI, E. J., WEIHE, S., & ARDUINO, P. (1997). Implicit integration of elasto-plastic consitutive models for frictional materials with highly non-linear hardening functions. *Mech.Coh.Frict.Mat.*, **2:** 1–29.

MASING, G. (1926). Eigenspannungen und Verfestigung beim Messing, In *Proc.2nd Int.Congr.Appl.Mech*, Zűrich.

MRÓZ, Z., NORRIS, V. A., & ZIENKIEWICZ, O. C. (1981). An isotropic, critical state model for soils subject to cyclic loading. *Geotechnique*, **31:** 451.

MRÓZ, Z. & ZIENKIEWICZ, O. C. (1984). Uniform Formulation of Consitutive Equations for Clays and Sands, In *Mechanical Engineering Materials*, pages 415–449, John Wiley, New York.

PASTOR, M., ZIENKIEWICZ, O. C., & CHAN, A. C. (1990). Theme/Feature paper: Generalized plasticity and the modelling of soil behavior. *Int.J.Num.Anal.Meth.Geom.*, **14:** 151–190.

SIMO, J. C. & TAYLOR, R. L. (1985). Consistent tangent operators for rate-independent elastoplasticity. *Comp.Meth.Appl.Mech.Engng.*, **48:** 101–118.

SUIKER, A. S. J. (1998). Fatigue behaviour of granular materials, part 2: Numerical formulation of the constitutive relations. Technical Report 7-98-119-3, Delft University of Technology.

WANG, W. M. (1997). *Stationary and propagative instabilities in metals - a computational point of view.* Dissertation, Delft University of Technology.

ADAPTIVE MESH SIMULATION OF PASSIVE EARTH PRESSURE FAILURE

M.A. Hicks
University of Manchester, Manchester, UK

ABSTRACT: Adaptive remeshing has been used in the finite element analysis of passive earth pressure failure. This has involved the updating of meshes at regular user-specified intervals, thereby enabling the progressive development of high shear strains to be more closely modelled. For a frictionless wall and elastic-perfectly plastic material behaviour, the computed results are in close agreement with analytical solutions. In the absence of strain-softening, the minimum element size has no influence, other than to control the shear band width. A criterion has been proposed for determining mesh-update increments automatically.

1 INTRODUCTION

Hicks and Mar (1994) have introduced an adaptive mesh refinement approach to analysing strain localisation problems in geomechanics. This is based on remeshing at regular intervals during loading, enabling accurate discretisation for those zones in which strain concentrations are developing. An important feature of this work is that, following each mesh-update, no re-analysis of previous increments is necessary. Also, the algorithm has been successfully used with complicated models of soil behaviour (e.g. Hicks (1997a,b)).

This paper continues the work of Hicks (1998) in analysing a drained passive earth pressure problem. Previously, the emphasis was on a detailed algorithm description and on comparing the responses of soils with different dilation angles. The present work investigates the influence of minimum element size. The concept of remeshing indicator has been introduced as a convenient criterion for determining mesh-update increments.

2 NUMERICAL ALGORITHM

Only brief algorithm details are given and these are confined to the use of simple elastoplastic models for drained computations. An extended description for such problems is given by Hicks (1998), while Hicks (1997a) describes the implementation for more complicated models and partially drained material response.

2.1 FINITE ELEMENT EQUATIONS

The global incremental equation is solved using a modified Newton-Raphson iteration scheme and takes the form

$$[KE]\{dr\}_n^m = \{df\}_n^{m-1} = \{df_{ext}\}_n + \{df_{err}\}_n + \{df_{body}\}_n^{m-1} \tag{2.1}$$

in which $[KE]$ is the symmetric elastic stiffness, $\{dr\}$ the incremental nodal displacements and $\{df\}$ the incremental nodal loads. Subscript and superscript refer to increment and iteration numbers, respectively, and iteration continues until the iterative change in $\{dr\}$ becomes sufficiently small.

$\{df\}$ comprises three components: $\{df_{ext}\}$, the incremental external loads, which remain constant from one iteration to the next; $\{df_{body}\}$, the incremental out-of-balance forces, which 'drive' the iteration process; and $\{df_{err}\}$, the unconverged out-of-balance forces from the end of the previous increment, which are re-applied in the following increment as an extra set of external loads.

$\{df_{body}\}$ is formed by integrating the incremental excess plastic stresses $\{d\sigma^p\}$ over the mesh domain Ω. Hence,

$$\{df_{body}\}_n^{m-1} = \int_\Omega [B]^T \{d\sigma^p\}_n^{m-1} d\Omega \tag{2.2}$$

in which $[B]$ contains the shape function derivatives and where $\{d\sigma^p\}$ is computed from

$$\{d\sigma^p\}_n^{m-1} = \{d\sigma^e\}_n^{m-1} - \{d\sigma^*\}_n^{m-1} \tag{2.3}$$

In this equation, $\{d\sigma^e\}$ are the incremental elastic stresses, while $\{d\sigma*\}$ are the smoothed approximation to the incremental elastoplastic stresses $\{d\sigma\}$. Note that, for adaptive mesh purposes, the algorithm is based on a smoothed stress-strain formulation and that the use of $\{d\sigma*\}$ in forming $\{df_{body}\}$ ensures compatibility between the smoothed and unsmoothed solutions.

Two steps are taken to increase the convergence rate for a given increment. Firstly, $\{df_{body}\}$ for the first iteration is taken to be that from the end of the previous increment (Smith and Griffiths, 1998), i.e.

$$\{df_{body}\}_n^0 = \{df_{body}\}_{n-1}^k \tag{2.4}$$

in which k is the number of iterations in the n-1th increment. Secondly, the use of $\{df_{err}\}$, defined by

$$\{df_{err}\}_n = \{df_{body}\}_{n-1}^k - \{df_{body}\}_{n-1}^{k-1} \tag{2.5}$$

means that larger convergence tolerances may be used (Hicks, 1995).

2.2 ADAPTIVE MESH FRAMEWORK

Mesh-updating takes place at regular intervals during an analysis, to account for new and/or changing regions of strain concentration. Following each mesh-update, internal variables (e.g. stresses, strains, etc.) are transferred from old to new meshes. This enables the analysis to continue without the need for re-analysing previous increments.

2.2.1 SMOOTHING

The smoothing of internal variables is required: (a) to enable mapping of variables in mesh-update increments; (b) for the estimation of discretisation errors. This has been achieved using a global projection smoothing formulation (e.g. Zienkiewicz and Zhu (1987)), i.e.

$$\int_\Omega [N^*]^T [N^*] d\Omega \{\bar{q}^*\} = \int_\Omega [N^*]^T \{q\} d\Omega \tag{2.6}$$

in which $[N^*]$ are the smoothing shape functions, $\{q\}$ the quantities to be smoothed and $\{\bar{q}^*\}$ the resulting smoothed quantities which are determined at the nodal points. For the present work, $\{q\}$ takes the form proposed by Hicks (1998) and comprises three vector components: $\{d\sigma\}$, the incremental elastoplastic stresses, which are smoothed for every iteration of every increment and used in equation (2.3); $\{d\varepsilon\}$, the incremental strains, which are smoothed at the end of every increment and used in equation (2.7); and $\{d\sigma^e\}$, the incremental elastic stresses, which are only needed for mesh-update increments. A smoothed approximation to this third component is needed immediately following each mesh-update, so that $\{df_{body}\}$ and $\{df_{err}\}$ can be re-computed for the new mesh at the start of the following increment, via equation (2.3).

2.2.2 ERROR ESTIMATION

Estimates of element discretisation error are computed using the error indicator proposed by Hicks and Mar (1994) and Hicks (1997a), i.e.

$$\left(\|e_{d\gamma}\|_{L_2}\right)_i = \left(\int_{\Omega_e} e_{d\gamma} e_{d\gamma} \, d\Omega_e\right)^{1/2} \tag{2.7}$$

in which Ω_e is the element domain and $(\|e_{d\gamma}\|_{L2})_i$ is the L_2 norm, for element i, of the error in the incremental shear strain invariant, $e_{d\gamma}=d\gamma^*-d\gamma$, where $d\gamma^*$ and $d\gamma$ are the shear strain invariants computed using the smoothed and unsmoothed incremental strain fields, respectively. Element error norms can be combined to give a global error norm, from which a global relative error η_{L2} may also be derived (Hicks (1998), Zienkiewicz and Zhu (1987)).

2.2.3 MESH GENERATION

New meshes are generated using an advancing front technique similar to that of Peraire et al. (1987), with the number and distribution of elements within the new mesh being based on the element error estimates computed using equation (2.7). The aims are: (a) to achieve the optimum mesh through a process of element error equalisation (Mar and Hicks, 1996); (b) to attain some user-specified global relative error $\bar{\eta}$.

The general refinement strategy is described by Hicks (1998) and based on the work of Zienkiewicz and Zhu (1987) and Zienkiewicz et al. (1988). In brief, new element sizes are governed by the element refinement indicator ζ, defined for element i as

$$\zeta_i = \frac{\left(\|e_{d\gamma}\|_{L_2}\right)_i}{\bar{e}_m} \tag{2.8}$$

Hence, the element error norm from equation (2.7) is compared with $\bar{e}_m$, the target element error norm, which is derived from $\bar{\eta}$ by assuming element error equalisation for an m-element mesh. Nodal spacings for the new mesh are defined relative to nodal points in the previous (i.e. background) mesh. Hence,

$$\delta_j = \frac{\bar{h}_j}{\bar{\zeta}_j^{1/p}} \tag{2.9}$$

in which δ_j is the new nodal spacing at the location of 'old' node j. Also defined at node j are: $\bar{h}$, a measure of old element size, found by averaging the lengths of element sides connected to node j; and $\bar{\zeta}^{1/p}$, the average of the element subdivision numbers for elements connected to node j, in which each subdivision number has been computed by raising equation (2.8) to the power $1/p$, where p is the polynomial order of the elements being used.

A converged mesh configuration is ensured by re-analysing each mesh-update increment using the iterative technique proposed by Hicks (1998). Although taking p=1, as has been done here, leads to over-refinement for the first iteration, convergence occurs in only three iterations. (Note that taking p=2 for 3-node triangles leads to a more realistic number of elements for the first iteration, due to this being based on comparing the squares of element lengths (i.e. their areas).)

3 PASSIVE EARTH PRESSURE EXAMPLE

Problem geometry and loading conditions are the same as in Hicks (1998). A smooth and rigid wall, 1m high, is translated into a drained uniform rectangular soil mass, 2m×5m, using 100 equal-sized displacement increments. Boundary conditions include a fixed mesh base and rollers allowing only vertical movement along the right-hand boundary, while the initial stresses have been defined by an effective unit weight, γ'=10kN/m^3, and by a coefficient of earth pressure at rest, K_0=1.0. The starting mesh comprises 640 equal-sized triangular elements (3-noded) and mesh-updating takes place at the end of every tenth increment, as well as just after the start of the analysis at the end of increment 2.

The soil has been modelled using an elastic-perfectly plastic Mohr-Coulomb idealisation. Material properties are as follows: Young's modulus, E'=10^{5}kPa; Poisson's ratio, ν'=0.3; friction angle, ϕ'=30°; cohesion, c'=0.0kPa; and dilation angle, ψ=10° (cf. Hicks (1998) in which dilation angles of 0, 20 and 30 degrees were also considered).

As in Hicks (1998), a graded maximum element size has been used, with δ_{max} varying linearly with distance from the wall: specifically, δ_{max} varies from 0.125m at the wall to 0.25m at the right hand boundary. However, for this investigation four minimum element sizes have been considered: i.e. δ_{min}=0.03, 0.04, 0.05 and 0.06 metres (cf. only 0.03m used previously). Again, a global relative aiming error of 7% has been used.

3.1 RESULTS

The results are summarised in Figures 1 to 4. Figure 1 illustrates the evolution of mesh configurations for a typical analysis, by showing the starting mesh and nine of the eleven mesh-updates. For reasons of clarity, this figure considers the case with the largest δ_{min}. It is seen that the mesh appearance stays fairly constant following localisation.

Figure 2 compares the load-displacement responses for the four analyses, by plotting mobilised earth pressure coefficient K_P (=F/F_0) against wall horizontal displacement Δ_H, non-dimensionalised with respect to wall height H. F and F_0 are the horizontal mobilised and initial loads acting on the wall, respectively, which have been computed by using the repeated trapezium rule to integrate the smoothed nodal stresses acting normal to the wall. Figure 2 shows that δ_{min} has no influence on load-displacement response, with each analysis computing a collapse load near to the Rankine limiting solution of K_P=tan^2(45°+ϕ'/2)=3.0.

Figure 3 shows contours of incremental shear strain invariant at the end of each analysis, demonstrating that δ_{min} also has no influence on shear band orientation: in each case there

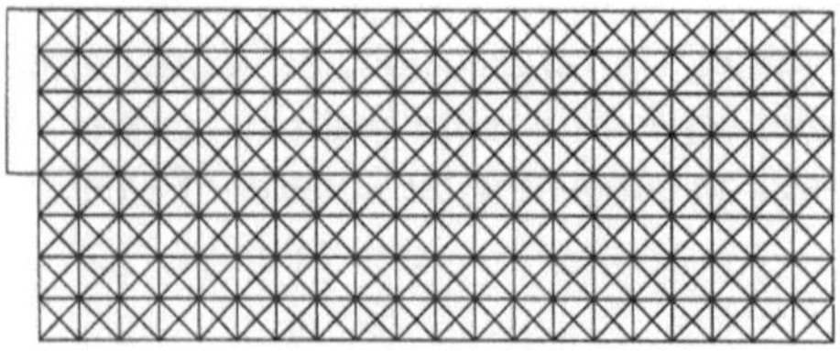

(a) starting mesh

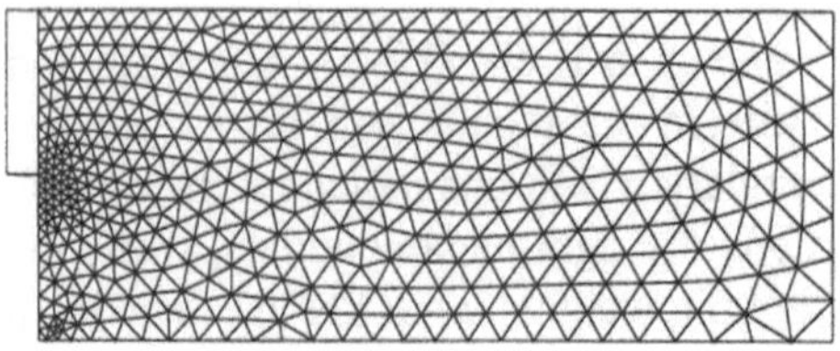

(b) 1st mesh-update

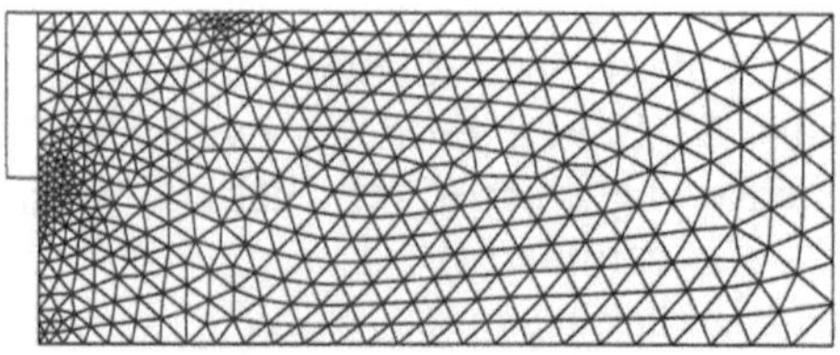

(c) 2nd mesh-update

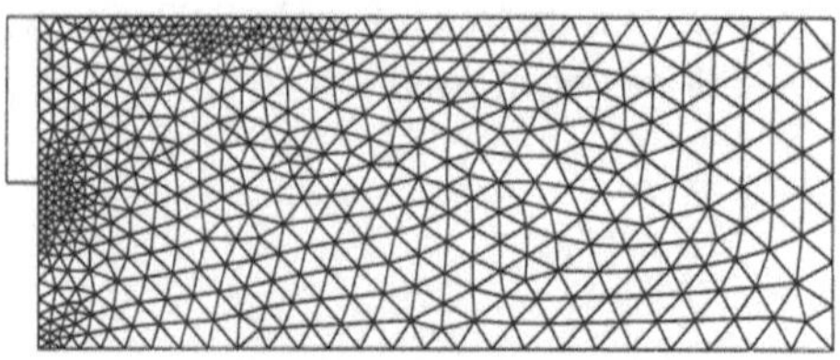

(d) 3rd mesh-update

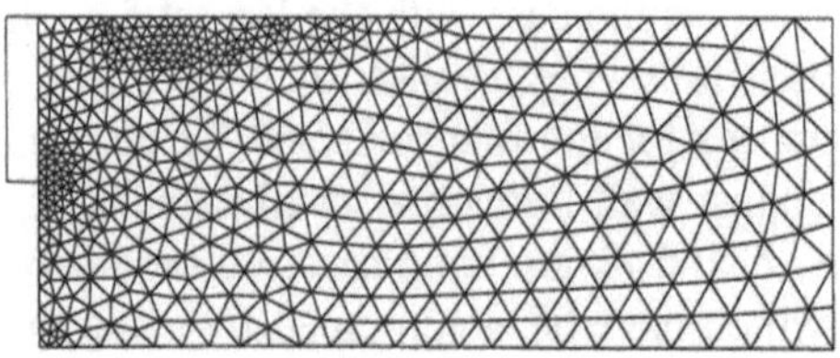

(e) 4th mesh-update

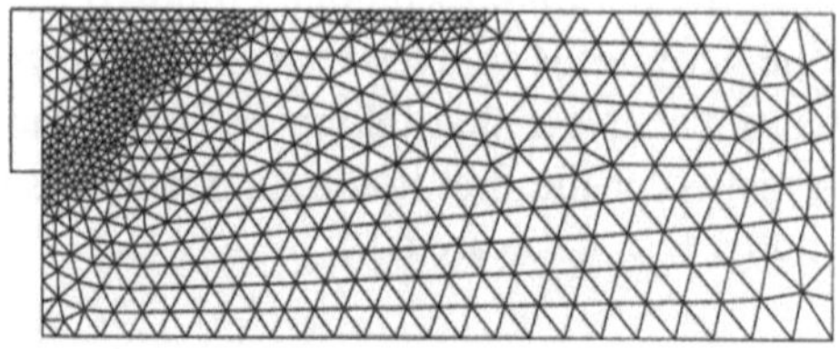

(f) 5th mesh-update

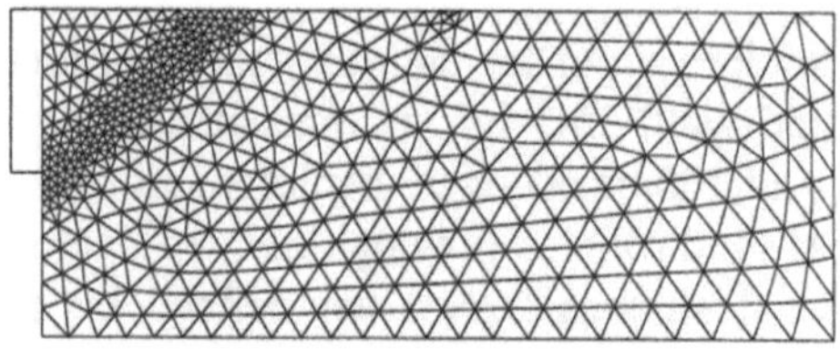

(g) 6th mesh-update

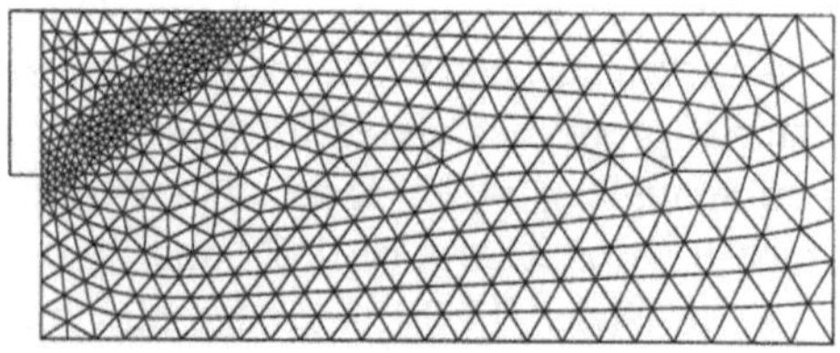

(h) 7th mesh-update

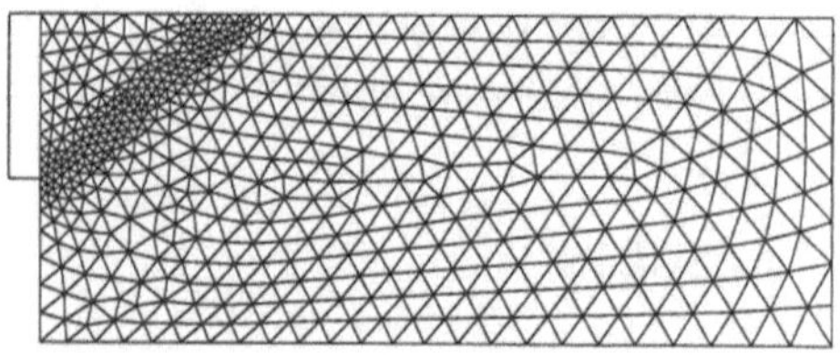

(i) 9th mesh-update

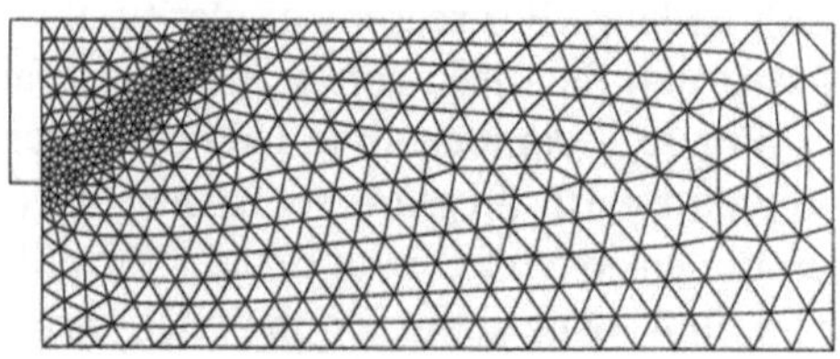

(j) 11th mesh-update

Fig. 1. Evolution of meshes for δ_{min}=0.06H

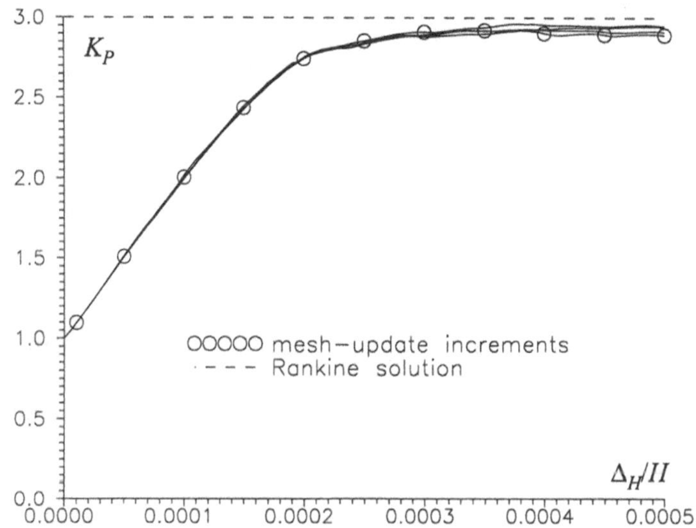

Fig. 2. Influence of minimum element size on load-displacement response

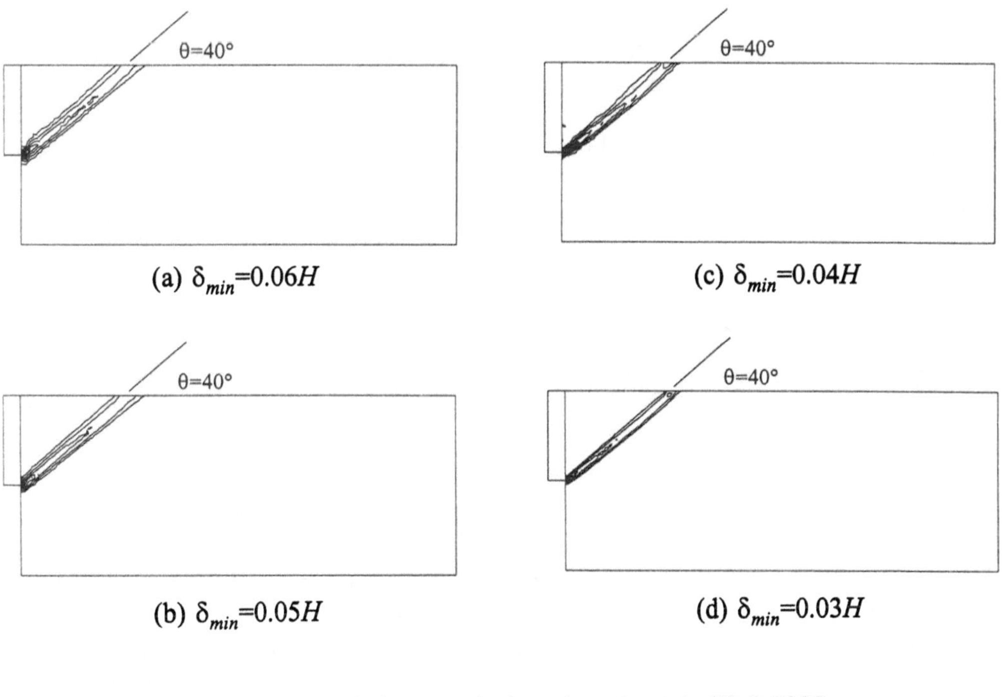

Fig. 3. Contours of incremental shear strain invariant dγ at Δ_H/H=0.0005

is good agreement with the Roscoe solution of θ=(45°-ψ/2)=40°. However, due to the absence of an internal length scale approach, shear band width is a function of minimum element size and, for the present work, is approximately equal to four times δ_{min}.

Figure 4 gives further details of algorithm performance by comparing various aspects of the δ_{min}=0.03 and 0.06 metre analyses. Specifically, as the displacement of the wall is increased, four quantities are compared: (a) mobilised earth pressure coefficient, K_P (as for Figure 2); (b) estimated global relative error, η_{L2}; (c) number of elements in the mesh, *ne*; (d) remeshing indicator, χ. In each figure, symbols are used to denote the mesh-update increments: i.e. circles and squares for the δ_{min}=0.03 and 0.06 metre cases, respectively.

Figure 4(b) shows that, for both minimum element sizes, and for the first three or four mesh-updates, mesh-updating results in a decrease in global error to approximately the aiming error $\bar{\eta}$ (shown by the horizontal broken line). Note that the error shown for a mesh-update increment tends to be an over-estimate, because it has been computed using the penultimate mesh generated during the mesh-update iteration process (see section 2.2.3): i.e. in the final iteration, but just before computation of the final mesh. Following localisation, however, the aiming error is clearly not attainable. The smaller minimum element size does lead to an increase in accuracy, but it is not certain whether decreasing the element size further would result in attainment of the aiming error, or whether the more severe localisation then being modelled would prevent this happening. Figure 4(b) underlines the difficulty in quantifying errors following localisation, thereby demonstrating why, in this instance, a remeshing criterion based only on η_{L2} would not be successful.

Figure 4(c) summarises the numbers of elements used in the two analyses. By comparing this figure with Figure 1, it is seen that, following localisation, successive meshes have similar numbers and distributions of elements. Therefore, despite the high errors computed for this stage of each analysis, there is little tendency for the mesh to change.

Figure 4(d) introduces the remeshing indicator, a quantity indicating when remeshing is required: the larger the value for χ, the greater the need for a new mesh. χ has been computed for each increment and is based, in part, on η_{L2}, via the vector of new nodal spacings $\{\delta_1\}$. However, it also takes account of the constraints imposed through specifying δ_{min} and δ_{max}. In equation form:

$$\chi=\frac{1}{nn}\sum_{i=1}^{nn}\sqrt{\left(\frac{\delta_1^i-\delta_0^i}{\delta_0^i}\right)^2}\ ;\qquad \delta_{min}\leq\delta_0^i\leq\delta_{max}\ ;\qquad \delta_{min}\leq\delta_1^i\leq\delta_{max} \tag{3.1}$$

in which *nn* is the number of nodes in the current mesh, $\{\delta_0\}$ are the nodal spacings corresponding to that mesh and superscript *i* denotes node number. Hence, χ is a measure of the tendency of a mesh to change, either with respect to number of elements, or distribution of elements, or both. By comparing Figures 1 and 4(d), it is seen that low values for χ correspond to minimal changes in mesh configuration, suggesting that this quantity may reasonably be used to automate the decision on when to perform mesh-updating.

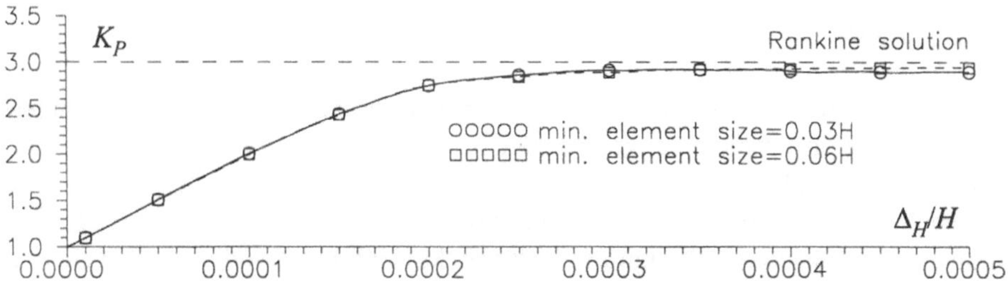

(a) mobilised earth pressure coefficient

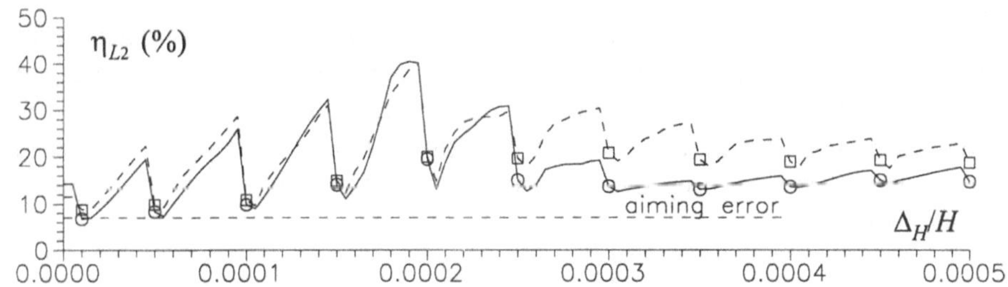

(b) global relative error

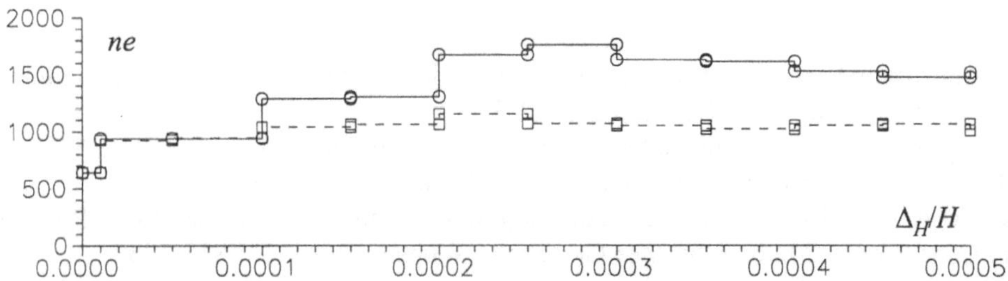

(c) number of elements in mesh

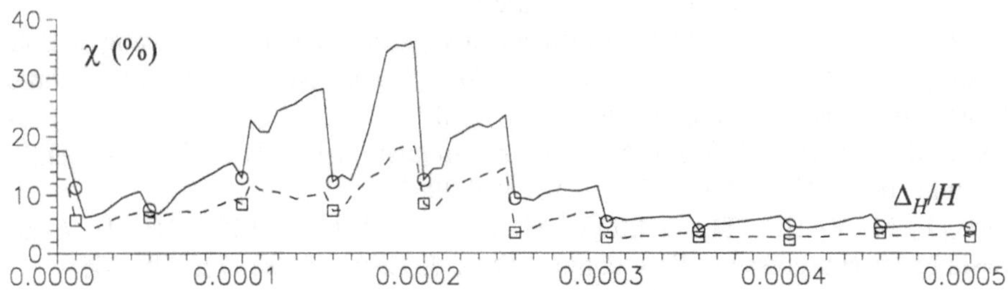

(d) remeshing indicator

Fig. 4. Variation of various quantities with respect to displacement

4 CONCLUSIONS

Accurate computations of collapse load and shear band orientation have been obtained for a simple deformation problem in geomechanics. In the absence of strain-softening, the minimum element size has no influence, other than to control the shear band width. For the 3-node elements used in this investigation the shear band width is approximately four times the minimum nodal spacing, although results by Hicks (1998) have indicated that shear band width is also a function of stress-strain response. Due to the absence of an internal length scale facility, the present formulation has no control over the number of elements traversing a band.

The results have also shown that it is difficult to quantify error estimates following localisation (see also Hicks and Mar (1994)), this being due, at least in part, to the constraints imposed through limiting element refinement. For this reason, global relative error provides a poor indication of when remeshing should take place. A remeshing indicator has therefore been proposed with a view to automating, rather than specifying, mesh-update increments.

REFERENCES

HICKS, M.A. (1995). MONICA—A computer algorithm for solving boundary value problems using the double-hardening constitutive law Monot: I. Algorithm development. Int. J. Num. Anal. Meth. Geomech., 19: 1-27.

HICKS, M.A. (1997a). Implementation of adaptive mesh refinement in MONICA. Proc. 6th Int. Symp. Num. Models Geomech., Montreal, Canada: 409-414.

HICKS, M.A. (1997b). An adaptive mesh study of localisation in a saturated soil. Proc. 9th Int. Conf. Comp. Meth. Advances Geomech., Wuhan, China, 3: 1853-1858.

HICKS, M.A. (1998). Error estimation and mesh refinement for computations of localisation in geomaterials. Proc. 4th Int. Workshop on Localization and Bifurcation Theory for Soils and Rocks, Gifu, Japan (1997). In press.

HICKS, M.A. & A. MAR (1994). Mesh adaptivity for geomaterials using a double-hardening constitutive law. Proc. 8th Int. Conf. Comp. Meth. Advances Geomech., Morgantown, West Virginia, U.S.A., 1: 587-592.

MAR, A. & M.A. HICKS (1996). A benchmark computational study of finite element error estimation. Int. J. Num. Meth. Eng., 39: 3969-3983.

PERAIRE, J., M. VAHDATI, K. MORGAN & O.C. ZIENKIEWICZ (1987). Adaptive remeshing for compressible flow computations. J. Comp. Phys., 72: 449-466.

SMITH, I.M. & D.V. GRIFFITHS (1998). Programming the finite element method (3rd ed.). Pub. Wiley.

ZIENKIEWICZ, O.C. & J.Z. ZHU (1987). A simple error estimator and adaptive procedure for practical engineering analysis. Int. J. Num. Meth. Eng., 24: 337-357.

ZIENKIEWICZ, O.C., Y.C. LIU & G.C. HUANG (1988). Error estimation and adaptivity in flow formulation for forming problems. Int. J. Num. Meth. Eng., 25: 23-42.

AN APPLICATION OF MESHLESS METHODS IN MODELING OF DAMAGED ROCK MASSES

D. Hoxha, F. Homand and M. Souley
École Nationale Supérieure de Géologie, Vandœuvre-lès-Nancy, France

ABSTRACT: A two-dimensional code coupling Element-Free Galerkin Method (EFGM) and Finite Elements Method (FEM) as described by Belytschko *et al.* (1995) is used to simulate the behavior of elasto-damaged materials. The EFGM is used in some regions of interest and the FEM in the other parts. Two damage models are implemented and tested. The first is a continuum damage model as described by Homand *et al.* (1998). The second is a micromechanical damage model based on the works of Kachanov (1982, 1994) and those of Nemat-Nasser & Obata (1988). Some simulations of laboratory tests using the micromechanical model are performed to check its applicability. Then the damage around a circular gallery was studied using the two models.

1. INTRODUCTION

The modeling of excavation damaged zone (EDZ), which may be formed around an underground opening excavated into brittle rocks, is a problem whose solution calls for experimental and *in situ* observations as well as the use of some appropriate calculation methods.

The observations have shown that a considerable number of defects of small sizes appears in rock matrix in function to its components even in the presence of macrocracks and fractures. The damage of rocks is essentially related to the development of such microcracks whose length is only some millimeters. Surely the other cracks of greater sizes are important but one could imagine the process of crack growth as a process of microcracks growth and coalescence.

In order to describe the mechanisms of crack initiation and growth, several micromechanical models have been proposed (Kachanov 1982, Nemat-Nasser and Obata 1988, Kemeny and Cook 1991). The micromechanical fracture models generalize the results of linear fracture mechanics established for a single crack in the set of cracks involved in a elementary representative volume (ERV). There are a few examples of the application of such models in the calculation of in situ structures because of the difficulties to implement such models in the actual running codes based on a space discretization.

The continuum damage models based on thermodynamic principles and using an internal damage variable, are successfully used to the damage modeling (Dragon et al 1994, Shao *et al.* 1997, Homand-Etienne *et al.* 1998, Souley *et al.* 1998). The principal advantage of continuum damage models is the simplicity of their implementation into the traditional codes. However it is difficult to catch and secure a physical meaning of some state variables in such models. The damage variable is considered to be calculated on some ERV, great enough to make a sense to the homogenization in the same time when a very refined mesh is necessary to be used in order to obtain the numerical stability. A solution to this contradict is the use of a non local constitutive law.

The meshless methods have known a progressive development during these last years, with constant increasing of calculating possibilities (Belytchko *et al.* 1996). As the bases of these methods are developed quickly, the first applications are appeared in many fields: static and dynamic fracture mechanics (Krysl and Belytschko1997), modeling of material interfaces, etc. What is more, it seems that these methods are well suited for the crack propagation analyses (Krysl and Belytschko 1997, Fleming 1997). In these applications the cracks are considered to be in a elastic matrix. The presence of microcracks could sensibly modify the overall mechanical behavior of the rock as well as the parameters of crack propagation. In such situations the cracks grow in close relation to the microcracks presence and interaction.

The objective of this paper is to check the applicability of Element Free Galerkin method, a meshless method, in the modeling of damaged rocks. We only consider microcracks and the proposed approach could be used with other methods then the meshless ones. However the use of meshless methods presents some advantages for the modeling of the interaction between microcracks and cracks (not discussed here). We present some bases of the method followed by a brief description of a micromechanical damage model. The continuum damage model used for the simulations is not presented here, but we refer the reader to Homand *et al.* (1998) for the details. Then we give some details of implementation and finally the simulation of the damage zone around a circular gallery is discussed.

2. THEORITICAL BACKGROUND

2.1 Element Free Galerkin Method (EFGM)

We present here only a very brief description of the method. The details could be found for example in Belytschko *et al.* (1996).

EFGM is a relatively recent approach proposed by Belytschko *et al.* (1994) as an improved version of diffuse element method introduced by Nayroles *et al.* (1992). In this method, the moving least square approximations is used to approach the unknown function u(x) by a ***n***-degrees polynom :

$$u_h(x) = \sum_{j=1}^{n} p_j(x)a_j(x) \tag{1}$$

where $p_j(x)$ are monomial basis functions, and $a_j(x)$ are their coefficients, which depend on the spatial coordinates x. The coefficients $a_j(x)$ are calculated by minimizing the difference between the local approximation and the function :

$$J = \sum_{I=1}^{m} w(x - x_I)\left[u_h(x, x_I) - u_I\right]^2 = \sum_{I=1}^{m} w(x - x_I)\left[p^T(x_I)a(x) - u_I\right]^2 \tag{2}$$

where $w(x - x_I)$ is a weighting function with a compact support an ***m*** is the number of nodes whose supports include x. To find the coefficients a(x) we obtain the extremum of J which gives a vectorial equation system :

$$A(x)a(x) = B(x)u \tag{3}$$

The matrix A and B are expressed as follows :

$$[A(x)]_{ij} = \sum_{q=1}^{m} w(x - x_q)p_i(x_q)p_j(x_q) \qquad [B(x)]_{ij} = w(x - x_j)p_i(x_j) \tag{4}$$

Finally, the approximation $u_h(x)$ could be expressed as

$$u_h(x) = \sum_{I=1}^{m} \phi_I(x)u_I \tag{5}$$

where the shape functions are given by:

$$\phi_I(x) = \sum_{j=1}^{n} p_j(x)\left[A(x)^{-1}.B(x)\right]_{jI} \tag{6}$$

The EFGM has no interpolating property so it has to use some special techniques to enforced the boundary conditions. A very attractive way is to use a coupling of FEM and EFGM as proposed by Belytschko *et al.* (1996). The FEM is used where some boundary conditions has to be enforced as well as in other regions with not particular interest. In the elements found between EFGM and FEM regions, some special shape functions are used making sure the continuity of the approximation.

In the case of the elasticity problem, u(x) to estimate is the displacement field and the weak form for Galerkin approximation could be written :

$$\int_{\Omega} \delta \underline{\underline{\varepsilon}} : \underline{\underline{\sigma}} \, d\Omega = \sum_{i} \int_{\Gamma_{t_i}} \delta \underline{u}_i . \underline{t}_i \, d\Gamma + \int_{\Omega} \delta \underline{u} . \underline{b} \, d\Omega \tag{7}$$

Since both FEM and EFGM regions are discretized by Galerkin approximations, the weak forms can be used directly in the coupled method to develop the discrete equations:

$$[K].\{u\} = \{f\} \tag{8}$$

where $K_{ij} = \int_{\Omega} B_I^T M B_J d\Omega \qquad B_I = \begin{bmatrix} \Phi_{I,x} & 0 \\ 0 & \Phi_{I,y} \\ \Phi_{I,y} & \Phi_{I,x} \end{bmatrix}$

In the case of a damaged material, stiffness matrix M is a function of crack geometry and applied stress, so the solution of discretized equations has to be performed iterratively. For the continuum damage model used in our simulations a damage tensor is calculated at each iteration and then the stiffness matrix is deduced using a potential characterizing the damage rock (Homand *et al.* (1998), Souley *et al* (1998)).

In the case of the micromechanical model the stiffness matrix at each step was written as:

$$M = M^0 + \Delta M \tag{9}$$

where M^0 is the initial elastic stiffness matrix and ΔM is the contribution of microcracks in the mechanical properties. ΔM is calculated by a direct scheme, knowing applied stress and strains induced by microcracks and supposing the existence of an potential for the damaged rock as proposed by Kachanov (1994).

2.2 Micromechanical model of crack initiation, growth and interaction

We consider an initial population of cracks described by some statistical distributions of their length, position and orientation, respectively $N(l)$, $N(X)$ and $N(\theta)$. In this notation $N(a)$ means the number of cracks with propriety a ranged between a and $a+da$. The initial distributions are deduced from microscopic observations. The other cracks with length above a critical user defined length could be studied using one of methods described by Fleming (1997), with some modifications concerning the interaction with microcracks array.

The interaction among the microcracks is considered by using the analyses proposed by Kachanov (1994). The distributed loading $t^j(\zeta^j)$ on the jth microcrack is presented as a sum of its average $<t^j>$ (uniform loading) and a nonuniform loading $t^j(\zeta^j) - <t^j>$. The key simplifying assumption is that the impact of the ith nonuniform loading on j^{th} microcrack is neglected. The problem could be solved as a linear set of equations for the unknown vectors $<t^j>$. This system could be written in a compact vectorial form (Kachanov 1994) :

$$(2\delta_{ik} - \Lambda^{ik}).<t^i> \; = \; t^{k(0)} \tag{10}$$

where summation over all microcrack numbers is assumed. Tensorial element Λ^{ik} represents the average loading generated along the kth microcrack line by the ith crack loaded by an uniform loading of arbitrary direction and unit intensity. The righthand part of eq.10 represents the remote conditions, while the lefthand part represents the intrinsic geometry of microcracks array.

Once the average loading are known the stress vectors on the i-th microcracks could be calculated very easy. Their normal and shear components are respectively :

$$p^i(\zeta) = p^{i(0)} + \underline{n}^i . \sum_{j \neq i} \left[< p^j > \sigma_n^j + < \tau^j > \sigma_\tau^j \right] . \underline{n}^i \tag{11.a}$$

$$\tau^i(\zeta) = \tau^{i(0)} + \underline{n}^i . \sum_{j \neq i} \left[< p^j > \sigma_n^j + < \tau^j > \sigma_\tau^j \right] . (\underline{\underline{I}} - \underline{n}^i . \underline{n}^i) \tag{11.b}$$

In these expressions the double underlined denotes a tensor of second order while $\underline{n}^i$ is the vector normal to the crack surface. σ_n^j and σ_τ^j are the "standard" stress fields generated by a crack loading of unit intensity (Kachanov 1994).

In the case of overall compressive stresses, it is difficult to analyze the growth and the deformations of microcracks. There are a very great number of proposed models mainly based on the sliding cracks (Kachanov 1982, Kemeny and Cook 1991, Nemat-Nasser and Obata 1988). We follow the scheme proposed by Nemmat-Nasser and Obata (1988) with some modifications for the crack interaction.

The stress intensity factor (SIF) for a winged crack is given by :

$$K_I = \frac{2\sqrt{c}}{\sqrt{\pi(l + l^*)}} f_{act} . \sin\Theta + \underline{n} . \underline{\underline{\sigma}} . \underline{n} \sqrt{\pi l} \tag{12}$$

where f_{act} is the driving force in the active flaw calculated as the shearing force corrected for the friction. In our analyses this force count also for the interaction of the microcracks, so the driving force on the crack line is expressed:

$$f_{act}^i = \int_{L_i} \left[\tau^i(\zeta) - p^i(\zeta) . \mu - C \right] d(\zeta) \tag{13}$$

The values of $\tau^i(\zeta)$ and $p^i(\zeta)$ are given by the expressions 11.a and 11.b, while μ and C are respectively the friction coefficient and cohesion of microcracks.

This active force makes possible the sliding and the opening of preexisting flaws. The relation between sliding $\boldsymbol{b}$ and the dilatancy $\boldsymbol{d}$ in the direction normal to the crack face has the following form:

$$d = d_0(1 - e^{-\xi b}) \tag{14}$$

The interaction of winged microcracks is analyzed by substitution of winged cracks with some equivalent linear cracks and by using the above mentioned scheme.

To count for the crack nucleation, the number of cracks in the volume has to be increased with the increase of deviatoric stress. A simple function of crack number verified on our laboratory observations is proposed:

$$N(\underline{\underline{\sigma}}) = N_0 e^{\omega(\frac{J_2^2}{I_1})} \tag{15}$$

Here ω is a parameter adjusted by microscopical observations and I_1 et J_2 represent respectively the first and the second invariants of stress tensor .

The overall strain induced by microcracks is the sum of the contribution of each microcrack. If these strains and the remote stress are known, then we could find the impact of this set of microcracks in the effective elastic properties.

Some examples of the application of this model on the conventional triaxial tests and lateral extension tests are presented in figures 1.a and 1.b.

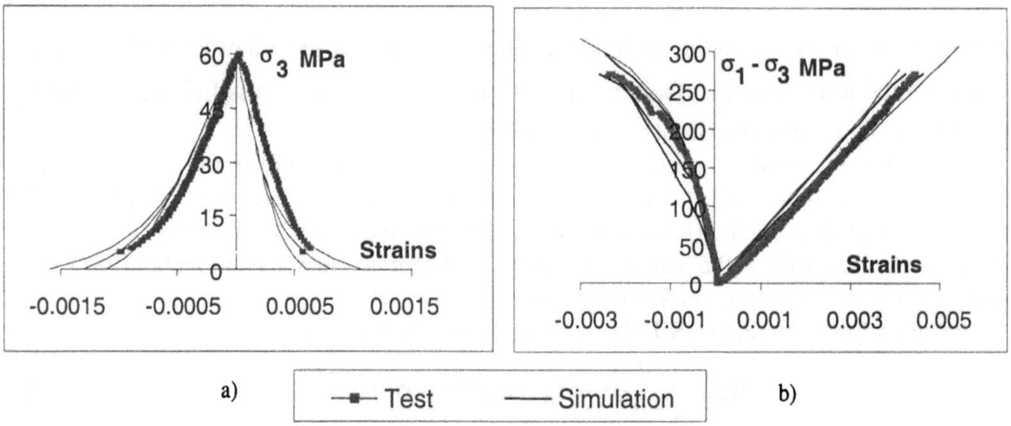

Fig. 1. Simulation of laboratory tests using the micromechanical damage model
a) triaxial compression test b) lateral extension test

3. IMPLEMENTATION PROCEDURE

A 2D code combining the FEM and EFGM was developed. The library of EFGM shape functions is ESFLib by Krysl (1996). Details of this library could be found in the same reference. To perform the calculations related to the microcracks, we have developed. an another library.

A random numbers generator was used as a base to produce user-chosen distribution of microcracks. Each crack is presented by its center coordinates, length and the angle with respect to x direction. At each EFGM node, a sample of cracks has to be generated by the program from a chosen distribution. The number of cracks to be generated would be sufficiently great in order to provide the numerical reproducibility. So the volume of the sample centered at any node must be calculated in order to ensure the minimal number of microcracks for a given microcrack density (number of cracks per unit volume). The same initial sample of microcracks is used for all nodes. The increase of cracks density is simulated not by the increase of number of cracks in a fixed volume but by the shrinkage of the volume in order to fit a density for a given maximal crack number.

Another generator was used to generate the FEM elements and EFGM nodes and EFGM background elements.

A very high time cost operation is the construction of the interaction matrix and the computation of the SIF for the equivalent linearly cracks. The integrals to be calculated are singular and to assure the necessary precision a coordinates transformation and a great number of Gauss points are necessary. One could speed up the calculation by an integration by parts who helps to separate the singular part in some elliptical functions whose values at Gauss points could be evaluated beforehand. In this case some derivatives of stress functions along the cracks are to be calculated, but the cost calculation time is minor.

Here is a simplified flow chart of the 2D code coupling EFGM and FEM :

1. Generation of FEM elements and EFGM nodes.
2. For a user-chosen crack distribution and a given density, estimate a minimal representative volume in order to provide a minimal number of microcracks. (for the micromechanical model only)
3.Generation of cracks in a volume sample. (for the micromechanical model only)
4. For each iteration:
   Estimate the stress in each EFM element and at each EFGM node
   For each EFGM node:
      Estimate the damage state variables
      Estimate the new mechanical properties in respect to the microcracks geometry
   Resolve the equation (8) for the current mechanical properties
   If the changes in mechanical properties are not significant compared to their previous values stop the process, otherwise go to another iteration.

4. APPLICATION ON THE SIMULATION OF DAMAGED ZONE AROUND A CIRCULAR GALLERY

The code is used to simulate the damage zone around a 3.5m diameter circular gallery supposed being excavated into a rock mass with an initial hydrostatic stress of 30 MPa. The elastic properties and microcrack characteristics for the micromechanical model are those used in the previous simulation of the conventional triaxial tests (Fig.1). The parameters of the continuum damage model as described by Homand *et al.* (1998) are determined on the same material and the simulation is carried out for the same gallery.

A general view of meshed zone is presented in figure 2.a.. The circular area between the radii 1.75 and 3.5m is filled with some EFGM nodes. There are as much as 270 nodes in this region. A band of FE elements is constructed around this region in order to enforce the boundary conditions.

The contours of stress principal difference for the micromechanical damage model are presented in figure 3.b in comparison to the same contours obtained using the continuum damage model (fig 3.a). We note the continuity of contours in the passage between FEM and EFGM for both models. The fluctuation of the contours far from the tunnel are explained by the fact that FE in this region are large and the variation of principal stress difference are small which put in difficulty our post processor.

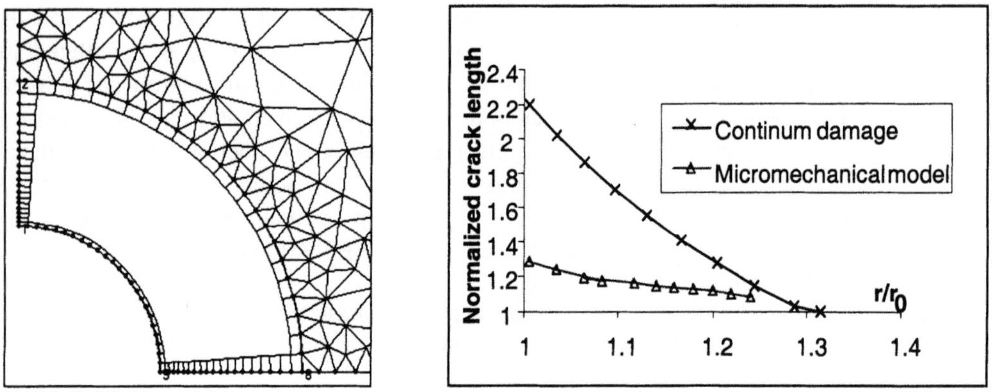

a) b)

Fig. 2. a) Close up of EFGM region
b) Variation of crack length in a radial direction

The maximal difference of principal stresses is just located around the tunnel, but this value is something smaller (55 MPa) in comparison to the theoretical elastic solution (60 MPa). This is explained by the unloading observed in the damaged rock.

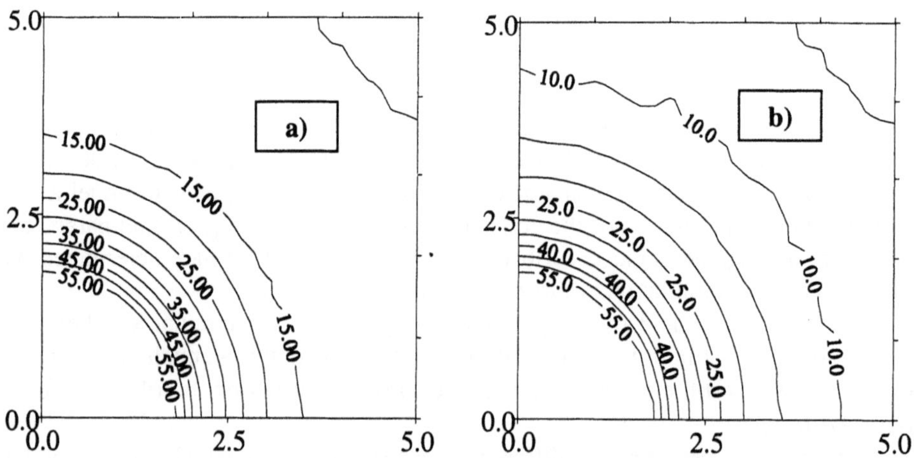

Fig.3. Contours of principal stress difference
a) continuum damage model b) micromechanical damage model

In the figure 2.b are presented the variation of normalized mean crack length along a radial direction. We note that the damage models discussed here refer to different level of rock structure. The one (continuum damage model) refers to changes in mesoscopic level and the other to the changes in microscopic level. For the continuum damage model the initial mean crack length is equal to the grain size (3 mm) while for the micromechanical model 0.06 mm. The mean crack length for the micromechanical model changes very slowly and this phenomena was observed also in some calculations using only FEM (not presented here). This result seems to be in good agreement with some observations which show that the growth of microcracks is accompanied with nucleation of smaller microcracks which makes the mean crack length in microscopic level to grow only moderately.

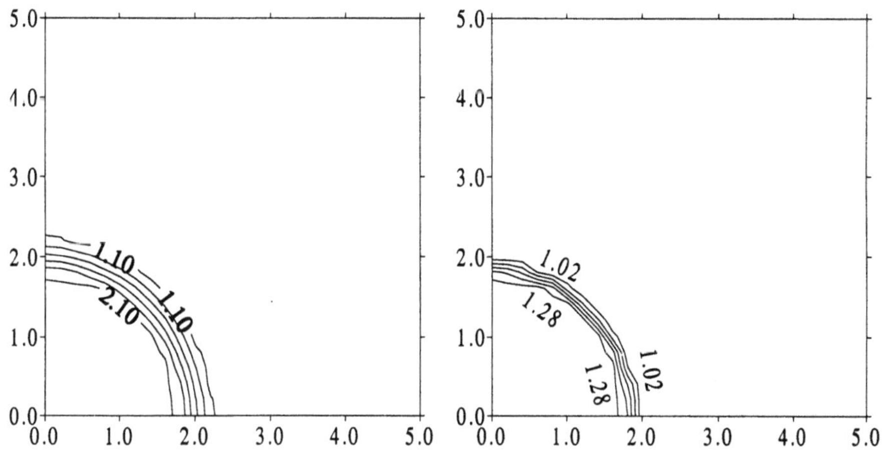

Fig. 4. Contours of normalized mean crack length around the tunnel
a) continuum damage model b)micromechanical damage model

The contours of normalized mean crack length around the tunnel are presented in figure 4. One could use these contours to estimate the extent of the damaged zone. For the continuum damage model (fig 4.a) the deep of damaged zone is as much as 51 cm. However for the micromechanical model the mean length of cracks could not be the only indicator of the damage. For this model we must also consider the density of microcracks, whose normalized value in the vicinity of tunnel is estimated 1.76. The extent of damage zone considering the crack density is about 1.2m while considering the mean crack length about 25 cm. So, the deep of 34 cm has to be considered only as a rough estimation of the damage zone considering the crack length and significant changes in crack density.

5. CONCLUSIONS

A 2D code was developed to check the applicability of meshless methods in damage rock modeling.

A continuum damage model and a micromechanical damage model are implemented and tested. It was found that meshless methods could be used successfully in the prediction of microcracks initiation and propagation. It would be of great interest to test these methods in the conditions of cracks-microcracks interaction.

REFERENCES :

BELYTSCHKO T., Y. KRONGAUZ, D. ORGAN, M. FLEMING AND P. KRYSL (1996) Meshless methods: An overview and recent developments, Computer Methods in Applied Mechanics and Engineering Vol. 139: pg 3-47.

BELYTSCHKO T., D. ORGAN AND Y KRONGAUZ (1995) A coupled finite element-element-free Galerkin method. Computational Mechanics , 17: 186-195

BELYTSCHKO T, Y. Y. LU AND L. GU (1994) Element Free Galerkin methods International Journal of Numerical Methods in Engineering, 37, 229-256.

DRAGON A., F. CORMERY, T. DÉSOYER, D. HALM (1994) Localized failure analysis using damage models, *in* Localization and bifurcation theory for soils and rocks, Chambon R., Desrues J. & Vardoulakis I. (eds), Balkema : 127-140

FLEMING A. (1997) The element-free Galerkin method for fatigue and quasi-static fracture. A dissertation submitted to the graduate school in partial fulfillment of the requirements for the degree Doctor of Philosophy, Evanston, Illinois

HOMAND-ETIENNE F., D. HOXHA AND J.F. SHAO (1998) A continuum damage constitutive law for brittle rocks, Computers and Geotechnics, 22 (2), 135-151.

KACHANOV M. (1982) A microcrack model of rock inelasticity. Part II : Propagation of microcracks. Mechanics of Materials, 1, 29-41.

KACHANOV M. (1994) Elastic solids with many cracks and related problems Advances in Applied Mechanics, Vol 30, 259-445

KEMENY J. and N.G.W. Cook (1991) Micromechanics of deformation in rocks, in Toughening mechanisms in quasi-brittle materials, S.P. Shah (ed), Kluwer Academic Publishers. 155-188

KRYSL P., T. BELYTSCHKO (1997) Application of the element free Galerkin method to the propagation of arbitrary 3-D cracks. Submitted to International Journal for Numerical Methods in Engineering.

NAYROLES B., G. TOUZOT AND P. VILLON (1992) Generalizing the finite element method: diffuse approximation and diffuse elements. Computational Mechanics, 10: 307-318.

NEMAT -NASSER S., M. OBATA (1988) A microcrack model of dilatancy in brittle materials, Journal of Applied Mechanics, Transaction of ASME Vol 55: 24-35

SHAO J.F., G. DUVEAU, N. HOTEIT , M. SIBAI , M. BART (1997) Time dependent continuous damage model for deformation and failure of brittle rock, International Journal of Rocks Mechanics and Mining Sciences, **34**(3-4), Paper No. 285

SOULEY M., D. HOXHA AND F. HOMAND (1998) Distinct element modeling of an underground excavation using a continuum damage model. International Journal of Rocks Mechanics and Mining Sciences, **35**(4-5), Paper No. 26

WHY WATER FLOW IS PREVENTED IN CLAY?

UNIFIED SIMULATION OF MOLECULAR DYNAMICS AND HOMOGENIZATION ANALYSIS FOR SEEPAGE PROBLEM

Y. Ichikawa
Nagoya University, Nagoya, Japan

K. Kawamura
Tokyo Institute of Technology, Tokyo, Japan

M. Nakano
Kobe University, Kobe, Japan

K. Kitayama
Tokyo Electric Power Co., Tokyo, Japan

H. Kawamura
Obayashi Co., Tokyo, Japan

ABSTRACT: There are two crucial issues for analyzing the behavior of micro-inhomogeneous materials: One is how to determine the characteristics of constituent components of the micro-continuum which are directly affected by their molecular movement, and another is how to relate the microscopic characteristics to the macroscopic behavior. Bentonite clay is a typical micro-inhomogeneous material. We proposed a unified method of Molecular Dynamics (MD) and Homogenization Analysis (HA) (henceforth, abbreviated as the *unified* MD/HA *method*) for treating these materials (Kawamura *et al.* 1997). The method can involve the molecular and macro-continuum behaviors in a seamless manner. That is, for determining the properties of each constituent component we apply the MD, then for relating the microscopic characteristics to the macroscopic behavior the HA is used. We here apply it to the water flow (i.e., seepage) problem in bentonite clay.

INTRODUCTION

The essential but yet unsolved problem in radioactive waste disposal is to establish a true physical-chemical procedure for analyzing the long-term behavior (more than 10,000 years) of barrier materials. We know any phenomenological theory such as a visco-elastic model is not applicable for this problem because we cannot provide long-term experimental data for the model. Note that the phenomenological model is essentially workable in the form of interpolation for the experimental data. In this sense the natural analogue is thought to be a prime tool for the long-term behavior problem at the current stage. But is it sufficient? Is there any alternative method, which supports the true physical-chemical process of the long-term behavior?

We proposed the unified MD/ HA method (Kawamura *et al.* 1997, Ichikawa *et al.* 1998) for treating micro-inhomogeneous materials (Fig.1). Based on a true physical-chemical process, this method provides an integrated procedure to simulate their molecular level behavior and the macro-continuum one in a seamless manner. Here the MD is used for defining micro-scale material properties, and the HA is used for extrapolating the micro-inhomogeneous continuum to the bulk-scale continuum.

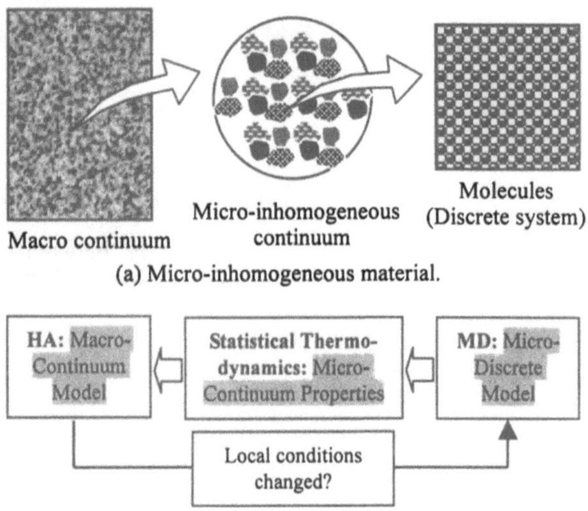

(a) Micro-inhomogeneous material.

(b) Schematic diagram of the unified MD-HA method.

Fig. 1 Concept of the unified MD/HA analysis.

Most of geomaterials belong to this micro-inhomogeneous category. We here apply the unified MD/HA procedure to the seepage problem of compacted bentonite clay, which is a main component of the engineered barrier system (EBS) of the Japanese concept of the high-level radioactive waste disposal. Note that the bentonite clay consists of lamellae of montmorillonite hydrate (i.e., montmorillonite minerals and interlamellar water) and macro-grains such as quartz, so the hydraulic conductivity is determined due to their spatial distribution and the water viscosity. The lowest permeability is undoubtedly attained if the bentonite totally consists of the montmorillonite hydrate, which we treat here as the extreme case.

MOLECULAR BEHAVIOR OF THE MONTMORILLONITE HYDRATE

A montmorillonite mineral is of lamellar shape with the size of about 100×100×1nm, and the montmorillonite hydrate consists of several lamellae of the montmorillonite minerals and interlamellar water. In the previous work (Kawamura *et al.* 1997) we showed that the shearing viscosity of the montmorillonite hydrate is so small that it does not affect its very long-term behavior. Since the hydraulic conductivity of the montmorillonite hydrate is extremely small, we can say that its long-term behavior is mainly governed by the seepage phenomena.

Now the unified MD/HA procedure applied to the seepage problem of the montmorillonite hydrate is given as the following three phases:

1) We develop a crystal model for Na-montmorillonite with surrounding water. After specifying the interatomic potential functions for each atom pairs we perform the MD computation.
2) By applying a standard procedure of statistical mechanics for the MD results we determine the bulk properties of the montmorillonite hydrate such as external/internal water viscosity at the neighborhood of the montmorillonite mineral.
3) In order to determine the macro-mechanics field we introduce a HA procedure (Sanchez-Palencia 1980) in which we use the former MD results (i.e., the water viscosity distribution).

In the MD calculation, motions of all the atoms are given by the Newton equation of motion. The force $\boldsymbol{F}_i$ is calculated from the potential function U_{ij} between two particles given by

$$\boldsymbol{F}_i = \sum_{j(i \neq j)} \boldsymbol{F}_{ij}; \qquad \boldsymbol{F}_{ij} = -\nabla U_{ij}. \tag{1}$$

The molecular formula of Na-montmorillonite hydrate with *n*-layers of interlamellar water is given by $Na_{1/3}Al_2[Si_{11/3}Al_{1/3}]O_{10}(OH)_2.nH_2O$. The interatomic potential for all atom-atom pairs is composed of the Coulomb, the short range repulsion, the van der Waals and the Morse terms. Some of these terms can be omitted due to the nature of the interaction.

$$\text{2-body term:}\quad U_{ij}(r_{ij}) = \frac{z_i z_j e^2}{4\pi\varepsilon_0 r_{ij}} + f_0(b_i + b_j)\exp[\frac{a_i + a_j - r_{ij}}{b_i + b_j}] - \frac{c_i c_j}{r_{ij}^6}$$
$$+D_{ij}[\exp\{-2\beta_{ij}(r_{ij} - r_{ij}^*)\} - 2\exp\{-\beta_{ij}(r_{ij} - r_{ij}^*)\}] \tag{2}$$

The 3-body term is added to the H-O-H interaction because of its sp^3 hybrid orbital:

$$\text{3-body term:}\quad U_{ijk}(\theta_{ijk}, r_{ji}, r_{jk}) = f_k[1 - \cos\{2(\theta_{ijk} - \theta_0)\}](k_i k_j)^{1/2} \tag{3}$$
$$k_i = \frac{1}{\exp\{g_r(r_{ij} - r_m)\} + 1}$$

Parameters $\{z, a, b, c\}$ and $\{D, \beta, r^*\}$ for the 2-body term, $\{f_k, \theta_0, g_r, r_m\}$ for the 3-body term are specified so as to reproduce structural and physical properties of several oxide crystals (quartz, corundum, feldspars, etc.; Kawamura 1992). Parameters for H_2O molecule including the 2-body and 3-body terms were taken from Kumagai *et al.* (1994).

The muscovite (K $Al_2(Si_3$ Al) O_{10} $(OH)_2$, monoclinic, $C_{2/c}$) structure is used to build the initial state of Na-montmorillonite. We employ an (NPT)-ensemble MD scheme under the condition of 300 K temperature and 0.1 MPa pressure. The equation of motion is integrated by the Verlet algorithm. The time increment is 0.4fs. The Ewald method is used to calculate the electrostatic energy and force in long range interaction.

A snap shot for the single-layer montmorillonite lamella system with external water (3,000 H_2O molecules) is shown in Fig. 2(a). Using the MD results the spatial distribution of diffusion coefficient in vicinity of the montmorillonite mineral is calculated, then under the Stokes-Einstein assumption the distribution of viscosity is obtained from the diffusion data as shown in Fig. 2(b). We found the structurally ordered water layer in contact with the clay-surface, which is called the "ice sheet". The thickness of the sheet is ca.0.5nm. Furthermore in the layer of 3 to 4nm thickness the viscosity is changed rapidly. We call such a water property the *iceberg effect*.

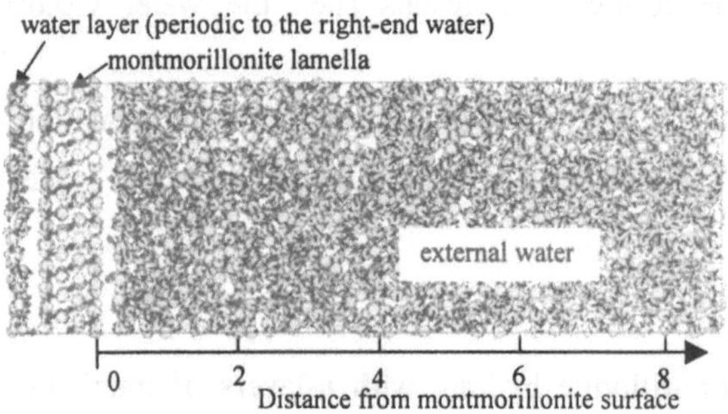

(a) Snap shot.

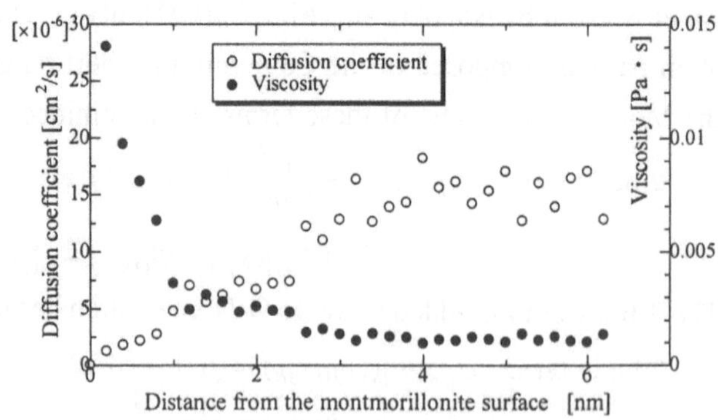

Fig 2. MD results for the montmorillonite hydrate with external water.

(b) Diffusivity and viscosity of the external water.

SEEPAGE PROBLEM FOR THE MONTMORILLONITE HYDRATE BY HA

The HA is a perturbation theory, which was developed for the behavior of a micro-inhomogeneous material with periodic microstructure (Fig. 3: Sanchez-Palencia 1980). We apply the HA to a seepage problem in bentonite with distributed water viscosity in vicinity of montmorillonite minerals. For this problem we start with the Navier-Stokes equation, and obtain a macroscopic seepage equation .

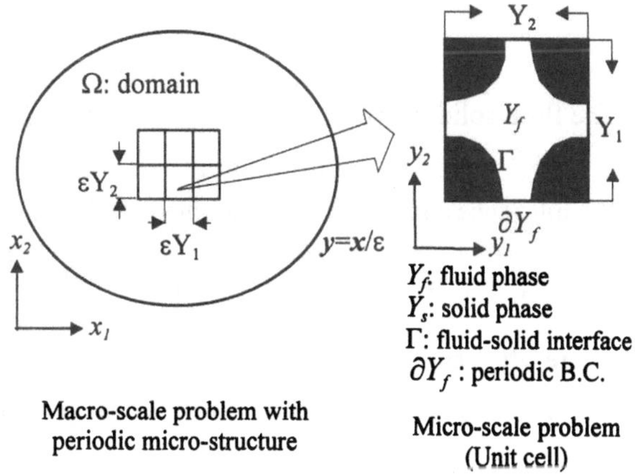

Fig. 3 Macro- and micro-scale problems in HA.

Let us introduce the local coordinate system $\boldsymbol{y}$, which is related to the global one $\boldsymbol{x}$ by $y=x/\varepsilon$. We treat the following incompressible Navier-Stokes flow field:

$$-\frac{\partial P^\varepsilon}{\partial x_i}+\eta\frac{\partial^2 V_i^\varepsilon}{\partial x_k \partial x_k}+F_i=0 \text{ in } \Omega_{\varepsilon f} \tag{4}$$

$$\frac{\partial V_i^\varepsilon}{\partial x_i}=0 \text{ in } \Omega_{\varepsilon f}, \quad V_i^\varepsilon=0 \text{ on } \partial\Omega_{\varepsilon f} \tag{5}$$

where V_i^ε is the velocity vector with the shearing viscosity η, P^ε the pressure, $\boldsymbol{F}_i$ the body force vector, and $\Omega_{\varepsilon f}$ the water flow region in the global coordinate system ($\partial\Omega_{\varepsilon f}$ its boundary).

Now we introduce an asymptotic expansion

$$\begin{aligned} V_i^\varepsilon(\boldsymbol{x}) &= \varepsilon^2 V_i^0(\boldsymbol{x},\boldsymbol{y})+\varepsilon^3 V_i^1(\boldsymbol{x},\boldsymbol{y})+\cdots , \\ P^\varepsilon(\boldsymbol{x}) &= P^0(\boldsymbol{x},\boldsymbol{y})+\varepsilon P^1(\boldsymbol{x},\boldsymbol{y})+\cdots, \end{aligned} \tag{6}$$

where $V_i^\alpha(\boldsymbol{x},\boldsymbol{y})$ and $P^\alpha(\boldsymbol{x},\boldsymbol{y})$ (α=0,1,...) are $\boldsymbol{Y}$-periodic functions such as $V_i^\alpha(\boldsymbol{x},\boldsymbol{y})=V_i^\alpha(\boldsymbol{x},\boldsymbol{y}+\boldsymbol{Y})$, $P^\alpha(\boldsymbol{x},\boldsymbol{y})=P^\alpha(\boldsymbol{x},\boldsymbol{y}+\boldsymbol{Y})$ with the size of a unit cell $\boldsymbol{Y}$. Next, because of $y=x/\varepsilon$, we change the differentiation as

$$\frac{\partial}{\partial x_i} \Longrightarrow \frac{\partial}{\partial x_i}+\frac{1}{\varepsilon}\frac{\partial}{\partial y_i}. \tag{7}$$

Substituting Eqns(6), (7) into Eqn(4) and taking $\varepsilon \rightarrow 0$ yield

$$\varepsilon^{-1}\text{-term: } \frac{\partial P^0}{\partial y_i}=0 \Longrightarrow P^0(\boldsymbol{x},\boldsymbol{y})=P^0(\boldsymbol{x}) \text{ in } Y_f \tag{8}$$

$$\varepsilon^0\text{-term: } -\frac{\partial P^1}{\partial y_i}+\eta\frac{\partial^2 V_i^0}{\partial y_k \partial y_k}=\frac{\partial P^0}{\partial x_i}-F_i \text{ in } Y_f. \tag{9}$$

Similarly Eqn(5) gives

$$\varepsilon^1\text{-term:}\quad \frac{\partial V_i^0}{\partial y_i} = 0 \quad \text{in } Y_f, \tag{10}$$

$$\varepsilon^2\text{-term:}\quad \frac{\partial V_i^0}{\partial x_i} + \frac{\partial V_i^1}{\partial y_j} = 0 \quad \text{in } Y_f. \tag{11}$$

The fluid-solid interface condition $(5)_2$ is written as

$$V_i^0 = 0, \quad V_i^1 = 0, \cdots \quad \text{on } \Gamma. \tag{12}$$

We introduce a separation of variables for Eqn(9) by

$$V_i^0 = \left(F_k(\boldsymbol{x}) - \frac{\partial P^0(\boldsymbol{x})}{\partial x_k}\right) v_i^k(\boldsymbol{y}), \qquad P^1 = \left(F_k(\boldsymbol{x}) - \frac{\partial P^0(\boldsymbol{x})}{\partial x_k}\right) p^k(\boldsymbol{y}), \tag{13}$$

where $v_i^k(y)$ and $p^k(y)$ (k=1,2,3) are called the characteristic velocity and the characteristic pressure, respectivelys. Then Eqn(9) is changed into a PDE of only y:

$$-\frac{\partial p^k}{\partial y_i} + \eta \frac{\partial^2 v_i^k}{\partial y_j \partial y_j} + \delta_{ik} = 0 \quad \text{in } Y_f. \tag{14}$$

In the similar manner the mass conservation law (10), the fluid-solid interface condition (12) and the periodic boundary conditions are written as

$$\frac{\partial v_i^k}{\partial y_i} = 0 \quad \text{in } Y_f, \qquad v_i^k = 0 \quad \text{on } \Gamma, \tag{15}$$

$$v_i^k(\boldsymbol{y}) = v_i^k(\boldsymbol{y} + \boldsymbol{Y}), \qquad p^k(\boldsymbol{y}) = p^k(\boldsymbol{y} + \boldsymbol{Y}) \quad \text{on } \partial Y_f. \tag{16}$$

Eqns(14)-(16) are called the *micro-scale equations system* [MiSE].

Now we introduce a mass averaging operation for Eqn$(13)_1$, and get the Darcy's law:

$$\widetilde{V}_i^0 = K_{ji}\left(F_j - \frac{\partial P^0}{\partial x_j}\right), \qquad K_{ji} = \widetilde{v}_i^j = \frac{1}{|\boldsymbol{Y}|}\int_{Y_f} \rho\, v_i^j \, dy \tag{17}$$

where $\widetilde{V}_i^0$ is the averaged mass velocity in the unit cell ($|\boldsymbol{Y}|$: volume of the unit cell), ρ the mass density of water which is assumed to be constant here because of its incompressibility. It can be shown that K_{ji} is symmetric and positive definite. The same averaging is applied to Eqn(11), then the second term vanishes because of the periodic condition of V_i^1, so we have the following *macro-scale equation* [MaSE], called the HA-seepage equation:

$$\frac{\partial \widetilde{V}_i^0}{\partial x_i} = 0 \quad \text{or} \quad \frac{\partial}{\partial x_i}\left\{K_{ji}\left(F_j - \frac{\partial P^0}{\partial x_j}\right)\right\} = 0 \quad \text{in } \Omega. \tag{18}$$

The true pressure P^ε and velocity V_i^ε are calculated in the first order approximation sense by Eqn(6) as

$$V_i^\varepsilon(\boldsymbol{x}) \simeq \varepsilon^2 V_i^0(\boldsymbol{x}, \boldsymbol{y}), \qquad P^\varepsilon(\boldsymbol{x}) \simeq P^0(\boldsymbol{x}). \tag{19}$$

The procedure to solve the total HA-seepage problem is summarized as follows: We first solve MiSE (14)-(16) and get v_i^k and p^k, then determine the Darcy's coefficient K_{ij} by Eqn$(17)_2$. Next by solving MaSE(18) we get the macro-pressure P^0. Substituting these into

Eqn(19) we can finally provide the true pressure and velocity fields.

In geotechnical engineering we usually use the same form with the HA-seepage equation (18) such as

$$\frac{\partial \widetilde{V}_i'}{\partial x_i} = 0; \quad \widetilde{V}_i' = K_{ji}' \left(F_j - \frac{\partial P}{\partial x_j} \right) \tag{20}$$

where $\widetilde{V}_i'$ is the mass velocity, and P the pore pressure. Compared this with Eqns(18)-(20), we understand the correspondence

$$\widetilde{V}_i' = \widetilde{V}_i^\varepsilon \simeq \varepsilon^2 \widetilde{V}_i^0, \tag{21}$$

so we have the following relationship between the HA-*permeability* K_{ij} and the conventional C-*permeability* K_{ij}':

$$K_{ij}' = \varepsilon^2 K_{ij}. \tag{22}$$

If the water viscosity η is constant, a plane Poiseuille-like flow shown in Fig. 4 we can solve MiSE as

$$v_1^1 = \frac{1}{8\eta}(d^2 - 4y_2^2), \quad p^1 = C_1, \quad p^2 = y_2 + C_2 \tag{23}$$

where C_1 and C_2 are constants. The HA-permeability is then calculated by Eqn$(17)_2$ as

$$K_{11} = \frac{\rho}{12\eta} \frac{d^3}{Y_1 Y_2}. \tag{24}$$

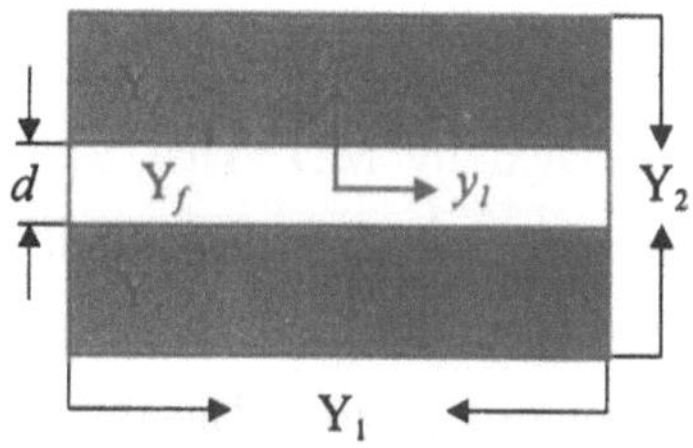

Fig. 4 Unit cell for plane flow.

Let us suppose that at the temperature 300K the water viscosity $\eta=0.8\times10^{-3}$[Pa s] and the mass density $\rho=0.99651$ [g/cm^3], and we assume $Y_1=1$, $Y_2=0.02$, $d=0.01$, $\varepsilon=100$[nm]$=10^{-5}$ [cm], so the C-permeability is calculated as $K_{11}'= 1.7517\times10^{-13}$[cm/s]. These conditions approximately correspond to the clay with its saturated wet density $\rho_w=1.76$. Note that the intrinsic mass density ρ_i of the montmorillonite mineral is obtained as 2.901[g/cm^3] by the MD calculation.

In the MD simulation we showed that in the vicinity of montmorillonite surface the motion of water molecules is strongly affected by the surface condition of the montmorillonite mineral, and water molecules are constrained on that surface. This is called the *iceberg effect*, which is quantitatively evaluated in our MD simulation. We account for this effect to estimate the Darcy's coefficients. That is, the problem which must be solved is given by MiSE(14)-(16). Then we employ the penalized weak formula for Eqns(14) and $(15)_1$ under the boundary conditions $(15)_2$ and (16):

$$\int_{Y_f} \eta \frac{\partial v_i^k}{\partial y_j} \frac{\partial w_i^k}{\partial y_j} dY + \frac{1}{\lambda} \int_{Y_f} \frac{\partial v_i^k}{\partial y_i} \frac{\partial w_j^k}{\partial y_j} dY - \int_{Y_f} \delta_{ik} w_i^k \, dY = 0 \quad \forall w_i^k \ (w_i^k = 0 \text{ on } \Gamma) \tag{25}$$

where λ ($0<\lambda<<1$) is the penalty coefficient. The finite element (FE) discretization procedure is same as for the classical incompressible Navier-Stokes problem, so details are omitted here.

We here show two cases of FE calculation by using the distributed viscosity η obtained by MD (see Fig. 2), and we compare them with the results of η=constant.

Case (1): As a FE simulation model of MiSE for the montmorillonite hydrate, we employ the same unit cell as presented in the previous section for the analytical solution method (Fig. 4) except that the FE model employs the distributed viscosity calculated by MD. That is, the viscosity at a Gaussian point of FE is specified by using the data shown in Fig. 2(b). The distance d between two montmorillonite minerals is chosen to be 8.37, 7, 6, 5, 4, 3, 2, 1, 0.6nm. A profile of the characteristic function v_1^1 is shown in Fig. 5, comparing with the constant-viscosity case given by Eqn(20). The calculated C-permeability transformed from the HA-permeability by Eqn(21) is given in Fig. 6. In Fig. 6 the results are also compared with he constant-viscosity case (solid dots in the figure). In Fig. 5 and Fig. 6 we can obviously find that the iceberg effect affects the bentonite permeability crucially.

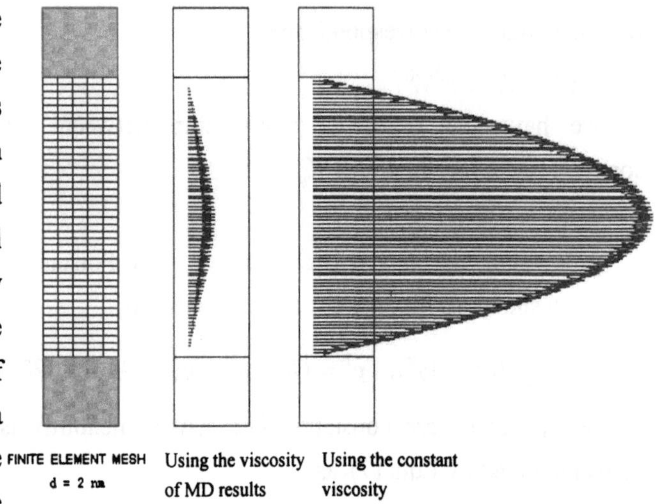

Fig. 5 Profiles of the characteristic velocity v_1^1 for Case (1).

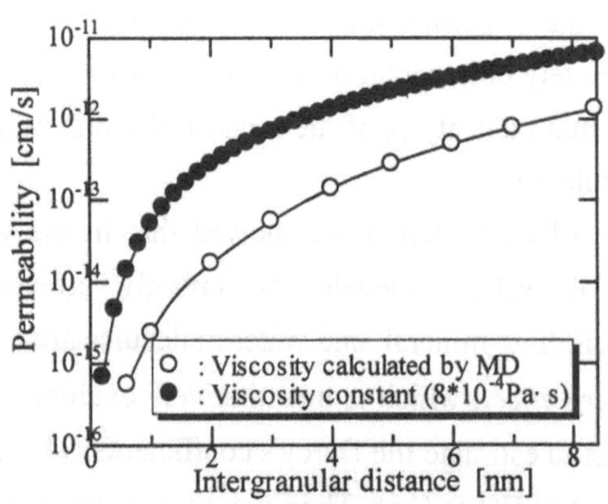

Fig. 6 Calculated permeability for Case (1).

Case (2): Next we treat a more realistic model for the compacted bentonite clay with the saturated density ρ_w=1.8[g/cm^3]. It is known that several lamellae of montmorillonite mineral are stacked in one group. We here assume that eight lamellae are stacked, and there exists n-molecules of water in the interlayer spaces (nH$_2$O-system). We calculated the cases of n=1,3,5. Since for each case the interlayer distance is given by the MD results as shown in Fig. 2, the thickness of the external water layer is obtained so as to be ρ_w=1.8[g/cm^3]. FE calculated profiles of the characteristic velocity $v_1^{\,1}$ are shown in Fig. 7. Note that the order of the characteristic velocity is different for each case of n. Consequently the C-permeability is changed as shown in Fig. 8.

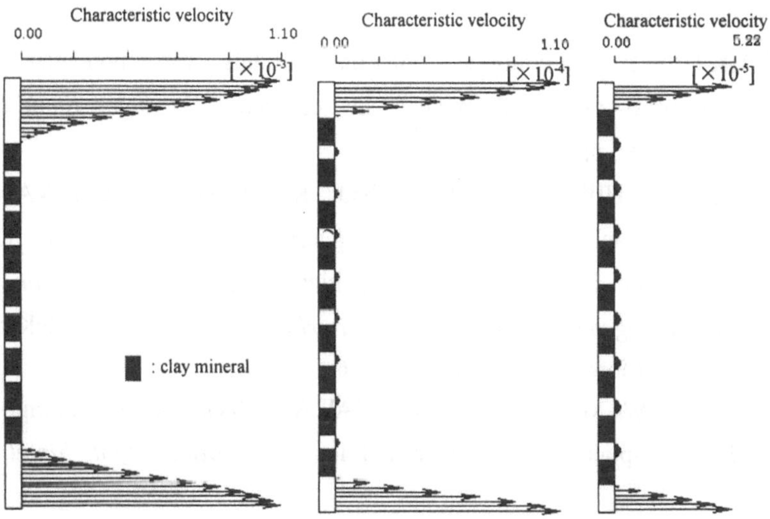

Fig. 7 Profiles of the characteristic velocity function $v^1{}_1$ for Case (2).

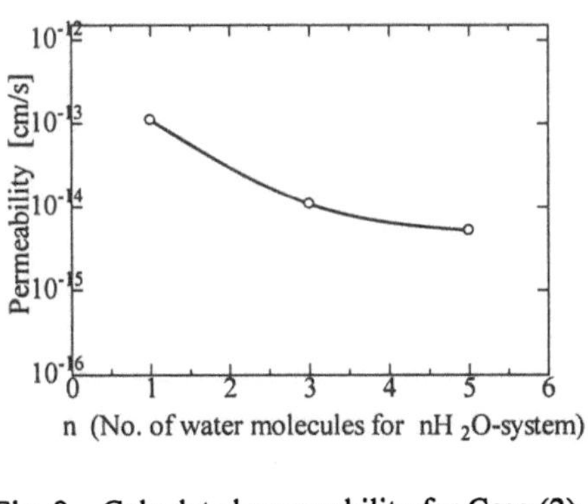

Fig. 8 Calculated permeability for Case (2).

CONCLUSIONS

We here presented a unified MD/HA method for analyzing the seepage problem of clay backfill. In the long-term analysis of backfill materials for radioactive waste management, it is important to establish a true physical-chemical procedure, and we believe the unified MD/HA method can provide such an integrated scheme.

REREFENCES

ICHIKAWA,Y., K.KAWAMURA, M.NAKANO, K.KITAYAMA & H.KAWAMURA (1998): "Unified molecular dynamics/ homogenization analysis for water flow in bentonite", *Proc. 1998 Int. High-Level Radioactive Waste Management Conf. (American Nuclear Soc.), Las Vegas, 1998.*

KAWAMURA,K.(1992); "Interatomic potential models for molecular dynamics simulations of multi-component oxides", in *Molecular Dynamics Simulations* (ed. F.Yonezawa), Springer, pp.88-97.

KAWAMURA,K., Y.ICHIKAWA, M.NAKANO, K.KITAYAMA & H.KAWAMURA (1997): "A new approach for predicting the long term behavior of bentonite: The unified method of molecular dynamics and homogenization analysis", *Scientific Basis for Nuclear Waste Management XXI., Davos (Swizerland) 1997*, ed. I.G. McKinley & C. McCombie, Material Research Soc., pp.359-366.

KUMAGAI,N., K.KAWAMURA & T.YOKOKAWA (1994); "An interatomic potential model for H_2O: Applications to water and ice polymorphs", *Mol. Simul.*, 12(3-6), pp.177-186.

SANCHEZ-PALENCIA,E. (1980); *Non-Homogeneous Media and Vibration Theory*, Springer-Verlag.

A THIN-LAYER ELEMENT FOR SEEPAGE ANALYSIS

J.M.M.C. Marques
University of Porto, Porto, Portugal

J.J.M.S. Domingues
Instituto Superior de Engenharia do Porto, Porto, Portugal

ABSTRACT

The presence of a very thin layer of material whose permeability differs significantly from that of the surrounding soil can impose the adoption of a very dense finite element mesh, because of the small element dimensions required for the modeling of the thin layer. This paper presents the formulation of an isoparametric thin-layer element for 2D analysis which can be employed with advantage for the modeling of such zones, allowing for greater economy in mesh development and analysis time, while at the same time providing accurate results. The proposed thin-layer element is derived from a quadrilateral element by making one dimension converge to zero and it can handle flow in both the normal and tangential directions.

1. INTRODUCTION

Joint/interface/thin-layer elements for mechanical problems have been proposed by various authors, such as Goodman (1968), Desai (1984) and Beer (1985), for modelling discontinuities arising in solid mechanics problems. Examples of application range from the fractured or jointed masses of rock engineering to the interfaces between dissimilar materials, as in soil-structure or soil-reinforcement interaction problems. The purpose of this paper is to present a zero thickness isoparametric element for the modelling of thin layers in groundwater flow or consolidation problems whose permeability contrasts with that of the adjacent materials. The element is able to handle flow in both the normal and tangential directions.

2. BASIC EQUATIONS

Let us concentrate for simplicity on the steady state analysis of 2D confined seepage problems. Assuming the validity of Darcy's law we have

$$\mathbf{v} = -\mathbf{k}\nabla\phi \tag{1}$$

where $\mathbf{v}$ is the discharge velocity, ϕ is the total head, ∇ is the differential operator $\nabla = \left\{\frac{\partial}{\partial x}, \frac{\partial}{\partial y}\right\}^T$ and $\mathbf{k}$ is the permeability matrix

$$\mathbf{k} = \begin{bmatrix} k_x & k_{xy} \\ k_{xy} & k_y \end{bmatrix} \tag{2}$$

Under steady state conditions the continuity equation imposes that

$$\nabla^T \mathbf{v} = Q \tag{3}$$

where Q=Q(x,y) is a specified inflow per unit volume. Substituting equation (1) into equation (3) we obtain the well known *quasi-harmonic equation*

$$\nabla^T \mathbf{k}\nabla\phi + Q = 0 \tag{4}$$

which governs seepage through the domain Ω, with essential boundary conditions $\phi = \bar{\phi}$ in Γ_ϕ and with natural boundary conditions $q_n = -\mathbf{n}^T \mathbf{k}\nabla\phi = \bar{q}_n$ in Γ_q, where the parts Γ_ϕ and Γ_q of the boundary Γ obey $\Gamma_\phi \cup \Gamma_q = \Gamma$ and $\Gamma_\phi \cap \Gamma_q = 0$.

Using the Galerkin weighted residual approach in the finite element solution of this equation, following for example Hinton and Owen (1979), we arrive at a linear equation system on the nodal values a_j of the total head

$$K_{ij}\, a_j = f_i \tag{5}$$

with

$$K_{ij} = \int_\Omega \nabla^T N_i\, \mathbf{k}\nabla N_j\, d\Omega \tag{6}$$

and

$$f_i = \int_\Omega Q N_i\, d\Omega - \int_{\Gamma_q} \bar{q}_n N_i\, d\Gamma + r_i \tag{7}$$

where

$$r_i = \int_{\Gamma_\phi} n^T k \nabla N^T a N_i\, d\Gamma \tag{8}$$

and N_i is a global vector of shape functions.

3. PLANE QUADRILATERAL ELEMENTS

Figure 1 depicts a plane isoparametric element showing the local coordinate axes with ξ and $\eta \in [-1,+1]$.

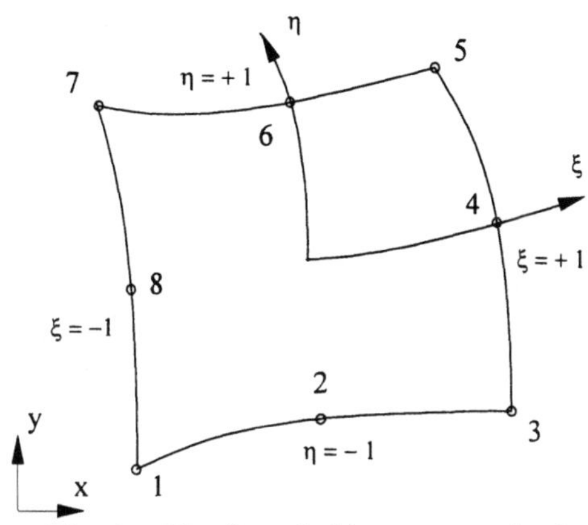

Fig. 1 – The 8-noded isoparametric element

The global coordinates of the n element nodes can be stored in a matrix **X**

$$\mathbf{X} = \begin{bmatrix} x_1 & x_2 & \cdots & x_n \\ y_1 & y_2 & \cdots & y_n \end{bmatrix}^T \tag{9}$$

Using the isoparametric concept the element geometry can be interpolated as

$$\mathbf{x} = \{x, y\}^T = \mathbf{X}^T \mathbf{N} \tag{10}$$

where

$$\mathbf{N} = \{N_1, N_2, \ldots, N_n\}^T \tag{11}$$

In order to compute the element "stiffness" matrix K the following transformations have to be performed in equation (6)

$$\nabla N_i(x,y) = \mathbf{J}^{-1}\,\hat{\nabla} N_i(\xi,\eta) \tag{12}$$

$$d\Omega = |\mathbf{J}|\, d\xi\, d\eta \tag{13}$$

where the differential operator $\hat{\nabla}$ is now $\hat{\nabla} = \left\{ \frac{\partial}{\partial \xi}, \frac{\partial}{\partial \eta} \right\}^T$ and **J** is the jacobian matrix

given by

$$\mathbf{J} = \begin{bmatrix} \frac{\partial x}{\partial \xi} & \frac{\partial y}{\partial \xi} \\ \frac{\partial x}{\partial \eta} & \frac{\partial y}{\partial \eta} \end{bmatrix} = \hat{\nabla}\,\mathbf{x}^T = \hat{\nabla}\,\mathbf{N}^T \mathbf{X} \tag{14}$$

Each element K_{ij} can now be computed as

$$K_{ij} = \int_{-1}^{+1}\int_{-1}^{+1} \hat{\nabla}^T \mathbf{N}_i\, (\mathbf{J}^{-1})^T\, \mathbf{k}\, \mathbf{J}^{-1}\, \hat{\nabla} \mathbf{N}_j\, |\mathbf{J}|\, d\xi\, d\eta \tag{15}$$

4. THIN-LAYER ELEMENT FORMULATION

The main objective of this element is to model the flow in thin layers whose permeability may contrast with that of the surrounding soil. Let us consider one such layer denoted by Ω_0 in Fig. 2, with boundaries Γ_1 e Γ_2.

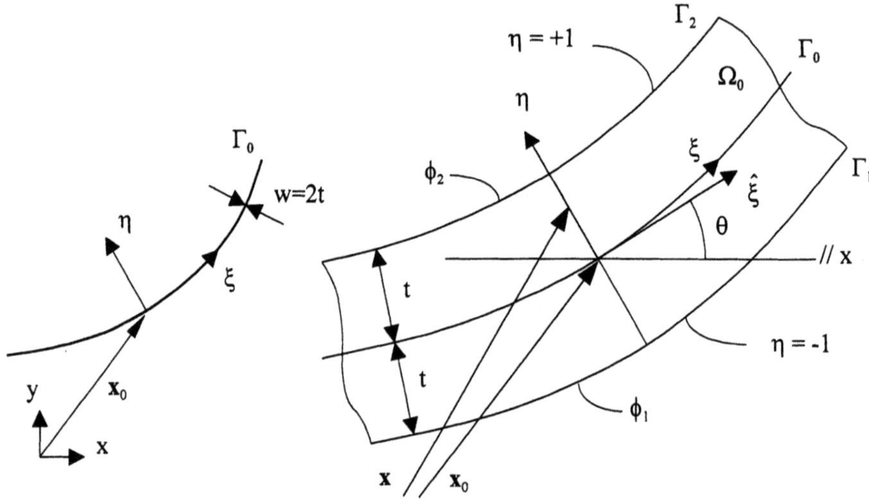

Fig.2 – Thin-layer geometry

Considering the local axes (ξ, η) along the centre line Γ_0, see Fig. 2, the position vector $\mathbf{x}$ of any point of Ω_0 can be obtained by

$$\mathbf{x} = \mathbf{x}_0 + \begin{Bmatrix} -m \\ l \end{Bmatrix} t\eta \tag{16}$$

where $\mathbf{x}_0 = \{x_o, y_o\}^T$ is the orthogonal projection of $\mathbf{x}$ onto the centre line Γ_0 and l and m are the components of the unit vector $\hat{\xi}$ tangent to Γ_0, that is,

$$\hat{\xi} = \begin{Bmatrix} l \\ m \end{Bmatrix} = \begin{Bmatrix} \cos\theta \\ \mathrm{sen}\,\theta \end{Bmatrix} = \frac{1}{\alpha} \begin{Bmatrix} \dfrac{\partial x_0}{\partial \xi} \\ \dfrac{\partial y_0}{\partial \xi} \end{Bmatrix} \tag{17}$$

with

$$\alpha = \sqrt{\left(\frac{\partial x_0}{\partial \xi}\right)^2 + \left(\frac{\partial y_0}{\partial \xi}\right)^2} \tag{18}$$

Note that in the above expressions x_0, y_0, α, l and m are functions of ξ only. Also note that definition (17) implies that

$$l^2 + m^2 = 1 \tag{19}$$

which, differentiated with respect to ξ, yields

$$l\frac{\partial l}{\partial \xi} + m\frac{\partial m}{\partial \xi} = 0 \tag{20}$$

The jacobian matrix associated with transformation (16) can be obtained according to the definition, equation (14). After differentiating and simplifying we get

$$\mathrm{J} = \begin{Bmatrix} \dfrac{\partial}{\partial \xi} \\ \dfrac{\partial}{\partial \eta} \end{Bmatrix} \mathbf{x}^{\mathrm{T}} = \begin{bmatrix} l\alpha - t\dfrac{\partial m}{\partial \xi}\eta & m\alpha + t\dfrac{\partial l}{\partial \xi}\eta \\ -t\,m & t\,l \end{bmatrix} \tag{21}$$

whose determinant is

$$|\mathbf{J}| = t(\alpha + t\beta\,\eta) \tag{22}$$

where

$$\beta = m\frac{\partial l}{\partial \xi} - l\frac{\partial m}{\partial \xi} \tag{23}$$

The jacobian matrix can be represented more compactly. Using expressions (19), (20) and (23) we conclude that

$$\frac{\partial l}{\partial \xi} = m\beta; \quad \frac{\partial m}{\partial \xi} = -l\beta \tag{24}$$

which can be substituted into expression (21) and, given equation (22) for the determinant, finally leads to

$$\mathbf{J} = \mathbf{G}\,\mathbf{T}^{\mathrm{T}} \tag{25}$$

where matrix **G** is given by

$$\mathbf{G} = \begin{bmatrix} \dfrac{|\mathbf{J}|}{t} & 0 \\ 0 & t \end{bmatrix} \tag{26}$$

and **T** is the coordinate transformation matrix

$$\mathbf{T} = \begin{bmatrix} l & -m \\ m & l \end{bmatrix} \tag{27}$$

Let us now consider the layer domain Ω_0 discretized into n-noded elements of constant width w=2t, see Fig.3. The nodes are located along the boundaries Γ_1 and Γ_2 so that each

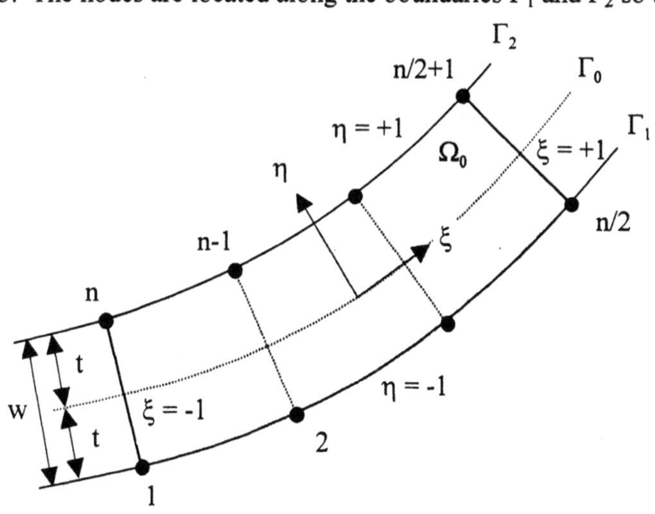

Fig. 3 – General n-noded thin-layer element topology

pair of nodes defines a normal to the centre line Γ_0. When the layer width tends to zero each nodal pair coincides in Γ_0.

Expression (15) previously obtained for plane isoparametric elements will now be the starting point for the K "stiffness" matrix of the isoparametric thin-layer element with zero thickness. A linear interpolation will be assumed across the element width and the integration with respect to η will be performed analytically, followed by the computation of the limit when $t \rightarrow 0$.

We will employ shape functions defined as

$$N_i(\xi,\eta) = \overline{N}_i(\xi)\,\overline{\overline{N}}_i(\eta) \tag{28}$$

where

$$\overline{\overline{N}}_i(\eta) = \frac{1}{2}(1+\eta\eta_i) \tag{29}$$

obviously with

$$\eta_i = \begin{cases} -1 & \text{if } i \in \Gamma_1 \\ +1 & \text{if } i \in \Gamma_2 \end{cases} \tag{30}$$

The interpolation functions $\overline{N}_i(\xi)$ for the 4- and 6-noded thin-layer elements are given in Fig.4.

4-noded elements

$\overline{N}_1 = (1-\xi)/2$

$\overline{N}_2 = (1+\xi)/2$

$\overline{N}_3 = \overline{N}_2$

$\overline{N}_4 = \overline{N}_1$

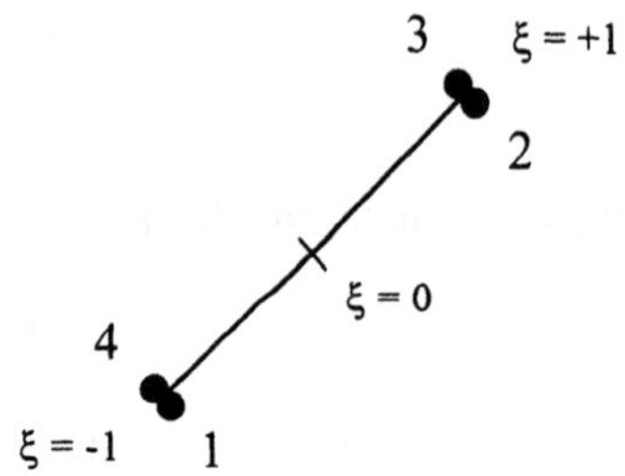

6-noded elements

$\overline{N}_1 = (\xi-1)\,\xi/2$

$\overline{N}_2 = (1-\xi^2)$

$\overline{N}_3 = (1+\xi)\,\xi/2$

$\overline{N}_4 = \overline{N}_3$

$\overline{N}_5 = \overline{N}_2$

$\overline{N}_6 = \overline{N}_1$

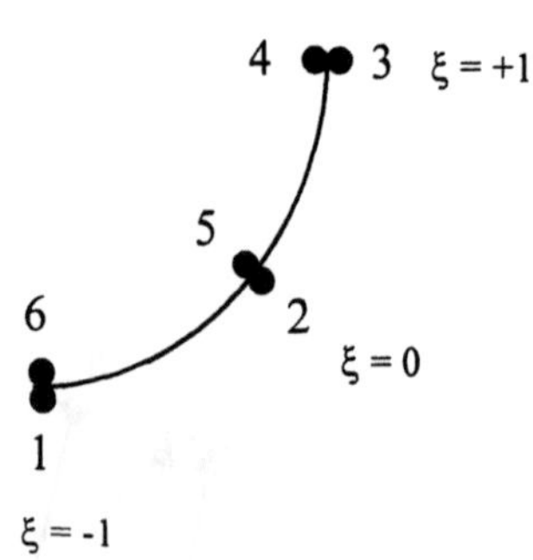

Fig.4 – Interpolation functions $\overline{N}_i(\xi)$ for 4- and 6-noded thin-layer elements

Differentiating the shape functions N_i with respect to ξ and η we have

$$\hat{\nabla} N_i = \begin{Bmatrix} \dfrac{\partial N_i}{\partial \xi} \\ \dfrac{\partial N_i}{\partial \eta} \end{Bmatrix} = \frac{1}{2} \begin{Bmatrix} (1+\eta\eta_i)\overline{N}_i' \\ \eta_i \overline{N}_i \end{Bmatrix} \tag{31}$$

with $\overline{N}_i' = \dfrac{\partial \overline{N}_i}{\partial \xi}$.

Considering that the principal directions of permeability coincide with the tangent and the normal to the centre line Γ_0, the permeability matrix $\mathbf{k}$ with respect to the global coordinates (x,y) can be obtained as

$$\mathbf{k} = \mathbf{T} \begin{bmatrix} k_\xi & 0 \\ 0 & k_\eta \end{bmatrix} \mathbf{T}^T \tag{32}$$

where $\mathbf{T}$ is given by (27). Then taking into account expressions (25) and (26) it is possible to compute the product $(\mathbf{J}^{-1})^T \mathbf{k} \mathbf{J}^{-1}$, which appears in equation (15) for K, as

$$\mathbf{J}^{-T} \mathbf{k} \mathbf{J}^{-1} = \mathbf{G}^{-T} \begin{bmatrix} k_\xi & 0 \\ 0 & k_\eta \end{bmatrix} \mathbf{G}^{-1} = \begin{bmatrix} \dfrac{t^2}{|\mathbf{J}|^2} k_\xi & 0 \\ 0 & \dfrac{1}{t^2} k_\eta \end{bmatrix} \tag{33}$$

Substituting (31) and (33) into (15), using (22) for $|\mathbf{J}|$ and working out the products we arrive at

$$K_{ij} = \frac{1}{4} \int_{-1}^{+1} \int_{-1}^{+1} \left[t k_\xi \frac{(1+\eta\eta_i)(1+\eta\eta_j)}{\alpha + t\beta\eta} \overline{N}_i' \overline{N}_j' + \frac{k_\eta}{t} \eta_i \eta_j (\alpha + t\beta\eta) \overline{N}_i \overline{N}_j \right] d\xi \, d\eta \tag{34}$$

Noting that the $\overline{N}_i$ and their derivatives $\overline{N}_i'$ depend solely on ξ, as well as the geometric parameters α e β, and assuming that the principal permeabilities k_ξ, k_η are constant along η, it is then possible to carry out separately the following integrations with respect to η:

$$I_\xi = \int_{-1}^{+1} \frac{(1+\eta\eta_i)(1+\eta\eta_j)}{\alpha + t\beta\eta} d\eta = \frac{(P\alpha^2 - S\alpha b + b^2)\ln\lambda + 2b(Sb - P\alpha)}{b^3} \tag{35}$$

$$I_\eta = \int_{-1}^{+1} (\alpha + t\beta\eta) d\eta = 2\alpha \tag{36}$$

with $P = \eta_i \eta_j$, $S = \eta_i + \eta_j$, $b = t\beta$ e $\lambda = (\alpha + b)/(\alpha - b)$. Expression (34) can then be written in the compact form

$$K_{ij} = \int_{-1}^{+1} \left(\frac{w}{8} k_\xi I_\xi \overline{N}_i' \overline{N}_j' + \frac{k_\eta}{w} \alpha P \overline{N}_i \overline{N}_j \right) d\xi \tag{37}$$

Let us now consider that the layer width w is very small. It can be proved that in the limit situation $w \to 0$ the value of I_ξ as given by (35) converges to a finite value. Using Maple V one gets

$$\lim_{w \to 0} I_\xi = \frac{2}{3} \frac{3P+1}{\alpha} \tag{38}$$

The terms $w k_\xi$ e k_η / w have physical meaning and a significant numerical value even when the layer width w is very small. Making

$$g = w k_\xi \tag{39}$$

and

$$h = \frac{k_\eta}{w} \tag{40}$$

taking expression (38) for the limit value of I_ξ, the expression for K can finally be written as

$$K_{ij} = \int_{-1}^{+1} \left(g \frac{3\eta_i \eta_j + 1}{12\alpha} \overline{N}_i' \overline{N}_j' + h\alpha\, \eta_i \eta_j \overline{N}_i \overline{N}_j \right) d\xi \tag{41}$$

5. TEST EXAMPLES

EXAMPLE 1. In the elementary one dimensional problem of Fig.5 a vertical thin-layer element is crossed by horizontal flow. The thin layer permeability is 100 times smaller than that of the adjacent regions. The total head profile shown in Fig. 5 coincides with the analytical solution.

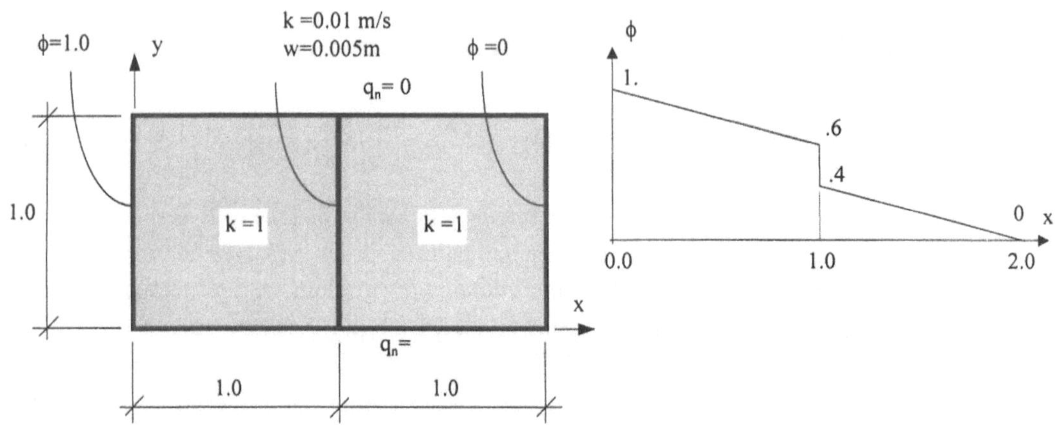

Fig. 5 – Mesh data and total head profile for Example 1

EXAMPLE 2. In the problem of Fig. 6, the thin-layer element is now parallel to the flow direction, has w=0.02m and is 100 times more permeable than the surrounding layers. The analytical solution is again trivial. The hydraulic gradient is identical for all three layers with a value of 0.5. The flow rate through the thin layer is 0.01 m^3/s, twice that corresponding to the adjacent layers.

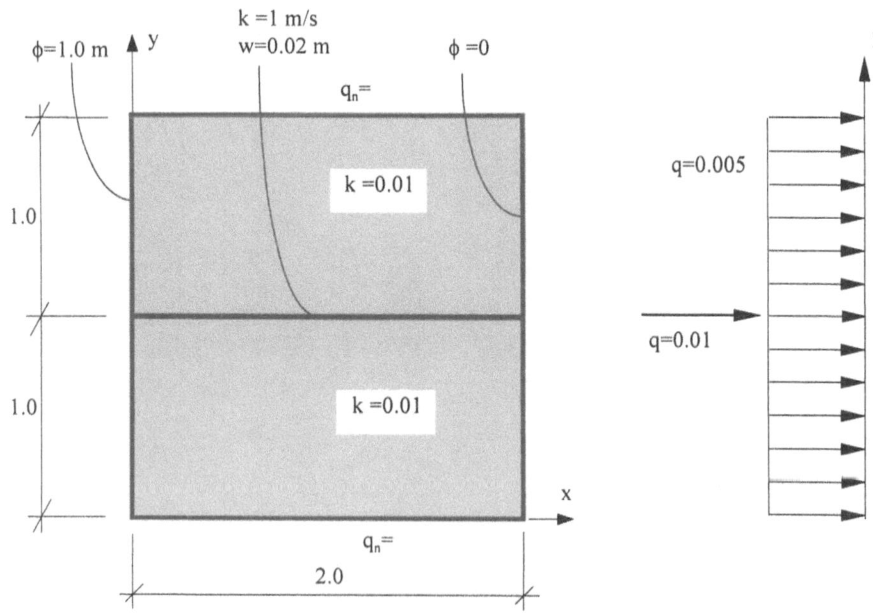

Fig. 6 – Mesh data and flow pattern for Example2

6. CONCLUSIONS

A thin-layer element has been presented for use in groundwater flow analysis, or indeed in other problems also governed by the quasi-harmonic equation, where strong field discontinuities may exist due to the presence of regions of small width with contrasting physical properties. The element is of zero-thickness type and can model flow in both the normal and tangential directions.

7. REFERENCES

BEER, G. (1985). An isoparametric joint/interface element for finite element analysis, Int. J. Num. Meth. Engng., 21, 585-600.

DESAI, C. S., M. M. ZAMAN, J. G. LIGHTNER & H. J. SIRIWARDANE (1984). Thin-layer element for interfaces and joints, Int. J. Num. Anal. Meth. Geomech., 8, 19-43.

GOODMAN, R. E., R. L. TAYLOR & T. L. BREKKE (1968). A model for the mechanics of jointed rock, J. Soil Mech. Found. Div. ASCE, 94, SM3, 637-659.

HINTON, E. & D. R. J. OWEN (1979). An Introduction to Finite Element Computations, Pineridge Press, Swansea.

A PROCEDURE FOR GENERATING WELL CONDITIONED TETRAHEDRAL MESHES

D.J. Naylor
University of Wales, Swansea, UK

ABSTRACT: On the premise that tetrahedra (particularly ten noded ones) make good finite elements and that they should be as near regular as possible a procedure for generating structured meshes of such elements is outlined. This involves two stages: First the definition, by means of macro elements, of a regular array of optimally shaped and suitably sized tetrahedra which fill the region. Then the distortion of the mesh thus obtained to better fit the boundaries. Examples are given which demonstrate the suitability of the procedure for, in particular, static analyses in geomechanics.

1 INTRODUCTION

Tetrahedra are not widely used in 3D finite element analysis. The reason it seems is not that they are inferior elements but simply that they are difficult to generate. Hexahedra can be more easily fitted together, particularly if the mesh boundaries are orthogonal as is often the case in geotechnical problems.

Why then tetrahedra? The answer is that they can be used with an integrating rule which is superior to the gauss rules available for the hexahedron. Attention here is restricted to the ten noded isoparametric tetrahedron. Their hexahedral counterpart is the 20 noded *brick* element. The problem with this element is that when used with *reduced* integration, ie a 2x2x2 gauss rule, it is under-integrated and under certain conditions (such as in zones of yield in elasto-plastic analysis) near mechanisms can develop. If *full* integration, ie a 3x3x3 gauss rule, is used the element becomes too stiff. Although there are ways round this such as the use of a 14 noded brick element as proposed by Smith (1991) the use of ten noded tetrahedra with four symmetrically placed integrating points (each one quarter of the way along a line from an apex to the centroid of the face opposite) provides an attractive alternative. The number of integrating points in relation to the number of degrees of freedom is about right, providing an element that is neither too soft nor too stiff. There is a close parallel with the 2D and axi-symmetric situations where the 8 noded quadrilateral can develop mechanisms with reduced integration but is too stiff and can sometimes even lock (Nagtegaal, *et al.*, 1974, Sloan and Randolf, 1982) but the 6 noded triangle with a three point integrating rule offers a good alternative. Naylor (1994) compared various rules for different triangular elements and concluded that this combination was one of the best.

Mesh generation can be *structured* or *unstructured*. Unstructured generation of four noded tetrahedral meshes is widely used in Computational Fluid Dynamic (CFD) applications. See eg Weatherill, *et al.* (1995). In this the mesh grows from the boundaries according to certain rules. It seems to be best suited to meshes in which there is just one material or in situations where mesh regeneration is required during the analysis. A problem with it is that there is little control over the shape of the tetrahedra and badly conditioned elements known as *slivers* (Cavendish, *et al.*, 1985) can form. These reduce the accuracy of the analysis. Structured mesh generation is probably more appropriate for most geotechnical analyses, particularly if more than one material type is involved. The generation procedure described here is structured.

A criterion for a good tetrahedral mesh is that the tetrahedra should be as near *regular* as possible. Unlike equilateral triangles regular tetrahedra cannot be fitted together *ad infinitum*. Interestingly Aristotle thought that they could and this error persisted for nearly 2000 years (Senechal, 1981)! Naylor (1998) investigated various assemblies of tetrahedra which would fill space and in doing so unearthed a particular shape of tetrahedra which was nearly regular and on its own could be fitted together indefinitely leaving no gaps. He found that this shape had been identified by Sommerville (1923) and was in fact the *only* shape of tetrahedron which on its own could fill space. An important conclusion was that an infinite mesh formed of these tetrahedra was the nearest one could possibly get to one composed of regular tetrahedra were this possible. Naylor called these tetrahedra *isotets*. Their four faces are identical isosceles triangles, each with two sides of length $\sqrt{3}/2$ times the base length.

Because of the geometric constraint of the boundaries an actual mesh cannot normally consist of isotets. By first subdividing the region into elements most of which are isotets and then distorting it to satisfy boundary and other constraints this ideal is approached in the procedure outlined here.

The procedure is described from the perspective of a user and examples are brought in at an early stage to communicate the basic idea. With this foundation some of the detail of how the actual elements can be numbered and their nodes numbered and co-ordinated is given. First a measure of the conditioning, which shows how an actual element compares with a regular tetrahedron, is described. Finally two examples of actual applications in geotechnical engineering are provided.

2 CONDITIONING MEASURE

There are a number of alternative measures which have in common that they range from zero, corresponding to a tetrahedron folded flat so that it has no volume, to 1 (or 100%), corresponding to a regular tetrahedron. Naylor (1998) compared five such measures and found that it was not critical which one was used as the hierarchy of a range of different tetrahedral shapes was the same. The measure preferred in that paper, designated the *Conditioning efficiency* (η), is used here. It is defined as the volume of the actual tetrahedron divided by the volume of the *regular* tetrahedron which has the same average edge length (e_q). Noting that the volume of a regular tetrahedron is $e_q^3/(6\sqrt{2})$:

$$\eta = \frac{6\sqrt{2}V}{e_q^{\ 3}} \quad \text{with} \quad e_q = \frac{1}{6}\sum_{i=1}^{6} e_i \tag{1}$$

in which V is the volume of the tetrahedron and e_i is the length of an edge i. In what follows (1) will be multiplied by 100 so that η is expressed a percentage.

η is 93.6% for an isotet.

3 MESH GENERATION - THE BASIC IDEA

The user must first conceive a uniformly spaced orthogonal grid which spans the region to be discretised. Element nodes will be at grid points initially defined by co-ordinates (i,j,k). The average size of the elements will be controlled by the number of grid subdivisions across the region so that there is an implied physical distance between grid points. In the first stage of the mesh generation physical locations are not explicit as the mesh is defined entirely by the integer coordinates. These define the *mapped* mesh. The second stage is to map (or strictly reverse-map) this on to the *actual* mesh in (x,y,z) space. In what follows the mapped mesh is viewed as a physical entity which is distorted to form the actual mesh.

Fig. 1 illustrates a mapped mesh. It is made up of macro elements the corners of which are assigned (i,j,k) co-ordinates. Interpreting the grid as occupying physical space this mesh will be of roughly the same shape as the actual mesh. Assignment of (x,y,z) co-ordinates to the macro element nodes maps the mesh on to real space so that the actual mesh fits an idealised boundary. This if it does not coincide with the actual boundary should be a reasonable approximation of it.

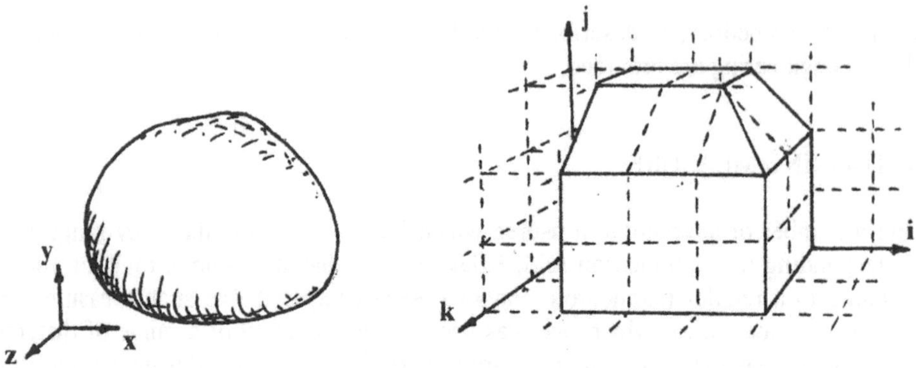

Fig. 1. A mapped mesh

Certain constraints must be observed in defining the macro element faces. The (i,j,k) co-ordinates must be chosen to define *plane* faces. If a macro element is to be divisible into isotets in the mapped mesh then there are just six possible planes in which the faces must lie. Pairs of these are mutually orthogonal, each with a normal vector perpendicular to one axis and at 45° to the other two. These planes are referred to as the *natural* planes. Two are illustrated in Fig. 1. It would be more convenient if in addition the mapped mesh faces could be orthogonal to the i, j and k axes. Accordingly this is permitted but it means that the ideal of a mesh entirely of isotets has to be compromised as these orthogonal planes cut through isotet elements sub-dividing them into either two or four. In addition to these nine constraints on the orientation of the macro element faces the (i,j,k) co-ordinates of corner nodes must be either all even or all odd. The reason for these constraints is clarified in section 4 below.

The macro elements are basically 20 noded isoparametric bricks. The assignment of (x,y,z) co-ordinates to the 12 edge nodes in addition to the corners allows the faces of the macro elements in the actual mesh to be non-planar, thus permitting a better fit to an irregular boundary. If the co-ordinates are not assigned the edge nodes default to the middle of the edges. Appropriate positioning of the edge nodes can also be used to refine the mesh across a macro element, thus causing the actual elements to vary in size across it.

The actual mesh (x,y,z) nodal co-ordinates are calculated from the nodal co-ordinates of the macro element in which they lie. The actual node macro element local co-ordinates (ξ,η,ζ) must first be calculated. The corresponding (x,y,z) can then be determined from the macro element shape functions in the normal way. An explicit procedure is used to calculate (ξ,η,ζ) for the actual nodes. There is not, however, space to go into it here.

To provide greater flexibility provision is made to 'collapse' the faces of the brick macro elements to convert them to wedges, pyramids or tetrahedra. This is achieved by assigning the same (i,j,k) co-ordinates to opposite nodes on a face to be collapsed.

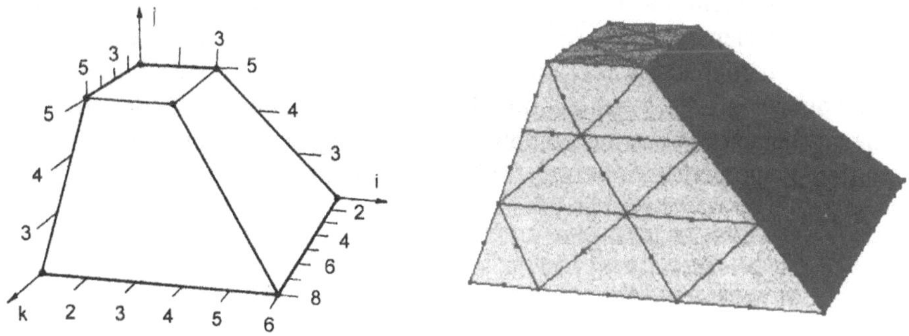

(a) Mapped mesh (b) Actual mesh

Fig. 2. Embankment end

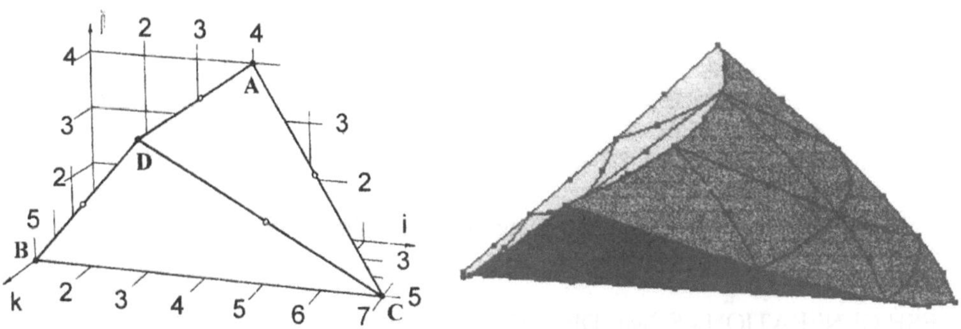

(a) Mapped mesh (b) Actual mesh

Fig. 3. Embankment dam

The examples in Fig.s 2 and 3 each involve just one macro element subdivided into respectively 128 and 27 actual elements. The corresponding numbers of nodes are 207 and 84 respectively. The macro element for the embankment end (Fig. 2) is a hexahedron. It has two natural planes and four orthogonal. It has a median conditioning efficiency (η) of 79%. The macro element for the embankment dam (Fig. 3) is a tetrahedron and all the faces lie in natural planes. Its median η is 70%. The relatively large number of elements in relation to the number of nodes in the first example is a result of the subdivision of the isotets on the four orthogonal boundaries. The lower conditioning efficiency for the dam reflects a greater distortion of the elements in the mapping process than in the other example. The curved crest and valley sides of the actual mesh in the dam are achieved by assigning non-central (x,y,z) co-ordinates to the edge macro nodes. In addition those along the edges AB, AC, DC and DB have been positioned to make the elements smaller near the valley floor.

Two important features are the grouping of macro elements into *zones* which can be used to separate different materials (concrete from soil, different strata, etc.) and the identification of macro element faces which are *internal*, ie are neither an interface between zones nor an external boundary. To minimise the sub-division of isotets those which are cut by internal boundaries are *not* sub-divided. They straddle the boundaries and may share two or four macro elements. While this complicates the coding both the number of elements and the proportion of sub-divided isotets are reduced thus improving the overall conditioning.

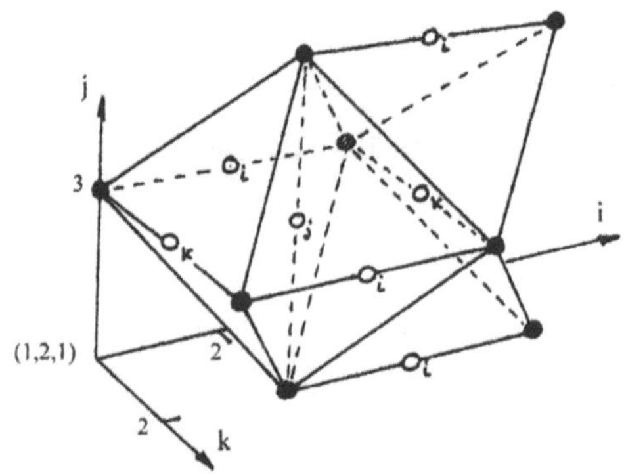

Fig. 4 A repeating assembly of six isotets.

4 MESH GENERATION - SOME DETAIL

To generate the mesh a loop is entered in the program in which an assembly of six isotets is 'stepped' through the grid. At each step it is ascertained (see below) which of the six isotets, or if they are sub-divided which parts of them, lie within the mesh boundaries. Fig. 4 shows one such assembly. Each isotet in it has a different orientation so that the assemblies are *repeating*, ie they fit together without gaps. (These six orientations may be compared with the 2D situation where just two orientations of equilateral triangles - rotated alternately 180° - are needed to fill space.) This loop is referred to as the *generation* loop.

In Fig. 4 the solid blobs represent corner nodes of isotet elements. They are referred to as *normal* corner nodes to distinguish them from the corner nodes of isotets subdivided by one or more orthogonal boundary. These are represented by the hollow blobs. The suffix identifies the orientation of the plane which cuts the isotet. Thus o_i is at a corner formed by an i = constant plane intersection. It can be seen from this figure that, as mentioned above, the (i,j,k) for the normal corner nodes must be all even or all odd. Thus the extreme left and top right solid blobs are at (1,3,1) and (4,4,2) respectively, whereas the bottom right mid-edge node o_i is at (3,2,2).

Before the generation loop is implemented node numbers are assigned to all the corner nodes including those of the sub-divided isotets. These are stored in an array which later will also contain the mid-edge node numbers. For it to be able to do this and also relate the nodes to the grid the (i,j,k) grid needs to be sub-divided to define an (i',j',k') grid in which

$$\left.\begin{aligned} i' &= 2i + 4 \\ j' &= 2j + 2 \\ k' &= 2k + 2 \end{aligned}\right\} \qquad (2)$$

The node numbers are then stored in an array *numnd*(i',j',k'). With this containing zeros for addresses which are not nodes and node numbers for those which are it can be used to find which isotets (or part-isotets) are to be included in the mesh in the generation loop. The reason for adding 2 or 4 to the right hand sides of equations (2) is to provide working space outside the limits of the mesh. Only positive i, j, and k values are allowed in the mesh definition and this ensures that i', j' and k' are also always positive. Mid-edge node numbers are added to *numnd* during the generation loop.

One interesting aspect of the coding is the determination of whether a grid point which is a potential node is within a macro element or on its face. If it is then it is indeed a node. This is done using *integer* vector algebra in (i',j',k') space, integer because it is both simpler and faster. In a loop through the faces of each macro element the scalar product of a vector normal to the face and a vector from the face to the grid point is calculated for each grid point. The sign of this product indicates on which side of the face the point lies or if zero shows that it lies in the plane of the face. It can then be determined from this set of scalar products whether or not the grid point is a node in that macro element.

5 PRACTICAL EXAMPLES

Figures 5 and 6 illustrate two actual problems. Figure 5 shows one half of a 6m long by 4m wide by 4m deep test pit used at the Building Research Establishment, England for tests on collapse settlement of a fill material. The three $1m^3$ zones illustrated in 5(b) by their absence contained a collapse settlement susceptible material. The surrounding fill was not susceptible to collapse. The tests involved flooding the pit after the material had first been placed under carefully controlled conditions (of water content and compaction). Stresses and movements in the fill material were measured The finite element modelling of these tests involves the placing of the fill material in five (finite element) layers and the simulation of the collapse associated with saturation. Figure 6 shows a finite element mesh of a large central core embankment dam in Southern Spain. The valley runs from left to right. The dark shaded overhang of the mesh shows how the dam spreads out above a steep sided central portion. This mesh is to be used to simulate the construction of this dam and the filling of the reservoir behind it. Again the simulation of collapse settlement is involved.

In these two examples the mesh is divided into zones, eight in the BRE test pit - one for each of the five layers and one for each zone of the collapse settlement susceptible material - and six for the dam. In the dam there are 3 finite element construction layers each of which is subdivided into central core and rockfill zones. There are more macro elements than zones, 70 being required for the test pit and 20 for the dam. The numbers of actual elements followed by the nodes in brackets are 1428 (2632) for the test pit and 1548 (2658) for the dam. The median conditioning efficiency for the test pit is 77% and that for the dam 66%. The poorer conditioning of the dam elements is due to the contorted shape of the valley.

An interesting feature of the mesh generation for the BRE test pit is the orientation of the mapped mesh at 45° to the i and k axes as shown in Fig. 5(a). This makes all the vertical faces natural planes thus considerably reducing the number of isotet intersections.

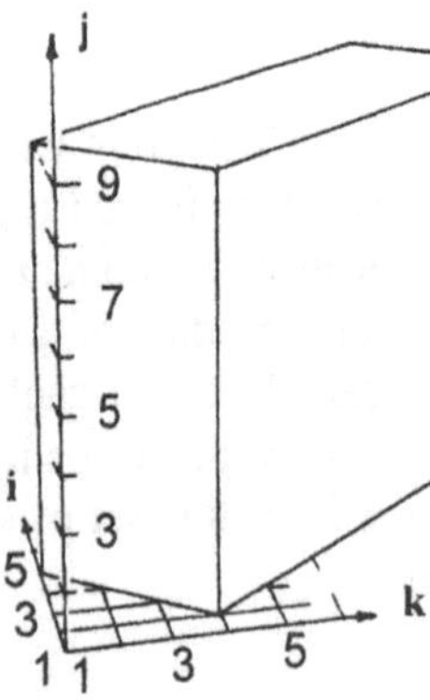

(a) Mapped mesh

(b) Actual mesh (viewed in opposite direction from (a))

Fig. 5. BRE's test pit for collapse settlement

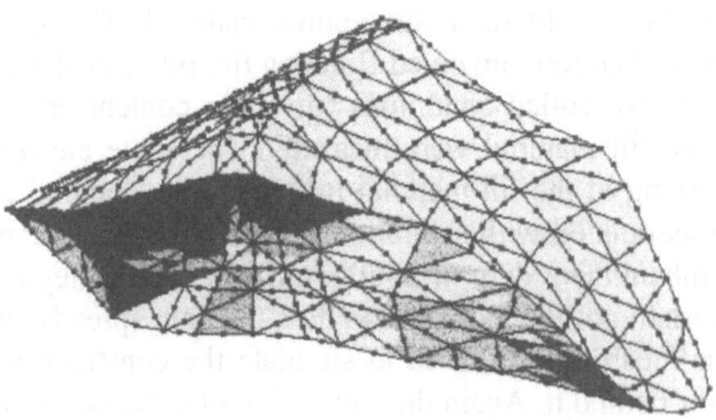

Fig. 6 Perspective of tetrahedral mesh for large central clay core dam

6 CONCLUSIONS

A case has been made for the use of structured meshes of ten noded tetrahedra for the 3D finite element analysis of at least some geotechnical problems. A generation procedure has been described based on an idea whereby a mesh composed of ideally shaped tetrahedra, but which only roughly matches the boundaries, is first generated. This mesh is then distorted to provide a good fit with minimal loss of quality. The procedure is new in that it is based on assemblies of the unique tetrahedron, the isotet, which has the closest possible shape to that of a regular tetrahedron but which can be fitted together to fill an indefinite space without gaps.

Two simple hypothetical examples have been used in the description of the algorithm and two actual examples have been used to illustrate its application. All of them relate to geotechnical structures. It is believed that this procedure is particularly appropriate for the generation of well conditioned meshes for problems of the type considered here.

REFERENCES

Cavendish, J.C., Field, D.A. and Frey, W.H. (1985). An approach to automatic three-dimensional finite element mesh generation. *Int. J. Numer. Methods in Engng*, **21**: 329-347.

Nagtegaal, J.C., Parks, D.M. and Rice, J.R. (1974). On numerically accurate finite element solutions in the fully plastic range. *Comp. Meth, Appl. Mech. Engng*, **4**: 153-177.

Naylor, D.J. (1998). Filling space with tetrahedra. Accepted for publication in *Int. J. Numer. Methods in Engng.*

Senechal, M. (1981). Which tetrahedra fill space?. *Mathematics Magazine*, **54**: 227-243.

Sloane, S.W. and Randolph, M.F. (1982). Numerical prediction of collapse loads using finite element methods. *Int. J. Numer. Anal. Methods in Geomech*, **6**: 47-76.

Smith, I.M. (1991). Are there any new elements? *The finite element method in the 1990's*. (Ed: Onate, Periaux, Samuelsson) Springer-Verlag. 109-118.

Sommerville, D.M.Y. (1923). Space-filling Tetrahedra in Euclidean Space. *Proc. Edinburgh Mathematical Soc.*. **41**: 49-57.

Weatherill, N.P., Hassan, O.and Marcum, D.L. (1995). Compressible Flowfield Solutions with Unstructured Grids Generated by Delaunay Triangulation. *AIAA Journal*, **33**, No. 7: 1196-1204.

BOUNDARY INTEGRAL METHODS IN ENGINEERING

A SALTWATER INTRUSION EXAMPLE

R.M. Pereira, E.V. Veiga and J.B. Martins
University of Minho, Braga, Portugal

1. ABSTRACT

In confined aquifers, in which its thickness and hydraulic conductivity is constant, and that there is no storing, with an essentially horizontal flux, the water flow is described by the Laplace equation.

In the saltwater intrusion problem presented here it is necessary to determine the separation line between freshwater and the saltwater. We will consider that both fluids are not miscible, and that the freshwater inflow is done with a constant flux, supposing that the level of the saltwater does not change with time.

In the simplified situation, we will not consider a zone of mixture of both saltwater and freshwater, and trough an iterative process supported in both the kinematic and static conditions that prevail in the interface, it is possible to determine its movement to an equilibrium configuration.

2. INTRODUCTION

The problem of saltwater intrusion is very important for hydraulic engineers, once one of their main tasks is to determine the good sites to build a pumping station , as well as to know how much freshwater can be pumped without damaging the aquifer.

Saltwater intrusion happens when to much freshwater is pumped from a coastal aquifer. If so, the piezometric head of the freshwater region drops, and, if the piezometric head of the saltwater region becomes greater than the one from the freshwater region, the interface between both regions advances, until a new equilibrium is reached.

Lots of people were interested in solving this problem, some examples are [Bear, Dagan, 64], [Sá da Costa, 79], [Cheng, Liu, Ligget, 81].

3. THE MODEL

The picture bellow represents the schematic description of our problem in a two dimensions situation.

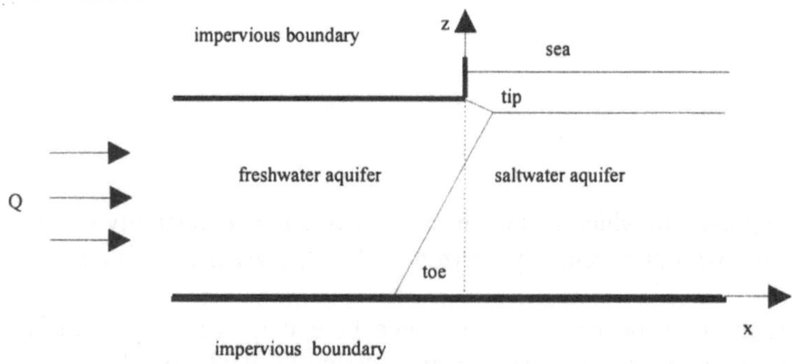

Fig. 1 - Saltwater intrusion problem

The equations that describe the flux in both regions, in a confined aquifer with constant width [Sá da Costa, 79] are:

$$\nabla.(K_i \nabla \phi_i) = S_i \frac{\partial \phi_i}{\partial t} (i = 1,2) \tag{1}$$

where, i represents the zone(i=1-freshwater zone, i=2-saltwater zone) , ϕ_i represents the piezometric head($\phi_i = z_i + \frac{p_i}{\rho_i g}$, p_i is the water pressure in the point z_i and g is the gravity acceleration), K_i is the permeability of the zone, e S_i is the storage coefficient due to the deformation of the mean. Considering that there is no storage, and that the hydraulic conductivity is constant then the equations above are reduced to the Laplace equations:

$$\nabla^2 \phi_i = 0 (i = 1,2) \tag{2}$$

3.1 Boundary conditions

There are two types of boundary - an exterior boundary and an interior one. In order to define them, let's consider picture 2 which is a simplification of picture 1.

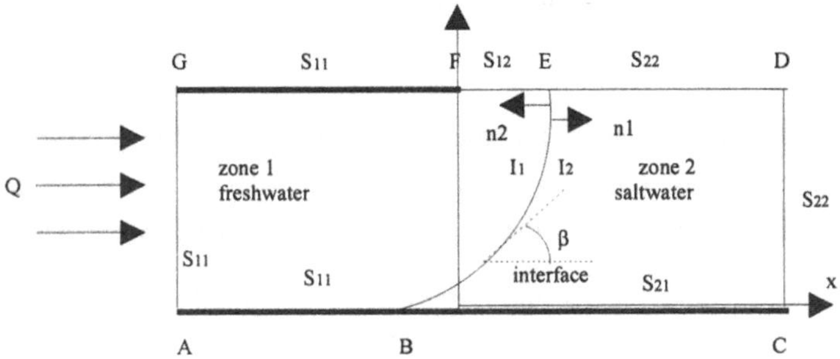

Fig. 2 - Mathematical model

In the outside boundary we have a set of surfaces that limit the freshwater aquifer(AB, EF, FG e GA) and another one that limit the saltwater aquifer (BC, CD e DE). DE is supposed to be far enough from BE. In each one of these surfaces, in order to specify our problem the following boundary conditions were considered:

i) $\phi = c$ in EF(boundary type S_{12}).

ii) $\phi = \frac{c}{s}$ in CD, DE(boundary type S_{22}).

iii) $q = -K\frac{\partial\phi}{\partial n} = 0$ in AB, FG(boundary type S_{11}) and BC(boundary type S_{21}) - impervious surfaces.

iv) $q = \bar{q}$ in GA (boundary type S_{11}).

where c is the piezometric head in EF, $s = \frac{\rho_2}{\rho_1}$ is the ratio between saltwater and freshwater densities, $q = -K\frac{\partial\phi}{\partial n}$ where K is the permeability.

In the interior boundary there are two types of boundary condition[Cheng, Liu, Ligget, 81] - kinematic and static which we will describe bellow.

The interface position is given at any time by:

$$x = \varsigma(z,t) \tag{3}$$

Once the interface is a mobile boundary, a kinematic condition prevails in it . This condition[Cheng, Liu, Ligget, 81] is given by:

$$\frac{\partial \varsigma}{\partial t} = \frac{K}{\sin\beta}\frac{\partial \phi_1}{\partial n_1} = -\frac{s.K}{\alpha}\frac{\partial \phi_2}{\partial n_2} \tag{4}$$

$$\beta = arcctg(\frac{\partial \varsigma}{\partial z})$$

where K is the permeability of the mean, ϕ_1 and ϕ_2 are the piezometric heads of the freshwater and saltwater zones, n_1 and n_2 are the normal vectors to the saltwater and freshwater interface surface, β is the angle that the interface makes with the xx axis, and $\alpha = \frac{\mu_1}{\mu_2}$, where μ_1 and μ_2 is the viscosity of the fluids in freshwater and saltwater zones(picture 2).

Another type of boundary condition is obtained because we know that the pressure must be the same in a point in the interface on both sides of the interface. By the definition of piezometric head we have:

$$\phi_1 = z + \frac{p_1}{\rho_1 g}; \phi_2 = z + \frac{p_2}{\rho_2 g} \tag{5}$$

solving in order to the variable z, and considering that $s = \frac{\rho_2}{\rho_1}$, we obtain the equation that defines the static boundary condition ($p_1 = p_2 = p$) in the interface:

$$z(s-1) = s\phi_2 - \phi_1 \tag{6}$$

3.2 Numerical model

The method we used to solve this problem was the well known Boundary Integral Method [Brebbia, Wrobel, Partridge, 92]. The boundary integral equation for each zone is

$$\int_s \left[\phi_i \frac{\partial}{\partial n}(\ln(r)) - \ln(r)\frac{\partial \phi_i}{\partial n}\right] ds = 0 (i = 1,2) \tag{7}$$

This equation can be written an algebraic equations system, in the variables ϕ and $\frac{\partial \phi_i}{\partial n}$.

$$H_i\phi_i = G_i \frac{\partial \phi_i}{\partial n} (i = 1,2) \tag{8}$$

after discretizing and appropriate treatment, H and G are the influence matrixes.

Both external and internal boundaries were discretized in a finite number of elements (line segments).

As we can see in picture 2, the boundary S_1 (boundary of the freshwater aquifer) is formed by S_{11} (part of the boundary where the flux $\frac{\partial \phi_{11}}{\partial n}$ is known), by S_{12} (part of the boundary where the piezometric head ϕ_{12} is known) and by I_1 (part of the boundary in the interface).In the zone 2, the same kind of reasoning was used. The algebraic system of equations (8) can be rewritten as:

$$\begin{aligned} H_{s_{11}}\phi_{s_{11}} + H_{s_{12}}\phi_{s_{12}} + H_{I_1}\phi_{I_1} &= G_{s_{11}}\frac{\partial \phi_{s_{11}}}{\partial n} + G_{s_{12}}\frac{\partial \phi_{s_{12}}}{\partial n} + G_{I_1}\frac{\partial \phi_{I_1}}{\partial n} \\ H_{s_{21}}\phi_{s_{21}} + H_{s_{22}}\phi_{s_{22}} + H_{I_2}\phi_{I_2} &= G_{s_{21}}\frac{\partial \phi_{s_{21}}}{\partial n} + G_{s_{22}}\frac{\partial \phi_{s_{22}}}{\partial n} + G_{I_2}\frac{\partial \phi_{I_2}}{\partial n} \end{aligned} \tag{9}$$

If we use the boundary conditions in the interface on (9), and, if we pass the unknowns to the LHS of (9), then, we can say:

$$\begin{bmatrix} H_{s_{11}} & -G_{s_{12}} & 0 & 0 & H_{I_1} & -G_{I_1} \\ 0 & 0 & H_{s_{21}} & -G_{s_{22}} & \frac{H_{I_2}}{s} & \frac{\alpha}{s}G_{I_2} \end{bmatrix} \begin{Bmatrix} \phi_{s_{11}} \\ \frac{\partial \phi_{s12}}{\partial n} \\ \phi_{s_{21}} \\ \frac{\partial \phi_{s_{22}}}{\partial n} \\ \phi_{I_1} \\ \frac{\partial \phi_{I_1}}{\partial n} \end{Bmatrix} = \begin{bmatrix} G_{s_{11}} & -H_{s_{12}} & 0 & 0 & 0 \\ 0 & 0 & G_{s_{21}} & -H_{s_{22}} & H_{I_2} \end{bmatrix} \begin{Bmatrix} \frac{\partial \phi_{s_{11}}}{\partial n} \\ \phi_{s_{12}} \\ \frac{\partial \phi_{s_{21}}}{\partial n} \\ \phi_{s_{22}} \\ (\frac{1}{s}-1)z_{I_2} \end{Bmatrix} \tag{10}$$

After solving this system, we obtain the values of the piezometric heads and fluxes in all the points of the boundary used to discretize it. The location of the interface, is then calculated using (4) written in a finite diference form:

$$\varsigma_{i+1} = \varsigma_i + \frac{k}{\sin\beta_i}\left(\theta\left(\frac{\partial\phi_{I_1}}{\partial n}\right)_i + (1-\theta)\left(\frac{\partial\phi_{I_1}}{\partial n}\right)_{i+1}\right)\Delta t \qquad (11)$$

where i represents the iteration, θ is a relaxation parameter used in the time discretization.

3.3 First example of saltwater intrusion and saltwater regression

The following example shows both saltwater intrusion and regression in an aquifer. Let's consider the following picture in which we have the problem specification. It has the boundary conditions and the number of points used in the discretization specified.

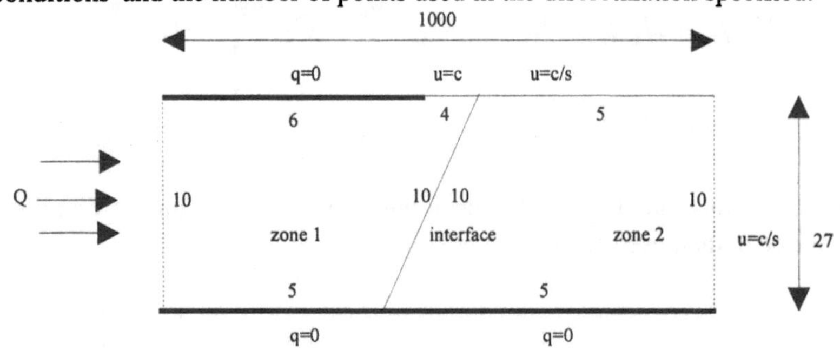

Fig. 3 - Schema of the problem to solve

We are considering that the aquifer has constant thickness, horizontal bottom and that the freshwater inflow is known ($Q = 93m^2 / s$). The value considered for the permeability was 69. With the inflow rate mentioned above, and considering an arbitrary initial configuration to the interface, we reach the initial state of equilibrium. In order to verify the arbitrary character of the initial configuration, several experiences were considered. Two of them can be seen in the following picture. In both of them we can see that the initial state of equilibrium is reached in the same place.

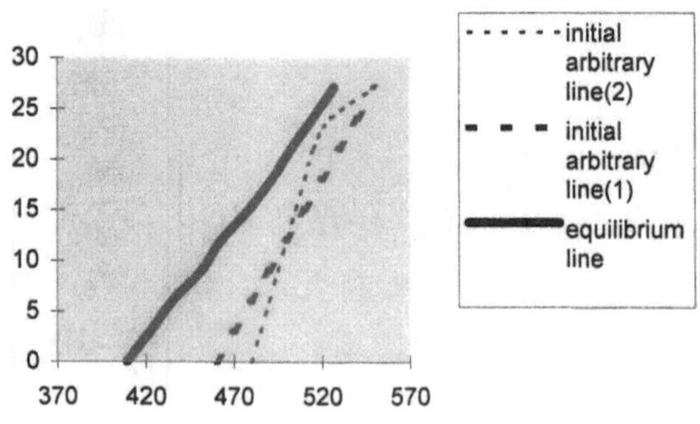

Fig. 4 - Initial state of equilibrium for the interface line.

Then, at t=0s, it was considered a sudden diminishing of the freshwater inflow($Q = 46m^2 / s$). The interface line moved into the freshwater aquifer way - saltwater intrusion phenomena.

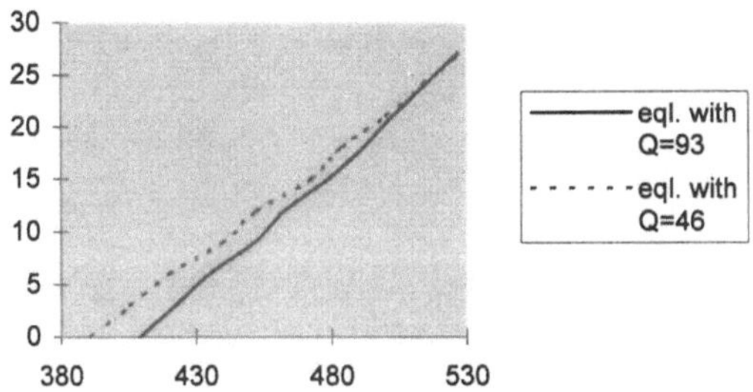

Fig. 5 -Saltwater intrusion results.

Then, with the purpose of having the inverse problem, we increased the freshwater inflow again($Q = 93m^2 / s$), obtaining the saltwater regression phenomena, as we can see in the following picture.

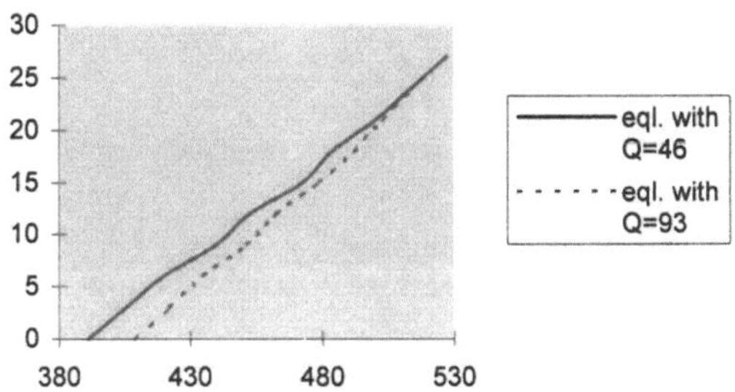

Fig. 6- Saltwater regression results.

One can see that considering we increase the freshwater inflow after reaching a equilibrium situation(with $Q = 46m^2 / s$) the interface line moves in the saltwater aquifer way. As the new value of freshwater inflow is the same as in the beginning of the example, the new equilibrium should be reached in the same place as the initial state of equilibrium line.(See Pic. 6) This is the saltwater regression phenomena.

3.4 Second Example

To see the effect of anisotropic permeability of the aquifer, in the saltwater intrusion phenomena, a second example was considered, where geometry, boundary and initial conditions are the same, but the aquifer has two stratum: the upper stratum with permeability K_u and a lower stratum with permeability K_l. In Fig.7 , $K_l < K_u$ (K_l=29; K_u=69). It is seen that the effect of having a lower permeability in the lower half of the aquifer, reduces the extent of salt intrusion. Comparing Fig. 8 with Fig. 5 it is seen the reverse, where the lower part of the aquifer is more permeable than the upper part ($K_l > K_u$ K_l=89; K_u=69). In this case, if we increse the permeability on the upper part of the aquifer the extent of salt intrusion is reduced.

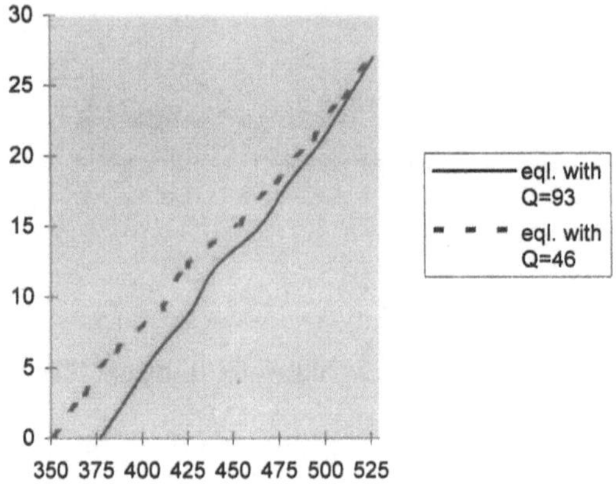

Fig 7.- Second Example($K_l < K_u$; K_l=29; K_u=69)

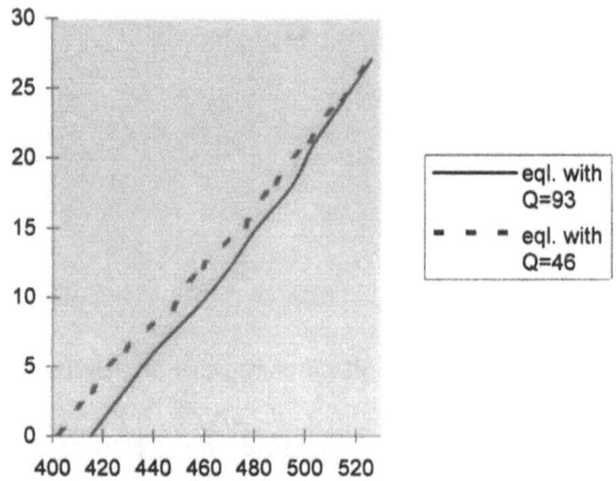

Fig 8.- Second Example($K_l > K_u$; K_l=89; K_u=69)

3.5 Problem discussion

The model is much simplified in the physical sense, since it considers no mixing zone in the interface between salt and freshwater. A better solution would be to build a model with the mixing zone. However,the problem migth become intratable, since the mathematical problem related to the simplified model is already strongly non linear. This happens mainly because of the factor $\frac{1}{\sin\beta}$ that enters in equation (4). The angle β (see Fig. 2) varies along the interface, and the interface changes with time.

Therefore, to reach an equilibrium position with physical meaning, it is needed to start with an arbitrary initial interface line in a position not far from the final one. Otherwise, the numerical solution may not converge to a mathematical result with physical meaning.

4. BIBLIOGRAPHY

[Bear, Dagan, 64] - Dagan B., Bear J., "Moving interface in coastal aquifers", Proceed. ASCE , Journal of hydraulics division", pp. 193-216, 1964.

[Brebbia, Wrobel, Partridge, 92] - Brebbia, Wrobel, Partridge, "The dual reciprocity boundary element method", Computational Mechanics Publications, 1992.

[Cheng, Liu, Ligget, 81] - Cheng, Liu, Ligget, "Boundary integral equation solution to moving interface between 2 fluids in porous media", 1981.

[Sá da Costa, 79] - Sá da Costa, Ph.D. thesis, "A numerical model for seawater intrusion in aquifers", MIT, 1979.

[Veiga, 85] - E. V. Veiga, PH.D. thesis, "O Método dos integrais de fronteira aplicado a alguns problemas de potencial escalar", Universidade do Minho, 1985.

APPLICATION OF ITERATIVE SOLVERS IN GEOMECHANICS WITH SPECIAL EMPHASIS ON PETROLEUM ENGINEERING

B. Plischke
ISAMGEO Engineering GmbH, Wiesbaden, Germany
V. Bulgakov
Moscow State University of Civil Engineering, Moscow, Russia

ABSTRACT:

For large 3D - finite element models, the use of iterative solvers can drastically reduce computation times compared to conventional direct solution methods. Some geotechnical applications require the simultaneous solution of unknown displacements and unknown pore pressure degrees of freedom. For these coupled problems, standard iterative solvers, which require a symmetric matrix with positive main diagonal members, cannot be used. Newly developed iterative solvers, capable to treat also coupled problems and optimized with respect to storage requirements, have been implemented into a 3D-finite element code. Their successful application to non-linear problems in rock mechanics related to petroleum engineering, which require large models (more than 80000 degrees of freedom), is described. Some limitations are discussed. Applicability in the field of civil engineering is evident.

1 INTRODUCTION

Numerical modeling is an indispensable tool for solving geomechanics problems in petroleum engineering. As a widely known example serves the platform sinking in some North Sea chalk fields. The finite element simulation of compaction and subsidence associated with the production of oil and gas serves for a forecast of future platform subsidence and provides a better understanding of the impact of the mechanical behaviour of the rock on the productivity of these fields. Such finite element models typically have dimensions of several kilometers. Nevertheless, a rather fine discretisation is necessary to model the variation of pore pressure and of porosity, which is decisive for the compaction behaviour of chalk. The greatest challenge for this kind of modeling is however not the size of these models, but the tedious calibration process. Rock parameters determined in laboratory tests on small scale samples from core material cannot be regarded as representative for the large scale rock mass behaviour. The stress strain model applicable for a full field compaction analysis must therefore be evaluated by means of a history match of field observations, e.g. measured platform subsidence or compaction measured directly by radioactive markers (Plischke, 1994). Additional sensitivity analyses are required to increase the reliability of the forecast model. Thus a great number of runs with large 3D-models is required. For some other FE-applications in petroleum engineering, typical dimensions are in the range of only a few centimeters, but nevertheless require sometimes complex 3D-models. This applies e.g. for the stability of perforations (Okland and Plischke, 1996), which are small openings in the steel casing of producing wells. They extend as cavities into the surrounding rock and enable the flow of oil or gas towards the well. If they fail, the well starts unwanted production of solids. For most of these applications, pore pressure changes must be considered in a coupled analysis, because the increase of effective stress as a consequence of pressure decline due to production is the governing rock loading mechanism.

For large 3D-models and pronounced non-linear stress strain behaviour, the computation time needed for the solution of the system of equations imposes limits for the refinement of models and the number of simulations to be run in parametric studies. This especially applies, when using direct equation solvers. For industrial applications, direct methods are often preferred because of their robustness and their predictable performance. Recent developments of efficient iterative approaches in combination with the availability of sufficient main memory in today's computers offer the opportunity to overcome the above mentioned limits. This paper describes the experience with new efficient iterative solvers, which can also be used for coupled analyses. It is evident that these approaches are also suitable for other applications in geotechnical engineering.

2 SOME BACKGROUND FOR ITERATIVE SOLUTION TECHNIQUES

Theoretically well founded iterative methods, which were extensively described in textbooks, suffer from slow convergence in realistic applications. In opposite, so called preconditioned Krylov subspace methods are not sufficiently theoretically grounded, but efficient for many practical applications. Formally, the preconditioning means that instead of the original equation system

$$\mathbf{A}\cdot\mathbf{x}=\mathbf{b}$$

we solve a new one

$$\mathbf{B}^{-1}\cdot\mathbf{A}\cdot\mathbf{x}=\mathbf{B}^{-1}\cdot\mathbf{b}$$

where **B** is a matrix (operator) called preconditioner, which is supposed to improve the properties of the original matrix **A** associated with the convergence of the iterative solution process. The true meaning of preconditioning applied to FE-systems is the reduction of the spectral condition number of the stiffness matrix. This applies only to positive definite, symmetric matrices, for which this number provides a rough theoretical convergence estimate. However, even for these matrices, a large spectral condition number does not necessarily mean bad convergence. The role of preconditioners is to provide a good clustering of eigenvalues rather than to reduce the condition number.

Usually the preconditioning matrix **B** is not explicitly inverted, but represented in a factorized form. The requirements for this matrix are quite evident. It should be

- "close" to **A**,
- readily represented in a factorized form, with each factor readily invertible, and sparse,
- the algorithm of constructing the preconditioner should be universal at least for a sufficiently wide class of considered problems.

The preconditioning is used in different iterative schemes. The schemes like CG (conjugate gradient), CGS (conjugate gradient squared), BiCGSTAB (bi-conjugate gradient stabilized) and GMRES (general minimum residual) belong to the class of so called Krylov subspace methods (Saad, 1996) which do not need the user intervention and proved to be efficient in many practical applications. The updates of solution $\mathbf{x}^k$ and residual $\mathbf{r}^k = \mathbf{b} - \mathbf{A}\cdot\mathbf{x}^k$ vectors along with some auxiliary vectors are involved in the iterative procedure. Practically all computational efforts are spent on actions of $\mathbf{B}^{-1}$ and **A** on a vector. The process is interrupted, when the relative norm of the residual vector $\|\mathbf{r}^k\|/\|\mathbf{r}^0\|$ is sufficiently small.

Our solver employs preconditioning based on incomplete factorization techniques usually called ILU (Incomplete LU) methods. This class of methods suggests a large variety of ILU preconditioning strategies and proved to be efficient in many applications. For problems with positive definite matrices (associated with standard elasticity problems) we use the modified version of the conjugate gradient method with the incomplete Cholesky factorization based preconditioner denoted by Modified ICCG (see e.g. Saad (1996) with other references).

The ICCG solver however cannot be used for coupled problems, which have negative entries on the matrix diagonal leading to indefinite matrices. For this purpose, the ILU-preconditioned iterative method for general purpose matrices has been implemented. The scheme of this method is based on the IKJ-variant of Gaussian elimination. It means that computed rows of the L (lower) and U (upper) factors are accessed successively. Let P be a zero static pattern of the incomplete factorization, i.e. a set of pairs of indexes (i,j) which correspond to those entries in factors, which are supposed to be zero. Then the general algorithm of the incomplete factorization can be sketched as following:

```
Do i=2,...,n
  Do k=1,...,i-1
    If ((i,k) doesn't belong to P) then
       A{i,k} := A{i,k}/A{k,k}
         Do j=k+1,...,n
           If((i,j) doesn't belong to P) A{i,j}:=A{i,j}-A{i,k}A{k,j}
         Enddo
    Endif
  Enddo
Enddo
```

Of course, implementation of this algorithm involves a special sparse storage of the original matrix and of obtained approximate factors. It also involves modifications associated with special matrix scaling, pivot improvement and other techniques. The original matrix **A** is in a sparse format, so that only nonzero entries of each row are stored. In addition, column numbers for each row entry and pointers to the beginning and to diagonal entries in a row are also stored. The same storage scheme is used for the factors. When P coincides with a zero pattern of **A**, the method is usually called ILU(0). A more sophisticated algorithm is employed, when the zero pattern P is not static, but worked out during the process of factorization. For this purpose, a special thresholding strategy for choosing appropriate matrix entries is employed.

In our experiments, different conjugate gradient-like iterative schemes for general matrices have been considered as candidates for an industrially used solver. It was found that for solving coupled problems the preconditioned BiCGSTAB external iterative scheme appeared to be more efficient than the preconditioned CGS scheme. The preconditioned GMRES scheme requires too much auxiliary memory for storing Krylov basis vectors.

3 IMPLEMENTATION INTO A FINITE ELEMENT CODE

Iterative solvers have been implemented into a three-dimensional finite element code, which offers a variety of non-linear stress strain models as well as specialized options for applications in petroleum engineering and other fields of geotechnical engineering. A rather conservative approach is chosen to model material non-linearities: For each load increment, the system stiffness matrix is composed from the elastoplastic stiffness matrices of individual elements, based on the previous state of stress. After evaluation of displacements and strains, stresses are evaluated and, if necessary, corrected back to the yield surface. Correction forces from an equilibrium check are applied in the next load increment.

For pure displacement unknowns with symmetric stiffness matrix, a solver of the ICCG type is implemented. Besides the compactly stored stiffness matrix **A**, a working array of similar size is required to keep the preconditioner. Two integer arrays store the row numbers of entries in **A** and the pointers to diagonal entries in of **A**. For coupled analysis, the solver KSNIS has been developed. It offers various options for tuning the efficient treatment of different problems with indefinite matrices. It requires approximately 1.5 times the storage for the ICCG solver and some additional working arrays.

4 PERFORMANCE OF THE ITERATIVE SOLVERS

As an example for the excellent performance of the iterative solver in case of a coupled analysis serves the model of a North Sea chalk reservoir, as shown on Fig. 1. It is made from 6580 isoparametric elements with 20 nodes, resulting in 82400 degrees of freedom. Pore pressure degrees of freedom are introduced for elements representing the rock over- and underlying the reservoir, which is composed of 13 layers. This large model mainly serves the purpose to verify that the mesh refinement of a coarser model of the same field

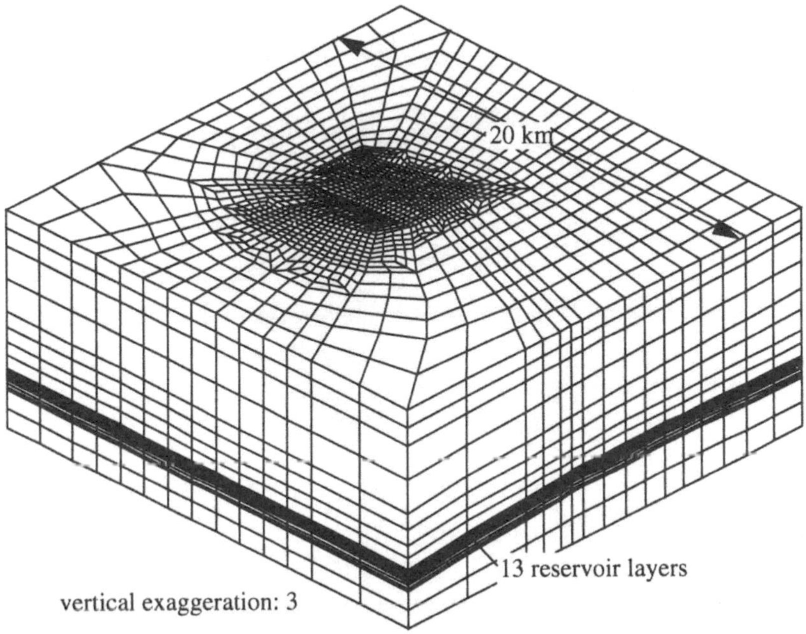

Fig. 1: Finite element mesh for analysis of compaction and subsidence of a North Sea chalk field. For this and the following figures, each 20-noded isoparametric element is subdivided into 8 elements with 8 nodes.

with approximately 60000 degrees of freedom is sufficient. On a HP 9000/735 workstation, one load step for the large model took about 34 CPU-minutes, from which about 20 minutes apply to the iterative solution routines. The required main memory was less than 300 MByte. No comparison run using the frontal solution technique was performed, as the estimated computation time was in the range of more than 16 CPU-hours per load step. The scratch file to be written in the course of the frontal solution would have reached a size of 3600 MByte.

For smaller models, the acceleration with respect to computation times needed for direct solution methods is generally less pronounced. Such a smaller model is shown on Fig. 2. It covers several gas reservoirs and the adjoining common aquifer in the northern part of the Adriatic Basin. The time history of pore pressure in the reservoirs and the adjoining aquifer due to envisaged gas production was evaluated elsewhere by means of a reservoir simulation model and imported into the finite element model for a forecast of subsidence. Pore pressure degrees of freedom are introduced in elements representing the soil and rock over- and underlying the reservoirs and the common aquifer in order to account for pore pressure changes induced by the deformation associated with the compaction of reservoirs and the adjoining aquifer. The number of degrees of freedom amounts to 32800. One load

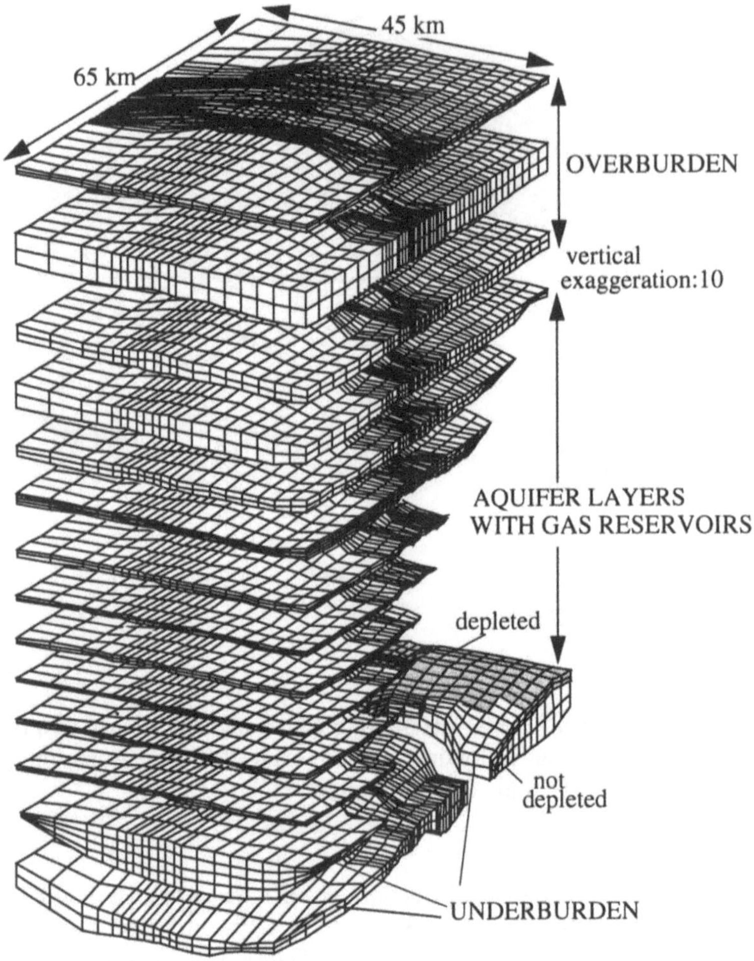

Fig. 2: Perspective view of the layers representing a finite element model for the analysis of subsidence due to envisaged gas production in the northern Adriatic Basin

step using the frontal solver took about 2 hours of CPU time, with a scratch file of 530 MBytes. The iterative solver took about 15 CPU-minutes per load step, which is an acceleration by a factor of 8. The required main memory was less than 115 MByte.
The implemented iterative solvers did not provide a solution for all analyzed cases. No or only very poor convergency was encountered in case of extremely unfavourable element geometries, for some coupled analyses with very small permeabilities and for structures with disconnected parts.

As an example for an unfavourable element geometry serves the analysis of a multilateral well junction, which models the branching off of a lateral well from an existing vertical main wellbore. Typically, kick-off angles are in the range of only 1 to 2 degrees. In order to facilitate the generation of the finite element model, the intersection of the two wells was modeled by increasing the intersection angle to 4 degrees. The analysis includes as intermediate construction stages the simulation of cutting the window into the casing of the main wellbore and the drilling of the lateral well. Fig. 3 shows a detail of the finite element mesh with a vertical geometry scale factor of 0.2. A rather unfavourable element geometry is inevitable when modeling the intersection with such a small angle under the pre-condition to keep the number of nodes within certain limits. The aspect ratio of element edges is rather high, and locally elements are strongly distorted in order to match simultaneously the shape of the main wellbore and of the lateral. Especially in case of a coupled analysis, which modeled pore pressure changes in the adjoining rock, no convergency could be achieved by means of the iterative solver. The impact of element geometry was demonstrated by analyzing the hypothetical case of a lateral branching off with an angle of 90 degrees. For this model, the element geometry is rather regular, while all other assumptions have not been changed. No convergency problems were encountered for this regular element geometry.

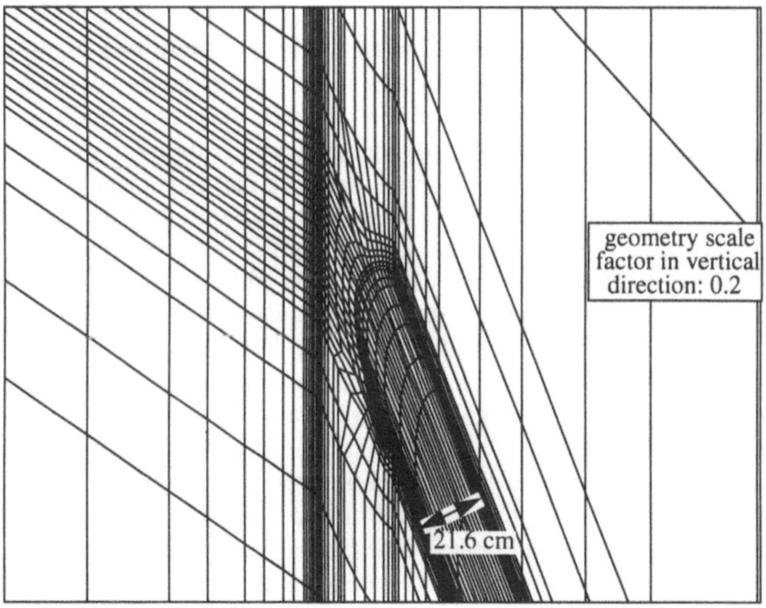

Fig. 3: Detail of the mesh for a multilateral well junction, showing the main bore and the lateral in a 90°- segment of the formation. Elements representing cement and casing are not shown.

For coupled analyses with pore pressure as additional degrees of freedom (consolidation), it can generally be stated that the convergency rate slows down with decreasing permeability. For some simulations modeling shales with permeabilities of less than 10^{-12} m/s, the iterative solvers either failed to converge or converged only rather slowly. Convergency was then improved either by some perturbation of the diagonal terms, when constructing the preconditioner or after some scaling of the system of equations. In this context, it should be recalled that the numerical solution of the consolidation problem generally results in some difficulties, if a too small time step is chosen for the time history integration scheme. Some authors, e.g. Britto and Gunn (1987), quote a criterion for the minimum time step Δt as a function of the consolidation coefficient c_v and the distance l of a node inside the continuum from a draining boundary with prescribed pore pressure:

$$\Delta t = l^2 / (12 \cdot c_v) \qquad (1)$$

If a smaller time step is chosen, even an unconditionally stable integration scheme results in a zigzag distribution of pore pressure versus the distance from the draining boundary. For a direct solution procedure as the frontal solution method, the non-appropriate choice of time steps results in some oscillations of evaluated pore pressure, which are of minor relevancy in most practical applications. However, the applied iterative solvers apparently fail to converge in case of a non-appropriate choice of time steps.

Poor convergency was also obtained in case of a very small rock compressibility, e.g. when simulating undrained behaviour by adding the compressibility of the pore fluid to the stiffness matrix without introduction of pore pressure degrees of freedom (equivalent to assuming a Poisson's ratio close to 0.5). Again, some perturbation of the diagonal terms during the construction of the preconditioner improved the convergency.

In the course of wellbore stability analyses and related problems, the following intermediate construction stage may be encountered: After the simulation of drilling the bore hole, elements representing the casing of the wellbore are added, while elements representing the cement, which fills the annulus between casing and bore hole wall, are to be introduced only in the next step of the analysis. For this intermediate stage, elements representing the casing do not share nodes with the remaining structure. Apparently, the applied iterative solver is not capable to arrive at a solution for such disconnected parts. Convergency is not reached.

The applied iterative solver never encountered convergency problems related to pronounced strain softening. Infinite boundary elements, which simulate zero displacements far from the finite part of the model, delay the convergency compared to finite boundaries. Furthermore it can be stated that the results obtained by the iterative solvers are in excellent agreement with results obtained by the frontal solution method, whenever convergency was reached.

A failure in convergence or slow convergence, as observed in some cases, may probably be overcome by using other iterative techniques. Suitable approaches are multi-level methods of the aggregation type (Bulgakov and Kuhn 1995) or of the ILU type exploiting reordering with an approximate Shur complement (see, e.g. Reusken 1997). However, these approaches require additional memory and operations. Thus they may be less efficient for the majority of "favourable" problems, which can successfully solved by the implemented solvers.

5 CONCLUSIONS

Iterative solvers offer an attractive alternative for large 3D-finite element models also in case of coupled analyses. They enable refined modeling and extensive parametric studies and sensitivity analyses, as mandatory for many applications. In some cases, the tremendous reduction in computation time compared to direct methods is linked with the need to test different options of the solver. Thus iterative solvers still require some experience of the user when solving industrial geomechanics problems.

As potential future applications, iterative solvers will enable a refined modeling of phenomena, which are currently only modeled by means of approximations. This may apply for example to the implementation of non-associated flow rules in plasticity. When using iterative solvers of the KSNIS type, non-symmetric matrices as a consequence of an exact treatment of non-associated plastic flow do not result in unacceptable long computation times. Another example is the more realistic analysis of the failure of granular materials by the formation of shear bands, for which Cosserat's theory provides a promising base. In case of three-dimensional problems, the increase of the number of degrees of freedom by the introduction of three additional rotational degrees of freedom, was considered as an obstacle for many practical applications, which can be overcome by the use of iterative solvers.

6 REFERENCES

BRITTO, A.M. & M.J. GUNN (1987). Critical State Soil Mechanics via Finite Elements. Ellis Horwood Ltd., Chichester.

BULGAKOV, V.E. & KUHN G. (1995). High-Performance Multi-Level Iterative Aggregation Solver for Large Finite-Element Structural Analysis Problems, Int. J. Numer. Meth. Eng., Vol. 38, 3529-3544.

OKLAND, D. & B. PLISCHKE (1996). Perforation Stability Analysis as a Tool for Predicting Sand Production: Application for the Brage Field. Proc. SPE - European Production Operations Conf., Stavanger, 1996.

PLISCHKE, B. (1994). Finite element analysis of compaction and subsidence - Experience gained from several chalk fields. Proc. EUROCK'94, Delft: 795 - 802.

REUSKEN, A. (1997). On the Approximate Cyclic Reduction Preconditioner. Bericht Nr. 144, Institut für Geometrie und Praktische Mathematik, Rheinisch-Westfälische Technische Hochschule Aachen.

SAAD, Y. (1996). Iterative methods for sparse linear systems. PWS Publishing Company, Boston.

INVESTIGATION ON FLOW PROBLEMS USING FRACTAL ANALYSIS

G. Telekes and Z. Czap
Technical University of Budapest, Budapest, Hungary

ABSTRACT

The fractal analysis is the result of recent developments in mathematics. The validity of the different mathematical methods depends on the models applied. The fractal models for granular soils are valid up to hundred times of the significant particle size, which is the scale of laboratory tests. In larger scale projects the methods of geostatistics should be applied, among them the fractal geometrical analysis is also available.

The fractal geometrical methods in Geotechnics are still under development. In the research presented association with fractal dimension and behaviour or physical parameters of the soil was investigated.

The paper introduces the basic ideas of fractals and its theoretical application for investigation on flow problems. The rules of flow in granular soils and the methods of its numerical modelling are also discussed.

1. THE NOTION OF DIMENSION

The fractals are not widely used in Geotechnics yet, so the basic idea needs to be introduced. At first, the definition of fractal dimension must be clarified. A common method of measuring a length, a surface area or a volume consists in encasing them with boxes whose length, surface area or volume is taken as a unit of measurement (Fig. 1). This is the principle that lies behind the use of multiple integration in calculating these quantities. If ε be the side (standard length) of a box and d its Euclidean dimension, the measurement obtained is

$$M = N \cdot \mu = N \cdot \varepsilon^{d}, \qquad (1)$$

where M is the measure, μ is the unit of measurement (length, surface area or volume) and N is a corresponding number.

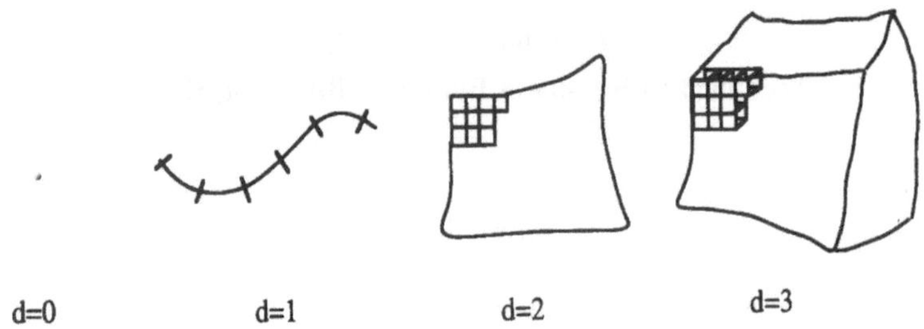

Fig. 1: Interpretation of dimension

This concept was extended to special objects where the dimension (Haussdorf-Besicovitch dimension) is not an integer number (Fig. 2). These fraction dimensional shapes are the fractals. The dimension is the tangent of the line from the diagram N and ε in a double logarithmic scale (Fig. 3). The connection between R (size of fractal) and N is similar but have opposite sense as between N and ε, with fractals made by growth pattern. Of course there are several other definitions for the fractals dimension.

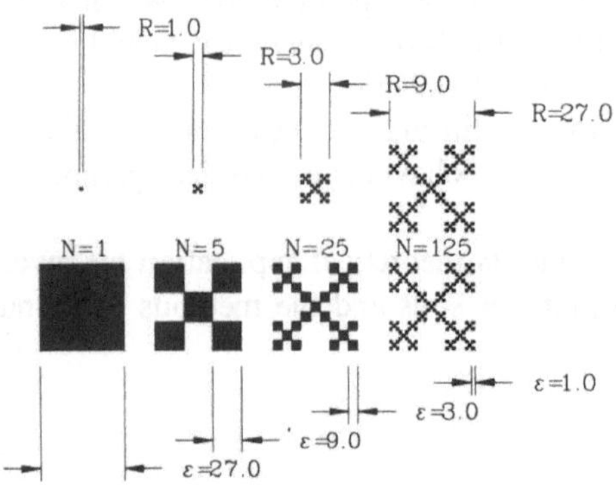

Fig. 2: Making fractals with growth pattern (upper) or erasion technique

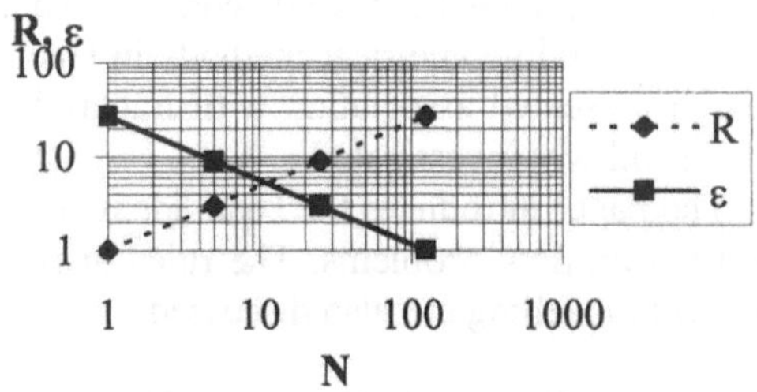

Fig. 3: Determination of fractal dimension

2. FRACTALS IN MATHEMATICS AND IN NATURE

In mathematics there are deterministic (Fig. 2) or random fractals (in Chapter 4). For the random fractals equation (1) is valid only asymptotically ($\varepsilon \to 0$, $N \to \infty$).

In Nature there are examples for both deterministic and random fractals. Leave, snowflake, cauliflower all are like deterministic fractals. There are examples for random fractals from any part of Nature from the Galaxy to the surface of sand particles.

There is limited validity of equation (1) to the fractals in the nature, for example the range for validity for a tree is from 1 cm to 10 m.

There are two ways of the formation of soil. One is the disintegration of rocks; the other is the sedimentation and cohesion of grains carried by wind or water. There are mathematical fractal analogies for both processes of formation of soil (Fig. 2), so it is plausible to apply the fractal geometric methods in Geotechnics.

The methods measuring the fractal dimension are based on equation (1). The test configuration measuring the porosity of soil is shown in Fig. 4. The soil sample is injected slowly with liquid Wood's metal (melting point 70°C) and cut into segments after getting cool. Image processing based on deviation of colours makes the determination of fractal dimension.

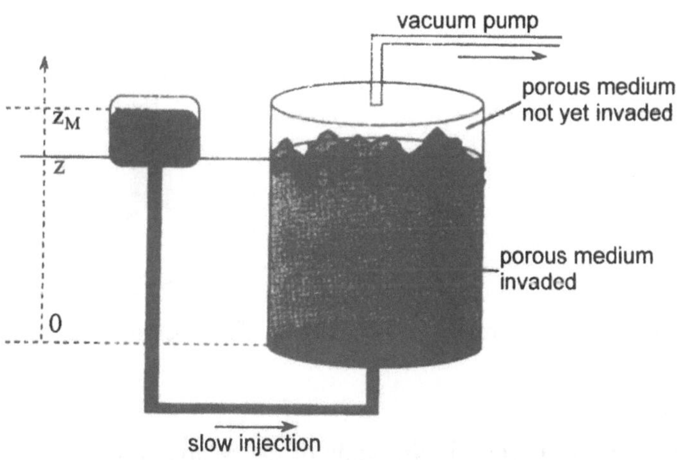

Fig. 4: Determination of fractal dimension on porous medium

3. RULES OF FLOW IN POROUS MEDIUM

Differentiation should be made between wetting and nonwetting fluids (including among fluids the air as well). Wetting is a relative concept and it depends on the interaction of fluids filling the pores of the soil. For example in the case of water and oil or water and air, water is the wetting material. The representative for the wetting capability is the angle θ

(Fig. 5, where i is the nonwetting, j is the wetting fluid), which is zero when the wetting is absolute. The other important representative is the interfacial tension (γ), from which the capillary pressure difference can be determined.

$$P_{cap} = \frac{2 \cdot \gamma \cdot \cos\theta}{r}. \quad (2)$$

The Q_S volume flux through a channel can be determined from the Poiseuille's law that is similar to the Darcy's law

$$Q_s = \frac{\pi \cdot r^4}{8 \cdot a \cdot \mu} \cdot \left(P_i - P_j - P_{cap}\right), \quad (3)$$

where μ is the effective viscosity (calculated from μ_1 and μ_2), a is the length of the channel. If $(P_i-P_j)>P_{cap}$, the nonwetting fluid expels the wetting one, in the opposite case the reverse occurs. If the radius of the channel

$$r < \frac{2 \cdot \gamma \cdot \cos\theta}{P_{inj}}, \quad (4)$$

where P_{inj} is the injection pressure of the nonwetting fluid, the injected fluid is not able to enter the channel (Fig. 6). The injected fluid therefore only enters those pores where the size of the narrow passage is adequate.

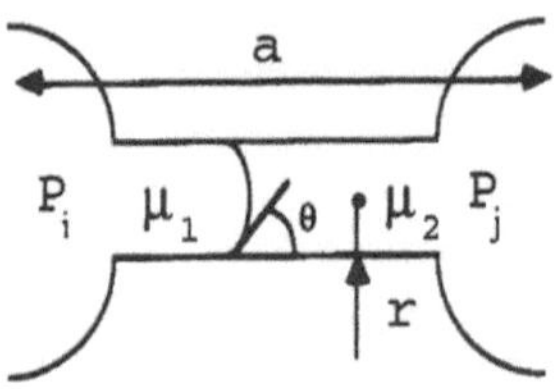

Fig. 5: Flow in narrow channel

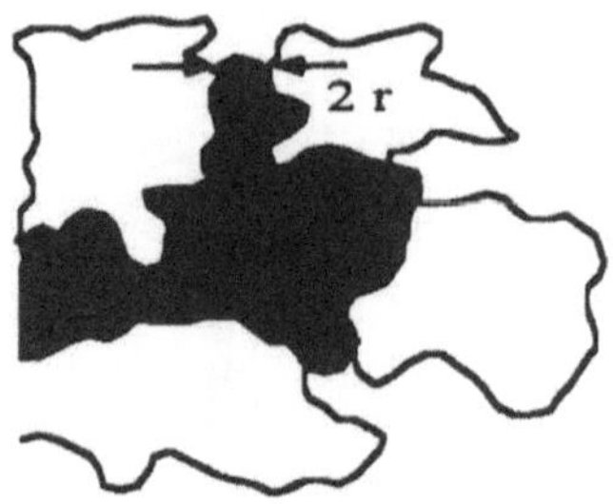

Fig. 6: The blocking effect of narrow passage

There are two important parameters in the study of the behaviour of the fluid. One is the number of capillarity

$$C = \frac{Q \cdot \mu_1}{A \cdot \gamma \cdot \cos\theta}, \quad (5)$$

(where A is the cross-sectional area of the channels and Q is the specific quantity of the injected fluid), which is the ratio between the viscous forces and capillary forces. The other one is the ratio of the viscosity of the fluids present,

$$M = \frac{\mu_1}{\mu_2}, \quad (6)$$

Numerical tests show that there are three typical behaviours of the injected fluid (Fig. 7). An interim zone separates the three categories. From viscous fingering to capillary fingering the transitions are shown in (Fig. 8).

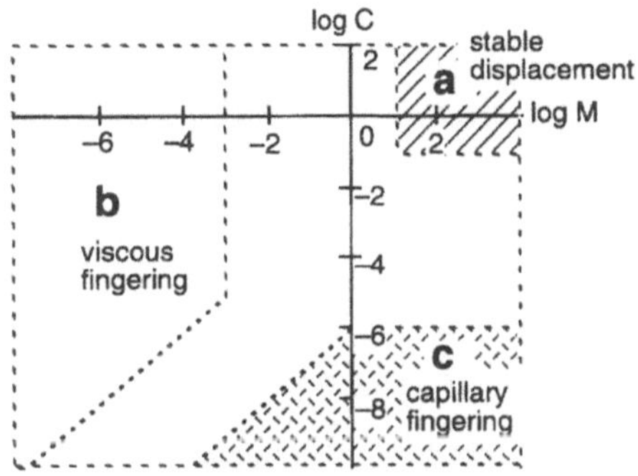

Fig. 7: Typical flow behaviours

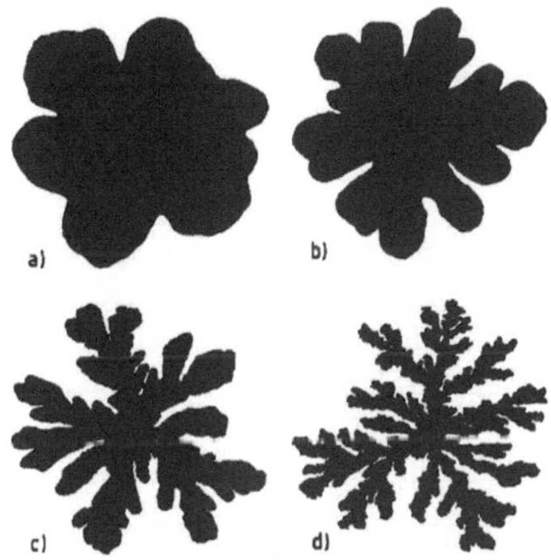

Fig. 8: Transitions from viscous fingering to capillary fingering

4. MODELLING BY NUMERICAL METHODS

The classical investigation on flow problems is made, using Darcy's law, by solving partial differential equations. As it was emphasised in Chapter 0, in the case of nonwetting fluid flow can be detected only in part of the channels and the injected nonwetting fluid is not able to expel the wetting one from all the pores.

In analysing the capillary fingering, one of the possibilities is the diffusion model. For the modelling making appropriate lattices is required. The reason for the imperfectly filled cross-sectional area is that the new particles stick with greater probability to the growing „fingers" tip, because in the course of meandering more particles can reach the tip than the side of the „fingers" (Fig. 9).

The other modelling possibilities is the growth model. The model simulates the proliferation of malignant cells (in black) starting from a single cell (the germ) by contamination of the closest neighbours. The reason for the closely filled cross-sectional area is that the new particles with greater probability stick to those points where the particles join on more sides to the others (Fig. 10).

Different fractal models were tried describing the based on the blocking effect shown in Fig. 6. Modelling flow in porous medium one of the possibilities is the diffusion model. For the modelling making appropriate lattices and an initial layer at the injected point is required. There are meandering particles in the model sticking to the initial layer or to the thicken layer because of diffusion. Fig. 11 shows a perpendicular sticking model, Fig. 12

shows a 45° sticking model. The diffusion model valid in laboratory condition to highly viscous liquid.

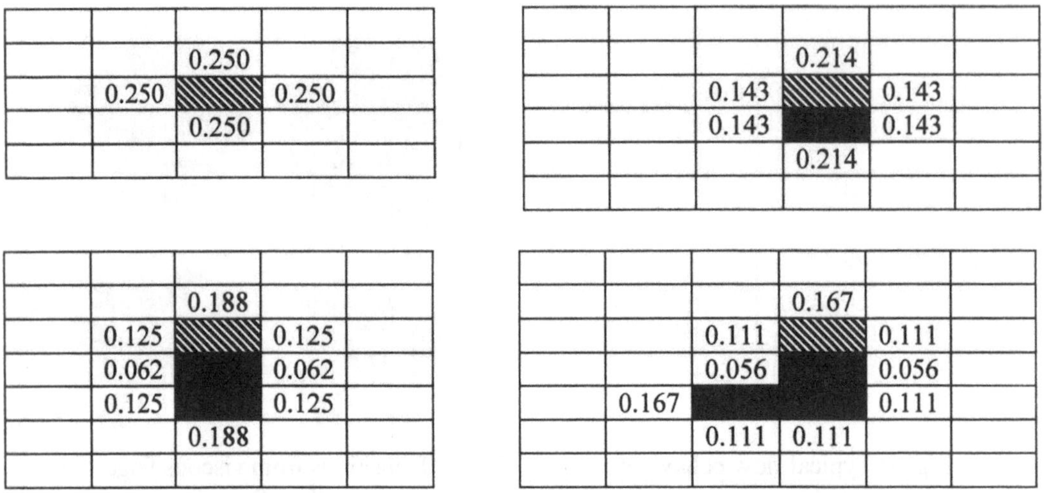

Fig. 9: The probability of cohesion of the new particles in the diffusion model

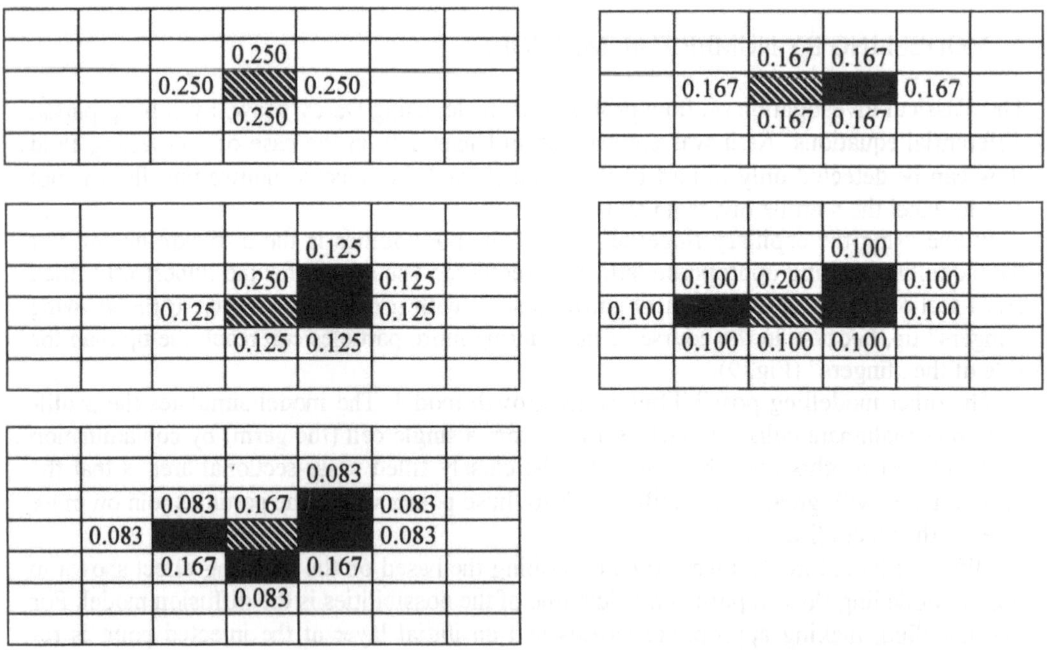

Fig. 10: The probability of cohesion of the new particles in the growth model

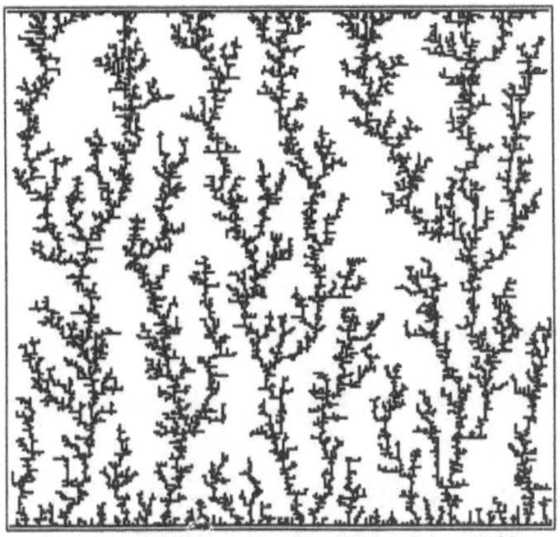

Fig. 11: Perpendicular sticking diffusion model (DLA1)

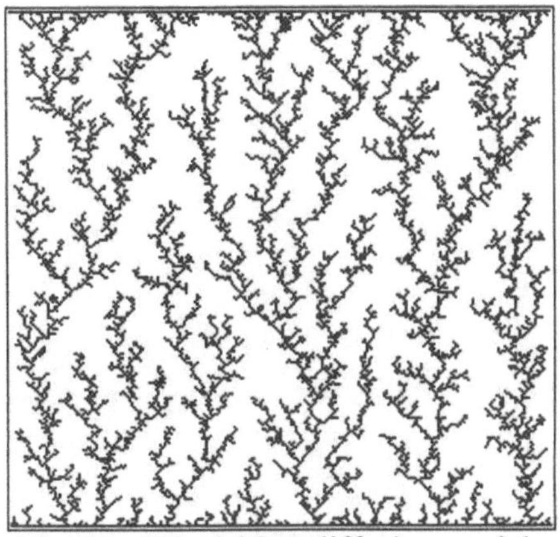

Fig. 12: 45° sticking diffusion model (DLA2)

The other modelling possibilities is the growth (Eden) model (Fig. 13). Than the meandering particles moves outward and sticking at the boundary. This model represents a very slow filling.

The third modelling possibility is the ballistic model. Than the meandering particles falls vertically and sticking at the boundary. Fig. 14 shows the vertical sticking, Fig. 15 the perpendicular sticking and Fig. 16 the 45° sticking ballistic model.

Fig. 13. The growth (Eden) model (E)

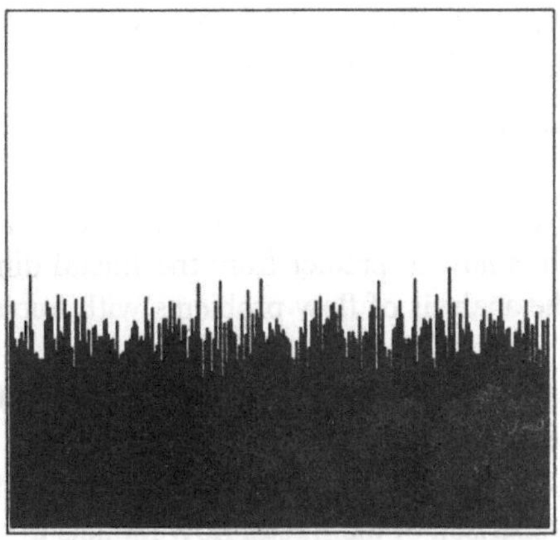

Fig. 14. The vertical sticking ballistic model (B1)

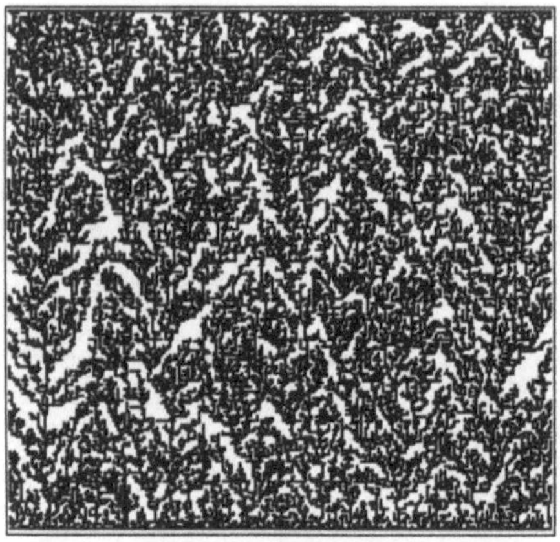
Fig. 15. The perpendicular sticking ballistic model (B2)

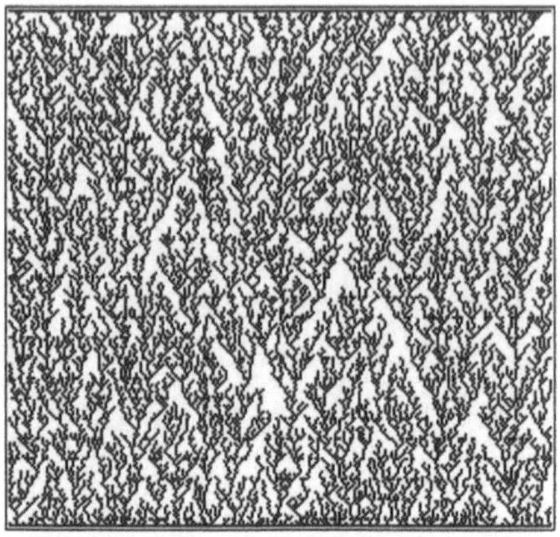
Fig. 16. The 45° sticking ballistic model (B3)

In soil mechanic applications the number of steps of the meandering particles analogous to the time, and the ratio of sticking particles analogous to the degree of saturation. The comparison of the presented flow models is shown in Fig. 17.

Of course in the laboratory modelling other formation or the combination of the presented mathematical models are also possible.

One very current test apparatus is the Hele-Shaw cell (Fig. 18). The gap between two glass-plates, which are slick or notched in one or more directions, at first is filled with viscous fluid, then air is injected through the centre of one of the glass-plates. A comparison between Hele-Shaw cell test (a) made by BAKUCZ (1996) and numerical tests (b) is shown in Fig. 19. The results of ballistic fractal modelling fit well to the tests result.

5. FIELDS OF APPLICATION

The fractal geometrical methods are yet in a developmental phase. The direction of the research is how to predict from the fractal dimension the typical behaviour of the materials. For the analysis of flow problems with fractal geometrical methods the research area could be:

- development of test methods for different types of soil in order to determine the fractal characteristics;
- to find relationship between the flow and fractal characteristics;
- comparison of the fractal flow models to continuum flow models.

In the investigation with contaminant soils and groundwater the main research area could be:

- development of the fractal characteristics for the distribution of contamination;
- application of parallel computing techniques for fractal modelling;

- comparison of the fractal and the continuum models for the distribution of contamination and for the spreading of contamination;
- development of fractal geometrical models of soil and groundwater cleaning technologies.

It is clear that there will be a wide scope of applications for the fractal geometrical methods in the near future. The present task is to increase the correctness and reliability of the fractal geometrical models.

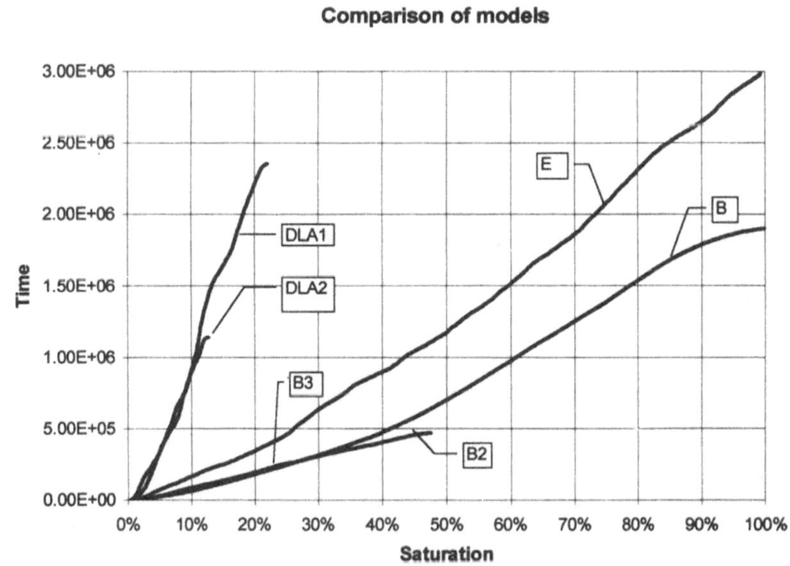

Fig. 17. Comparison of the presented numerical models

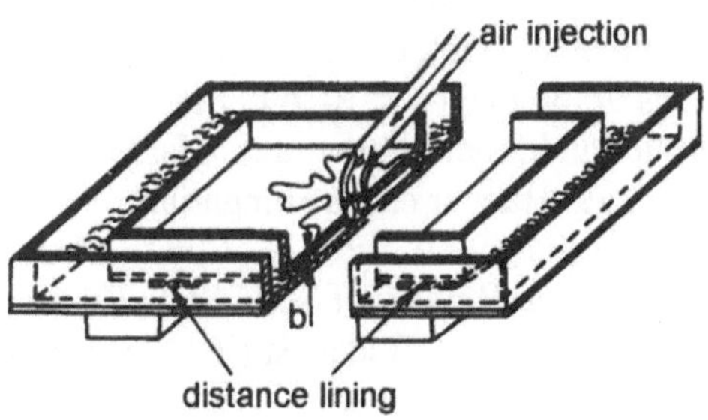

Fig. 18: Hele-Shaw cell

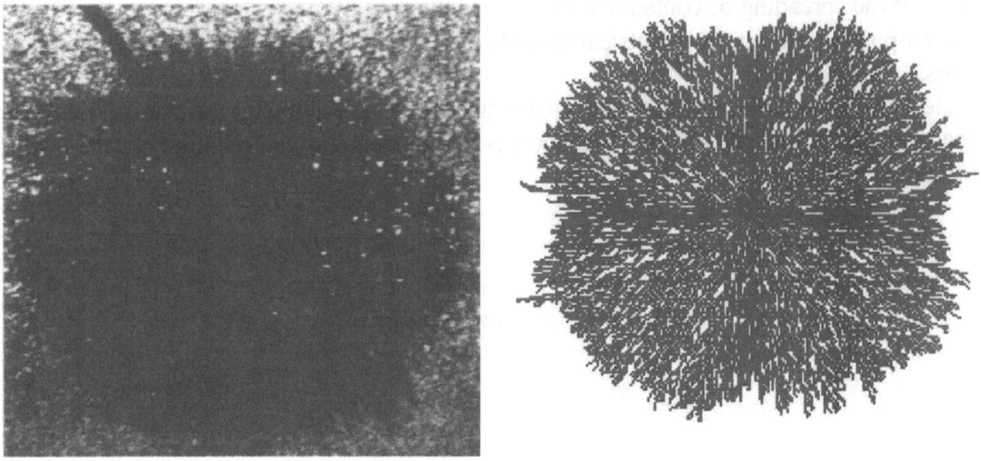

a) b)

Fig. 19: A Hele-Shaw cell test (a) and its ballistic fractal model (b)

ACKNOWLEDGEMENTS

The present study was supported by the Hungarian National Scientific and Research Foundation (OTKA # T 022138 and T 020446) and by the COPERNICUS (PARCOMPWASTE).

REFERENCES

BAKUCZ, P. (1996). The hydrodinamical fractal dispersion. Hidrológiai Közlöny, Vol. 76. No. 5. Budapest (in Hungarian)

BARABÁSI, A. L. - STANLEY, H. E. (1995). Fractal Concepts in Surface Growth. Cambridge University Press

CZAP, Z. - TELEKES, G. (1997). Potential Uses of Fractal Analysis on Flow Problems in Unsaturated Soils. Environmental Geotechnics – Design Parameters for Computing Applications, the Proceedings of the Lectures, REAT, Prague

GOUYET, J-F. (1996): Physics and Fractal Structures, Masson, Springer-Verlag

Physics (1993). SH atlas. Springer Hungarica (in Hungarian)

STAUFFER, D. - STANLEY, H. E. (1990). From Newton to Mandelbrot. Springer-Verlag

VICSEK, T. (1990). Fractal growing. Review of Solid State Research 22. Publishing House of the Hungarian Academy of Sciences. Budapest (in Hungarian)

A PARALLEL COMPUTING SOLUTION OF THE FINITE ELEMENT ANALYSIS OF THERMO/HYDRO/MECHANICAL BEHAVIOUR OF UNSATURATED SOIL

H.R. Thomas, A.D. Jefferson and E.E. Nkposong
University of Wales, Cardiff, UK

ABSTRACT

This paper presents a parallel iterative solution or systems which incorporate the thermo/hydro/mechanical behaviour of unsaturated soil. The approach adopts an iterative solution of the governing equations for the fully coupled analysis in which the primary variables are pore water pressure, pore air pressure, temperature and displacement. The current approach extends this work to problems of greater complexity via a finite element scheme. The developed parallel code was applied to the simulation of a thermo-hydro-mechanical experiment for use in the performance assessment of deep geological repositories for the disposal of nuclear waste. Good speed-ups were achieved, ranging from 1.85 on two processors to 4.71 on 9 processor

1. INTRODUCTION

Despite the tremendous advancement of computers in the last decade the solution of large sparse non-linear systems can still be prohibitive in terms of time and storage requirements. An example of this is the prediction of fully coupled heat and moisture transfer in a deformable unsaturated soil. This problem is of particular importance in predicting the long-term performance of engineered buffer material used as a sealant for nuclear waste, in deep geological repositories.

The underlying physical phenomena are complex and lead to a complex mathematical description of the problem, via a coupled set of partial differential equations, defining up to

6 unknown variables. The complexity of the work is such that the geometry of the domain is very large with real solutions requiring long periods of time to complete. As a result the problems are computationally very demanding. One of the key components, therefore in current state-of-the-art engineering design practice for such systems is the use of computer software to perform the required analysis. To date, such computational work performed almost entirely in sequential mode and not in a parallel environment.

The development of high performance parallel computers in recent years has, for the first time, provided the potential for realistic solutions to these multi-dimensional problems. Furthermore the development of multi-processor operating systems for desktop computers such as Microsoft Windows NT means that it is now possible to exploit the benefits of parallelisation in a typical design office environment.

The aim of this paper is to present a parallel iterative algorithm for the fully coupled thermo/hydro/mechanical behaviour of unsaturated soil. An important feature of this algorithm is its portability to other parallel platforms such as mentioned above, which will enhance the attractiveness of the method outside the academic community. Previous work on parallel solutions have been achieved for coupled flow problems (Thomas and Li 1997) using a finite difference approach. The current approach extends this work to problems of greater complexity via a finite element scheme.

The work presented here was programmed in FORTRAN. All computations were performed on a PARAMID parallel computer at the University of Wales, Cardiff (UWC). The Multiple Instruction Multiple Data (MIMD) PARAMID parallel computer with distributed memory architecture, is based on the Transtech parallel system. The system has 48 processors, each consisting of a 100 MFLOP Intel i860-XP running at 50 MHz.

A number of parallel numerical tests were undertaken to verify the correctness and efficiency of the model proposed. The analysis of a laboratory experiment is used to assess the computational efficiency of the new parallel code and to validate the results against experimental data.

2. THEORY OF HEAT, MASS AND AIR TRANSFER IN UNSATURATED SOIL

The behaviour of unsaturated soil consisting of solid, liquid and gas phases can be described mathematically as a set of non-linear differential equations. The formulation employed here has been presented elsewhere in considerable detail, (Thomas and King 1991, Thomas and Sansom 1995, Thomas and He 1995, Thomas and He 1998), and therefore only a brief description of the salient points will be given here.

2.1 Moisture Transfer

The moisture transfer equation, accommodating both liquid and vapour phases is expressed in the form

$$(n\rho_l S_l)_{,t} + (n\rho_v S_a)_{,t} + (\rho_l (V_l)_j)_{,j} + (\rho_l (V_v)_j)_{,j} + (\rho_v (V_a)_j)_{,j} = 0 \tag{1}$$

where n denotes the porosity, ρ the density, S the degree of saturation, t the time and V the velocity. The subscripts *l*, *a* and *v* refer to liquid, air and water vapour respectively. The subscript j, ranging from 1 to 3, is tensor index.

2.2 Dry air transfer

Using Henry's law to take account of dissolved air in the pore water, conservation of dry air flow dictates that

$$[n\rho_{da}(S_a + H_c S_l)]_{,t} + [\rho_{da}((V_a)_j + H_c(V_l)_j)]_{,j} = 0 \quad (2)$$

where H_c is Henry's coefficient of solubility, and, ρ_{da} is the density of dry air.

2.3 Temperature

From consideration of conservation of energy, the governing equation for heat transfer can be written as

$$\Phi_{,t} = Q_{i,i} \quad (3)$$

where Q, the global heat flux per unit volume and Φ, the heat capacity of the soil per unit volume are defined as;

$$\Phi = n(\rho_l S_l C_{pl} + \rho_v S_a C_{pv} + \rho_a S_a C_{pa})(T - T_r) + (1-n)\rho_s C_{ps}(T - T_r) + n\rho_v S_a L$$

$$Q_j = -\lambda_T T_{,j} + (\rho_l (V_v)_j + \rho_v (V_a)_j)L + (C_{pl}\rho_l (V_l)_j + C_{pv}\rho_l (V_v)_j + C_{pv}\rho_v (V_a)_j + C_{pa}\rho_a (V_a)_j)(T - T_r)$$

In the above λ_T is the intrinsic thermal conductivity of the soil, C_{pl}, C_{pv}, C_{pa} and C_{ps} are the specific heat capacities of pore liquid, water vapour, pore air and the solid particles respectively, T and T_r are the temperature and reference temperature respectively,ρs is the density of the solid particles, and L is the latent heat of vaporisation.

2.4 Stress equilibrium equation and constitutive relationship.

The total strain can be written in an incremental form as

$$d\varepsilon = d(\varepsilon^{\sigma}) + d(\varepsilon^{s}) + d(\varepsilon^{T}) \quad (4)$$

where the superscripts σ, s and T refer to net mean stress, suction and temperature respectively.

The stress strain relationship can be expressed as

$$d\sigma'' = \mathbf{D}(d\underline{\varepsilon} - d\underline{\varepsilon}_s - d\underline{\varepsilon}_T) \quad (5)$$

where σ'' is the net stress and **D** is the elastic matrix. An elasto plastic constitutive relationship is used in the model (Thomas and He 1998).

2.5 Finite element formulation

By exploiting the internal relations between variables, and employing the Galerkin method, the finite element representation of the governing equations can be expressed in terms of the primary variables, i.e. $u_l, T, u_a, \underline{u}$, as follows;

$$\begin{bmatrix} \mathbf{K}_{ll} & \mathbf{K}_{lT} & \mathbf{K}_{la} & \mathbf{0} \\ \mathbf{K}_{Tl} & \mathbf{K}_{TT} & \mathbf{K}_{Ta} & \mathbf{0} \\ \mathbf{K}_{al} & \mathbf{0} & \mathbf{K}_{aa} & \mathbf{0} \\ \mathbf{0} & \mathbf{0} & \mathbf{0} & \mathbf{0} \end{bmatrix} \begin{Bmatrix} u_l \\ T \\ u_a \\ \underline{u} \end{Bmatrix} + \begin{bmatrix} \mathbf{C}_{ll} & \mathbf{C}_{lT} & \mathbf{C}_{la} & \mathbf{C}_{Tu} \\ \mathbf{C}_{Tl} & \mathbf{C}_{TT} & \mathbf{C}_{Ta} & \mathbf{C}_{Tu} \\ \mathbf{C}_{al} & \mathbf{C}_{Tl} & \mathbf{C}_{aa} & \mathbf{C}_{Tu} \\ \mathbf{C}_{ul} & \mathbf{C}_{uT} & \mathbf{C}_{ua} & \mathbf{C}_{uul} \end{bmatrix} \begin{Bmatrix} \dot{u}_l \\ \dot{T} \\ \dot{u}_a \\ \dot{\underline{u}} \end{Bmatrix} = \begin{Bmatrix} f_l \\ f_T \\ f_a \\ f_u \end{Bmatrix} \quad (6)$$

where u_l represents pore water pressure, T the temperature, u_a the pore air pressure, and $\underline{u}$ the displacement vector, K_{ij} and C_{ij} represent the corresponding matrices of the governing equations, (i,j=l, T, a, u), $\langle\cdot\rangle$ refers to the derivative with respect to time, $\underline{u}$ is the vector of displacement.

Equation (6) can be rewritten in a simplified notation as;

$$K\{\varphi\}+C\left\{\frac{\partial\varphi}{\partial t}\right\}=\{R\} \tag{7}$$

where φ refers to the global unknowns, that is $\{u_l \quad T \quad u_a \quad \underline{u}\}^{\mathrm{T}}$

To solve equation (7), one useful method for transient problems is direct temporal integration. A general form of two level difference method is employed to discretise the governing equations in the time domain (Cook, 1981). Applying this method to equation (7) gives the form;

$$\chi K\{\varphi^{n+1}\}+C\{\varphi^{n+1}\}/\Delta t=(1-\chi)\{R^n\}+\chi\{R^{n+1}\}-(1-\chi)K^n\{\varphi^n\}+C\{\varphi^n\}/\Delta t \tag{8}$$

where χ is an integration factor, which defines the required time interval ($\chi\in[0,1]$).

Rewriting equation (8) in an alternative notation yields

$$A\{\varphi^{n+1}\}=\{F^{n+1}\} \tag{9}$$

where

$$A=\chi K+C/\Delta t$$

$$F^{n+1}=(1-\chi)\{R^n\}+\chi\{R^{n+1}\}-(1-\chi)K^n\{\varphi^n\}+C\{\varphi^n\}/\Delta t$$

3. ITERATIVE SOLVERS

The Conjugate Gradient class of solvers form the basis for the parallel solution algorithm presented in this paper. In recent years it has been shown that in certain cases iterative solvers are as efficient as direct solvers for large scale problems on single processor machines. Furthermore iterative solvers are more naturally parallel than direct solution methods, which enhances their portability to multi processor machines.

A number of parallel algorithms have been developed for the solution of a general system of equations. It is important to emphasise that these algorithms do not fully take account of the characteristics of the PARAMID i.e. high processor speed to communication ratio. A new parallel algorithm has therefore been developed to reduce communications to a minimum and, where possible, overlap these communications with computation, while at the same time maintaining general communication efficiency. This algorithm should be efficient not only on the PARAMID but on other platforms which have high processor speed to communication ratios such as a network of high performance PC's.

The highly non-linear fully coupled thermo/hydro/mechanical problems under consideration produce an ill-conditioned asymmetric system of equations and hence an asymmetric solver has been used. The Bi-Conjugate Gradient (BiCG) algorithm was adopted for the solver. It has the same basic structure as the standard Conjugate Gradient

(CG) algorithm, but utilises two sets of iterates, one based on the standard matrix and the other on its transpose. Its efficiency, in terms of the number of iterations to achieve convergence is comparable with other state of the art asymmetric CG solvers (Barret et al, 1995). Diagonal preconditioning is used for the present work because it naturally lends itself to parallelisation, however, further research is being carried out to obtain more effective parallel preconditioners.

The non-linear system of equations presented in equation (9) is the basis from which the vector of unknown's φ^{n+1} are found. The parallel algorithm presented addresses the problems associated with the storage of a large system of equations by using a Compressed Storage Scheme (CRS) to store only the non-zero elements of the stiffness matrix A. In this method, the matrix is stored as a one-dimensional array of non-zero numbers along with two pointer arrays which give the column number and an array which, for each row, gives the entry number of the last entry of the row.

Advantage is taken of the structure of the finite element discretisation to distribute the assembly and storage of the stiffness matrix between processors. A set of rows of the stiffness matrix are associated with each processor by assigning a cluster of nodes to each processor. Elemental information is then obtained from the associated nodes. The advantage of this approach is that it can be employed on regular or irregular grids because nodal assignment to processors is independent of the geometry. Global assembly of the stiffness matrix is never carried out because each processor assembles and stores a local sub-matrix, with only those element which contribute to the nodes assigned to a processor being evaluated for that processor. The CRS scheme is then utilised to compact the sparse local sub-matrix, storing only the non-zero values in a dense vector. This was of particular importance in the present work because there is limited memory on each processor of the PARAMID.

Although the CRS scheme is entirely general and gives minimum storage it however, requires an addressing step for each matrix-vector operation, which reduces the efficiency of the scheme for vector processors. However, by packing two vectors in advance of the computation and thereby undertaking all addressing before the scalar multiplication's, the scheme was efficient and made good use of the vector processors. As part of the assembly process additional pointer arrays are established which list the components of the search vector to be sent to and required from each processor.

Table 1 shows the new parallel BiCG algorithm which minimises communications and overlap them with computations where possible. Convergence is governed by the scalar product of the preconditioned residual and residual vectors, normalised by the scalar product of the preconditioned force and force vectors. The use of the scalar product of the residual and preconditioned residual as a stopping criteria follows (Axelsson 1994) and the normalisation serves to provide a dimensionless measure which gives a consistent level of accuracy between different analyses. This scalar product is equal to the scalar product of the initial displacement and force vectors i.e.

$tn = (\mathbf{M}^{-1} \cdot \mathbf{f}) \cdot \mathbf{f} = \varphi_0 \cdot \mathbf{f}$.

$\mathbf{M} = \mathrm{diag}(\mathbf{A})$	Set diagonal preconditioning matrix
$\varphi_0 = \mathbf{M}^{-1}\ \mathbf{f}$	Calculate initial estimate of the unknowns.
$\varphi_o^{\,lf} = \phi_f^{\,lf}$	Set the prescribed variables Calculate the initial residuals, ***in parts***
$\mathbf{r}_0 = \mathbf{f} - \mathbf{A}\,\varphi_o,\ \bar{\mathbf{r}}_0 = \mathbf{r}_0$	Zero the residual at fixed degrees of freedom
$\mathrm{r}_0^{\,\mathrm{lf}} = 0$	Precondition the initial residuals, ***in parts***
$\mathbf{z}_0 = \mathbf{M}^{-1}\ \mathbf{r}_0, \bar{\mathbf{z}}_0 = \mathbf{M}^{-1}\ \bar{\mathbf{r}}_0$	Compute the inner product ***in parts.*** Send receive and add. Set initial search vectors to residual vectors, ***in parts*** send Mes 1
$\rho_0 = \bar{\mathbf{r}}_0 \cdot \mathbf{z}_0$	Set pre-conditioned right hand side norm ***in full***
$\mathbf{p}_0 = \mathbf{z}_0, \bar{\mathbf{p}}_0 = \bar{\mathbf{z}}_0$	Enter iteration loop until convergence achieved
$tn = \varphi_0 \cdot \mathbf{f}$	Compute mat-vect. products, in parts receiving Mes 1 ***in parts***
for i = 1, 2 , ...	
$\mathbf{q}_{i-1} = \mathbf{A}\,\mathbf{p}_{i-1},$	Compute inner product, ***in parts*** and send Mes 2
$\bar{\mathbf{q}}_{i-1} = \mathbf{A}^{\mathrm{T}}\ \bar{\mathbf{p}}_{i-1}$	Receive Mes 2, sum γ and calculate α
$\gamma_{i-1} = \mathbf{q}_{i-1} \cdot \bar{\mathbf{p}}_{i-1}$	Computer residuals ***in parts***
$\alpha_{i-1} = \rho_{i-1} / \gamma_{i-1}$	Set residual for fixed degrees of freedom to zero
$\mathbf{r}_i = \mathbf{r}_{i-1} - \alpha_{i-1}\,\mathbf{q}_{i-1}$	Precondition the residual, ***in parts***
$\bar{\mathbf{r}}_i = \bar{\mathbf{r}}_{i-1} - \alpha_{i-1}\,\bar{\mathbf{q}}_{i-1}$	Compute inner product ***in parts*** and send Mes 3
$\mathrm{r}_i^{\,\mathrm{lf}} = 0$	Update vector of unknowns, ***in parts***
$\mathbf{z}_i = \mathbf{M}^{-1}\,\mathbf{r}_i, \bar{\mathbf{z}}_i = \mathbf{M}^{-1}\,\bar{\mathbf{r}}_i$	Rec. Mes 3 and compute ρ , then form the scaling factor β
$\rho_i = \bar{\mathbf{r}}_i \cdot \mathbf{z}_i$	Update search vectors ***in parts***, sending Mes 1
$\varphi_i = \varphi_{i-1} + \alpha_{i-1}\,\mathbf{p}_{i-1}$	
$\beta_i = \rho_i / \rho_{i-1}$	Compute the error norm. Check for convergence
$\mathbf{p}_i = \mathbf{z}_i + \beta_i\,\mathbf{p}_{i-1}$	
$\bar{\mathbf{p}}_i = \bar{\mathbf{z}}_i + \beta_i\,\bar{\mathbf{p}}_{i-1}$	Parts of Mes. 1 (i.e p) need clearing from the message buffer Send unknowns to all processors
$\mathrm{enm} = \rho_i / \mathrm{tn}$	Compute 'reactions' for fixed degrees of freedom
If enm < tol exit loop	
Close loop	
Clear mes. buffer	
Send and recv. φ	
$rf^{\,lf} = \mathbf{A} \cdot \varphi^{lf}$	

Table 1. Parallel preconditioned BiCG algorithm

It was found that the use of this norm, which has the dimensions of work (for the deformation components), produced greater consistency (in terms of accuracy) between different problems than did a norm of the right-hand-side vector alone. This is because both the unknowns and right-hand-side terms can vary by many orders of magnitude for the different phases but that the products of unknown and right-hand-side vectors for the phases vary much less.

4. APPLICATION

A fully coupled thermo/hydraulic/mechanical laboratory based experiment has been used to investigate the performance of the parallel finite element code for large scale simulations. The experiment was performed on a Boom clay sample, uniaxially compacted within a stainless steel cell 146 mm high and 150 mm in diameter at an initial water content of 14.8 % to a dry density of 1.65 g/cm^3. A diagrammatic representation of the thermo/hydraulic cell is shown in Figure 1. The base-plate of the cell has two hydration channels, which allow drainage or hydration of the sample via a porous plate. The device is composed of an electrical resistance, which is regulated electronically by a thermocouple. The incorporated electrical resistance configuration is known as a 'punctual heater' and is housed in a copper cell, which is 10 mm in diameter and projects 15 mm into the soil sample.

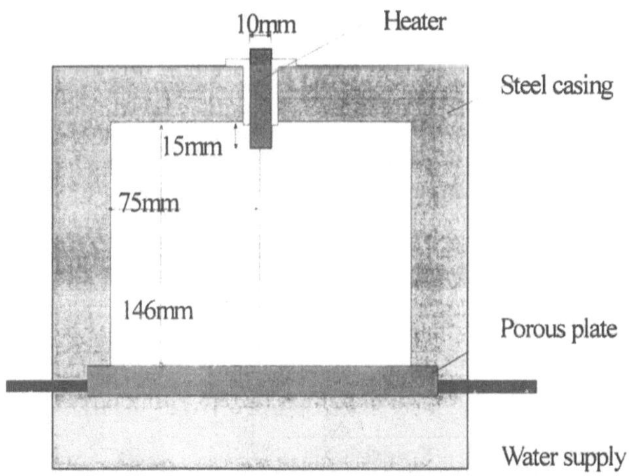

Fig. 1. Schematic diagram of the thermo/hydraulic cell.

Nine thermocouples placed at various locations measure the temperature variation within the soil sample.

The temperature at the nodes of the heater was fixed at 100 °C. The temperature along the outer boundaries was fixed at 28 °C due to the thermo-shower. Initial conditions for the simulation were estimated from measured experimental data as follows:

Degree of saturation $S_l = 0.5$, Void ratio $e = 0.72$, Temperature $T = 20$ °C. Initial deformation was taken as zero.

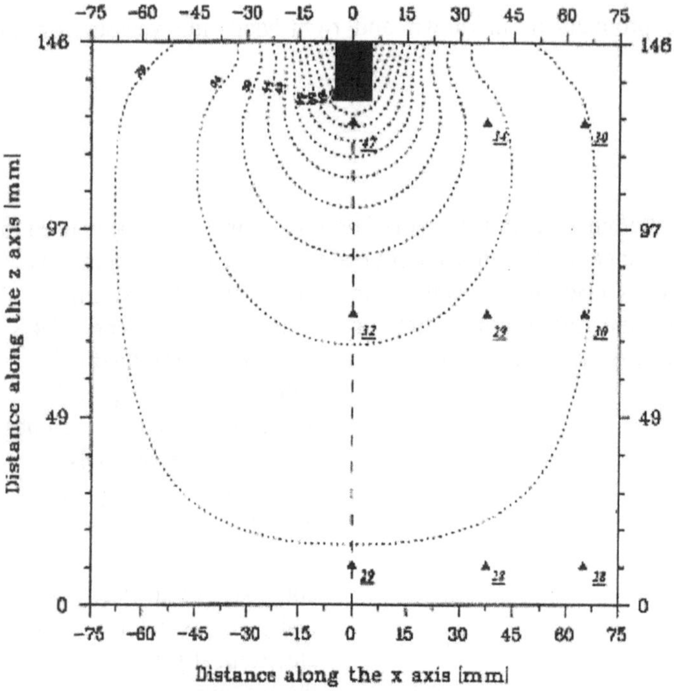

Fig. 2 Plot of temperature.

Advantage was taken of the symmetry and only half the sample was modelled. A mesh 146 mm high and 75 mm wide, with 364 8-node isoparametric elements was chosen to represent the test sample.

The results obtained for the steady state distributions of temperature within the soil sample is presented in the form of contour plots in Figure 2. The Speedup and Efficiency are shown in Table 2.

No. of Proc.	Time (%)	Efficiency (%)	Speed-Up
1	591.72		
2	319.85	92.50	1.85
3	239.67	82.30	2.47
4	194.66	75.99	3.04
6	152.38	64.72	3.88
9	125.55	52.37	4.71

Table 2. PARAMID results for 5865 degree of freedom fully coupled problem.

Speedup is defined as the time on one processor divided by the time on *n* processors and Efficiency, expressed in percentage terms, is Speedup divided by the number of processors.

5. CONCLUSION

A parallel iterative solution of a fully coupled thermo/hydro/mechanical simulation of the behaviour of unsaturated soil has been presented. The parallel version of COMPASS is based on iterative solution algorithm namely the Bi-Conjugate Gradient (BiCG). The results demonstrate the achievements of the developed parallel codes in predicting the complex interactions which are computationally very demanding.

The improvements have been achieved in the parallel codes because they efficiently manage the memory storage requirements generated by the huge system of linear equations which arise from these problems and reduce the overall solution times greatly, and enable the results to be obtained in a reasonable time. Also, the effective division of the computation between processors leads to large speed-ups in solution.

The improved computational performance achieved using the parallel codes provides a greatly enhanced capability for modelling partially saturated media used as a sealant in the disposal of nuclear waste.

6. REFERENCES

AXELSSON, O. (1989) "On a class of preconditioned iterative methods on parallel computers", *Int. J. Num Meth. in Eng*, **27**, 637-654 (1989).

AXELSSON, O. (1994) *Iterative solution methods,* Cambridge University Press.

BARRETT, R., BERRY, M., CHAN., T., DEMMEL, J., DONATO, J. DONGARRA, J., EIJKOUT, V., POZO, R., ROMINE C.,AND VAN DER VORST, H. (1995) T*emplates for the solution of linear building blocks for iterative methods.*

THOMAS, H. R AND HE, Y.(1995) "Analysis of coupled heat, moisture and air transfer in a deformable unsaturated soil." *Geotechnique*, 45(4): 677-689.

THOMAS, H. R AND HE, Y."Modelling the behaviour of unsaturated soil using an elasto plastic constitutive relationship." Accepted for publication in *Geotechnique* - 1998.

THOMAS, H. R, HE, Y, ONOFREI, C (1998) "An examination of the validation of a model of the hydro/thermo/mechanical behaviour of engineered clay barriers." *Int. J for Numerical and Analytical Methods in Geomechanics*, 22: 49-71.

THOMAS, H. R. AND LI, C.L.W. (1991). "A parallel computing solution of coupled flow in soil." *ASCE J. Comp. Civil Eng.*, 5: 428-443.

THOMAS, H. R. AND LI, C.L.W. (1995). "Modelling transient heat and moisture transfer in unsaturated soil using a parallel computing approach." *Int. J For Numerical and Analytical Methods in Geomechnics,* 19: 345-366.

THOMAS H R AND LI C.L.W (1997) Modelling heat and moisture transfer in unsaturated soil using a finite difference self implicit method on parallel computers. *Int. J For Numerical and Analytical Methods in Geomechnics*, 21(6): 409-419.

VILLAR, M. V.& P.L.MARTIN (1993). "Suction controlled oedometric tests in montmorillonite clay." *Proc. 29th Annual Conference of the Engineering Group of the Geological Society of London*, 337-342.

results demonstrate the achievements of the [illegible] complex [illegible] computationally very [illegible]

The [illegible] have [illegible] [illegible] [illegible] which are [illegible] problems and reduce the [illegible] and [illegible] the [illegible] [illegible] when processes [illegible]

The [illegible] [illegible] [illegible] [illegible]

REFERENCES

[illegible]

[illegible]

[illegible] MARTIN [illegible] [illegible]

[illegible]

[illegible]

[illegible] [illegible] (1997) "Modelling heat and moisture transfer in [illegible] [illegible] in parallel computers", Int. J. for Numerical and Analytical Methods in Geomechanics, 21(5): 41-[illegible]

[illegible] MARTIN (1994) "Suction controlled oedometric tests in [illegible]", [illegible] Society of Canada [illegible]

THERMOPOROELASTIC ANALYSIS OF A DEEP CIRCULAR TUNNEL

EFFECT OF HETEROGENEITY

G. Thouvenin and A. Giraud
École Nationale Supérieure de Géologie, Vandœuvre-lès-Nancy, France

ABSTRACT: Semi-analytical solutions are presented for the behavior of a cylindrical cavity excavated in a porous rock mass. Two homogeneous media constitute the rock mass: a 'skin' surrounding the cavity and the porous rock mass. The behavior of the rock mass is thermoporoelastic and contrasts of mechanical and hydraulical properties between the two media are taken into account to model approximately degradation and damage due to the opening of the cavity. General solutions are given for an isotropic saturated porous medium with only one space coordinate, the radius of the cavity. Results are presented, using realistic data, to evaluate temperature, pore pressure, effective stress distributions and radial displacement for two materials representative of a clay and a shale.

1. INTRODUCTION

The problem of cavity excavation in porous media such as deep clayey formations may involves strong coupling effects between hydraulical, thermal and mechanical behaviors. As it depends on the medium characteristics, two different ones are study and it can be yet interesting to study the effects of discontinuity between two different media: indeed, the excavation may be follow by a hydromechanical degradation of a small thickness of medium around the cavity.

As analytical solutions can only be obtained on problems characterized by simple geometries and boundary conditions, such ones have been developed to solve more complex and realistic problems because of the development of symbolic computation. They are very useful because they allowed the validation of numerical code and parametric studies can be easily performed thanks to speed calculation.

Semi-analytical methods have been developed by various authors under the hypotheses of linear behavior, using Fourier or Hankel transforms to reduce a two or three dimensional problem to a one dimensional problem (Booker & Small, 1987; Small & Booker, 1986. In this paper the porous rock mass is considered semi-infinite and only one spatial coordinate is necessary. Furthermore, the time-dependent problem is carried out by applying Laplace transforms and a numerical method, the Talbot's algorithm (Talbot 1979) modified by Piessens (Piessens et al. 1983) is then performed to invert the transformed solutions. Semi-analytic solutions have been yet performed to simulate the thermohydromechanical behavior of a multi-layered porous rock mass containing a waste disposal (Giraud & Thouvenin, 1997; Giraud et al, 1998).

2. GOVERNING EQUATIONS AND SOLUTION SCHEME

2.1. Governing equations

In a thermoporoelastic theory, neglecting heat convection, the three fully coupled linear field equations to solve can be written as follows (see Coussy 1995, index (i) denotes the medium):

Fluid diffusion equation:

$$\lambda_H^{(i)} \nabla^2 P^{(i)} = \frac{1}{M^{(i)}} \frac{\partial P^{(i)}}{\partial t} - 3\alpha_m^{(i)} \frac{\partial T^{(i)}}{\partial t} + b^{(i)} \frac{\partial \varepsilon^{(i)}}{\partial t} \tag{1}$$

Thermal diffusion equation:

$$\lambda_T^{(i)} \nabla^2 T^{(i)} = -3\alpha_m^{(i)} T_0 \frac{\partial P^{(i)}}{\partial t} + C^{(i)} \frac{\partial T^{(i)}}{\partial t} + 3\alpha_0^{(i)} K_0^{(i)} T_0 \frac{\partial \varepsilon^{(i)}}{\partial t} \tag{2}$$

Navier equation:

$$\left(\lambda_0^{(i)} + 2G^{(i)}\right)\nabla \varepsilon^{(i)} = b^{(i)} \nabla P^{(i)} + 3\alpha_0^{(i)} K_0^{(i)} \nabla T^{(i)} \tag{3}$$

Where $\varepsilon = \mathrm{Tr}(\varepsilon)$ denotes the volumetric strain, P is the excess pore pressure, T is the excess temperature. The linear isotropic thermoporoelastic model contains 9 independent coefficients. G is the shear modulus, K_0 is the bulk modulus of drained solid, b is the Biot dimensionless coefficient of effective stress, M is the Biot bulk modulus of the fluid, λ_0 is the drained Lamé coefficient, λ_H denotes the ratio between the permeability k_w (ms^{-1}) and the unit weight of the fluid γ_w ($\lambda_H = k_w/\gamma_w$), α_0 and α are respectively the drained and undrained expansion coefficients of matrix particles ($\alpha K = \alpha_0 K_0 + \alpha_m b M$, where $K = K_0 + b^2 M$ denotes the bulk modulus of the undrained solid), λ_T is thermal conductivity of the porous medium, C is the volumetric specific heat and T_0 is a reference temperature. Contrasts of all the thermohydromechanical properties have been taken into account.

2.2. Problem definition

We consider the one-dimensional problem (characterized by the radial coordinate denoted r) of a medium represented by a skin denoted medium (1) ($r_i \leq r \leq r_e$) surrounding a cylindrical cavity and the rock mass denoted medium (2) ($r_e \leq r \leq \infty$).

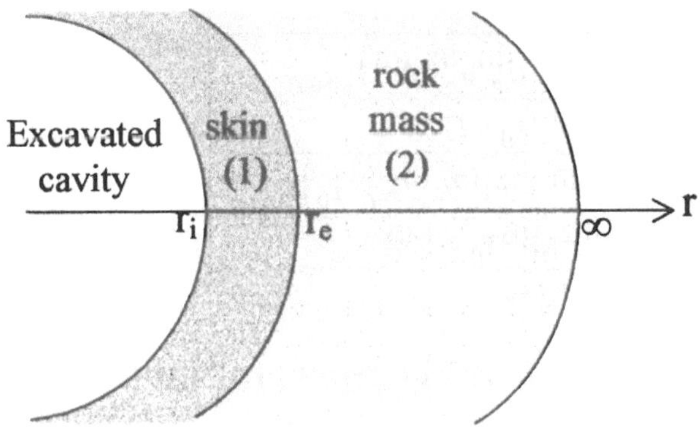

Fig. 1: Geometry

In the initial state, the rock mass is submitted to uniform hydrostatic stress, pore pressure and temperature fields respectively denoted σ_0, P_0 and T_0. And at time t = 0 the cylindrical cavity is instantaneously excavated.

The problem satisfies at t > 0 the following boundary conditions:

Internal boundary ($r = r_i$):

$$\sigma_{rr}(r = r_i, t) = -\varpi, \qquad P(r = r_i, t) = 0, \qquad T(r = r_i, t) = T_i \tag{4}$$

Infinite boundary ($r \to \infty$):

$$\sigma_{rr}(r \to \infty, t) = \sigma_0, \qquad P(r \to \infty, t) = P_0, \qquad T(r \to \infty, t) = T_0 \tag{5}$$

Continuities of displacement, pore pressure, temperature, fluid mass flux and heat flux are imposed at the interface between the layers ($r = r_e$).

2.3. Instantaneous response

The medium is instantaneously loaded on its internal boundary ($r = r_i$) by a mechanical pressure ϖ ($\sigma_{rr} = -\varpi < 0$). Its instantaneous response is undrained and adiabatic, so the fluid mass supply and the volume heat rate are zero. Classical elastic solutions with undrained parameters (λ, μ) gives the instantaneous response (results are shown in Table 1).

$\varepsilon^{0+(i)}$	$2\alpha^{(i)}$
$\sigma_{rr}^{0+(i)}$	$2\left(\lambda^{(i)}+G^{(i)}\right)\alpha^{(i)}-2G^{(i)}\dfrac{\beta^{(i)}}{r^2}$
$T^{0+(i)}$	$T_0+\dfrac{3T_0\left(\alpha_0^{(i)}K_0^{(i)}+\alpha_m^{(i)}b^{(i)}M^{(i)}\right)}{9\alpha_m^{(i)2}M^{(i)}T_0-C^{(i)}}\varepsilon^{0+(i)}$
$p^{0+(i)}$	$P_0+\dfrac{M^{(i)}\left(b^{(i)}C^{(i)}+9\alpha_0^{(i)}\alpha_m^{(i)}K_0^{(i)}T_0\right)}{9\alpha_m^{(i)2}M^{(i)}T_0-C^{(i)}}\varepsilon^{0+(i)}$

Table 1: Instantaneous response

where $\delta = 2G^{(1)}\left(\lambda^{(1)}+G^{(1)}+G^{(2)}\right)r_e^2-2\left(\lambda^{(1)}+G^{(1)}\right)\left(G^{(1)}-G^{(2)}\right)r_i^2$

$$\alpha^{(1)}=\frac{\left(G^{(1)}-G^{(2)}\right)r_i^2\varpi}{\delta},\beta^{(1)}=\frac{\left(\lambda^{(1)}+G^{(1)}+G^{(2)}\right)r_i^2r_e^2\varpi}{\delta},\alpha^{(2)}=0,\beta^{(2)}=\frac{\left(\lambda^{(1)}+2G^{(1)}\right)r_i^2r_e^2\varpi}{\delta}$$

If the shear modulus is the same in the "skin" and the rock mass, both coefficients $\alpha^{(1)}$ and $\alpha^{(2)}$ are equal to zero and the instantaneous response of volumetric strain and pore pressure do not change the initial state. At the opposite, a discontinuity of shear modulus between the two media yields a discontinuity of all the hydromechanical variables. It can be noticed that this discontinuity can be taken into account thanks to the semi-analytical approach whereas exact calculation seems to be impossible.

2.4. Solution scheme

The field equations to solve are constituted by two coupled diffusion equations and the Navier equation. As we consider one-dimensional problem, the displacement field is non-rotational, and besides, as the medium is semi-infinite, the Navier equation can be simply integrated as follows in one dimensional problems (f is an integration constant, function of time):

$$\left(\lambda_0^{(i)}+2G^{(i)}\right)\varepsilon^{(i)}-b^{(i)}\left(P^{(i)}-P_0\right)-3\alpha_0^{(i)}K_0^{(i)}\left(T^{(i)}-T_0\right)=f(t) \qquad (6)$$

After introducing the Navier equation into the diffusion equations, the system to solve is constituted by two coupled diffusion equations (coefficients of matrix $\delta^{(i)}$, $\chi_T^{(i)}$ and $\chi_P^{(i)}$ are given in appendix):

$$\frac{\partial}{\partial t}\begin{bmatrix} P^{(i)} \\ T^{(i)} \end{bmatrix} = \delta^{(i)}.\nabla^2 \begin{bmatrix} P^{(i)} \\ T^{(i)} \end{bmatrix} + \begin{bmatrix} \chi_P^{(i)} \\ \chi_T^{(i)} \end{bmatrix} \frac{\partial f^{(i)}}{\partial t} \tag{7}$$

Applying Laplace transform (a superior bar indicates that a Laplace transform has been applied to a field quantity):

$$\overline{X}(r,s) = \int_0^\infty X(x,t) e^{-st} dt \tag{8}$$

The system can be written:

$$\begin{bmatrix} \overline{P}^{(i)} \\ \overline{T}^{(i)} \end{bmatrix} = \frac{1}{s}\delta^{(i)}.\nabla^2 \begin{bmatrix} \overline{P}^{(i)} \\ \overline{T}^{(i)} \end{bmatrix} + \begin{bmatrix} \chi_P^{(i)} \\ \chi_T^{(i)} \end{bmatrix} \overline{f}^{(i)} + \frac{1}{s}\begin{bmatrix} P^{0+(i)} - \chi_P^{(i)} f^{0+(i)} \\ T^{0+(i)} - \chi_T^{(i)} f^{0+(i)} \end{bmatrix} \tag{9}$$

$P^{0+(i)}$, $T^{0+(i)}$ and $f^{0+(i)}$ represent respectively instantaneous response of pore pressure, temperature and integration constant for each medium (i) to instantaneous mechanical loading.

Introducing intermediate variables:

$$\overline{T}'^{(i)} = \overline{T}^{(i)} - \frac{T^{0+(i)}}{s} - \chi_T^{(i)}\left[\overline{f}^{(i)} - \frac{f^{0+(i)}}{s}\right] \quad \text{and} \quad \overline{P}'^{(i)} = \overline{P}^{(i)} - \frac{P^{0+(i)}}{s} - \chi_P^{(i)}\left[\overline{f}^{(i)} - \frac{f^{0+(i)}}{s}\right] \tag{10}$$

yields:

$$\begin{bmatrix} \overline{P}'^{(i)} \\ \overline{T}'^{(i)} \end{bmatrix} = \frac{1}{s}\delta^{(i)}\nabla^2 \begin{bmatrix} \overline{P}'^{(i)} \\ \overline{T}'^{(i)} \end{bmatrix} \tag{11}$$

and the solutions of the coupled system can then be expressed as a linear combination of two eigenfunctions $\overline{v_i}$:

$$\begin{bmatrix} \overline{P}'^{(i)} \\ \overline{T}'^{(i)} \end{bmatrix} = \theta^{(i)}.\begin{bmatrix} \overline{v_1}^{(i)} \\ \overline{v_2}^{(i)} \end{bmatrix} \qquad \text{where } \theta^{(i)} = \begin{bmatrix} \theta_{11}^{(i)} & \theta_{21}^{(i)} \\ \theta_{12}^{(i)} & \theta_{22}^{(i)} \end{bmatrix} \tag{12}$$

where column (k) of the matrix $\theta^{(i)}$ contains the two components of the eigenvector associated to the eigenvalue (k) of the matrix $\delta^{(i)}$. Eigenfunctions are solutions of a classical uncoupled diffusion problem:

$$\nabla^2 \overline{v}_j^{(i)} - \left(\omega_k^{(i)}\right)^2 \overline{v}_j^{(i)} = 0 \tag{13}$$

and can be expressed as:

$$\overline{v}_j^{(i)} = A_j^{(i)}(s) I_0\left(\omega_j^{(i)} r\right) + B_j^{(i)}(s) K_0\left(\omega_j^{(i)} r\right) \qquad \text{where} \quad \omega_j^{(i)} = \sqrt{\frac{s}{\delta_j^{(i)}}} \tag{14}$$

where $\delta_j^{(i)}$ is an eigenvalue of the matrix $\delta^{(i)}$, I_0 and K_0 represent modified Bessel functions of order zero.

Four constants $A_j^{(i)}(s)$, $B_j^{(i)}(s)$ and one integration constant $f^{(i)}(s)$ have to be determined for each medium by boundary and continuity conditions. In fact, a system of ten equations needs to be solved: exact calculations in Laplace transform space are performed using Mathematica, so analytic expressions can be obtained for the ten constants thanks to development of symbolic calculation. To inverse these expressions and return in time space, Talbot's algorithm modified by Piessens (Piessens et al. 1983) is used.

3. NUMERICAL RESULTS AND DISCUSSION

The parameters of the porous media have been chosen to represent an excavation of a deep tunnel ($r_i = 1$ m, $r_e = 2$ m) in clayey media. Two media are taken into account (table 2): the first one corresponds to a shale of the Paris sedimentary basin (ANDRA, 1996) and the second one is a plastic clay representative of Boom clay (Belgium, Giraud et al, 1998). The study is performed with the following constant properties: $\lambda_T = 1.7$ Wm^{-1}K^{-1}, $C = 2.85.10^6$ Jm^{-3}K^{-1}, $\alpha_0 = 10^{-5}$ K^{-1} and for each medium, we consider a degradation of the hydromechanical properties in the skin around the cavity: K_0 and G are 50% lower and λ_H is multiplied by a factor 10.

	Paris basin shale		Boom clay	
	Skin	Rock mass	Skin	Rock mass
Drained bulk modulus K_0 (MPa)	970	1940	50	100
Shear modulus G (MPa)	445	890	30	60
Biot coefficient b	0.82	0.64	1	1
Biot bulk modulus M (MPa)	4650	5500	5500	5500
Porosity ϕ	0.2	0.2	0.4	0.4
Undrained coeff. of thermal dilatation α (K^{-1})	$2.44.10^{-5}$	$1.95.10^{-5}$	$4.43.10^{-5}$	$3.54.10^{-5}$
Permeabilities k_w (ms^{-1})	4.10^{-11}	4.10^{-12}	4.10^{-11}	4.10^{-12}

Table 2: Thermohydromechanical parameters of the two materials.

All the results are presented as distributions of variables for different times, from 1 to 10^7 days with an increasing rate equal to 10. The arrow indicates their growing evolution.

The equation of heat diffusion can be uncoupled for problems involving low gradients of fluid flow in low permeability media (see Giraud et al., 1998). In this paper, permeabilities of all the layers are lower than 10^{-9} ms^{-1} and the figure 2 shows that coupling effects in heat equation are negligible. Besides, the evolution is similar for both considered media because the two materials have the same thermal properties.

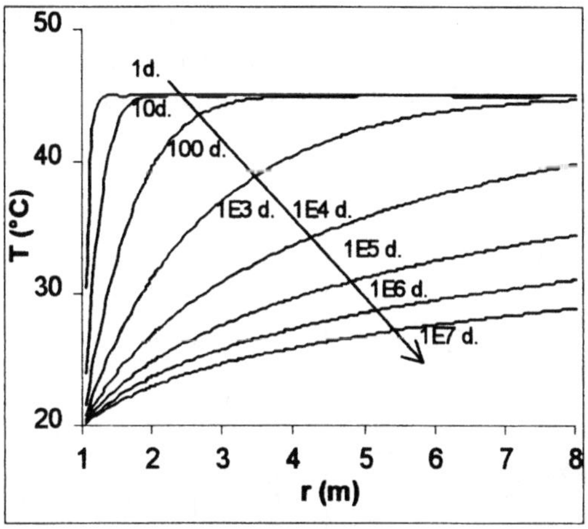

Fig.2: Temperature distribution

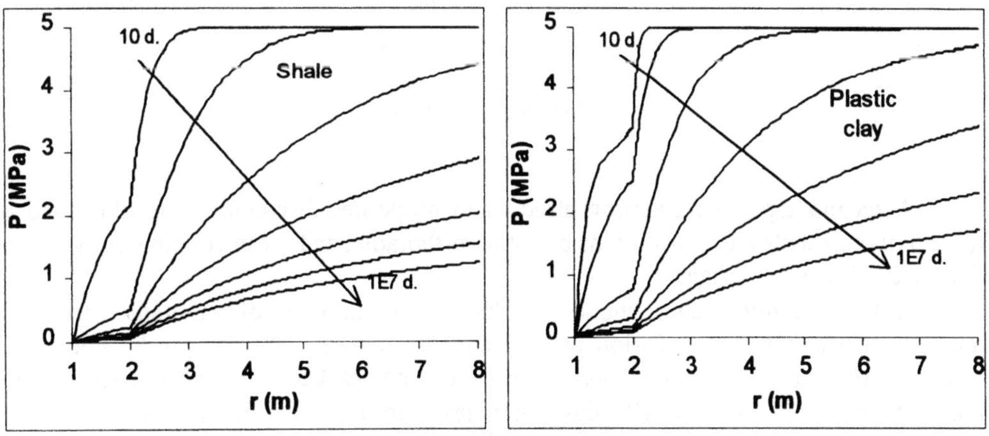

Fig.3: Pore pressure distribution

The influence of hydromechanical coupling effects is very significant on the distribution of pore pressure (figure 3). Indeed, the pore pressure dissipation is faster in the shale and the pressure reaches values close to zero in the skin nearly ten times sooner than in plastic clay. Besides, as would be expected, the heterogeneity of permeability yields a strong discontinuity of pore pressure whatever the material.

The kinetics of thermal and hydraulical diffusion processes is of the same order in the rock mass. This can be explained by similar values of characteristic diffusion times or diffusivities (for example eigenvalues of matrix diffusion δ for the plastic clay are C_1 = 5.96 10^{-7} m^2s^{-1} and C_2 = 3.54 10^{-7} m^2s^{-1}).

In order to illustrate the influence of coupling effects and hydromechanical properties discontinuity, we have plotted in figure 4 the displacement distribution for both materials. Negative values indicate a tensile phenomenon: the steady flow is reached nearly instantaneously in the skin whereas it increases with time in the rock mass. The evolution is the same in both materials, but due to the strong difference of shear modulus between them, displacement values are lower in the shale with a similar ratio.

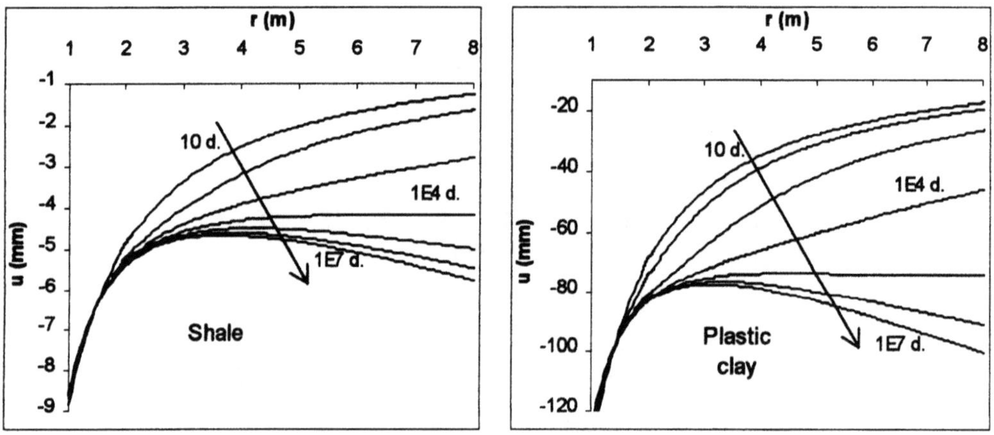

Fig.4: Displacement distribution

A second aspect of coupling effects and properties discontinuity is illustrated on figure 5: it represents the distributions of radial, orthoradial and mean effective stresses, and of the second stress invariant.

A discontinuity of shear modulus induces a discontinuity of orthoradial stress $\sigma_{\theta\theta}$. Coupling effects between diffusion process and mechanical behavior can be noticed on radial effective stress : a peak of pore pressure can be reached in the rock mass near the skin at times t = 1, 10 and 100 days. Besides significant discontinuities of total and effective stresses at the interface are not realistic compared, for example, to elastoplastic calculations. At last, a comparison with isothermal calculation shows that effective stresses are mainly affected by hydraulical loading (because of the "relatively" low thermal loading taken into account for the numerical application).

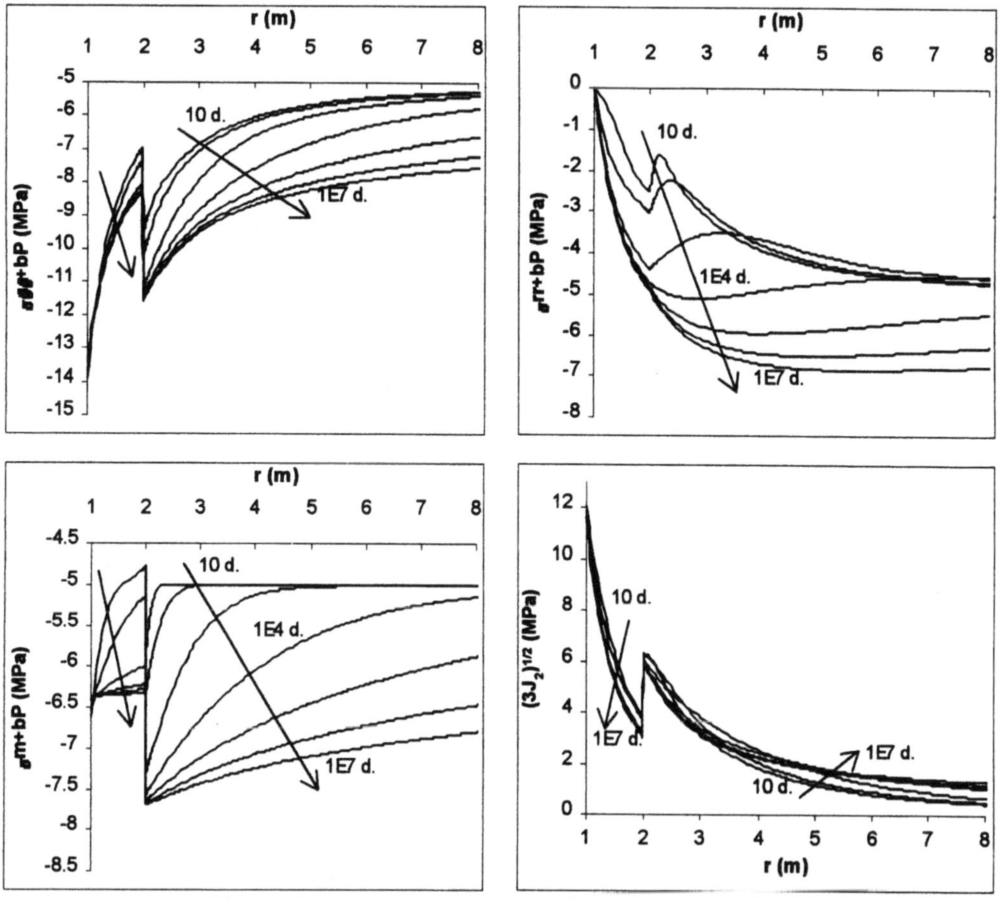

Fig.5: Effective stresses distributions for plastic clay

4. CONCLUSION

A simplified preliminary approach to model the degradation around a deep excavated cavity has been presented for two different materials using semi-analytic method. Even if the stresses distributions present discontinuities at the interface which are not realistic, this study gives order of magnitude of stresses, pore pressure and kinetics of diffusion processes, and allows the estimation of the importance of coupling effects. Besides we can insist on the importance of semi-analytical methods: they are very useful because they provide fast speed calculation, contrarily to finite elements method, and they give punctual calculation. They can then be used for repetitive calculation, like in parametric studies, or to validate numerical codes.

REFERENCES

ANDRA (1996). *Laboratoires souterrains d'études géologiques. Résultats des travaux de reconnaissance géologique*. Report DG/RP/96-01.

Booker, J. R. & Small J. C. (1987). A method of computing the consolidation behavior of layered soils using direct numerical inversion of Laplace transforms. *International Journal for Numerical and Analytical Methods in Geomechanics*, **11**, 363-380.

Coussy, O. (1995). *Mechanics of porous continua*.. John Wiley and Sons.

Giraud, A., Homand, F. & Rousset, G. (1998). Thermoelastic and thermoplastic response of a double-layer porous space containing a decaying heat source, *International Journal for Numerical and Analytical Methods in Geomechanics*, **22**, 133-149.

Giraud, A. & Thouvenin, G. (1997). Numerical simulation of a multi-layered porous rock mass containing a waste disposal, *Int. J. Rock Mech. & Min. Sci*, Vol. **34**, 3-4.

Piessens, R. et al (1983). *Quadpack :a subroutine package for automatic integration*, Springer-Verlag.

Small, J. C. & Booker, J. R. (1986). The behavior of layered soil or rock containing a decaying heat source. *International Journal for Numerical and Analytical Methods in Geomechanics*, **10**, 501-519.

Talbot, A. (1979). The accurate numerical inversion of Laplace transforms', *J. Inst. Maths Applics*, vol. **23**, 97-120.

APPENDIX

$$\delta^{(i)} = \frac{1}{\alpha_1^{(i)}\beta_2^{(i)} - \alpha_2^{(i)}\beta_1^{(i)}}\begin{bmatrix} \beta_2^{(i)} & -\alpha_2^{(i)} \\ -\beta_1^{(i)} & \alpha_1^{(i)} \end{bmatrix} \qquad \begin{bmatrix} \chi_P^{(i)} \\ \chi_T^{(i)} \end{bmatrix} = \frac{1}{\alpha_1^{(i)}\beta_2^{(i)} - \alpha_2^{(i)}\beta_1^{(i)}}\begin{bmatrix} \alpha_1^{(i)}\beta_3^{(i)} - \alpha_3^{(i)}\beta_1^{(i)} \\ \alpha_2^{(i)}\beta_3^{(i)} - \alpha_3^{(i)}\beta_2^{(i)} \end{bmatrix}$$

where $$\alpha_1^{(i)} = \frac{1}{\lambda_H^{(i)}}\left[\frac{1}{M^{(i)}} + \frac{b^{(i)^2}}{\lambda_0^{(i)} + 2\mu^{(i)}}\right], \qquad \beta_1^{(i)} = \frac{T_0}{\lambda_T^{(i)}}\left[\frac{3\alpha_0^{(i)}K_0^{(i)}b^{(i)}}{\lambda_0^{(i)} + 2\mu^{(i)}} - 3\alpha_m^{(i)}\right]$$

and $$\alpha_2^{(i)} = \frac{1}{\lambda_H^{(i)}}\left[\frac{3\alpha_0^{(i)}K_0^{(i)}b^{(i)}}{\lambda_0^{(i)} + 2\mu^{(i)}} - 3\alpha_m^{(i)}\right], \qquad \beta_2^{(i)} = \frac{1}{\lambda_T^{(i)}}\left[C + \frac{9\alpha_0^{(i)^2}K_0^{(i)^2}T_0}{\lambda_0^{(i)} + 2\mu^{(i)}}\right]$$

and $$\alpha_3^{(i)} = \frac{b^{(i)}}{\lambda_H^{(i)}\left(\lambda_0^{(i)} + 2\mu^{(i)}\right)}, \qquad \beta_3^{(i)} = \frac{3\alpha_0^{(i)}K_0^{(i)}T_0}{\lambda_T^{(i)}\left(\lambda_0^{(i)} + 2\mu^{(i)}\right)}$$

NUMERICAL ANALYSIS OF PRESSUREMETER TEST IN SOIL

R. Zentar, G. Moulin and P.Y. Hicher
École Centrale de Nantes, Nantes, France

ABSTRACT : In this paper, finite elements analysis is used to simulate pressuremeter tests. In order to discuss the effect of the finite length of the pressuremeter, the test is treated as a two dimensional axisymetric problem. The soil is modelled as an undrained elastic-perfectly plastic material, using linear isotropic elasticity and a Tresca yield criterion for plasticity.

1- INTRODUCTION :

The pressuremeter test is essentially an expansion test of a cylindrical cavity. The basic idea behind the pressuremeter test is the expansion of cylindrical cavity formed in the ground in order to measure a relationship between pressure and deformation for the soil (Louis Menard 1975).
The concept of pressuremeter testing has led to the development over the years of many different devices falling mainly into two categories :

- Those installed in pre-bored holes, the Menard type (MPM).
- Those forming their own hole with minimal disturbance, the self-boring types (SBP).

The expansion of a cylindrical cavity can be analysed rigorously by existing theories which can take even very complex soil properties into account.
In the analysis of a pressuremeter test it is usual to assume that the expansion process takes place under conditions of axial symmetry and of plane strain in the vertical direction. In

other words, the finite length of the pressuremeter is ignored, and the expansion curve is assumed to be the same as would occur for an infinitely long pressuremeter.
In this paper finite element analysis (FEA) is used to simulate pressuremeter tests.
In the first time, the expansion problem of the cylindrical cavity is studied as infinite cavity expansion. In other words with assuming that the expansion process takes place under conditions of axial symmetry and of plane strain in the vertical direction. The results of the finite element analysis is compared with the analytical solution for expansion of a cylindrical cavity developed by Monnet (1995).
In the second time, the expansion of pressuremeter of finite length was treated as a two-dimensional axisymmetric problem. Two discretisation are assumed in order to take in account various L/D ratios. The L/D ratio of self-boring pressuremeter is close to 2.3, those of the Menard pressuremeter is close to 6.6 (Waschkowski, 1976).
Throughout this paper the soil is modelled as an undrained (and therefore incompressible) elastic-perfectly plastic material, using linear isotropic elasticity and a Tresca yield criterion for plasticity. An associated flow rule is also assumed. The model is therefore completely characterised in terms of the values of G and c_u. The analysis is in terms of total stresses.

2- ANALYSIS OF THE PRESSUREMETER TEST :

The analysis of undrained expansion of cylindrical pressuremeter in clay is described by several authors (Bishop and al 1945, Menard 1957, Gibson and Anderson 1961, Salençon 1966, Vesic 1972, Prevost and Hoëg 1975, Monnet 1995). Only the important results of analytical solution developed by Monnet are reproduced here for completeness.
The pressuremeter is assumed to expand under conditions of axial symmetry and plane strain in the axial direction, so that all variables are functions of radius only, reducing the problem to one of a single dimension.
Monnet in his development take into account the vertical stress effect. He shows that two different plastic zones may develop around the pressuremeter. The first one is linked to the difference between the circumferencial stress and the radial stress. The second one is linked to the difference between vertical stress and the circumferencial stress. The relationship which distinguish between the two behaviour is function of the initial state stress and the undrained strength.
Monnet developed a relationship between stress and strain for each case.
In this study the initial state of stress and the undrained strength is chosen in order to develop only one zone as :

$$c_u > (1-K_0)\sigma_{v0} \qquad (1)$$

The relationship derived by Monnet for the expansion problem of cylindrical cavity in elastic-perfectly plastic medium, in the case of one plastic zone, is as follows :

$$\frac{u_a}{a} = \frac{c_u}{2\mu} e^{\left(\frac{\Delta P}{c_u} - 1\right)} \qquad (2)$$

Where :
- a : initial radius
- c_u : undrained strength
- ΔP : pressure increment ($P - P_0$)
- μ : shear Modulus

3- FINITE ELEMENT ANALYSIS :

The analysis of a pressuremeter test was studied by use of finite element program CESAR-LCPC developed at the LCPC (Laboratoire Centrale des Ponts et Chausses). Eight nodded quadrilateral elements were used, with nine Gauss points for integrating of the stiffness matrix. The non-linear cavity expansion problem was solved by use of the initial stress approach.

Soil behaviour around the probe was analysed using three different discretisations. In the first step, cavity expansion was modelled as an axisymmetric problem under plane strain condition in the vertical direction (model 1), using a mesh of 535 nodes and 144 eight-node isoparametric elements (fig1). This simulation will be compared with the analytical solution developed under the same hypothesis.

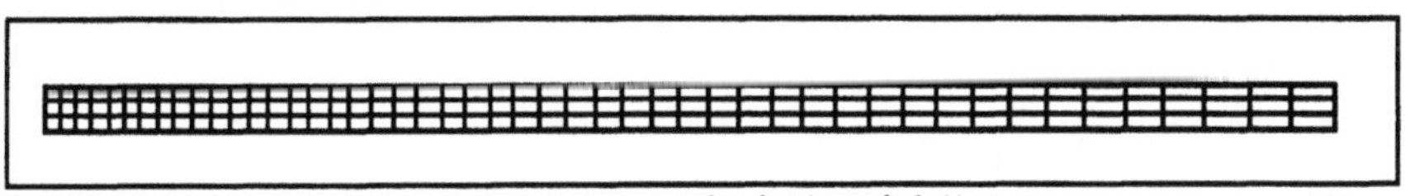

Fig. 1 - FE mesh for pressuremeter analysis (model 1).

In the second step, The analysis of a pressuremeter of finite length was treated as a two-dimensional axisymmetric problem. To take in account various L/D ratios, two meshes were used.

In the case of Menard pressuremeter (model 2), the mesh was composed of 6959 nodes and 2256 elements of the same kinds as before. The symmetry of the pressuremeter was exploited to reduce the size of the meshes for the analysis. A typical mesh is shown in figure 2. It consists of a rectangular grid of element, 14 in the radial direction and 13 in the axial direction. In the case of the self boring pressuremeter (model 3), the mesh was composed of 5645 nodes and 1824 elements of the same kinds as before. A typical mesh is shown in figure 3. It consists of a rectangular grid of element, 14 in the radial direction and 13 in the axial direction.

This discretisation neglects the finite length of the device below the membrane. The effect of the pressuremeter membrane is not take in account.
The loading was applied by specifying increments or normal (radial) pressure along the portion of the cylindrical cavity in contact with the pressuremeter membrane. Level of each increment is 1 kPa.
All tests are carried out with the initial state of stress of $\sigma_{v0} = 50$ kPa and $K_0 = 0.6$.
The elastic parameters for this study are chosen as G=3355.70 kPa (with E=10000 kPa and ν=0.49). The plastic one is c_u=30 kPa.

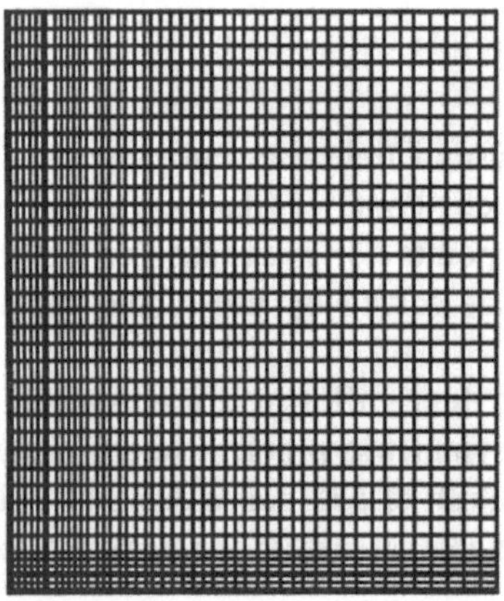

Fig 2 : FE mesh for pressuremeter analysis (model 2, L/D=2.3)

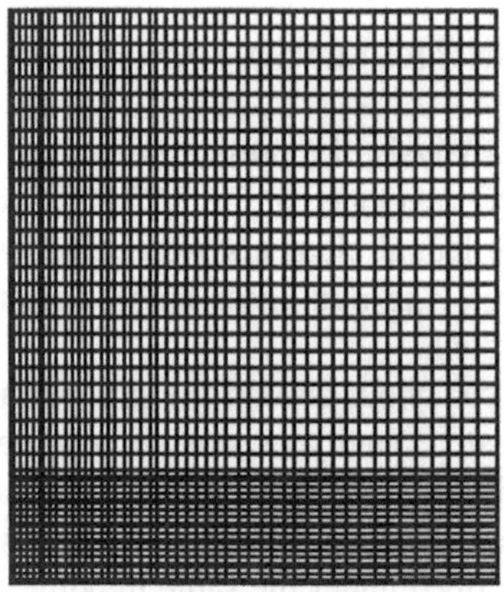

Fig 3 : FE mesh for pressuremeter analysis (model 2, L/D=6.6)

4- RESULTS AND INTERPRETATION :

Figure 4 shows a comparison of the results of the finite element analysis (model 1) with the analytical solution presented by Monnet (with G= 3355,7 and c_u =30 kPa).
In the first we can see the good agreement between the two response. The Curve from the finite element analysis is analysed as the result of real test with theory of cylindrical expansion cavity in infinite medium under plane strain condition.
The simulated tests were interpreted as follows :

The measured shear modulus was determined from the slope of the pressure-expansion curve up to the theoretical yield point $P=P_0+c_u$.

The measured undrained shear strength was found by fitting a straight line to the plot of $\Delta P/c_u$ against $\ln(\varepsilon=ua/a)$ for the section of the plastic expansion curve from $\varepsilon=c_u/2G$ to ε= 5% according to the relationship presented by Monnet.
The parameters derived from this analysis are postponed in table 1. It is to note that The shear modulus derived is less than the used one by a factor of 0.96. One part of this factor

is due to the geometrical model used (external to internal radius ratio) the other part is due to numerical calculating accuracy. The undrained strength is derived with an accuracy less than 2 %. However the difference are small and in the identifying problem this accuracy is sufficient and more accuracy is illusory.

parameters	reference set	from FEA	apparent to reference parameter
G	3355,70	3212,84	0,96
c_u	30	29,67	0,99

Table 1 : reference set and apparent parameters derived from FEA.

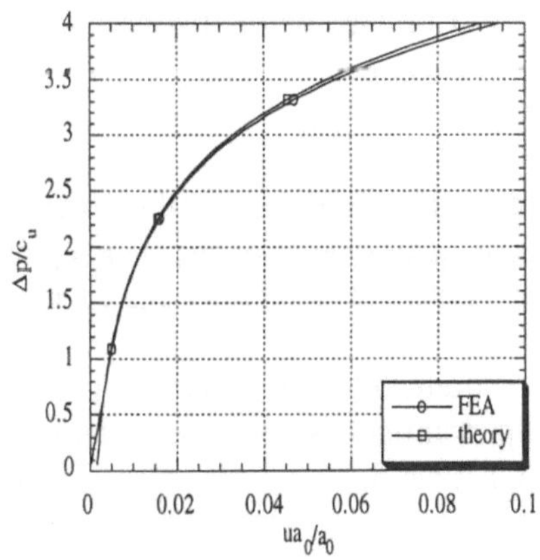

Fig. 4 -Comparison of pressuremeter curves.

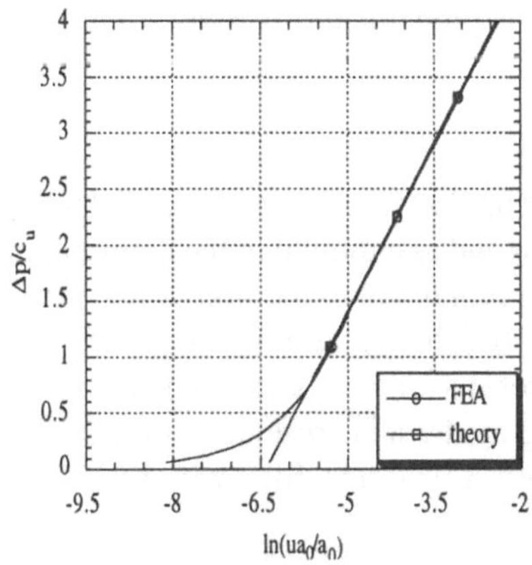

Fig. 5 -Comparison of pressuremeter curves : logarithmic time scale.

The finite element analysis of the pressuremeter tests as the bidimentional problem is analysed both as an expansion of the cavity in an elastic medium and in an elastic-perfectly plastic medium.

The result of numerical calculations for the expansion of cylindrical cavity with various L/D ratios in elastic medium is studied according to the radial and vertical strain induced around of the probe. The pressure increment applied is equal to 100 kPa.

Figure 6, where abscissa axis presents the radial (ua) and vertical (va) strain relating to the initial radius (a).The ordinate axis presents vertical distance (z) from the membrane centre to half-height of the probe (c), shows that the plane strain condition is not respected. The vertical strain, for z/c=1, represents 1/5 of the radial strain. The cylindrical expansion seems for the Menard pressuremeter (L/D=6,6) proved. For the self-boring pressuremeter (L/D=2,3) the cylindrical expansion seems proved only for 2/3 of probe height.

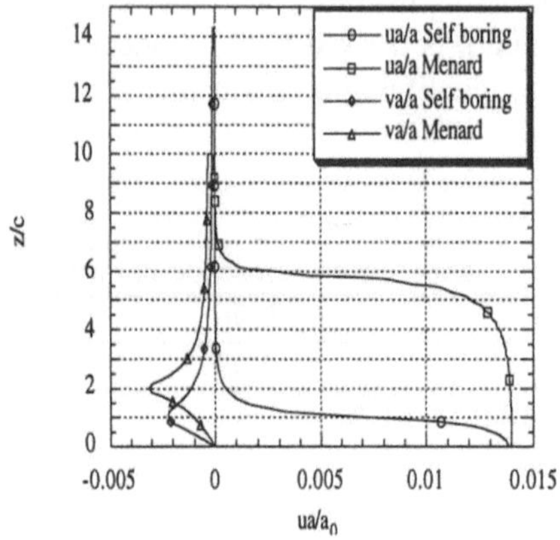

Fig. 6 - Comparison of strains around the probe.

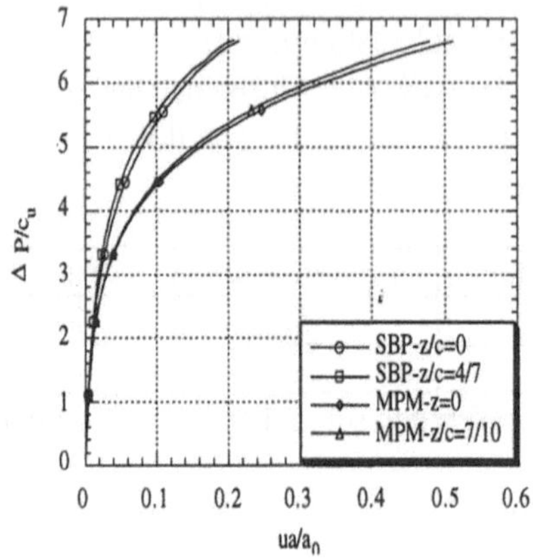

Fig. 7 - Comparison of pressuremeter curves.

Knowing that for real test the radial strain are calculated by measurements of the volumeter change and assuming that the deformation is cylindrical, an additional error can be integrated if this cylindrical expansion is not proved. The level of this error is directly linked to the strain distribution around the probe.

The results of the finite element analysis of the pressuremeter test in elastic-perfectly plastic medium is presented in figure 7. The relationship strain-stress are presented for two different L/D ratios. SBP with L/D=2,3 and MPM with L/D=6,6.

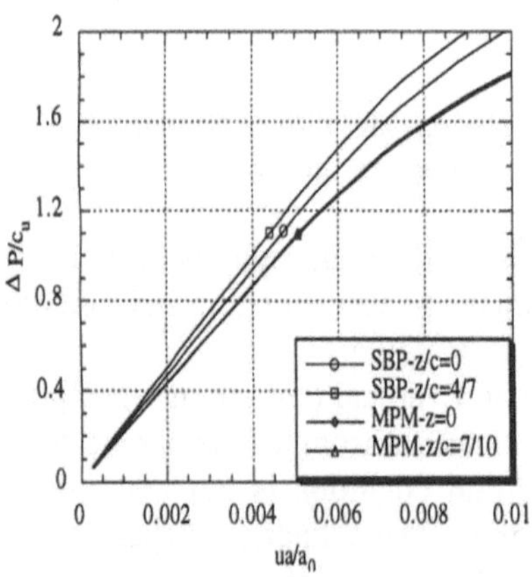

Fig. 8 - Zoom of elastic range.

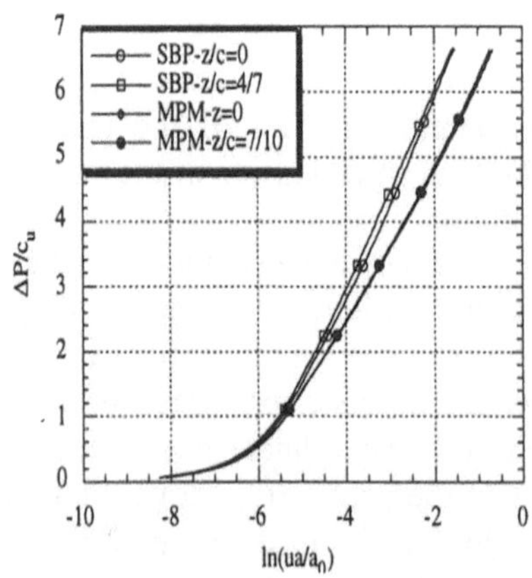

Fig. 9 - Comparison of pressuremeter curves : logarithmic time scale.

The response at two different levels are presented. At the centre of the probe (z/c=0) and at nearly 2/3 of probe height. It is to note that for this different level the curves are similar. Figure 7 shows that conventional limit pressure derived from Menard pressuremeter (for $\Delta V=V_0$) is reached for 20 % of radial strain in the case of the self-boring pressuremeter. Zoom of elastic range is presented in figure 8. For MPM the response for different level are similar due as explain before to the cylindrical expansion. The SBP shows stiffer response than the MPM. The shear modulus derived from this tests are postponed in table 2.

parameters	reference set	from FEA at centre	apparent to reference parameter
G	3355,70		
MPM		3191	0,95
SBP		3488	1,04
c_u	30		
MPM		32,88	1,096
SBP		38,16	1,272

Table 2 : reference set and apparent parameters derived from FEA in elastic-perfectly plastic medium.

The undrained shear strength was found by fitting a straight line to the plot of $\Delta P/c_u$ against $\ln(\varepsilon=ua/a)$ for the section of the plastic expansion curve as shown in figure 9.
It is to note that the undrained shear strength derived from the SBP is too higher by a factor of 1,272.

CONCLUSION :

The expansion problem of the cylindrical cavity is studied as infinite cavity expansion. The results of the finite element analysis is compared with the analytical solution for expansion of a cylindrical cavity developed by Monnet (1995). This comparison shows good agreement between the solutions.
In the second time, the expansion of pressuremeter of finite length was treated as a two-dimensional axisymmetric problem. Two discretisation are assumed in order to take in account various L/D ratios.
The test with the L/D ratio of self-boring pressuremeter close to 2.3, shows stiffer behaviour and the shear modulus derived is too higher by a factor of 1,04. The undrained shear strength is too higher by a factor of 1,272.
The test with the L/D ratio of 6,6 (MPM) shows that the cylindrical expansion is proved and the derived parameters are in more agreement with the reference set.

REFERENCES :

Bishop R.F., Hill R., and Mott N.F., (1945), « The theory of identation and hardness tests », Proc. Physical Society, 57, N° 321, pp. 147-159.
Gibson R.E. and Anderson W.F. (1961), « In-situ measurement of soil properties with the pressuremeter », Civil Enginneering and Public Works Review, Vol. 56, N°658, pp. 615-618.
Menard L. (1957), « Mesure in situ des propriétés physiques des sols », Annales des P. et C., N°3, Mai-Juin, pp. 356-377.
Monnet J. (1995), « Analyse théorique et expérimentale de l'équilibre élasto-plastique d'un sol cohérent autour du pressiomètre », The pressuremeter and its new avenues, Balkema, Rotterdam, pp 193-200.
Salencon J. (1966), « Expansion quasi statique d'une cavité à symétrie sphérique ou cylindrique dans un milieu élastoplastique », Annales des Ponts et Chaussées III.
Prevost J.H. and Hoëg K. (1975), « Analysis of pressuremeter in strain-softening soil », Journal of the Geotechnical Engineering Division ASCE, N° GT 8, pp 717-732.
Vesic A. (1972), « Expansion of cavities in an infinite soil mass », Jour. of Soil Mech. and Found. Div., ASCE, Vol. 98, N° SM 3, March, pp. 265-290.
Waschkowski E. (1976), « Contribution à l'étude du comportement du sol autour d'une sonde pressiométrique et d'un pieu chargé verticalement », Mémoir de fin d'étude, CNAM.

4. MISCELLANEOUS

SOME EXPERIENCES OF MODELLING TUNNELLING IN SOFT GROUND USING THREE-DIMENSIONAL FINITE ELEMENTS

C.E. Augarde, H.J. Burd and G.T. Houlsby
University of Oxford, Oxford, UK

ABSTRACT: A three-dimensional finite element model has been developed at Oxford University to study the effects of subsidence from soft ground tunnelling on adjacent surface structures. Simulation of excavation and the ground loss associated with tunnelling are incorporated in the model. Surface buildings are also included, as groups of interconnected two-dimensional façades composed of an elastic no tension material, to model masonry. This paper describes the development, implementation and performance of procedures to model the tunnelling processes. A description is also given of the methods used to generate the finite element meshes and to post-process the data.

1 INTRODUCTION

The construction of a tunnel in soft ground usually leads to subsidence of the ground surface. The size and shape of the settlement trough at a *greenfield* site may be estimated by a well-tested semi-empirical approach, described in many references (e.g. Mair *et al.*, 1996). The presence of a surface structure usually changes the settlement profile, due to the interaction between the ground and the building. If differential settlements are significant they may damage the building. It is important, therefore, that reliable methods for settlement and damage prediction are available if urban tunnelling schemes are to be successfully promoted.

Numerical methods, principally the finite element method, have been applied for some years to this problem, although most studies have been two-dimensional and have not included a realistic model of a building (e.g. Potts & Addenbrooke, 1997). It is thought that realistic modelling of this problem can be achieved only with three-dimensional models. In order to respond to this need, research has been underway at Oxford University since 1993 to develop a three-dimensional numerical model for the prediction of the effects of soft ground tunnelling on surface structures.

A paper presented at the last in this series of conferences described the proposed composition of a three-dimensional finite element model of tunnelling (Burd *et al.*, 1994). The model has now been implemented and is described in detail in Augarde (1997) and Liu (1997). Some initial results using the model are given in Burd *et al.* (1998). This paper describes some of the challenges met during its development and subsequent use and discusses some of the issues that face developers of complex numerical models of this type. The paper is concerned particularly with the procedures adopted to model tunnel installation.

2 DESCRIPTION OF THE NUMERICAL MODEL

The numerical model uses finite elements to represent the ground, a tunnel and a surface structure. A typical problem is shown in Figure 1, where a straight circular tunnel of 5m diameter, with its axis at a depth of 10m, is constructed beneath four connected façades of a large masonry building of plan dimensions 10m by 20m. The building is unsymmetrical with respect to the tunnel centreline. The tunnel is assumed to be installed with a ground loss of 2%.

A non-linear, elasto-plastic material formulation is used to model an overconsolidated clay deposit with increasing stiffness and strength with depth. The building is constructed from masonry, modelled as an elastic no tension material. Tunnel linings are linear elastic. Tunnel installation is simulated in the model in discrete stages, each of which consists of the simultaneous removal of soil and activation of shell elements to simulate the liner. Four stages of tunnel installation are used in the analysis shown in Figure 1. Elements that are to be removed in the current stage are ignored when the structure stiffness is formed. Nodal loads are imposed on elements surrounding the excavated block to render the exposed faces free of surface tractions (Brown & Booker, 1985). The calculation of these loads is described in Augarde *et al.* (1995).

An important requirement of a finite element model of tunnelling is the ability to model ground loss. Ground loss arises in practice from two sources; radial movement of the soil around the tunnel liner (tail loss) and inward movement of the soil at the tunnel heading (face loss). The amount of ground loss that occurs in practice is determined mainly by the installation method and the quality of the construction procedures. In this model, the excavated face of the tunnel is unsupported, thus permitting movement to represent face loss. Tail loss is simulated by uniform shrinkage of the lining elements in a plane normal to the tunnelling centreline. This occurs at the same time as excavation and lining activation.

While large masonry buildings are three-dimensional structures, their most significant feature, with respect to settlement damage, is the in-plane response of the main façades. The façades are represented by meshes of six-noded triangular plane stress elements, each

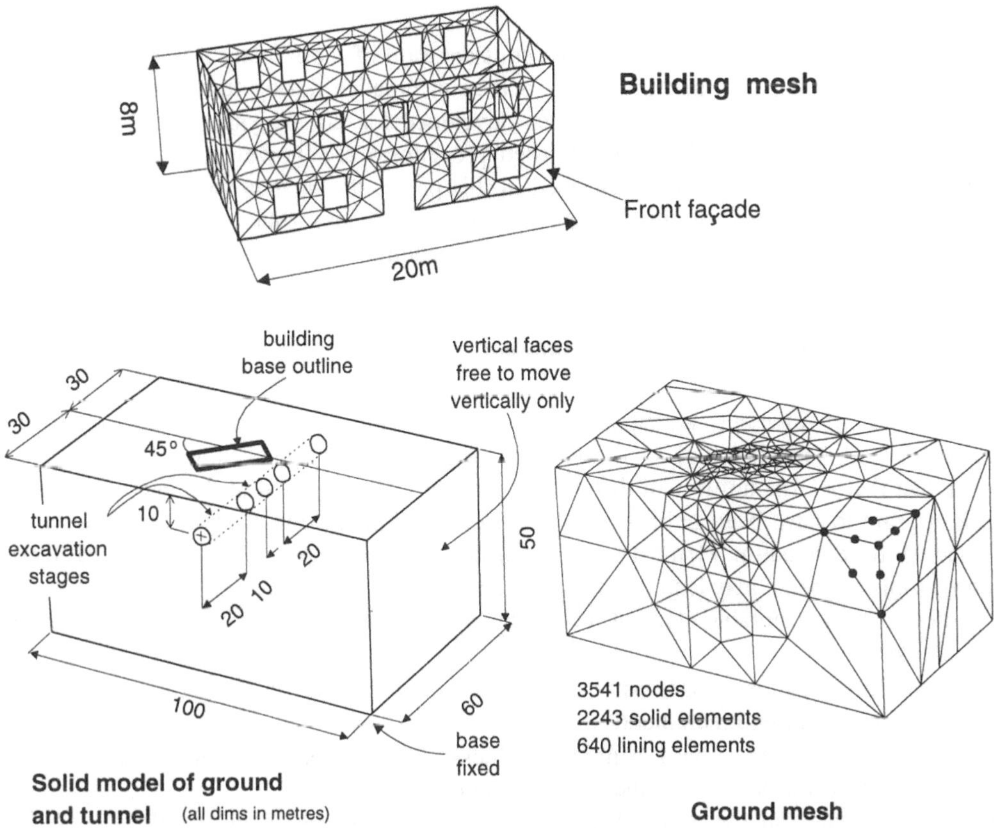

Fig. 1; Example problem

façade having its own local co-ordinate system. The façade meshes are joined to each other, and to the surface of the ground mesh, using a novel system of tie elements that implement the kinematic constraints linking each local two-dimensional co-ordinate system with the three-dimensional global system (Liu, 1997).

The various components of the model described above have been implemented in OXFEM, which is a finite element program written in Fortran 90 and developed at Oxford University. Non-linearity is accommodated in an incremental scheme, the Modified Euler method. Loads are applied in steps, followed by an equilibrium check to determine the out-of-balance force at each nodal degree-of-freedom. This out-of-balance force is then corrected in the next step. Incremental techniques such as this require the user to specify the numbers of steps, a decision that requires some experience of the non-linearities involved.

3 NUMERICAL MODEL OF TUNNEL INSTALLATION

A numerical model of tunnelling requires consideration of the structural behaviour of the liner, and the ground loss that invariably occurs in practice during tunnel installation. A variety of modelling procedures have been developed for 2D analysis (e.g. Rowe, 1983; Potts & Addenbrooke, 1997); no established procedures are, however, available for 3D analysis. In the numerical model described in this paper, the liner is modelled using shell elements, and the effect of one component of ground loss (tail loss) is included by numerical shrinking of the liner elements. These procedures are discussed below.

3.1 The choice of lining element formulation

The choices of element formulations for the ground and building are relatively straightforward. The choice of a suitable shell element for the tunnel lining is complicated by the fact that a large number of different approaches are proposed in the literature. Little work appears to have been published, however, on the behaviour of shell elements when connected to continuum elements. It is possible, for example, that the use of these incompatible elements may lead to numerical problems.

In determining a suitable element a choice has to be made from the two main classes of shell element:

- true curved elements based on classical shell theory or derived from the degeneration of a solid continuum element,
- faceted elements, where bending stiffness is attributed to a plate element and membrane stiffness to a plane stress, continuum element.

Many formulations in the first category have been proposed. Yang *et al.* (1990), for example, reviewed over 280 publications relating mainly to curved shell elements. Degenerated solid shell elements appear to be more commonly used than those based directly on shell theory, and are available in various commercial packages. Faceted formulations are usually less complicated both to code and to operate, since membrane and bending effects are uncoupled. This has the advantage that separate formulations for bending and membrane actions can be adopted (Hughes *et al.*, 1995).

The element used in this project was a faceted shell with displacement, but no rotational, degrees of freedom. The formulation was developed by Phaal & Calladine (1992) and is novel in that bending stiffness is provided by a six-noded plate element having four flat triangular facets while membrane stiffness is provided by three-noded triangular elements which overlay each facet. Thin plate elements must fulfil the Poisson-Kirchoff requirement (i.e. continuity of displacement gradient) and this is met in conventional plate formulations by the inclusion of rotational degrees of freedom at nodes, in addition to the translational degrees of freedom. In Phaal & Calladine's shell element, the facets of adjacent plate component elements overlap to achieve this continuity without the need for rotational degrees of freedom (Figure 2). This unconventional feature leads to many difficulties in the inclusion of this element in a standard finite element program.

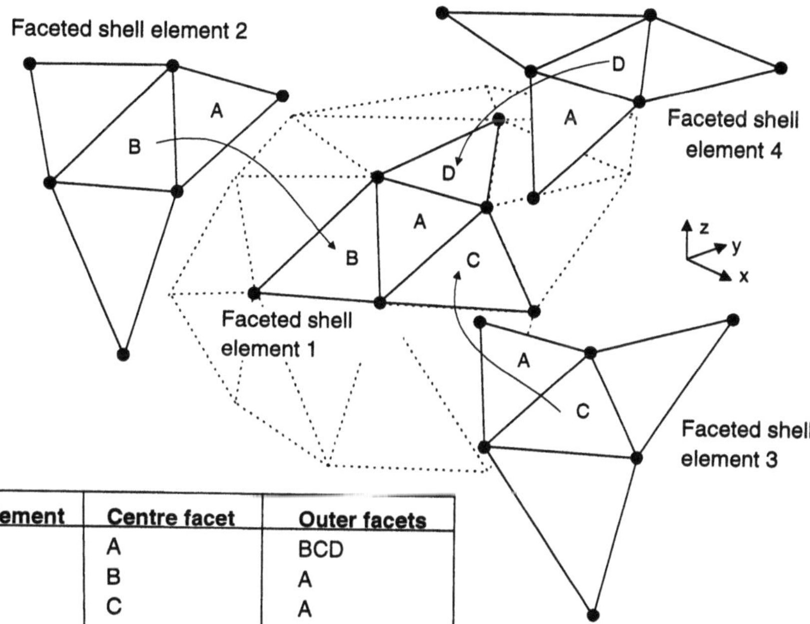

Element	Centre facet	Outer facets
1	A	BCD
2	B	A
3	C	A
4	D	A

Fig. 2; Overlapping shell elements

This element was chosen because it appeared to offer reasonable behaviour and accuracy without the need for rotational degrees of freedom at the element nodes. The lack of rotational degrees of freedom was thought to have three advantages:

- The number of global degrees of freedom in the mesh is less than that required for conventional shell formulations,
- all nodes in the model all have the same number of degrees of freedom. This makes the element more convenient to implement.
- There is no need to consider the in-plane rotational degree of freedom (the drilling degree-of-freedom).

The Phaal & Calladine (1992) formulation is, at first sight, highly complex. The formulation is presented in a general form, and further work is needed to cast it in a form that may be implemented in a finite element program. The original formulation was described in the context of the analysis of linear shell problems. The model described in this paper is non-linear, however, and this required some minor further development of the shell element formulation.

The linear elastic material model for the tunnel lining requires three properties to be specified: Young's modulus, E, Poisson's ratio, ν, and thickness, t. These parameters are used to derive the flexural stiffness of the plate component of the shell, D where

$$D = \frac{Et^3}{12(1-\nu^2)} \tag{1}$$

and the membrane stiffness component as Et (per unit length of lining). The elements performed reasonably well when used to model tunnel linings. There are, however some unresolved difficulties using the elements in combination with the non-linear elasto-plastic material model for the ground. It is occasionally found that the analysis fails to converge with the out-of-balance force at the equilibrium check (see Section 2) increasing with each step. The problem may be avoided by increasing the flexural stiffness, D, but this leads to a tunnel liner that is substantially stiffer than would be expected in practice. While the need to impose unrealistic properties on the liner is unfortunate, since it is difficult then to interpret the stress resultants, it does not unduly affect the surface response of the model, which is of primary interest in this study.

3.2 Modelling ground loss in 3D

Procedures for modelling ground loss in 2D finite element analysis are well established. Potts & Addenbrooke (1997), for example, use a technique in which the unloading caused by soil removal is carried out in a series of small increments. At the end of each increment the amount of ground loss is recorded; when the ground loss reaches the specified value then the calculation is terminated. A more complex procedure is described by Rowe (1983). In this procedure separate meshes of beam elements to model the lining and continuum elements to model the soil are used. The soil mesh contains a zone of elements that are removed to simulate excavation and the lining elements are activated to simulate the installation of the lining. The mesh of lining elements is separated from the surrounding ground by a specified gap. After the soil has been excavated and the liner elements activated, the ground closes onto the lining mesh as the soil around the tunnel unloads. Numerical procedures are used to model the contact between the soil and liner after the gap between the two meshes closes.

The above methods would be complex to implement in a 3D analysis; this is particularly the case for the gap parameter approach of Rowe (1983). A third approach was therefore selected in which the elements within the tunnel are removed (to model excavation) and, simultaneously, lining elements are activated to model the liner. At the end of this procedure the lining elements are subjected to a uniform hoop shrinkage to develop the required amount of ground loss. The hoop shrinkage is achieved by the application of a suitable set of radial forces within the tunnel liner. The process leads to fictitious stresses within the liner, but the liner is elastic and so this does not affect the way in which the ground and the liner interact. An important feature of this finite element model is that the tunnel is installed in separate stages. During any stage of tunnel installation (except the first one) it is important to ensure that the applied numerical shrinkage does not cause additional shrinkage in the liner installed during the previous stage. This was achieved by constraining to zero the shrinkage deformation applied at the tunnel heading (except for the final

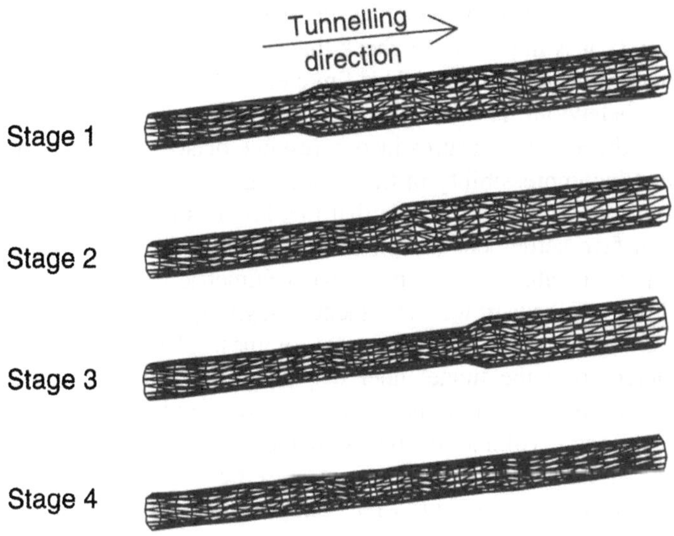

Fig. 3; Lining shrinkage for each tunnelling stage

tunnelling stage). The nodal forces associated with these constraints were then removed during the shrinkage phase of the next tunnelling stage. This was a somewhat artificial, but necessary, device to achieve satisfactory modelling of volume loss for incremental tunnel construction in a 3D model.

Figure 3 shows the lining elements at each stage of the analysis associated with the mesh in Figure 1. The shrinkage at each stage, which is displayed to an exaggerated scale, follows that specified by the user.

4 PERFORMING ANALYSES WITH THE MODEL

4.1 Preparation of input data and interpretation of output

The meshes used with the numerical model are unstructured, to permit localised refinement around the tunnel and below the building. This feature precludes hand generation, and an automated method must be used. Unstructured mesh generation has been the subject of considerable research in other areas, notably computational fluid dynamics, and guidance is available from the literature in this area. It is recommended, however, that a commercial package is used, since the effort involved in writing a robust generator for three-dimensional unstructured meshes is considerable. The package used in this research is I-DEAS, primarily a solid modelling package but with excellent mesh generation tools.

It is vital that the output from the analysis is checked carefully. This is a complex task, however, because of the large amount of computed data. Apart from checking that the nodal forces are in equilibrium, which is an integral part of the solution phase of the analysis (Section 2), it is important to check that the tunnel installation procedures produce the specified amount of ground loss. This finite element model is based on the assumption of undrained soil behaviour, and therefore that the soil is incompressible. Although the soil model adopted in the analysis ensures incompressible plastic deformation, it is not possible to specify perfect incompressibility in the elastic region. (In these calculations a Poisson's ratio of 0.49 was adopted). It is possible that this loss of incompressibility may lead to the volume of the surface settlement trough being less than the total volume of ground loss generated by tunnel installation. To check that volumetric deformation of the soil did not reduce the volume of the surface settlement trough by an appreciable amount, careful comparisons were made between these two volumes. The ground loss achieved in the tunnel is computed from the tunnel liner displacements. The procedure is based on the computation of the volume of a mesh of tetrahedra within the tunnel formed by 3D triangulation. The total volume of the settlement trough at the ground surface was computed by numerical integration of the surface settlements. It was found that the ground loss could be controlled to within 0.1%, and that the volume of the surface trough correlated very closely with the ground loss.

The final stage of an analysis is the post-processing of output data into a useful format. The large volume of data precludes any hand manipulation and the options are either in-house or commercially available software. In this research, use is made of both approaches with I-DEAS and an in-house program, 2CAN, used for generation of contours of surface settlements. 2CAN is also used to process and display the cracking of building façades. This is a feature not available in I-DEAS, and was one of the specific post-processing tools to be written for this project. Figure 4 shows the predicted crack pattern for each tunnel stage for the front façade of the building in the analysis of Figure 1. The intensity and inclination of cracks in the masonry model can be obtained from the OXFEM output. These data are displayed as collections of lines, at the given inclination, centred on the element integration point. The number of lines indicates the magnitude of the crack strain. (Crack strain is closely associated with principal strain normal to the crack and provides a convenient measure of the intensity of cracking, Liu, [1997]). This display gives an immediate indication of the damaged areas of the building façades.

Figure 4 indicates that the location of the most severe damage to the front façade changes as the tunnel approaches and passes beneath the building. At stage 1, the tunnel heading is closest to the large door opening in the façade, and transverse settlements cause cracking adjacent to and above this region. There is little effect in the façade directly ahead of the tunnel at this stage. The greatest damage is predicted at the end of stage 2, when the tunnel heading is situated between the front and side façades. At this point, the building is subjected to the combined effects of differential settlements along the tunnel axis and in a transverse direction. Much of the cracking seen at stage 1 has closed up by stage 2 due to the changing settlement profile along the base of the façade. Stages 3 and 4 are similar since the tunnel heading has progressed beyond the building. The differential settlements along the tunnel axis have reduced, leaving slight damage from transverse differential settlements.

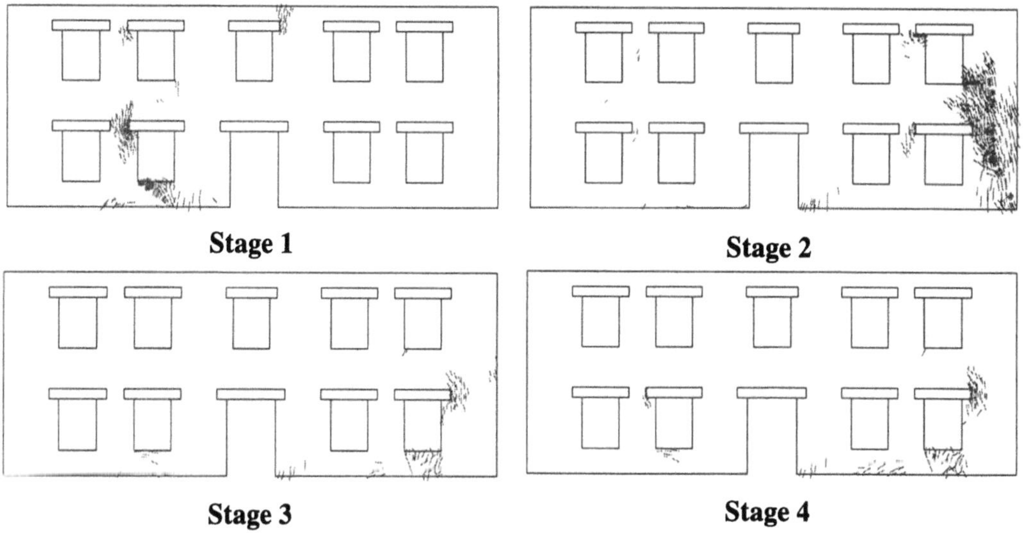

Fig. 4; Predicted crack patterns in the front façade of the building analysed using the model shown in Figure 1

5 ANALYSIS COSTS

Typical hardware used to run analyses with the model described in this paper is currently a Sun Microsystems UltraSparc 2 machine with processors running at 200MHz and 256Mb of RAM. Runs of the analysis shown in Figure 1 with this hardware are completed in approximately five days. During solution, virtual memory use rises to approximately 250Mb, which can be accommodated entirely within the RAM, provided other processes are small or prevented from running via a queue system. While this level of hardware configuration is of a high specification, it is not excessively expensive, considering the value of the information it is able to provide. Equally, while runtimes are long they are not excessively so, considering the complexity of the model.

6 FUTURE DEVELOPMENTS

Further work is underway to increase the scope of the model and to improve its efficiency. In particular, the modelling of compensation grouting, using interface finite elements, is under development. It is also intended to develop an effective stress model for soil in order to study the effect of consolidation settlements. The behaviour of the shell elements in the model has prompted an investigation of other methods of modelling volume loss. It appears possible to use thin continuum elements with high stiffness for the lining, which should remove any difficulties associated with element non-compatibility.

A parallel project began in 1996 to compare predictions from the model with field data from tunnelling schemes in London. This, and further studies, should serve to validate the model and its components and indicate areas for improvement. Work has recently begun on porting the code to run on OSCAR, the Oxford Supercomputer where significant reductions in solution times are anticipated.

7 ACKNOWLEDGEMENTS

Financial support for the development of the numerical model, between 1994-7, was provided by EPSRC.

REFERENCES

Augarde, C.E. (1997). *Numerical Modelling of tunnelling processes for assessment of damage to structures.* D.Phil. Thesis, University of Oxford.

Augarde, C.E., H.J.Burd & G.T.Houlsby (1995). A three-dimensional finite element model of tunnelling. *Numerical models in Geomechanics, Proc. NUMOG V,* 457-462.

Brown, P. and J.Booker (1985). Finite element analysis of excavation. *Computers and Geotechnics* 1(3):207-220.

Burd, H.J., G.T.Houlsby, L.Chow, C.E.Augarde & G.Liu (1994). Analysis of settlement damage to masonry structures. *Proc. 3rd European Conf. Numerical Methods in Geotechnical Engineering* (ed. I.M. Smith), 203-208.

Burd, H.J., G.T.Houlsby, C.E.Augarde & G.Liu (1998). Prediction of tunnel-induced settlement damage to masonry structures. Oxford University Dept of Engineering Science Report OUEL 2162/98.

Hughes, T., A.Masud & I.Harari (1995). Numerical assessment of some membrane elements with drilling degrees of freedom. *Computers and Structures* 55: 297-314.

Liu, G. (1997). *Numerical modelling of damage to masonry buildings due to tunnelling.* D.Phil. Thesis, University of Oxford.

Mair, R.J., R.N.Taylor & J.B.Burland (1996). Prediction of ground movements and assessment of risk of building damage due to bored tunnelling, in *Geotechnical aspects of underground construction in soft ground* (ed. R.N.Taylor), 713-718.

Phaal, R. & C.R.Calladine (1992). A simple class of finite elements for plate and shell problems. II: An element for thin shells, with only translational degrees of freedom. *Int. J. Num. Meth. Eng.* 35: 979-996.

Potts, D.M. & T.I.Addenbrooke (1997). A structure's influence on tunnelling-induced ground movements. *Proc. ICE, Geotech. Engng*, 125:109-125.

Rowe, R.K., K.Y.Lo & G.J.Kack (1983). A method of estimating surface settlement above tunnels constructed in soft ground. *Can. Geotech. J.*, 20:11-22.

Yang, H., S. Saigal & D.Liaw (1990). Advances of thin shell finite elements and some applications – version 1. *Computers and Structures* 35(4):481-504.

NUMERICAL MODELLING OF ROCK MASS STRESS STRAIN CHANGES CAUSED BY UNDERGROUND EXCAVATION IN THE BOR COPPER MINE

D. Divac and D. Vuckovic
Institute for the Development of Water Resources "Jaroslav Cerni", Beograd, Yugoslavia

M. Zivkovic and N. Grujovic
Faculty of Mechanical Engineering, Kragujevac, Yugoslavia

ABSTRACT

Extensive investigations have brought up the fact that there are still considerable amounts of ore reserves of copper mine "Bor" (at Yugoslavia), located at the slope and under the abandoned open pit mine. This paper presents the methodology of modeling the process of inducing mass deficit, the process of caving and corresponding stress-strain changes in the surrounding rock mass. The rock mass of the investigated area is composed of: extrusive magmatic rocks, pyroclastites, volcanic sediment rocks (the Bor pelites) and conglomerates. The problem of determining parameters of a rock mass constitutive relations in a large scale has been analyzed. The finite element method (FEM) is adopted as a numerical procedure. The constitutive relations of the rock strata are modeled as elastic and plastic-brittle behaviour. The effects of strain softening around the ground openings are taken into account. Key problems of the model are caving moment criterion and size of the caved zone. Assuming that physical instability coincides with numerical divergence, the criterion of the moment of caving has been introduced as initial divergence of the solution of the stress-strain problem. The results show the development process of the zones of instability as a function of the progress of the exploitation. The proposed numerical procedure combined with in-situ investigation works has many advantages over the usual empirical procedures, and can be a valuable tool to mining engineers assessing the short and long term stability of rock mass caused by underground excavition at open pit slope.

1. INTRODUCTION

Abundant ore resources of the Bor copper mine are located in the slope under the toe of the abandoned open pit mine bluff. The abandoned open pit mine is situated in the very city of Bor. A road, a hydrotransportion pipeline, industrial and housing facilities, etc., are in its immediate vicinity. The open pit exploitation of the ore-bearing rock is not rational due to the need for extensive surface excavation. The caving method to be applied to the ore body together with the upper rock mass (hard stiff andesites) implies the rock mass undercutting and falling. The cave excavation under so complicated conditions induces significant stress-strain changes in the rock mass. The complexity of the problem is enhanced by the possibility of a large potential energy accumulation within the rock mass and its sudden release - the phenomenon known as "rock burst". On the other hand, the rock mass caving may cause the open pit uncontrolled slope slides which might endanger the structures built in the surface area.

The paper presents the methodology of modeling the rock mass deficiency induction and caving processes, and the corresponding stress-strain changes in the surrounding rock mass due to the neglected underground excavation. The model development aims were:

- to determine the terrain surface deformations for the improved exploitation;
- to determine the unstable zones within the rock mass for the proper location of the capital storage;
- to analyze the possibility of a rock burst.

The whole process of inducing the rock mass deficiency by mining works, and the stress-strain response of the surrounding rock medium provoking the instability, were modeled using the data on real rock mass physical-mechanical properties.

2. THE TILVA ROŠ ORE BODY GEOLOGICAL COMPOSITION

The terrain (-150 a.s.l.) is composed of the Cretaceous rock mass and only partly covered by artificial materials (embankment - disposal site). The oldest creations are represented by extrusive magmatic rocks, originating from the Turon Period (Volcanic Phase I) when the hornblende stockwork composed of hornblende andesite and hornblende biotite-andesite (tymocite) was formed. During the Volcanic Phase I period up to the Volcanic Phase II, explosive volcanic eruptions of the viscous lava were taking place and forming pyroclastitic rocks, most often englomerates and breccia. That caused deep hydrotermal changes in the existing andesite rock mass and the formation of various forms. The volcanic processes were completed in the period of Senon, while the creations of the Phase II are now known as the Bor Pelites. Marine and submarine sediments, marl and slate were formed during the settling phases. The Bor Conglomerates deposited during the Mastricht-Damme period. The Bor Fault dividing volcanites form conglomerates is characteristic for the investigated area. The elements of the slope are approximately 220/80, while the original route was passing through the central part of the investigated area. Now, therefore, due to the W-E and NW-SE directions lateral faulting, the Bor Fault route does not spread continuously, but is diverted eastwards for about 500-600 m. The fault zone is 5-10 m wide and backfield with tectonic clays. Beside the fault, the investigation site rock mass is divided by fractures (tensional and shear), i.e. by numerous fraction systems.

3. GEOTECHNICAL ZONING

The geotechnical media zoning was performed according to the fissuring and the rock mass freshness within lithologic structural units. The fracture and fracture walls properties, as well as the weathering degree, were analyzed in detail. The monolith physical and mechanical properties were statistically processed using the database formed for each lithologic unit. The following geotechnical media (quasihomogenous zones) were determined:

- lower fresh andesite α
- upper fresh andesite α'
- lower chlorinated andesite α_h
- upper chlorinated andesite α_h'
- lower kaolinated andesite α_k
- upper kaolinated andesite α_k'
- upper silificated andesite α_s'
- pyroclastite θ
- lower conglomerate K
- upper conglomerate K'
- ore

Since the data on the rock mass fissuring and freshness were collected only for several open pit mine horizons, the geotechnical zones boundaries are based on the observed rock mass history. This cause a lot of approximation, but such kind of zoning are still more suitable than the lithologic-structural averaging. The design indicators for defining the geotechnical media boundaries were: the rock mass exposure to alteration factors and the high stress zones. In order to further apply the geologic research results, the geotechnical media geometrization has consciously neglected many gradual boundaries and its own penetrations into some of the media, so that the investigated area became homogenous regarding the exhibition of the defined geotechnical media.

The rock mass was classified by dedicating the rating points to each quasuhomogenous zone and using two approaches; the Bieinawski approach from 1976. (RMR) and the Barton from 1974 (Q). A special attention was directed to the rock mass rating definition for the "residual" conditions. The "residual" conditions denote a rock mass state after the brittle plastic deformations (followed by an increased fissuring) caused by the stress state reaching the fracture criterion. The quasihomogenous zones fracture porosity was quantified for peak and residual conditions and according to the number of fracture families, frequency and aperture. The value of the parameters was equal to the value of the parameters used for the quasihomogenous zones qualification.

4. ROCK MASS CONSTITUTIONAL MODEL

An integral, reduced data interpretation was made using the available documentation. The development of the rock mass constitutional model represents the result of the above investigation phase. The quasihomogenous zones properties were quantified using deformabilty and strength parameters. A special attention was paid to the scale effects quantification when determining the rock mass behaviour parameters. The constitutional model adoption starting premises were:

- a rock mass, although essentially a discontinual medium, could be approximated by a quasicontinuum within the scale of a problem;
- a constitutional model should describe the most important properties of a fissured rock mass: the elsto-plastic behaviour with strain hardening, i.e. softening, including the dilatation behaviour;
- the stress-strain behaviour of a rock mass is influenced by all rock mass structural elements (monolith, fissures, fractures up to the discontinuity of the investigated area order of magnitude);

The constitutional model of a rock mass behaviour is represented by the critical state concept. The material behaviour according to this model is determined by the boundary state line in the stress area, the initial flow surface and the hardening, i.e, softening law. The boundary state line separates the hardening zone within which the material is deformed by the volume decrease, and the softening zone within which the material is deformed by the volume increase.

The behaviour of the material modeled using the critical state concept, is defined as follows:

- the flow law: $d\varepsilon_{ij} = d + d\lambda \cdot (\partial F/\partial\sigma_{ij})$
- the flow condition: $F(\sigma_{ij}) = M^2 \cdot p^2 - M^2 \cdot p_0 \cdot p + q^2 = 0$
- the hardening/softening law: $d\varepsilon_v^p = \alpha/(1+e_0) \cdot (dp_0/p_0) = 0$

where:

σ_{ij}	stress tensor
ε_{ij}	strain tensor
$p = I_1/3$	I_1 - stress tensor first invariant
$q = (3I_{2d})^{0.5}$	I_{2d} - stress tensor deviation part second invariant
ε_v^p	plastic volume strain
M	parameter (defines stress state line position)
p_0	parameter (defines initial flow condition)
α	parameter (defines hardening/softening speed)
e_0	parameter (initial porosity)

The elastic behaviour parameters were derived indirectly, on the basis of known empirical correlations with elastic waves velocities and the rock mass rating correlatons. The adopted principle is more suitable than using the sample test results because of the problems with the scale effects quantification. The quasihomogenous zones initial flow condition parameters of the fissured rock masses, for the adopted critical state concept, were determined using laboratory tests (shear strength tests, singleaxial pressure test, the Brazilian Test), and the data on the rock mass RMR and Q classifications rating obtained using the Hoek-Brown fracture condition. The initial fracture porosity was immediately defined using the EG mapping data. The softening law parameter was determined using the residual hardness criterion and the corresponding changes in the fracture porosity.

5. STRESS INTEGRATION NUMERICAL PROCEDURE FOR CRITICAL STATE MATERIAL MODELLING

The stress integration procedure during the time step is presented by an algorithm in Table 1. A more detailed description is given in [7], [8], [9].

Table 1. Stress Integration Basic Steps

---

1. Check for elastic solution

$$^{t+\Delta t}\sigma_m^E = {}^t c_m \, {}^{t+\Delta t}e''_m \qquad \text{(mean elastic stress)}$$

$$^{t+\Delta t}S_{ij}^E = \frac{1}{a_E} {}^{t+\Delta t}e''_{ij} \qquad \text{(elastic deviatoric stress)}$$

$^t c_m$ and a_E are elastic constants.

If $^{t+\Delta t}f_y(^{t+\Delta t}\sigma_m^E, {}^{t+\Delta t}S_{ij}^E) \le 0$, the deformation is elastic - next load step calculation.

---

2. The strain is plastic, assume material hardening.

Start without mean plastic strain increment, $\Delta e_m^P = 0$

$$^{t+\Delta t}\sigma_m^{(k)} = {}^t c_m (^{t+\Delta t}e''_m - \Delta e_m^{P(k)}) \;, \qquad ^{t+\Delta t}p_0^{(k)} = {}^t p_0 e^{-3\frac{1+^{t+\Delta t}e}{k_s}\Delta e_m^{P(k)}}$$

$$\Delta\lambda^{(k)} = \frac{3\Delta e_m^{P(k)}}{2^{t+\Delta t}\sigma^{(k)} - {}^{t+\Delta t}p_o^{(k)} - T} \quad \text{parameter of plastic strain increment}$$

$$f_y^{(k)} = f_y(^{t+\Delta t}\sigma_m^{(k)}, {}^{t+\Delta t}p_0^{(k)}, \Delta\lambda^{(k)})$$

$$\Delta e_m^{P(k+1)} = \Delta e_m^{P(k)} - \frac{f_y^{(k)}}{\partial f_y^{(k)} / \partial(\Delta e_m^P)}$$

Iterate according to $\Delta e_m^{P(k+1)}$ until the next condition is satisfied

$$^{t+\Delta t}f_y = \left(^{t+\Delta t}\sigma_m - \frac{^{t+\Delta t}p_0 - T}{2}\right)^2 - \left(\frac{^{t+\Delta t}p_0 - T}{2}\right)^2 + \frac{3d^2}{2M^2\left(a_E + \frac{3\Delta\lambda}{M^2}\right)} = 0$$

$$\Delta e_m^P = \Delta e_m^{P(k)}$$

The final values determination $^{t+\Delta t}S_{ij}$, $^{t+\Delta t}e_{ij}'^P$. The hardening parameters correction.

---

3. Critical State. Determination of $\Delta\lambda$

$$\Delta\lambda = \frac{M^2}{3}\left(\frac{\sqrt{2}\,^{t+\Delta t}d}{2M\left|^{t+\Delta t}\sigma_m - T\right|} - a_E\right)$$

The final values determination. $^{t+\Delta t}S_{ij}$, i $^{t+\Delta t}e_{ij}'^P$ .The hardening parameters correction.

The applied model basic parameters meaning for the elastoplastic analysis is shown in Fig. 1.

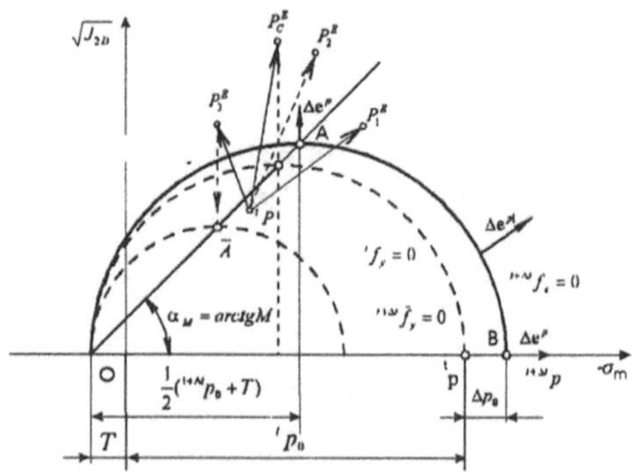

Fig. 1: Critical State Model

6. STRESS-STRAIN PROCESSES FEM MODEL

The stress-strain process modeling is based on the finite element method application. A unique network of 8-noded 3D elements was developed for all computations including the total of 5669 nodes and 4835 elements (Fig. 2). The boundary conditions were treated in the usual way: following the forces along the model contour which corresponds to the surface of the terrain and following the movements (the prevented movements perpendicular to the contour) along the side and the lower boundaries of the model.

The stress-strain analysis of the exploitation process was impeded due the static system

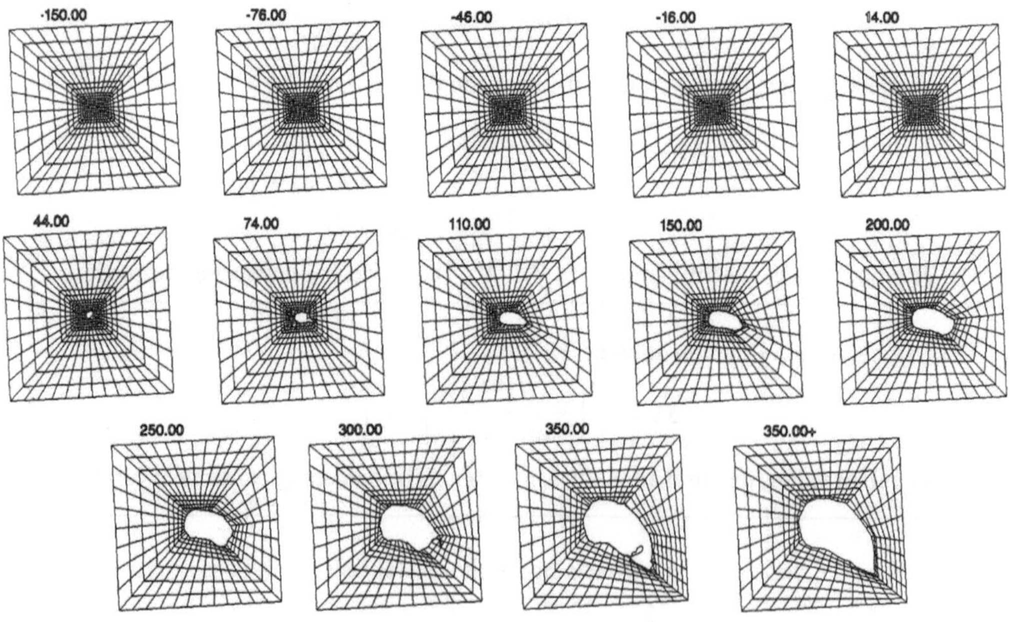

Fig. 2: Finite elements mesh

change caused by the exploitation improvement. The change occurred as the consequence of a gradual mass decrease and the surrounding rock medium loss of stability and fall. Two characteristic processes were noticed:

- the process of an underground opening formation due to the excavation of a subetage
- the rock mass overthrowing process due to the loss of stability

Both characteristic processes are followed by simultaneous changes of the static system and the rock mass behaviour model. Therefore, each step forward must be followed by two computations quantifying the changes due to the characteristic processes (the first one under the static system conditions with an underground opening and the second one under the static system conditions and the filled in underground opening and changed characteristics of the material in the overthrowing zone). The key problems of the model are the overthrowing moment criterion and the overthrowing zone size quantification.

Using the principle of the physical instability coinciding with the numerical divergence, the overthrowing moment criterion was defined, i.e. a numerical procedure which could be shortly described with follows iteration scheme:

$$ {}^{(i)}\varepsilon_n^{(j-1)} = {}^{(i)}B_{mn}^{(j-1)} \cdot {}^{(i)}u_m^{(j-1)} $$

$$ {}^{(i)}C_{mn}^{(j-1)} = \partial\, {}^{(i)}\sigma_m^{(j-1)} / \partial\, {}^{(i)}\varepsilon_n^{(j-1)} $$

$$ [\int {}^{(i)}B_{mk}^{(j-1)}\, {}^{(i)}C_{kl}^{(j-1)}\, {}^{(i)}B_{ln}^{(j-1)}] \cdot [{}^{(i)}u_n^{(j)} - {}^{(i-1)}u_n^{(j)}] = {}^{(i)}F_m - \int {}^{(i)}B_{mn}^{(j-1)}\, {}^{(i)}\sigma_n^{(j-1)} $$

$$ {}^{(i)}\Delta E^{(j)} = [{}^{(i)}F_m^{(j)} - \int {}^{(i)}B_{mn}^{(j-1)}\, {}^{(i)}\sigma_n^{(j-1)}] \cdot [{}^{(i)}u_m^{(j)} - {}^{(i-1)}u_m^{(j)}] $$

Overthrowing criterion:

$$ j \rightarrow \infty \quad \Rightarrow \quad {}^{(i)}\Delta E^{(j)} \neq 0 \qquad \text{- overthrowing} $$

Where:

$(i) = 1,2,3...$	exploitation step (excavation step)
$(j) = 1,2,3...$	step computational iteration
${}^{(i)}\sigma_m^{(j)}$	vector shaped stress tensor
${}^{(i)}\varepsilon_n^{(j)}$	vector shaped strain tensor
${}^{(i)}u_m^{(j)}$	vector of displacements
${}^{(i)}F_m$	external forces vector
${}^{(i)}x_m^{(j)}$	arbitrary spatial point coordinates
${}^{(i)}B_{mn}^{(j)}$	stress-displacements transformation matrix
${}^{(i)}C_{mn}^{(j)}$	constitutive matrix
${}^{(i)}\Delta E^{(j)}$	unbalanced energy

The procedure is repeating for each excavation increment.

All stress-strain analyses of the problem described were performed using the PAK computing program [10].

The results of the stress-strain modeling of the exploitation process were interpreted as a function of the stated aims. By using the presented procedure for the adopted model, immediate computing results are obtained for the whole observed area. The results are given in a graphical form showing the instability zones development as a function of the exploitation progress and the spatial position of the observed point. It is therefore possible to immediately compute the costs as a function of time arising due to caving and massive deformations at the surface of the terrain, and to determine the exploitation impact on the capital storage. Some of the results are presented in Fig. 3 and Fig. 4.

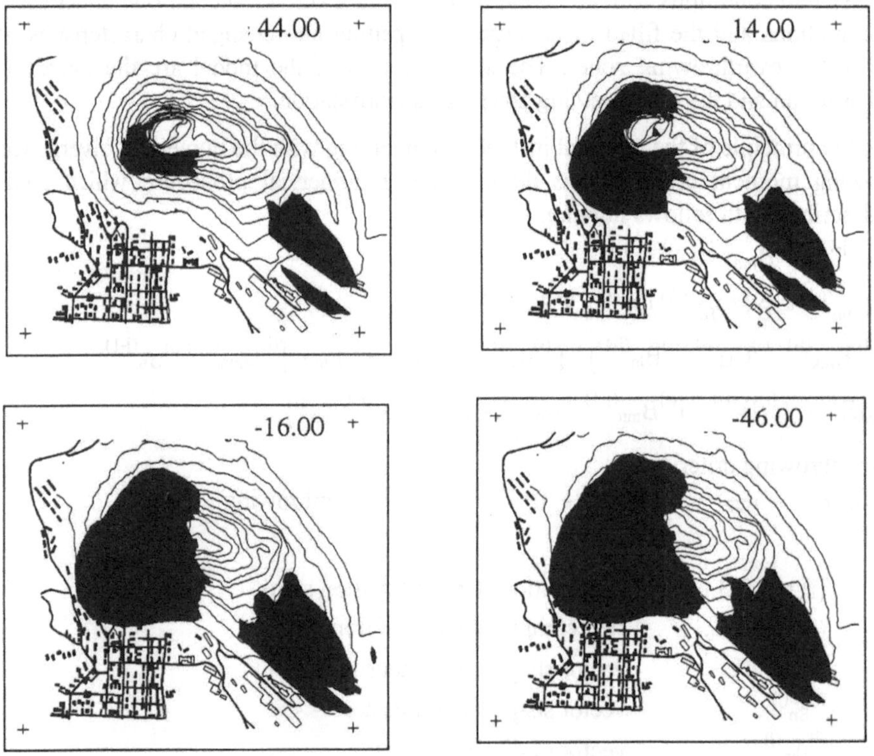

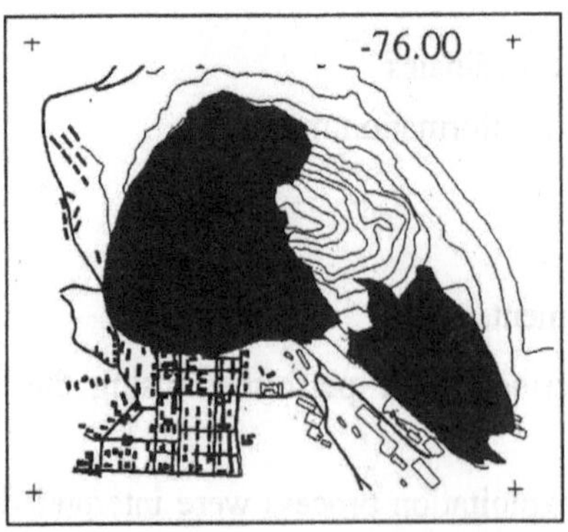

Fig. 3: Development of instability zones - horizontal plan

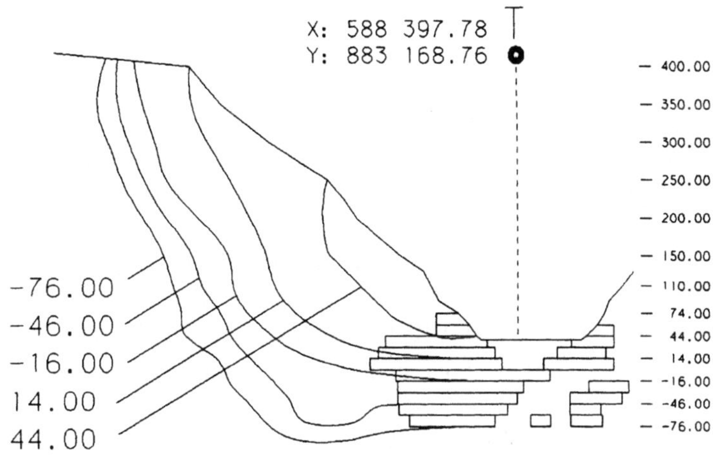

Fig. 4: Development of instability zones - vertical cross section

Beside the instability zones determination, another important issue is the exploitation method applicability in present circumstances. As to the rock burst occurrence possibility, characteristic is the energy released by overthrowing and sliding. A computation was therefore made to quantify the part of the energy that is gradually, continuously released due to the development of fissures and fractures and uniform caving processes, and the part which could be released in a form of sudden changes and unforeseeable cavings.

7. CONCLUDING REMARKS

The rock mass deficiency formation process, caving process, and stress-strain changes in the surrounding rock mass due to planned underground excavations were modeled on the basis of the real physical-mechanical rock properties. The research, analyses and computations carried out were multidisciplinary and based upon available data.

It is necessary to confirm the obtained results by appropriate observations and by intended measuring. The program applicability to modified input parameters enables the use of the observed and measured data for the "calibration" of the model in further phases.

The applied method of the stress-strain processes modeling and the rock mass instability zones prediction has many advantages when compared to traditional procedures mostly based upon experience. It enables grasping of real geomorphological and geomechanical conditions, especially important is the connection between the computation and investigations and the time development process simulation.

This method proved applicability in solving other similar problems (The Soko coal mine) [4]. Having in mind the potentials offered by this method, we suggest its use for solving problems of other layers planned to be exploited. However, a wise and intended data collection through investigations and observations is essential.

LITERATURE:

1. BARTON N., LIEN R., LUNDE J. (1974). Engineering Classification of Rock Masses for the Desing of Tunnel Support. Rock Mechanics, Vol. 6, No. 4.
2. BIENIAWSKI Z.T. (1979). The Geomhechanics Classification in Rock Engineering Application. Proc. 4th Congress of the Int. Society for Rock Mechanics, Motreux, Volum 2, PP 41-48.
3. DESAI C.S., CHRISTIAN J.T. (1977). Constitutive Laws for Geologic Media. Numerical Methods in Geotechnical Engineering, McGraw-Hill inc., USA.
4. DIVAC D., VUČKOVIĆ D., MAŠALA S., DENIĆ D. (1996). A new method of modelling subsidance above underground mining excavation. News Journal ISRM, Vol. 2, N 2.
5. DIVAC D., VUČKOVIĆ D., MAŠALA S., DENIĆ D. (1996). A New Method of Rock Mass Stress-Strain Modelling for Mining Works Impact Analysis in the Bor Mine Tilva Ros Ore Body. 11th Symp. on Hydro and Eng. Geol., Budva.
6. HOEK E. (1994). Strength of rock and rock masses. News Journal ISRM, Vol. 2, N 2.
7. KOJIC M. (1993). General Concept of Constitutive Relations at Nonelastic Material Strain Implicite Iteration. Ce. for Sci.res. od SANU and Univ. of Kragijevac.
8. KOJIĆ M., SLAVKOVIĆ R., GRUJOVIĆ N., VUKIĆEVIĆ M. (1994). Implicit stress integration algorithm for the Cam-clay material. Teorijska i primenjena mehanika 20, pp 95-118.
9. KOJIĆ M. (1997). Computational Procedures in Inelastic Analysis of Solid and Structures. Cen. for Sci. Res. of SANU and Univ. of Kragujevac.
10. KOJIC M, SLAVKOVIC R, ZIVKOVIC M., GRUJOVIC N. (1995). PAK - General Program for Linear and Nonlinear Structural Analyses and Heath Transfer. Mechanical Engeenering Faculty, Univ. of Kragujevac.
11. The Bor Copper Mine Documentation on Open Pit Mine Investigations (papers from various fields prepared during last 30 years).

OBJECT-ORIENTED ANALYSIS OF A FINITE ELEMENT FRAMEWORK FOR GEOTECHNICAL APPLICATIONS

P. Fritz and X. Zheng
Swiss Federal Institute of Technology, Zurich, Switzerland

SUMMARY

Up till now several object-oriented Finite Element (FE) frameworks have been presented which are partially extendable. However, the extendability is limited to a few specific directions, e.g. the introduction of new element types or solving strategies. Much less support is available for task control, creation of new material models, configurable field variable types or extensions of the analysis model. No framework is available which is especially designed to cover the problems encountered when dealing with geotechnical engineering. IMAGINE tries to close these gaps.

The nucleus of the Finite Element framework presented relies on abstractions, which include common concepts accepted in mathematics, mechanics, engineering and interactive visualization techniques, and serve as the fundamental object-oriented framework of classes for finite element applications and task management. The aim of the framework was not to include as many features as possible (e.g. a variety of types of finite elements or material laws), but to provide a sophisticated and robust foundation which may be used in the future due to its inherent capabilities of simple maintenance, adaptability to new resources and extendability.

Keywords: object-oriented programming, OOP, finite elements, framework, toolkit, C++, geotechnical engineering.

1 Introduction

Since many years, professionals have been aware of the so-called software crisis (Wirfs et al., 1990). The main problem of large software systems is the great effort necessary for the continuous support of software, i.e.

- maintenance, e.g. debugging,
- adaptation to new resources (e.g. compilers or hardware), and
- extensions.

This is particularly true for FE programs. An increasing demand on new analytic capabilities, constitutive models and easy-to-handle graphical user interfaces requires the introduction of new design methodologies. However, as a matter of fact, current textbooks on the Finite Element Method (FEM) are still very similar to those of two decades ago. This is also illustrated by 60% of all codes of engineering software being written in FORTRAN, about 6% in C and 0% in object-oriented C++ (Smith, 1994). Only a handful of research projects have been undertaken in the last few years to overcome these limitations.

Typically, development of finite element codes has been started by research organizations. Then the codes have been transferred to private companies, where they have been extended and enhanced. The researchers have been left with a multitude of rather incomplete code fragments, each tailored to a specific topic of interest. Due to advances in programming and because of the conventional, inflexible design, scientists usually could not rely on existing code and had to start practically from scratch for each new research project which extended the current model. Furthermore, for scientific consulting work or for teaching purposes, they used some third party software which was, at least for geotechnical engineering, not one hundred per cent tailored to the specific problem to be solved, but was much easier to handle thanks to an elaborate, comfortable user interface.

Therefore, both research organizations and practicing engineers would benefit from a Finite Element framework[3] for geotechnical applications which should satisfy the following requirements:

- simple software support, i.e. easy maintenance, adaptation to new resources and extendability,
- reusability, implying understandability (readability, traceability, learnability), homogeneity and portability,
- reliability, which designates the probability with which software does not cause the failure of a system, or lead to incorrect results,
- software security, which means the degree to which software protects itself from unauthorized actions,
- software safety, i.e. the probability that even an unintended usage does not lead to a mishap (hazard).

As discussed by Fritz and Zheng (1998), these requirements are met by Object-Oriented Programming (OOP) in an exemplary way. Therefore, Microsoft (MS) Visual C++ has been chosen as programming language, together with the MS Foundation Class (MFC) library for

[3] A framework is a collection of classes that provide a set of services for a particular domain (Booch, 1994), e.g. for buiding compilers.

the user interface (Microsoft Visual C++, 1993). As development environment MS-Windows 4 and NT are used.

The life cycle of software viewed in the light of the object-oriented paradigm comprises
- Object-oriented requirement analysis (OOA)
- Object-oriented design (OOD)
- Object-oriented programming (OOP)
- Testing
- Maintenance.

The analysis phase is the main topic of this report. It outlines the external characteristics of the system which are visible to the user (programmer) and the key abstractions of the framework. Additionally, sources of information are identified which may help to solve the task in hand, and an overall description of the whole system is given.

A more detailed report dealing with these topics and also about the design and the implementation of IMAGINE are given elsewhere (Fritz and Zheng, 1998).

2 Aim of the Framework

The objectives of the FE framework IMAGINE (Integrated Modeling and Analysis in Geotechnics by finite Elements) are twofold:
- provide scientists in the field of geotechnical engineering with the means to develop specific applications
- development of an example application for geomechanics for a de facto proof of the applicability of the framework. Furthermore, this application might be used by students, scientists and practicing engineers.

The first aim requires as mentioned above
- extendability (with regard e.g. to statical or physical modeling, algorithms and task management),
- understandability from the programmer's point of view,
- maintainability,

and the second aim
- a basic set of the most frequently used features,
- a satisfying user interface.

It must be realized in advance that it is not aimed to develop a universal framework which may be used for all kinds of FE application. An example of a universal FE system is NASTRAN which required hundreds of man-years for development, further hundreds for maintenance, and still cannot satisfy the specific needs of each engineer.

3 State of the Art

Recently OOP has been applied to FEM by various authors. Although the specific interests have been different, the common basis has been to model more closely the concepts of interest.

In a first approach several authors discussed the advantages and applicability of OOP for the field of structural engineering (Fenves (1989), Miller (1988, 1989)). A sophisticated framework (implemented with the Common Lisp Object System) has been presented by Rehak (1986), Baugh and Rehak (1987 and 1989), and Rehak and Baugh (1988). This work may be considered as the definitive proof of the concept of OOP for an FE framework. Forde (1989) compared a conventional numerical analysis program with its object-oriented counterpart, and Dubois (1992) compared two implementations of the same FE program written in two different languages, one version in Smalltalk and the other in C++.

Frameworks which have been designed to be extendable have been presented by Langtangen (1993) and Sims (1994). However, both frameworks would require major enhancements to be applicable as an FE building instrument.

Summarizing it can be said that several frameworks have been presented which are more or less extendable. However, the extendability is limited to a few specific directions, e.g. the introduction of new element types or solving strategies. Much less support is available for task control, creation of new material models or extending the analysis model. No object-oriented framework is available which is especially designed to cover the problems encountered in dealing with geotechnical engineering. IMAGINE tries to close these gaps.

4 Object-oriented Analysis

Architecture of IMAGINE

When developing a framework for the FE domain, two issues will be of central importance: to design an architecture which is adaptable to future requirements, and a data organization which is easy to use, guarantees data integrity and consistency, and avoids redundancy. Without object-oriented programming, it is extremely difficult to satisfy these aims. But even when basing on OOP it is largely up to the design how far one goes to reach these aims. OOP employed in a narrow sense may just help to write conventional programs in a better style. At the other extreme, i.e. when OOP is exploited to its full extent, even a large program system may be developed which is still perfectly lucid and manageable. The price for this is a big expense in its basic design. IMAGINE's approach is more oriented towards the second way. About 80% of the effort has been invested in a comprehensive class design, and only about 20% in developing a simple application in the sense of a "proof of applicability".

The first design decision was to completely separate everything which concerns the FE method from the items which concern managing an analysis. Accordingly, the class architecture is based on two subsystems: one for the *project and resource management*, and the other for *FE modeling* (Fig. 1). Each subsystem includes several main abstractions (class hierarchies) which are used to model different concepts, each for its own domain of competence.

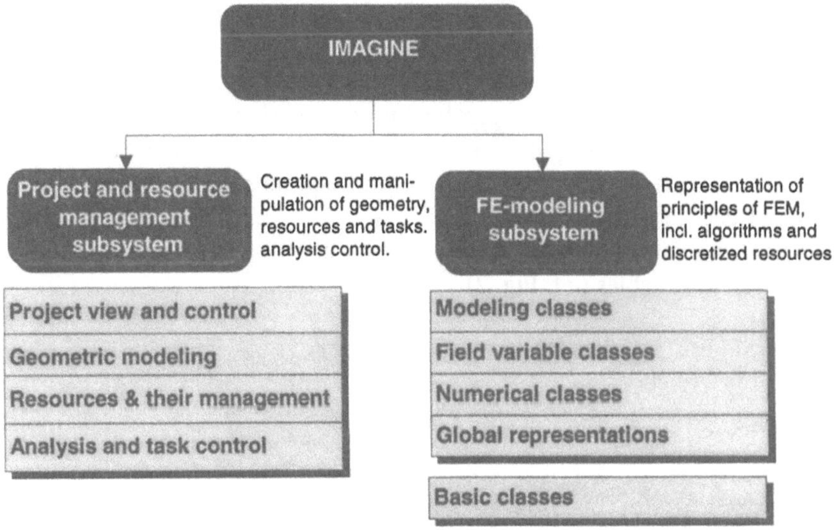

Fig. 1 General class architecture

The ***FE modeling subsystem*** includes the computational abstractions of the principles of the finite element method:

- modeling classes, e.g. FE types, nodes with information about geometry and element connectivity,
- field variable classes, e.g. stress and strain, degrees of freedom,
- numerical classes, e.g. templates of shape functions,
- global representations, e.g. the global discretized FE system, the equation solvers, and
- basic classes, vectors and matrices, points, and coordinate systems.

The *FE modeling subsystem* provides the necessary classes or data types to build a comprehensive representation of the underlying FE domain. However, the creation and manipulation of the representation has to be done at another level. In IMAGINE this domain is realized in the ***Project and resource management subsystem***. This subsystem is considered to be equally important as the *FE modeling subsystem*. It includes

- the project view and control, which is responsible at the highest level for the views and the control of the project,
- the geometric modeling, which is the intermediary between the geometry as defined by the solid modeler and the one used by the *FE modeling* subsystem,
- the resources and their management, typically applied to zones or regions, and
- the analysis and task control, where the individual computing tasks are defined and the actual computation is executed.

FE modeling subsystem

The basic abstractions of the kernel domains of a Finite Element analysis (FEA) model in a direct way the individual parts of an analysis. An FE analysis includes establishing the local element stiffness matrices and assembling the global one, applying the boundary conditions (loads and constraints), solving the system of equations, and computing secondary field variables (e.g. stresses and strains) from the primary ones (e.g. displacements). The prime aim of the framework IMAGINE is to be open for future extensions. Therefore, it is most important to introduce abstractions which separate the individual concerns as far as possible. Thereby abstractions may include both structural aspects and algorithmic aspects.

The two most important class hierarchies of the *FE modeling subsystem* are the *Modeling Classes*, which model the components of the FE system, and the *Field Variable Classes*, which contain all physical variables (c.f. Fig. 2).

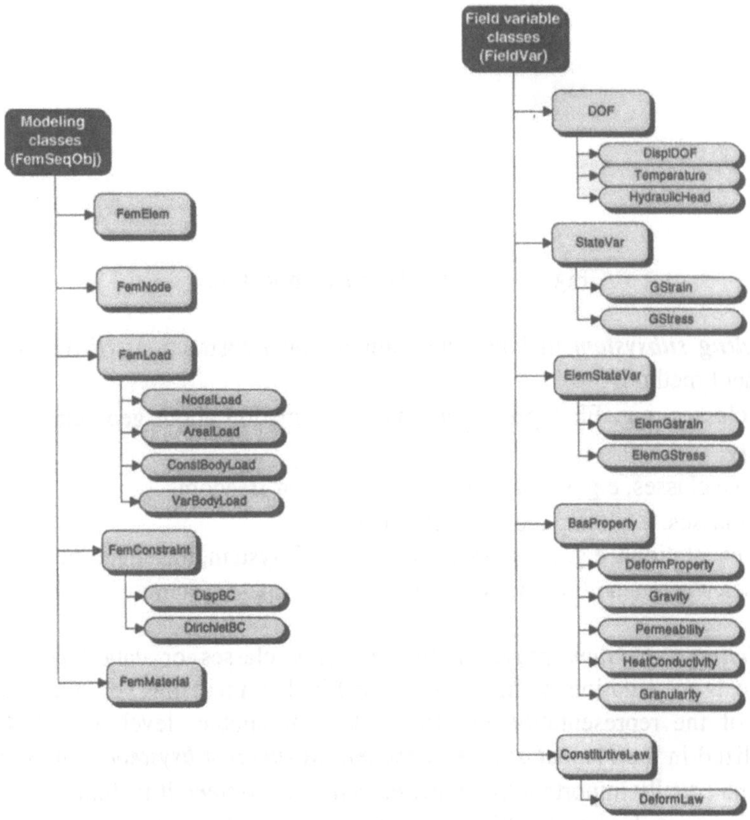

Fig. 2 FE subsystem: modeling and field variable classes

The abstraction which probably influences the quality of an FE framework design to the greatest extent is the one for a finite element (**FemElem**[4]). Here the designer has much freedom and even greater responsibility to separate levels of concern. Elements may be

[4] In this report all class names used by IMAGINE are written in bold characters.

characterized by their geometric shape (bar, triangle, etc.), by their mechanical behavior (bar, beam, etc.), the field variables acting in them (displacements, stresses, water pressure, etc.), and the governing material law. In conventional programs often individual combinations of these characteristics form different element types. IMAGINE adopts another approach. This is indicated by the fact, that currently **FemElem** does not have any subclasses at all. The **FemElem** class itself is directly responsible just to provide geometric and topologic information. All other characteristics are covered by element-independent abstractions. If required, the corresponding objects may be linked dynamically to **FemElem**, i.e. it serves as a sort of container. **FemElem** just assembles the information from these objects, e.g. to form the material matrix, without being aware of their type. In this way **FemElem** is completely open for integrating future extensions, e.g. the introduction of a new type of analysis or field variable, or a new shape function.

As for the element also the abstraction for a material **FemMaterial** has no subclasses. This is achieved by extensive separation of levels of concern. Analogously to the element, **FemMaterial** serves as a container, but not for field variables and numerical objects, but for constitutive laws and their properties. I.e. constitutive laws are situated for **FemMaterial** at the same level of abstraction as field variables for **FemElem**. They may be linked to **FemMaterial** dynamically. If a material behavior is path-dependent gathering the actual data is delegated to the project control. Thanks to this design the introduction of a new material law should be possible without changing the underlying concept.

The physical field variables are derived from the common base class **FieldVar**. One important role of **FieldVar** is to cover the association requirements of its subclasses by means of type identifiers. Primary variables (e.g. displacements) are abstracted in the degree of freedom class **DOF**. Secondary, i.e. derived field variables (e.g. strains and stresses) are contained in the state variable classes.
The **DOF** class has more content than just the "number of degrees of freedom". **DOF** and its subclasses contain information about how many unknowns associated with a certain type of field variable apply in a **FemNode**, the state of these unknowns (free or constrained) and their indices in the discretized FE system. A local coordinate system will be provided wherever it is necessary since in practice the **DOF** objects sometimes may only be defined relative to a local coordinate system (e.g. for constraints applied at an inclined support).

The discretized representation of the physical structure is abstracted in a fundamental class at the *global level* (c.f. Fig. 3) named **FemStruct.** It contains the individual components which form an FE system at the local level. The most important ones are abstractions for the finite elements, the loads and the material law. However, these abstractions are own independent classes. **FemStruct** does not need to know e.g. the type of an element, it just assembles them and puts them in a global frame. Its main responsibilities are topologic information about elements and global assignment of degrees of freedom.
The second abstraction with global level concern is the class **FemSolver**. It assembles the global equation system as required by the actual solver type, and solves it. It is the responsibility of the project control to choose and assign an adequate type of solver. The *numerical classes* are primarily responsible to provide integration support at the element level.

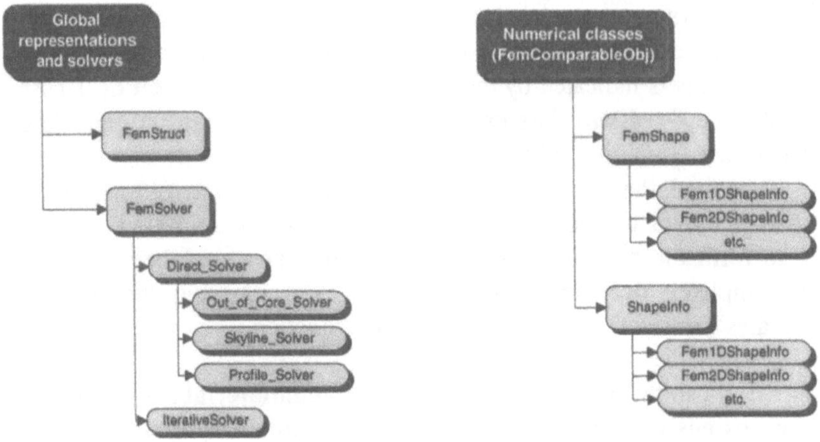

Fig. 3 FE subsystem: global representations and numerical classses

Project and resource management subsystem

In conventional programming, the FE kernel is usually considered to be the most important part of an FE system, whereas its management is condensed to a sequence of instructions specifying when and what to compute. Thereby it is often overlooked that the FE kernel takes over responsibilities of the management system, e.g. the definition of the computational tasks.

Following the object-oriented approach with its separation of concerns, the management of the projects and its resources should be completely separated from the concerns of the FE kernel system. In this way it may easily be seen that the project management has far more responsibilities than usually assumed. As a fact, it is at least as important as the FE kernel itself, which may be illustrated e.g. through its greater size in terms of lines of code.

Managing the project and its resources includes geometric modeling, defining the resources, defining the computational tasks, and dispatching them for execution (Fig. 4). All these items refer principally to a higher level of abstraction, i.e. they do not refer to the level of discretized finite elements. E.g. a load may be defined for a geometric zone. Its assignment to individual finite elements is done later by the FE kernel before solving the system of equations. I.e. the project management includes all definition of resources and their management, including the definition and dispatching of computational tasks, but excluding all definitions and manipulations at the level of the finite elements themselves. Only preparation of the mesh is also included here.

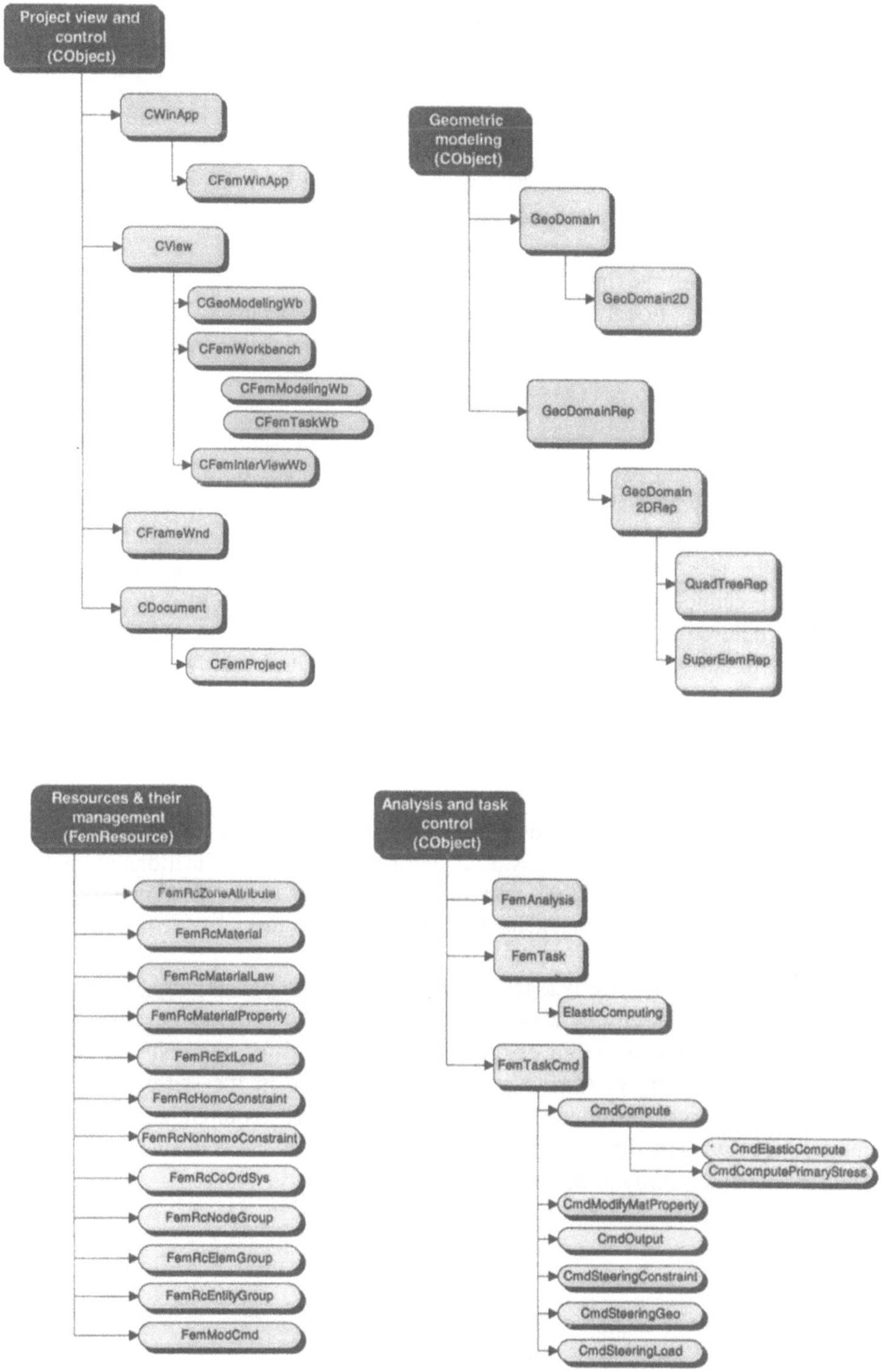

Fig. 4 Project and resource management subsystem

In somewhat greater detail the project management has four responsibilities:

- Project control at the uppermost level. This includes steering the user interface.
- Geometric modeling: definition of the project geometry in terms of a solid model. Conversion to data structures suitable for meshing. Actual meshing.
- Resources: definition of all resources, e.g. loads, material laws, geometric subregions.
- Analysis: definition of computational tasks, execution of the analysis.

The strict separation of the individual levels of concern facilitates inclusion of future extensions. E.g. computational tasks are treated like other resources. They are completely separated from the analysis. This may facilitate using artificial intelligence techniques for the definition of these tasks.

Thanks to OOP the project management is much less rigid than its conventional counterpart. OOP favors "non-anticipation" by its design. An object is a self-contained conceptual entity, on which all foreseen methods may be employed at any time. Before it exists it cannot be manipulated. Translated to the project management this means that the sequence how the responsibilities are fulfilled is of subordinate importance. If an object does not exist, it cannot be acted upon. E.g. assigning a distributed load to element nodes cannot be done before the mesh has been created. If a method needs more information to be executed, this will automatically be created, and thanks to OOP mostly without any special provisions from the programmer's side.

5 Conclusions

The analysis of the object-oriented Finite Element framework IMAGINE has been presented which may be used as a robust foundation for building applications in the field of geotechnical engineering. IMAGINE aims to be extendable in terms of material models, types of physical variables and management of computational tasks.

It has been recognized that there is a great difference between developing a framework just as a proof of the applicability of a specific concept, and to provide a framework which should be usable and extendable for engineering practice and for research work. The main aim of the IMAGINE project was to provide a framework in the sense of a numerical infrastructure, on which developers can build to integrate required extensions with a minimum of development time and side effects. This aim of the framework has certainly been fulfilled. It is now ready to develop applications upon it.

References

Baugh John W., Rehak Daniel R. (1987): Object-oriented design of Finite Element programs. *Report Dept. Of Civil Engng., Carnegie-Mellon University, Pittsburg.*

Baugh John W., Rehak Daniel R. (1989): Computational Abstractions for Finite Element Programming. *Report 89-182. Dept. Of Civil Engng., Carnegie-Mellon University, Pittsburg.*

Dubois-Pelerin Yves-Dominique (1992): Object-oriented finite elements: programming concepts and implementation. *Diss. No. 1026, Swiss Federal Institute of Technology Lausanne.*

Fenves Gregory L. (1989): Object-oriented models for engineering data. *Computing in civil engineering, Atlanta.*

Forde B., Foschi R., Stiemer S. (1989): Object-Oriented Finite Element Analysis. *Computers & Structures, v.34, no.3.*

Fritz P., Zheng X. (1998): A Finite Element Framework for Geotechnical Applications based on Object-Oriented Programming. *Report Swiss Federal Institute of Technology, Zürich.* http://www.igt.ethz.ch/www/search/publication.asp?publication=145.

Microsoft Visual C++ (1993): Microsoft Visual C++, Programmers Guides. *Microsoft Corporation.*

Miller Greogry R. (1988): A LISP-based object-oriented approach to structural analysis. *Engineering with Computers, v.4, no.1/2.*

Miller Gregory R. (1989): An object-oriented approach to structural analysis and design. *Civil-Comp Press, AI techn. and applications for civil and structural engineering, Sept. 1989.*

Rehak Daniel R. (1986): Artificial Intelligence Based Techniques for Finite Element Program Development. *Reliability of methods for Engineering Analysis. Pineridge Press.*

Rehak Daniel R., Baugh John W. (1988): Alternative programming techniques for finite element program development. *Report Dept. Of Civil Engngn., Carnegie-Mellon University, Pittsburg.*

Sims John Michael (1994): An object-oriented development system for finite element analysis. *Dissertation no.9424155 of the Arizona State University, U.S.A. Dissertation Abstracts International. Volume: 55-04, Section: B, page: 1564.*

Smith I.M. (1994): The advantages of Fortran 90 in computational geomechanics. *Computer Methods and Advances in Geomechanics, Balkema, Rotterdam.*

Wirfs-Brock Rebecca, Wilkerson Brian, Wiener Lauren (1990): Designing object-oriented software. *Prentice Hall Inc., New Jersey.*

SLOPE STABILITY OF EMBANKMENTS ON SOFT CLAY

F.F. Martins, C.F.S. Vieira and J.B. Martins
University of Minho, Azurém, Guimarães, Portugal

ABSTRACT: Embankments on soft clay are studied from the point of view of slope stability.

Comparative analyses of the results obtained using the traditional slope stability and using partial factors of safety according to the Cases A, B and C from Eurocode 7 (EC7) for situations of rapid loading are performed.

These analyses are done using the Fellenius and the Bishop rigid-plastic methods. The results obtained are compared with those obtained by the elasto-plastic Finite Element Method (FEM).

A parametric study to see the influence of the undrained strength of the clay in the maximum allowable height of the embankment is done. The traditional case and the case C of the EC7 are compared.

It is also analysed the influence of the strength of the embankment in the slope stability safety factor.

1. INTRODUCTION

The construction of embankments on soft clay is a subject very hard to deal with, due to the low strength, the high deformability and the low permeability of this material. The low strength limits the height of the embankment that it is possible to place quickly with acceptable safety in relation to the general sliding.

As these problems are very important, the study of embankments on soft soil has been discussed in many international conferences by many authors (Leitão Borges, 1995).

The purpose of this work is to compare the factors of safety using the usual rigid-plastic methods of Fellenius and Bishop and the elasto-plastic finite element method for the traditional case of slope stability analysis and the new partial safety factors analysis according to Eurocode 7 (EC7) (ENV 1997 – 1: 1994).

To perform the finite element analysis it is used the professional version 6.10 of the PLAXIS program where triangular 15 node elements and the Mohr-Coulomb yield criterion are selected.

2. EXAMPLE

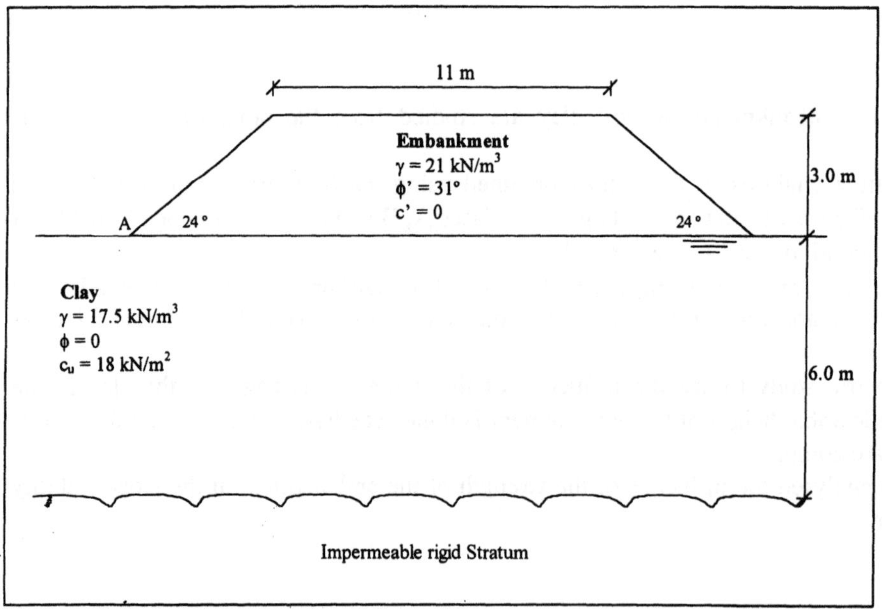

Fig. 1 – Embankment on soft clay

Consider the fig. 1, which represents an embankment with 3 m of height, 11 m of width at the top and a slope inclination of 24°. The thickness of the saturated clay is 6 m. An impermeable rigid stratum is below the clay layer.

The characteristics of the soils are the following:

Embankment: $\gamma = 21$ kN/m^3
$\phi' = 31°$
$c' = 0$
$E = 100000$ kN/m^2
$\nu = 0.3$

Clay: $\gamma = 17.5$ kN/m^3
$\phi_u = 0$
$c_u = 18$ kN/m^2
$E = 10000$ kN/m^2
$\nu = 0.495$

It is considered that the undrained strength of the clay does not change with depth. This is not true, mainly with normally consolidated clays. However, the main aim of this paper is only to discuss the application of EC7 to this kind of problems.

The finite element analysis is performed using a mesh of 58 triangular elements of 15 nodes.

Eight situations are considered here to obtain the factor of safety. In the first four situations it is not considered surcharge at the top of the embankment.

These four situations are the following:

1. Traditional analysis using the characteristic values of the soil properties,
2. Analysis using partial factors of safety for case A - EC7 ($\gamma_{m\phi'}=1.1$; $\gamma_{mcu}=1.2$; $\gamma_g=1.0$),
3. Analysis using partial factors of safety for case B - EC7 ($\gamma_{m\phi'}=1.0$; $\gamma_{mcu}=1.0$; $\gamma_g=1.35$),
4. Analysis using partial factors of safety for case C - EC7 ($\gamma_{m\phi'}=1.25$; $\gamma_{mcu}=1.4$; $\gamma_g=1.0$).

The last four situations are identical to the first four situations with a surcharge of 10 kN/m^2 applied at the top of the embankment.

The minima factors of safety obtained are presented in table 1.

	Without Surcharge			With Surcharge (10kN/m^2)		
	Fellenius	Bishop	FEM	Fellenius	Bishop	FEM
Traditional	1.49	1.57	1.57	1.29	1.36	1.38
Case A	1.24	1.32	1.32	1.08	1.14	1.16
Case B	1.11	1.18	1.19	1.00	1.05	1.08
Case C	1.07	1.13	1.13	0.92	0.97	<1*

*Failure occurs with the self-weight of the embankment and the surcharge

Table 1 - Obtained values for minimum F_s

The circular sliding surfaces for the minima factors of safety obtain in situations 1 and 4 are presented in fig. 2 and 5.

The scaled field displacements for situations 1 and 4 are presented in fig. 3 and 6 and the variation of the total displacement of node A with the factor of safety for situations 1 and 4 is shown in fig. 4 and 7.

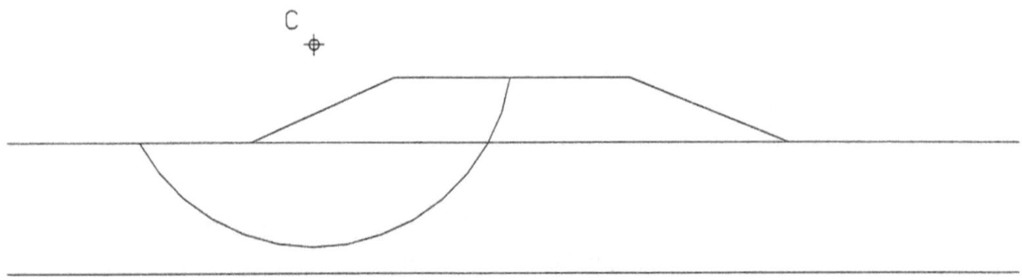

Fig. 2 – Sliding surface for the minimum factor of safety obtained in situation 1

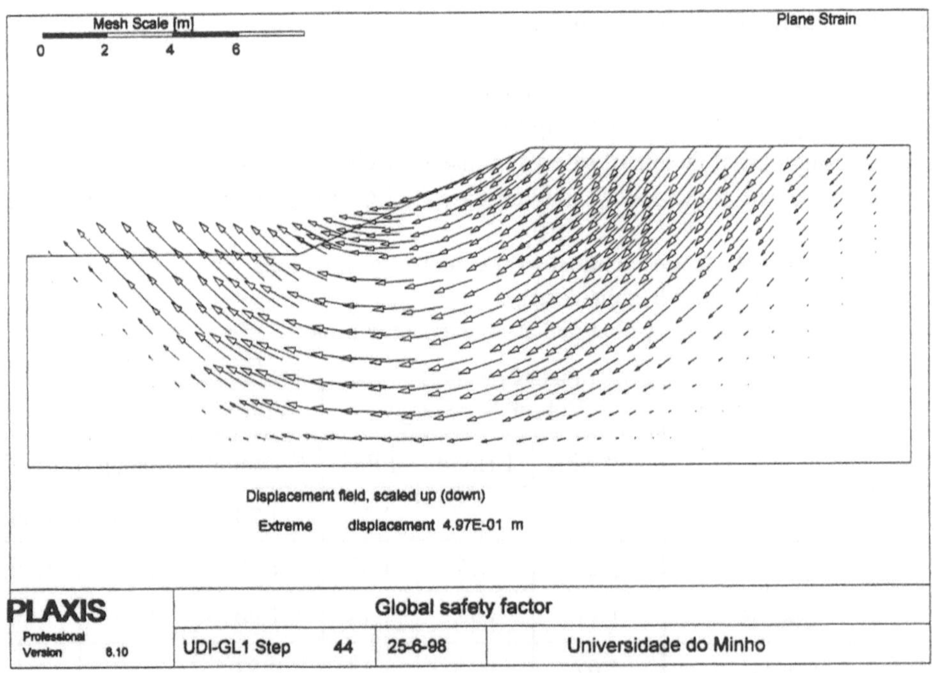

Fig. 3 – Scaled field displacements for situation number 1

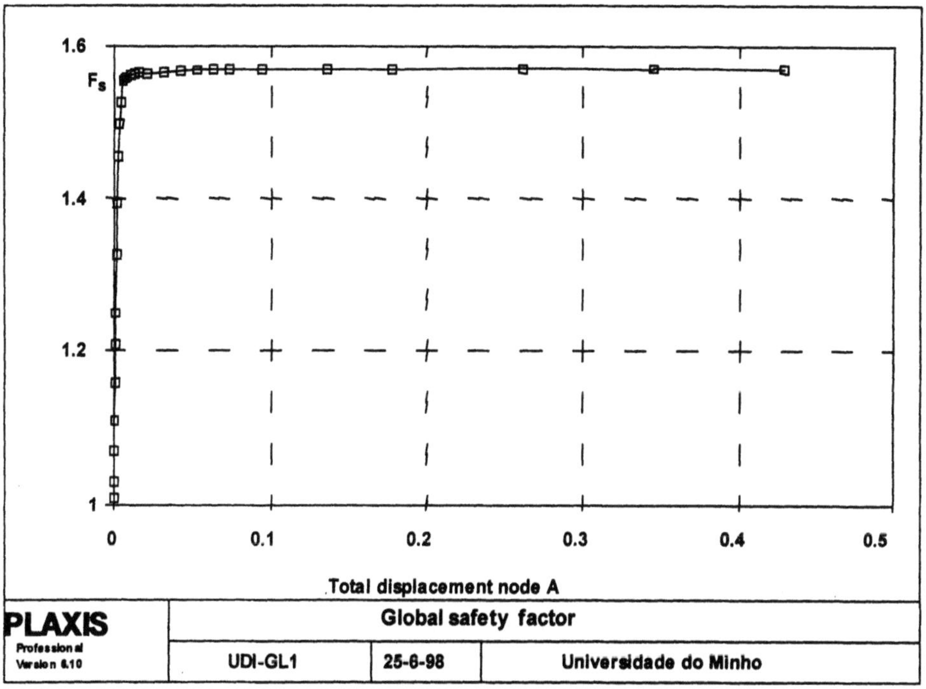

Fig. 4 – Factor of safety versus total displacement node A for situation number 1

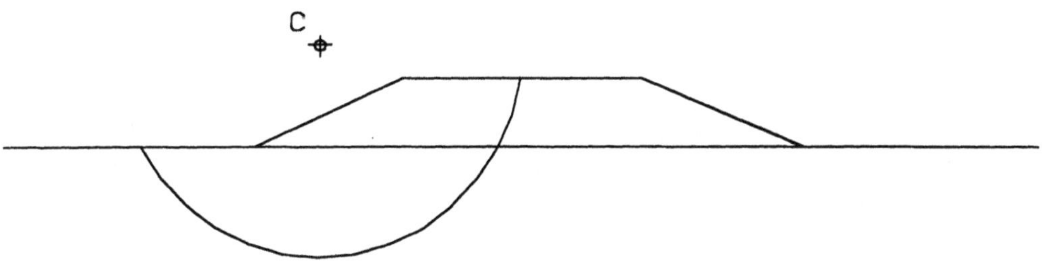

Fig. 5 – Sliding surface for the minimum factor of safety obtained in situation 4

It can be seen in table 1 that the factor of safety obtained for case C is less than those obtained for cases A and B in all situations. This is in accordance to the EC7, which refers that "Case C is generally critical in cases such as slope stability problems, where there is no strength of structural elements involved".

The results obtained using the FEM are closer those obtained using the Bishop method than those obtained using the Fellenius method.

As the factor of safety for the traditional case (situation 1) is greater than 1.5 the embankment can be considered stable. The same conclusion can be obtained in relation to case C (situation 4) where the factor of safety is greater than 1.0.

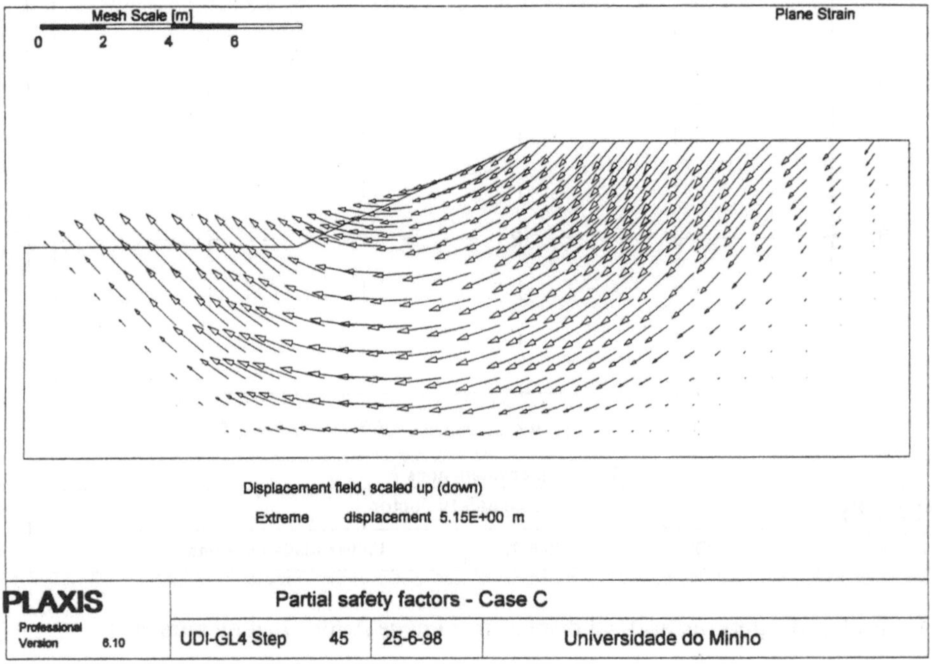

Fig. 6 – Scaled field displacements for situation number 4

If a surcharge of 10 kN/m^2 is applied on the top of the embankment the factor of safety obtained in the traditional case is less than 1.5 and in case C is less than 1.0. Therefore, it can be conclude that the safety is not satisfied in both cases.

3. PARAMETRIC STUDIES. INFLUENCE OF THE STRENGH PARAMETERS.

As we have seen before, case C is the most unfavourable. Therefore, in these parametric studies the comparison between this case and the traditional case is done. These studies are performed using the Bishop method.

It is analysed the stability of an embankment on soft clay changing the undrained strength of the clay between 12 kN/m^2 and 30 kN/m^2. For each undrained strength it is computed the maximum height of the embankment that ensures its stability. It is assumed that a factor of safety equal 1.5 and 1.0 is enough to ensure the stability of the embankment for the traditional case and the case C, respectively.

The characteristics of the embankment are the same used in the example of fig. 1. In relation to the clay layer only the undrained strength changes.

The slope inclination and the width of the top of the embankment remain unchanged in relation to the example of fig. 1.

The results are presented in table 2.

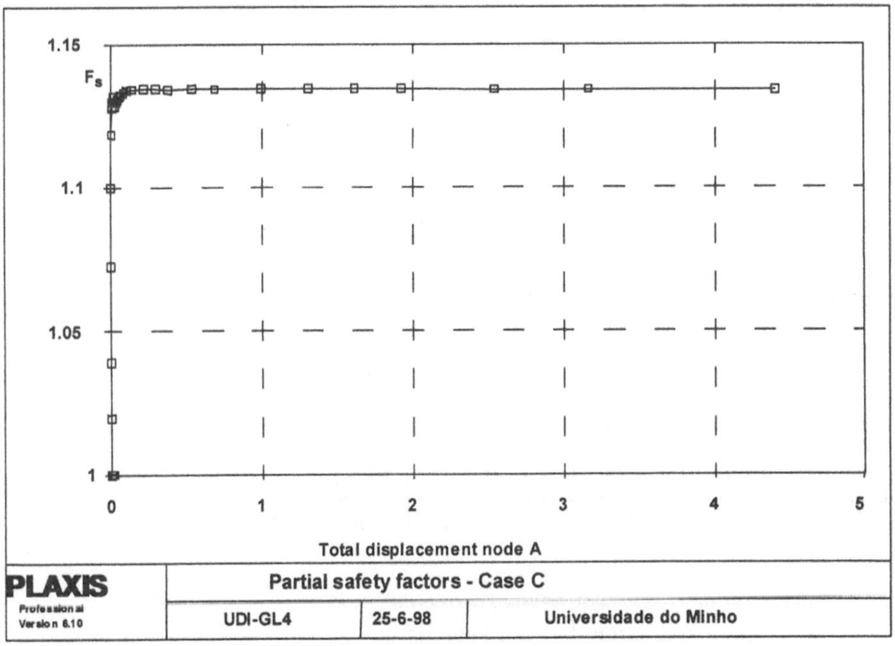

Fig. 7 – Factor of safety versus total displacement node A for situation number 4

		c_{uk} (kN/m^2)				
		12	15	20	25	30
H (m)	Traditional	2.1	2.6	3.5	4.35	5.45
	EC7 – Caso C	2.25	2.8	3.8	4.9	5.95
Change (%)		7.1	7,7	8,6	12.6	9,2

Table 2 - Height of the embankment

It is possible to see in this table the increase in the height of the embankment when it is used the EC7 case in relation to the traditional case.

It can be conclude that when using the case C the embankment height is allowed to increase more than 7% in relation to that obtained when using the traditional case. These values are higher than that obtained by Maranha das Neves (1994) that found an increase of 7%.

To analyse the influence of the angle of shearing resistance ϕ' of the embankment in its stability the embankment of the example of fig.1 is used. It is only changed ϕ' between 29° and 35°.

The results obtained for traditional case and case C with and without surcharge are presented in tables 3 and 4.

	Without surcharge		With surcharge (10 kN/m^2)	
ϕ'_k	Fellenius	Bishop	Fellenius	Bishop
29°	1.49	1.57	1.29	1.36
31°	1.49	1.57	1.29	1.36
33°	1.49	1.58	1.29	1.36
35°	1.49	1.58	1.29	1.37

Table 3 – Minimum Fs for different values of ϕ'_k - Traditional case

	Without surcharge		With surcharge (10 kN/m^2)	
ϕ'_k	Fellenius	Bishop	Fellenius	Bishop
29°	1.06	1.13	0.92	0.97
31°	1.07	1.13	0.92	0.97
33°	1.07	1.13	0.93	0.98
35°	1.07	1.14	0.93	0.98

Table 4 – Minimum Fs for different values of ϕ'_k – Case C

It can be seen that ϕ'_k has no influence in the results.

4. CONCLUSIONS

In the example of fig.1, there is a fair agreement between the factors of safety computed using the FEM and those computed using the Bishop method. In relation to the Fellenius method there are some differences.

The values obtained for case C are more unfavourable than those obtained for case A and B. This is in accordance with comments of the EC7.

In relation to the parametric studies, it is seen that the results are not influenced by the variation of the angle of shearing resistance ϕ' of the embankment but there is a significant influence of the undrained strength of the clay.

The maximum allowable height of the embankment is higher for the case C of the EC7 than for the traditional case. The difference is more than 7 %.

ACKNOWLEDGEMENTS

This work has been partly financed by the R/D project "Praxis/2/2.1/CEG/42/94" financed by PRAXIS XXI (National Science and Technology Foundation – FCT).

REFERENCES

ENV 1997-1: 1994 – Eurocode 7 – Geotechnical Design – Part 1: General rules. CEN.

Leitão Borges, J. M. (1995). Aterros sobre solos moles reforçados com Geossintéticos. Análise e Dimensionamento. Dissertação para Doutoramento em Engenharia Civil, FEUP,Porto.

Maranha das Neves, E. (1994). Estados limites e Segurança em Geotecnia. Geotecnia, N.° 72, Lisboa.

RESULTS FROM TWO GEOTECHNICAL BENCHMARK PROBLEMS

H.F. Schweiger
Technical University Graz, Graz, Austria

ABSTRACT: One of the aims of the working group 1.6 "Numerical Methods in Geotechnics" of the German Society for Geotechnics is to provide recommendations for numerical analyses in geotechnical engineering. In addition benchmark examples are specified and the results obtained by various users employing different software are compared. So far two simple examples, namely a tunnel excavation and a deep excavation problem, have been set up and results have been discussed at a workshop. This paper summarizes the most important aspects of this exercise. It shows that despite the simplicity of the examples and the rather strict specifications significant differences in the results are obtained even in cases where the same software has been utilized by different users.

1 INTRODUCTION

Numerical modelling has gained increasing importance in solving practical problems in geotechnical engineering. Enhancements in hard- and software have made it feasible to perform complex numerical analyses with relatively little effort. Although in general desirable there are some drawbacks associated with these developments. Unexperienced users with little theoretical background may be tempted to perform complex nonlinear analyses without appreciating that even with sophisticated constitutive laws results will often be more of qualitative nature than quantitative because of uncertainties in input parameters and e.g. modelling of construction sequences. Furthermore less experience in applying numerical methods is available as compared to classical calculation procedures and a good deal of engineering judgement is often necessary for correct interpretation of results. So far relatively little attention has been paid in the literature on validation and reliability of numerical models in general and on specific software in particular although some attempts have been made (e.g. Schweiger 1991).

In order to provide some guidance on application of numerical models in practical geotechnical engineering the Working Group 1.6 "Numerical Methods in Geotechnics" of the German Society for Geotechnics has published recommendations for numerical simulations in tunnelling (Meissner 1996) and is working on recommendations for deep excavations to be published in the near future. In addition example problems should demonstrate the influence of various modelling assumptions e.g. the type of constitutive law employed on the results. So far two (simple) benchmark examples have been specified, namely a tunnel excavation and a deep excavation problem. They have been sent to various university institutes, consulting companies and software developers. Some of the most significant results of these comparative calculations will be presented in the following.

2 A NOTE ON SPECIFICATIONS FOR BENCHMARK EXAMPLES

Arguments for developing benchmark tests have been briefly discussed above. Keeping in mind the purpose of benchmarking from a practical point of view the following requirements for benchmark examples can be postulated:

- No analytical solution available (any code solves these problems).
- Actual practical problem should be addressed, simplified in such a way that the solution can be obtained with reasonable computational effort.
- No calibration of laboratory tests (this is done extensively in research and is of minor interest for engineers in practice).
- Examples should be preferably set in such a way that - in addition to global results - specific aspects can be checked (e.g. handling of initial stresses, dilation behaviour, excavation procedures etc.).

Once a series of examples has been designed and solutions are available they could serve as:

- Check of commercial codes.
- Learning aids for young geotechnical engineers to become familiar with numerical analysis.
- Verification examples for proving competence in numerical analysis of geotechnical problems.

In addition these examples will identify limitations of the present state of the art in numerical modelling in practice, provide the possibility to show alternative modelling assumptions and highlight the importance of appropriate constitutive models.

3 EXAMPLE NO. 1 - TUNNEL EXCAVATION

3.1 Specification

Fig. 1 depicts the geometry of the example and Table 1 lists the material parameters given. Additional specifications are as follows:

- plane strain
- linear elastic - perfectly plastic analysis with Mohr-Coulomb failure criterion
- perfect bonding shotcrete/ground
- shotcrete lining: beam or continuum elements, 2 rows of elements over cross section if continuum elements with quadratic shape function are used
- to account for deformations occuring ahead of the face (prerelaxation) the load reduction method or a similar approach should be used

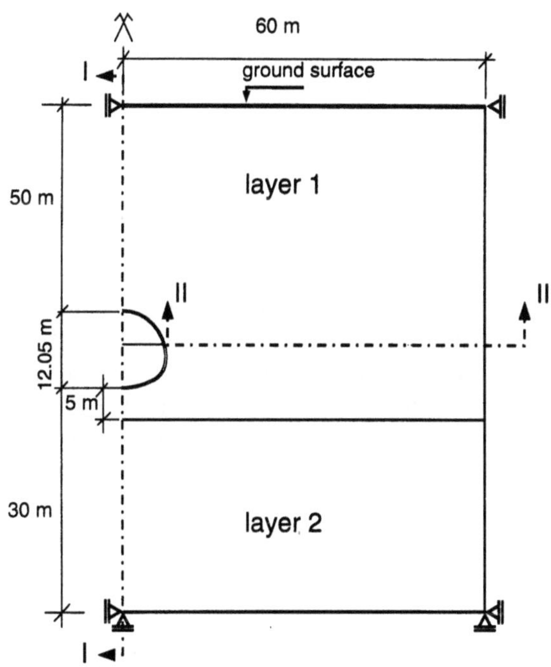

Fig.1 Geometry of example "tunnel excavation"

	E (kN/m²)	ν	φ (°)	c = (kN/m²)	K_o	γ (kN/m³)
Layer 1	50 000	0.3	28	20	0.5	21
Layer 2	200 000	0.25	40	50	0.6	23

Shotcrete (d = 25 cm): linear elastic $E_1 = 5\,000 \times 10^3$ kN/m²
$E_2 = 15\,000 \times 10^3$ kN/m²
$\nu = 0.15$

Table 1 Material parameters for example "tunnel excavation"

The following computational steps had to be performed:

Initial stress state is given by $\sigma_v = \gamma H$, $\sigma_h = K_o\gamma H$ (deformations = 0)
The prerelaxation factors given are vaild for the load reduction method.

I. Excavation sequence top heading followed by bench
construction stage 1: prerelaxation top heading 40%
construction stage 2: excavation top heading, $\text{shotcrete}_{\text{top heading}}$ with ($E = E_1$)
construction stage 3: prerelaxation bench 30%, $\text{shotcrete}_{\text{top heading}}$ with ($E = E_2$)
construction stage 4: excavation bench, $\text{shotcrete}_{\text{bench}}$ with ($E = E_1$)
II. Excavation in one step
construction stage 1: prerelaxation full cross section 40%
construction stage 2: excavation full cross section, shotcrete with ($E = E_2$)

3.2 Selected Results

3.2.1 Excavation in one step

In the following some of the most interesting results are presented. In Fig. 2 surface settlements obtained from 10 different analyses are compared. 50 % of the calculations give 5.2 or 5.3 cm as maximum settlement and most of the others are in a range of approximately 20 % difference. However, calculation TL1A and TL10 show significantly lower settlements. The reason in both cases was that it was not possible to apply the load reduction method correctly and therefore other methods have been used. TL10 used the stiffness reduction method and it is difficult to match these two methods (Schweiger 1997). TL9 also employed the stiffness reduction method but obtained larger settlements which is rather unusual.

Fig. 3 shows calculated normal forces and bending moments in the shotcrete lining. Reasonable agreement is observed for normal forces with the exception of TL1 and TL10 but a wider scatter is obtained for bending moments. Fig. 3 indicates also the location of maximum bending moments and the significant differences in magnitude (approximately 300 %) and location is obvious. Even if TL1, TL9 and TL10 are excluded because they did not refer exactly to the specification the variation is still 70 %. Unfortunately it was not possible from the information available to identify the reasons for these discrepancies.

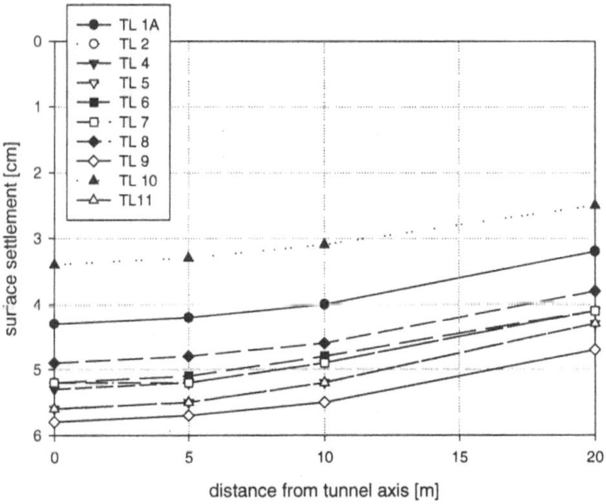

Fig. 2 Surface settlements for excavation in one step

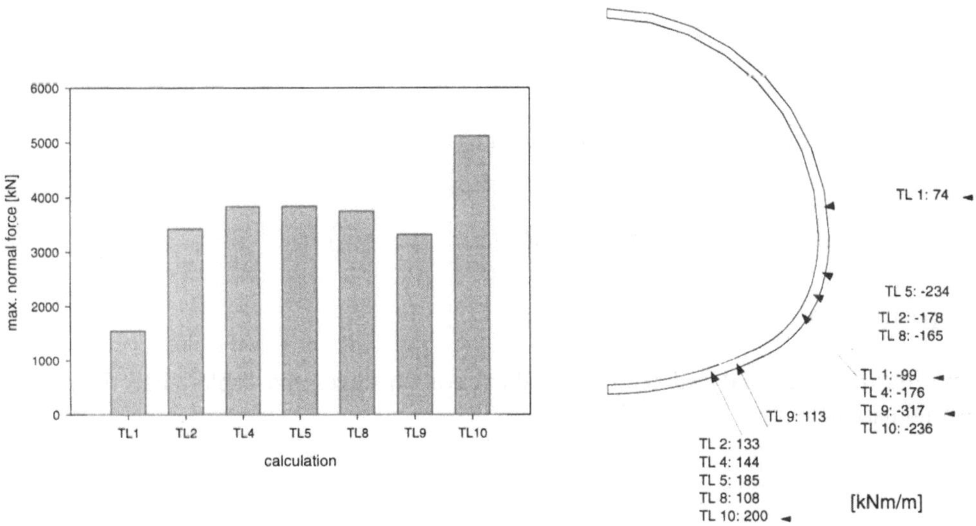

Fig. 3 Comparison of maximum normal forces and bending moments in lining

3.2.2 Excavation in two steps

Simulation of tunnel excavation in two steps with top heading and bench/invert included crucial construction stages namley the stages 2 and 3 because no temporary top heading invert has been specified (which probably would have been made in practice under the given conditions). Thus these stages were numerically at the edge of failure and it was interesting to see that some analyses introduced additional measures in order to account for longitudinal arching, some indicated high residual forces thus no equilibrium could be obtained and some just indicated large displacements. So the wide range of calculated surface settlements (Fig. 4) is not surprising.

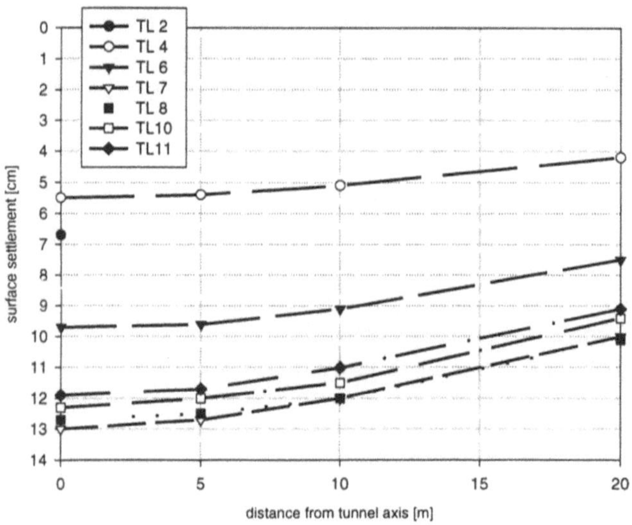

Fig. 4 Surface settlements for excavation in one step – construction stage 4

4 EXAMPLE NO. 2 - DEEP EXCAVATION

4.1 Specification

Fig. 5 illustrates the analysed geometry and excavation stages and Table 2 lists the relevant material parameters.

Additional specifications:

- plane strain
- linear elastic - perfectly plastic analysis with Mohr-Coulomb failure criterion

- perfect bonding diaphragm wall/ground
- struts may be modelled as rigid (i.e. horizontal degree of freedom fixed)
- influence of diaphragm wall construction may be neglected, i.e. initial stresses without wall, then wall "wished-in-place"
- diaphragm wall modelling: beam or continuum elements, 2 rows of elements over cross section if continuum elements with quadratic shape function are used

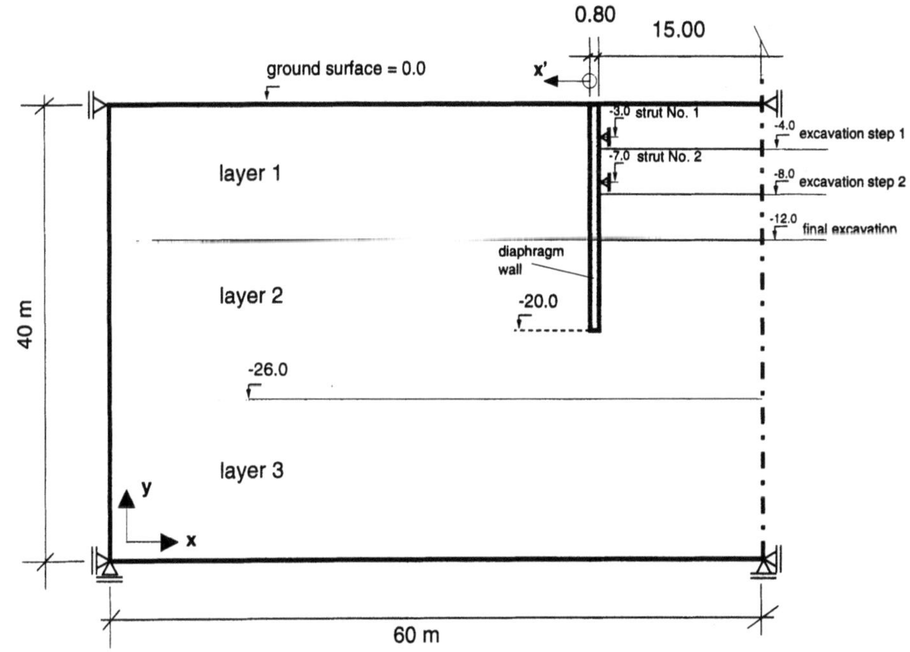

Fig. 5 Geometry of example "deep excavation"

	E (kN/m^2)	ν	φ (°)	c = (kN/m^2)	K_o	γ (kN/m^3)
Layer 1	20 000	0.3	35	2.0	0.5	21
Layer 2	12 000	0.4	26	10.0	0.65	19
Layer 3	80 000	0.4	26	10.0	0.65	19

Diaphragm wall (d = 80 cm): linear elastic $E = 21\,000 \times 10^3\ kN/m^2$
$\nu = 0.15$
$\gamma = 22\ kN/m^3$

Table 2 Material parameters for example "deep excavation"

The following computational steps had to be performed:

Initial stress state is given by $\sigma_v = \gamma H$, $\sigma_h = K_o \gamma H$
(deformations = 0, then wall "wished-in-place")

construction stage 1: excavation step 1 (to level -4.0 m)
construction stage 2: excavation step 2 (to level -8.0 m), 1. strut (in -3.0 m) active
construction stage 3: final excavation (to level -12.0 m), 1. and 2. strut (in -7.0 m) active

4.2 Comparison of results

It is worth mentioning that 5 out of the 12 calculations which have been submitted for comparison have been made with the same programme but different users. Fig. 6 compares surface displacements for construction stage 1 and the wide scatter of results is obvious.

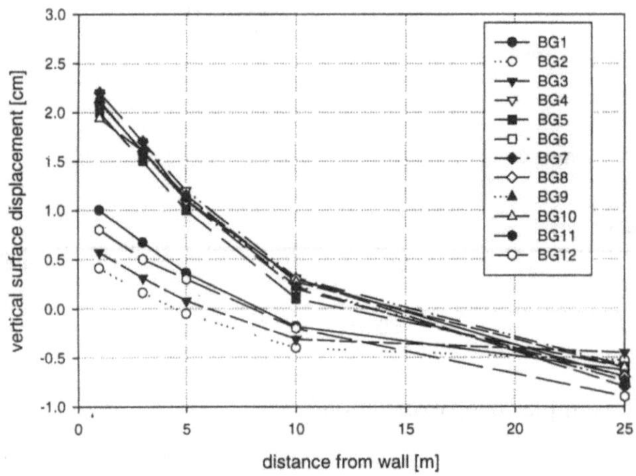

Fig. 6 Vertical displacement of surface behind wall – construction stage 1

It is apparent from Fig. 6 that elasto-perfectly plastic constitutive models are not well suited for analysing the displacement pattern around deep excavations especially for the surface behind the wall because the heave predicted is certainly not realistic. However it was not the aim at this stage to compare results with actual field observation but merely to see what differences are obtained when using slightly different modelling assumptions within a rather tight problem specification. The lower values for the heave from calculations BG1 and BG2 may be explained from the use of interface elements which have been used despite the specification did not forsee them. The results of BG3 and BG12 could not be explained in detail. There were indications though that for the particular programme a significant difference in vertical displacements was observed whether beam or continuum elements have been used for modelling the diaphragm wall. This emphasizes the significant influence of different modelling assumptions and the need for evaluating the validity of these models under certain conditions. It may be worth mentioning thart this effect was not observed to that extent in other programmes used.

For the final excavation step (Fig. 7) again a significant scatter of results has been observed.

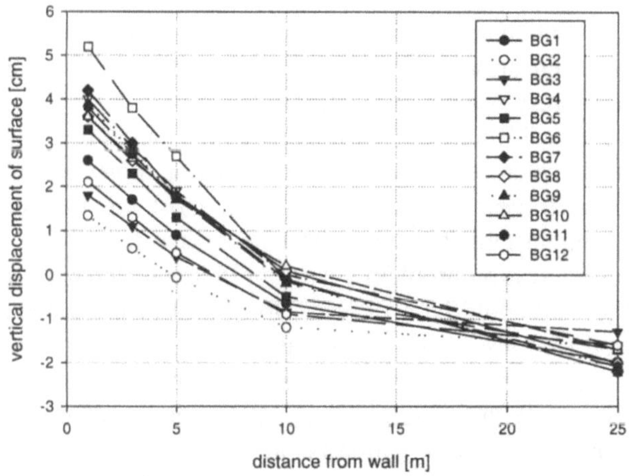

Fig. 7 Vertical displacement of surface behind wall – construction stage 3

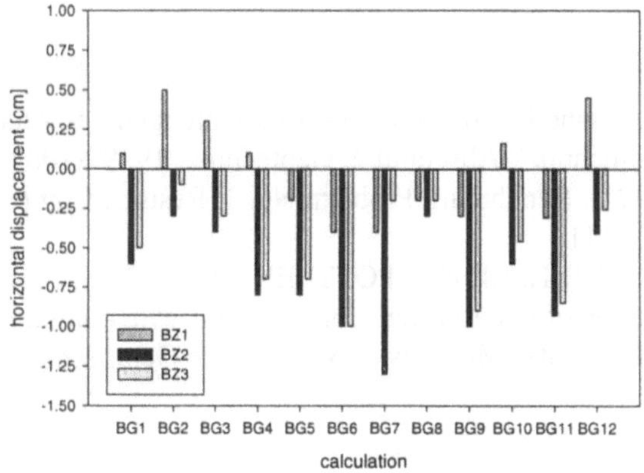

Fig. 8 Horizontal displacement of head of wall – excavation step 1

It is interesting to compare the horizontal displacement of the head of the wall for excavation step 1. Only 50 % of the analyses predict displacements towards the excavation (+ve in Fig. 8) whereas the other 50 % predict movements towards the soil which seems to be not very realistic for a cantilever situation. The horizontal displacements of the bottom of the wall, the heave inside the excavation and earth pressures distribution did not show significant differences. Calculated bending moments varied within 30 % and strut forces for excavation step 2 varied from 155 to 232 kN/m.

5 CONCLUSION

Some results from two benchmark examples suggested for comparing results obtained from various analyses have been presented. Although quite tight specification have been given significant differences in results have been observed.

For the tunnel problem difficulties in simulating the so-called prerelaxation have caused significant differences in calculated displacements. Bending moments also scatter in a wide range whereas normal forces are comparable for most of the submitted analyses. Difficulties in identifying unstable behaviour became apparent.

In case of the deep excavation problem again significant differences in displacements of the surface behind the wall could be observed. Horizontal displacements of the wall head for the first excavation step also differ significantly. Bending moments differ as well but not dramatically. Earth pressure distributions agree very well.

It could be clearly shown that employing a simple elastic-pefectly plastic constitutive model is not able to predict the deformation behaviour of complex geotechnical structures involving excavation not even to the accuracy required from a practical point of view. The series of examples will be extended and some more sophisticated specifications will be elaborated.

REFERENCES

MEISSNER H. (1996). Tunnelbau unter Tage – Empfehlungen des Arbeitskreises 1.6 "Numerik in der Geotechnik", Abschnitt 2. Geotechnik, 19: 99-108.

SCHWEIGER H.F. (1991). Benchmark Problem No. 1: Results. Computers and Geotechnics, 11: 331-341.

SCHWEIGER H.F., H.SCHULLER & R.PÖTTLER (1997). Some remarks on 2-D-models for numerical simulation of underground constructions with complex cross sections. Proc. 9th Int. Conf. Computer Methods & Advances in Geomechanics, Wuhan, 2: 1303-1308.

NUMERICAL MODELS FOR THE DESIGN OF SHALLOW FOUNDATIONS FOR LOW-RISE BUILDINGS

H.D. Skinner, M.S. Crilly and J.A. Charles
Building Research Establishment Ltd., Watford, London, UK

ABSTRACT: For lightly loaded shallow foundations, most problems are due to ground movements caused by factors not related to the weight of the building. Current methods for calculating ground movements associated with changes in the moisture content of partially saturated soils rely on largely empirical techniques and are only capable of providing an estimate of the likely movements. Particular difficulties involved in modelling the behaviour of shrinkable soils for shallow foundations are generally related to boundary conditions. The use of more realistic models for soil behaviour, together with the appropriate laboratory and field testing, could put design and analysis of shallow foundations on to a more rational and scientific basis, permitting the quantitative assessment of the relative merits of a number of different foundation solutions.

1. INTRODUCTION

The development of soil mechanics as a rational and coherent scientific discipline in the earlier part of the twentieth century was largely associated with deep foundations for heavy civil engineering structures. The subject therefore was concerned with the behaviour of soil at depth under high stresses and usually in a saturated state. This might have seemed sensible on the premise that deep foundations for heavy structures presented the major challenge. Harding and Glossop (1951), describing the influence of soil mechanics on foundation construction, reflected this view: "Since the construction of shallow foundations is generally a simple matter, this paper deals chiefly with deep foundations, such as may be required for heavy structures on sites where ground of adequate bearing capacity occurs at some depth below the surface; ..."

In reality it is usually easier to predict behaviour at high stresses than at low stresses and certainly far simpler to predict the behaviour of saturated soil at depth than partially saturated soil close to the ground surface. Cooling and Ward in 1948 had already recognised the problem of shallow foundations on shrinkable clays: "In the last five years a very large number of examples of building damage due to foundation movement has been encountered in which the effect of the weight of the building on the ground is negligible. Broadly speaking the damage arises from movement of the ground below the level of the foundations due to either climatic changes or the action of heat from industrial buildings."

It might be argued that low-rise buildings are less important because the value of the structures is much less. While this is true of buildings viewed individually, it is not the case when low-rise buildings are considered as a generic entity. In some years the value of insurance claims for subsidence and heave damage in the United Kingdom has exceeded £500 million.

Some of the problems have been appreciated for many years and advice has long been available. Cooling (1951) quoted a recommendation by Rankine (1862) that foundation depths should be such that the building "rests on earth below the reach of the disintegrating effects of frost and drought". Rankine suggested that minimum foundation depths in Britain should be 0.9m for sand and 1.2m for clay. Cooling commented that economic pressures had led to a tendency to cut down on depth, even in areas of highly shrinkable clay with consequent damage to houses in these areas. This led to a study of the problem by the Building Research Establishment. Field measurements have been carried out at many sites and guidance on shallow foundations on difficult soils has been given. Some consideration is given here to the application of numerical models to the design of shallow foundations.

2. SHALLOW FOUNDATION DESIGN

The typical model for assessing foundation behaviour is concerned with the application of a load to the ground and determining whether or not the allowable bearing capacity of the ground will be exceeded or intolerable settlement will be caused. Consolidation settlement of clay soils has often been calculated using a numerical model based on an elastic stress distribution to determine the additional stress due to the applied load and compressibilities determined in oedometer tests. Soil-structure interaction is important and the relative stiffnesses of the foundation and the soil will modify the pattern of movements, although this is rarely calculated explicitly.

However, this type of problem is not the dominant one for low-rise building foundations. Low-rise buildings typically apply relatively small stress increments to soil close to the ground surface. The stress distribution under footings subjected to a load of 50kN/m run (a typical load due to a party wall) is illustrated in Figure 1. With a footing width of 0.5m, the increase in stress in the ground decays to half the applied stress at 0.5m below footing level and is only one fifth of the applied stress at 1.5m depth. At this depth the increase in stress due to the applied load is only about 50% of the original overburden stress. Figure 2 shows

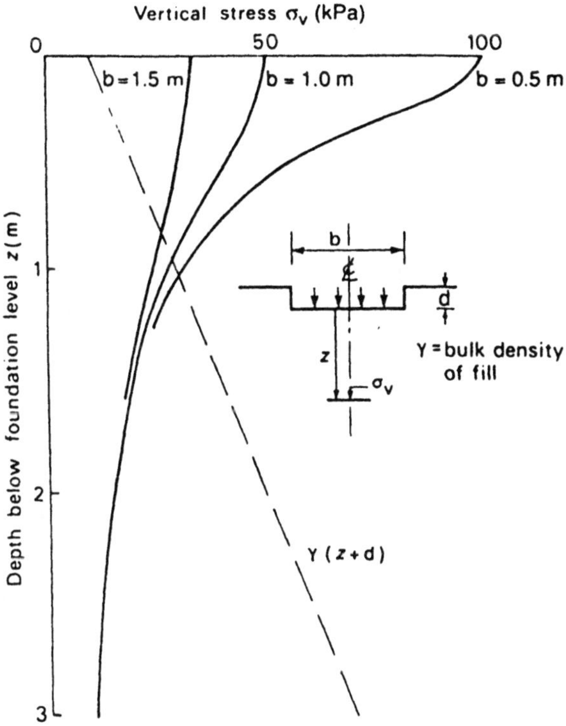

Fig. 1 Vertical stress (σ_v) on centre-line of strip footings of different widths with net load of 50kN/m run

how the maximum settlement of around 10mm, which might have been expected for a 0.5m width footing at 1m depth in London Clay due to a loading of 50kN/m run, compares with the measured seasonal settlements at foundation depth near trees. The seasonal movements were up to five times the calculated load induced settlements. For lightly loaded shallow foundations, most problems are due to ground movements caused by factors not related to the weight of the building. In these situations the building is subjected to imposed deformations, which may originate close to the ground surface or may be deep-seated.

Volumetric deformations in clays result from moisture content changes. If the water content of some clay soils is reduced below its 'natural' water content, the clay will shrink and if the water content is then increased the clay will swell. These changes in moisture content may be effected by seasonal changes or by vegetation or they may be a result of human activity. Deep-seated ground movements may result from mining or tunnelling whilst some phenomena such as collapse compression on inundation or biodegradation can produce effects at both shallow and large depths. Depending on the depth at which the ground movement occurs, the imposed deformation on the foundation will be affected by arching in the soil and by soil-structure interaction. An analysis of the foundation response to such imposed deformations will require an estimate of the likely movements due to each action.

The designer may vary the foundation type, width, depth and stiffness in order to obtain the most cost-effective design which fulfils the required serviceability and safety criteria. There are a number of sources of information and guidance on foundation design but there are few references to the specific problems associated with partially saturated soils. Driscoll et al (1996) described some of the potential hazards and included a number of foundation solutions suitable for low-rise construction on both natural and filled ground.

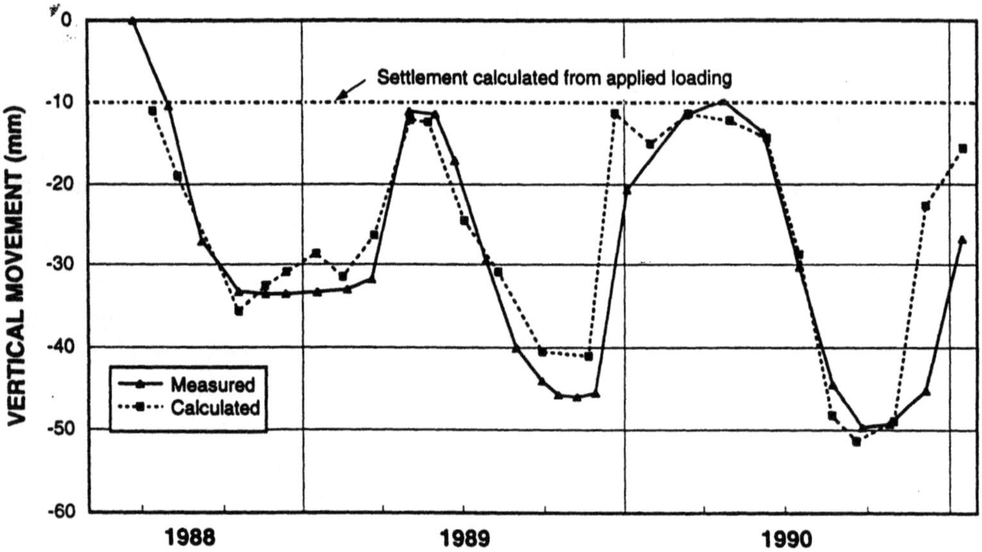

Fig. 2 Measured and calculated movements at 1m below ground level on London Clay near trees.

In order to obtain the most economic and safe design the designer must be able to predict with reasonable confidence the behaviour of the foundation and the soil, with respect to foundation deformation due to loading and other effects. Current methods for calculating the ground movements associated with changes in the moisture content of partially saturated soils rely on largely empirical techniques and are only capable of providing an estimate of the likely movements. In part this is due to the difficulty in measuring the changes in suction which result from the action of vegetation, re-hydration or inundation and which are a major factor in the determination of the ground deformations.

3. PARTIALLY SATURATED SOILS

Shallow foundations for low-rise buildings are rarely on saturated ground and an understanding of the behaviour of partially saturated soils is fundamental to the design of such foundations. In recent years major developments have taken place in the study of partially saturated soils and this interest was reflected in the First International Conference on Unsaturated Soils held in Paris in September 1995. Advanced testing techniques and numerical models have been developed. However, it may be questioned whether any of these developments, as yet, can be usefully applied to the shallow foundations of low-rise buildings.

Suction is a vitally important parameter in developing an understanding of the behaviour of partially saturated soils. Recent developments in the measurement of total and matric suction have led to a number of more accurate and reliable devices. Some of these have been used in engineering situations to provide a measure of suction changes, but generally these applications have been associated with the stability of embankment slopes and the monitoring of pavements and not in the monitoring of foundations.

There have also been major advances in the laboratory testing of partially saturated soils. Testing with suction measurement and control is possible for a wide range of stress paths but remains time-consuming and expensive. Only recently have tests with suction measurement been reported which match the flow conditions to be found in the field. If the latest constitutive models for partially saturated soils are to be validated for a wider range of soils, and extended to other stress paths, then many more test programmes on partially saturated soils will be necessary.

4. NUMERICAL MODELS

There has been significant progress made in the theoretical description of the mechanical and hydro-mechanical behaviour of partially saturated soils in recent years. Elastic models are easy to use and both linear elastic and non-linear elastic models have been formulated for partially saturated soils. Such models assume that the behaviour of the soil is reversible, and no account is taken of irreversible (plastic) strains. Elastic models should only be used when monotonic changes in either stress or water content are involved and the soil parameters should be measured in the laboratory along appropriate stress paths and over suitable strain ranges.

Elasto-plastic models include all constitutive formulations which allow the separation of strains which are elastic or recoverable and plastic or non-recoverable. This implies the existence of a yield criterion which defines the transition between elastic strains, for stress increments which lie within the yield surface, and plastic strains on reaching the defined surface whose location may be fixed in stress space or may be dependent on some past history. Recent developments in numerical modelling include elasto-plastic critical state model formulations which describe a number of important features of partially saturated soil behaviour, for example Alonso et al (1990). Such models predict that yield may occur either due to increased loading or due to a reduction in suction (collapse compression) and replicate path dependent behaviour of the soil following a number of loading and suction changes. If loading takes place along a path in which suction is not controlled, a further relationship is required in order to relate changes in water content to sample volume change and changes of suction to water content.

Wheeler (1996) suggested the use of a second volumetric state variable, specific water volume in addition to specific volume, in order to properly incorporate water content variations in the elasto-plastic framework. This type of approach appears to have greater potential than the use of more complex stress state variables.

All of the elasto-plastic models proposed are based on isotropic hardening, whereas there is evidence which suggests that the structure built up during deposition or compaction can have a significant effect on the initial suction and initial degree of expansion of the yield surface. It is not clear how much effect compaction has on the model parameters. In common with all complex soil models, whether saturated or partially saturated, the experimental determination of the soil parameters is difficult and time consuming.

5. MOVEMENT OF WATER

In 1939, Terzaghi pointed out that: "... in engineering practice difficulties with soils are almost exclusively due not to the soils themselves but to the water contained in their voids. On a planet without any water there would be no need for soil mechanics."

In the application of advanced soil models to the design of shallow foundations it is important to give due consideration to a number of water related effects. It is necessary to model not only pore pressures but also the flow of water through partially saturated soils, drying and wetting cycles and collapse compression.

The theoretical modelling of water flow through partially saturated soils is based on the same principles as saturated soils. The water is assumed to be incompressible and continuity and Darcy's Law are applied. The main difference is that the water storage capacity and the permeability of a partially saturated soil are dependant on the suction. The simplest case is where the flow of air is ignored. The analysis of water flow through partially saturated soil can be utilised in the prediction of deformations which may result from changes in the moisture content of either shrinking and swelling or collapsible soils. A solution can be obtained by first analysing the flow regime and then the subsequent deformations, but, as the flow properties are dependant on the soil state, a calculation in which the deformations and flow are coupled is more appropriate.

Experimental and field data from drying and wetting cycles show that a significant irreversible volumetric strain can be induced on the first wetting of samples which contain active clay minerals. The models described previously cannot model this behaviour as swelling and reloading within the yield surface follow elastic paths. Gens and Alonso (1992) suggested a modified form of elasto-plastic model, which distinguished between a saturated microstructure of individual clay packets and an unsaturated macrostructure formed by much larger inter-packet voids. The model defined behaviour at the macrostructural level, behaviour at the microstructural level and the coupling between the two levels. The model permits the qualitative explanation of features of partially saturated expansive soil behaviour which differ from non-expansive soils and in particular the pattern of swelling and shrinking over a number of cycles of suction. However, the possibility of collapse of the macrostructure which may accompany the swelling of the microstructure is not included in the model. Very little data on expansive clays in different stress paths is available to validate many of the key assumptions of this model and more data will be required before it could be used with confidence. Likely difficulties in obtaining the model parameters, in particular separating the strains associated with each level of the structure and in deriving the coupling relationships, and the extra complexity in the model itself mean that this type of model will be more difficult to use and formulate in numerical applications than those for non-expansive partially saturated soils.

As already discussed, significant advances have been made in constitutive models for shrinking and swelling clays over recent years. However, the particular difficulties involved in modelling the behaviour of such soils for shallow foundations are generally related to boundary conditions. In the UK most problems associated with shrinkable clays are related to desiccation induced by trees; the behaviour of tree roots and the factors that influence their growth are many, complex, poorly understood and the subject of some controversy. The selection of appropriate boundary conditions to model the magnitude of ground movement that might occur as a tree grows to maturity is therefore somewhat subjective, but is likely to be more important than the precise details of the soil constitutive model. The formation of desiccation cracks in shrinkable clays also gives rise to significant difficulties in analysis. The opening and closing of such cracks, in particular during the swelling process, gives rise to significant differences between anticipated and actual ground movements. Typically the results of heave potential calculations are divided by 3 to obtain more realistic predictions (Crilly et al, 1992). Figure 2 shows the measured seasonal movements at 1m below ground level compared with movements calculated from measured water content changes, assuming no change in degree of saturation and one dimensional behaviour. In order to replicate the observed behaviour the calculated movements were divided by 4! Whilst some of this difference can be attributed to the assumption of unchanging saturation, cracking is considered to be the most important factor. Blight (1997) discusses the use of the Griffith failure criterion for the onset of cracking; again, the use of a realistic crack formation and growth model is probably more important at present than the details of the soil constitutive model.

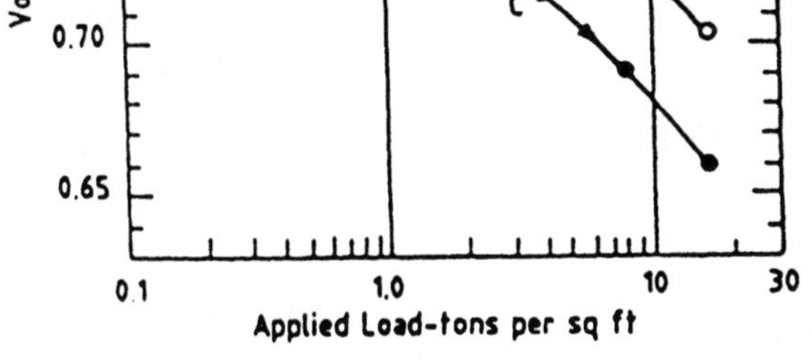

Fig. 3 Effect of saturation of a silt sample at constant axial stress and constant volume in a one dimensional compression test (after Burland, 1965).

Nobari and Duncan (1972) described a procedure for simulating collapse compression in rockfill dams based on experimental research and observations. The method was based on the observation that if one of two initially unsaturated identical samples is saturated and then loaded and the other is first loaded then saturated, the final stress-strain point is virtually the same. After saturation at constant stress, further loading causes the stress-strain curve to revert to that for a sample saturated at zero stress. This is illustrated under one dimensional loading in Figure 3 (after Burland, 1965).

When modelling collapse compression in dams during first filling Nobari and Duncan first calculated the stress state prior to filling. The stress change on wetting at constant strain was found from the stress difference between loading a dry and a saturated sample to the same strain. The stress change was applied to the stress state in the dam, resulting in a hypothetical stress distribution which did not satisfy equilibrium. The out-of-balance stress was re-applied, with the material model appropriate to the saturated material, in order to restore equilibrium and collapse strains resulted. Naylor et al (1989) developed a procedure for finding this stress change without reference to any particular constitutive law and implemented a collapse algorithm in a finite element program.

A similar technique has been applied by Skinner et al (1998) in two dimensions and and by Naylor (1997) in three dimensions to a large scale test involving collapse of limited zones of fill but has not yet been applied to foundations. The testing indicated that significant stress redistribution and arching occurred during collapse of a limited zone and the deformation patterns were highly dependent on the depth of the zone below ground surface. Approximate deformation moduli for the fill in compacted and saturated states were found from oedometer tests and used to model the collapse of a limited zone of material at a depth of 0.5m below ground surface (Figure 4). The measured surface settlements were not particularly well represented by a numerical analysis using an isotropic linear elastic-plastic model.

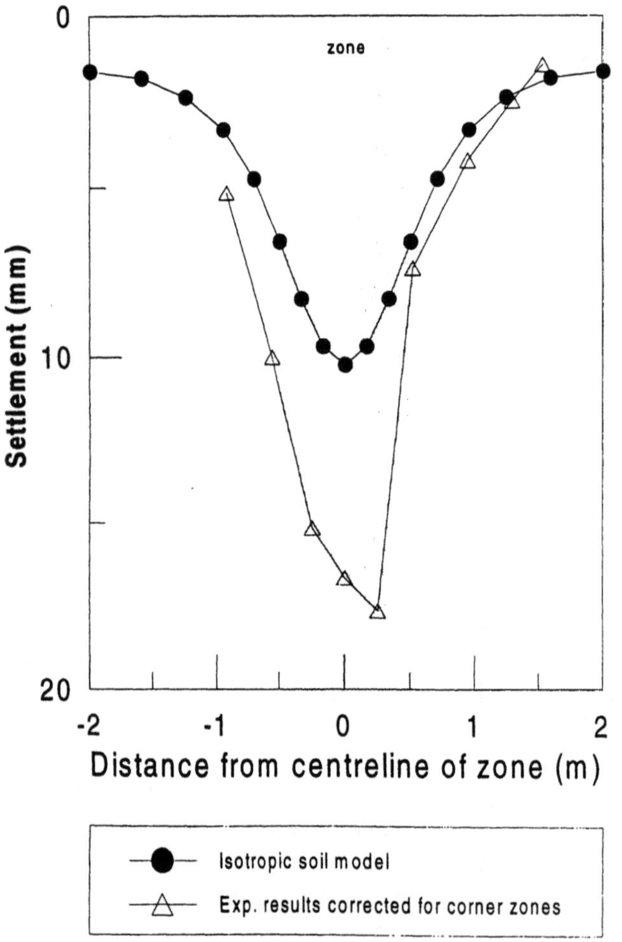

Fig. 4 Surface settlements – comparison of experimental and numerical results for 0.5m deep zone.

The application of the collapse algorithm is not dependent on the choice of model - the pre-inundation and post-inundation stress strain behaviour is represented by the same model but with different parameters - and hence the partially saturated nature of the soil is not addressed explicitly in the modelling. Large scale testing has shown that significant arching can occur if limited zones of soil suffer collapse and has highlighted the importance of modelling the behaviour of the partially saturated soil surrounding any collapsing zones.

6. CONCLUSIONS AND RECOMMENDATIONS

Current methods for the design and analysis of shallow foundations on partially saturated soils do not necessarily fully address the risks posed by foundation movement not caused by the applied loading. The use of more realistic models for soil behaviour, together with the appropriate laboratory and field testing, could put design and analysis of shallow foundations on to a more rational and scientific basis. The greater use of numerical techniques would allow soil-structure interaction to be taken into account and permit the quantitative assessment of the relative merits of a number of different foundation solutions for any given problem. Hence cost-effective and innovative foundations could be more readily identified and adopted whilst inappropriate foundation solutions could be avoided. Where foundations are already suffering distress, the use of improved soil models and appropriate testing techniques will permit a more rational assessment of the likely extent of further damage and lead to better targeted remedial action.

Although there has been progress in the modelling of the mechanical and hydro-mechanical behaviour of partially saturated soils in recent years, and the latest models permit the non-linear and stress path dependant aspects of the mechanical behaviour to be replicated, few models incorporate the changes in water content and suction which occur in general loading paths and, in particular, the hydraulic hysteresis which is demonstrated by most partially saturated soils. Further work is required on appropriate boundary conditions for shrinking and swelling clay problems. The models have in general been formulated to model element tests for soils with no anisotropy and must be validated against the behaviour of a wider range of 'real' soils and fills both in the laboratory and the field. The use of such models in the design or analysis of shallow foundations could provide an improvement over the current, largely empirical, techniques.

7. REFERENCES

ALONSO E.E., GENS A. AND JOSA A. (1990). A constitutive model for partially saturated soils. Geotechnique, 40: 405-430.

BLIGHT G. E. (1997). Thirty-seventh Rankine Lecture: Interactions between the atmosphere and the Earth. Geotechnique, 47: 715-767.

BURLAND J. B. (1965). Some aspects of the mechanical behaviour of partly saturated soils. Proceedings of the International Research and Engineering Conference on Expansive Clay Soils, Australia. Butterworth, 1: 270-278.

COOLING L. F. (1951). Some foundation problems in Great Britain. Proceedings of Building Research Congress, London, September, Division 1, Part III: 157-164.

COOLING L. F. AND WARD W. H. (1948). Some examples of foundation movements due to causes other than structural loads. Proceedings of 2nd International Conference on Soil Mechanics and Foundation Engineering, Rotterdam, 2: 162-167.

CRILLY M. S., DRISCOLL R. M. C. AND CHANDLER R. J. (1992). Seasonal ground and water movement observations from an expansive clay site in the UK. Proceedings of the 7th International Conference on Expansive Soils, Dallas, 1: 313-318.

DRISCOLL R. M. C., CRILLY M. S. AND BUTCHER A. P. (1996). Foundations for low-rise buildings. The Structural Engineer, vol. 74, 11/4: 179-185.

GENS A. AND ALONSO E.E. (1992). A framework for the behaviour of unsaturated expansive clays. Canadian Geotechnical Journal, 29: 1013-1031.

HARDING H. J. B. AND GLOSSOP R. (1951). The influence of modern soil studies on the construction of foundations. Proceedings of Building Research Congress, London, September, Division 1, Part III: 146-156.

NAYLOR D.J. (1997). Private communication.

NAYLOR D.J., TONG S-L. AND SHAHKARAMI A.A. (1989). Numerical modelling of saturation shrinkage. Proceedings of the International Symposium on Numerical Models in Geomechanics, Niagara Falls, Canada, 636-647.

NOBARI E. S. AND DUNCAN J. M. (1972). Movements in dams due to reservoir filling. Proceedings of the American Society of Civil Engineers Special Conference on Performance of Earth and Earth-supported structures. Purdue University, Indiana, 797-815.

RANKINE W. J. M. (1862). Civil Engineering, p 380. 26th edition. Charles Griffin and Company, London.

SKINNER H.D., CHARLES J.A. AND WATTS K.S. (1998). Ground deformations and stress redistribution due to a reduction in volume of zones of soil at depth. Accepted for publication in Geotechnique.

TERZAGHI K. (1939). Soil mechanics - a new chapter in engineering science. Journal of the Institution of Civil Engineers, 12: 106-141.

WHEELER S.J. (1996). Inclusion of specific water volume within an elasto-plastic model for unsaturated soil. Canadian Geotechnical Journal, 33: 42-57.

ARTIFICIAL NEURAL NETWORK ANALYSIS FOR THE EVALUATION OF SLOPE STABILITY

W. Xu
Unniversity of Hydraulic and Electric Engineering, Yichang, Hubei, China
J.-F. Shao
Université des Sciences et Technologies de Lille, Villeneuve d'Ascq, France

ABSTRACT The knowledge data-base which consist of 80 practical case problems for predicting slope stability is developed based on the Statistical Artificial Neural Network method. The forecasting results for the slope stability related to permanent shiplock rock slope engineering in Three Gorges Project have illustracted in details. It is shown that application of Statistical ANNs methods for the prediction of stability of slope engineering is realiable and practical.

KEYWORDS slope engineering; artificial neural network; stability; Three Gorges Project

1 Statistical Artificial Neural Networks and its Graphics

Artificial Neural Networks(ANN), neurocomputing or brainlike computation is based on the wishful hope taht we can reproduce at least some of flexibility and power of human brain by artificial means. Artificial Neural Networks consist of many simple computing element-generally simple nonlinear summing junctions - connected together by connections of varying strength, a gross abstraction of brain, which consist of very large numbers of far móre complex neurons connected together with far more complex and far more structured couplings.

The statistical artificial neural networks for prediction some geotechnical problem is developed by authors these years. As a self-organized forwardfeed artificial neural networks it is based on the statistical priciple familiar to self-study nector networks. The major advance of this networks is to solve problem rapidly and to simplify large problems.

The graphics of the statistical artificial neural networks is shown as Figure1.

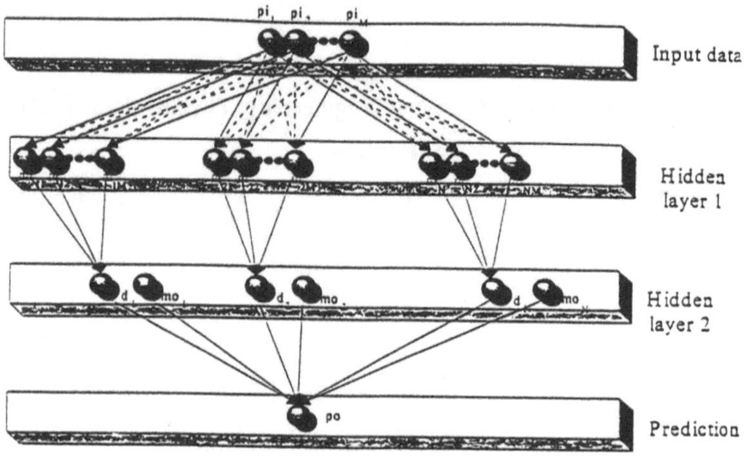

Fig.1 Graphics of Statistical Artificial Neural Networks

Notation in Figure1 have the following meaning:

p: prediction vector,
m: model vector,
i: indicates the neuron, belonging to the input variable,
o: indicates the neuron, belonging to the output variable,
N: number of model vectors.
M: number of input varieable of the phenomenon.

Different statistical methods demand the selection of shape of the function which best suits the discription of the geotechnical problem. The coefficients of empirical/regression equations are then determined with the least squares method. In such a way we try to describe the prblem with some advanced presumed empirical law. While the data are usually incomplete the selected law is fitted very good to the available data and usually fails when more data is acquired. The available database are not representative in most of the practical cases and therefore the automatic modeling of the phenomenon is very appropriate when new data are obtained. Compared to the parametric methods the algoritmic model prepared by our software does not need any advanced presumed law. The training phase here is very quickly and corresponds to the presentation of the model vectors. The prediction phase corresponds to the calculation of values of processing elements and to the calculation of the unknown output values.

Therefore as with all ANNs processing system it's clear that the reliability of the solution, and the furthering of the efficiency of the model strongly depends on the quality of the data base, and on the representativity of the samples. Representativity of the samples means an approximately uniform distribution of the samples in the prpblem space and, of course, a sufficient number of those samples.

Incidentally in our case we have to get several measures of rock mechanics with regulartime to form a data base before starting the modeling process. If the datas are noisy which will inevitably occur in such a field it is not necessary to collect much more data. in fact the noise depends on the way of modelling a geotechinical problem. Noisy data do not have an effect on the predicted result but will on the estimated error. As a result the forecasted output will present quite the similar gap as the initial database.

The major attribute that distinguishes the software here from other Artificial Neural Networks is the analysis speed. We use an algorithm, which does not require any learning phase, and than the answers about prediction can obtained almost immediatelly.

One of the features of the application is that the software uses only one coefficient, which has major influence on the results. The coefficient which is called penalty coefficient must be defined. It has an indirect relation to the learing error in classical ANNs. Knowing the right value for this coefficient let us to solve the entire problem.

Another feature is the ability to dynamic change the knowledge base. When the data are changes or new data are added to the data base the software adjusts the model automatically. This means that we can add some new data to neural network / or remove old ones, add some additional variables / or remove old ones and still get answers right away. It means there is no time consuming learning phase.

2 Knowledge - base building for rock slope stability evaluation

The model vectors for rock slope stability prediction is listed as the following.

$$\{mv\} = \{\gamma, c, \varphi, \phi, H, \upsilon; F, S\}$$

For each model there are six variables as the input values and two output varieties. The input varieties are:

- γ: rock unit weight,
- c: cohesion,
- φ: :angle of internal friction,
- ϕ: :angle of slope,
- H: height of slope,
- υ: ratio of pore pressure.

The output varieties are:

- F: safety index of slope,
- S: conditon of slope stability.

The slope is considered as stable if the safety index of slope is bigger than 1.20. If the safety index is smaller than 1.20 the slope is considered as failure. For the first situation the conditon of slope stability S is defined as 1.0 and for the second situation the S is 0.

After detailed collection and analysis of 80 practical case study related to the slope stability which are covered not only in china but also in world the knowledge-base for prediction slope stability with the methods ANNs is illustracted as Table1.

Table1 Practical slope stability case knowledge data-base

No.	γ KN/m^3	C (KPa)	φ:	ϕ:	H (m)	υ: ratio pore pressure	F	S
1	18.86	26.34	15	35	8.23	0.00	1.11	0
2	18.84	14.36	25	20	30.50	0.00	1.88	1
3	18.84	57.46	20	20	30.50	0.00	2.05	1
4	28.84	29.42	35	35	100.0	0.00	1.78	1
5	28.84	39.23	38	35	100.0	0.00	1.99	1
6	14.8	0	17	20	50.0	0.00	1.13	0
7	14.8	0	17	20	50.0	0.00	1.13	0
8	14	11.97	26	30	88.00	0.00	1.02	0
9	25	120	45	53	120.0	0.00	1.30	1
10	26	150.0	45	50	200.0	0.00	1.20	0
11	18.5	25	0	30	6.00	0.00	1.09	0
12	18.5	12	0	30	6.00	0.00	0.78	0
13	22.4	10	35	30	10.00	0.00	2.00	1
14	21.4	10	30.34	30	20.00	0.00	1.70	1
15	20.41	24.9	13	22	10.67	0.35	1.40	1
16	19.63	11.97	20	22	12.19	0.41	1.35	1
17	21.82	8.62	32	28	12.80	0.49	1.03	0
18	20.41	33.52	11	16	45.72	0.20	1.28	1
19	18.84	0	20	20	61.00	0.50	1.03	1
20	18.84	15.32	30	25	10.67	0.38	1.63	0
21	21.43	0	20	20	61.00	0.50	1.03	0
22	19.06	11.71	28	35	21.00	0.11	1.09	0
23	18.84	14.36	25	20	30.50	0.45	1.11	0
24	24.51	6.94	30	31	76.80	0.38	1.01	0
25	14	11.97	26	30	88.00	0.45	0.63	0
26	18	24	30.15	45	20.00	0.12	1.12	0
27	23	0	20	20	100.00	0.30	1.20	0
28	22.4	100	45	45	15.00	0.25	1.80	1
29	22.4	10	36	45	10.00	0.40	0.90	0
30	20	20	36	45	50.00	0.50	0.83	0
31	20	0	36	45	50.00	0.25	0.79	0
32	20	0	36	45	50.00	0.50	0.67	0
33	22	0	40	33	8.00	0.35	1.45	1
34	24	0	40	33	8.00	0.30	1.58	1
35	20	0	24.5	20	28.00	0.35	1.37	1

36	27	40	35	43	420.00	0.25	1.15	0
37	27	50	40	42	407.00	0.25	1.44	1
38	27	35	35	42	359.00	0.25	1.27	1
39	27	37.5	35	37.8	320.00	0.25	1.24	1
40	27	32	33	42.6	301.00	0.25	1.30	1
41	27	32	33	42.4	289.00	0.25	1.30	1
42	27.3	14	31	41	110.00	0.25	1.25	1
43	27.3	31.5	29.7	41	135.00	0.25	1.25	1
44	27.3	16.8	28	50	90.50	0.25	1.25	1
45	27.3	26	31	50	92.00	0.25	1.25	1
46	27.3	10	39	41	511.0	0.25	1.43	1
47	27.3	10	39	40	470.0	0.25	1.42	1
48	25	46	35	47	443.0	0.25	1.28	1
49	25	46	35	44	435.0	0.25	1.37	1
50	25	46	35	46	432.0	0.00	1.23	1
51	26	150	45	30	200.0	0.00	1.20	0
52	18.5	25	0	30	6.00	0.00	1.09	0
53	18.5	12	0	30	6.00	0.00	0.78	0
54	22.4	10	35	30	10.00	0.00	2.00	1
55	21.4	10	30.34	30	20.00	0.00	1.70	1
56	22	20	36	45	50.00	0.00	1.02	0
57	22	0	36	45	50.00	0.00	0.89	0
58	12	0	30	35	4.00	0.00	1.46	1
59	12	0	30	45	8.00	0.00	0.80	0
60	12	0	30	35	4.00	0.00	1.44	1
61	31.3	68	37	49	200.0	0.25	1.20	0
62	20	20	36	45	50.00	0.25	0.96	0
63	27	40	35	47.1	292.0	0.25	1.15	0
64	25	46	35	50	284.0	0.25	1.34	1
65	31.3	68	37	46	366.0	0.25	1.20	0
66	25	55	36	44.5	299.0	0.25	1.55	1
67	27.3	10	39	40	480.0	0.25	1.45	1
68	25	46	35	46	393.0	0.25	1.31	1
69	25	48	40	49	330.0	0.25	1.49	1
70	31.3	18.6	37	47	305.0	0.25	1.20	0
71	25	55	36	45.5	299.0	0.25	1.52	1
72	31.3	68	37	47	213.0	0.25	1.20	0
73	22	20	36	45	50.00	0.00	1.02	0
74	22	0	36	45	50.00	0.00	0.89	0
75	12	0	30	35	4.00	0.00	1.46	1
76	12	0	30	45	8.00	0.00	0.80	0
77	23.47	0	32	37	214.0	0.00	1.08	0
78	12	0	30	45	8.00	0.00	0.80	0
79	25.14	23.94	20	65	30.50	0.25	1.36	1
80	16.5	11.49	00	303	3.66	0.00	1.00	0

It seems very reasonable because the model vector is resumed from practical problem related to the nature slope/landslide or slope related to some practical rock engineering.

3 Case study for the slope stability prediction related to TGP

The Three Gorges Hydropower Project (TGP) is the largest engineering project not only in china but also in world. The permanent shiplock rock slope engineering is the key part related to TGP. A lot of geotechnical problems faced to civil engineer for the design and construction of 185m high and deep rock slope engineering. Amone then the stability and deformation of rock slope are the fundamental problems have to be answered succefully by the scientists.

Based on the understanding of the geological condition and construction methods of the rock slope engineering related to the permanent shiplock engineering TGP the followed slope section case study is carried out with the ANNs method which is developed in this paper.

Critical problem here for the calculation is to choice the optimal value of penalty coefficient. The best way to find the optimal value for penalty coefficients is to run filtration and verification and to observe the Root Mean Square (RMS) error. The filtration means the value of each output variable for each model vector from the database is predicted on the basis of all models vectors from the database. The verification refers two thirds of available data is used for the preparation of the model and the last third is used for the verification. The value, which will give the smallest verification RMS error, will be usually very close to the optimal penalty coefficient. The most important thing is also to find the type of penalty coefficient and his value, which minimize the verification Root Mean Square global error.

The error distribution for each output variable will be listed as the following two types of error distribution. One is defined as square root of sum squares (error power square) divided by the number of model vectors (N):

$$G_j = \sqrt{\frac{1}{N}\sum_{i=1}^{N}\left(E_{ij}\right)^2}$$

The another one is the global error of distribution for all the output. It is defined similarly, the expression under the square root is divided with the number of all output variables (M), as well:

$$G = \sqrt{\frac{1}{N * M}\sum_{j=1}^{M}\sum_{i=1}^{N}\left(E_{ij}\right)^2}$$

The results of the penalty coefficient optimization related problem of some parts of slope stability in the permanent shiplock engineering TGP are listed as following:

Table2 Penalty coefficient optimization

Normalization method		Statistical	Regular
Static Penalty Coefficient	Optimal penalty coefficient	0.3639	0.0272
	Global Filtrate. RMS error	0.19311	0.03874
	Global Verify. RMS error	0.58736	0.20973
Dynamic Penalty Coefficient	Optimal penalty coefficient	0.1185	0.0594
	Global Filtrate. RMS error	0.17372	0.036863
	Global Verify. RMS error	0.058899	0.20863
Nearest Penalty Coefficient	Optimal penalty coefficient	0.3129	0.1025
	Global Filtrate. RMS error	0.1067	0.027786
	Global Verify. RMS error	0.57769	0.23347

Finally it has chosen the dynamic penalty coefficient and the regular normalization method, because it is the combination which minimizes the global RMS Verification error. With this coefficient the distribution error is shown as Figure2.

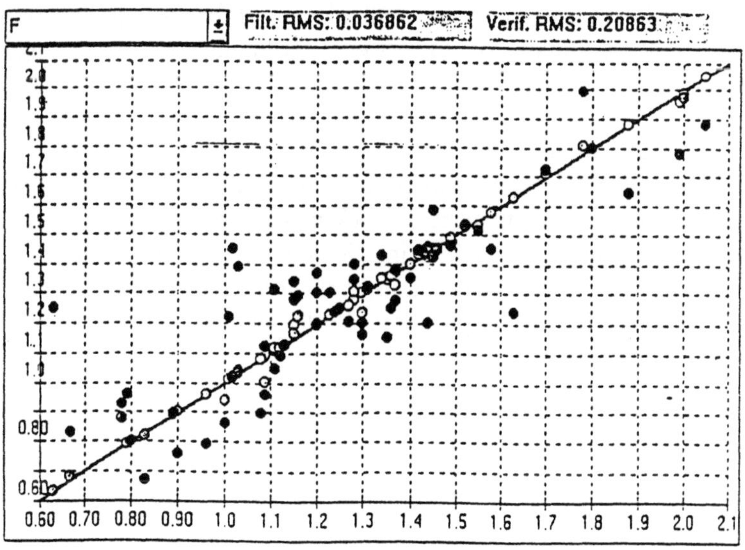

Figure2 Error distribution for the safety index

With this model we have gotten following results for the slope stability prediction of several section parts of permanent shiplock rock slope engineering as listed in Table3. The predicted results are satisfied and reasonable compared with the practical situation of slope engineering in TGP.

Table3 Predicted Results Related TGP Based on Knowledge Data-base

No.	γ KN/m^3	C (KPa)	φ:	ϕ:	H (m)	υ: ratio pore pressure	F	S
1	26.49	150	33	45	73.00	0.15	1.23	1
2	26.7	150	33	50	130.00	0.25	1.80	1
3	26.89	150	33	52	120.00	0.25	1.80	1
4	26.57	300	38.7	45.3	80.00	0.15	1.20	0
5	26.78	300	38.7	54	155.00	0.25	1.20	0
6	26.81	200	35	58	138.00	0.25	1.20	1
7	26.43	50	26.6	40	92.20	0.15	1.25	1
8	26.69	50	26.6	50	170.00	0.25	1.25	1
9	26.81	60	28.8	59	108.00	0.25	1.25	1

This work is a part of results in subject NSFC59493600 which is jointed supported by the Nature Science Foundation of China(NSFC) and China Yangtze Three Gorges Project Development Corporation.

REFERENCES

Weiya XU and J.-F. SHAO, The thoery and methodology of feedback progressive design for rock slope engineerig. Journal of University of Hydraulic and Electric Engineering, Vol.20, 1997, p.1-8.

Weiya XU and J.-F. SHAO, Application of ANNs in the rock slope engineering. Journal of University of Hydraulic and Electric Engineering, Vol.20, 1997, p.9-20

Weiya XU and Shijing WANG, Some main problem of feedback design based on pregeological forecasting for permanent shiplock rock mass slope speed construction Three Gorges Project, Journal of Engineering Geology, Vol.5, 1994,p. 36-44.

Weiya XU, Evolutionary model building and its application in rock mechanics. OPTIMIZATION-techniques and application, World Scientific Publishing, 1995

5. LATE PAPERS

STRESS-STRAIN BEHAVIOUR OF COHESIONLESS SOILS
EXPERIMENTS, THEORY AND NUMERICAL COMPUTATIONS

W. Ehlers and H. Müllerschön
Stuttgart University, Stuttgart, Germany

ABSTRACT: Cohesionless soils as for example sand show a very complex stress-strain behaviour depending on the stress state and the load history. To capture this material behaviour and, furthermore, to describe cohesionless soils under saturated and unsaturated (empty) conditions, an elasto-plastic model is used and presented in the framework of the Theory of Porous Media (TPM) defined by the elements mixture theory and concept of volume fractions. Within the constitutive equations of the elasto-plastic model, the elastic response is described by a materially non-linear elasticity law. Plastic deformations are considered in the context of a single-surface yield function with isotropic hardening properties. Non-associated flow is realized with an additional plastic potential function. Isotropic behaviour is assumed and the study is restricted to small strains. Parameter identification for the described model is shown for dense Berlin Sand on the basis of triaxial tests. The presented model is implemented into the finite element code PANDAS*, numerical examples will demonstrate the capability of the formulation.

1. INTRODUCTION

Basically, the description of saturated, unsaturated and empty porous solid materials (e. g. frictional materials) can be carried out by use of the well-founded framework of the Theory of Porous Media (TPM). The TPM approach is based on a homogenization process to handle the microscopically inhomogeneous porous solid material with a macroscopic continuum mechanical description, *cf.* e. g. the work by TRUESDELL & TOUPIN (1960), BOWEN (1976, 1980), DE BOER *et al.* (1991) and EHLERS (1993).

*Porous media Adaptive Nonlinear finite element solver based on Differential Algebraic Systems

To describe the stress-strain relation of a granular soil matrix, an elasto-plastic model seems to be suitable. Basis of an elasto-plastic theory is the assumption that the total deformation consists of the sum of a recoverable elastic part and an irreversible plastic part. Furthermore, experimental results indicate that the elastic parameters, as e. g. the shear modulus or the YOUNG's modulus, are functions of the stress state, respectively of the deformation of the structure. Nevertheless, a thermodynamically consistent elasticity law can be formulated without any restriction to the existence of a hyperelastic potential (LORET 1985).

Plastic strains are described within a rate-independent plasticity approach. The presented contribution proceeds from a single-surface yield function, a non-associated flow rule based on an additional plastic potential function and an isotropic work-hardening formulation. The evolution of the yield surface is defined by rate equations for two parameters included in the yield criterion. Although the model is rather complex, the included parameters can be identified on the basis of the results of conventional triaxial compression and extension tests.

For the numerical treatment of the problem within the finite element method (FEM), the weak formulation of the field equations is used. The elasto-plastic model is implemented in the numerical code PANDAS to calculate the stress-strain relations of the soil matrix under saturated and empty conditions. Finally, the capability of the model is demonstrated by numerical examples.

2. FIELD EQUATIONS FOR A TWO-PHASE MODEL

The description of coupled deformation-flow processes in porous materials requires the consideration of a multi-phase problem within the Theory of Porous Media proceeding from the classical mixture theory and the concept of volume fractions.

In the following, a materially incompressible two-phase model is considered. Thereby, the constituents φ^α are a solid phase ($\alpha = S$), desribed through an elasto-plastic model, and a viscous fluid phase ($\alpha = F$). The inclusion of material incompressibility of a constituent implies incompressibility in the material's micro range, which does not lead to macroscopic incompressibility. Thus, even if the realistic density $\rho^{\alpha R}$ of a constituent φ^α is constant, the partial densities $\rho^\alpha = n^\alpha \rho^{\alpha R}$ can still change through changes in the volume fractions n^α. The model incorporates three independent fields, the solid displacement $\mathbf{u}_S$, the seepage velocity $\mathbf{w}_F$ and the effective fluid pressure p. The corresponding three equations for quasi-static considerations (negligible accelerations),

$$-\frac{k^F}{\gamma^{FR}}(\operatorname{grad} p - \rho^{FR}\,\mathbf{b}) = n^F\,\mathbf{w}_F\,, \qquad (\text{DARCY's law}) \qquad (1)$$

$$\operatorname{div}(\mathbf{T}^S_E - p\,\mathbf{I}) + (n^S\rho^{SR} + n^F\rho^{FR})\,\mathbf{b} = \mathbf{0}\,, \qquad (2)$$

$$\operatorname{div}\left[(\mathbf{u}_S)'_S + n^F\,\mathbf{w}_F\right] = 0\,, \qquad (3)$$

can be obtained from the kinematics, the balance relations and the constitutive equations described in EHLERS (1993) and DIEBELS *et al.* (1998). Therein, k^F is the DARCY permeability parameter, $\mathbf{b}$ the body force (gravity), γ^{FR} the effective specific weight and $(\mathbf{u}_S)'_S$ the time derivative of the solid motion. The solid extra stress tensor $\mathbf{T}_E^S$ represents the effective stress of the soil matrix.

3. ELASTIC STRAINS

Elastic strains can be isolated through unload-reload cycles in triaxial tests. In case of moderate unloading, hysteresis effects can be neglected and intermittent strains are assumed to be entirely recoverable and hence elastic (Fig. 1). A special property of granular materials is the dependency of the elastic stiffness of the stress state, respectively of the deformation state of the structure. To take into account this phenomenon, the elastic parameters vary with the elastic volumetric strain, and furthermore consider the plastic volumetric strain and the point of compaction as governing parameters.

In particular, hydrostatic tests show a non-linear behaviour towards the point of compaction, compare Fig. 2, defined by a critical packing density, where an increase of the isotropic load cannot cause a further increase of the volumetric strains, unless there is grains chrunching.

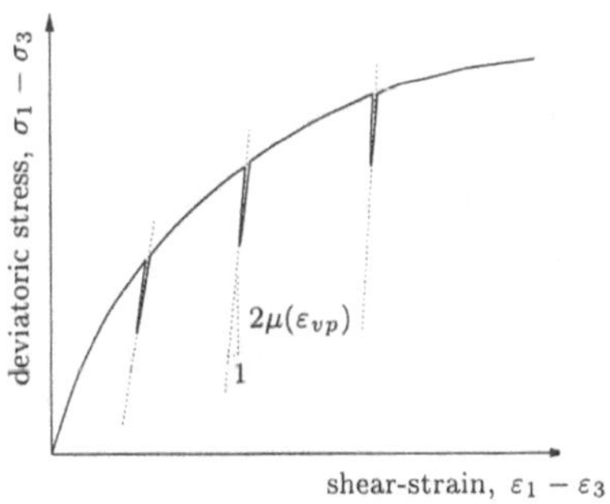

Fig. 1: Unload-reload cycles for triaxial compression, $\sigma_3 =$ const. (qualitatively)

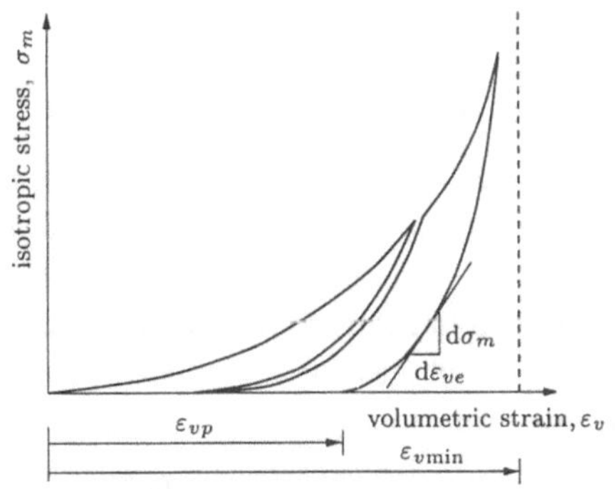

Fig. 2: Unload-reload cycles in isotropic compression test (qualitatively)

To describe the elastic behaviour, it is assumed that

$$\mathbf{T}_E^S = 2\mu\, \boldsymbol{\varepsilon}_e^D + k\,(\boldsymbol{\varepsilon}_e \cdot \mathbf{I})\mathbf{I} \tag{4}$$

be a modification of HOOKE's law. The tensor $\boldsymbol{\varepsilon}_e^D = \boldsymbol{\varepsilon}_e - \frac{1}{3}(\boldsymbol{\varepsilon}_e \cdot \mathbf{I})\mathbf{I}$ is the deviatoric part of the elastic strain tensor $\boldsymbol{\varepsilon}_e$. To consider the variation of the shear strength, the shear modulus μ is introduced as a linear function of the plastic volumetric strain $\varepsilon_{vp} = \boldsymbol{\varepsilon}_p \cdot \mathbf{I}$:

$$\mu = \mu(\varepsilon_{vp}) = \mu_0\, |\varepsilon_{vp}|. \tag{5}$$

Therein, ε_{vp} must be understand as a parameter and thus, μ is a constant at any given state of the plastic deformation. The bulk modulus k not only depends on ε_{vp} as a parameter of the plastic deformation, but is also a function of the elastic volumetric strains $\varepsilon_{ve} = \boldsymbol{\varepsilon}_e \cdot \mathbf{I}$. Finally, the point of compaction must be included by the parameter $\varepsilon_{v\,\min}$ describing the maximum of the volume contraction. Thus,

$$\begin{aligned} k &= k\,(\varepsilon_{ve}; \varepsilon_{vp}, \varepsilon_{v\,\min}) \\ &= k_1 + k_2 \ln\left[1 + \varepsilon_{ve}\left(\varepsilon_{ve} + \varepsilon_{vp} - \varepsilon_{v\,\min} + \frac{1}{\varepsilon_{vp} - \varepsilon_{v\,\min}}\right)\right] \varepsilon_{ve} \quad . \end{aligned} \tag{6}$$

k_1 and k_2 are constants to be fitted at unload-reload cycles in isotropic compression tests. Parameters for the considered dense sand are given in Section 5, Table 1.

4. PLASTIC SOIL PROPERTIES

4.1 CONSTITUTIVE EQUATIONS

In order to bound the elastic domain, a *single-surface yield function* (Ehlers 1995) exhibiting a closed and smooth shape in the principal stress space is introduced:

$$F(\mathbf{T}_E^S, \boldsymbol{q}^d, \boldsymbol{q}^h) = \sqrt{\mathrm{II}_D \left(1 + \gamma\, \mathrm{III}_D/(\mathrm{II}_D)^{3/2}\right)^m + \tfrac{1}{2}\alpha \mathrm{I}^2 + \delta^2 \mathrm{I}^4} + \beta \mathrm{I} + \epsilon \mathrm{I}^2 - \kappa = 0\,, \tag{7}$$

$\mathrm{I}, \mathrm{II}_D, \mathrm{III}_D$: Invariants of the extra stress tensor $\mathbf{T}_E^S$,

$\left.\begin{aligned} \boldsymbol{q}^d &= (\gamma, m)^T \\ \boldsymbol{q}^h &= (\alpha, \beta, \delta, \epsilon, \kappa)^T \end{aligned}\right\}$: Parameters associated to the hydrostatic and the deviatoric plane.

The additional index $(...)_D$ characterizes the deviatoric stress invariants. The material parameters $\boldsymbol{q}^h$ govern the effect of *isotropic work-hardening* processes, realized through the rate equations

$$(\boldsymbol{q}^h)'_S = C_i\,(\overset{\star}{\boldsymbol{q}}{}^h - \boldsymbol{q}^h)\,(W_p)'_S\,, \tag{8}$$

with the boundary conditions

$$\boldsymbol{q}^h(W_p = 0) = \boldsymbol{q}_0^h > \overset{\star}{\boldsymbol{q}}{}^h \quad \text{and} \quad \boldsymbol{q}^h(W_p \to \infty) = \boldsymbol{q}_\infty^h = \overset{\star}{\boldsymbol{q}}{}^h\,. \tag{9}$$

W_p is the plastic work per unit volume accumulated during the deformation process, C_i are constants to fit equation (8) to experimental results, $\boldsymbol{q}_0^h$ is the parameter set of the initial yield surface and $\overset{\star}{\boldsymbol{q}}{}^h$ comprises the parameters of the limiting yield surface, respectively of the failure surface. Reformulation of equation (7) with the Reuss variables r (yield radius) and Θ (Lode angle) leads to

$$F(\mathrm{I},\, \mathrm{II}_D,\, \mathrm{III}_D;\, \mathbf{q}^d, \mathbf{q}^h) \;\Rightarrow\; F(\mathrm{I},\, r,\, \Theta;\, \mathbf{q}^d, \mathbf{q}^h). \tag{10}$$

Thus, the yield function can be given by

$$r(\Theta,\mathrm{I}) = \underbrace{\sqrt{2}\left[\left(\epsilon^2-\delta^2\right)\mathrm{I}^4 + 2\beta\epsilon\mathrm{I}^3 + \left(\beta^2-\frac{1}{2}\alpha-2\epsilon\kappa\right)\mathrm{I}^2 - 2\beta\kappa\mathrm{I} + \kappa^2\right]^{1/2}}_{F_h(\mathrm{I})} \cdot$$

$$\cdot \underbrace{\left[1+\frac{2}{\sqrt{27}}\gamma\sin(3\Theta)\right]^{-m/2}}_{F_d(\Theta)} . \qquad (11)$$

Therein, $r(\Theta,\mathrm{I})$ is the yield radius given as the product of two independent functions, $F_h(\mathrm{I})$ and $F_d(\Theta)$, where $F_h(\mathrm{I})$ represents the hydrostatic shape function with $\Theta = 0°$, while $F_d(\Theta)$ describes the shape of the yield condition in the deviatoric plane.

In general, soils are frictional materials and the description of plastic flow requires a *non-associated flow rule.* Therefore, an additional plastic potential (EHLERS & MAHNKOPF 1998)

$$G(\mathbf{T}_E^S, \boldsymbol{q}^d, \boldsymbol{q}^h) = \sqrt{\psi_1 \mathrm{II}_D + \frac{1}{2}\alpha\,\mathrm{I}^2 + \delta^2\,\mathrm{I}^4} + \psi_2\beta\,\mathrm{I} + \epsilon\,\mathrm{I}^2 \qquad (12)$$

is defined. The parameters ψ_1 and ψ_2 specify the deviation of the shape of the plastic potential, respective of the shape of the yield surface in the hydrostatic plane. In the deviatoric plane, coaxial flow perpendicular to the hydrostatic axis is assumed. Given equation (12), the flow rule

$$(\boldsymbol{\varepsilon}_p)_S' = \Lambda\,\frac{\partial G}{\partial \mathbf{T}_E^S} \qquad (13)$$

holds, where Λ is the usual plastic multiplier.

4.2 PARAMETER IDENTIFICATION – DEVIATORIC PLANE

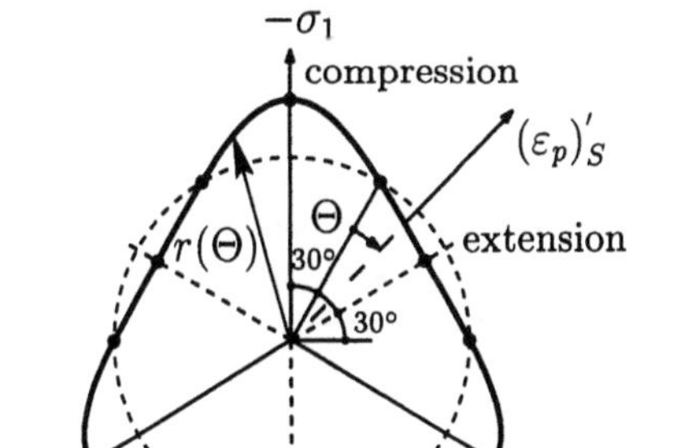

Fig. 3: Octahedral plane; cross-section perpendicular to the hydrostatic axis

Cubical triaxial tests, performed for example by YAMADA & ISHIHARA (1979), indicate that the deviatoric contours of the plastic work (equipotential surfaces of W_p) have circular shapes at low stress levels and gradually change into rounded triangular shapes when the stress level increases up to failure, compare Fig. 3. The formulation of the yield function in (7) allows the description of this property for a suitable choice of γ and m. The ratio of the compression and the extension radii, r_c and r_e, in the failure state, provides the first condition for the determination of γ and m.

In addition, the restriction to convexity evaluated at the point of extension delivers the second condition for these parameters. Change of the shape of the yield curve is realized through a linear increase of γ from $\gamma = 0$ (circular) to γ_{max} (rounded triangular shape) according to the deviatoric stress level.

4.3 PARAMETER IDENTIFICATION – HYDROSTATIC PLANE

In equation (11), F_h describes the contour of the yield curve in the hydrostatic plane for $\Theta = 0°$. The parameters $\boldsymbol{q}^h$ specify the shape of the curve. For sand as a cohesionless material, κ and α are zero. Thus, there are three remaining parameters β, δ and ϵ to determine the yield criterion.

Since yielding is a continous process for frictional materials, there may not be a distinct point of initial yielding on a stress-strain curve. The determination of yield surfaces have been studied in detail since the plasticity theory has been found effective in modeling deformation of frictional materials. In the presented model, the method of identification of yield points and yield surfaces in stress space is based on the assumption that yield surfaces are equivalent to plastic work contours as was suggested by LADE & KIM (1988). For each stage of several triaxial compression tests and of an isotropic compression test, the plastic work W_p is calculated. A family of yield contours is approximated through an optimization process to stress points with the same values of W_p, compare Fig. 5.

The parameter β is directly connected to the slope of the yield surface at the origin of the stress space (I = 0). Comparison with the friction angle ϕ of the well-known MOHR COULOMB theory leads to the relation $\beta = \frac{1}{3}\sin\phi$ for the failure state. The evaluation of ϕ for compression tests with six different values of the confining pressure is shown in Fig. 4. An increase of the friction angle with a decrease of applied confining pressure is observed. For I = 0, $\phi = 48.6°$ is assumed, which corresponds to $\beta = 0.25$. During hardening and hence for the above mentioned optimization, β is kept constant at this value.

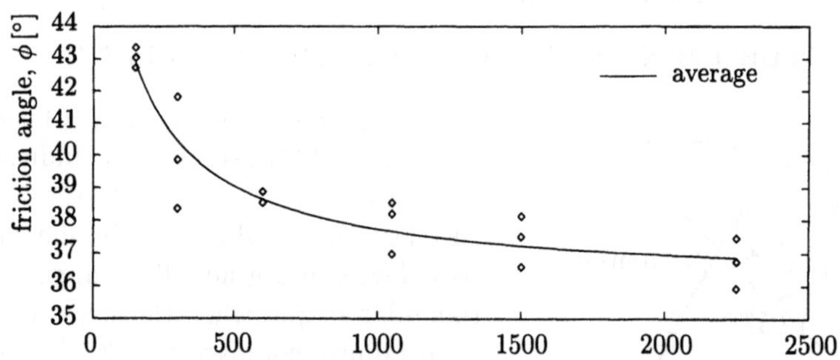

Fig. 4: Friction angle vs. confining pressure observed in triaxial compression tests on dense Berlin sand

Hence, δ and ϵ are the hardening parameters, which are evaluated according to the hardening law in equation (8). Fig. 5 shows a choice of adapted yield curves for different values of W_p. Since there are two hardening parameters for the description of the hydrostatic yield function, a non-similar evolution of the shape of the yield curve is possible and therefore a more flexible fit to the experimental results is feasible.

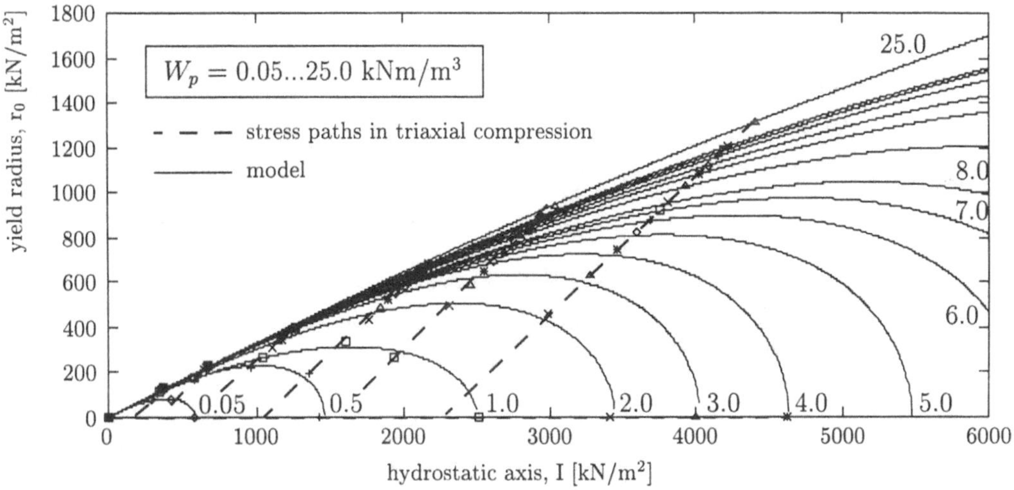

Fig. 5: Contours of constant plastic work and exp. data points shown on ($\Theta = 0°$)-plane

A failure surface is enclosed in the model defined through $\overset{\star}{\delta}$ and $\overset{\star}{\epsilon}$, which are the limit values for the evolution of δ and ϵ, compare Fig. 6. In hydrostatic direction, the expansion of the yield surface for sand is very sensitive, e. g. increments of volumetric plastic strains are rather small for high stress levels. In combination with the presented elasticity law in Section 3, the increments of total volumetric strains converge to very small values, which makes sense for granular materials. Following a triaxial compression stress path leads to a critical point of deviatoric stress level, where the yield surface no more expands and hence ideal plasticity takes place.

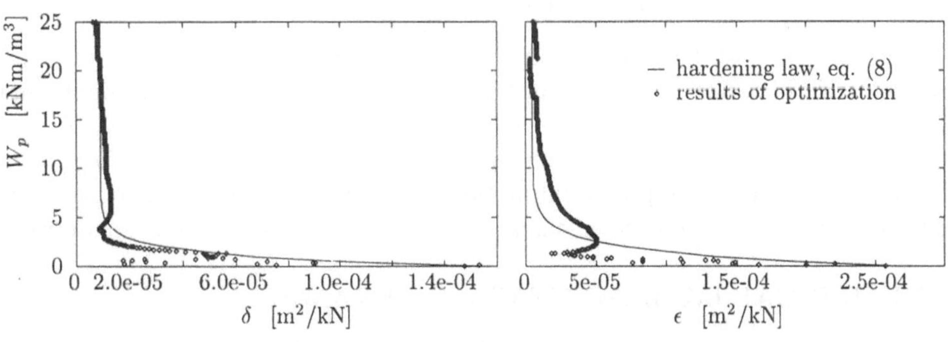

Fig. 6: Development of the parameters δ and ϵ during hardening

As mentioned above, non-associated flow can be realized through an additional plastic potential G (eq. (12)). The parameters ψ_1 and ψ_2 specify the projection of the flow direction in the hydrostatic plane. Determination of these parameters is done through a fit to volume change curves of triaxial compression tests.

5. PARAMETER VALUES FOR BERLIN SAND

Table 1 shows the parameter values for the presented elasto-plastic model determined in triaxial tests on dense Berlin sand.

Elasticity			
Parameter	Symbol	Value	Unit
Elasticity law	μ_0	$1.07 \cdot 10^8$	kN/m^2
	k_1	$6.11 \cdot 10^6$	kN/m^2
	k_2	$3.86 \cdot 10^4$	kN/m^2
	$\varepsilon_{v\,\min}$	-0.019	–
Plasticity			
Yield and hardening functions	α	0	–
	κ	0	kN/m^2
	β	0.25	–
	$\overset{\star}{\delta}$	$8.81 \cdot 10^{-6}$	m^2/kN
	C_δ	0.81	m^2/kN
	δ_o	$1.48 \cdot 10^{-4}$	m^2/kN
	$\overset{\star}{\epsilon}$	$5.00 \cdot 10^{-6}$	m^2/kN
	C_ϵ	0.60	m^2/kN
	ϵ_o	$2.58 \cdot 10^{-4}$	m^2/kN
	γ	0.569	–
	m	1.664	–
Plastic potential	ψ_1	0.342	–
	ψ_2	0.832	–

Table 1: Material parameters for the elasto-plastic model

The characteristics of the sand are summerized as follows: coefficient of uniformity: 2.1; specific weight of grains: $\rho^{SR} = 2.653$ g/cm^3; maximum void ratio: 0.817; minimum void ratio: 0.481. All tests are realized on samples with a bulk density of $\rho_d = n_{0S}^S\, \rho^{SR} = 1.71$ g/cm^3. This density corresponds to an initial volume fraction of the solid of $n_{0S}^S = 0.644$. Viscous effects within the pore fluid motion are included by the DARCY permeability coefficient k^F. For the considered sand, $k^F = 2.5 \times 10^{-4}$ [m/s] is an adequate value.

6. NUMERICAL IMPLEMENTATION – EXAMPLE

The presented elasto-plastic model is implemented in the finite element code PANDAS. For the numerical treatment of the problem, the weak formulation of the field equations (1-3) is used. The incompressibility conditions for both constituents, fluid and solid, lead to a system of differential-algebraic equations (DAE) of first order in time. This DAE system is handled by a diagonally implicit RUNGE-KUTTA (DIRK) time integration scheme (DIEBELS *et al.* 1998). Time discretization of the elasto-plastic equations, described in Section 4, leads to a system of nonlinear equations. For the solution, a NEWTON procedure is applied in each time step. A consistent linearization of the nonlinear equations guarantees quadratic convergence within the NEWTON procedure.

Model validation – triaxial tests:

Predictions for conventional triaxial compression tests on Berlin sand were calculated from the model and compared with experimental results of three similar tests in Fig. 7. In these tests, the confining pessure was kept constant at $\sigma_3 = 500$ kN/m^2, while the major principle compressive stress σ_1 was increased to peak failure. The model predictions capture the observed stress-strain relations accurately.

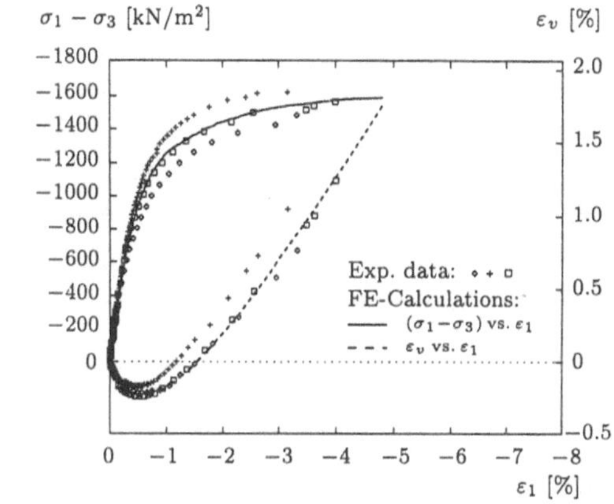

Fig 7: Comparison with exp. results

Coupled fluid-solid problem:

This example handles a fluid-saturated sand region. Load is applied by a time dependent external force on a rigid strip footing. The boundary beside the footing is perfectly drained, the other boundaries are rigid and undrained. Fig. 8 shows the contours of isolevels for the effective fluid pressure p at a given time. The arrows display the seepage velocity of the pore fluid. The problem is calculated for different values of the DARCY permeability coefficient k^F. The load is increased linearly at a load-velocity of 1 kN/m^2 per second. The load-settlement and time-settlement curves (Fig. 9) demonstrate the influence of the variation of k^F on the displacement of the footing. The viscocity of the fluid leads to a time-dependent deformation. In case of a lower value of k^F, the settlement is larger, since fluid pressure decreases very fast and the soil matrix has to carry the greater part of the load. In Fig. 8b, the load is kept constant at 500 s, after some time the curves converge to the final settlement.

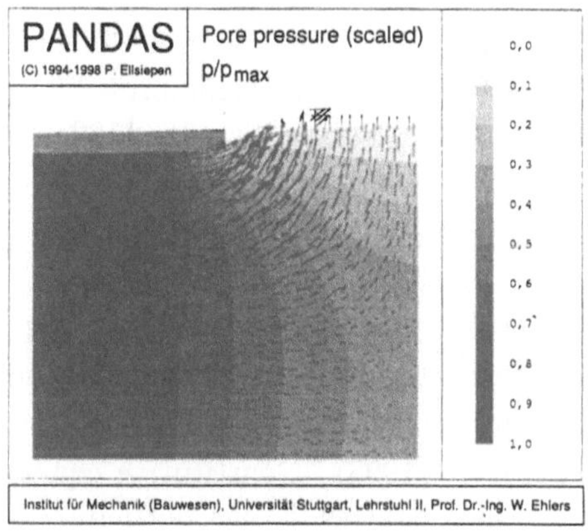

Fig 8: Consolidation problem for fluid-saturated sand

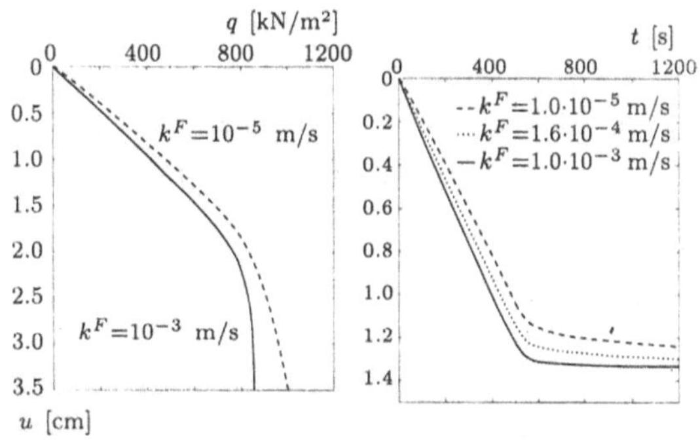

Fig 9: Load-settlement/Time-settlement curve

7. CONCLUSIONS

In the present article, it was shown that a unified approach can be used to describe the stress-strain behaviour of frictional materials. A combination of elasticity and plasticity theories may provide a reasonable and consistent description of this behaviour. In addition to a non-linear elasticity law, the components of this framework are the single-surface yield function, the plastic potential, the flow rule and the work-hardening law.

The elasto-plastic model within a multiphase theory is implemented in the FE-Code PANDAS. Validation of the model was achieved by comparison of predicted and experimental stress-strain curves.

REFERENCES

DE BOER R., W. EHLERS, S. KOWALSKI & J. PLISCHKA (1991). Porous Media – A Survey of Different Approaches. Forschungsberichte aus dem Fachbereich Bauwesen, Heft 54, Universität-GH-Essen.

BOWEN R. M. (1976). Theory of mixtures. in A. C. Eringen (ed.): Continuum Physics, Vol. III. Academic Press, New York: 1-127.

BOWEN R. M. (1980). Incompressible porous media models by use of the theory of mixtures. Int. J. Engng. Sci. 18: 1129-1148.

DIEBELS S., P. ELLSIEPEN & W. EHLERS (1998). Error-controlled RUNGE-KUTTA Time Integration of a Viscoplastic Hybrid Two-phase Model. Technische Mechanik, submitted 4/1998.

DIEBELS S., P. ELLSIEPEN & W. EHLERS (1998). A two phase-model for viscoplastic geomaterials. Proceedings of the International Symposium on Dynamics of Continua, Bad Honnef 1996, appears in 1998.

EHLERS W. (1993). Constitutive equations for granular materials in geomechanical context. In K. Hutter (ed.): Continuum Mechanics in Environmental Sciences and Geophysics, CISM Courses and Lecture Notes No. 337. Springer-Verlag, Berlin: 313-402.

EHLERS W. (1995). A single surface yield function for geomaterials. Arch. Appl. Mech. 65: 63-76.

EHLERS W. & D. MAHNKOPF (1998). Elstoplastizität und Lokalisierung poröser Medien bei finiten Deformationen. ZAMM 98, submitted 5/1998.

LADE P. V. & M. K. KIM (1988). Single Hardening Constitutive Model for Frictional Materials, Part II. Computers and Geotechnics, 6: 13-29.

LORET B. (1985). On the choice of elastic parameters for sand. Int. J. Num. Anal. Methods Geomech., 9: 285-292.

TRUESDELL C. & R. A. TOUPIN (1960). The classical field theories. In S. Flügge (ed.): Handbuch der Physik, Band III/1, Springer-Verlag, Berlin: 226-902.

YAMADA Y. & K. ISHIHARA (1979). Anisotropic Deformation Characteristics of Sand under Three Dimensional Stress Conditions. Soils and Foundations, JSSMFE, 19:79-94.

UNSATURATED BEHAVIOUR OF EARTHFILL DAMS

A. Modaressi and A. Michali
École Centrale de Paris, Châtenay-Malabry, France

ABSTRACT

The computation and safety analysis of earth dams requires a good knowledge of soil behaviour and a global understanding of the phenomena involved. For example, pore pressure distribution within earth dam, presence of the air, etc, should be analysed through out the dam life. A unified approach covering the dam behaviour and different phenomena encountered during construction, consolidation an impounding is considered. This approach covers both mechanical and hydraulic behaviour of multiphase soil. The role of the introduced parameters is demonstrated through a parametric study of a homogeneous earth dam (La Ganne). Laboratory tests performed on the material of this dam have been used to calibrate the model parameters, which have then been used in a Finite Element computation to estimate the behaviour of the dam. These results are compared to the measured quantities obtained by the dam monitoring.

1. INTRODUCTION

The multiphase behaviour of non saturated soil either on the mechanical or hydraulic point of view has been extensively studied in the past decades. The stress-strain behaviour of unsatutrated soil has been the subject of numerous experimental and theoretical investigations. A number of constitutive relationships such as the so-called state surface approach [Bishop and Blight (1963), Matyas and Radhakrishna (1968), Fredlund (1979), Lloret and Alonso (1985)] or more recently elastoplastic type models, have been proposed [Alonso et al. (1990), Kohgo et al (1993), Modaressi and Aboubekr (1994)]. The important feature differentiating these models is the choice of total or generalised effective stresses to express the behaviour of the unsaturated soil. In this way, combining the momentum equilibrium and the soil-water mass balance equation, a consistent and unified approach can be applied to simulate both saturated and unsaturated soil behaviour [Aboubekr (1995)]. The aim of this work is to demonstrate the applicability of such approach to embankment dams and to illustrate the role of the compressibility of the unsaturated soil on the dam behaviour.

In addition, as several phenomena influence this compressibility, the parameters characterising this aspect of the behaviour and their influence on the overall behaviour of the dam are demonstrated through a parametric study.

La Ganne embankment dam has been used for this analysis as several laboratory tests

[Poulain (1993)] on its constitutive material are available and pore pressure and stress gauges have been installed to monitor its behaviour. It is a 14m high homogeneous dam made of alluvium clay with a horizontal and vertical drain at the downstream. The computation has included the three phases of construction, consolidation and impounding of the reservoir.

2. GOVERNING EQUATIONS

2.1. Effective stress in partially saturated soil

Later works show that Terzaghi's principle is suitable for both saturated soil with negative pore pressure as well as quasi-saturated materials in which water is the continuous phase ($S > 95\%$). When the degree of saturation (S) decreases, air enters the pores and water forms menisci at the grain contact points. These menisci apply intergranular attracting forces whose overall action is identified by capillary stress that when deduced from total stresses gives effective stresses. Based on experimental observations, and the micromechanical model proposed by [Biarez et al. (1993)], the capillary pressure π_c which is a hyperbolic function of suction is introduced. It varies from the air entry pore pressure to an asymptotic value which can be computed as a function of the void ratio, surface tension and the mean grain size characterised by D_{10} parameter [Modaressi et al. (1995)].

In this approach, it is assumed that this generalised effective stress governs the deformation of the skeleton. It can be given as:

$$\boldsymbol{\sigma}' = \boldsymbol{\sigma} - w\boldsymbol{I} \quad [1]$$

Where $\boldsymbol{\sigma}'$ is the generalised effective stress tensor, $\boldsymbol{\sigma}$ the total stress tensor an $\boldsymbol{I}$ the unity matrix. w is defined as:

	$w = -u_w$			$w = \pi_c$	[2]
if	$u_w \geq u_{wd}$	in a drying path	if	$u_w < u_{wd}$	in a drying path
or	$u_w \geq u_{wr}$	in a wetting path	or	$u_w < u_{wr}$	in a wetting path

Where u_{wd} is the air-entry suction corresponding to the shrinkage limit of the material and u_{wr} is the resaturation suction which marks the transition from unsaturated domain to the saturated one in a wetting path.

Though the assumption of $w = S\, u_w$ usually used for partially saturated soil might be acceptable for specific cases, its generalisation to all degrees of saturations is not correct.

2.2. Balance equations

The momentum conservation coupled with the mass balance equation in which Darcy's law generalised to unsaturated soil is introduced result into the following hydromachanical system of equations:

$$Div\, \boldsymbol{\sigma}' + \boldsymbol{grad}\, w + \rho\, \boldsymbol{g} = 0 \quad [3]$$

$$div\,\dot{u} + c\,\dot{p} - div\left[\frac{K}{\mu}\left(grad p + \rho_l g.x\right)\right] = 0 \quad [4]$$

With

$$\rho = Sn\,\rho_w + (1-n)\,\rho_s \quad [5]$$

In the above equations u is the displacement vector of the solid skeleton, p is the pore pressure in the saturated soil and suction in the unsaturated soil, n is the porosity, ρ_i is the mass density of the i phase (either water or skeleton) and $\boldsymbol{g}$ represents the gravity vector. K is the permeability matrix, μ the viscosity of the fluid and c is the storage term which is given by $$c = \frac{n}{K_w} + \frac{dS(p)}{dp}$$ [6]

where K_w is the water compressibility. dS/dp takes into account the variation of degree of saturation with suction. As it will be recalled later, this quantity plays a key role in the variation of pore pressure. This system of equations can be solved for given initial and boundary conditions which apply on both mechanical and hydraulic variables. The numerical resolution of the above problem is done by the Finite Element code GEFDYN [Aubry and Modaressi (1996)].

2.3 Constitutive relations

Both the mechanical and hydraulic behaviour of the unsaturated soil have to be specified. As far as the mechanical behaviour is concerned an elastoplastic constitutive model developed in terms of effective stress for dry or saturated soil [Aubry et al 1980, Hujeux 1985], has been extended to unsaturated soils [Modaressi and Aboubekr 1994] using a capillary hardening. The proposed model is based on the decomposition of the irreversible phenomena into four coupled elementary mechanisms: one isotropic and three deviatoric. Each deviatoric mechanism related to a plane, has its own hardening parameters corresponding to distortion in this plane and follows a *Coulomb*-type failure criterion and its yield function is that of *Cam Clay*. The model is essentially governed by the influence of suction variations on the void ration. Suction variations induce an evolution of the consolidation (critical) pressure. This aspect is explicitly taken into account in the isotropic mechanism. From an elastoplastic point of view it is equivalent to an increase of the elastic domain taken into account by introducing a cohesion or in other words by increasing the overconsolidation pressure due to the capillary forces. Thus, the isotropic yield function will be written as:

$$f_i = p' - r^{iso}\, d\, p_c^{ns}$$ [7]

with:

p' the effective mean stress in the unsaturated domain estimated using relation [1].

p_c^{ns} the critical pressure which is a function of volumetric plastic strain, with β the plastic compressibility modulus, p_{c0} the critical pressure corresponding to the initial state and $R(\pi_c)$ a function of the capillary pressure which controls the extent of the elastic domain due to desaturation, its expression is given as: $p_c^{ns} = p_{c0} e^{\beta \varepsilon_v^p} - R(\pi_c)$ [8]

As an initial approximation, we assume R to be linear function of π_c: $R(\pi_c) = k' \pi_c$

r^{iso} the mobilization ratio of the isotropic mechanism during monotonous loading. It evolves from a minimum which defines the elastic domain to unity at the perfectly plastic state. Its evolution is controlled by variations of the total plastic volumetric strain ε_v^p.

d a parameter denoting the distance between the virgin consolidation line and the critical state line in *e-Lnp'* plane, where e is the void ratio.

The deviatoric yield function corresponding to the kth-plane is proposed as:

$$f_k = q_k - p'_k \sin\phi \left(1 - b \operatorname{Log} \frac{p'}{p_c^{ns}}\right) r_k$$ [9]

where p'_k and q_k are respectively the *reduced* effective mean and deviatoric stresses corresponding to the k_{th}-mechanism. ϕ is the friction angle of the material, and *rk* the degree of shear mobilisation of the k_{th}-mechanism defined as the mobilised friction over maximum friction that may be mobilised. *b* is a model parameter which controls the shape of the yield function (i.g. *b* = 0 Mohr-Coulomb, *b* = 1 Cam Clay). The friction angle is supposed not to change due to non-saturation whose influence is taken into account by p_c^{ns} through which the isotropic and deviatoric mechanisms are coupled. For details concerning hardening rules see [Hujeux (1985)]. As far as the hydraulic behaviour is considered, the Darcy's law is generalised to the case of flow in unsaturated soil. The variation of the permeability coefficient with void ratio as well as the degree of saturation are to be taken into account. An important quantity of empirical relations are available in the literature. In absence of experimental data, we have chosen the one proposed by Van Genuchten [Van Genuchten (1980)] which is largely used in dam engineering [Ozanam-Hautefeuille (1988)]. Thus the coefficient of permeability is decomposed in two terms, one taking into account the role of the geometry of pores (**K'**), and the other the influence of saturation variation k_r Thus:

$$\mathbf{K} = k_r\,(S)\,\mathbf{K}'(n) \qquad [10]$$

$$\mathbf{K}'\,(n) = \mathbf{K}'\,(n_0)\,\frac{n^3}{n_0^3}\,\frac{(1-n_0)^2}{(1-n)^2} \qquad k_r(S) = \frac{(S-S_r)^3}{(1-S_r)^3} \qquad [11]$$

where n_0 is the reference porosity for which the permeability is equal to K'_0. S_r is the minimum degree of saturation which can be reached in a given soil. The other variable which should be characterised is the degree of saturation. In compacted soil, it is very difficult to saturate completely the material and gas and air remain trapped in the pore water. So the variation of degree of saturation with respect to water pressure exhibits the form given in Figure 1 on which two domains can be distinguished. Usually hysteresis, non considered in this study, is observed. In one zone, the degree of saturation decreases with the positive pore pressure following a smooth curve. In this domain, no menisci is formed and the non saturation of the material is due to the dissolution of air and gas bubbles in water characterised by Boyle's and Henry's laws. The second domain corresponds to pore pressures less than the air-entry pressure. As both these phenomena are characterised by the term dS/dp in the mass balance equation, we have decided to take them into account in the same manner. For this reason, we have slightly modified the Van Genuchten relation introducing the pore pressure at saturation p_{sat}. We have thus the following relation with α and p_{sat} as the model parameter:

$$S = S_r + \frac{1-S_r}{\sqrt{1+[\alpha\frac{p-p_{sat}}{\rho_w g}]^2}} \qquad [12]$$

3. NUMERICAL SIMULATIONS OF THE LABORATORY TESTS

A very detailed experimental study concerning the embankment material is made [Poulain (1993)], specially undrained isotropic triaxials tests are performed in order to study the compressibility of the non-saturated material and determine the Skempton's B factor. The inconsistency of the results let us suppose that the specimen were not sampled in a homogeneous material. Nevertheless we could set up a unique set of parameters from the experimental results using correlations and parameters already obtained for other similar materials.

Triaxial undrained tests allow us to determine the friction angle ϕ. The identification of the elastic parameter E is conditioned by the existence of triaxial or shear tests for which the strain measurement is performed in the strain interval $[10^{-5}, 10^{-4}]$, where the behaviour is really elastic. Calculations already performed [Michali (1994)] proved that an E modulus equal to 100 MPa is appropriate for the embankment material.

In addition to the compressibility parameters E and β, the two important model parameters, which influence the compressibility of the soil material, are p_{c0} and the α of the Van Genuchten model. In order to identify these parameters we have simulated a number of isotropic compression undrained tests. Several initial degree of saturation have been studied (86.6%, 89.7%, 90.7%, 93%, 95% and 96%).

Two sets of the Van Genuchten model [Van Genuchten (1980)] parameters, equation [12], are tested:

set 1: $\alpha = 0.03$ and $p_{sat} = 150$ kPa set 2: $\alpha = 0.075$ and $p_{sat} = 100$ kPa

The retention curves and the dS/dp term which represents the compressibility are shown in Figures 1 and 2.

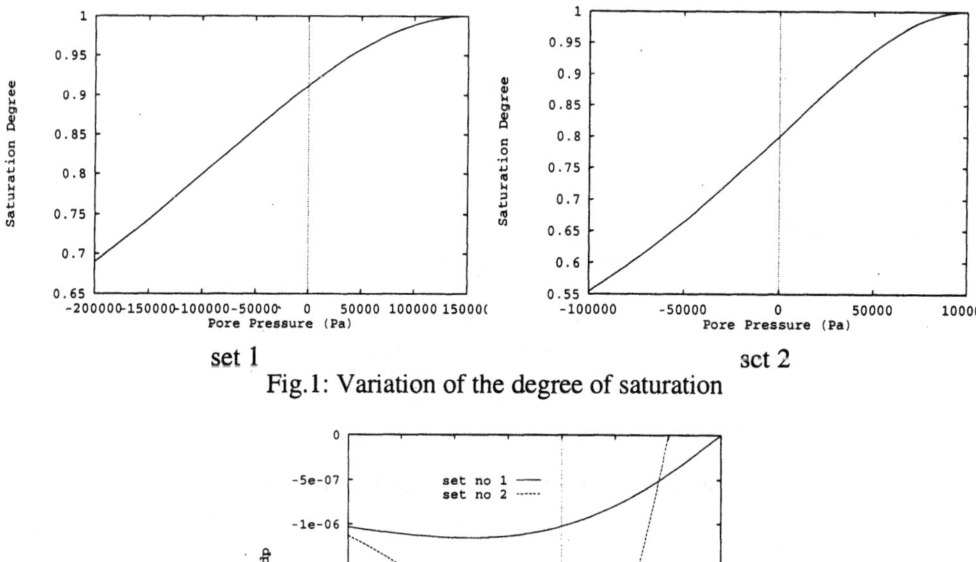

set 1 sct 2

Fig.1: Variation of the degree of saturation

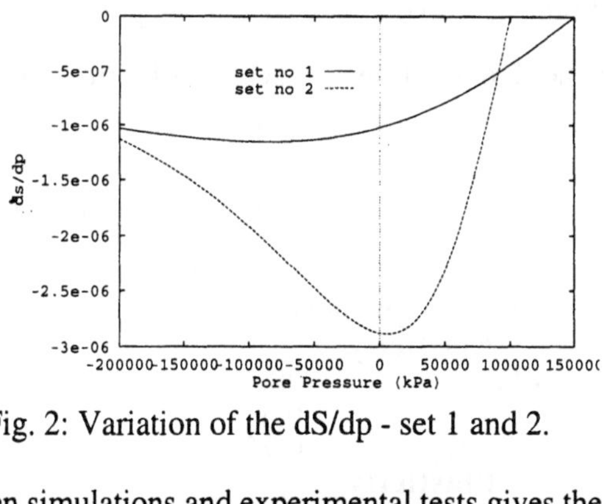

Fig. 2: Variation of the dS/dp - set 1 and 2.

The comparison between simulations and experimental tests gives the most appropriate value for the p_{c0} parameter. Ensuing the degree of saturation of the material we can note that a value of 60 or 90 kPa seems correct; the stronger the saturation degree the weaker the parameter p_{c0}. This observation is in harmony with the unsaturated soil's nature. The value of the plastic compressibility β=8 turns out to be quite satisfactory for the performed simulations.

We have noted that the set 1 generates more pore pressure than the set 2. In the Figures 3 and 4 we can see two of the performed simulations for the isotropic undrained compression.

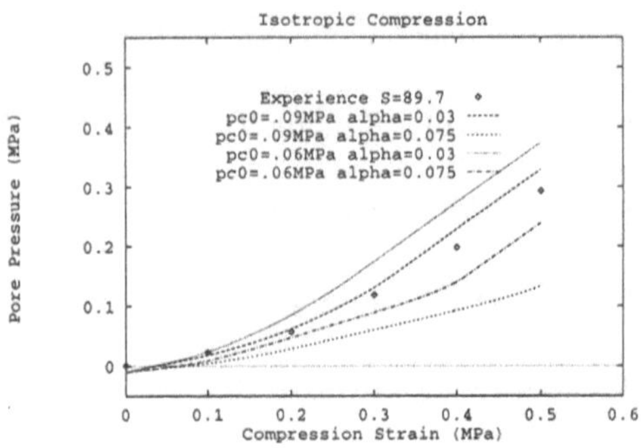

Fig. 3: Test P10046 - S_0=89.7%.

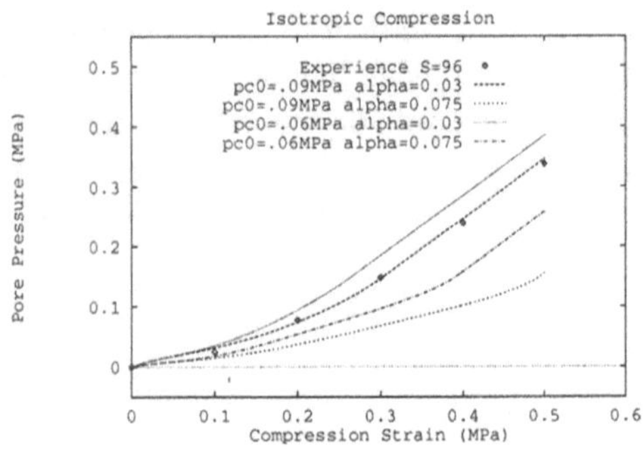

Fig. 4: T10006 and S_0=96%

4.MODELING OF THE LA GANNE EMBANKMENT

The used mesh (Figure 5) comported 441 Q4 and T3 elements with 505 nodes. A 2D plane strain computation has been performed modelling the three periods of construction by layers, consolidation and step by step impounding the reservoir one after the other.

The boundary conditions are the ones classically used. Only because of the high permeability of the drains, atmospheric pressure has been prescribed on the corresponding nodes. For the layered construction phase, the pore pressures are initialised considering a degree of saturation of 90% for the constructed layers' material. The results obtained at the end of each phase, compose the initial values of the next phase computation.

In order to bring to the fore the role of some of the model parameters, which are quite important for the embankment construction, we have performed a parametric study. The parameters that have not been changed during the study are given.

Initial values and physical properties		**Hydraulic properties**	
n_0	0.38	K_w (GPa)	2222
S_0	0.90	S_r	0
ρ_s(Kg/m^3)	2700	$K'_h(n_0)\ \rho_w\ g/\mu$ (m/s)	10^{-9}
ρ_w (Kg/m^3)	1000	$K'_v(n_0)\ \rho_w\ g/\mu$ (m/s)	10^{-8}

Mechanical properties

Elasticity

E_{ref} (MPa)	286
ν	0.3
n (non linearity)	0.75

Plasticity

φ (°)=ψ (°)	29	a1	0.001
b	0.9	a2	0.0008

r^{ela}	0.1	r^{hys}	0.3
r^{mob}	0.5	$r4^{iso}$	0.005
d	2.0	c	0.1
ccyc	0.05		
	Non saturation		
k'	0.5	D_{10}	$0.3.10^{-6}$

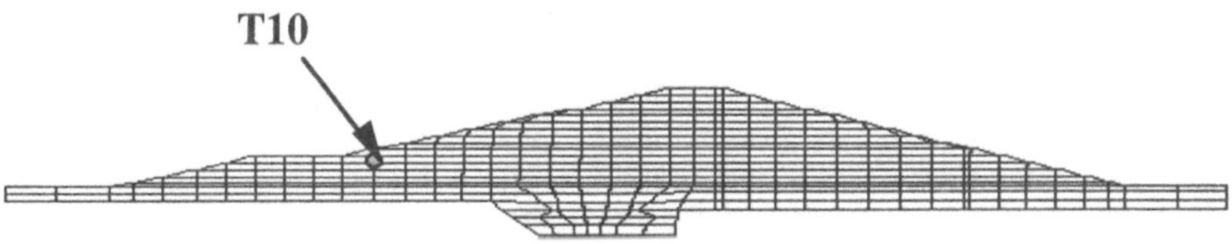

Fig. 5: The used mesh and the control point T10.

Four calculations are performed. We give the set of parameters on which the parametric study is performed. The case D has not been continued until the water loading.

case	α	psat (kPa)	β	p_{c0} (kPa)
A	0.03	150	8	60
B	0.03	150	8	90
D	0.075	100	8	60
E	0.03	150	14	350

5. ANALYSIS OF THE RESULTS

Settlement analysis:
Maximum and minimum settlements and horizontal displacements as well as the relative displacements with respect to the end of construction for both consolidation and impounding phases are given:

	Case A		Case B		Case D		Case E	
(mm)	H	V	H	V	H	V	H	V
settlement after construction	-46 +20	-410 +43	-27 +38	-210 +3	-81 +170	-570 +140	-9 +10	-47 +3
relative displacement after consolidation	-34 +66	-138 +75	-33 +4	-5 +74			-2 +2	-0,2 +3
relative displacement after impounding	-20 +14	-18 +64	0 7	-10 +1			0 +4	-4 +0,2

One can see that the maximum of the settlement is obtained in the case D where the pore pressures are very low. This is quite reasonable given that the effective stress values are thehighest possible. Among the cases studied with the same values of (α, p_{sat}) the one generating higher pore pressures is the case A where p_{c0} has its weakest value. In this case the maximum settlement is obtained. This points out that the consolidation phase play an important role on the structure's behaviour and shows the importance of the materials porosity.

In all cases, given the embankment's form, it is deformed towards the downstream. It should be noticed that only in the case of important consolidation, embankment's horizontal displacement, after construction remain symmetrical.

Pore pressure analysis:
The pore pressure evolution analysis on point T10 (Figure 6) shows that the maximum pore pressure is obtained in case A where a weak initial consolidation pressure is applied. If one chooses a higher consolidation pressure (case E), one achieves 20% decrease of the pressure. Case A gives the best results with respect to the experimental measurements. Case D gives the lowest pressure. This is due to the form of the (dS/dp - p) curve considered with parameter α=0.075 which results into higher storage quantity a lower compressibility of equivalent water-gaz fluid. One can note that for cases B and E there is no great difference for the corresponding pressures at point T10.

The computed pore pressure at the point T10 is 20 kPa at the end of the construction for the case A while for the cases B and E the pore pressure is equal to 15 kPa. Nevertheless the pore pressure is stabilised at the value of 30 kPa for all the cases except the case E where the pressures are smaller.

Horizontal and vertical stresses analysis:
The stress analysis at the T10 point give very interesting results. We bring to the fore the role of the consolidation pressure to the stress distribution in the embankment. We can note that the vertical effective and total stresses are identical for all the studied cases (Figures 7 and 8). In the case E, while the pore pressures are weak, the effective stresses are stronger.

The horizontal stresses are influenced by the p_{c0} value. One can see that for the case A, where the p_{c0} is the smallest, the total horizontal stress is maximum for the point T10.

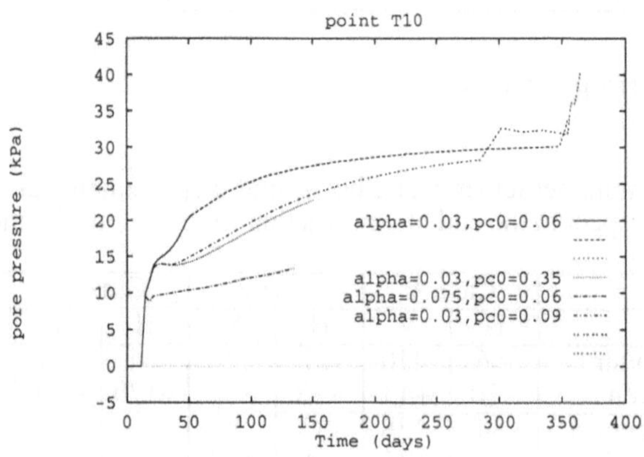

Fig. 6: Pore pressure comparison for 400 days in the point T10.

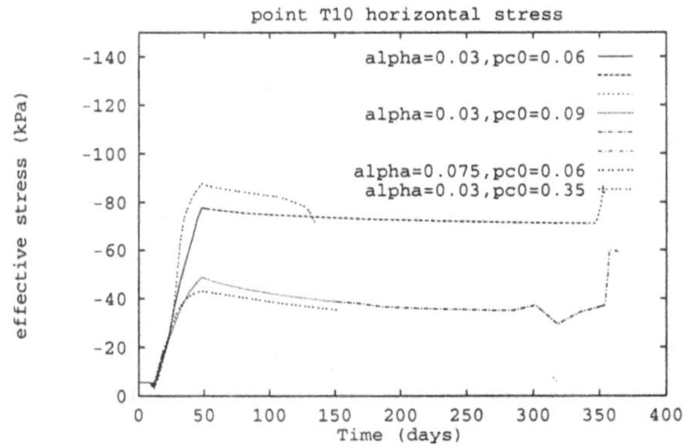

Fig. 7: Horizontal effective stress comparison for 400 days in the point T10.

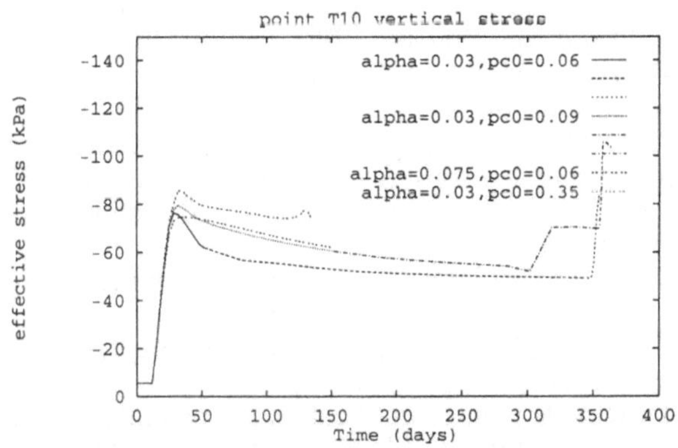

Fig. 8: Vertical effective stress comparison for 400 days in the point T10.

6. CONCLUSIONS

The essential phenomena governing the behaviour of partially saturated soil is presented. Both mechanical and hydraulic constitutive models necessary to describe such behaviour are presented. The validity of the approach is illustrated modelling the homogeneous La Ganne earthfill dam, during its construction by layers phase, its consolidation and the reservoir impoundment.

The role of the key factors such as the initial consolidation or density, controling the pore water pressure and consequently the settlement of the dam has been illustrated by a parametric study. Both the skeleton compressibility (β) and the storage capacity ($S(p)$) govern the compressibility of the multiphase soil and the generation of pore pressure.

Other interesting information on the stability of the dam were obtained which are not presented here due to lack of space.

7. ACKNOWLEDGMENTS

The authors wish to thank EDF-CNEH for their financial and material aid.

8. REFERENCES

ABOUBEKR N. (1995). Modélisation du comportement mécanique et hydraulique des sols partiellement saturés. Thèse de Doctorat, École Centrale de Paris.
ALONSO E.E., GENS A. and JOSA A.(1990). A constitutive model for partially saturated soils. Géotechnique 40: 405-430.
AUBRY D., HUJEUX J.C., LASSOUDIERE F. & MEIMON Y. (1982), A double memory model with multiple mechanisms for cyclic soil behaviour, Int. Symp. Num. Mod. Geomech., Balkema, pp 3-13.
AUBRY D., MODARESSI A (1996). Manuel Scientifique GEFDYN.
BIAREZ J., FLEUREAU J.M. &TAIBI S. (1993), Constitutive model for unsaturated granular media, Proceedings of the Second International Conference on Micromechanics of granular Media. Birmingham.
BISHOP and BLIGHT (1963). Some aspects of effective stress in saturated and partially saturated soils. Géotechnique, n° 3, vol. XIII, pp. 177-197.
HUJEUX J.C (1985), Une loi de comportement pour le chargement cyclique des sols, Génie Prarasismique, Ed. V. Davidivici, Presses ENPC, pp. 287-302.
KOHGO Y., NAKANO M.& MIYAZAKI T. (1993a), Theoretical aspects of constitutive modelling for unsaturated soils, Soils and foundations, Vol. 33, No 4, 49-63.
KOHGO Y., NAKANO M.& MIYAZAKI T. (1993b). Verification of the generalized elastoplastic model for unsaturated soils, Soils and foundations, Vol. 33, No 4, 64-73.
FRENDLUND D.G (1979). Appropriate concepts and technology for saturated soils. Canad. Geotech. J., 16, 121-139.
MATYAS E.L.and RADHAKRISHNA H.S. (1968). Volume change characteristics of partially saturated soils. Geotechnique, 18, 432-448.
LLORET A. and ALONSO E.E (1985). State surfaces for partially saturated soils. 11th I.C.S.M.F.E. Vol.2, San Francisco, pp. 557-562.
MICHALI A. (1994). Méthode pour l'identification des paramètres d'une loi élastoplastique à partir d'essais de laboratoire et in-situ-Modélisation numérique du tassement d'un remblai sur sol compressible. Thèse de Doctorat, École Centrale de Paris.
MODARESSI A. and ABOUBEKR N. (1994). Constitutive model for unsaturated soils: Validation on a silty material. Numerical Methods in Geotechnical Engineering, Proceedings of theThird ECONMIG, I.M. Smith (ed.), Balkema.
MODARESSI A., ABOUBEKR N. & FRY J.J. (1995). Unified approach to model partially saturated and saturated soil. Third European Conference on Numerical Methods in Geomechanics, Manchester, Balkema eds., pp. 91-96.
OZANAM-HAUTEFEUILLE O. (1988). Modélisation numérique des sols élastoplastiques non saturés-Application aux barrages en remblais. Thèse de Doctorat, ECP.
POULAIN D. (1993). Pressions interstitielles de construction dans les barrages homogènes en matériaux argileux humides. Thèse de Doctorat, Université de Bordeaux I.
ROSCOE K.H. and BURLAND I.B. (1968). On the generalised stress-strain behaviour of wet clay, Engineering Plasticity, Cambridge, pp. 535-609.
TAIBI S. (1994). Comportement mécanique et hydraulique des sols soumis à une pression interstitielle négative. Etude expérimentale et modélisation. Thèse de Doctorat, ECP.
VAN GENUCHTEN R. (1980). A closed form equation for predicting the hydraulic conductivity of unsaturated soils. Soil Science AM. Soc., 44, 892-898.

A NUMERICAL CODE FOR COMPUTATION OF TRANSPORT PHENOMENA COUPLED WITH MECHANICS IN POROUS MEDIA

J. Gaombalet and K. Su
École Polytechnique, Palaiseau, France

ABSTRACT : *This paper describes the numerical code ANTHYC developed at G.3S for modeling of transport phenomena in porous media. Using a method derived from the Finite Volume method, the code forces the local conservation of transfer and mechanical processes and leads to a very good accuracy. The theoretical development of the method is described, and a number of examples are given.*

1. INTRODUCTION

A new numerical code, ANTHYC, has been developed by G.3S. It allows the modelling of various coupled effects between mechanical behavior and transport phenomena in porous media. This code uses a method based on Finite Volume Method. this method, much used in Fluids Mechanics, is very suited to the study of porous media where transfer processes are involved. The code has been developed with a general purpose, so that every coupled term can be taken into account, and may be non linear. Moreover, material properties can be anisotropic.

The present version of the code has been fully validated and used for coupled thermo-hydro-chemico-mechanical modeling in two dimensions in saturated porous media. A version treating multiphase flow is being tested for the study of unsaturated rocks.

In this paper, after a general presentation of ANTHYC, a few examples of computations performed using the code for its validation are given.

2. CALCULATION METHOD

2.1. Finite Volume Schemes

A first version of ANTHYC, devoted to specific cases, was based on a Finite Difference method. This method, very simple in principle, could not be generalized easily, especially from a geometrical point of vue. This explains ANTHYC evolution towards a method derivated from Finite Volume methods.

Finite Volume schemes are often used to solve numerically conservation laws in gas dymamics. These methods are not so far, somehow, from Finite Difference methods, but they are based on a different approach.

The basic idea of Finite Volumes methods is the following : let us consider a diffusion equation written in a conservative form :

$$\frac{\partial W}{\partial t} + \mathrm{div}[F(W)] = 0 \tag{1}$$

where W is a conservative variable (or a set of conservative variables).

Equation (1) traduces in fact a conservation law : for each fixed volume $V \subset \Re^3$, one can deduce from (1) (using Green's Formula) that :

$$\frac{d}{dt}\left(\int_V W\right) + \int_{\partial V} F(W).n = 0 \tag{2}$$

where ∂V denotes the frontier of the domain V, and n its the outward unit normal vector. The equation (2) expresses that the variation of the integral $\int_V W$ is due to the the flux $\mathcal{F} =$ F(W).n on the boundary.

In addition to (2), it is possible to write equations in conservative form (in form of flux on the contour) for mobile volumes V(t) as :

$$\frac{d}{dt}\left(\int_{V(t)} W\right) + \int_{\partial V(t)} \psi.n = 0 \tag{2'}$$

The principle, in Finite Volume schemes, consists of the direct discretization in the form (2) or (2'), which guarantees conservation properties for the volume V.

2.2. General method

Discretization principle, used in ANTHYC, is based on the idea expressed above : to ensure conservation properties locally, thanks to the formulation in form of flux on the contour. The method used in Anthyc is actually a combined method Finite Volume/Subdomain collocation. This may appear in a simple way through a diffusion equation (which can be, for instance, thermal diffusion).

Let us consider the following Cauchy's problem and boundary conditions for a domain Ω :

$$\begin{cases} \dfrac{\partial X}{\partial t} + \mathrm{div}(F(X)) + Q = 0 & \text{on } \Omega \quad \text{(a)} \\ A(X) = f & \text{on } \partial\Omega \quad \text{(b)} \\ X(M, t = 0) = X^0(M) & \text{on } \Omega \quad \text{(c)} \end{cases} \tag{3}$$

where X is a scalar variable, F designates a flow and Q a source term.

The integral form of (3-a) is written as follow :

For each weighting function φ, we have :

$$P = \int_\Omega [\frac{\partial X}{\partial t} + \mathrm{div}(F(X)) + Q].\varphi = 0 \tag{4}$$

a) <u>Weak formulation</u>

For space discretization, a Finite Element method is used. Let Ω_h be the approximation of the domain Ω and D_h the discretization by Finite Elements of Ω_h.

A mesh is built, defining, around each node M_i, a small volume C_i called <u>basic cell</u> or <u>Finite Volume</u>, as shown in figure 1. The association of all the cells is a new partition of Ω_h :

$$\Omega_h = \bigcup_{i=1}^{N_d} C_i$$

We seek a closed form solution $X_h = \sum_i N_i X_i$ for X solution of (4) [P = 0] by a subdomain collocation, choosing C_j as the subdomains. The weak form obtained thus is :

$$\int_{\Omega_h} \frac{\partial X_h}{\partial t}.\varphi_j + \int_{\Omega_h} \mathrm{div}(F(X_h)).\varphi_j + \int_{\Omega_h} Q.\varphi_j = 0 \ , \ \forall j \in \{1, N_d\} \tag{5}$$

where φ_j are defined by :

$$\varphi_j = \begin{cases} 1 \text{ if } M \in C_j \\ 0 \text{ if } M \notin C_j \end{cases} \tag{6}$$

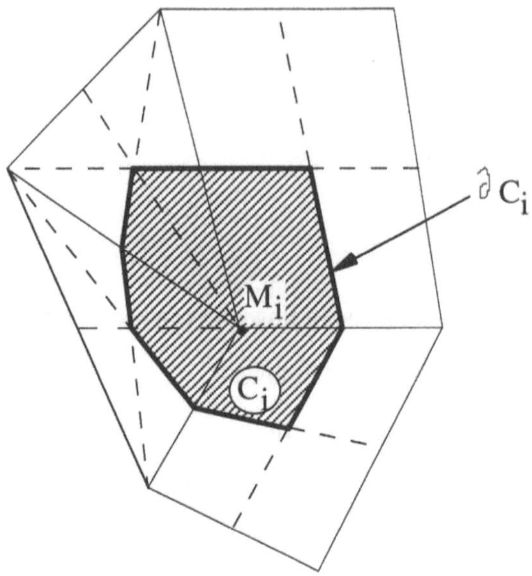

Fig.1 : *A basic cell around the node M_i when a first order approximation is adopted using finite elements.*

This formulation (5) can also be written as a simple integration of (3-a) on each cell C_j :

$$\int_{C_j} \frac{\partial X_h}{\partial t} + \int_{C_j} Q + \int_{C_j} \mathrm{div}(F(X_h)) = 0 \ , \ \forall j \in \{1, N_d\} \tag{7}$$

The final formulation is obtained, using Green's Formula :

$$\int_{C_j} \frac{\partial X_h}{\partial t} + \int_{C_j} Q + \int_{\partial C_j} F(X_h).n = 0 \ , \ \forall j \in \{1, N_d\} \tag{8}$$

b) Time integration

When discretized, the problem (8) leads to a first order differential equation in time, which can be linear or non linear :

$$[C].\frac{dU(t)}{dt} + [K].U(t) = S(t) \tag{9}$$

To the system (9) must be added the initial values of U. It is solved by a direct implicit/explicit integration method (a parameter allows to choose the resolution method).

In case of non linear problems, the choice of a non-explicit resolution naturally leads to that of a non linear problem at each time step. A simple method of substitution is used to solve it.

2.3. Method for Mechanical equilibrium

Through the last exemple, the resolution of a transfer problem was described. The same principle is applied for Mechanics : the method consists in maintening local equilibrium for each cell C_j :

$$\int_{C_j} \operatorname{div}(\underline{\underline{\sigma}}) + \rho \underline{f} = 0 \tag{10}$$

Thus, Green's Formula is used again :

$$\int_{\partial C_j} \underline{\underline{\sigma}}.\underline{n} + \int_{C_j} \rho \underline{f} = 0 \tag{11}$$

Equations (11) express that "the sum of forces applied on each cell C_j is zero".

3. FORMULATION AND POSSIBILITIES

Coupling transport and mechanical behavior in porous media implies the silmutaneous resolution of equilibrium equations and those of each transfer phenomenon, considering initial and boundary values. Transfer phenomena may be various (see for instance Coussy 1991, Schrefler 1995, Jouanna 1995, Lassabatère 1994, Bear *and al.* 1991, Alonso *and al.* 1990) : multiphase flow, thermal and soluted mass transfer, electrical transfer (Casagrande 1952), etc ...

All these transfer phenomena are in principle fully coupled. They are also coupled with mechanical behavior via volumetric strain.

3.1. About transport equations

Whatever may be the transfer phenomena taken into account, and after the choice of main unknowns, one has to solve a set of coupled transport equations, also coupled with Mechanics.

This led us to introduce in the code some generic transport equation in porous media, which allows :

- any choice of transport variables ;
- any coupling between these variables, and with mechanics ;
- anisotropic coefficients ;

Let us consider a problem where N transfer variables are involved, written $X_1, .., X_N$. Generic equation for the variable i can be written as follows :

$$\begin{aligned} \operatorname{div}\left(\underline{\underline{\lambda}}^1_{j\to i}.\underline{\operatorname{grad}}(X_j) + \underline{\lambda}^2_{j\to i}.X_j\right) + \underline{\lambda}^3_{j\to i}.\underline{\operatorname{grad}}(X_j) + \lambda^4_{j\to i}.(X_j - \overline{X}_j) + \lambda^5_{j\to i} \\ = \lambda^6_{j\to i}\frac{\partial}{\partial t}(\lambda^7_{j\to i}.X_j) + \underline{\underline{\delta}}^1_{k\to i} : \frac{\partial}{\partial t}(\delta^2_{k\to i}\underline{\underline{Y}}_k) \end{aligned} \tag{12}$$

- $\underline{\underline{Y}}_k$ are strain tensors (total, elastic or irreversible strain), and terms $\underline{\underline{\delta}}^1_{k\to i}$ and $\delta^2_{k\to i}$ represent action of mechanics on the the equation i (for total strain, one can choose a direct coupling or, as it is the case for elastic or irreversible strain, take the results of the last iteration ...) ;

- scalars, vectors or tensors " $\lambda^k_{j\to i}$" are coefficients bringing into play action of field j on equation i in what a so-called " k - type coupling " ;

- scalars $\overline{X}_j$ are reference values (solubility, etc ...).

Material properties may be anisotropic. They may be also non linear : some functions are integrated in the code, but any other non linearity can be induced through users subroutines (the large use of "commons" allows to reach easily any variable at any time). Moreover, the code allows to take into account not only diffusive but also convective terms.

3.2. Mechanical aspect

For Mechanics, constituve equations of the solid phase (the skeleton for porous media) are introduced using the concepts of total stress $\underline{\underline{\sigma}}$ or effective stress $\underline{\underline{\sigma}}^{eff}$:

$$\underline{\underline{\sigma}}^{eff} = \underline{\underline{\sigma}} + \underline{\underline{B}}_i P_i \tag{13}$$

$\underline{\underline{B}}_i$ is a Biot tensor [1],
P_i is the partial pressure of the fluid i.

Making the assumption of infinitesimal displacements, the strain tensor $\underline{\underline{\varepsilon}}$ is taken as the sum of an elastic part $\underline{\underline{\varepsilon}}^e$, a plastic part $\underline{\underline{\varepsilon}}^p$, a viscoplastic one $\underline{\underline{\varepsilon}}^{vp}$, and $\underline{\underline{\varepsilon}}^i$ for strains due to variation of transfer variable i (save pressures) : let, $\underline{\underline{\varepsilon}}^i = \underline{\underline{\alpha}}^i(X_i - X_i^0)$ where X_i^0 is the initial value of the field i.

Mechanical evolution of the porous medium is quasi-static and governed by the traditionnal following equations and data :

$$\begin{cases} \text{- équilibrium equations} \\ \text{- constitutive laws} \\ \text{- initial and boundary values} \end{cases}$$

In the present version of the code, the mechanical behavior may be elastic, plastic and/or viscoplastic. Irreversible part can be treated in terms of total or effective stresses (plastic effective stress, viscoplastic effective stress). Parameters related to plasticity or viscoplasticity can depend on any variable (indeed, in agreement with second principle of thermodynamics). Thus, mechanical behavior can be strongly non linear and coupled with transfer phenomena. Moreover, properties (Stifness tensor, thermal expansion tensor, ...) can be anisotropic.

Apart from those already introduced in the code, any other constitutive law, for instance related to damage, can be easily introduced by a user procedure.

3.3. Possibilities

Anthyc allows, thanks to its "generic" formulation (12), to treat a wide range of coupled transport-mechanical problems in porous media. These couplings are direct for diagonal and non diagonal terms. Material properties can be anisotropic and non linear. Indeed, heterogeneities due to the presence of different materials can be taken into account. The code has been designed with a concern of user-friendliness, facing this great number of parameters and possibilities.

In the version already in use, the code can solve problems in the framework of Thermodynamics of Open Continuum Media in transient or steady state regimes, concerning hydraulic, thermal or chemical transport fully coupled together and also with mechanics.

4. Examples of Validation

Many comparisons have been done between ANTHYC's results and analytical solutions or numerical solutions obtained using reliable codes. The results show a good agreement with theoritical calculations and those of other other codes. Moreover, the tests have been performed for each coupled term introduced. A few simple but significant examples are given here.

4.1. Plasticity

Let us consider a cylindrical gallery in a homogeneous and isotropic medium, the behavior of which is elastoplastic.with perfect plasticity obeying to Tresca's criterion. The rock mass is assumed to be infinite. The model is 1D axisymmetric, and the massif is assimilated as a hollow cylinder, infinitely long, which internal and external radii are R_i = 1m and R_e = 230m respectively. Analytical solutions for this problem are given by Bernaud (1991).

Mechanical properties :
- Young's modulus : E = 1430MPa ;
- Poisson's ratio : $\nu = 0.25$;
- cohesion : C = 0.56 MPa ;
- initial isotropic stress σ^0= -4.5MPa ;
- boundary conditions : radial stress σ^i= -1.5MPa on the cavity wall (r = R_i) and $\sigma^e = \sigma^0$= -4.5MPa at outer boundary r = R_e .

Figure 2 shows that ANTHYC's solution and analytical solution are the same.

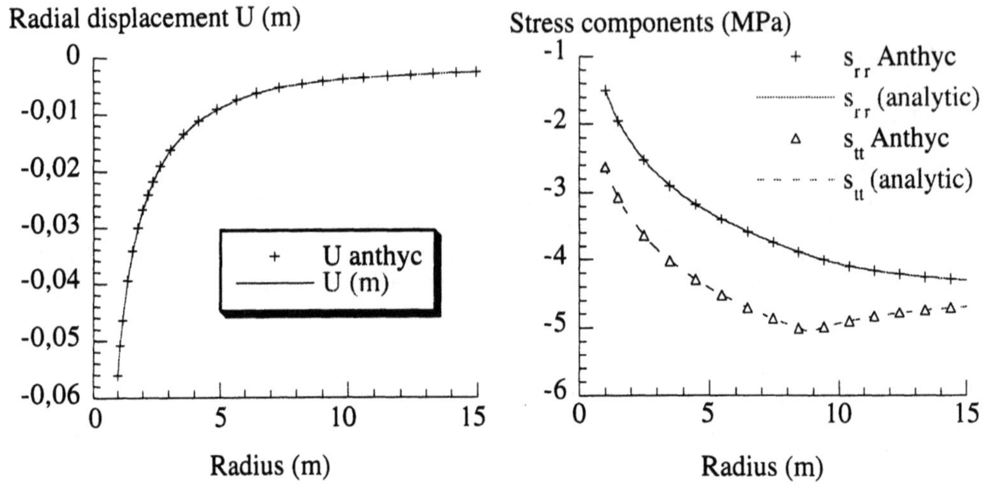

Fig.2 : *Radial displacement, radial and orthoradial stress versus radius for a Tresca perfectly elastoplastic material. Comparison between numerical and analytical solution.*

4.2. Viscoplasticity

Again in axisymmetric condition, let us consider a viscoplastic behavior obeying to Norton-Hoff's law (without yield limit, as it is supposed for rock salt). Viscoplastic potential is given as follows :

$$\Omega(\underline{\underline{\sigma}}) = \frac{a\sigma_0}{n+1}\left(\frac{\sigma_{eq}}{\sigma_0}\right)^{n+1} \qquad (\dot{\underline{\underline{\varepsilon}}}^{vp} = \frac{\partial\Omega}{\partial\underline{\underline{\sigma}}})$$

where : σ_{eq} is the deviatoric equivalent stress : $\sigma_{eq} = \sqrt{3J_2} = \sqrt{\frac{3}{2}\underline{\underline{s}}:\underline{\underline{s}}}$ (with $\underline{\underline{s}}$: deviatoric stress), σ_0 is a normalization stress, and a, n are parameters of the model.

If q_0 and q(t) denote respectively the initial value and the value at the time t of σ_{eq} on the boreholewall, it is known that the ratio $S(t) = \frac{q(t)}{q_0}$ tends towards a limit value $S_\infty = \frac{1}{n}$ when $t \to \infty$ (Pouya 1991). A closed form solution of S(t) is given by the following expression :

$$\hat{S}(t^* = \frac{t}{\tau}) = \frac{1+(n-1)\left(\frac{1}{1+(n-1)t^*}\right)^{\frac{1}{n-1}}}{n}$$

where τ is given by : $\tau = \frac{\sigma_0}{aE}\left(\frac{\sigma_0}{q_0}\right)^{(n-1)}$ (E designates the Young's modulus).

Since no analytical solution is available for this problem, we consider for validation the good agreement between the curve $\widehat{S}(t)$ and that of ANTHYC (figure 3).

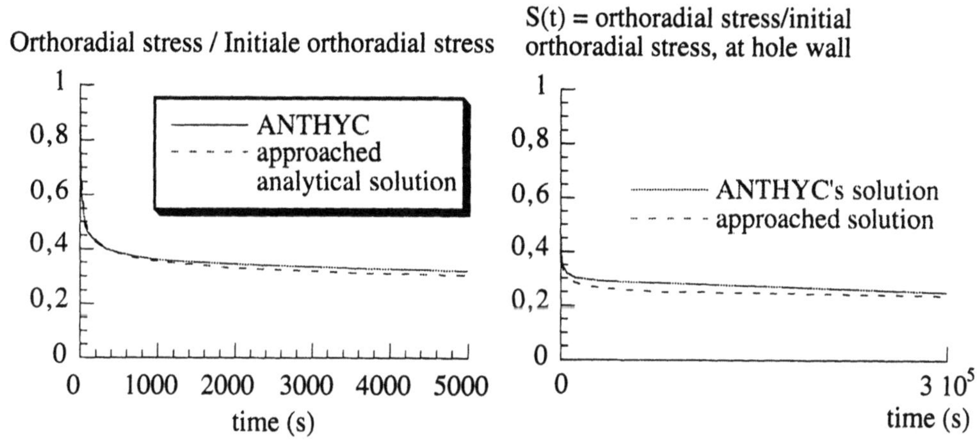

Fig.3 : *Comparison between numerical evolution of* S(t) *(hollow cylinder) and closed form solution* $\widehat{S}(t)$ *for two time intervals (n=5).*

4.3. Thermal diffusion

The third example concerns thermal diffusion in spherical symmetry conditions. Let us consider a hollow sphere. The internal radius is 0,25m, the external radius is 10m., Furthermore, let us suppose an interface with a radius of 2m. The initial state of the whole is defined by a temperature of 30°C. Then, a temperature of 100°C is applied at the internal face.

Spheres 1 et 2 separated by the interface have, as thermal conductivity, respectively λ_1=5.734W/m/K et λ_2 = 1.47W/m/K. The steady stated and the transient evolutions are considered. For the transient calculation, the specific heat depends on temperature, with the same law for both spheres ($\rho c = 1.8705.10^6 + 3.8772.10^2.\theta$, where θ is the temperature in °C, and $[\rho c] = J.m^{-3}.K^{-1}$). The steady state evolution is calculated analytically, since the problem is 1D ; the equation becomes :

$$\Delta T = \frac{1}{r^2}\frac{\partial}{\partial r}(r^2 \frac{\partial T}{\partial r}) = 0 \ .$$

Results are given on figure 4. One can observe that ANTHYC's steady stated calculation is in good agreement with the analytical solution. Moreover, the transient calculation tends towards the permanent solution when t tends towards infinity.

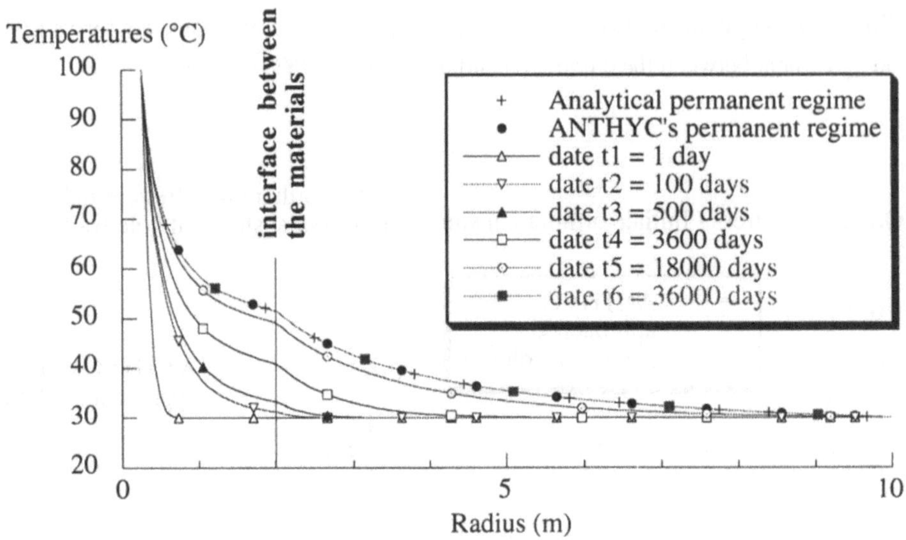

Fig.4 : *Temperature versus radius at different times for the transient calculation (ANTHYC), and for steady state evolution calculated using Anthyc and analytically.*

4.4. An example of application

The use of ANTHYC has become very common at G.3S, for the calculations on underground storage in France and in other european countries.

In order to show the interest of a coupled approach, and at the same time the use of ANTHYC for the modeling of underground repositories, an example of calculation in poroplasticity is selected here. We consider a set of circular parallel galleries, the diameter of which is 7m, separated by a distance of 70m.

To investigate the influence of an extended plastic zone arround the galleries, we suppose that the galleries are 690m deep (initial vertical stress equal to the overburden weight) and we assume a significant anisotropy of initial stress tensor $\underline{\underline{\sigma}}^0$ (the ratio horizontal stress components over vertical stress is taken equal to K_0=1,5). Furthermore, a standard plastic behavior is considered. The plastic criterion used is parabolic in the plane "mean stress, deviatoric equivalent stress", and leads to a plastic dilatancy.

The main caracteristics of the porous medium are the following (those of a hard clay):

intrinsic permeability	$k = 10^{-21}\ m^2$
porosity	$\phi = 9\%$
Biot's coefficient	$b = 0{,}4$
Biot's modulus	$M = 12700MPa$
drained Bulk modulus	$K_o = 5600MPa$
drained Poisson's ratio	$\nu_o = 0{,}25$

The digging of galleries is supposed to be instantaneous. No wall support is used. The calculation is made considering transient evolution. Besides, plane strain condition is assumed, with only one gallery (figure 6) : interaction between galleries is assured by symmetrical boundary conditions. Two calculations are made :

• considering only hydraulic flow uncoupled with mechanical behavior (reference calculation),

• in poroplasticity.

The results are illustrated on figures 6 and 7, where two phenomena are visible :

1) Due to initial stress anisotropy, the initial plastic zone (figure 6) is not symmetrical around the gallery. Its extension is about 0.8 times the gallery's radius.

2) this dilatant plastic zone leads to a sudden drop in interstitial pressure in the gallery nearby (figure 7). Its extension naturally depends on the hydromechanical properties. In the studied case, high negative pressures are obtained and denote that rock mass is not saturated around the galleries in a relatively extended zone. This zone tends to become progressivelly saturated. This resaturation leads to a pressure build-up, so that mean effective stress decreases (in absolute value). As deviatoric effective stress remains the same, this leads to an additional plastification, resulting to slow down resaturation ...

Such a phenomenon can not be explained without taking into account hydromecanical coupled phenomena. The appropriate modelling of this problem requires to use unsatured media formalism, justifying the present development of the code ANTHYC.

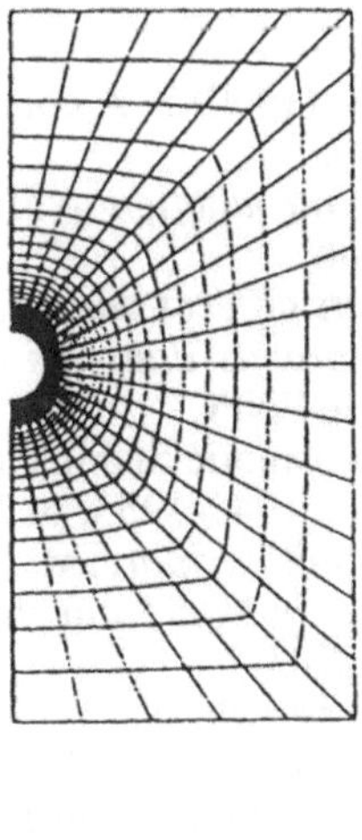

Fig.6 : *Initial plastic zone extent*

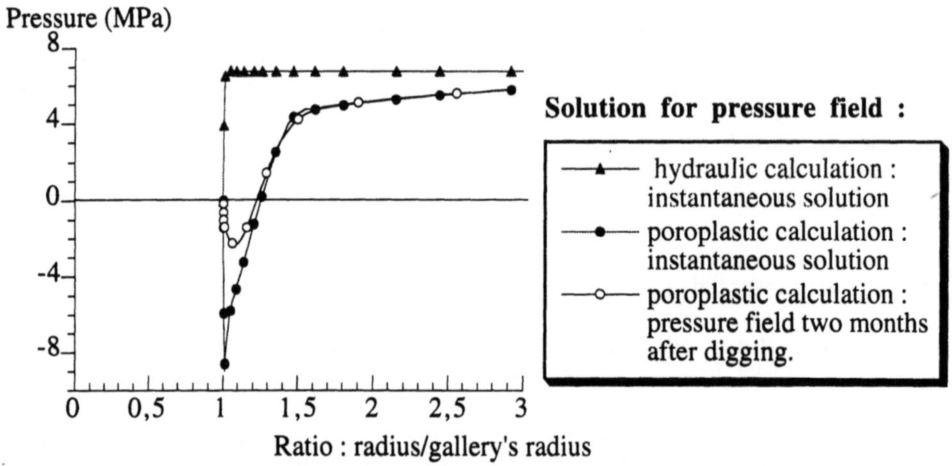

Fig.7 : *Pressure versus radius immediatly after digging for both the hydraulic and coupled hydro-mechanical calculations*

6. Closing remarks

G.3S now has at its disposal a calculation code, validated and already used to model transfer-mechanical coupled effects in rock masses. The code is structured in quite a general way, so that we can easily introduce any new transfer equation. In the same time, the code is still flexible, and user-friendly for a common use in the Groupement. It is possible to perform large, nonlinear simulations involving various transfer phenomena fully coupled with mechanics.

References

ALONSO E.E., GENS A. & JOSA A. (1990) ; "A constitutive model for partially saturated soils", Geotechnique,40 : 405-430.

BEAR J. & BACHMAT Y. (1991) ; *Introduction to Modelling of Transport Phenomena in Porous Media* , ISBN 0-7923-1106.

BENAMAR I. (1996) : *Etude des effets différés dans les tunnels profonds* ; Thèse ENPC.

BERNAUD D. (1991) ; *Tunnels profonds dans les milieux viscoplastiques : approche expérimentale et numérique* , Thèse ENPC.

CASAGRANDE L. (1952) ; "Electro-osmotic Stabilisation of Soils. Boston Society of Civil Engineers. Contrib. Soil. Mechanic 1941-1953, p. 285.

COUSSY O. (1991) ; *Mécanique des milieux poreux* , Éd. Technip. ISBN : 2-7108-0595-2.

JOUANNA P. (1995) ; *Modern Issues in Non-Saturated Soils*, pp.128-159, CISM Courses & Lectures n°357, ISBN 3-211-82783-8.

LASSABATERE T. (1994) ; *Couplages hydromécaniques en milieu poreux non saturé avec changement de phase : application au retrait de dessication* , Thèse ENPC.

POUYA A. (1991) : *Comportement rhéologique du sel gemme. Application à l'étude des excavations souterraines* ; Thèse ENPC.

SCHREFLER A. (1995) ; "Coupled heat, water and gas flow in deformable porous media", International Journal for Numerical Methods in Fluids, Vol. 20, 969-987 (1995)

DETERMINATION OF ROCK MASS STRENGTH PROPERTIES BY HOMOGENIZATION

A. Pouya, M. Ghoreychi
École Polytechnique, Palaiseau, France

ABSTRACT : A method for determining fractured rock masse properties is given on the basis of a homogenisation approach. The homogenized constitutive model of the rock mass considered as a heterogeneous medium constituted of intact rock and of fractures, is studied numerically using Finite Element Method. The constitutive model of the rock masse, elastic properties, ultimate strength and hardening law, were determined assuming linear elasticity and Mohr-Coulomb strength criterion for intact rock and for fractures.

INTRODUCTION

Mechanical behaviour of fractured rock masses depend on the properties of the intact rock and of the fractures. The properties of the intact rock determined by laboratory experiments overestimate the strength of the rock mass for large scales and underestimate its deformability. Rock mass classification methods give very empirical estimations of mechanical properties of fractured rock masses (Barton *et al.* 1974, Bieniawski 1976). Rigorous estimations of these properties can be obtained by semi analytical calculations for the cases where the fractures are parallel to one or two different directions (Bekaert et Maghous 1996, de Buhan et Maghous

1997). In this paper, an intermediate approach is given for determining fractured rock mass properties by numerical homogenisation using Finite Elements. The method was first used by F. Coste (1997) to determine non linear elastic properties of rock masses. In the present paper this method is extended beyond the elastic domain as to determine the strength properties.

1) NUMERICAL METHOD

Rock mass is considered as a heterogeneous medium constituted by intact rock and by different families of fractures having random size and orientation distributions and specific mechanical behaviour (Figure 1).The homogenized strain and stress ε^{hom} and σ^{hom} of this medium are defined as the average values of the local strain and stress distributions ε and σ in the medium. The homogenized behaviour of this medium, i.e., the relations between ε^{hom} and σ^{hom}, is determined by numerical calculation. To do this, displacements or forces corresponding to homogeneous macroscopic strains or stresses are prescribed on the boundary of a sufficiently large domain (Representative Elementary Volume) and strains and stresses are calculated using Finite Element Method. The average values of strains and stresses in the domain are then calculated. The calculations are carried out under plane stresses hypothesis because in this case it is easier to determine the strength properties in the considered plane.

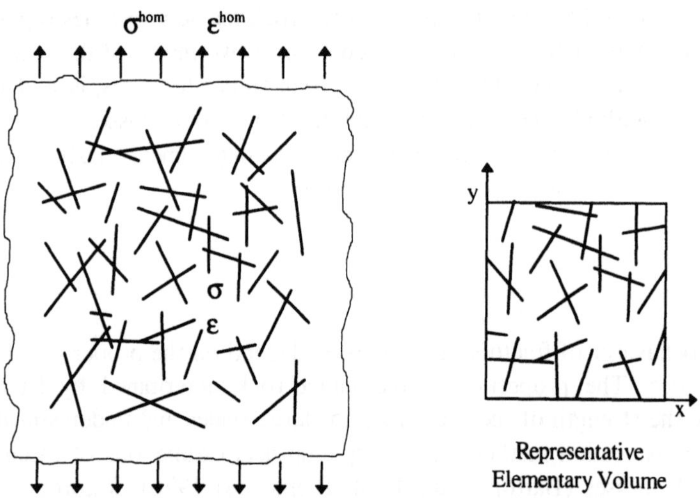

Fig. 1 : The fractured rock mass considered as a heterogeneous medium constituted of intact rock and of fractures

2) FRACTURING MODEL

The fractures size, orientation, density and aperture introduced in this model have been deduced from some geological data corresponding to a granitic fractured rock mass. In the 2D representation of the rock mass studied here, the fractures have a size varying from 20cm to 30m and a slope varying from 0 to 90°. The statistical distribution of the fractures size is assumed to obey to an exponential law with an average value of 10m. The orientation obeys a normal distribution law with an average value of 60° (sub-vertical fractures) and a variance of 10°. The density of fractures is uniform and equal to 0.4 m^{-2}.

3) CONSTITUTIVE MODEL OF INTACT ROCK AND OF FRACTURES

The elastic behaviour of the intact rock is linear and isotropic in the elastic stage with Young modulus E = 72000 MPa, Poisson ratio $\nu = 0.25$. The strength properties are assumed to be given by the Mohr-Coulomb criterion with Cohesion C = 17 MPa, internal friction angle ϕ = 57°. The fractures behaviour is assumed to be also linear in the elastic stage and obeying to Mohr-Coulomb criterion cohesion c = 1.51 MPa and internal friction angle φ = 27°. These values for the fractures have been measured at laboratory on natural filled fractures. These simple models allow to compare easily the rock mass behaviour to its constituents (intact rock and fractures) behaviour. To simplify the calculation the elastic-brittle behaviour of the intact rock and of the fractures is replaced by an elastic-perfectly plastic behaviour.

4) NUMERICAL LOAD EXPERIMENTS

A square domain of the fractured medium is considered (figure 2) and different loads are simulated by prescribing forces and displacements on the boundary of this domain :

1) Compression in x direction with different values of the lateral stress σ_{yy}
2) Compression in y direction with different values of the lateral stress σ_{xx}
3) Shear in xy direction with different values of the mean stress $\sigma_m = \sigma_{xx} = \sigma_{yy}$

Homogenized stress and strain are calculated for each load case by the method presented above. To simplify the notations, the homogenized stress and strain are in the following denoted by σ and ε. Stress-strain curves obtained for different load cases are given in the Figure 2.

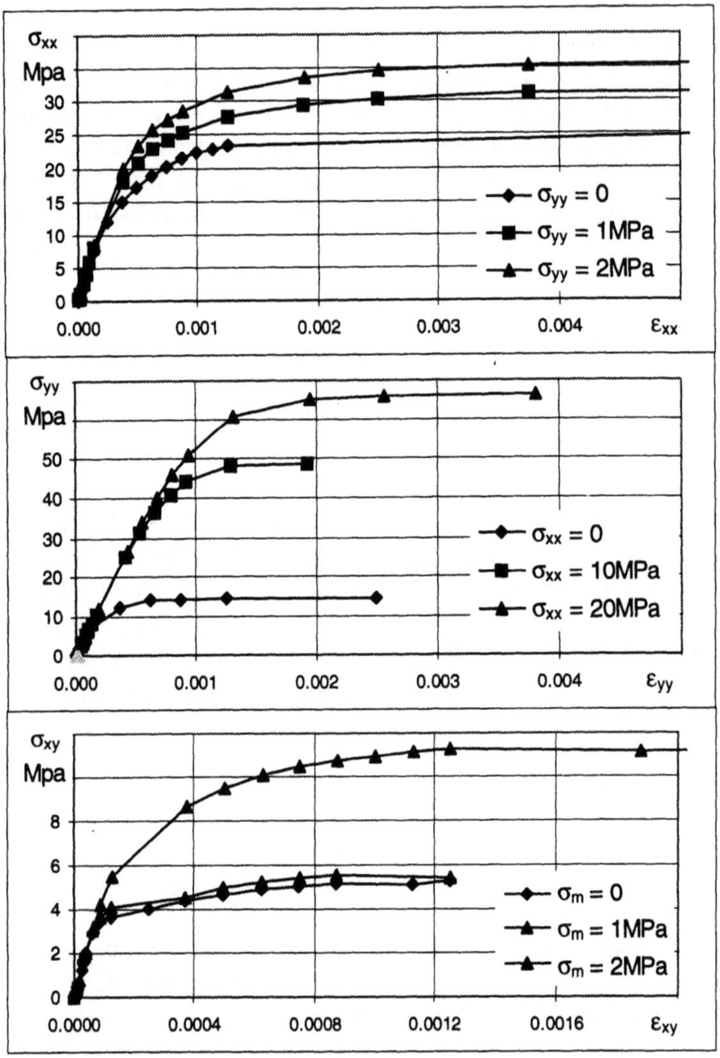

Fig. 2 : Stress-strain curves of the homogenized fractured rock mass for different loadings

5) CONSTITUTIVE MODELLING

The stress-strain curves given in the figure 2 can be considered as some "experimental data" from which the constitutive model of the rock mass must be deduced. These data must be completed, as for as real experimental data, by hypotheses concerning the form of the constitutive model.

The analysis of these results shows up three different stages of deformation. The stress-strain curves begin by a linear elastic stage. During the following stage, the slope of the curve decreases but the stress is always increasing (positive hardening stage). The curves then tend to a horizontal asymptote corresponding to a maximum value of the stress which can not be exceeded (asymptotic perfect plastic behaviour). In the following, will be analyzed first the results corresponding to the elastic stage, then those characterizing the ultimate plastic behaviour, and finally those corresponding to the transient hardening stage.

5.1) ELASTIC STAGE

During the elastic stage, stress-strain curves have a linear trend. This stage extends up to the strain values of about 10^{-5}. Ratios obtained for different variables during this stage are given in the table 1. This table shows that when compressive load in xx or yy directions is applied, very small values are obtained for $\varepsilon_{xy}/\varepsilon_{xx}$ or $\varepsilon_{xy}/\varepsilon_{yy}$ compared to the values of $\varepsilon_{yy}/\varepsilon_{xx}$ or $\varepsilon_{xx}/\varepsilon_{yy}$. The shear strain ε_{xy} is so negligible in these cases. Moreover, elastic moduli in x and y directions have very close values. For these reasons, the elastic behaviour can be assumed, with a good approximation, to be isotropic. The average values of Young modulus and Poisson ratio E as given by the table 1 are the following :

Young modulus:	$E = 62.6$ GPa
Poisson ratio :	$\nu = 0.267$

With this values on can calculate the shear modulus $G = E/2(1+\nu) = 24.7$ GPa. This value corresponds very accurately to that obtained numerically in the last line of the table 1 : $G = \sigma_{xy}/2\varepsilon_{xy} = 24{,}7$ GPa. This is a supplementary confirmation of the hypothesis of linear and isotropic elastic behaviour.

Load direction xx	$\sigma_{xx} / \varepsilon_{xx}$ (GPa)	$\varepsilon_{yy} / \varepsilon_{xx}$	$\varepsilon_{xy} / \varepsilon_{xx}$
	63.2	0.269	0.015
Load direction yy	$\sigma_{yy} / \varepsilon_{yy}$ (GPa)	$\varepsilon_{xx} / \varepsilon_{yy}$	$\varepsilon_{xy} / \varepsilon_{yy}$
	62.0	0.264	0.005
Load direction xy	$\sigma_{xy} / \varepsilon_{xy}$ (GPa)	$\varepsilon_{xx} / \varepsilon_{xy}$	$\varepsilon_{yy} / \varepsilon_{xy}$
	49.4	-0.014	-0.03

Table 1 : Different ratios obtained during the elastic stage

5.2) INITIAL ELASTIC LIMIT

Determination of the initial elastic limit is important because it characterizes the beginning of damage which has important effects on the permeability of the rock mass. The initial elastic limit in different directions can be deduced from the numerical curves or can be determined by theoretical calculations with some simplification hypotheses. This second method is easier and gives an idea of the global form of the domain and will be adopted in the following.

The elastic limit corresponds to the limit of the macroscopic (homogenized) stress for which one of the constituents of the heterogeneous medium attains its elastic limit. Because the cohesion of the intact rock is very greater than the cohesion of the fractures (C=17MPa, c=1.51Mpa) and because the fractures lay in all the directions, the fractures attains their elastic limit before the intact rock.

Consider now a fracture with an angle θ with respect to x direction, with normal and tangent unit vectors $\vec{t}=(\cos\theta, \sin\theta)$ and $\vec{n}=(-\sin\theta, \cos\theta)$. Assume that the shear and normal stresses on this fracture are given by the projections of the homogenized stress tensor σ on the plane of this fracture. This simplification hypothesis is equivalent to neglect the interaction between fractures during the initial elastic stage and also to assume that they have an infinite size. The shear and normal stresses τ and σ_n on the fracture are respectively given by $\sigma_n = \vec{n}.\sigma.\vec{n}$ and $\tau = \vec{t}.\sigma.\vec{n}$. The elastic domain for the fracture is given by (compressive stress computed negatively) :

$$|\tau| < c - \sigma_n \, tg\varphi$$

This inequality leads to following condition for the homogenized stress σ :

$$|\vec{t}.\sigma.\vec{n}| < c - \vec{n}.\sigma.\vec{n} \, tg\varphi \quad (1)$$

This defines a Mohr-Coulomb criterion with cohesion c and internal friction angle φ. So, the initial elastic domain of the homogenized rock mass is given by the Mohr-Coulomb criterion with the same cohesion and internal friction angle than the fractures.

5.3) ASYMPTOTIC STAGE - STRENGH CRITERION

As shown in the figure 2, stress-strain curves tend to a horizontal asymptote. The stress corresponding to this asymptote represents the ultimate strength of the rock mass in the corresponding direction. This ultimate strength depends on the lateral stress for compressive loads or on the mean stress for shear loads. These results show up an important anisotropy of the plastic behaviour of the rock mass. The strength in x direction is in fact about 2 times greater than in y direction. The ultimate stress versus lateral or mean stress has been plotted in the figure 3. This figure shows a linear dependence of the strength with respect to the mean stress.

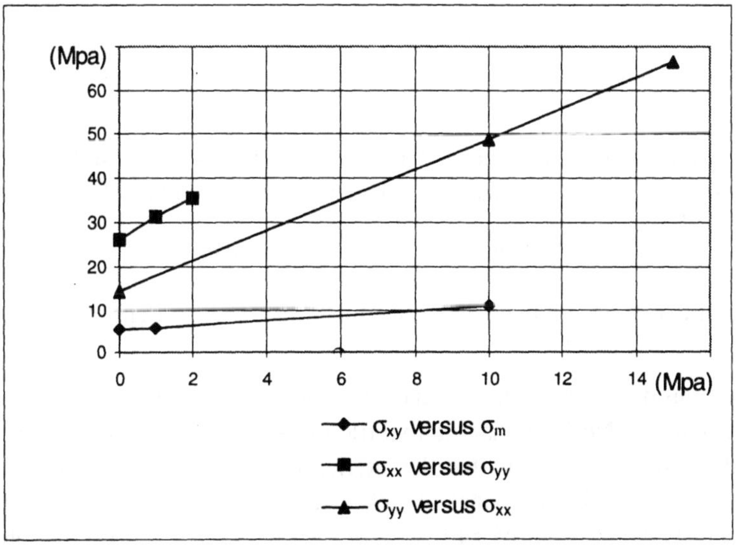

Figure 3 : Effect of the lateral or of the mean stress on the strength in different directions

These results allow to deduce an internal friction angle ϕ^{hom} and a cohesion C^{hom} for each load direction. The values obtained for these parameters are given in the table 2 :

Load direction	C^{hom} (MPa)	ϕ^{hom}
xx	6.02	40,5°
yy	3.35	39,4°
xy	5.02	31,7°

Table 2 : Homogenized Cohesion and internal friction angle in different directions

One can notice that values obtained for homogenized cohesion and internal friction angle are intermediate between those of intact rock (C =17 MPa, ϕ = 57°), and of fractures (c=1.51 MPa, φ=27°).

These results will be used for elaborate a model of the ultimate strength of the rock mass. This model can be written as $f(\sigma) \leq 0$ and the function f must be determined using the above results.

Under the plane stresses hypothesis, the only components of the stress σ are σ_{xx} , σ_{yy} and σ_{xy}. Let us define the mean stress σ_m and a deviatoric stress s in this plane defined by :

$$\sigma_m = \frac{1}{2}(\sigma_{xx}+\sigma_{yy}) \qquad s = \frac{1}{2}(\sigma_{xx}-\sigma_{yy}) \qquad (2)$$

The function $f(\sigma_{m,}\, s, \sigma_{xy})$ will be determined in such a way so that f =0 give the straight lines of the figure 3.

Let us first consider the case $\sigma_{xy} = 0$ and try to determine the boundary of the domain $f \leq 0$ in (s,σ_m) plane (figure 4). Noticing that internal friction angle have very close values in the table 2 for x and y directions, we assume that the internal friction angle in the plan (s,σ_m) is equal to the average value of 40°. The two lines of table 2 for x and y directions define then two straight lines in (s,σ_m) plane with angles of ±40° with respect to σ_m axis (notice that changing x in y is equivalent to changing s in -s and so φ in -φ and c in -c). These lines pass through the points B=(0, 6.02) and C(0, -3.35) ant intersect with together at the point A(5.59, 1.335). These lines delimit the strength domain in (σ_m, s) plane.

In the same way one can determine the strength domain in (σ_m, σ_{xy}) plane.

Consider now the plane (s, σ_{xy}) for a given value of σ_m (Figure 4). For this given value of σ_m, one can identify three points of the boundary of the strength domain verifying $f(\sigma)=0$ by the data of the table 2 : two points with $\sigma_{xy} = 0$ and s given by the straight lines of figure 3 and one point with s=0 and $\sigma_{xy} = 5.02-\sigma_m tg(31.7°)$. The dimensions of the strength domain in this plane vary linearly with σ_m with different rates in different directions (angle 40° for s direction and 31.7° for σ_{xy} direction). The simplest form allowing to satisfy all these conditions is an ellipse passing through the three points identified above and the centre of which has the same s than the point A, s=5.59.

The elliptic form of the domain in (σ_{xy}, s) plane and the linear variation of its diameters with σ_m can be given with a function f having the following expression :

$$f(\sigma) = f(\sigma_m, s, \sigma_{xy}) = (a\,s-d)^2 + (b\,\sigma_{xy} -d)^2 -(-\sigma_m + g)^2 -d^2 \leq 0 \qquad (3)$$

where a,b,d and g are four parameters to be determined.

Supposing $\sigma_{xy}=0$, the expression (3) leads to $f(\sigma_m, s, 0) = (a\,s-d)^2 -(-\sigma_m + g)^2 = 0$. Comparing this expression to the equation of the two straight lines of the figure 3 or the data of the table 2, one can identify : a = cotg (40°), d = 1.59 MPa, g = 5.59 MPa.

In the same way, supposing s=0, the expression (3) leads to : $f(\sigma_m, 0, \sigma_{xy}) = (b\,\sigma_{xy} -d)^2 -(-\sigma_m + g)^2 = 0$. Comparing this expression to the data of the table 3, one can identify : b = cotg(31.7°). This expression leads to two expressions of linear variation of σ_{xy} as a function of σ_m which can be written : $\sigma_{xy} = -\sigma_m\,tg(31.7°)+ 4.43$MPa and $-\sigma_{xy} = -\sigma_m\,tg(31.7°)- 2.47$ MPa. The cohesion c for the first expression is equal to 4.43 MPa instead of the value 5.02 MPa given in the table 2. This approximation is necessary if one wants to maintain the simple form of the strength criterion given by the expression 3. The cohesion c of the second linear expression of variation of σ_{xy} as a function of σ_m is equal to 2.47 MPa. This value has to be verified by further numerical calculations (simulation of shear load in the -xy direction).

Inside the elliptic strength domain one finds the initial elastic domain which corresponds to a circle centred on the origin and with a radius :

$$R = c\cos\varphi - \sigma_m \sin\varphi \qquad (4)$$

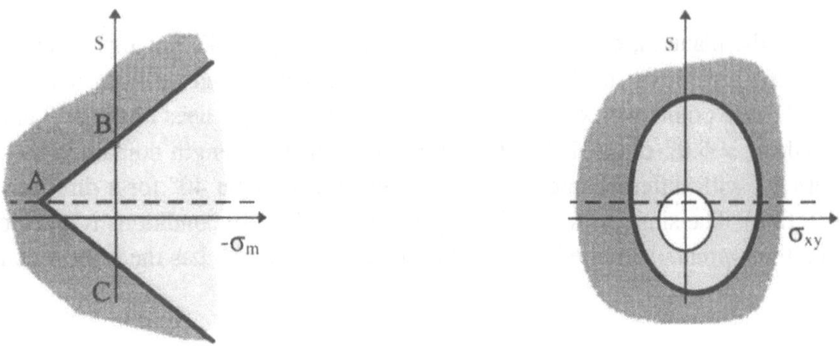

Fig.4 : Strength domain in (σ_m, s) plane is bounded by two straight lines making an angle of ±40° with $-\sigma_m$ axis. Strength domain in (σ_{xy}, s) plane is an ellipse containing the initial elastic domain (circle centred on the origin).

5.4) HARDENING STAGE

During the transient hardening stage of the strain-stress curves, the representative stress point in (σ_{xy}, s) plane goes on a path beginning at the initial elastic domain boundary and tending (asymptotic behaviour) to the ultimate strength boundary. The elastic domain varies during this stage (hardening effect). Noticing that the initial elastic domain can also be described by the same expression (3) with different values of parameters (a, b, d, g), one can assume that the elastic domain is at every stage described given by the same expression (3) with varying parameters (a, b, d, g). The evolution of theses parameters must be deduced from the stress-strain curves of the figure (2).

Let denote (a_0, b_0, d_0, g_0) the initial values of parameters (a, b, d, g). These values can be deduced from (4) and one finds : $a_0 = b_0 = 1/\sin(27°) = 2.20$, $d_0 = 0$ and $g_0 = 1.51 \operatorname{cotg}(27°) = 2.96$ MPa. The final values of these parameters giving the ultimate strength domain, denoted by (a_∞, b_∞, d_∞, g_∞) have been identified here above : $a_\infty = \operatorname{cotg}(40°)$, $b_\infty = \operatorname{cotg}(31.7°)$, $d_\infty = 1.59$ MPa, $g_\infty = 5.59$ MPa. These parameters are assumed to vary between their initial and final values as functions of plastic strain. More precisely, these parameters are assumed to be given by expressions such as :

$$a = a_0 + (a_\infty - a_0)\,(1 - e^{-\beta\xi}) \qquad (5)$$

in which the hardening parameter ξ is defined by $\dot{\xi} = \|\dot{\varepsilon}^p\|$ and β is a constant parameter. This constant has to be determined by fitting the stress-strain curves in the transient stage. The fitting of different stress-strain curves corresponding to different load directions, lead to different values of β varying between 547 to 2447. A unique value of β = 1000 for all the directions has been considered to simplify the model and to reduce the number of its parameters. The result of curve fitting with this unique value of β = 1000 has given sufficiently good agreement with numerical curves.

6) CONSTITUTIVE MODEL

The homogenized rock mass behaviour is finally given by the following constitutive model :

Yield Criterion :

$$f(\sigma, \xi) = f(\sigma_{m,} s, \sigma_{xy}, \xi) = (a\, s - d)^2 + (b\, \sigma_{xy} - d)^2 - (-\sigma_m + g)^2 - d^2 \leq 0$$

Hardening law :

$$\dot{a} = \beta\,(a_\infty - a_0)\; e^{-\beta\xi}\, \dot{\xi}$$

$$\dot{b} = \beta\,(b_\infty - b_0)\; e^{-\beta\xi}\, \dot{\xi}$$

$$\dot{d} = \beta\,(d_\infty - d_0)\; e^{-\beta\xi}\, \dot{\xi}$$

$$\dot{g} = \beta\,(g_\infty - g_0)\; e^{-\beta\xi}\, \dot{\xi}$$

$$\dot{\xi} = \|\dot{\varepsilon}^p\|$$

Initial state : $\xi = 0$, $a = a_0$, $b = b_0$, $d = d_0$, $g = g_0$

Numerical Values : $\beta = 1000$

$a_0 = 2.20$, $b_0 = 2.20$, $d_0 = 0$, $g_0 = 2.96$ MPa

$a_\infty = 1.19$, $b_\infty = 1.62$, $d_\infty = 1.59$MPa, $g_\infty = 5.59$ MPa

7) CONCLUSION AND PERSPECTIVES

The constitutive model determined here for rock mass can be easily introduced in numerical codes for calculation of stress and strain fields around underground openings. Some hypotheses adopted in this model are been verifying by complementary calculations. An important question is the 3D extension of this 2D model with requires complementary work.

More generally, through the studied case, this work has presented a method of determination of rock mass properties by numerical homogenization. The results of this method, involving of course some simplification hypotheses, are very more accurate than those given by commonly used « rock mass classification» methods. An interesting perspective and application of the present work is, in fact, to determine the strength properties of typical cases of rock masses considered in classification tables by numerical calculations instead of empirical estimations. This represents of course an important work requiring a quantity of long calculations. But the results will be of great interest for engineering applications.

REFERENCES

BARTON N., LIEN R. & LUNDE J. (1974). " Engineering classification of rock masses for the design of tunnel support ", Rock Mechanics, vol. 6, n° 4, pp. 189-236.

BEKAERT A. & MAGHOUS S. (1996). " Three-dimensional yield strength properties of jointed rock mass as a homogenized medium ", Mechanics of cohesive frictional materials, vol. 1, 1-24.

BIENIAWSKI Z.T. (1976). " Rock mass classifications in rock engineering ", Proceedings of the symposium on Exploration for Rock Engineering, Johannesburg November 1976.

DE BUHAN P. & MAGHOUS S. (1997). " Comportement élastique non linéaire macroscopique d'un matériau comportement un réseau de joints ", C.R. Acad. Sci. Paris, t. 324, Série II b, p. 209-218.

COSTE F. (1997). " Comportement Thermo-Hydro-Mécanique des massifs rocheux fracturés ", Thèse de Doctorat, Ecole Nationale des Ponts et Chaussées.

FINITE ELEMENT ANALYSIS OF TRANSIENT PULSE METHOD FOR PERMEABILITY MEASUREMENTS

Ö. Aydan
Tokai University, Shimizu, Japan

ABSTRACT : There are various methods to measure the permeability of rocks in literature. Among these methods, *Transient Pulse Method* proposed by Brace et al. (1968) has been widely used for rocks with low permeability in recent years. When the experimental relations of elapsed time with pressure difference between at the ends of a specimen are used to obtain the permeability of rocks, the compressibility of intrinsic rock material and fluid is not taken into account. Therefore, it is natural to expect that computed permeability values may be different from actual values. In this article, the author proposes a finite element method to model the transient pulse method by taking into account the compressibility of intrinsic rock material and fluid and investigate its effect on the computed permeability values through a series of parametric analyses and simulations of permeability tests.

1 INTRODUCTION

It is well known that ground water adversely affects the stability of excavations as well as the working conditions (Aydan & Ersen 1983). In recent years, studies on the disposal of radioactive wastes in rockmass have been increasing in the countries where a great proportion of energy production is based on nuclear power plants. In case of any accidence during disposing these wastes in rock mass, the transportation of the radioactive wastes to the ecological environment strictly depends upon the seepage characteristics of rock mass.

(Brace vd., 1968, Gale 1990, Tsang & Witherspoon 1981). Furthermore, the productivity of wells in petroleum industry also depends upon the seepage characteristics of rock mass.

To measure the seepage characteristics of rock masses, numerous methods are proposed, and the preference to select an appropriate method is generally associated with the expected permeability values of rock masses (Aydan et al. 1997). When the expected permeability of rock mass is relatively large, constant head or falling head permeability tests, which are widely used in soil mechanics, are employed. On the other hand, the transient pulse test proposed by Brace et al. (1968) is used when the permeability of rock masses is relatively small. The permeability is obtained from time-pressure difference relation observed in this test. Aydan et al. (1967) extended this method to axisymmetric radial flow tests. The compressibility of solid and fluid phases of samples is not taken into account in interpreting the test results in this method. It is natural to expect that the permeability values may be different if the compressibility of the samples is taken into account. In the first part of this article, the author presents the governing equation of the fluid flow in porous rocks and then describes its finite element formulation. In the remaining part, a series of parametric studies is performed to illustrate the effect of sample compressibility on permeability results and permeability tests are simulated and results are discussed.

2 GOVERNING EQUATIONS

In the transient pulse test method proposed by Brace et al. (1968), it is assumed that two reservoirs with given pressures and volumes are attached to the ends of a sample, and a sudden pressure difference is caused at one of the reservoirs to initiate a longitudinal flow through the sample. When the pressure difference between two reservoirs disappears, the test is presumed to be terminated. Figure 1 illustrates the fundamental concept of one dimensional longitudinal and radial flow tests.

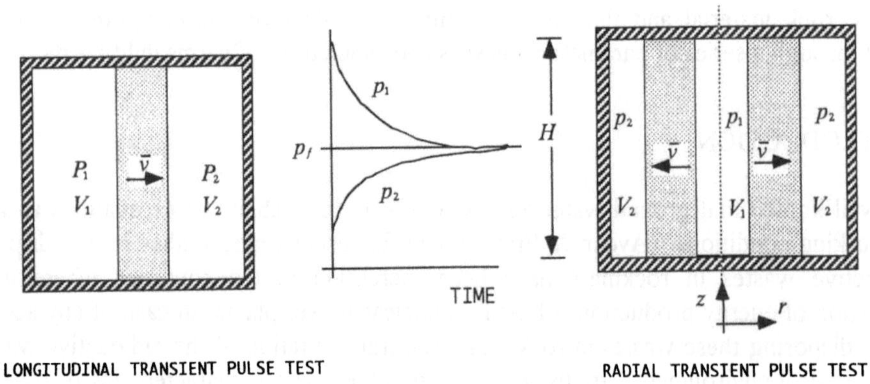

Figure 1 The fundamental concept of one dimensional longitudinal and radial flow tests.

2. 1 Governing Equations of Fluid Flow through A Sample

The mass conservation law for fluid flowing through the pores a sample may be given in the following form with the use of the mixture theory and assuming that a coordinate system fixed to the solid phase:

$$\frac{\partial(\phi\rho_f)}{\partial t} = -\nabla\cdot(\phi\mathbf{q}_f),\quad \nabla = \frac{\partial}{\partial x_i}\mathbf{e}_i,\ i = 1,3 \tag{1}$$

where ρ_f : fluid density, ϕ : porosity, $\mathbf{q}_f$: fluid flux. One may write the following relation for fluid flux in terms of relative velocity $\mathbf{v}_r$ of fluid and the velocity $\mathbf{v}_s$ of solid phase as

$$\phi\mathbf{q}_f = \rho_f(\mathbf{v}_r + \phi\mathbf{v}_s) \tag{2}$$

Let us assume that the flow of fluid obeys the D'Arcy's law. Thus we have the following

$$\mathbf{v}_r = -\frac{k}{\eta}\nabla p \tag{3}$$

where k is permeability, η is viscosity of fluid. Inserting Eqs. (3) and (2) into Eq. (1) yields the following

$$\frac{\partial(\phi\rho_f)}{\partial t} = \nabla\cdot(\rho_f(\frac{k}{\eta}\nabla p - \phi\mathbf{v}_s)) \tag{4}$$

The material derivative operator according to Eulerian description may be written as (Eringen 1980):

$$\frac{d_s}{dt} = \frac{\partial}{\partial t} + \mathbf{v}_s\cdot\nabla(\ ) \tag{5}$$

Introducing this operator into Eq. (4), we have the following relation

$$\frac{d_s\phi}{dt} + \frac{\phi}{\rho_f}\frac{d_s\rho_f}{dt} = \nabla\cdot(\frac{k}{\eta}\nabla p) - \frac{1}{\rho_f}\nabla\cdot(\phi\mathbf{v}_s) \tag{6}$$

The following constitutive relations are assumed to hold among porosity, fluid and solid densities and pressure (Zimmerman et al. 1986)

$$\frac{d_s\phi}{dt} = (C_b - (1+\phi)C_s)\frac{d_s p}{dt},\quad \frac{1}{\rho_f}\frac{d_s\rho_f}{dt} = C_f\frac{d_s p}{dt} \tag{7}$$

If the velocity of solid phase is assumed to be small so that it can be neglected, Eq. (6) takes the following form with the use of Eq. (7):

$$\beta\frac{\partial p}{\partial t} = \nabla\cdot(\nabla p) \tag{8}$$

where

$$\beta = \left[(C_b - C_s) + \phi(C_f - C_s)\right]\frac{\eta}{k}$$

Eq. (8) can be re-written for one dimensional longitudinal flow as:

$$\beta\frac{\partial p}{\partial t} = \frac{\partial^2 p}{\partial x^2} \tag{9}$$

Similarly Eq. (8) can be also written for axisymmetric radial flow as:

$$\beta \frac{\partial p}{\partial t} = \frac{1}{r}\frac{\partial}{\partial r}\left(r\frac{\partial p}{\partial r}\right) \tag{10}$$

2. 2 Governing Equations of Fluid in Reservoirs Attached to A Sample

Using the mass conservation law and the constitutive relation between pressure and fluid density, the velocities v_1, v_2 of fluid contained in reservoirs numbered (1) and (2) and attached to the ends of a sample can be written as:

$$v_1 = -C_f V_1 \frac{\partial p_1}{\partial t}, \; v_2 = -C_f V_2 \frac{\partial p_2}{\partial t} \tag{11}$$

where V_1 and V_2 are volumes of reservoirs, and p_1 and p_2 are pressures acting on fluid reservoirs.

3 FINITE ELEMENT FORMULATION

In this section the finite element formulation of governing equation (9) of the transient pulse test method is given.

3.1 Longitudinal Flow

For incremental variation δp, the integral form of Eq. (9) can be written as:

$$\int_{x=a}^{b} \delta p \beta \frac{\partial p}{\partial t} dx = \int_{x=a}^{b} \delta p \frac{\partial^2 p}{\partial x^2} dx \tag{12}$$

Applying integration by parts to the equation above yields the weak form of the governing equation as

$$\int_{x=a}^{b} \delta p \beta \frac{\partial p}{\partial t} dx + \int_{x=a}^{b} \frac{\partial \delta p}{\partial x}\frac{\partial p}{\partial x} dx = \delta p \hat{t}\Big|_{x=a}^{b}, \; \hat{t} = \frac{\partial p}{\partial x} n \tag{13}$$

Let us assume that the domain is discretised into M elements. Since Eq. (13) is valid for the total domain, it must also hold element-wise. Let us further assume that pressure in an element is interpolated as given below:

$$p = N_i P_i + N_j P_j \quad \text{veya} \quad p = [N]\{P(t)\} \tag{14}$$

where

$$[N] = \begin{bmatrix} N_i & N_j \end{bmatrix}, \; \{P\}^T = \{P_i \quad P_j\}, \; N_i = \frac{x_j - x}{L}, N_j = \frac{x - x_i}{L}, L = x_j - x_i$$

Thus Eq. (13) can be written for a typical element with the use of Eq. (14) as

$$\int_{x=x_i}^{x_j} \beta [N]^T [N]\{\dot{P}\} dx + \int_{x=x_i}^{x_j} [B]^T [B]\{P\} dx = [N]^T \{\hat{t}\}\Big|_{x=x_i}^{x_j} \tag{15}$$

where

$$B_i = \frac{-1}{L}, B_j = \frac{1}{L}$$

The equation above may be re-written as:

$$[M]^e\{\dot{P}\}^e + [K]^e\{P\}^e = \{F\}^e \tag{16}$$

where

$$[M]^e = \beta \int_{x=x_i}^{x_j} [N]^T[N]dx \ , \ [K]^e = \int [B]^T[B]dx, \ \{F\}^e = [N]^T \hat{t}\Big|_{x=x_i}^{x_j}$$

Eq. (16) given for a typical element may be transformed to the equation below for the total domain as

$$[M]\{\dot{P}\} + [K]\{P\} = \{F\} \tag{17}$$

With the use of the θ method, Eq. (17) takes the following form for time step t_{n+1} as

$$[M^*]\{P\}_{n+1} = \{F^*\}_{n+1} \tag{18}$$

where

$$[M^*] = \left(\frac{1}{\Delta t}[M] + \theta[K]\right),$$

$$\{F^*\}_n = \left(\frac{1}{\Delta t}[M] - (1-\theta)[K]\right)\{P\}_n + \theta\{F\}_{n+1} + (1-\theta)\{F\}_n$$

As understood from Eq. (11), the boundary conditions to solve Eq. (18) will vary with time. With use of D'Arcy's law, the velocities at the ends of a sample must be continuous in relation with those of the reservoirs. Therefore, one may write the followings:

$$q_1 = -\frac{k}{\eta}\frac{1}{A}\left(\frac{\partial p}{\partial x}n\right)_{x=a}, \ q_2 = -\frac{k}{\eta}\frac{1}{A}\left(\frac{\partial p}{\partial x}n\right)_{x=b} \tag{19}$$

where A is sample cross section, and $n_{x=a} = -1$, $n_{x=b} = 1$. From the Taylor expansion of Eq. (11) for time step t_n for the both ends of the sample, one can obtain the followings:

$$P_1^{n+1} = P_1^n - \frac{\Delta t}{C_f V_1} q_1, \quad P_2^{n+1} = P_2^n - \frac{\Delta t}{C_f V_2} q_2 \tag{20}$$

Thus the boundary conditions of Eq. (15), which change with time, can be replaced with the equivalent boundary conditions given by Eq. (20) and the resulting simultaneous equation system can be solved.

3.2 Radial Flow

The finite element formulation of Eq. (10) for radial flow can be obtained in a similar manner. Since the formulation would be the same except the specific forms of matrices and vectors, which will be different as a result of the geometry of the domain, it is not presented herein.

4 APPLICATIONS AND DISCUSSIONS

In this section, the following two cases are analysed using the finite element method presented in the preceeding section:

- *Case 1:* The compressibilities of rock and fluid are constant, and porosity is variable
- *Case 2:* The compressibility of rock is variable, and the compressibility of fluid and porosity are constant

4.1 The Effect of Porosity

In order to investigate the effect of porosity, the compressibilities of solid phase and fluid phase are assumed to be constant while porosity is varied. The effect of variation of porosity enters the equation system through constant β. Bulk compressibility C_b is generally obtained from experiments. We can also make an estimation from the limiting values of the bulk compressibility from the compressibility of a sample when porosity has values of 0 and 1. When the porosity has a value of 0, the bulk compressibility must be equal to that of solid phase. On the other hand, when the porosity has a value of 1, then the bulk compressibility must be equal to that of fluid phase. From these requirements it simply implies that the bulk compressibility is twice that of solid phase. In each computation, the value of constant β is computed using this requirement. Figures 2 and 3 show computed results.

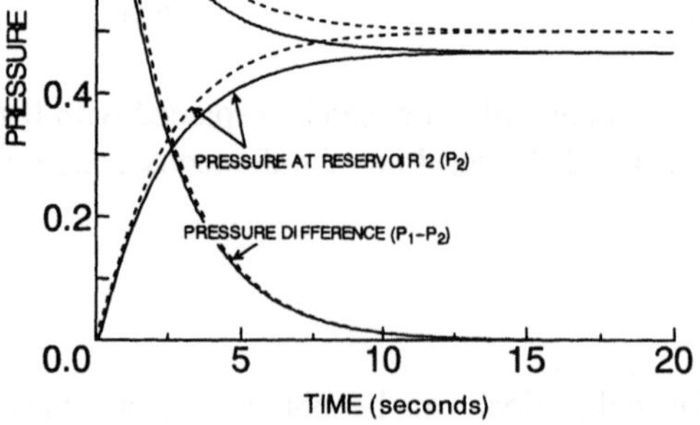

Figure 2: Variation of pressure difference with time (porosity $\phi = 0.01$)

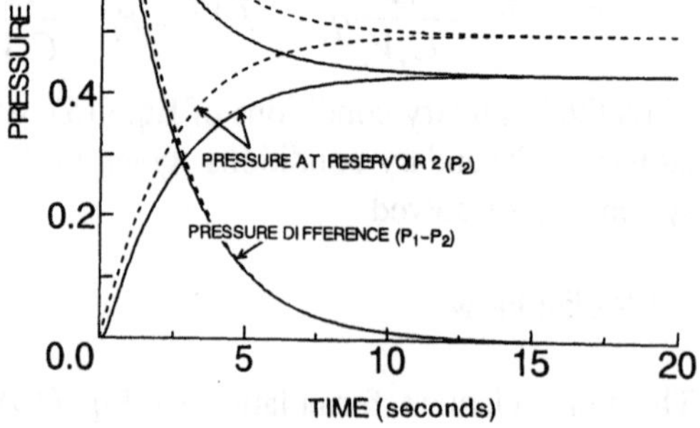

Figure 3: Variation of pressure difference with time (porosity $\phi = 0.20$)

In the same figures, relations used in interpreting transient pulse tests are also included for comparisons. As seen from these figures, the computed relations become markedly different from the conventionally used relations as the value of porosity increases. From this observation it can be said the relations used in interpreting transient pulse test results may be only valid when porosity is very small. Otherwise, the estimated permeability values will be greater than the actual ones.

4.2 The Effect of Compressibility of Solid Phase

In order to investigate the effect of the compressibility of solid phase, the compressibility of fluid phase and porosity are assumed to be constant while the compressibility of solid phase is varied. The value of constant β is computed using the procedure described in the previous case. Figures 4 and 5 show computed results. In the same figures, relations used in interpreting transient pulse tests are also included for comparisons. As also seen from these figures, the computed relations become markedly different from the conventionally used relations as the value of solid phase compressibility increases. From this observation it can also be said the relations used in interpreting transient pulse test results are only valid when porosity and the compressibility of solid phase are very small . Otherwise, the estimated permeability values will be greater than the actual ones.

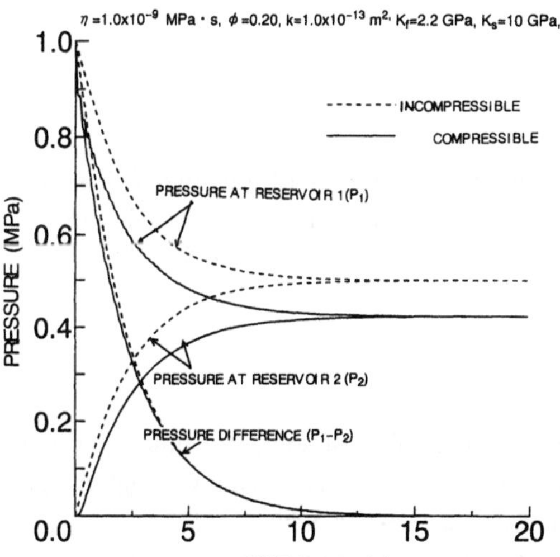

Figure 4: Variation of pressure difference with time (compressibility of solid phase C_s=0.1 GPa^{-1})

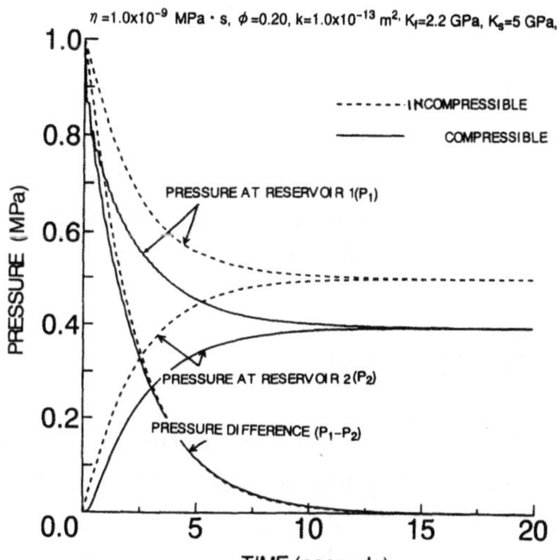

Figure 5: Variation of pressure difference with time (compressibility of solid phase C_s=0.2 GPa^{-1})

In the next series of computations, the effect of the rigidity of solid phase is investigated when and the value of porosity is small. The compressibility of solid phase was assumed to have the values of 0.0 and 0.02 and the value of porosity was taken as 0.01. The value of constant β is computed using the procedure described in the previous case. Figures 6 and 7 show computed results. In the same figures, relations used in interpreting transient pulse tests are also included for comparisons. As also seen from these figures, the computed relations become very close to the conventionally used relations as the solid phase becomes rigid.

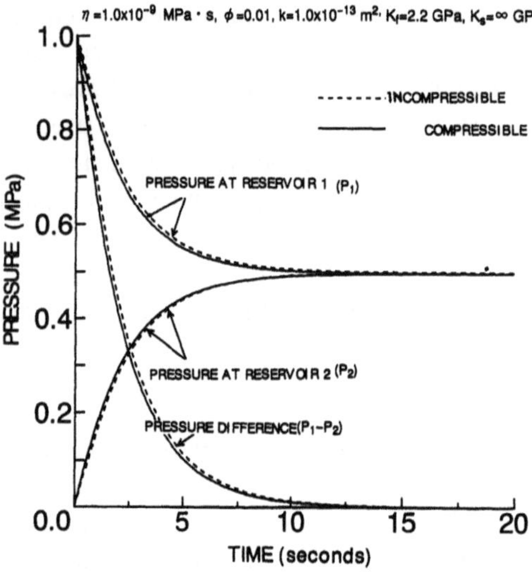

Figure 6: Variation of pressure difference with time (compressibility of solid phase C_s=0.00 GPa^{-1})

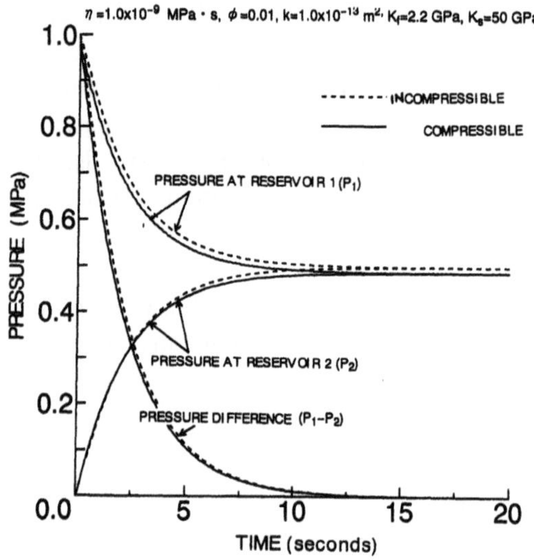

Figure 7: Variation of pressure difference with time (compressibility of solid phase C_s=0.02 GPa^{-1})

5 CONCLUSIONS

In this study, the effect of consideration of the compressibility of solid and fluid phases on the permeability values obtained from the transient pulse method has been investigated using a finite element method. Two series of parametric analyses, in which porosity and solid phase compressibility were varied, were carried out and these results are presented. From the results of parametric studies, it has become clear that the consideration of the total compressibility of samples has a remarkable effect on permeability values and it was shown that the actual permeability of samples must be less than that calculated from the conventional procedure in interpreting the transient pulse test results. If the porosity and the compressibility of solid phase of rock are small, then the computed results will be similar to that computed from the conventional procedure. Nevertheless, it can be said that the relations used in interpreting transient pulse test results can be used to estimate an initial value in the back analysis for obtaining the actual permeability values.

REFERENCES

AYDAN, Ö. and A. ERSEN (1983). Ground freezing method and its application (in Turkish). *Madencilik*, **22(2)**, 33-44.

AYDAN, Ö., H. ÜÇPIRTI and R. ULUSAY (1997). Theoretical formulation of D'Arcy's law for fluid flow through porous and/or jointed rock and its validity (in Turkish). *Bulletin of Rock Mechanics*, Ankara, **13**, 1-18.

AYDAN, Ö., H. ÜÇPIRTI and N. TÜRK (1997). Theory of laboratory methods for measuring permeability of rocks and tests (in Turkish). *Bulletin of Rock Mechanics*, Ankara, **13**, 19-36.

BRACE, W.F., J.B. WALSH, J.B. and W.T. FRANGOS (1968). Permeability of granite under high pressure. *Journal of Geophysical Research*, **73(6)**, 2225-2236.

ERİNGEN, A. C. (1980). Mechanics of Continua. R.E. Krieger Pub. Co., 2nd Ed., New York.

GALE, J. (1990). Hydraulic behaviour of rock joints. *International Symposium on Rock Joints*, Loen, 351-362.

TSANG, Y.W. P.A. WITHERSPOON (1981). Hydro-mechanical behaviour of a deformable rock fracture subject to normal stress. *J. Geophys. Res.*, **86(B10)**, 9287-9298.

ZIMMERMAN, R.W. , W.H. SOMERTON, M.S. KING (1986). Compressibility of porous rocks. *J. Geophys. Res.*, **91(B12)**, 12765-12777.

DEFORMATIONAL ANALYSIS OF GEOMATERIALS CONSIDERING STRAIN-INDUCED DAMAGE

S. Sakurai and C.F.A. Akayuli
Kobe University, Kobe, Japan

ABSTRACT : In this paper a constitutive equation is proposed, which is derived on the basis of experimental results of simple shear tests. In the constitutive equation shear modulus decreases with increase of maximum shear strain, while Young's modulus is kept constant, so that the material damages caused by shear strain can be taken into account. In order to demonstrate the applicability of the constitutive equation to engineering practices both tunnel and land-slide problems are solved.

1. INTRODUCTION

It is well recognized that deformational behavior of geostructures like tunnels, shafts, slopes, retaining walls, etc., observed during construction often differs from the one predicted by numerical analysis. This difference mainly comes from the difficulty in the determination of input data such as geological conditions, characteristics of joints, material properties, underground water conditions, etc., which are used in numerical analysis.

In order to overcome these difficulties, observational methods are often adopted. In the method, observation and field measurements are carried out during the construction and the adequacy of design and construction methods are assessed, and if necessary, they are modified in order to achieve rational design and constructions. This method sounds very impressive, however, the method cannot function effectively unless the measurement results are interpreted adequately. One of the most powerful tools for interpreting field measurements is back analysis, which can assess the input data (design parameters) used in the design analysis, properly.

Various back analysis procedures have already been developed and used in practice. However, non-linear back analysis is not so easy to use so it is hardly applicable to practice. This difficulty is caused by a complex constitutive equation. In this paper a simple constitutive equation is described which may be easily adopted in back analysis. Deformational analysis for tunnels and slopes are demonstrated by using the proposed constitutive equation.

2. ANISOTROPIC DAMAGE OF GEOMATERIALS

In order to clarify the failure mechanism of geomaterials simple shear tests were carried out on sand under an unsaturated condition. During the shear tests a vertical force was applied by loading and unloading while shear strain was kept constant during vertical loading and unloading. The vertical loading and unloading tests were done several times at different levels of shear strain. Experimental results are shown in Figs. 1 and 2. Fig. 1 shows the relationship between shear stress and shear strain, and Fig. 2 shows the relationship between normal stress and normal strain in the vertical direction which were obtained under different shear strain levels. It can be seen from Fig. 2 that the stress-strain relationships are almost identical for different levels of shear strain. This means that the normal stress and normal strain relationship is independent of shear strain so that Young's modulus is constant no matter how large the shear strain is. Concerning the shear modulus, however, it is obvious from Fig. 1 that the shear modulus decreases with increase of shear strain because it is defined as the slope of the curve. In this paper the secant shear modulus is adopted.

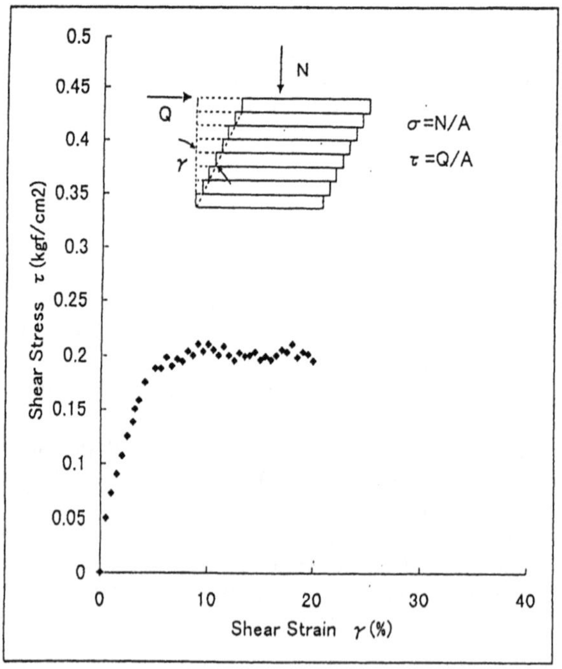

Fig. 1 Shear stress - shear strain relationship in a simple shear test.

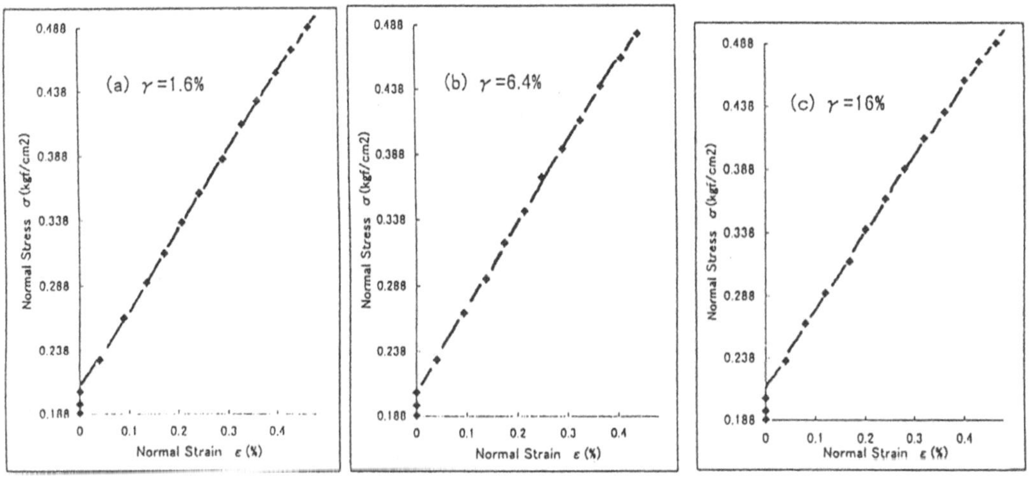

Fig. 2 Normal stress versus normal strain recorded during a simple shear test.

The shear modulus and Young's modulus are plotted in relation with shear strain as shown in Figs. 3 and 4, respectively. The ratio of the shear modulus to Young's modulus is defined as the parameter *m*. The parameter *m* is also plotted in relation with shear strain as shown in Fig. 5. Thus, the parameter *m* is expressed as a function of maximum shear strain in the following general form.

$$m = f(\gamma \max) \tag{1}$$

The author defines the anisotropic damage parameter *d* as follows:

$$d = \frac{1}{2(1+\nu)} - m \tag{2}$$

where ν is Poisson's ratio.

3. CONSTITUTIVE EQUATION

Considering the results of the above described simple shear tests, the following constitutive equation can be derived as a relationship between stress and strain in a two-dimensional plane strain condition:

$$\begin{Bmatrix} \sigma_{x'} \\ \sigma_{y'} \\ \tau_{x'y'} \end{Bmatrix} = [D'] \begin{Bmatrix} \varepsilon_{x'} \\ \varepsilon_{y'} \\ \gamma_{x'y'} \end{Bmatrix} \tag{3}$$

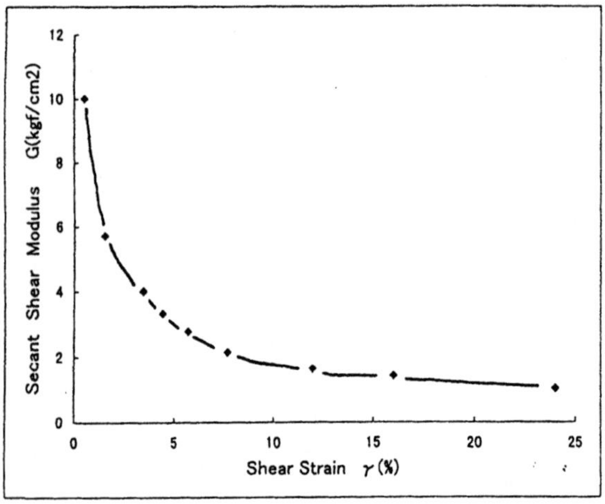

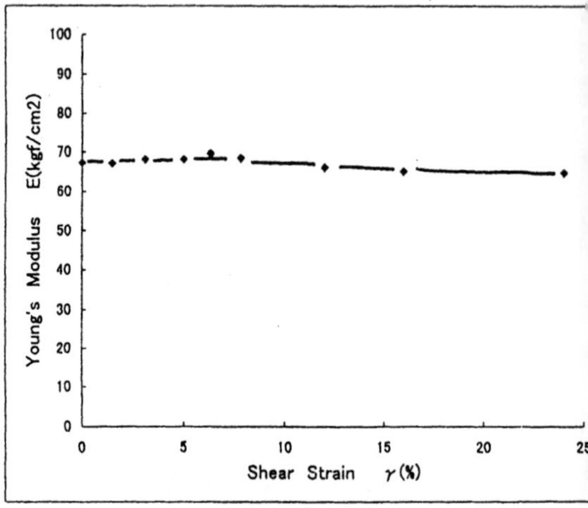

Fig.3 Shear modulus versus shear strain. Fig.4 Young's modulus versus shear strain.

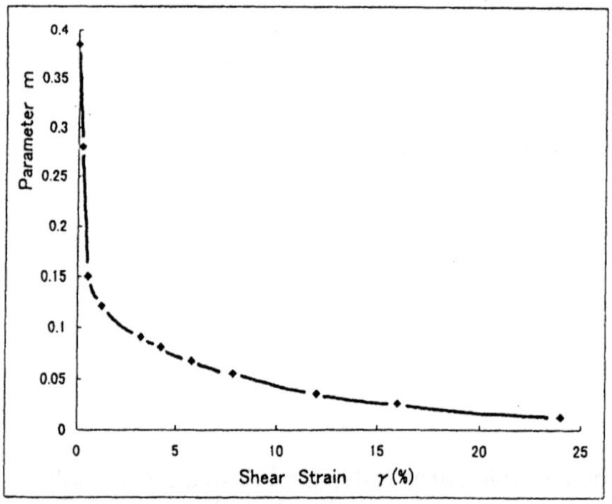

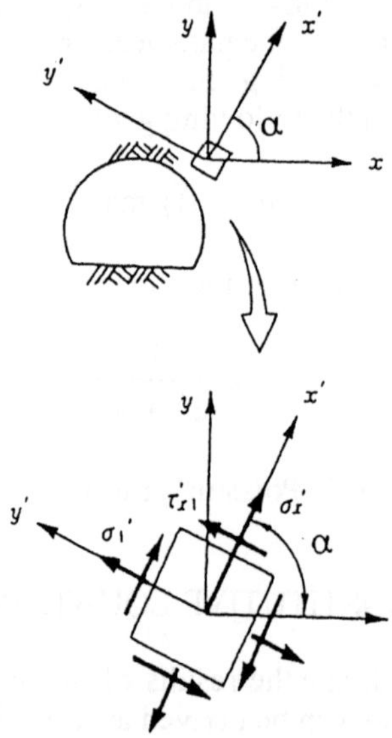

Fig.5 Parameter m ($= G/E$) versus shear strain. Fig.6 Local and global coordinates.

where

$$[D'] = \frac{E}{1-\nu-2\nu^2}\begin{bmatrix} 1-\nu & \nu & 0 \\ \nu & 1-\nu & 0 \\ 0 & 0 & m(1-\nu-2\nu^2) \end{bmatrix} \quad (4)$$

Eq. (3) is valid in the local coordinate system. The x'-axis is parallel to a potential slip plane where the stress state is situated on the Mohr-Coulomb's failure criterion. Eq. (4) can be transformed to global coordinates as follows:

$$[D] = [T][D'][T]^T \quad (5)$$

where $[T]$ is a transformation matrix.

It should be noted that there exists, in general, two potential slip planes at the given point within a material under a stressed condition as shown in Fig. 6. It is seen from Fig. 7 that one is located at the angle of α_1 counted counter-clockwise from the direction of minimum principal stress (compression is positive), and the other is at the angle of α_2. The angle of α_1 and α_2 are determined for geomaterials which obey the Mohr-Coulomb yielding criterion as follows:

$$\alpha_1 = 45^\circ + \frac{\phi}{2} \quad , \quad \alpha_2 = 135^\circ - \frac{\phi}{2} \quad (6)$$

where ϕ is the internal friction angle.

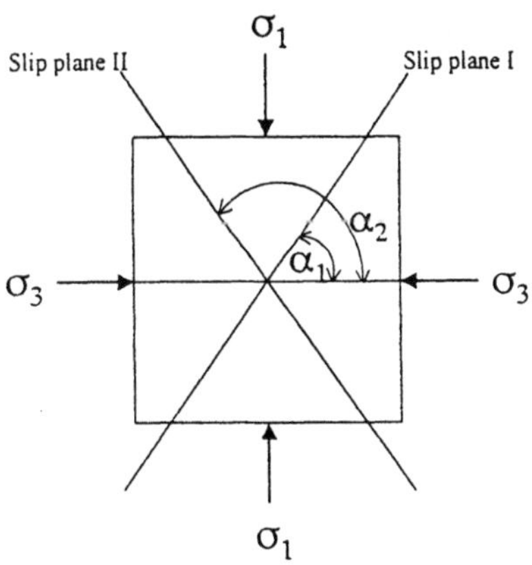

Fig. 7 Conjugate slip planes under a triaxial compressive stress condition.

The stress-strain relationship for the global coordinate system is then given in the following form:

$$\{\sigma\} = [D]\{\varepsilon\} \quad (7)$$

where

$$[D] = \left[\frac{1}{2}\left([D_1]^{-1} + [D_2]^{-1}\right)\right]^{-1} \quad (8)$$

$[D_1]$ and $[D_2]$ are the matrices defined for two conjugate potential slip planes, which can be calculated by Eq. (4) for slip planes I and II, respectively. If the parameter m is obtained as a function of shear strain, in addition to Young's modulus and Poisson's ratio, all the elements of the stress matrix can be determined.

4. APPLICATION OF THE PROPOSED STRESS-STRAIN RELATIONSHIP TO ENGINEERING PRACTICE

4.1 *Tunnel*

Two parallel tunnels excavated in sandy ground are considered as an example problem. First, in order to investigate the mechanism of deformation occurring around the tunnels, physical model tests were carried out in a two dimensional fashion.

In the physical model tests thousands of aluminum pellets were piled up to simulate a sandy ground. Two parallel tunnels (diameter of each tunnel was 15 cm) were set in the medium of piled aluminum pellets, by applying air pressure on the inner surface of the tunnels which was equivalent to overburden pressure. The excavation of the tunnels was simulated by reducing the air pressure gradually until a collapse occurred. Since the details of the tests have been published elsewhere (Akayuli, et. al., 1996), only some of the test results are shown here.

Fig. 8 indicates the maximum shear strain distribution which was recorded just before the collapse occurred. It is seen from this figure that the maximum shear strain distribution tends to concentrate in the region above the tunnels, and shear bands start to be formed. This is known as a strain localization phenomenon.

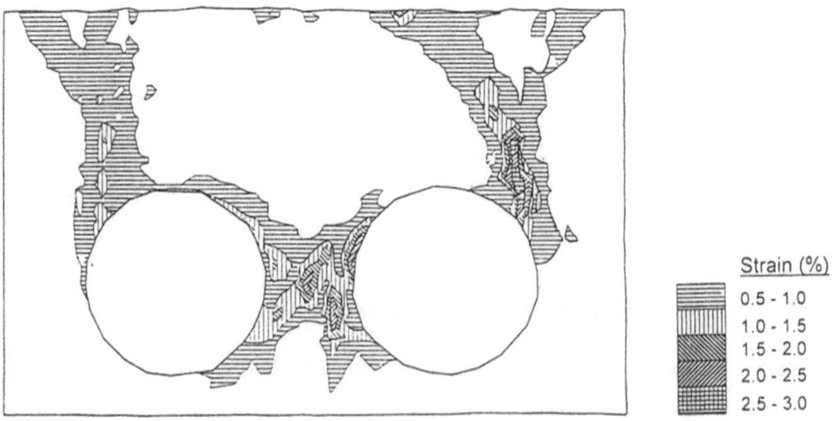

Fig. 8 Maximum shear strain distribution around the two parallel tunnels. (Experiments)

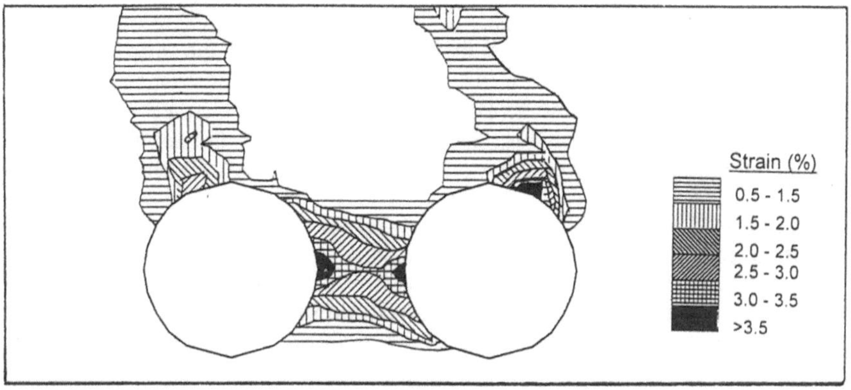

Fig. 9 Maximum shear strain distribution around the two parallel tunnels. (Numerical analyses)

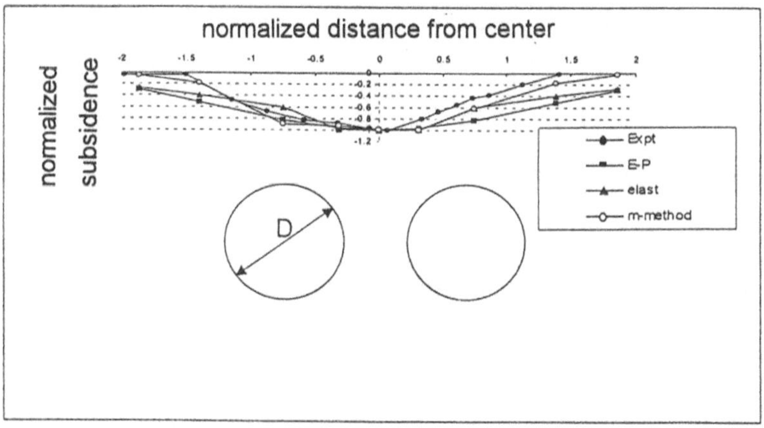

Fig.10 Comparison of numerical results for subsidence calculated by three different models, i.e. (1) elastic, (2) elasto-plastic, and (3) strain-induced damage, considering the parameter *m*.

The deformational behavior of two parallel tunnels in the physical model tests is now simulated numerically by FEM, where the stress-strain relationship given in Eq. (7) is used. The input data adopted in this calculation are shown in Table-1.

Unit weight	w	$1.8 kNm^{-3}$
Young's modulus	E	540.5kPa
Poisson's ratio	ν	0.33
Internal friction angle	ϕ	28.8°

Table 1. Input data used in back analysis

The parameter *m* is back-calculated so as to obtain the maximum shear strain distribution similar to that of the physical model test. When the appropriate value of the parameter *m* is determined, finite element analysis is then carried out and the maximum shear strain distribution is obtained as shown in Fig. 9. The surface subsidence calculated by this method was plotted on the same axes as the subsidence calculated by the elastic and elastoplastic analyses as shown in Fig. 10. It can be seen that the subsidence calculated by employing the parameter *m* showed close agreement with the experimental results.

Comparing Fig. 8 with Fig. 9, it is seen that there is a good agreement. This means that the stress-strain relationship derived on the basis of shear tests of sand can be used to simulate the deformational behavior of the physical model consisting of aluminum pellets. Therefore, it may be concluded that the proposed stress-strain relationship introducing the anisotropic damage parameter can be used for analyzing the deformational behavior of granular materials like various types of soils.

4.2 *Slope*

Land-slide is a time-dependent deformational behavior of slopes. Therefore, in order to predict a land-slide, the time-dependent characteristics of the ground materials must be investigated. However, the time-dependent characteristics are so complex that it is a very difficult task to predict the behavior of land-slides.

To overcome this difficulty, displacement measurements are often carried out for monitoring the stability of land-slides. In assessing the measurement results, back-analysis is a powerful tool which can be used to back-calculate the mechanical properties of the ground materials. In the back analysis of land-slides, the stress-strain relationship given by Eq. (3) is proposed.

The local coordinate x'-axis is taken parallel to the sliding surface. The parameter *m* is then back-calculated from measured displacements. It is assumed that land-slides occur due to the reduction of the value of the parameter *m*. Since the parameter *m* decreases with increase of shear strain, the reduction of *m* causes the displacement increment of land-slides(see Fig.11).

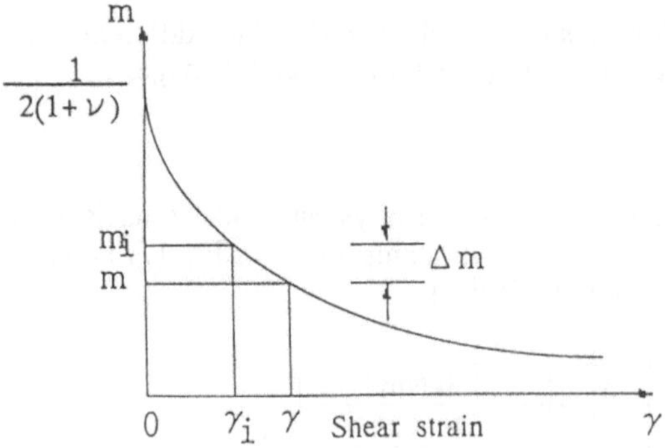

Fig.11 Relationship between increments of parameter m and shear strain.

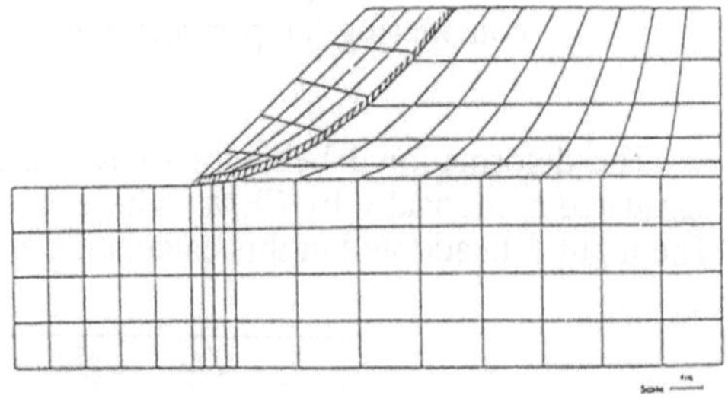

Fig.12 Finite element mesh.

The displacement increment $\Delta\delta$ is then related to the reduction of m, say Δm, in the following equation:

$$[K]\{\Delta\delta\} = [\Delta P] \tag{9}$$

where

$$\{\Delta P\} = \int [B]^T [D][\Delta C]\{\sigma\} dV \tag{10}$$

$[K]$ is a stiffness matrix, $[B]$ is a matrix relating strain occurring in an element to displacements at nodal points of the element. $[D]$ is the elasticity matrix for isotropic elastic materials. $\{\sigma\}$ is stress in damaged regions where the anisotropic damage parameter d increases so that the parameter m decreases as a function of shear strain. $[\Delta C]$ is given for a two-dimensional case in the following form:

$$[\Delta C] = \begin{bmatrix} 0 & 0 & 0 \\ 0 & 0 & 0 \\ 0 & 0 & \dfrac{\Delta m}{m \cdot m_i} \end{bmatrix} \tag{11}$$

In back analysis m_i and m may be determined so as to minimize the following equation:

$$\varepsilon = \frac{\sum_{i=1}^{N} \left(u_i^m - u_i^c\right)^2}{\sum_{i=1}^{N} u_i^m} \tag{12}$$

where u_i^m and u_i^c are measured and computed displacements at measuring point i, respectively, and N is the total number of measuring points.

In order to demonstrate the applicability of the proposed calculation procedure, an example problem is solved. In this example problem the reduction of parameter m is given in a zone where sliding occurs. Finite element mesh is given in Fig. 12. The input data used in the analysis is shown in Table -2. The displacement vectors are calculated and shown in Fig.13.

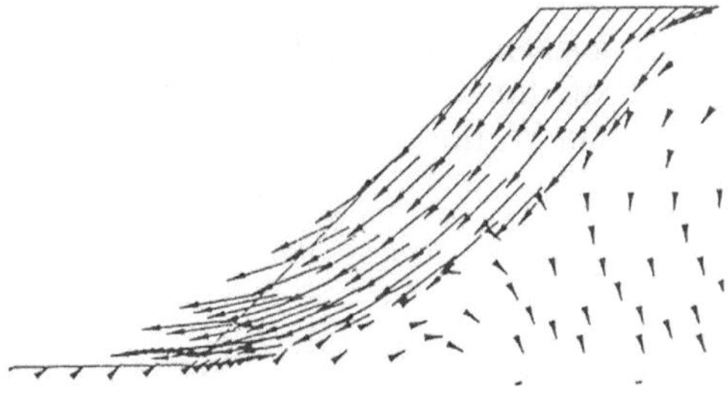

Fig.13 Displacement vectors obtained by numerical analysis.

Young's modulus	E	100MPa
Poisson's ratio	ν	0.3
Anisotropic damage	m_i	0.10
Parameter	m	0.05

Table 2. Input data for slope problem

5. CONCLUSIONS

In this paper the anisotropic damage parameter has been described by showing the results which were obtained by simple shear tests in the laboratory. On the basis of the results of experiments, the stress-strain relationship was introduced.

In order to demonstrate the applicability of the proposed stress-strain relationship, the example problems such as tunnels and a land-slide have been solved, and they demonstrate that the stress-strain relationship should be well applicable to engineering practices.

ACKNOWLEDGMENT

The authors acknowledge the contributions of Mr. K. Hamada to the calculations of the slope problem used in this paper.

REFERENCES

AKAYULI C.F.A., T. MARUOKA, S. SAKURAI & S. AKUTAGAWA (1996). Stress distribution in a pillar of underground openings. Research Report of Construction Engineering Research Institute, No.38-A, pp15-31.

SAKURAI S. & K. HAMADA (1997). Assessment of the stability of slopes (keynote paper), Proc. 9th Int. Conf. on Computer Methods & Advances in Geomechanics, Wuhan, 1: 57-60.

SAKURAI S., A. HIRAOKA & K. HORI (1998). Strain- induced damage of rocks. Proc. 3rd Int. Conf. on Mechanics of Jointed and Faulted Rock (MJFR-3), Wien.

SAKURAI S.(1998). Back analysis based on field measurements, with particular reference to underground openings. Prof. Int. Conf. on Geomechanics/Ground Control in the Mining and Underground Construction, Wollongong.

3-D GEOTECHNICAL MODEL FOR THE NORTH/SOUTHLINE IN AMSTERDAM

B.F.J. van Dijk and F.J. Kaalberg
Design Office North/Southline, Amsterdam, The Netherlands

ABSTRACT:

Under the historic inner city of Amsterdam 3,8 km of the new metro-line is planned to be constructed by means of Tunnel Boring Machine (TBM) tunnelling. The specific environmental conditions in Amsterdam require an outstanding TBM design and process control. During the design process 3D numerical modelling has been and will be used to determine the effects of tunnelling.

Specific research topics for which 3D Finite Element Method (FEM) calculations have been carried out are:

- determination of soil settlements due to tunnelling;
- optimizing the design of a new TBM concept;
- design of the tunnel lining;
- the development of an interactive boring control system; and
- determination of deformation behaviour of adjacent buildings.

For the first four purposes a 3D geotechnical model was made. In this model a number of new numerical ideas have been implemented. The TBM has been modelled in detail. The borefront support and the tail void grouting are modelled as active pressure controlled liquids, that cause (readable) reaction forces on the shield and the jacks in the model. The first results of the model will be presented in this paper.

1. INTRODUCTION

The North/Southline will form an important extension of the current metro-net in Amsterdam. The metro-line will connect the inner city, the Pijp and Amsterdam North with the growing business centres at the southern side of Amsterdam. The new metro-line will both improve the accessibility and habitability of the centre of the city and form an important connection with the business area in the south of Amsterdam.

The first phase of the North/Southline is the construction of the main part between station Buikslotermeer in the north of Amsterdam and Zuid/WTC (World Trade Centre) in the south of Amsterdam. The section is about 9,5 km and contains 7 new metro-stations, of which 5 will be constructed underground. About 6 km of the metro-line will be constructed underground, of which about 3,8 km will be constructed by means of TBM tunnelling. An overview of the metro-line is depicted in figure 1. A geotechnical profile of the boring track can be seen in figure 2.

Fig. 1. Track of the North/Southline.

During the design process a 3D geotechnical model was developed with the FEM program DIANA. The model has been used and will be used to determine the effects of tunnelling in Amsterdam. The specific research topics for which the model has been and will be used are:

- determination of soil settlements due to tunnelling;
- determination of deformation behaviour of adjacent buildings;
- optimizing the design of a new TBM concept;
- design of the tunnel lining; and
- an important tool for an interactive boring control system (IBCS).

2. SPECIFIC BOUNDARY CONDITIONS FOR TUNNELLING IN AMSTERDAM

The conditions for the TBM can be derived from the integrated design aspects regarding alignment, geotechics, foundation technics and constructions.

- From a legal and administrative point of view the track must follow the pattern of the streets as much as possible. Therefore at several places the track has a small radius (minimal radius approximately 190 m). The streets are relatively narrow causing both tubes to lay close to each other (regular distance between the two tubes is 0,5D and for part of the track even above each other).

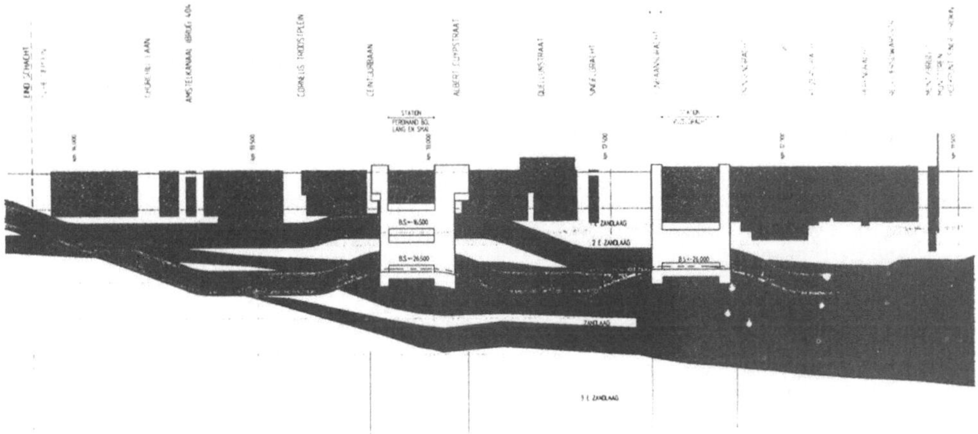

Fig. 2. Geotechnical profile of the boring track.

- The TBM goes partly through soft soil layers with a high watertable. The soil layers exist of sand, soft clay and on top the Holocene toplayers containing peat and mixtures of sand and clay. The TBM bores through regular changing geological conditions. The geological length profile and vertical alignment of the bore track of the North/Southline are shown in figure 2.
- The TBM goes partly through soft soil layers with a high watertable. The soil layers exist of sand, soft clay and on top the Holocene toplayers containing peat and mixtures of sand and clay. The TBM bores through regular changing geological conditions. The geological length profile and vertical alignment of the bore track of the North/Southline are shown in figure 2.
- The existing buildings along the track are all founded on 14 to 20 m long wooden and concrete piles. The settlement pattern caused by the boring process on the foundation level of the piles is decisive for the prediction of the possible damage to buildings. Most adjacent buildings have a great historical value and in other cases demolishing of buildings is prohibited. Therefore potential damage has to be minimized.

3. INTERACTION BETWEEN TBM, SUBSOIL AND ADJACENT BUILDINGS

On the one hand, in soft subsoil the TBM has a function of creating a clearance for the tunnel by means of supportive digging of the borefront, while on the other hand, the created borehole has to be temporarily supported until the tunnel segments have been placed. The separate components of this support (borefront, shieldjacket, transition zone shield/tunnel segments) cause changes in the stress situation in the subsoil resulting in short and long term settlements.

The longitudinal settlement wave caused by tunnelling has a decisive influence on the deformation behaviour of the adjacent buildings, especially in deformation sensitive soil conditions. This is particularly the case for the North/Southline project because the development of this settlement wave has to be considered at the level of the pile toe instead of the ground level. The relatively small distance between the TBM and the pile toe reduces the levelling effect of the intermediate soil layers that is often found elsewhere. Therefore minimization of all the separate contributions to the development of the settlement wave must be pursued. The separate contributions to the total deformations were analyzed.

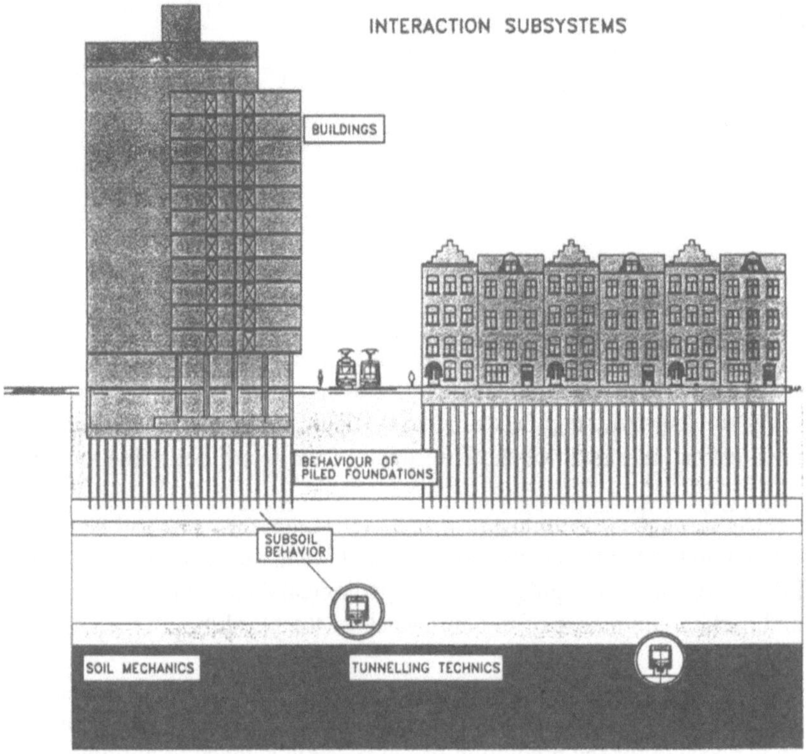

Fig. 3. Cross section F. Bolstreet.

In front of the TBM, the soil will start to settle due to the pressure that is applied at the boring front.

The rigid shieldjacket only accomplishes a passive support of the soil. To activate the support a certain amount of soil deformation will take place. This is amplified by the light conic shape of the shield.

The tail void has to be filled with grout. When injected the grout is liquid. Until the grout has hardened, the soil is supported by the liquid grout. While hardening, the TBM advances forwards and excess water flows from the grout into the soil. These processes influence the pressure in the tail void. The soil has to make an equilibrium with the liquid grout resulting in soil deformations.

After the grout has hardened, it has to support the uprising force of the tunnel lining. Again the soil has to make an equilibrium with the grout and tunnel lining, resulting in soil deformations.

Especially in soft soils, long term effects such as creep and consolidation will result in time dependent settlements of the soil.

4. MINIMIZING POSSIBLE DAMAGE AND SETTLEMENTS

To take into account all the aspects mentioned above, the problem has to be regarded from several angles. Extensive research is and has been carried out on the field of settlement predictions, pile behaviour, definition of damage classifications and soil improvement.

FEM calculations have been carried out together with TNO Building and Constructions Research to investigate the deformation of adjacent constructions due to soil settlements that will be caused by usage of the present-day tunnelling technics (see figure 4). Then a Geographical Information System (GIS) has been applied to visualize extensive parameter studies in an efficient way.

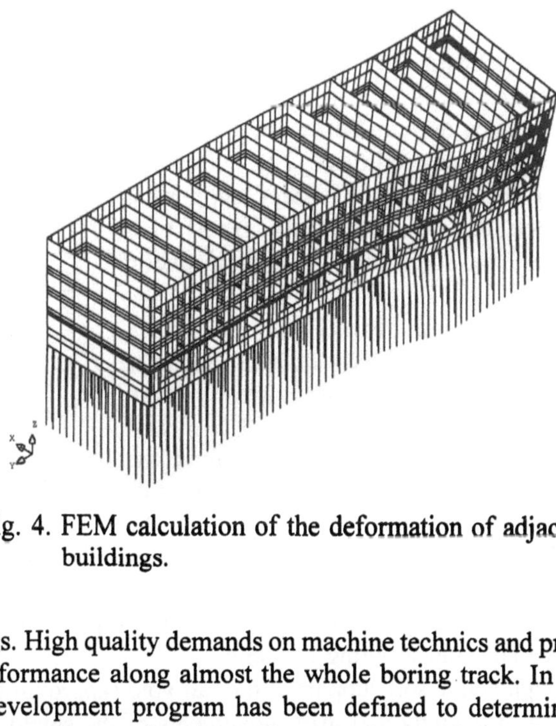

Fig. 4. FEM calculation of the deformation of adjacent buildings.

So far this research to the Amsterdam conditions has shown that the use of a conventional boring process may lead to unacceptable settlements and damage to buildings. High quality demands on machine technics and process control can improve the TBM performance along almost the whole boring track. In order to minimize possible damage a development program has been defined to determine the innovation potential of the TBM regarding optimized settlement characteristics. At those locations where, even with an optimized TBM concept and TBM control, the expected settlements may still lead to unacceptable possible damage, extra measures such as improvement of the soil characteristics will be taken to further reduce settlements.

5. INNOVATION POTENTIAL FOR THE TBM CONCEPT

The development program concentrates on the one hand, to the development of adjusted machine technical concepts, and on the other hand, to the development of a new settlement oriented control of the TBM.

Based on the specific boundary conditions in Amsterdam mentioned before the following requirements were drawn for the TBM:

- to develop a TBM concept that can change to another supporting principle by simple exchange of components during the boring process and therefore can adjust to changing geological conditions;
- to develop a TBM concept that through a smaller length will supersede the soil less in curves, reduce the passive support of the soil and that is less top heavy; and
- to improve the tail void grouting concept to a supporting system that actively supports the borehole and thus realizes a better balance with the soil stress and minimizes the deformations of the tunnel elements.

These requirements resulted in the determination of following technical innovation potential:

- An adjustment of the shape of the immersed wall:
  This creates the conditions for innovative converting possibilities. The discharge device of all sorts of front supporting concepts can be changed by means of a standardized connection under atmospheric pressure.
- A reduced length of the TBM:
  Conventional machines of this diameter (approximately 6,5 m) are normally about 7,5 m long. When the bulkhead is moved forwards it becomes possible to move the end of the main jacks more to the front of the machine. The total length of the machine will thus be reduced to approximately 5,7 m and the centre of gravity of the machine is less excentric. By means of a changed tail void concept it becomes possible to reduce the length of the passive (shield) support even further, to approximately 3,5 m!
- Active tail void grouting concept:
  Through active soil support in the tail void (approximately 1,2 m in length) not only the effects of conicity and drift movements of the shield are reduced, but at the same time the soil relaxation effects can be balanced by a liquid pressure medium. The optimal consistency of the tail void grouting material is still being researched, with due observance of unplanned problem situations.

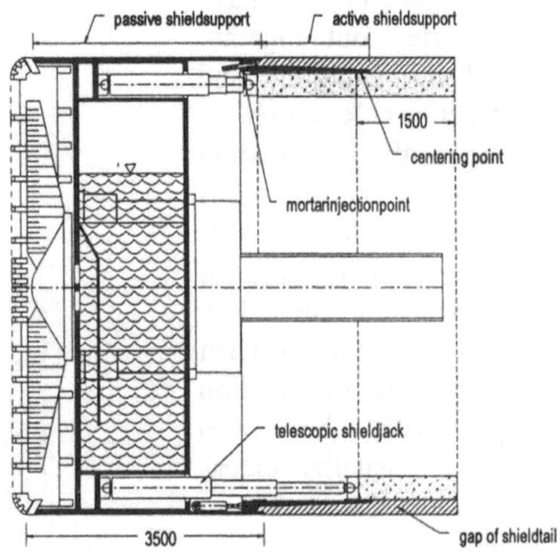

Fig. 5. Vario compact-shield.

The technical innovation potentials are shown in figure 5.

6. 3D DIANA CALCULATIONS

6.1 General

So far, designing practice has been to use numerical FEM soil-models to model the soil and to use framework models to model structures such as the lining of a tunnel.

In reality both settlements of the soil and deformations of structures influence each other. Therefore, when calculated separately, the interaction between the soil and structures can only be modelled through iterations. In DIANA, both the soil and structures can be modelled in a single model. Thus, one is able to model the soil, the TBM, the lining and adjacent structures in a single calculation including the interaction between them. The only limit is the capacity of the computer that is used.

Presently, at the Design Office North/Southline a number of soil-structure problems have been modelled regarding the interaction between soil, TBM and lining. Due to the complexity of the problem and the capacity of the present computers, including the behaviour of adjacent structures in the same model is not possible yet.

6.2 TBM Concept

To verify the innovation potential of the TBM, mentioned above, 3D models have been made at different cross sections of the Amsterdam soil (see figure 6). In these models the TBM has been modelled in detail. All the main jacks were implemented. Also the working chamber , the bulkhead and the articulated tail of the TBM were modelled, thus making it possible to verify for example the thickness and type of material needed in order to obtain a rigid shield. A model detail of the TBM is shown in figure 7.

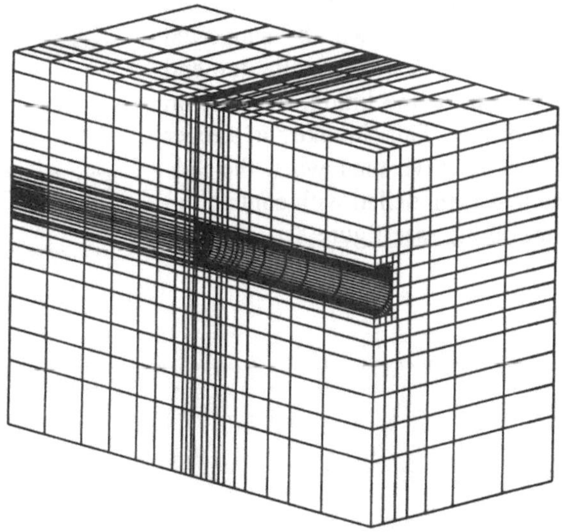

Fig. 6. 3D FEM model of tunnel and subsoil.

6.3 3D Pressure elements

The borefront support and the tail void grouting are modelled in an innovative way by using active pressure controlled liquids that balance the soil pressure. At the same time the pressure at the front mitiates reaction forces on the shield and the jacks in the model. This innovative technique (pressure support of the soil, instead of using a predefined volume loss by contraction of the lining) is also currently under research in the L520 committee for the Centre for Underground Research (CUR) in the Netherlands. The distinguishing aspect of the models used for the North/Southline is that the liquid pressure is modelled with 3D pressure elements. Using pressure controlled liquids initially caused numerical problems at the transition of the shield and the tail void that were solved with the use of interface elements.

When the volume loss is modelled by centric contraction of the lining, the soil follows this shrinkage. By using concentric contraction, the deformation of the soil at the sides is therefore the same as on top and at the bottom of the lining.

However, in reality, the grout in the tail void is liquid, directly after injection. When using regular grout, the grout remains liquid for a few hours. This phenomena is simulated in the liquid pressure support model. In the model the grout has to make an equilibrium with the soil. In virgin (Dutch) soil, the horizontal soil pressure is normally lower than the vertical soil pressure. Therefore with a hydrostatical distribution of the grout pressure, the soil will be supported more at the sides, than at the top and the bottom of the lining, to make an equilibrium. This invokes a steeper settlement curve than when the volume loss is modelled with contraction of the lining (see figure 8.).

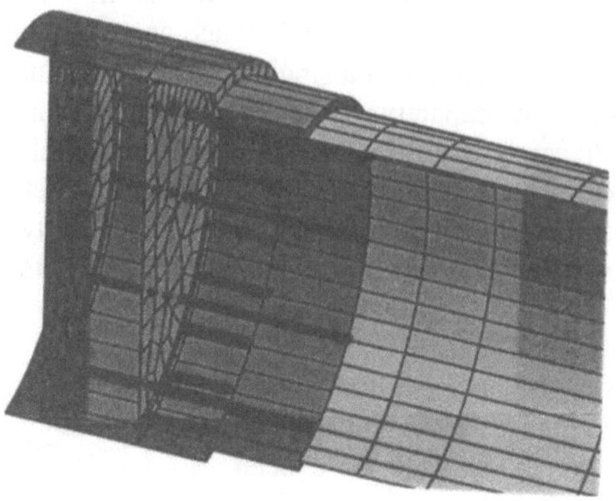

Fig. 7. 3D FEM detail of TBM.

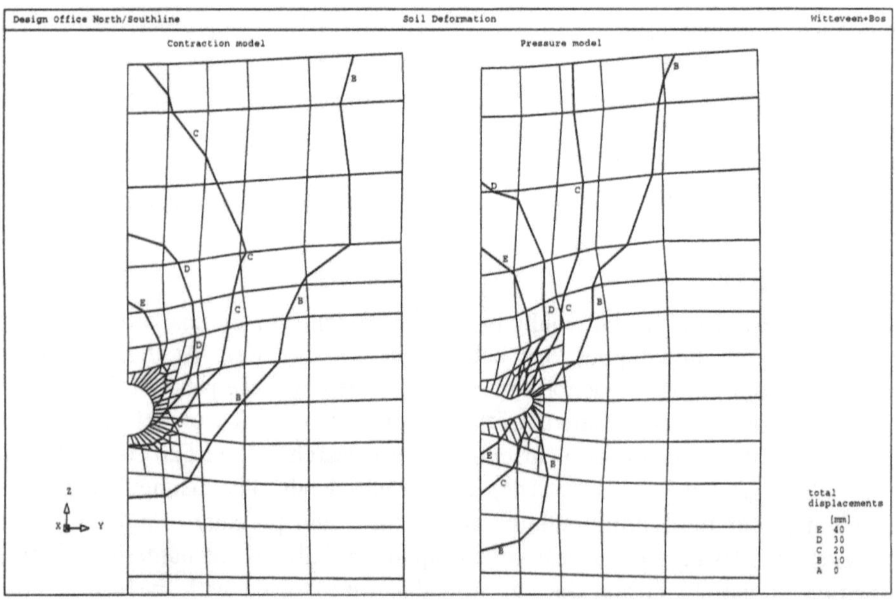

Fig 8. Deformation of the soil when using the contraction (left) and pressure (right) model, with the same volume loss.

6.4 Calibration

The model has been calibrated with measurements carried out at the testproject of the 2nd Heinenoord-tunnel near Rotterdam (the Netherlands). At the location of the 2nd Heinenoordtunnel a Test Pile Project (TPP) was initiated by the design office of the North/Southline. For this location another location specific 3D FEM model was made, using the techniques described above (see figure 8). Basically by adjusting the tail void grout pressure in the model, the measured displacements could be reproduced with the model (see figure 9). In the same figure the results for a 3D contraction model are shown. The settlement curve calculated with the contraction model, perpendicular to the axis of the tunnel, at surface level, is not as steep as the measured data.

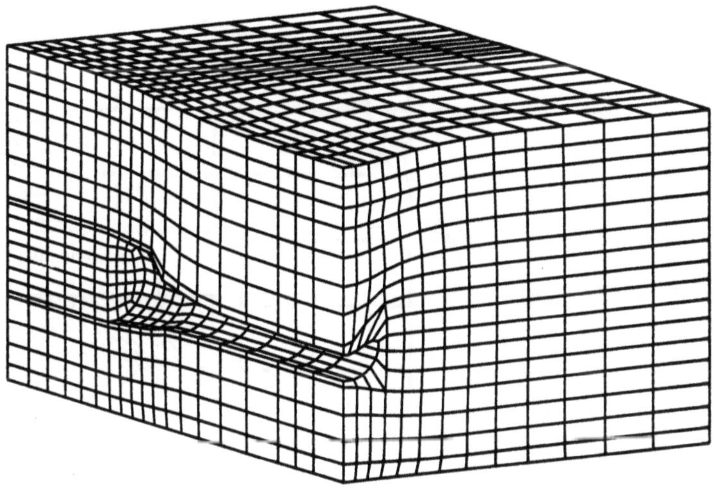

Fig. 9. 3D FEM model of Second Heinenoordtunnel

It appears that the pressure model is very sensitive to variations in tail void grout pressure. It also showed that measurements of the grout pressure (measured near the grout pumps) that are carried out at the 2nd Heinenoord can not be correlated to the pressure in the tail void. This most probably has to do with the leakage of excess water from the grout into the soil, which makes it difficult to validate the model. In order to validate the model and to obtain more insight in the rheologic processes in the tail void, additional measurements of the tail void pressure itself are very important. Measurements of deformations of the surface and the subsoil at the 2nd Heinenoord tunnel have confirmed that there is a strong relation between settlements of the soil and injected grout volume and pressure.

6.5 Settlement predictions for the North/Southline

In this stage of the design phase the model is used for 3D settlement predictions at specific sensitive locations. Due to the fact that for the Amsterdam conditions, the longitudinal trough is determining for the possible damage to buildings, the use of 3D models is necessary.

In a later phase, calculations will be made for different cross sections of the track of the North/Southline with the 3D model. The calculated deformations of the soil at the pile toe levels and the surface level are imported in a GIS model of the track mentioned before. For each building along the track, potential damage is then calculated. Thus, it was visualized where mitigating measures have to be taken along the track to avoid possible damage. The effect of mitigating measures is to be calculated with an adapted 3D FEM model.

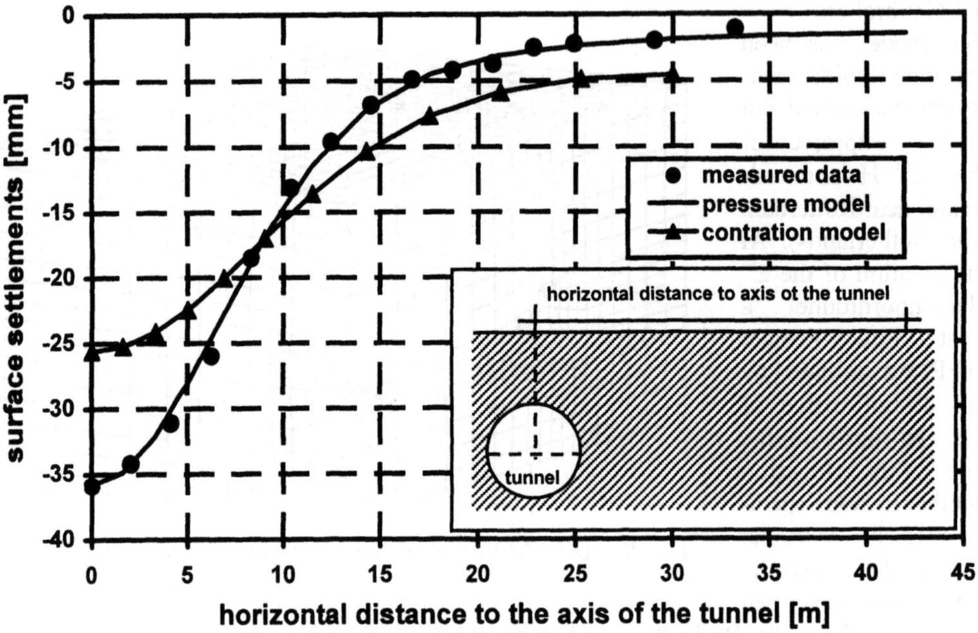

Fig. 10. Calibration of the 3D FEM model at the Second Heinenoordtunnel.

6.6 3D Calculations for the design of the lining

3D calculations for the design of the lining have been carried out. The basis of this model is the same as for the 3D model mentioned above. In the model for the lining, two tunnels, each consisting of two rings, have been modelled. The segments of the rings and the axial and radial joints have been modelled together with the soil. The rings are placed in the model sequentially. Thus making it possible to calculate the interaction between the two tunnels and soil. This interaction is simulated by using the same liquid pressure support method.

7. INTERACTIVE BORING CONTROL SYSTEM (IBCS)

To secure that the interactions with the surrounding are kept within critical boundaries (with a technically optimal TBM) it is necessary to use a systematic process control. The North/South metro-line project demands a process control of the machine with observance of the soil/structure interactions due to the strict settlement conditions set by the historical buildings of the city. Therefore during the construction an Interactive Boring Control System (IBCS) will be used that will be developed further, in the design phase (Figure 11).

The distinguishing characteristic for the IBCS concept, as opposed to well-known process control systems, is the fact that the machine is not controlled solely on the basis of tunnel and machine data, but rather, using settlement simultaneously as a criterion. The actual boring process is performed virtually for this, immediately providing information about

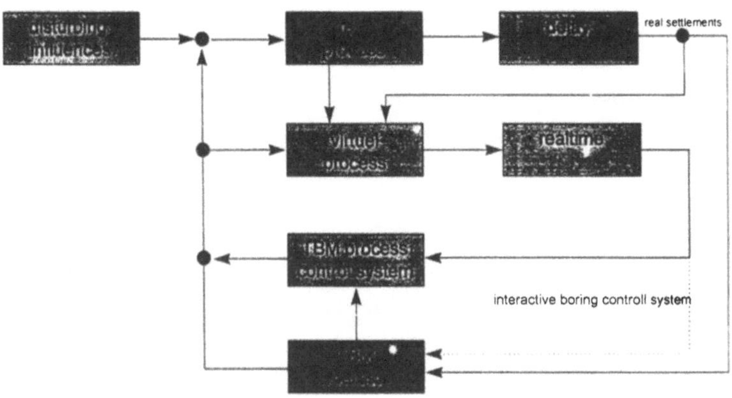

Fig. 11. Interactive Boring Control System.

settlement. This information is then verified and actualized permanently by an expert-team on the interactive basis of monitoring measurements of the surroundings (soil and buildings).

Using the above mentioned 3D FEM computer model, in which the decisive process parameters for the TBM are modeled, the motion of the TBM can be simulated. By adjusting the process parameters to the monitoring data by the process control system, the process sequence can be successively optimized with regard to settlement generation. The active pressure-controlled liquids exert their reaction forces (yielded from calculations) on the shield and the cylinders. In the 3D model propagation of the force effects can be tracked further mathematically to the surrounding soil. The 3D computer model thus serves to combine settlement predictions, monitoring data and process rules, yielding a closed, time-filled information circuit for the interactive system.

In the current phase of development, the model same is used for substantiating the TBM concept that has been developed, and for settlement predictions. Calibration and validation of the model is performed by a comparison of the results of the test pile field at the 2nd Heinenoord tunnel.

8. CONCLUDING REMARKS

3D FEM analysis has proven to be a powerful tool in various research topics for the design office North/Southline. It has been used to solve complex interactive design problems between the soil and structures. In the near future 3D geotechnical FEM models will be used in the final design of the TBM, and the lining, to predict settlements and to estimate the effect of mitigating measures. In the construction phase, the models will play an important role in the Interactive Boring Control System, to simulate the effects of the boring process.

Design Office North/Southline,
Amsterdam

DETERMINING RELATIVE DENSITY OF SANDS FROM CPT USING FUZZY MEASURE THEORY

T. Chi and E. Dembicki
Technical University of Gdansk, Gdansk, Poland

ABSTRACT

In this study, we develop a model based on fuzzy measure theory for aggregating some of the available correlations between relative density, R_d, and cone penetration test (CPT) data. In which, three interval numbers according to three empirical levels, low, medium, and high of compressibility measured by friction ratio of sands are selected. A degree of belief of these levels are determined based on "difference" measure of the given or actual compressibility and the numbers that represent the three predefined levels of compressibility. From that, the relative density R_d is determined based on the base correlations using fuzzy measures and fuzzy integral. Other method, for comparing purpose, using fuzzy set theory is presented also. Obtained results from different methods are compared each other.

INTRODUCTION

As well know, relative density (R_d) is an important parameter in geotechnical engineering. On the one hand, it indicates the state of denseness of a sandy soil. On the other hand, it is used to estimate other engineering properties of soil. Several empirical correlations between R_d and CPT data are available in the literature. They usually relate the cone-tip resistance (q_c) to R_d with consideration of effective overburden stress (σ'_v) and soil compressibility. For all sands, however, no single correlation seems to be able to predict correctly R_d. For example, the correlation proposed by Villet et.al. [9] is able to predict reliably R_d from q_c for sands of low compressibility. The correlation defined by Schmertmann's [8] is more applicable to sands of high compressibility, while the correlation defined by Baldi et al [1] was developed for sands of medium compressibility. These seem, in this study, to be three base correlations.

As, the compressibility of sands is not a well-determined parameter. Then fuzzy measure theory, a proven tool for dealing with imprecise parameters, is utilized in formulation of the new approach. A model to interpolate R_d from three base correlations is developed. Note that, fuzzy measures and fuzzy integral are presented in reference [10]. In this study, we represent rather how we can realize these measures and fuzzy integral numerically.

CORRELATION OF RELATIVE DENSITY WITH CONE-TIP RESISTANCE

In this study, the correlation curves by Schmertmann [8] on Hilton sand, by Baldi et al. [1]

on Ticino sand, and by Villet and Mitchell [9] on Monterey sand are seen as base correlations for sand of high, medium, and low compressibility, respectively. A general correlation has the following form (established by Kulhawy et.al. [6])

$$R_d^2 = \left(\frac{1}{Q_f}\right)\left[\frac{q_c / p_a}{(\sigma'_v / p_a)^{0.5}}\right] \tag{1.1}$$

Where p_a denotes atmospheric pressure (about 100 kPa); Q_f is an empirical constant. Regression analyses performed by Kulhawy et. al. [6], with some adjustments to account for compressibility effects, yields Q_f values of 332, 305, and 278 for NC sand of low, medium and high compressibility, respectively. Note that the over-consolidation ratio (OCR) is set to 1 in this study of normally consolidated (NC) sands. To deal with NC sands, the friction ratio (r) defined as sleeve friction (f_s) divided by cone-tip resistance (q_c) (Juang et. al. [4]) is used to characterize the sand compressibility. A degree of expert's belief (subjective evaluation) of this value is represented by a belief function τ, which is defined as:

$$\tau : R \rightarrow [0,1] \tag{1.2}$$

where, $R = \{r_x\}, x = 1,2,\ldots,n$

$$\tau_x = \tau(r_x), r_x \in R \tag{1.3}$$

DETERMINING RELATIVE DENSITY

The following interval numbers are used to represent the three compressibility qualifiers: low, L = [0, 0.3], medium, M = [0.3, 0.7], and high, H = [0.7, 1.0]. These numbers are subjectively established based on the current knowledge of the friction ratio-compressibility relationship, which are shown in Fig. 1

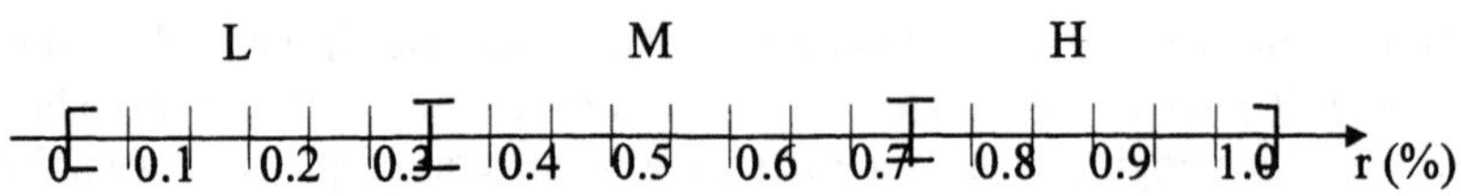

Fig. 1 Interval numbers that represent (low, medium, high) levels of compressibility of a sand

A degree of truth/belief of each predefined level of compressibility (low, medium, high) in respect of the given or actual compressibility, r_a, is determined based on a "difference" measure of r_a and the numbers represented the low, medium, high levels of compressibility respectively. In this example, difference of k_j-level compressibility and number represented the given or actual compressibility, r_a, ($diff(k_j)_{r_a}$) is defined as follows:

$$diff\left(k_j\right)_{r_a} = \left| r_a - r^*_{k_j} \mid \tau\left(r^*_{k_j}\right) = maximum \right| ; r^*_{k_j} \in k_j \subset R \tag{1.4a}$$

This is distance between number, r_a, represented the actual compressibility and number r^*_{kj}, $r^*_{kj} \in k_j$, $k_j \subset R$, R = {L, M, H}, which represents the most probable value so that $\tau(r^*_{kj})$ reaches the maximum value for sand of low, medium and high compressibility. This distance is used as a means of measuring how close the actual compressibility is to each of the predefined levels of compressibility. Note that for subjective evaluation: $\tau(r_1) = \tau(r_2) = \ldots = \tau(r_n)$, ($r_x \in k_j$, x = 1, 2, …, n) then:

$$diff(k_j)_{r_a} = 0 \mid \tau(r_1) = \tau(r_2) = \ldots = \tau(r_n); \forall r_x, r_a \in k_j, x = 1,2,\ldots,n \tag{1.4b}$$

Smaller distance indicated a higher degree of similarity, and the level of compressibility corresponding to a higher is assigned a greater degree of truth/belief, which is represented as truth/belief function τ:

$$\tau(k_j) = 1 - diff(k_j)_{ra}. \tag{1.5}$$

Where, j = 1, 2, 3 denotes index according to levels of compressibility (L, M, H).

Now, we can construct the τ_λ-fuzzy measure for all subsets of set R, R = {L, M, H} as follows: in first, from eq. 1.4b and eq. 1.5 we have;

$$\tau(R) = \tau(L \cup M \cup H) = 1 - diff(R)_{r_a} = 1 - 0 = 1$$

Using cq. 1.6 we have:

$$\tau(R) = \tau(\{L \cup M \cup H\}) = 1 = \frac{1}{\lambda}\left\{\prod_{i=1}^{n}\left[1 + \lambda.\tau(\{k_j\})\right] - 1\right\} \tag{1.6}$$

From that, the value λ can be found from the equation:

$$\lambda + 1 = \left\{\prod_{i=1}^{n}\left[1 + \lambda.\tau(\{k_j\})\right]\right\} \tag{1.7}$$

And,

$$\tau(k_i \cup k_j) = \tau(k_i) + \tau(k_j) + \lambda.\tau(k_i).\tau(k_j) for: k_i, k_j \in R, k_i \cap k_j = \varnothing \tag{1.8}$$

On the other hand, in practice, the relative densities ($R_d(k_1)$, $R_d(k_2)$, $R_d(k_3)$) are determined from the correlations defined for sands of low, medium and high compressibility, respectively. Scores of these are regarded as a measurable function $f_{R_d}(k_j)$ defined on (R, $\wp$(R)) such that $f_{R_d}(k_j) \in [0, 1]$ for each $k_i \in R$, where $\wp$(R) is power set of R. Then, relative density (R_d) value for common sands will be determined by fuzzy integral of a function f_{R_d} with respect to the τ_λ-fuzzy measures as follows:

$$R_d = \oint_R f_{R_d} d\tau = \sup_{\alpha \in [0,\infty]} \left[\alpha \wedge \tau(R \cap F_\alpha)\right], R = \{L, M, H\} \tag{1.9}$$

where, '$\wedge$' denotes 'min' operation, F_α is α-cut of f_{R_d} which is defined by:

$$F_\alpha = \{k \mid f_{R_d}(k) \geq \alpha\} \tag{1.10}$$

7.1.3 DATA AND RESULTS

Data from two sites, the Hunter's point, California site (Dimillio et al. [2] and the Texas A&M University National Geotechnical Experimentation Site (Gibbens and Briaud [3]), are used to illustrate the new approach.

1. *Texas A&M University site.*

The soil from the ground surface to the depth of 11 m is predominantly fine sands. Below the sand layers is very hard plastic clay extended to a depth of at least 33 m. A water table is at a depth of about 4.9 m from the ground surface. Laboratory tests of the sand samples yield a dry unit weight of 14.7 kN/m^3, a natural unit weight of 15.5 kN/m^3, and a natural moisture content of about 5%. Four CPT sounding are obtained (Gibbens and Briaud 1994) at this site. Consolidated drained triaxial tests, borehole shear tests, and density tests, among others are conducted with samples taken from various depth, particularly at the depths of 0.6 and 3.0 m.. Since R_d values are only reported in the literature for sands at the depths of 0.6 and 3.0 m, data at other depths are not considered here. Examination of soil strata obtained from boring logs indicates that at the depth of 0.6 m, the soil is fine sand, while the interpretation of CPT soundings suggests otherwise. Thus, for purpose of verifying the proposed model, the CPT data at the depth of 3.0 m is considered to be most appropriate. The data from four CPT sounding along with other pertinent data used for determining relative density are listed in the tab. 1.

CPT nr.	Depth (m)	σ'_v (kPa)	q_c (kPa)	f_s (kPa)	r_a (%)
1	3.0	46.5	6260	38	0.61
2	3.0	46.5	6590	35	0.53
5	3.0	46.5	6590	43	0.65
6	3.0	46.5	6190	37	0.60

Table 1. CPT data from Texat A&M University site

According to Robertson and Campanella [7], the value, r, increases with increasing sand compressibility. For most NC sands, the r^*_m value for medium compressibility is about 0.5%, but the r^*_L value is almost 0 for sands of low compressibility and r^*_H = 1% (or more) for sands of high compressibility. Difference of low level compressibility (L) and the number represented the actual compressibility, r_a, r_a = 0.61% for example is determined by using eq.1.4a, as follows:

$$diff(L)_{0.61} = 0.61 - 0.0 = 0.61,$$

Next, from eq. 1.5 we have

$$\tau(L) = 1 - 0.61 = 0.39$$

Repeating this process for all other levels of compressibility we obtain $\tau(M) = 0.89$, and $\tau(H) = 0.61$. From that we can construct the τ_λ-fuzzy measures for other subsets of set R, R = {L, M, H} as follows: in first, λ parameter is determined by using eq. 1.7 as follows:

$$\lambda + 1 = (1 + 0.39\lambda).(1 + 0.89\lambda).(1 + 0.61\lambda)$$

The solution of this equation according to the unique root greater than -1 is λ, λ = -0.99. And the τ_λ-fuzzy measures for other subsets of set R are defined by eq. 1.8 as follows: for example

$$\tau(L \cup M) = 0.39 + 0.89 - 0.99.(0.39).0.89 = 0.93$$

By similar way, we can obtain τ(L, H) = 0.76; τ(M, H) = 0.96. Finally, value R_d is calculated by equations (1.9) and (1.10), for example, for value r_a = 0.61 and various values of $R_d(k_j)$ given in tab. 2:

$$R_d = \oint_R f_{R_d}\, d\tau = [0.526 \wedge \tau(L \cup M \cup H)] \vee [0.549 \wedge \tau(M \cup H)] \vee [0.575 \wedge \tau(H)]$$
$$= (0.526 \wedge 1) \vee (0.549 \wedge 0.96) \vee (0.575 \wedge 0.61) = 0.575.$$

where '∧' and '∨' denote 'min' and 'max' operations respectively. The above process is repeated for data from other CPTs. The complete results are shown in table 2, in which, R_d values is determined by the proposed model using fuzzy measures and fuzzy integral, R'_d values are the results obtained from Juang et.al [4].

CPT nr.	r_a (%)	$R_d(L)$ -	$R_d(M)$ -	$R_d(H)$ -	τ(L) -	τ(M) -	τ(H) -	τ(L,M) -	τ(L,H) -	τ(M,H) -	R_d (%)	R'_d (%)
1	0.61	.526	.549	.575	0.39	0.89	0.61	0.93	0.76	0.96	**57.5**	**55**
2	0.53	.539	.562	.589	0.57	0.97	0.53	0.99	0.80	0.98	**56.2**	**56**
5	0.65	.539	.562	.589	0.35	0.85	0.65	0.92	0.78	0.97	**58.9**	**57**
6	0.60	.523	.545	.571	0.40	0.90	0.60	0.96	0.77	0.98	**57.1**	**55**

Table 2. Determination of relative density, R_d, by τ_λ-fuzzy measure and fuzzy integral and R'_d by Juang's et.al. method - Texas A&M University site.

Thus, R_d values determined by the proposed model in all four CPTs examined here agree well with the results obtained by Juang et al. [4].

2. *Hunter's point, California site.*

Details on the Hunter's point, California site are described in a Federal Highway Administration (FHwA) reported by Dimillio et al. [2], which documents results of a pile prediction event. The stop soil of the site is a man-made fill, a mixture of gravel, sand, and clay of about 1.5 m thick. Below that, the soil is clean sand. The ground-water table is about 2 m below the ground surface. The average moisture content is 23% and the average saturated unit weight is about 19 kN/m^3. The CPT data at various depths along with other pertinent data are listed in the table 3.

SPT Nr.	Depth (m)	σ'_v (kPa)	q_c (kPa)	f_s (kPa)	r_a (%)
15	3.0	58.0	4790	17	0.35
7	4.5	67.0	4070	9	0.22
12	6.0	81.0	5030	3	0.06
18	7.5	95.0	6700	11	0.16
25	9.0	110	8860	21	0.24

Table 3. CPT data from Hunter's point site

By same ways represented above, the predicted R_d values obtained from the proposed model using λ-fuzzy measures and fuzzy integral are listed in table 4 along with pertinent intermediate data and R'_d values obtained from Juang et. al. [4].

Depth (m)	r_a (%)	R_d(L) -	R_d(M) -	R_d(H) -	τ(L) -	τ(M) -	τ(H) -	τ(L,M) -	τ(L,H) -	τ(M,H) -	R_d (%)	R'_d (%)
3.0	0.35	.435	.454	.475	0.65	0.85	0.35	0.97	0.78	0.92	**45.4**	**45**
4.4	0.22	.387	.404	.423	0.78	0.72	0.22	0.95	0.83	0.78	**40.2**	**40**
6.0	0.06	.410	.428	.448	0.94	0.56	0.06	1.00	0.97	0.60	**42.8**	**41**
7.5	0.16	.455	.475	.500	0.84	0.66	0.16	0.97	0.94	0.84	**47.5**	**47**
9.0	0.24	.504	.526	.551	0.76	0.74	0.24	0.98	0.83	0.81	**52.6**	**52**

Table 4. Determination of relative density by τ_λ-fuzzy measure and fuzzy integral - Hunter's point site.

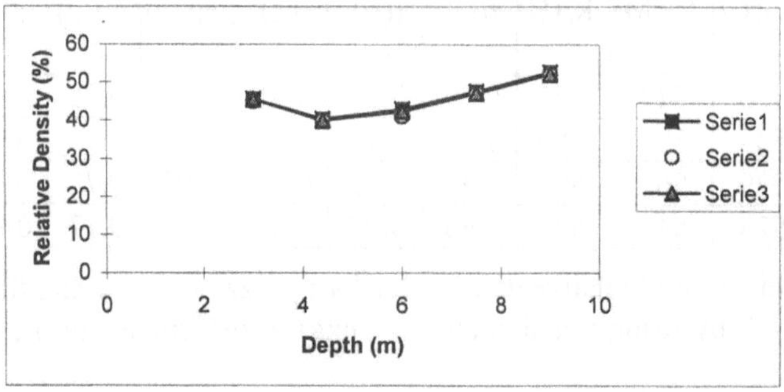

Fig. 2 Differences of the results obtained from proposed method (series 1), Juang's et.al. method (series 2) and Kulhawy's et. al. method (series 3)

Differences of results obtained from three different methods are graphically represented in Fig. 2. It is indicated that the R_d values determined by the using fuzzy measures and fuzzy integral in all of five CPTs examined here agree very well with the results obtained by Juang's et. al. method and Kulhawy's et. al. method.

DETERMINING RELATIVE DENSITY OF SANDS FROM CPT USING FUZZY SETS

In this example, three fuzzy subsets are used to represent the three compressibility qualifiers: low, medium, high. They are expressed in the form of membership functions, which are established based on engineering judgment as shown in Fig.3. In which, the triangular fuzzy number 'Medium' (M), for example, is expressed as:

$$M = (m - u, m, m + v)$$

Where, m, u, v, are mode, difference between the mode and the lower bound and difference between the mode and the upper bound of the fuzzy number M respectively. It is shown in Fig. 3. For sand of lower compressibility, the mode coincides with the lower bound and for high compressibility the mode coincides with the upper bound.

To represent the fuzziness in characterizing the sand compressibility and to handle this fuzziness in the computation process for predicting R_d, a fuzzy number $\mathbb{A}$ shown in Fig.3 is determined by fuzzyfication of the given value r_a.

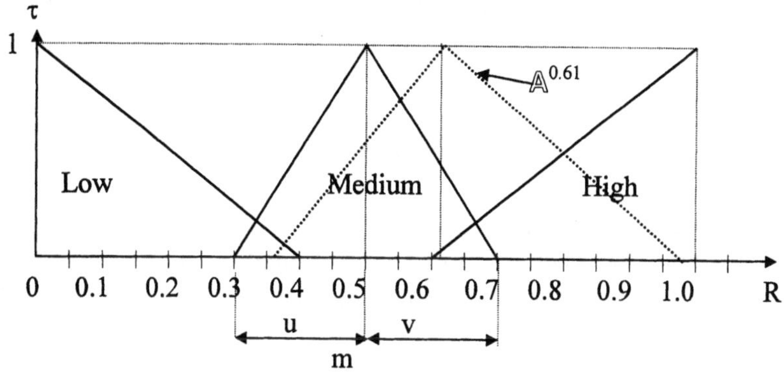

Fig. 3. Fuzzy numbers that represent three levels of compressibility qualifiers

FUZZYFICATION OF THE GIVEN VALUES

As, the measured r_a value is not perfect in characterizing the compressibility. We want to attach the given value with a some degree of fuzziness as follows: in this example, the measured r_a value is taken as the mode (m) of a fuzzy number $\mathbb{A}$, and its spread parameters u and v, is assumed as second-order polynomials of r_a. It is described as follows (see Juang et al. 1996):

$$u = y_1 r_a^2 + y_2 r_a + y_3 \tag{1.11}$$

$$v = y_4 r_a^2 + y_5 r_a + y_6 \tag{1.12}$$

The next step is to associate these polynomials with the established fuzzy numbers (L, M, H) that represent the three levels of compressibility so that the undetermined coefficients y_i, i = 0, 1, 2, ..., 6 can be determined. With this association, three equations can be established from (1.11) and the three other from (1.12). Thus, six equations are available for solving the six unknown coefficients. Solving these equations simultaneously yields the following results: y = (0, 0.4, 0, -0.8, 0.4, 0.4). i.e., eq. (1.11) and eq. (1.12) are completely defined. From that, we can "fuzzify" the given value, $r_a = 0.61$, for example, as follows:

$$m\left(\mathbb{A}^{0.61}\right) = 0.61$$

$$u\left(\mathbb{A}^{0.61}\right) = 0 + 0.4(0.61) + 0 = 0{,}244$$

$$v\left(\mathbb{A}^{0.61}\right) = -0.8\left(0.61^2\right) + 0.4(0.61) + 0.4 = 0.348$$

The lower bound ($l(\mathbb{A}^{0.61})$) and upper bound ($up(\mathbb{A}^{0.61})$) are calculated as follows:

$$l\left(\mathbb{A}^{0.61}\right) = m\left(\mathbb{A}^{0.61}\right) - u = 0.61 - 0.244 = 0.366$$

$$up\left(\mathbb{A}^{0.61}\right) = m\left(\mathbb{A}^{0.61}\right) + v = 0.61 + 0.348 = 0.958$$

Thus, fuzzy number $\mathbb{A}^{0.61}$ is completely defined and expressed as (0.366, 0.61, 0.958) see Fig. 3. It represents the fuzzified values of r_a conforming to the current understanding of the three levels of compressibility.

DETERMINING RELATIVE DENSITY

A degree of truth/belief of each predefined level of compressibility (low, medium, high) in respect of the given or actual compressibility, $\mathbb{A}$, is determined based on "difference" of $\mathbb{A}$ and fuzzy numbers (L, M, H). In this example, difference of k_j-level compressibility and $\mathbb{A}$ (diff(k_j)) is defined as follows. Let $k_1 = L = \mathbb{A}1$, for example

$$diff(\mathbb{A}1)_{\mathbb{A}} = \mathbb{A}1/\mathbb{A}$$

Using arithmetic operations on fuzzy numbers (see Klir et. al. [5]). This technique employs the extension principle, by which operations on real numbers are extended to operations on fuzzy numbers. The fuzzy number $diff(\mathbb{A}1)_{\mathbb{A}}$ is defined as follows: in first, using the α-cut concept of fuzzy set, we have:

$$^{\alpha}(\mathbb{A}1/\mathbb{A}) = {}^{\alpha}\mathbb{A}1/{}^{\alpha}\mathbb{A}$$

we can defined the all of values $^{\alpha}\mathbb{A}1$ and $^{\alpha}\mathbb{A}$ as follows:

$$^{0.2}\mathbb{A}1^R = 0.4-(0.2.(0.4)) = 0.32;\quad ^{0.4}\mathbb{A}1^R = 0.4-(0.4.(0.4)) = 0.24$$
$$^{0.6}\mathbb{A}1^R = 0.4-(0.6(0.4)) = 0.16;\quad ^{0.8}\mathbb{A}1^R = 0.4-(0.8.(0.4)) = 0.08;$$
$$^{0.2}\mathbb{A}^L = 0.366+(0.2.(0.244)) = 0.41;\quad ^{0.4}\mathbb{A}^L = 0.366+(0.4.(0.244)) = 0.46$$
$$^{0.6}\mathbb{A}^L = 0.366+(0.6.(0.244)) = 0.51;\quad ^{0.8}\mathbb{A}^L = 0.366+(0.8.(0.244)) = 0.56$$
$$^{0.2}\mathbb{A}^R = 0.958-(0.2.(0.348)) = 0.89;\quad ^{0.4}\mathbb{A}^R = 0.958-(0.4.(0.348)) = 0.82$$
$$^{0.6}\mathbb{A}^R = 0.958-(0.6.(0.348)) = 0.75;\quad ^{0.8}\mathbb{A}^R = 0.958-(0.8.(0.348)) = 0.68$$

From that, we can calculate:

$$^{0}(\mathbb{A}1/\mathbb{A}) = \left[min\left(\frac{0}{0.366},\frac{0}{1.0},\frac{0.4}{0.366},\frac{0.4}{1.0}\right), max\left(\frac{0}{0.366},\frac{0}{1.0},\frac{0.4}{0.366},\frac{0.4}{1.0}\right)\right] = [0,1.09]$$

Repeating of this process, we can obtain the fuzzy number ($\mathbb{A}1/\mathbb{A}$) as shown in Fig. 4.

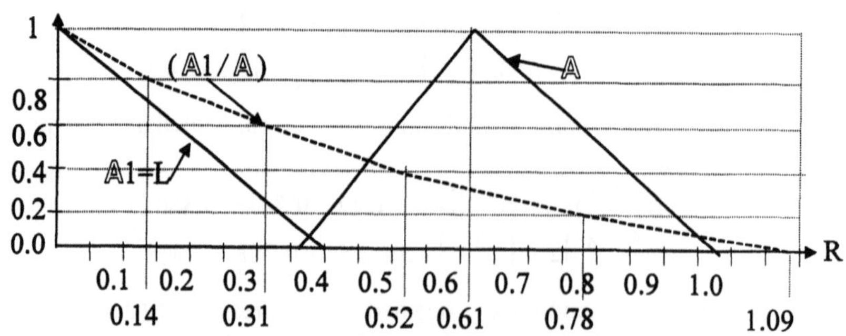

Fig. 4. Fuzzy number that represent the difference of the fuzzy numbers $\mathbb{A}1$ and $\mathbb{A}$

In next step, the resulting fuzzy number ($\mathbb{A}1/\mathbb{A}$) is *defuzzificated,* and the resulting real number which described the difference of two fuzzy numbers, $diff(\mathbb{A}1)_{\mathbb{A}}$ is obtained as follows:

$$diff(\mathbb{A}1)_{\mathbb{A}} = \frac{0.78x0.2 + 0.52x0.4 + 0.31x0.6 + 0.14x0.8}{0.2 + 0.4 + 0.6 + 0.8} = 0.33$$

As, smaller difference indicates a higher similarity according to higher degree of belief τ. Thus, we have

$$sim_{\mathbb{A}1\mathbb{A}} = |1 - diff(\mathbb{A}1)_{\mathbb{A}}| = |1 - 0.33| = 0.67$$

$$\tau_L = \tau_{\mathbb{A}1} = sim_{\mathbb{A}1\mathbb{A}} = 0.67$$

Repeating of the above operations with fuzzy numbers "medium" (M) and "high" (H) leads to results: $\tau_M = 0.15$ and $\tau_H = 0.28$. Continuously, the relative density R_d obtained from the three base correlations are aggregated as follows:

$$R_d = R_d(L).\tau_L + R_d(M).\tau_M + R_d(H).\tau_H$$

where $R_d(L)$, $R_d(M)$, $R_d(H)$ denote relative densities determined from the correlations defined for sands of low, medium, and high compressibility, respectively. Final result is shown in the table 5.

CPT nr.	Depth (m)	r_a (%)	$R_d(L)$ (%)	$R_d(M)$ (%)	$R_d(H)$ (%)	R_d (by fuzzy measure theory) (%)	R_d (by fuzzy set theory) (%)
1	3	0.61	52.6	54.9	57.5	57.5	59.5

Table 5. Comparison of Relative Density obtained by fuzzy measure and fuzzy set theory.

The R_d value of the sand at the depth of 3 m obtained from proposed method using fuzzy measures and fuzzy integral at this site is 57.5%. It is similar to the value determined by the method based on fuzzy set theory.

The new model for predicting R_d is formulated based on a set of three correlations that are believed to be applicable to sands of low, medium, and high compressibility. In this model, interval numbers and fuzzy numbers are generally established based on an empirical approach using expertise. While the fuzzy measure theory or fuzzy set concepts are employed, the entire procedure leading to the prediction of R_d is unambiguous and the results can easily be obtained by no difficult calculation.

The using of the interval numbers (1) or fuzzy numbers (2) (fuzzy subsets) depends on context of practical problems and the nature and quality of given information. For example, when observed parameters x_i and differences between them are small numbers: ($\mathbf{x}$ = 0.12, 0.15, 0.19) and they have same 'degree of belief', we can select (1). Whereas own soil types, gravel, sandy-gravel, silty-gravel, gravely-sand, sand etc. have gradual character. Then for their description we will select (2).

However, it is worthy to note that in the comparing of the fuzzy concepts using fuzzy arithmetic, conclusions (in this case; i.e., there are different methods for this purpose) are formally not well agree with our intuition (degree of similarity of two fuzzy subsets). Whereas, results of proposed method using fuzzy measures and fuzzy integral are clearer. Besides, the using of fuzzy integral allows us to deal with problems such as fusion of subjective evaluations from diverse sources, combination of subjective information and synthetic evaluation of objects in terms of multiple quality factors also. This problem will be addressed in another paper.

REFERENCES

[1] Baldi G., Bellotti R. Ghionna V., Jamiolkowski M. and Pasqualini E. (1982), Design Parameters for Sands from CPT. Procc. 2nd Eur. Sym. On Penetration Test. A.A. Balkema, Rotterdam, The Netherlands, Vol. 2 pp. 425-432

[2] Dimillio A.F., Briaud J.L., O'Neill M.W. and Geo-Resources Consultants (1987), Pile Group Prediction Symposium: Summary Vol. 1, Sandy Soil. Rep. No. FHwA-ts-87-221, Fed. Hwy Admin, Washington D.C.

[3] Gibbens R. and Briaud J. L. (1994). Data and Prediction Request for the Spread Footing Prediction Event. Predicted and Measured Behavior of Five Spread Footing on Sand. ASCE Geotech. Spec. Publ. No. 41 ASCE N.Y. pp 11-85.

[4] Juang C. H., Huang X. H. Holtz R. D. and Chen J. W. (1996), Determining Relative Density of Sands from CPT Using Fuzzy Sets. J. of Geo. Eng. Vol. 122, pp. 1-5.

[5] Klir G. J. and Yuan B. (1995). Fuzzy Set and Fuzzy Logic. Theory and Application, Prentice Hall PTR, Upper Saddle River, N.Y.

[6] Kulhawy F. H. and Mayne P. W. (1991). Relative Density, SPT and CPT Interrelationships. Proc. Ist Int. Symp. On Calibration Chamber Testing, Elsevier, N.Y. pp.197-211.

[7] Robertson P. K. and Campanella R. (1985). Liquefaction Potential of Sands Using the CPT. J. Geotech. Engrg, ASCE Vol. 109(3), pp. 458-482.

[8] Schmertmann J. H. (1978). Guidelines for Cone Penetration Test, Performance and Design. Rep. FHwA-TS-78-209, Fed. Hwy. Admin Washington, D.C

[9] Vilet W. C. and Mitchel J. K. (1981). Cone Resistance, Relative Density and Friction Angle. Procc. Session on Cone Penetration Testing and Experience, ASCE National Convention, G.N. Norris and R. D. Holtz eds., ASCE New York pp. 178-208

[10] Wang Z. and Klir G. J. (1992). Fuzzy Measure Theory. Plenum Press, N.Y. and London.

AUTHORS' INDEX

A. Aalto - Helsinki University of Technology, Espoo, Finland

C.F.A. Akayuli - Kobe University, Kobe, Japan

I. Al Abram - U.R.G.C. Géotechnique, Institut National des Sciences Appliquées de Lyon, Villeurbanne, France

L. Andresen - Norwegian Geotechnical Institute, Oslo, Norway

A. Asaoka - Nagoya University, Nagoya, Japan

C. Athanasiu - NOTEBY AS Consulting Engineers, Oslo, Norway

C.E. Augarde - University of Oxford, Oxford, UK

E. Ausilio - University of Calabria, Cosenza, Italy

Ö. Aydan - Tokai University, Shimizu, Japan

K. Bagi - Technical University of Budapest, Budapest, Hungary

B. Baharom - City University, London, UK

K.J. Bakker - Delft University of Technology, Delft, The Netherlands and Public Works Department "Rijkswaterstaat", Utrecht, The Netherlands

S. Belkhir - Le Havre University, Le Havre, France

N. Bergersen - NOTEBY AS Consulting Engineers, Oslo, Norway

F. Bernier - SCK/CEN, Mol, Belgium

C.B.M. Blom - Holland Railconsult, Utrecht, The Netherlands

P. Bonnier - Plaxis BV, Rhoon, The Netherlands

I. Bojtár - Technical University of Budapest, Budapest, Hungary

V. Bulgakov - Moscow State University of Civil Engineering, Moscow, Russia

H.J. Burd - University of Oxford, Oxford, UK

E. Cabella - University of Genoa, Genoa, Italy

P. Chambon - École Centrale de Nantes, Nantes, France

J.A. Charles - Building Research Establishment Ltd., Watford, London, UK

T. Chi - Technical University of Gdansk, Gdansk, Poland

A. Cividini - Polytechnic of Milan, Milan, Italy

E. Conte - University of Calabria, Cosenza, Italy

M.S. Crilly - Building Research Establishment Ltd., Watford, London, UK

Z. Czap - Technical University of Budapest, Budapest, Hungary

P. Dangla - Laboratoire Central des Ponts et Chaussées, Paris, France

R. De Borst - Delft University of Technology, Delft, The Netherlands

V. De Gennaro - CERMES, École Nationale des Ponts et Chaussées-LCPC, Paris, France

E. Dembicki - Technical University of Gdansk, Gdansk, Poland

D. Dias - U.R.G.C. Géotechnique, Institut National des Sciences Appliquées de Lyon, Villeurbanne, France

B.F.J. van Dijk - Design Office North/Southline, Amsterdam, The Netherlands

D. Divac - Institute for the Development of Water Resources "Jaroslav Cerni", Beograd, Yugoslavia

J.M. Dluzewski - Warsaw University of Technology, Warsaw, Poland

J.J.M.S. Domingues - Instituto Superior de Engenharia do Porto, Porto, Portugal

H.C.W. Duurland - Holland Railconsult, Utrecht, The Netherlands

A.-M. Duval - CETE Méditerranée, L.R.P.C., Nice, France

W. Ehlers - Stuttgart University, Stuttgart, Germany

M. El Yazidi - Institut National des Sciences Appliquées, Rennes, France

Erizal - Ehime University, Matsuyama, Ehime, Japan

G.S.K. Fernando - Nagoya University, Nagoya, Japan

A.O. Ficchì - Civil Engineer, Catanzaro, Italy

M.G. Freiseder - Technical University Graz, Graz, Austria

P. Fritz - Swiss Federal Institute of Technology, Zurich, Switzerland

D. Gallipoli - University of Glasgow, Glasgow, UK

J. Gaombalet - École Polytechnique, Palaiseau, France

B. Gatmiri - CERMES, École Nationale des Ponts et Chaussées, Paris, France

M. Ghoreychi - École Polytechnique, Palaiseau, France

D.R. Gill - Trinity College, University of Dublin, Dublin, Ireland

G. Gioda - Polytechnic of Milan, Milan, Italy

A. Giraud - École Nationale Supérieure de Géologie, Vandœuvre-lès-Nancy, France

N. Grujovic - Faculty of Mechanical Engineering, Kragujevac, Yugoslavia

R.-M. Günther - Institut für Gebirgsmechanik, Leipzig, Germany

T. Guttormsen - SAGA PETROLEUM ASA, Oslo, Norway

S.K. Haigh - Cambridge University, Cambridge, UK

W. Hamadeh - BCEOM, Montpellier, France

O.M. Heeres - Delft University of Technology, Delft, The Netherlands

M. Hellou - Institut National des Sciences Appliquées, Rennes, France

P.Y. Hicher - École Centrale de Nantes, Nantes, France

M.A. Hicks - University of Manchester, Manchester, UK

C.C. Hird - University of Sheffield, Sheffield, UK

F. Homand - École Nationale Supérieure de Géologie, Vandœuvre-lès-Nancy, France

G.T. Houlsby - University of Oxford, Oxford, UK

D. Hoxha - École Nationale Supérieure de Géologie, Vandœuvre-lès-Nancy, France

Y. Ichikawa - Nagoya University, Nagoya, Japan

H.P. Jostad - Norwegian Geotechnical Institute, Oslo, Norway

P.S. Jovanovic - Holland Railconsult, Utrecht, The Netherlands

F.J. Kaalberg - Design Office North/Southline, Amsterdam, The Netherlands

K. Kaneda - Nagoya University, Nagoya, Japan

M. Karstunen - University of Glasgow, Glasgow, UK

R. Kastner - U.R.G.C. Géotechnique, Institut National des Sciences Appliquées de Lyon, Villeurbanne, France

H. Kawamura - Obayashi Co., Tokyo, Japan

K. Kawamura - Tokyo Institute of Technology, Tokyo, Japan

K. Kitayama - Tokyo Electric Power Co., Tokyo, Japan

H. Konietzky - ITASCA Consultants GmbH, Gelsenkirchen, Germany

J. Kos - Czech Technical University, Prague, Czech Republic

P. Koskela - University of Oulu, Oulu, Finland

P. Kuklík - Czech Technical University, Prague, Czech Republic

T. Kyoya - Tohoku University, Sendai, Japan

A.D. Jefferson - University of Wales, Cardiff, UK

B. Jenab - CERMES, École Nationale des Ponts et Chaussées, Paris, France

C. Jommi - Polytechnic of Milan, Milan, Italy

T. Länsivaara - Norwegian University of Science and Technology, Trondheim, Norway

B.M. Lehane - Trinity College, University of Dublin, Dublin, Ireland

D. Levacher - Le Havre University, Le Havre, France

M. Lojander - Helsinki University of Technology, Espoo, Finland

S.P.G. Madabhushi - Cambridge University, Cambridge, UK

J. Mares - Czech Technical University, Prague, Czech Republic

J.M.M.C. Marques - University of Porto, Porto, Portugal

J. Martinez - Institut National des Sciences Appliquées, Rennes, France

F.F. Martins - University of Minho, Azurém, Guimarães, Portugal

J.B. Martins - University of Minho, Azurém, Guimarães, Portugal

P. Mathieu - U.R.G.C. Géotechnique, Institut National des Sciences Appliquées de Lyon, Villeurbanne, France

J.R. McDougall - Napier University, Edinburgh, UK

S. Mezazigh - Le Havre University, Le Havre, France
A. Michali - École Centrale de Paris, Châtenay-Malabry, France
A. Modaressi - École Centrale de Paris, Châtenay-Malabry, France
G. Moulin - École Centrale de Nantes, Nantes, France
H. Müllerschön - Stuttgart University, Stuttgart, Germany
A. Näätänen - Helsinki University of Technology, Espoo, Finland
M. Nakano - Nagoya University, Nagoya, Japan
M. Nakano - Kobe University, Kobe, Japan
D.J. Naylor - University of Wales, Swansea, UK
K. Nesnas - Napier University, Edinburgh, UK
E.E. Nkposong - University of Wales, Cardiff, UK
T. Noda - Nagoya University, Nagoya, Japan
S. Nordal - Norwegian University of Science and Technology, Trondheim, Norway
G.P.C. Oosterhout - TNO-Bouw, Delft, The Netherlands
G.N. Pande - University of Wales, Swansea, UK
R. Passalacqua - University of Genoa, Genoa, Italy
R.M. Pereira - University of Minho, Braga, Portugal
N.D. Pierpoint - Mott MacDonald, London, UK
B. Plischke - ISAMGEO Engineering GmbH, Wiesbaden, Germany
A. Pouya - École Polytechnique, Palaiseau, France
I.C. Pyrah - Napier University, Edinburgh, UK
O. Ravaska - University of Oulu, Oulu, Finland
S.W. Rees - University of Wales, Cardiff, UK
R. Rekonen - Helsinki University of Technology, Espoo, Finland
Y. Riou - École Centrale de Nantes, Nantes, France
P. Royis - École Nationale des Travaux Publics de l'Etat, Vaulx-en-Velin, France
C. Sagaseta - University of Cantabria, Santander, Spain
T. Sakai - Ehime University, Matsuyama, Ehime, Japan
S. Sakurai - Kobe University, Kobe, Japan
K. Salzer - Institut für Gebirgsmechanik, Leipzig, Germany
H.F. Schweiger - Technical University Graz, Graz, Austria
J.F. Seignol - École Nationale des Travaux Publics de l'Etat, Vaulx-en-Velin, France
M. Sejnoha - Czech Technical University, Prague, Czech Republic
J.-F. Semblat - Laboratoire Central des Ponts et Chaussées, Paris, France

J.-F. Shao - Université des Sciences et Technologies de Lille, Villeneuve d'Ascq, France
H.D. Skinner - Building Research Establishment Ltd., Watford, London, UK
I.M. Smith - University of Manchester, Manchester, UK
M. Souley - École Nationale Supérieure de Géologie, Vandœuvre-lès-Nancy, France
S.E. Stallebrass - City University, London, UK
D. Sterpi - Polytechnic of Milan, Milan, Italy
K. Su - École Polytechnique, Palaiseau, France
K. Sugawara - Kumamoto University, Kumamoto, Japan
G. Swoboda - University of Innsbruck, Innsbruck, Austria
T. Tanaka - Meiji University, Kawasaki, Kanagawa, Japan
G. Telekes - Technical University of Budapest, Budapest, Hungary
H.R. Thomas - University of Wales, Cardiff, UK
G. Thouvenin - École Nationale Supérieure de Géologie, Vandœuvre-lès-Nancy, France
N. Tokashiki - Ryukyu University, Okinawa, Japan
J. Vaunat - Polytechnic University of Catalunya, Barcelona, Spain
E.V. Veiga - University of Minho, Braga, Portugal
P. Vepsäläinen - Helsinki University of Technology, Espoo, Finland
P.A. Vermeer - Stuttgart University, Stuttgart, Germany
C.F.S. Vieira - University of Minho, Azurém, Guimarães, Portugal
C. Vogt - Stuttgart University, Stuttgart, Germany
D. Vuckovic - Institute for the Development of Water Resources "Jaroslav Cerni", Beograd, Yugoslavia
S.J. Wheeler - University of Glasgow, Glasgow, UK
E. Yamada - Nagoya University, Nagoya, Japan
W. Xu - Unniversity of Hydraulic and Electric Engineering, Yichang, Hubei, China
M. Zaki - Minia University, Minia, Egypt
R. Zentar - École Centrale de Nantes, Nantes, France
X. Zheng - Swiss Federal Institute of Technology, Zurich, Switzerland
M. Zivkovic - Faculty of Mechanical Engineering, Kragujevac, Yugoslavia

MIX
Papier aus verantwortungsvollen Quellen
Paper from responsible sources
FSC® C105338

If you have any concerns about our products,
you can contact us on
ProductSafety@springernature.com

In case Publisher is established outside the EU,
the EU authorized representative is:
Springer Nature Customer Service Center GmbH
Europaplatz 3, 69115 Heidelberg, Germany

Printed by Libri Plureos GmbH
in Hamburg, Germany